Institute of Physics Publishing

Inhaled Particles X
2008

Journal of Physics: Conference Series Vol. 151

September 23-25, 2008
Manchester, UK

Printed from e-media with permission by:

Curran Associates, Inc.
57 Morehouse Lane
Red Hook, NY 12571
www.proceedings.com

ISBN: 978-1-60560-959-1

Some format issues inherent in the e-media version may also appear in this print version.

Institute of Physics Publishing
Institute of Physics Publishing
Dirac House, Temple Back
Bristol BS1 6BE UK

Phone: 44 1 17 929 7481
Fax: 44 1 17 920 0979

techtracking@iop.org

TABLE OF CONTENTS

Inhaled Particles X: 23-25 September 2008, Sheffield UK.

Editors: Lee Kenny[1] and Fintan Hurley[2]

1 Health and Safety Executive, 241 Glossop Road, Sheffield S10 2GW,UK

2 Institute of Occupational Medicine, Edinburgh, EH14 4AP, UK

Lee.kenny@hse.gsi.gov.uk

Abstract: Multi-disciplinary research is vital to the science of particle-mediated lung disease and a distinctive feature of the Inhaled Particles Conference series. Continuing this tradition, Inhaled Particles X brought together world-renowned and up-and-coming researchers from a wide range of specialist disciplines, but with a common interest and purpose: to understand better the nature of inhalable particles and their effects once inhaled, for the protection of workers' and public health. IPX was an integrative Conference, in three ways. First, as usual with Inhaled Particles, IPX welcomed the full range of disciplines and scientists concerned with protection of health from inhaled particles. This included engineers concerned with dust control; scientists characterising and modelling emissions; exposure measurement, in the field and in the laboratory; experts in particle deposition and clearance; toxicologists discovering mechanisms of damage; epidemiologists developing and applying methods to link exposure with risk; and those concerned with policy – with setting standards, with protection of individuals. Secondly, IPX was concerned with particles in the workplace, in outdoor air, and in the general indoor environment; and tried to find common threads and encourage cross-over of ideas between these three broad fields. And thirdly, IPX integrated experience and perspectives internationally, including issues such as silicosis and pneumoconiosis that are now unfashionable as research topics in the developed Western economies but still constitute a major public health risk internationally. Contributors to IPX were invited to submit extended abstracts for publication in these Proceedings; in addition, all authors were asked to prepare a short abstract. These short abstracts (including those attached to extended abstracts and papers) are collected and published together in this Conference Overview, to maintain the integrity of the Proceedings as a record of the Conference as a whole.

1. Introduction

Inhaled Particles X is the tenth conference in the Inhaled Particles series, organised by the British Occupational Hygiene Society (BOHS) in conjunction with The Aerosol Society of Britain and Ireland. The first Inhaled Particles conference was held in Oxford, UK, in 1960.

2. Scientific Committee

The core UK Scientific Committee for Inhaled Particles X included:
Fintan Hurley - IOM (Institute of Occupational Medicine), Edinburgh (Chair)
Professor Raymond Agius – University of Manchester
Professor Ross Anderson – St. George's Hospital, London
Professor Jon Ayres – University of Aberdeen
Dr. John Cherrie – IOM and University of Aberdeen
Professor Ken Donaldson – University of Edinburgh
Dr. Sarah Dunnett – Loughborough University
Dr. Paul Harrison – IEH (Institute of Environment and Health), Cranfield University

Professor Roy Harrison – University of Birmingham
Dr. Lee Kenny – HSE (Health and Safety Executive), Sheffield
Professor Bob Maynard – HPA (Health Protection Agency), Chilton
Professor Anthony Seaton – IOM and University of Aberdeen
Dr. Sean Semple - University of Aberdeen
These were be advised and supported by International experts associated with the key conference sessions.

3. Conference themes

The remit of Inhaled Particles is very wide and it aims to embrace a wide range of research and policy communities, which overlap one another only partially. In developing themes for the conference one useful perspective is to consider *methodologies* (e.g. control, instrumentation / measurement; modelling of personal exposures; deposition and clearance of inhaled particles; fate / mechanisms of disease; epidemiology; health impacts and links with policy). Another is to look at *substantive topics*. The following topical questions were posed to IPX contributors in order to stimulate contributions taking a broad perspective of the field:

3.1. Cardiovascular disease:

What is the role of inhaled particles in the development of cardiovascular disease and/or the triggering of heart attacks?
Given the results on outdoor air pollution, is it to be expected that (i) occupational exposure and (ii) exposure indoors also lead to cardiovascular disease?
Is this an issue for combustion particles only or for other particles also?
Does outdoor air pollution affect the cardiovascular health only of people who are not fit enough to work; and if so, what does this imply for life expectancy?

3.2. Exposure to asbestos and other soil contaminants:

As well as the traditional research base on occupational exposure to asbestos, there are active areas of research in terms of (i) exposure within buildings and (ii) outdoor exposure, e.g. from contaminated soil. What can we gain from looking at these issues together?
What about other soil contaminants, for example, particles acting as vectors for persistent organic compounds of low volatility?

3.3. Nano-particles:

Engineered nano-particles are a major focus of attention worldwide.
What can we learn from studies of 'incidental' nano-particles, e.g. from diesel traffic? Are results likely to be transferable?
What is new with regard to mechanisms?
What characteristics, other than size (aerodynamic diameter) are likely to be important – Shape? Surface area? Surface properties? Solubility?

3.4. Environmental Tobacco Smoke

The recent ban on smoking, in Scotland and in several other countries, has been accompanied by research to evaluate its effects – on attitudes, on indoor air in public buildings, on tobacco use in the home, on public health, on health of bar workers.
What have we learned from these studies?
To what extent is smoking cessation a good 'model' for reduction of air pollution, in terms of the time-course of recovery?

3.5. Developing countries:

Traditional dusty diseases are less important now in many industrialised (post-industrial?) countries. Partly this is because of the decline of heavy industry in these countries; partly it is because of better controls on dust exposure at work. Internationally, however, exposure to silica, to coalmine dust, to asbestos and other fibres remains a major problem for worker's health.

Journal of Physics: Conference Series **151** (2009) 011001 doi:10.1088/1742-6596/151/1/011001

What is the state-of-play internationally – With incidence and prevalence of disease? With exposure control? With surveillance? What are examples of good practice?
To what extent is the existing very substantial research base on these dust-related diseases still relevant to current conditions worldwide? What new or different issues are now important?

3.6. Domestic exposure to biomass fuels:
This is a major issue internationally. What do we know about the combustion particles responsible? What can be done to reduce exposure, while maintaining people's capacity to get energy for warmth and cooking? What does it take for these measures to be implemented widely?

The contributions collected in these Proceedings address these topical questions and raise many more. Several papers highlight the ongoing burden of exposure to inhaled particles worldwide and the continuing damage to health that this causes. For some highly toxic materials such as asbestos, the legacy of former industrial use continues to damage health even where manufacturing of asbestos products has long ceased. This experience reinforces the need to understand and assess the potential future health risks posed by novel industrial materials, such as the engineered nanoparticles now finding a wide range of commercial uses, a major topic at this conference.

Inhaled Particles X demonstrates that there is still a need for basic research to understand the mechanisms that underlie the health impacts of particle inhalation, though much progress has been made in this field since Inhaled Particles IX. A multidisciplinary conference such as IPX brings together those working in fundamental and applied sciences; it links historical perspectives with forward-looking contributions, and it enables us to transfer knowledge and experience between the developed and the developing world.

We hope that these IPX Proceedings will prove a worthy contribution to the long and distinguished tradition of Inhaled Particles conferences. The **Conference Overview** summarises the conference programme and incorporates links to Short Abstracts and Extended Abstracts of the papers presented.

Journal of Physics: Conference Series **151** (2009) 011002 doi:10.1088/1742-6596/151/1/011002

Predicting nanoparticle toxicity: Significance of Dose- and Response-Metric

G Oberdörster[1], E K Rushton[1], J Jiang[2], S S. Leonard[3], S Eberly[1], V Castranova[3], P Biswas[2], A Elder[1], X Han 1, R Gelein[1] and J Finkelstein[1]

1 University of Rochester, Rochester, New York, USA;

2 Washington University in St. Louis, St. Louis, MO, USA;

3 National Institute for Occupational Safety and Health, Morgantown, WV, USA

E-mail: gunter_oberdorster@urmc.rochester.edu

Abstract: The multitude of existing and continuously created new nanoparticles (NPs) represents a huge challenge for assessing potential risks they may pose to humans and the environment. Even for characterizing the hazard of NPs represents a major problem, there is no generally agreed upon validated protocol for predicting a NP hazard. Thus, although concepts specific to nanotoxicology have been well advanced, there is still no consensus about the usefulness of *in vitro* tests for predicting *in vivo* toxicity of NPs. Many high dose studies, *in vitro* and *in vivo*, suggest that NPs can induce significant toxicity, yet comparative studies have shown that *in vitro* and *in vivo* results do not correlate well and can show largely different toxicity rankings. One reason for this seemingly poor correlation is that applied doses and resultant shapes of dose-response curves are different for *in vitro* and *in vivo* assays, which makes it difficult to decide which point on a dose-response curve should be used for comparing results between assays. Thus, the question is, can corresponding equivalent doses on the *in vivo* and *in vitro* dose-response relationships be identified? We hypothesize that the steepest slope of a dose-response curve (mathematically the 1st derivative) identifies this dose, it reflects, by definition, the maximum response per unit of the dose, but it is not the maximum response. We suggest to select the corresponding *in vivo* and *in vitro* responses observed at this steepest point for comparing results of different assays, the response-metric being the response per unit dose. Doses below this point and doses above this point result in a lower response per unit dose, they are in the flat sections of the dose-response. We reason that this steepest section reflects doses that have saturated biotransformation processes, binding sites or defense mechanisms, and it identifies matching conditions between assays. Responses at extremely high doses can be very misleading, do not correlate between different assays, and –if used for comparison– *in vitro* assays will then erroneously be judged as not being predictive for *in vivo* responses. We suggest further that NP surface area should be used as the most appropriate dosemetric. In a proof-of-principle multi-assay study we tested this concept, using 8 different well-characterized NPs with diverse physico-chemical characteristics, ranging from more benign TiO_2 and gold NPs to very reactive copper NPs. Endpoints evaluated in these assays related to induction of oxidative stress and included: two cell-free assays (DCFH oxidation to measure ROS generation; ESR signal to measure radical generation potential); two cellular assays (A549Luc1 cell line indicating IL-8 production; ESR in presence of alveolar macrophages) and one *in vivo* assay in rats for determining a pulmonary inflammatory response (lung neutrophil response after intratracheal instillation of the 8 NPs). The most important aspect of these studies was to express and compare responses in each of these assays based on the concept of selecting the maximum response per unit dose (expressed as NP surface). We found that responses in all non-*in vivo* assays correlated significantly with *in vivo* results ($R \geq 0.76$), with the two cellular assays correlating best ($R \geq 0.88$). We conclude that this concept of maximum response per unit dose (response-metric) allows for a meaningful toxicity testing of NPs through the selection of an appropriate dose and that it has a high predictive value for *in vitro* to *in vivo* extrapolation. This concept provides also a novel scheme for toxicological characterization and classification of NPs by categorizing different NP classes based on the reactivity or effects per unit surface area of NPs.

Supported by Grants: DoD MURI (AFOSR) FA9550-04-1-0430; EHS Center (NIEHS) P30 ESO1247.

Inhaled Particles X, (23–25 September 2008, Manchester) IOP Publishing
Journal of Physics: Conference Series **151** (2009) 011002 doi:10.1088/1742-6596/151/1/011002

Application of a dynamic population-based model for evaluation of exposure reduction strategies in the baking industry

Tim Meijster[1,2], Nick Warren[3], Dick Heederik[2], Erik Tielemans[1]

[1] TNO Quality of Life, Business unit Quality and Safety, Zeist, The Netherlands

[2] Utrecht University, Institute of Risk Assessment Sciences, Division of Environmental Epidemiology, Utrecht, The Netherlands

[3] Health and Safety Laboratory, Harpur Hill, Buxton, Derbyshire, United Kingdom

Tim.meijster@tno.nl

Abstract. Recently a dynamic population model was developed that simulates a population of bakery workers longitudinally through time and tracks the development of work-related sensitisation and respiratory symptoms in each worker. Input for this model comes from cross-sectional and longitudinal epidemiological studies which allowed estimation of exposure response relationships and disease transition probabilities This model allows us to study the development of diseases and transitions between disease states over time in relation to determinants of disease including flour dust and/or allergen exposure. Furthermore it enables more realistic modelling of the health impact of different intervention strategies at the workplace (e.g. changes in exposure may take several years to impact on ill-health and often occur as a gradual trend). A large dataset of individual full-shift exposure measurements and real-time exposure measurements were used to obtain detailed insight into the effectiveness of control measures and other determinants of exposure. Given this information a population wide reduction of the median exposure with 50% was evaluated in this paper.

Journal of Physics: Conference Series **151** (2009) 011002 doi:10.1088/1742-6596/151/1/011002

GIS-based models for ambient PM exposure and health impact assessment for the UK

John R Stedman, Susannah Grice, Andrew Kent and Sally Cooke

AEA Energy & Environment, The Gemini Building, Fermi Avenue, Harwell

Didcot, Oxfordshire, OX11 0QR

John.stedman@aeat.co.uk

Abstract. GIS-based models have been developed to map ambient PM_{10} and $PM_{2.5}$ mass concentrations across the UK. The resulting maps are used for the assessments of air quality required by the EU ambient air quality directives, health impact assessment and the development of UK air quality policy. Maps are presented for 2006 along with projections to 2020. The largest single contribution to the UK population-weighted mean annual mean background concentrations of PM_{10} in 2006 is estimated to be from secondary PM (43%), followed by the contribution from primary PM (24%). Concentrations are predicted to decline by 15% for PM_{10} and 13% for $PM_{2.5}$ over the period from 2006 to 2020. The extent of exceedence of the 24-hour limit value is predicted to decline from 1.9% to 0.1% of urban major roads over the same period. The potential health benefits of reductions in ambient PM are large. A reduction in concentration of 0.93 $\mu g\ m^{-3}$ as a result of a possible package of measures has been estimated within the UK Air Quality Strategy to result in a reduction in life years lost of approximately 2 – 4 million over a period of 100 years.

Inhaled Particles X, (23–25 September 2008, Manchester) IOP Publishing
Journal of Physics: Conference Series **151** (2009) 011002 doi:10.1088/1742-6596/151/1/011002

Aerosol exposure: concepts, criteria, standards and applications

James H. Vincent

Department of Environmental Health Sciences, School of Public Health, University of Michigan, Ann Arbor, MI 48109, U.S.A.

jhv@umich.edu

Abstract. This paper places *Inhaled Particles X* in the context of the whole sequence of such symposia, going back to the first one in 1961. It draws together some of the essential principles that have been learned since that earlier meeting about the nature of exposure and exposure assessment and thus provides a framework by which to integrate the new knowledge presented at this latest one. In the process, the importance of understanding the formal definition of aerosol exposure is stressed, including the distinction between exposure intensity and exposure history, and how that relates to some measure of cumulative dose which, in turn, may be linked with knowledge about intrinsic toxicity, etc. This then leads to a definition of exposure standards, and the important ingredients of criteria, sampling and limit values. A summary is provided of the current set of particle size-selective criteria that have been widely agreed in the international occupational and environmental health community. Some ideas are presented about how this set might be expanded for certain applications, the important case of ultrafine aerosols being one of them.

Journal of Physics: Conference Series **151** (2009) 011002 doi:10.1088/1742-6596/151/1/011002

MATCH Project – The case of personal exposure to atmospheric polycyclic aromatic hydrocarbons

Noel Aquilina, Juana Maria Delgado Saborit, Steve Baker, Claire Meddings, Adrienne Wynn, Stuart Harrad and Roy M. Harrison

Division of Environmental Health and Risk Management, School of Geography, Earth and Environmental Sciences, University of Birmingham, Birmingham, B15 2TT

Abstract. The project MATCH (**M**easurement and Modelling of Exposure to **A**ir **T**oxic **C**oncentrations for **H**ealth Effect Studies) aims to quantify the levels and range of individual personal exposures to air toxics, namely a group of 16 particle-phase polycyclic aromatic hydrocarbons (PAH) and a group of 15 gas-phase volatile organic compounds (VOC). This work comprises results from a personal exposure (PE) sampling campaign carried out in London, Birmingham and Wales, UK, during the period 2005-2007. A group of 100, adult and healthy non-smokers were chosen. To characterize better the exposure patterns of this group to PAH, home microenvironment (ME) sampling was carried out together with other microenvironments to reflect as much as possible the various lifestyles of the group. Although the principal sources of PAHs are traffic emissions and other combustion processes, less is known about indoor exposure, especially to environmental tobacco smoke (ETS) in homes and other environments like pubs and restaurants. From experiments with ETS in a controlled environment it has been shown that semi-volatile PAH are persistent in air for more than 2 hours. On the other hand high molecular weight particle-associated PAH show high concentrations only within the first hour, and correlate well with the VOC ETS marker 3-ethenylpyridine. In England, on 1st July 2007 a smoking ban was enforced in all public places. Sampling of PAH, 1,3-butadiene and particulate matter concentrations of size ranges $PM_{1.0}$, $PM_{2.5}$ and $PM_{7.0}$ was carried out in a typical pub microenvironment, before and after the ban. Reductions in concentrations for all the compounds measured were observed immediately after the ban. Results of this study will be presented

Journal of Physics: Conference Series **151** (2009) 011002 doi:10.1088/1742-6596/151/1/011002

Occupational Exposure to Polycyclic Aromatic Hydrocarbons in Wood Dust

C K Huynh, P Schüpfer and P Boiteux

Institute for Work and Health, rue du Bugnon 21, CH-1005 Lausanne, Switzerland

chuynh@hospvd.ch

Abstract. Sino-nasal cancer (SNC) represents approximately 3% of Oto-Rhino-Laryngology (ORL) cancers. Adenocarcinoma SNC is an acknowledged occupational disease affecting certain specialized workers such as joiners and cabinetmakers. The high proportion of woodworkers contracting a SNC, subjected to an estimated risk 50 to 100 times higher than that affecting the general population, has suggested various study paths to possible causes such as tannin in hardwood, formaldehyde in plywood and benzo(a)pyrene produced by wood when overheated by cutting tools. It is acknowledged that tannin does not cause cancer to workers exposed to tea dust. Apart from being an irritant, formaldehyde is also classified as carcinogenic. The path involving carcinogenic Polycyclic Aromatic Hydrocarbons (PAHs) emitted by overheated wood is attractive. In this study, we measured the particle size and PAHs content in dust emitted by the processing of wood in an experimental chamber, and in field situation. Quantification of 16 PAHs is carried out by capillary GC-ion trap Mass Spectrometric analysis (GC-MS). The materials tested are rough fir tree, oak, impregnated polyurethane (PU) oak. The wood dust contains carcinogenic PAHs at the level of $\mu g.g^{-1}$ or ppm. During sanding operations, the PU varnish-impregnated wood produces 100 times more PAHs in dust than the unfinished wood.

Inhaled Particles X, (23–25 September 2008, Manchester) IOP Publishing
Journal of Physics: Conference Series **151** (2009) 011002 doi:10.1088/1742-6596/151/1/011002

Exposure to respirable crystalline silica in South African farm workers

Andrew Swanepoel[1,2], David Rees[1,2], Kevin Renton[2], Hans Kromhout[3]

[1] University of the Witwatersrand, School of Public Health, Johannesburg, South Africa.

[2] National Institute for Occupational Health, Johannesburg, South Africa.

[3] Environmental Epidemiology Division, Institute for Risk Assessment Sciences, University of Utrecht.

Abstract Although listed in some publications as an activity associated with silica (quartz) exposure, agriculture is not widely recognized as an industry with a potential for silica associated diseases. Because so many people work in agriculture; and because silica exposure and silicosis are associated with serious diseases such as tuberculosis (TB), particular in those immunological compromised by the Human immunodeficiency virus (HIV), silica exposure in agriculture is potentially very important. But in South Africa (SA) very little is known about silica exposure in this industry. The objectives of this project are: (a) to measure inhalable and respirable dust and its quartz content on two typical sandy soil farms in the Free State province of SA for all major tasks done on the farms; and (b) to characterise the mineralogy soil type of these farms. Two typical farms in the sandy soil region of the Free State province were studied. The potential health effects faced by these farm workers from exposure to respirable crystalline silica are discussed.

Journal of Physics: Conference Series **151** (2009) 011002 doi:10.1088/1742-6596/151/1/011002

Indoor concentrations of fine particles and particle-bound PAHs in Gothenburg, Sweden

S Johannesson[1], K Bergemalm-Rynell[1], B Strandberg[1], G Sällsten[1]

[1] Department of Occupational and Environmental Medicine, Sahlgrenska University Hospital and Academy, Gothenburg, Sweden

E-mail: sandra.johannesson@amm.gu.se

Abstract. Fine particles are formed in a variety of processes, both natural and anthropogenic. Epidemiological studies have shown an association between exposure to particulate matter and adverse health effects. Airborne particles contain a variety of compounds, including polycyclic aromatic hydrocarbons (PAHs), and several of the PAHs are known or suspected carcinogens. In this study, stationary measurements of $PM_{2.5}$ were performed in the residences of 20 study participants along with simultaneous monitoring at an urban background site. The collected particle mass was then analyzed for its content of some particle-bound PAHs using GC-MS. The median level of $PM_{2.5}$ indoors was 7.3 $\mu g/m^3$ and in urban background 5.3 $\mu g/m^3$. For benzo(a)pyrene (B(a)P) the corresponding results were 10 pg/m^3 and 35 pg/m^3, respectively. There were significant correlations between indoor and ambient levels for both $PM_{2.5}$ (r_s=0.58, p=0.02) and B(a)P (r_s=0.67, p=0.007). No significant correlation was, however, found between the concentration of $PM_{2.5}$ and the associated levels of the investigated PAH compounds. This finding implies that exposure to B(a)P or other particle-bound PAH components needs to be separately assessed.

Inhaled Particles X, (23–25 September 2008, Manchester) IOP Publishing
Journal of Physics: Conference Series **151** (2009) 011002 doi:10.1088/1742-6596/151/1/011002

Mesothelioma in Qingdao, PRC (2000 – 2007)

Arthur L. Frank [1], Pang Zengchang [2], Zhang Huaqiang [3], Zhang Yun [2]

[1] Drexel University School of Public Health, Philadelphia, PA 19102

[2] Qingdao Center for Disease Control and Prevention, Qingdao, PRC

[3] Qingdao Institute of Women's and Children's Health Care, Qingdao, PRC

E-mail: alf13@drexel.edu

Abstract. The city of Qingdao, PRC has been the site of two asbestos product facilities that operated for almost fifty years, as well as a shipyard. Because of a new computerized data collection system for death certificates, almost all 48,000 yearly deaths from a population base of 7.5 million are now recorded with cause of death and "usual occupation". All mesothelioma deaths from 2000 through 2007 are reviewed and the unusual finding is that of a predominance of cases in females. The issues of competing causes of death and potential underreporting are discussed.

Heterogeneity of exposure and attribution of mesothelioma: Trends and strategies in two American counties.

B W Case[1,2] **and J L Abraham**[3]

[1] Department of Pathology and School of Environment, McGill University, Montréal, Québec, Canada.

[2] Professeur invité, Epidemiology & Biostatistics Unit, INRS-Institut Armand-Frappier, Université du Québec, Laval, Québec, Canada

[3] Department of Pathology, SUNY Upstate Medical University, Syracuse, NY 13210, USA.

bruce.case@mcgill.ca

Abstract. As mesothelioma risk has begun to decline in the United States, two trends are gaining relative importance. "Legacy" exposures causing this disease are most important in locales having past asbestos industry, shipyards, and/or local distribution of asbestos amphibole-containing material as a result. "Future" exposures are of particular concern in relation to so-called "naturally occurring asbestos" (NOA) areas which include unequivocally asbestiform amphibole. In this paper, Jefferson Parish, Louisiana is used as an example of the first trend, and El Dorado County, California as an example of the second. Available tumor registry, epidemiology, historical and mineralogical data, and lung-retained fibre content are used as indicators of disease and exposure. Jefferson Parish, LA was chosen as the prototype of "legacy" exposures on the basis of historical evidence of asbestos plants with known mesotheliomas in the workforce, known shipyards in the same area, EPA records of distribution of crocidolite-containing scrap to and remediation of over 1400 properties, NIOSH published data on mesothelioma by county, and exposure data including lung-retained fibre analyses in victims, where available. El Dorado, CA was chosen as the prototype of NOA amphibole exposures on the basis of tumor registry data, activity-based EPA sampling data in one area, and lung-retained fibre analyses in area pets, and future risk assessment based on tremolite-specific modelling in Libby, Montana and elsewhere. As expected, the legacy exposure area was high in mesothelioma incidence and mortality. Lung-retained fibre content confirms crocidolite exposures in exposed plant-workers and those exposed to crocidolite-containing scrap, and amosite in shipyard workers. In contrast, to date, cancer registry data in the NOA-amphibole ("future") county does not show a clear increase in incidence or mortality, but grouped county data from the area show a shift in higher incidence rates to the NOA areas and away from California "legacy" (e.g., shipyard) areas from 1988-2005. EPA active sampling has confirmed excess tremolite/ actinolite fibre(s) in air, although there is debate about its nature and the appropriateness of the area sampled. Lung-retained fibre in local pets shows unequivocally elevated asbestiform tremolite/ actinolite in areas thought to be most affected, but numbers are small. Future risk is expected to rise due to a vastly increased population base coupled with exposures potentially created by related construction activities. Although legacy exposures are producing smaller numbers of cases with time, they continue to occur at high rates, and new sources of legacy exposure are being discovered in highly localized "hotspots". Differential exposure sources remain a problem in attribution, but continued remediation seems the best strategy for prevention. In the "future" risk county and surrounding areas, incidence trends are less clear, but again highly localized exposures as opposed to broad areas seem important. Activity-based air sampling; targeted soil samples, and lung-retained fibre analyses may be useful in defining areas of highest future risk and potential prevention.

Trends in pneumoconiosis and other lung diseases, as reported to a UK-based surveillance scheme for work-related ill-health

S Turner[1], R McNamee[2], M Carder[1], R Agius[1]

[1]Occupational and Environmental Health Research Group, University of Manchester, UK

[2]Biostatistics, Health Methodology Research Group, University of Manchester, UK

E-mail: susan.m.turner@manchester.ac.uk

Abstract: The changing nature of industries associated with exposure to hazardous dusts in manufacture or in use, as well as better control methods, might be expected to be associated with a reduction in incidence of pneumoconiosis and other lung diseases. Data collected by the University of Manchester's ODIN/THOR network on work-related ill-health in the UK (as diagnosed by specialist physicians) can be used to estimate time trends in the lung diseases reported to the surveillance schemes. Reporters of work-related lung diseases in THOR (previously ODIN) mainly comprise two groups, namely clinical specialists in respiratory medicine and occupational physicians. These reporters return information on work-related cases of ill-health using postal reporting cards or an on-line web form. 'Report cards' are returned even if no new cases are seen, with responses recorded each month i.e. whether a card is returned and number of cases returned. Probabilities of a non response and, for returned cards, of a 'zero' return were modelled as a function of calendar time and/or membership time using 2-level logistic models. Annual change in disease incidence (all work-related respiratory disease and specific diagnoses) was estimated using 2-level Poisson models controlling for reporter characteristics, season, and whether or not a first report. The impact of membership time on reporting was also investigated. Case reports include information on patient demographics, diagnoses, industry, occupation, and suspected agents/exposures. These case details are coded and analysed using SPSS. Annual change in incidence of all work-related respiratory disease reported by specialist chest physicians (1999-2006) was -1.7% (95% CI: -3.1%, -0.2%). Specific diagnoses reported by chest physicians showed that the annual change in incidence for asthma was -3.1% (95% CI: -5.8%, -0.4%), for mesothelioma was -4.1% (95% CI: -6.7%, -1.5%), for benign pleural plaques was +1.1 (95% CI; -1.0%, +3.2%), and for pneumoconiosis was -2.6 (95% CI: -6.6, +1.5) over the same time period. Occupational physicians' reporting showed a change in incidence of -6.1% (95% CI: -11.6%, -0.4%) for all respiratory disease, and -8.4% (95% CI: -15.3%, -0.9%) for asthma. Given variation between reporter groups, and according to model assumptions, time trends from surveillance data need to be interpreted with caution, but may have some place in planning interventions aimed at improving the health of a workforce. Further work to investigate case details (such as suspected agent/exposures) should also add to this knowledge base.

Journal of Physics: Conference Series **151** (2009) 011002 doi:10.1088/1742-6596/151/1/011002

Respirable crystalline silica – a failure to control exposure!

J R Cain MSc MBA MFOH CBiol FIBiol Dip Occ Hyg

HM Regional Specialist Inspector (Occupational Hygiene), Health and Safety Executive, Marshalls Mill, Marshall Street, Leeds LS11 9YJ

John.cain@hse.gsi.gov.uk

Abstract: Several sites were visited to monitor stonemason exposure to respirable crystalline silica (RCS), inhalable dust and respirable dust. At all sites, exposure to RCS exceeded the Workplace Exposure Limit of 0.1 mg/m^3 8-hour TWA. There was therefore a continuing high risk of workers developing silicosis unless the appropriate measures were instigated to prevent or control exposure. Exposure control was ineffective at all sites e.g. water wall extraction systems were not well designed. There was evidence that foreign workers were at a greater exposure risk. But even with appropriate controls to mitigate exposure to RCS it may not be possible to sustain exposure to below 0.1 mg/m^3 8-hour TWA without on-going HSE intervention.

Inhaled Particles X, (23–25 September 2008, Manchester) IOP Publishing
Journal of Physics: Conference Series **151** (2009) 011002 doi:10.1088/1742-6596/151/1/011002

Rat- and Human-based Risk Estimates of Lung Cancer from Occupational Exposure to Poorly-Soluble Particles: A Quantitative Evaluation

E D Kuempel,[1] R J Smith,[1] D A Dankovic,[1] LT Stayner[2]

[1]National Institute for Occupational Safety and Health, Education and Information Division, Risk Evaluation Branch, Cincinnati, Ohio, USA

[2] University of Chicago School of Public Health, Division of Epidemiology and Biostatistics, Chicago, Illinois, USA

ekuempel@cdc.gov

Abstract. In risk assessment there is a need for quantitative evaluation of the capability of animal models to predict disease risks in humans. In this paper, we compare the rat- and human-based excess risk estimates for lung cancer from working lifetime exposures to inhaled poorly-soluble particles. The particles evaluated include those for which long-term dose-response data are available in both species, i.e., coal dust, carbon black, titanium dioxide, silica, and diesel exhaust particulate. The excess risk estimates derived from the rat data were generally lower than those derived from the human studies, and none of the rat- and human-based risk estimates were significantly different (all p-values$\geq$0.05). Residual uncertainty in whether the rat-based risk estimates would over- or under-predict the true excess risks of lung cancer from inhaled poorly-soluble particles in humans is due in part to the low power of the available human studies, limited particle size exposure data for humans, and ambiguity about the best animal models and extrapolation methods.

Journal of Physics: Conference Series **151** (2009) 011002 doi:10.1088/1742-6596/151/1/011002

A case cross-over analysis of acute coronary syndrome and short term exposure to ambient air pollution

G Bero Bedada[1], M Carder[1], A Hirst[1], J P New[2], J M Gibson[2], R Agius[1]

1 Occupational & Environmental Health Research Group, School of Translational Medicine, Faculty of Medical and Human Sciences, University of Manchester, UK

2 Department of Diabetes and Endocrinology, Hope Hospital, Salford, UK

Abstract: Epidemiologic research suggests that short term exposure to ambient particulate and gaseous pollution contributes to pulmonary and cardiovascular disease, including ischaemic cardiac events. This study was undertaken to assess the effects of short-term exposure to ambient particles and gases on myocardial damage as evidenced by cardiac troponin T levels in patients admitted to hospital for acute coronary syndrome (ACS). A case-crossover study was conducted to study the association between short term changes in particulate and gaseous pollutant concentrations and ischaemic cardiac events. Daily emergency hospital admission data for the period from July 2000 to December 2006 for actual or suspected ischaemic events, and measured cardiac troponin T (TnT) levels were obtained from Hope Hospital, Salford. Ambient PM_{10} and gaseous pollutant (SO_2, CO, Ozone, NO_2 and NO) concentration measurements were obtained from monitoring stations in Greater Manchester. There were 28,687 admissions out of which 1004 (3.5%) were for ACS with clinical myocardial infarction (TnT $\geq$ 1ng/dl). The majority of the TnT measurements (79%) were below the detection level. Both particles and gaseous pollutants were associated with admission for ACS after adjustment was made for meteorological confounders and long term trend. Using individual monitors, the highest estimated odds ratio associated with an interquartile increase in NO_2 was 1.18 (95% CI: 1.02 -1.37). Similarly, an interquartile increment in PM_{10} was associated with a 17% increase hospital admission for ACS (95% CI: 1.04-1.31). Most consistent results were observed in lags of 1 day, only PM_{10} and CO showed a delayed effect at lag of 4 days. In contrast, no associations between hospitalization for ACS and ozone and SO_2 were found. The results of this study present evidences of the deleterious role of short term exposure to PM_{10}, CO, NO and NO_2 on the myocardium as shown by elevated troponin T levels in patients admitted for ACS with clinical myocardial infarction.

Journal of Physics: Conference Series **151** (2009) 011002 doi:10.1088/1742-6596/151/1/011002

Occupational characteristics of respiratory cancer patients exposed to asbestos in Lithuania

R Petrauskaitė Everatt[1,3], G Smolianskienė[1,5], A Tossavainen[2], S Cicėnas[3,4] and R Jankauskas[1]

[1] Institute of Hygiene, Etmonu 3/6, LT-01129 Vilnius, Lithuania

[2] Finnish Institute of Occupational Health, Topeliuksenkatu 41a A, FI-00250 Helsinki, Finland

[3] Institute of Oncology, Vilnius University, Santariškių 1, LT-08660 Vilnius, Lithuania

[4] Institute of Rehabilitation, Sport Medicine and Nursing, Antakalnio 57, LT-10207, Vilnius, Lithuania

E-mail: grazina.smolianskiene@dmc.lt

Key words: asbestos; exposure assessment.

Abstract. Objective: To assess characteristics of asbestos exposure in respiratory cancer patients in Lithuania. **Methods.** Information on occupational exposure to asbestos was collected by personal interviews and occupational characteristics were evaluated among 183 lung cancer and mesothelioma patients with cumulative asbestos exposure ≥0.01 fibre years hospitalized at the Institute of Oncology, Vilnius. Additionally, some results of workplace air measurements were reviewed. **Results.** Cases with estimated cumulative exposure ≥5 fibre years had worked mainly in the construction industry (49%), installation and maintenance (13%), foundry and metal products manufacturing (6%), heating trades and boilerhouses (6%) as fitters/maintenance technicians, construction workers, welders, electricians or foremen. Typical asbestos materials used by the patients were asbestos powder, asbestos cement sheets and pipes, asbestos cord, brake and clutch linings. Patients were exposed to asbestos when insulating boilers, furnaces, pipes in power stations, industrial facilities, ships, locomotives, buildings, while covering and repairing roofs, at the asbestos cement plant or unloading asbestos products. Most patients with estimated cumulative exposure of ≥0.01-4.9 fibre years worked as lorry, bus or tractor drivers and motor vehicle mechanics. In 2002-2007 workplace air asbestos concentrations exceeded the limit value of 0.1 f/cm^3 in 11 samples out of 208 measurements. **Conclusion.** The results of this study indicate that since the 1960s occupational exposure to chrysotile asbestos was extensive in Lithuania.

Inhaled Particles X, (23–25 September 2008, Manchester) IOP Publishing
Journal of Physics: Conference Series **151** (2009) 011002 doi:10.1088/1742-6596/151/1/011002

Investigating the reliability and validity of diagnosing occupational asthma (OA) by reporters within The Health and Occupation Reporting (THOR) network

Turner S[1], McNamee R[2], Carder M[1], Agius R[1], Bradshaw L[3], Curran A[3], Francis M[3], Fishwick D[3]

[1] Occupational and Environmental Health Research Group, University of Manchester, UK

[2] Biostatistics Group, University of Manchester, UK

[3] Centre for Workplace Health, Health and Safety Laboratory, Buxton, UK

mailto:susan.m.turner@manchester.ac.uk

Abstract: THOR is a UK-wide voluntary surveillance scheme receiving case reports of occupational ill-health from physicians. Cases are reported if the clinician believes that the condition was caused by occupational factors, however the degree of reliability and validity that THOR reporters assign to a diagnosis of OA has not been studied. This study aimed to investigate the reliability and validity of assigning a diagnosis of OA by occupational and respiratory physicians reporting to THOR. In phase 1 of this study, 19 "possible" case histories of OA were sent to 51 occupational physicians (OPs) and 53 respiratory physicians (RPs) who were active participants in THOR. Each physician was asked to determine the certainty (0-100%) that (s)/he would assign to an OA diagnosis for 4 case histories. For phase 2, each physician was sent results of investigations for 2 of the 4 cases, and again asked to determine the certainty that (s)/he would assign to an OA diagnosis for these 2 cases. Data were analysed using SPSS v 14.0. Kappa coefficients (yes/no classifications) and interclass correlation coefficients (continuous scores) were calculated as statistical measures of reliability for OA diagnosis; a score of 0 = no better than chance and 1 = perfect agreement between raters. Comparisons were made using tests based on multilevel models, taking into account correlations between ratings of the same case. The participation rate was 87/104 (83.7%). The range of probabilities of OA assigned by physicians was wide, ranging from 0-100%. Mean overall scores from phase 1 were 52.1% for OPs and 50.0% for RPs; the difference (OPs' mean – RPs' mean) not being statistically significant (difference 2.1%; 95% CI: -2.6, 6.8 p=0.37). In phase 2, mean overall scores for OPs and RPs were 46.1% and 41.5% respectively, the difference again not being statistically significant (difference 4.6%; 95%CI: -3.5, 12.5 p=0.27). Further analysis is in progress to investigate the observation that, using case history information alone, OPs were more likely to decide on an OA diagnosis than RPs, and that there was a higher level of agreement (reliability) among OPs than RPs. In phase 2 of the study, RPs appear more likely to assign ratings at the extremes of the range, and the reliability for OPs' and RPs' ratings was similar. Patterns of reporting to THOR may differ within and between clinical specialties, perhaps caused by variations in physicians' case loads/mixes. However, such differences are not necessarily reflected in reliability or validity of OA diagnosis.

Inhaled Particles X, (23–25 September 2008, Manchester) IOP Publishing
Journal of Physics: Conference Series **151** (2009) 011002 doi:10.1088/1742-6596/151/1/011002

Orthotic dust: An occupational hazard for podiatrists?

V Gunn[1], N A McLarnon[1], J G Burrow[1] and M Hepher[2]

[1]The Division of Podiatric Medicine and Surgery, School of Health and Social Care, Glasgow Caledonian University, 70 Cowcaddens Road, Glasgow G4 0BA

[2] School of the Built and Natural Environment, Glasgow Caledonian University, 70 Cowcaddens Road, Glasgow G4 0BA

N.A.McLarnon@gcal.ac.uk

Abstract. This was a unique study, the aim of which was to determine whether podiatrists are at risk from exposure to dust produced during orthotic/insole manufacture. To date, analysis of dust produced during podiatric procedures has focussed mainly upon the reduction of thickened (onychauxic/onychogryphotic) toenails. Two common orthotic manufacturing materials: Blue Ethyl Vinyl Acetate (EVA) (Algeos Ltd, UK) and Northene (North Sea Plastics, Glasgow) were bevelled using both coarse and fine sanding discs and analysed for the presence of inhalable and respirable dust particles. Current and previous literature identified that exposure to inhalable and respirable dust particles may lead to a number of health problems. Therefore, it is important to determine the nature of dust produced during orthotic manufacture and the potential associated health risks for podiatric practitioners, orthotic technicians or Prosthetists and Orthotists. The results obtained establish the presence of inhalable and respirable dust from both orthotic materials sampled. Additionally, the fine sanding of Northene produced the greatest amount of dust, whereas coarse reduction of blue EVA produced the least amount of dust overall. EVA on a coarse sand produced a mean of 0.514 mg/m3, EVA on a fine sand created 0.700 mg/m3. Northene on a coarse reduction produced 0.8164 mg/m3 and finally Northene on a fine sanding produced the greatest amount of dust overall at 0.8256 mg/m3. All samples were bevelled/sanded for exactly four minutes. Dust found to be 4 µm or smaller is classed as respirable dust with an ability to penetrate deep into the lungs. Northene dust on a coarse bevel was analysed using electron microscopy and was found to produce dust as small as 2.274 µm in length and 926 nm in width. This evidence suggests that respirable Northene dust produces minute sized dust particles allowing it to penetrate even deeper into the lung tissue. The results of this study demonstrate that the grinding of orthotic materials such as Blue EVA and Northene may present an occupational hazard to podiatrists through the production of dust particles that may become lodged in the respiratory tract or be deposited on the skin or eyes and that risk reduction strategies need to be employed by practitioners using these materials to ensure their health and safety.

Medical Geology: Dust Exposure and Potential Health Risks in the Middle East

M B Lyles[1], H L Fredrickson[2], A J Bednar[3], H B Fannin[4], D Griffin[5] and T M Sobecki[6]

1. US Navy Bureau of Medicine & Surgery, 2300 E Street, NW; Washington, DC 20372.

2. US Environmental Protection Agency, 26 West Martin Luther King Blvd, Cincinnati, OH 45268.

3. US Army Corp of Engineers, 3909 Halls Ferry Road, Vicksburg, MS 39180-6199.

4. Murray State University, Department of Chemistry, Murray KY 42071.

5. US Geological Survey, 2010 Levy Avenue, Suite 100 Tallahassee, FL 32310.

6. US Army Corp of Engineers, 72 Lyme Road, Hanover, NH 03755

mark.lyles@med.navy.mil

Abstract. In the Middle East, dust & sand storms are a persistent problem & can deliver significant amounts of micro-particulates via inhalation into the mouth, nasal pharynx, & lungs due to the fine size and abundance of these micro-particulates. The chronic & acute health risks of this dust inhalation have not been well studied nor has the dust been effectively characterized as to its chemical composition, mineral content, or microbial flora. Scientific experiments were designed to study the Kuwaiti & Iraqi dust as to its physical, chemical, & biological characteristics for its potential to cause adverse health effects. First, dust samples from different locations were collected & processed & exposure data collected. Initial chemical & physical characterization of each sample including particle size distribution & inorganic analysis was conducted, followed by characterization of biologic flora of the dust, including bacteria, fungi and viruses. Initial data indicates that the mineralized dust is composed of $CaCO_3$ in a coating over a matrix of metallic silicate crystals containing a variety of trace metals constituting ~3 % of the PM10 by weight, of which ~1% is bioavailable Al & Fe each. Microbial analysis reveals a significant biodiversity of bacterial, fungi, and viruses of which ~25 % are known pathogens. The level of total suspended particle mass along with environmental & physiological conditions present constitute an excessive exposure to micro-particulates including PM 2.5 & the potential for long-term adverse health effects. These data suggest that the level of dust exposure coupled with the microbial & metal content could constitute a significant health risk. When taken with other existing work suggest that further research is warranted to provide insight into potential human health risks both acute and chronic.

Journal of Physics: Conference Series **151** (2009) 011002 doi:10.1088/1742-6596/151/1/011002

Development of an advanced exposure assessment tool: Advanced REACH Tool (ART)

E Tielemans[1], N Warren[2], T Schneider[3], M van Tongeren[4], M Tischer[5], P Ritchie[4], W Fransman[1], H Kromhout[6], J Schinkel[1], J Hemmen[1], J Cherrie[4]

1 TNO Quality of Life, The Netherlands

2 HSL, United Kingdom

3 NRCWE, Denmark

4 IOM, United Kingdom

5 BAuA, Germany

6 IRAS, Utrecht University, The Netherlands

Erik.Tielemans@tno.nl

Abstract. The REACH process requires a tiered exposure assessment to effectively cope with the broad range of exposure scenarios. Currently, a higher tier model generating realistic exposure estimates is missing. A new advanced exposure assessment tool is under development. The new framework incorporates both a mechanistic model and an empirical part with information from an exposure database. Both parts are to be combined using a Bayesian process in order to produce realistic exposure estimates. The approach facilitates the inclusion of any new data that become available in the future or during the risk assessment process. The proposal consists of five strongly interrelated Work Packages: 1) Development of mechanistic model, 2) Bayesian framework, 3) Database development, 4) Software development, 5) Testing and validation. The advanced REACH tool will be completed by 2010 and will be made freely available to stakeholders, probably as web-based tool.

Development of an Electrostatic Aerosol In Vitro Exposure System (EAVES 2) for Inhaled Particle Toxicology

John Volckens[1], Glen Walters[2], Lisa Dailey[3], and Robert Devlin[3]

1 Colorado State University, Fort Collins CO USA

2 University of North Carolina, Chapel Hill NC USA

3 U.S. EPA, Chapel Hill NC USA

john.volckens@colostate.edu

Abstract. The objective of this work is to develop a more physiologically and environmentally relevant model of air pollutant deposition to the human lung in vitro¬. Our central hypothesis is that the toxicology of inhaled particulate matter depends strongly on both the particulate dispersion state and the mode of delivery to cells. Our in vitro cell model employs a combination of unipolar charging and electrostatic force to deposit particles onto cells grown at an air-liquid interface in a heated, humidified exposure chamber. In this manner, we may sustain exposures from periods of minutes to hours. We show that cells exposed in this manner to concentrated, coarse ambient particulate matter express increased levels of inflammatory markers (mRNA) at both 1 and 24 hours post exposure and relative to controls exposed to particle-free air. More importantly, these effects are seen at doses (characterized by deposited mass per tissue area) that are an order of magnitude lower than doses applied using traditional in vitro methods (i.e., delivery of filter-extracted PM via syringe spike). Particle collection efficiency has been characterized as a function of charging current (1 – 10 μA), electric field strength (0.5 – 5 kV/cm), and flowrate (1 – 5 L/min). The device is portable and may be transported to workplace and field locations. Model application, advantages, and limitations are also discussed.

Journal of Physics: Conference Series **151** (2009) 011002	doi:10.1088/1742-6596/151/1/011002

Lung Deposition Sampler for Inhalable Aerosol

Philip Clark, Kirsten Koehler, and John Volckens

Colorado State University, Fort Collins CO USA

john.volckens@colostate.edu

Abstract. The goal of this research is to develop a more powerful exposure assessment tool that directly measures the deposition of inhaled particles within the human lung. Inhaled particles do not always deposit in the lung. Therefore, a device that estimates the size-specific deposition of inhaled particles will improve the power of risk assessment by providing a more physiologically-relevant measure of dose, and hence, inhalation hazard in the workplace. We have developed a model to predict particle penetration through porous foam media as a function of foam characteristics, particle size, and flowrate. This semi-empirical model, originally developed by Vincent et al. and later augmented by Kenny et al., utilizes the three particulate collection mechanisms: diffusion, inertia, and settling. The current model was developed from a non-linear regression of particle-foam deposition efficiency data generated in the laboratory. Four foam types of differing thickness (5, 10, 20 mm) and porosity (45, 60, 80, 100 PPI) were tested, along with both solid (NaCl) and liquid (oil) particles ranging in size from 0.02 to 10 μm. Inversion of the model indicates that a foam sampler may be developed with deposition characteristics that approximate the ICRP lung-deposition curve. We will discuss model development, uncertainty, limitations, and application.

Human *in vitro* method for testing the toxicity of manufactured nanoparticles.

Jane Green[1], Angela Curtis[1], Minnamari Vippola[2], Gareth S Evans[1] & Rosemary M Gibson[1]

1 Centre for Interdisciplinary NanoResearch at the Health & Safety Laboratory, Harpur Hill, Buxton, SK17 9JN;

2 Finnish Institute for Occupational Health, Helsinki, Finland.

mailto:Rosemary.Gibson@hsl.gov.uk

Abstract. Nanotechnology has the potential to bring enormous benefits to many aspects of society. Nanoparticles may however present potential health and safety hazards to workers, and the special properties conferred by nanoscale may alter the behaviour and toxicity compared to larger forms of the same material. A cost effective "first stage" screen for potential human health effects based on *in vitro* toxicity assays has been developed at the Centre for Interdisciplinary NanoResearch (CiNR) at the Health & Safety Laboratory (HSL). This screen uses human skin and respiratory cell lines and *ex vivo* human white blood cells. The assays are being used to evaluate cell activation, inflammation and cell death in response to titanium dioxide nanoparticles. Whilst many *in vitro* tests focus on a single cell type, this assay investigates the effects of nanoparticles on a cell mixture, and the responses reflect interactions between different cells as well as those of individual cells. Titanium dioxide nanoparticles (anatase: 25 nm; rutile: 10x40 nm, Sigma-Aldrich) were examined by transmission electron microscopic (EM) and energy dispersive x-ray (EDX) analyses. EDX revealed that the rutile nanoparticles (not the anatase) were coated with silicon dioxide. The cytotoxicity of the nanoparticles on A549 human respiratory cells and HaCaT human skin cells was analysed by staining treated cells with the fluorescent dyes Hoescht 33342 and propidium iodide after incubation for 24 hours. Over a concentration range of 0.003-3 mg/ml, both nanoparticles induced dose-dependent cell death (up to 10% and 20% necrosis with the highest doses tested on A549 and HaCaT cells respectively). In *ex vivo* human white blood cells, the rutile TiO_2 nanoparticles induced slightly more necrosis than the anatase nanoparticles. The rutile TiO_2 nanoparticles also dose-dependently activated the white blood cells to a considerably greater extent than the anatase form, as indicated by myeloperoxidase release from the cells after incubation for 4 hours. These results suggest that different nanoparticle forms of titanium dioxide may have different effects on human respiratory, skin and white blood cells, although the effects of the coating of the nanoparticles require further study. Furthermore, the results suggest that this screening assay may represent a sensitive approach for analysis of the potential human health effects of nanoparticles.

Journal of Physics: Conference Series **151** (2009) 011002
doi:10.1088/1742-6596/151/1/011002

Tungsten Oxide Fiber Dissolution and Persistence in Artificial Human Lung Fluids

A B Stefaniak and M Chirila

National Institute for Occupational Safety and Health, Morgantown, WV, USA, 26505

E-mail: AStefaniak@cdc.gov

Abstract. Tungsten is a dense metal that is used in a range of industrial applications, including non-sag wire for light bulb filaments. During the conversion of tungsten oxide powder into tungsten metal powder for use as filaments, aerosols may be generated which contain tungsten sub-oxide particles having fiber morphology. To evaluate whether these fibers pose a yet unrecognized inhalation hazard due in part to their biodurability, we characterized the physicochemical properties and measured relative dissolution of fiber-containing ($WO_{2.81}$, $WO_{2.66}$, $WO_{2.51}$) and isometric-shaped ($WO_{3.00}$, $WO_{2.98}$) powders in artificial lung fluids. Raman spectroscopy results present a shift in the main frequencies for tungsten oxide samples that were sonicated in surfactant, confirming a decrease in the size of the crystalline domains by de-agglomeration. Geometric mean fiber aspect ratios were 8.3 ($WO_{2.81}$), 7.9 ($WO_{2.66}$), and 6.9 ($WO_{2.51}$). In artificial extracellular lung fluid, alkylbenzyldimethylammonium chloride (ABDC), added to prevent mold growth during experiments, inhibited ($p < 0.05$) dissolution of $WO_{2.98}$, $WO_{2.81}$, and $WO_{2.66}$. Less ($p < 0.05$) of the fibrous $WO_{2.66}$ and $WO_{2.51}$ dissolved relative to W metal; however, biodurability was only modestly greater than W metal. These data are useful for understanding the inhalation dosimetry of fibrous and non-fibrous forms of tungsten oxide materials.

Journal of Physics: Conference Series **151** (2009) 011002	doi:10.1088/1742-6596/151/1/011002

Investigations on the Effects of Short Term Diesel Exhaust Inhalation in Rat Brain

D van Berlo[1], C Albrecht[1], A M Knaapen[2*], F R Cassee[3], M E Gerlofs-Nijland[3], N Palomero-Gallagher[4,5], H J Bidmon[4], F J van Schooten[2], J Krutmann[1] and R P F Schins[1].

[1]Institut für Umweltmedizinische Forschung (IUF) at the Heinrich Heine University Duesseldorf, Germany;

[2]Nutrition and Toxicology Research Institute Maastricht (NUTRIM), Department of Health Risk Analysis and Toxicology, University of Maastricht, The Netherlands, *current address: Department of Toxicology and Drug Disposition, Schering-Plough, Oss, The Netherlands;

[3]National Institute for Public Health and the Environment (RIVM), Bilthoven, The Netherlands;

[4]C. & O. Vogt Institute for Brain Research, Heinrich-Heine-University, Germany;

[5]Institut fur Medizin, Forschungszentrum Julich, Julich, Germany.

roel.schins@uni-duesseldorf.de

Abstract. Combustion-derived nanoparticles (CDNP), such as diesel exhaust particles, have been implicated in the adverse health effects of particulate air pollution. Recent studies indicate that CDNP may translocate to the brain, either along the olfactory nerve upon nasal deposition, or via translocation from the alveoli into the blood and across the blood-brain barrier. However, the functional consequences of such translocation processes for the brain largely remain to be determined. The aim of our study was to investigate whether short-term diesel exhaust exposure elicits oxidative stress, inflammation and xenobiotic metabolizing enzyme responses in the rat brain. Male Fischer F344 rats were exposed for 2 hours (nose-only) to 1.9mg/m3 diluted diesel exhaust from the idling (1500rpm) engine type F3M2011 (Deutz Ag, Köln, Germany) with a primary particle size of 65 nm. Animals were sacrificed at post-exposure time intervals between 4 and 72 hours, subsequently brains were removed and dissected into six regions, i.e. pituitary gland, hypothalamus, olfactory bulb, olfactory tubercles, cerebral cortex and cerebellum. Total RNA was extracted for qRT-PCR determination of the expression of Heme Oxygenase-1 (HO-1), inducible Nitric Oxide Synthase (iNOS), Cyclooxygenase-2 (COX-2), and Cytochrome P450 1A1 (CYP1a1). Ongoing investigations revealed significant enhancement of the expression of HO-1, COX-2 and CYP1a1 in specific brain sections at 4 and 18 hours, suggesting that short-term diesel exhaust exposure might elicit gene expression changes in rat brain. Whether these effects result from direct actions of the particulate and/or volatile component of diesel exhaust in the brain, or via indirect mechanisms remains to be elucidated.

Journal of Physics: Conference Series **151** (2009) 011002 doi:10.1088/1742-6596/151/1/011002

Genomic instability in quartz dust exposed rat lungs: Is inflammation responsible?

C Albrecht[1], AM Knaapen[2*], G Cakmak Demircigil[3], Erdem Coskun[3], FJ van Schooten[2], PJA Borm[4] and RPF Schins[1]

1 Institut für Umweltmedizinische Forschung (IUF) at the Heinrich Heine University Duesseldorf, Germany.

2 Nutrition and Toxicology Research Institute Maastricht (NUTRIM), Department of Health Risk Analysis and Toxicology, University of Maastricht, The Netherlands (*current address: Department of Toxicology and Drug Disposition, Schering-Plough, 5340 BH Oss, The Netherlands)

3 Gazi University, Faculty of Pharmacy, Dept. of Toxicology, Ankara, Turkey

4 Centre of Expertise in Life Sciences (Cel), Hogeschool Zuyd, Heerlen, Netherlands

e-mail: catrin.albrecht@uni-duesseldorf.de

Abstract. Exposure to quartz dusts has been associated with lung cancer and fibrosis. Although the responsible mechanisms are not completely understood, progressive inflammation with associated induction of persistent oxidative stress has been discussed as a key event for these diseases. Previously we have evaluated the kinetics of pulmonary inflammation in the rat model following a single intratracheal instillation of 2mg DQ12 quartz, either in its native form or upon its surface modification with polyvinylpyridine-N-oxide or aluminium lactate. This model has been applied now to evaluate the role of inflammation in the kinetics of induction of DNA damage and response at 3, 7, 28, and 90 days after treatment. Bronchoalveolar lavage (BAL) cell counts and differentials as well as BAL fluid myeloperoxidase activity were used as markers of inflammation. Whole lung homogenate was investigated to determine the induction of the oxidative and pre-mutagenic DNA lesion 8-hydroxy-2-deoxy-guanosine (8-OHdG) by HPLC/ECD, while mRNA and protein expression of oxidative stress and DNA damage response genes including hemeoxygenase-1 (HO-1) and apurinic/apyrimidinic endonuclease (APE/Ref-1) were evaluated using Western blotting and real time PCR. Isolated lung epithelial cells from the treated rats were used for DNA strand breakage analysis using the alkaline comet assay as well as for micronucleus scoring in May-Gruenwald-Giemsa stained cytospin preparations. In the rats that were treated with quartz, no increased 8-OHdG levels were observed, despite the presence of a marked and persistent inflammation. However, DNA strand breakage in the lung epithelial cells of the quartz treated rats was significantly enhanced at 3 days, but not at 28 days. Moreover, significantly enhanced micronucleus frequencies were observed for all four time points investigated. In the animals that were treated with the PVNO modified quartz, micronuclei scores did not differ from controls, while in those treated with the aluminium coated quartz intermediate effects were found. These findings were in line with the kinetics of inflammation and epithelial proliferation in the rat lungs for the different treatments. Notably, a highly significant correlation was observed between neutrophil numbers and micronucleus frequencies, indicative for a role of inflammation in eliciting genomic instability in target cells of quartz-induced carcinogenesis. Our ongoing investigations focus on the evaluation of the causality between both in relation to quartz exposure.

Pyrogenic effect of respirable road dust particles

Umesh Jayawardena, Linda Tollemark, Christer Tagesson and Per Leanderson

Occupational and Environmental Medicine, University Hospital, S-581 85 Linköping, Sweden

E-mail: per.leanderson@lio.se

Abstract. Because pyrogenic (fever-inducing) compounds on ambient particles may play an important role for particle toxicity, simple methods to measure pyrogens on particles are needed. Here we have used a modified *in vitro* pyrogen test (IPT) to study the release of interleukin 1β (IL-1β) in whole human blood exposed to respirable road-dust particles (RRDP). Road dusts were collected from the roadside at six different streets in three Swedish cities and particles with a diameter less than 10 µm (RRDP) were prepared by a water sedimentation procedure followed by lyophilisation. RRDP (200 µl of 1 – 10^6 ng/ml) were mixed with 50 µl whole blood and incubated at 37 °C overnight before IL-1β was analysed with chemiluminescence ELISA in 384-well plates. Endotoxin (lipopolysaccharide from Salmonella minnesota), zymosan B and Curdlan (β-1,3-glucan) were used as positive controls. All RRDP samples had a pyrogenic effect and the most active sample produced 1.6 times more IL-1β than the least active. This formation was of the same magnitude as in samples with 10 ng LPS/ml and was larger than that evoked by zymosan B and Curdlan (by mass basis). The method was sensitive enough to determine formation of IL-1β in mixtures with 10 ng RRDP/ml or 0.01 ng LPS/ml. The endotoxin inhibitor, polymyxin B (10 µg/ml), strongly reduced the RRDP-induced formation of IL-1β at 1 µg RRDP/ml (around 80 % inhibition), but had only marginal or no effects at higher RRDP-concentrations (10 and 100 µg /ml). In summary, all RRDP tested had a clear pyrogen effect in this *in vitro* model. Endotoxin on the particles but also other factors contributed to the pyrogenic effect. As opposed to the *limulus* amebocyte lysate (LAL) assay (which measures endotoxin alone), IPT measures a broad range of pyrogens that may be present on particulate matter. The IPT method thus affords a simple, sensitive and quantitative determination of the total pyrogenic potential of ambient particles.

Inhaled Particles X, (23–25 September 2008, Manchester) IOP Publishing
Journal of Physics: Conference Series **151** (2009) 011002 doi:10.1088/1742-6596/151/1/011002

Effective biological dose from occupational exposure during nanoparticle synthesis

Evangelia Demou[1], Lang Tran[2] and Christos Housiadas[3]

[1] ETH Zurich, Institute of Environmental Engineering, Zurich, Switzerland
[2] Institute of Occupational Medicine, Edinburgh, UK
[3] "Demokritos" National Centre for Scientific Research, Athens, Greece

Email: evangelia.demou@ifu.baug.ethz.ch

Abstract. Nanomaterial and nanotechnology safety require the characterization of occupational exposure levels for completing a risk assessment. However, equally important is the estimation of the effective internal dose via lung deposition, transport and clearance mechanisms. An integrated source-to-biological dose assessment study is presented using real monitoring data collected during nanoparticle synthesis. Experimental monitoring data of airborne exposure levels during nanoparticle synthesis of $CaSO_4$ and $BiPO_4$ nanoparticles in a research laboratory is coupled with a human lung transport and deposition model, which solves in an Eulerian framework the general dynamic equation for polydisperse aerosols using particle specific physical-chemical properties. Subsequently, the lung deposition model is coupled with a mathematical particle clearance model providing the effective biological dose as well as the time course of the biological dose build-up after exposure. The results for the example of $BiPO_4$ demonstrate that even short exposures throughout the day can lead to particle doses of 1.10·E+08 #/(kg-bw·8h-shift), with the majority accumulating in the pulmonary region. Clearance of particles is slow and is not completed within a working shift following a 1 hour exposure. It mostly occurs via macrophage activity in the alveolar region, with small amounts transported to the interstitium and less to the lymph nodes.

Journal of Physics: Conference Series **151** (2009) 011002 doi:10.1088/1742-6596/151/1/011002

Simultaneous on-line size and chemical analysis of gas phase and particulate phase of mainstream tobacco smoke

J McAughey[1], T Adam[2], C McGrath[1], C Mocker[2,3] and R Zimmermann[2,3,4]

1 British American Tobacco, Group R&D Centre, Southampton SO15 8TL, UK

2 Institute of Ecological Chemistry, Helmholtz Zentrum München, Germany

3 Research Center for Environmental Health, D-85764 Neuherberg, Germany3Institute of Chemistry, University of Rostock, D-18051 Rostock, Germany

4 Bavarian Institute of Applied Environmental Research & Technology GmbH, D-86167, Augsburg, Germany

E-mail: john_mcaughey@bat.com

Abstract. Tobacco smoke is a complex and dynamic physical and chemical matrix in which about 4800 components have been identified. It is known that deposition efficiencies of smoke particles in the lung in the lung (60-80%) are greater than expected for smoke particles of 150–250 nm count median diameter (CMD). Various mechanisms have been put forward to explain this enhanced deposition pattern, including coagulation, hygroscopic growth, condensation and evaporation, changes in composition, or changes in inhalation behaviour. This paper represents one of three studies seeking to better quantify smoke chemistry, inhalation behaviour and cumulative particle growth. This information will improve dosimetry estimates in quantitative risk assessment tools as part of a harm reduction process. In this study smoke particle size and chemistry were measured simultaneously in real-time using electrical mobility spectrometry and soft-ionisation, time-of-flight mass spectrometry respectively. Qualitative puff-by-puff resolved yields of three selected compounds (acetaldehyde, phenol, and styrene) are shown and compared with particle number and count median diameter from different smoking intensities and filter ventilation. Yields of chemical analysis, particle diameter and concentration are in good agreement with the intensity of the smoking regime and the dilution of smoke by filter ventilation.

Journal of Physics: Conference Series **151** (2009) 011002 doi:10.1088/1742-6596/151/1/011002

Real-time measurement of inhaled and exhaled cigarette smoke: Implications for dose

Conor McGrath[1], Nigel Warren[1], Philip Biggs[1] and John McAughey[1]

[1] British American Tobacco, Group R&D Centre, Southampton, SO15 8TL, UK

conor_mcgrath@bat.com

Abstract. Inhalation of tobacco smoke aerosol is a two-step process involving puffing followed by inhalation. Measured smoke deposition efficiencies in the lung (20-70%) are greater than expected for smoke particles of 150 – 250 nm count median diameter (CMD). Various mechanisms have been put forward to explain this enhanced deposition pattern, including coagulation, hygroscopic growth, condensation and evaporation, changes in composition, or changes in inhalation behaviour. This paper represents one of a series of studies seeking to better quantify smoke chemistry, inhalation behaviour and cumulative particle growth. The studies have been conducted to better understand smoke dosimetry and links to disease as part of a wider programme defining risk and potential harm reduction. In this study, the average CMD of inhaled smoke was 160 nm while the average CMD of exhaled smoke was 239 nm with an average growth factor of 1.5.

Journal of Physics: Conference Series **151** (2009) 011002 doi:10.1088/1742-6596/151/1/011002

Puffing and inhalation behaviour in cigarette smoking: Implications for particle diameter and dose

Colin Dickens, Conor McGrath, Nigel Warren, Philip Biggs and John McAughey

British American Tobacco, Group R&D Centre, Southampton, SO15 8TL, UK

colin_dickens@bat.com

Abstract. Inhalation of tobacco smoke aerosol is a two-step process involving puffing followed by inhalation. Measured smoke deposition efficiencies in the lung (20-70%) are greater than expected for smoke particles of diameter 150 – 250 nm CMD. Various mechanisms have been put forward to explain this enhanced deposition pattern, including coagulation, hygroscopic growth, condensation and evaporation, changes in composition, or changes in inhalation behaviour. This paper represents one of a series of studies seeking to better quantify smoke chemistry, inhalation behaviour and cumulative particle growth. The studies have been conducted to better understand smoke dosimetry and links to disease as part of a wider programme defining risk and potential harm reduction. In this study, it was noted that particle deposition increased with increasing inhalation depth, and that smoke inhalation volumes were generally greater than normal tidal breathing volumes. A weak association was observed between particle diameter and puff flow, but no strong association between particle diameter and retention efficiency.

Inhaled Particles X, (23–25 September 2008, Manchester) IOP Publishing
Journal of Physics: Conference Series **151** (2009) 011002 doi:10.1088/1742-6596/151/1/011002

Has the introduction of smokefree legislation had an impact on the second hand smoke levels in the homes of non-smokers who live with smokers? Results of a study in Northern Ireland

M van Tongeren[1], K S Galea[1], B G Miller [1], S Semple[1,2], A Apsley[1,2], L MacCalman[1], N McCay[3], G Gilmore[3], D Gossrau-Breen[3] and J W Cherrie[1]

[1] Institute of Occupational Medicine, Edinburgh, EH14 4AP, UK.

[2] University of Aberdeen, Department of Environmental & Occupational Medicine, Foresterhill Road, Aberdeen, UK.

[3] Health Promotion Agency for Northern Ireland, 18 Ormeau Avenue, Belfast, UK.

E-mail: martie.van.tongeren@iom-world.org

Abstract. Legislation to restrict smoking in public places and workplaces was introduced in Northern Ireland in April 2007. Some concerns were raised that this could lead to displacement of smoking from enclosed public places into the homes. We carried out a study among non-smokers who live with smokers to investigate self-reported second hand smoke (SHS) levels and residential airborne nicotine levels before and after the introduction of legislation. Self-reported domestic SHS exposure was reduced from approximately 3 hours in Phase 1 to under 2 hours in Phase 2. However, the average airborne nicotine levels were generally higher in Phase 2 compared to Phase 1. This difference was predominantly observed in households with relatively low nicotine levels in Phase 1. The increase was still present after taking into account the effects of smoking practices and therefore cannot be explained entirely by any changes in residential smoking patterns. Other factors such as differences in ventilation due to outdoor weather conditions may play an important role. Further studies are required to determine any long term trends in the residential SHS exposures.

Journal of Physics: Conference Series **151** (2009) 011002 doi:10.1088/1742-6596/151/1/011002

Regional lung deposition of aged and diluted sidestream tobacco smoke

W Hofmann[1], R Winkler-Heil[1] and J McAughey[2]

[1]Division of Physics and Biophysics, Department of Materials Engineering and Physics, University of Salzburg, Hellbrunner Str. 34, 5020 Salzburg, Austria
[2]British American Tobacco, Group R&D Centre, Southampton SO15 8TL, UK

E-mail: Werner.Hofmann@sbg.ac.at

Abstract. Since aged and diluted smoke particles are in general smaller and more stable than mainstream tobacco smoke, it should be possible to model their deposition on the basis of their measured particle diameters. However in practice, measured deposition values are consistently greater than those predicted by deposition models. Thus the primary objective of this study was to compare theoretical predictions obtained by the Monte Carlo code IDEAL with two human deposition studies to attempt to reconcile these differences. In the first study, male and female volunteers inhaled aged and diluted sidestream tobacco smoke at two steady-state concentrations under normal tidal breathing conditions. In the second study, male volunteers inhaled aged and diluted sidestream smoke labelled with ^{212}Pb to fixed inhalation patterns. Median particle diameters in the two studies were 125 nm (CMD) and 210 nm (AMD), respectively. Experimental data on total deposition were consistently higher than the corresponding theoretical predictions, exhibiting significant inter-subject variations. However, measured and calculated regional deposition data are quite similar to each other, except for the extra-thoracic region. This discrepancy suggests that either the initial particle diameter decreases upon inspiration and/or additional deposition mechanisms are operating in the case of tobacco smoke particles.

Journal of Physics: Conference Series **151** (2009) 011002 doi:10.1088/1742-6596/151/1/011002

UK smoke-free legislation: Changes in $PM_{2.5}$ concentrations in bars in Scotland, England and Wales

Sean Semple[1,2], Audrey Naji[1], Martie van Tongeren[2], Karen Galea[2], Laura MacCalman[2], Ivan Gee[3], Odette Parry[4] and Jon Ayres[1]

[1] Department of Environmental & Occupational Medicine, Liberty Safe Work Research Centre, University of Aberdeen, Foresterhill Road, Aberdeen AB25 2ZP, UK

[2] Institute of Occupational Medicine, Research Park North, Riccarton, Edinburgh EH14 4AP, U.K.

[3] Centre for Public Health, Liverpool John Moores University, Kingsway House, Hatton Gardens, Liverpool, L3 2AJ

[4] Social Inclusion Research Unit (SIRU), Glyndwr University, Mold Rd, Wrexham, L11 2AW.

E-mail : sean.semple@abdn.ac.uk

Abstract. Comprehensive smoke-free legislation was introduced in Scotland in 2006 with Wales and England following in 2007. Prior to implementation, workers in the hospitality sector were known to have among the highest exposures to secondhand smoke (SHS) of any occupational group. As part of three studies to evaluate the introduction of the legislation in each country, $PM_{2.5}$ as a marker of SHS concentrations was measured covertly for 30 minute periods before smoke-free legislation was introduced, again at 1-2 months post-ban (not Wales) and then, again, at 12-months post-baseline (not Scotland). Measurements were carried out with a TSI Personal Aerosol Sampler and time-weighted averages calculated. In addition, in Scotland and England a small number of overt measurements were carried out to assess bar workers' full-shift personal exposures to $PM_{2.5}$. Postcode data were also used to derive indicators of socio-economic status of the bar location. $PM_{2.5}$ levels in bars prior to introduction of the smoke-free legislation were highest in Scotland (median 197 $\mu g/m^3$), followed by Wales (median 184 $\mu g/m^3$) and England (median 92 $\mu g/m^3$). All three countries experienced a substantial and consistent reduction in $PM_{2.5}$ concentrations with the median size of the reduction ranging from 84% to 93% depending on the follow-up phase and country. Personal exposure reductions were also within this range. There was no evidence of significant differences in SHS levels by the deprivation score of the postcode of the bar.

Diesel exhaust particulate material expression of in vitro genotoxic activities when dispersed into a phospholipid component of lung surfactant

Shi X C, Keane M J, Ong T M, Harrison, J C, Slaven J E, Bugarski AD, Gautam M, Wallace W E.

US Centers for Disease Control and Prevention, National Institute for Occupational Safety and Health, Health Effects Laboratory Division, Morgantown, WV, USA

mjk3@cdc.gov

Abstract. Bacterial mutagenicity and mammalian cell chromosomal and DNA damage in vitro assays were performed on a diesel exhaust particulate material (DPM) standard in two preparations: as an organic solvent extract, and as an aqueous dispersion in a simulated pulmonary surfactant. U.S. National Institute for Standards and Technology DPM SRM 2975 expressed mutagenic activity in the *Salmonella* reversion assay, and for in vitro genotoxicity to mammalian cells as micronucleus induction and as DNA damage in both preparations: as an acetone extract of the DPM mixed into dimethylsulfoxide, and as a mixture of whole DPM in a dispersion of dipalmitoyl phosphatidyl choline. Dispersion in surfactant was used to model the conditioning of DPM depositing on the deep respiratory airways of the lung. DPM solid residue after acetone extraction was inactive when assayed as a surfactant dispersion in the micronucleus induction assay, as was surfactant dispersion of a respirable particulate carbon black. In general, a given mass of the DPM in surfactant dispersion expressed greater activity than the solvent extract of an equal mass of DPM.

IOP Publishing
doi:10.1088/1742-6596/151/1/011002

Quartz-Containing Ceramic Dusts: *In Vitro* Screening of the Cytotoxic, Genotoxic and Pro-Inflammatory Potential of 5 Factory Samples

C Ziemann[1], P Jackson[2], R Brown[3], G Attik[4], B H Rihn[4], O Creutzenberg[1]

[1]Fraunhofer Institute of Toxicology and Experimental Medicine, Hannover, Germany
[2]CERAM Research Ltd., Stoke-on-Trent, UK
[3]TOXSERVICES, Stretton, UK
[4]Nancy-University, Faculté de Pharmacie, Nancy, France

E-mail: christina.ziemann@item.fraunhofer.de

Abstract. Inhalation of some respirable crystalline silica (MMAD ≤ approx. 4 μm) leads to inflammatory and malignant diseases. Comprehensive physicochemical/biological data and suitable *in vitro*/*in vivo* methods may distinguish between more or less harmful quartz-varieties. Within the European Collective Research Project SILICERAM an *in vitro* screening battery was established to evaluate cytotoxicity (LDH-release, MTT-assay), genotoxicity (Comet-assay) and pro-inflammatory potential (PGE_2-liberation, TNF-α mRNA expression) of 5 respirable quartz-containing dusts from ceramic plants: brickwork (BR: 7.8% quartz), tableware granulate/cast (TG/TC: 5.8%/3.1%), tiles (TI: 8.1%), refractory (RF: 3.7%). DQ12 (87% α-quartz) and Al_2O_3 were used as particulate positive and negative controls, respectively. Primary rat alveolar macrophages and the macrophage cell line NR8383 served as model systems. Aluminium lactate was used as inhibitor of biologically active silica, enabling differentiation of silica- and non-specific toxicity. At 200μg/cm² (2h) the dusts did not alter significantly LDH-release (except TC), whereas the MTT-assay demonstrated the mainly quartz-independent rank order: DQ12>RF>TG>TI≥BR>TC>Al_2O_3. DNA-damage was maximal for BR and TI followed by DQ12>TG>TC>RF>Al_2O_3. All dusts induced PGE_2-liberation (DQ12>BR>TC>TG>TI>RF>Al_2O_3) at 50μg/cm² (4h), but TNF-α mRNA (10μg/cm², 24h) was only increased by DQ12, TG (quartz-dependently), and TC. In conclusion, these *in vitro* tests were an adequate approach to screen the toxic potential of quartz-containing ceramic dusts, but the quartz-content was too low to differentiate the various quartz-varieties.

Journal of Physics: Conference Series **151** (2009) 011002 doi:10.1088/1742-6596/151/1/011002

Cytokine secretion pattern in multiple cellular co-cultures exposed to carbon nanotubes

Ernesto Alfaro-Moreno[1,2], Benoit Nemery[1], Peter Hoet[1]

1) Lung Toxicology Unit, K.U. Leuven, Leuven, Belgium.

2) Investigación Básica, Instituto Nacional de Cancerología, México.

Abstract. To understand the cellular mechanisms related to particulate matter (PM) toxicity, *in vitro* models have proof to be very useful. The use of single cell cultures, or even co-cultures of two different cells may overlook the important interactions with other cell types and the communication with distant cells, such as the possible cross talk between the lung epithelium with the vascular endothelium. In the present proposal, we are discussing the cellular effects of carbon nanotubes on multiple cell cultures that have been previously used successfully to evaluate urban particulate matter. Four different human cell lines are used in this study, pneumocytes (A549), macrophages (THP-1 cells differentiated with PMA), mast cells (HMC-1) and endothelial cells (EAHY926). The cytotoxicity of the carbon nanotubes is tested in the different cell lines and the co-cultures. The pneumocytes-macrophages-mast cells (tricultures) are co-cultured in a ration 10:2:1, to mimic the lung physiology. We expose the tricultures to 0, 1, 3, 10, 30 and 100 μg/cm^2 of carbon nanotubes. After 12 h of exposure, an insert with confluent endothelial cells is added, and 12 h later, the supernatants are collected for further cytokine analysis. The cytokines are analysed first with a commercial cytokine array system, and those cytokines showing changes larger than 50 % in comparison with controls will be quantified by Facs analysis using a beads array method. In order to evaluate the role of different cytokines in the cellular cross talk, antibodies against different cytokines will be added to evaluate changes in the patter of secretion of the other cytokines…

Indoor air pollution in Malawi: Use of biomass fuels produces high particulate matter and carbon monoxide concentrations in both urban and rural homes

Duncan G Fullerton[1,2], Sean Semple[3] , Francis Kalambo[1] , George Henderson[3] Aryo Suseno[1,2], Rose Malamba[1], Jon Ayres[3], Stephen B Gordon[2]

1. Malawi-Liverpool-Wellcome Clinical Research Laboratories, Universities of Malawi and Liverpool (UK), P.O. Box 30096, Blantyre, Malawi.

2. Liverpool School of Tropical Medicine, Pembroke Place, Liverpool L3 5QA, U.K.

3. Department of Environmental & Occupational Medicine, Liberty Safe Work Research Centre, University of Aberdeen, Foresterhill Road, Aberdeen AB25 2ZP, UK

E mail duncan.fullerton@liv.ac.uk

Abstract. Air pollution from biomass fuels in Africa is likely to be a significant cause of mortality and morbidity both in adults and children. We measured the indoor air quality in 61 homes (31 rural / 31 urban) in Malawi using 4 different devices (sidepak, UCB, gravimetric and carbon monoxide monitor) over 24 hrs. A questionnaire of cooking and lighting fuel type was also administered. Respirable dust levels in both the urban and rural environment were high with a mean gravimetric device TWA level of 226μg/m^3. Levels of particulate matter (measured using a sidepak photometric device and a UCB monitor) were >250 μg/m^3 for over 1 hour per day in over 50% of rural homes compared to 22% of urban homes. Carbon monoxide levels in the urban homes were significantly higher than in rural homes ($p<0.001$). We conclude that indoor air levels in Malawian homes are high in both rural and urban environments and justify further investigation. Interventions should be sought to reduce exposure to concentrations less harmful to health.

Airborne asbestos concentrations in system built schools

Garry Burdett, Steve Cottrell and Catherine Taylor

Health and Safety Laboratory, Harpur Hill, Buxton, Derbyshire, UK. SK17 9JN

Garry.burdett@hsl.gov.uk

Abstract. Enclosure and encapsulation of asbestos containing materials is often used to minimise the risk of damage and the potential for airborne fibre release. Therefore there was concern when it was reported that significant levels of airborne fibre release had been measured from the enclosed columns of a system built school. The release was found to occur, when as part of the deliberate disturbance, the steel casings enclosing the structural support columns and asbestos insulation were struck firmly a number of times. The measured phase contrast microscopy concentrations inside the enclosure suggested a significant release of airborne fibres could sometimes occur. HSL investigations of similar columns gave a range of values, with the highest being 0.44 f/ml by phase contrast microscopy (PCM). The fibres released were shown to be predominantly amosite fibres, when analysed by analytical transmission electron microscopy (TEM). An airborne release of this level from an enclosed source of asbestos was unexpected and of particular concern as this is one of the recommended methods of control. The escape of fibres into the rooms was found to be mainly due to gaps in the joins of the column casing. Although other potential routes of release were also present from the top and bottom of the cladding. This presentation summarises the PCM and TEM results, from air samples taken to assess, remediate and control airborne asbestos emissions from the columns. The results were used to assess the effectiveness of the remediation before and after the rooms were reoccupied. Even partial remediation by the filling and taping of the visible gaps in the column casing was shown to be effective in minimising exposures to occupants. Several schools and ex-school buildings were monitored by PCM and TEM before, during, and after remediation over an extended period of time to assess airborne fibre and asbestos concentrations during normal occupation.

Journal of Physics: Conference Series **151** (2009) 011002
doi:10.1088/1742-6596/151/1/011002

Emphysema and Pulmonary Impairment in Coal Miners: Quantitative Relationship with Dust Exposure and Cigarette Smoking

E D Kuempel, Ph.D.[1]; V Vallyathan, Ph.D.[2]; F H Y. Green, M.D.[3]

1 National Institute for Occupational Safety and Health, Education and Information Division, Risk Evaluation Branch, Cincinnati, Ohio, USA

2 National Institute for Occupational Safety and Health, Health Effects Laboratory Division, Pathology and Physiology Research Branch, Morgantown, West Virginia, USA

3 Department of Pathology, Faculty of Medicine, University of Calgary, Calgary, Alberta, Canada

E-mail: ekuempel@cdc.gov

Abstract. Coal miners have been shown to be at increased risk of developing chronic obstructive pulmonary diseases including emphysema. The objective of this study was to determine whether lifetime cumulative exposure to respirable coal mine dust is a significant predictor of developing emphysema at a clinically-relevant level of severity by the end of life, after controlling for cigarette smoking and other covariates. Clinically-relevant emphysema severity was determined from the association between individuals' lung function during life (forced expiratory volume in one second, FEV_1, as a percentage of predicted normal values) and emphysema severity at autopsy (as the proportion of lung tissue affected). In a logistic regression model, cumulative exposure to respirable coal mine dust was a statistically significant predictor of developing clinically-relevant emphysema severity, among both ever-smokers and never-smokers. The odds ratio for developing emphysema associated with FEV_1 <80% at the cohort mean cumulative coal dust exposure (87 mg/m^3 x yr) was 2.30 (1.46-3.64, 95% confidence limits), and at the cohort mean cigarette smoking (among smokers: 42 pack-years) was 1.95 (1.39-2.79).

High Temporal Resolution Measurements of Roadside Particle Size Distributions and Their Implications for Exposure

A S Tomlin, D T Young, J J N Lingard and E L Agius

Energy and Resources Research Institute, University of Leeds, Leeds, LS2 9JT, UK

A.S.Tomlin@leeds.ac.uk

Abstract High time resolution, size segregated studies of ultrafine particle number concentrations (PNC) were performed at two UK roadside locations using a DMS500 particle spectrometer including sampling with and without a thermodenuder. Intercomparisons between the DMS500, an SMPS and a CPC, indicate that the SMPS can underestimate PNC due to the transient nature of excursions in concentrations and the longer scan time of the instrument. Short duration transient periods of high number concentrations made considerable contributions to hourly average PNC. The transient nature of changes in PNC potentially affects the type of exposure experienced in the roadside environment, as well as indoor outdoor exchange. The transient shifts were dominated by Aitken mode particles during high traffic periods, with the Aitken mode comprised of internally mixed particles with a low volatility core (thought to be primarily soot) and a surface of condensed and absorbed material. Transient periods of high nucleation mode PNC were also found, but were relatively more important during night-time and afternoon periods, due to secondary formation processes and lower concentrations of pre-existing particles. The thermodenuder studies indicated the presence of highly volatile liquid droplets in the nucleation mode which were lost on heating. Transient events and complexities in particle mixing state within the ultrafine fraction, coupled with discrepancies between instruments, make the potential adoption of a total number based air quality standard more complicated than first appears. Nucleation mode particle numbers are highly dependant on atmospheric and traffic conditions as well as pre-existing particle surface area and may be toxicologically less significant if they are predominantly liquid droplets. Aitken mode particles have been shown in previous studies to have higher toxicological significance due to their chemical composition. Their hourly average PNC correlated well with $PM_{2.5}$ and estimated PM_1 with lower significance of equivalent correlation of nucleation mode particles.

Inhaled Particles X, (23–25 September 2008, Manchester) IOP Publishing
Journal of Physics: Conference Series **151** (2009) 011002 doi:10.1088/1742-6596/151/1/011002

Home outdoor models for traffic-related air pollutants do not represent personal exposure measurements in Southern California

R Ducret-Stich[1], R J Delfino[2], T Tjoa[2], A Gemperli[1], A Ineichen[1], J Wu[2], H C Phuleria[1], L-J S Liu1[3]

[1] Institute of Social and Preventive Medicine, University of Basel, Switzerland

[2] Department of Epidemiology, School of Medicine, University of California, Irvine, CA, USA

[3] Department of Environmental and Occupational Health Sciences, University of Washington, Seattle, WA, USA

E-mail: regina.ducret@unibas.ch

Abstract. Background: Recent studies have used measurements or estimates of traffic-related air pollutants at home or school locations to link associations between exposure and health. However, little is known about the validity of these outdoor concentrations as an estimate for personal exposure to traffic. **Method:** This paper compares modelled outdoor concentrations at home with personal exposure to traffic air pollution of 63 children in two areas in Los Angeles in 2003/2004. Exposure monitoring consisted of sixteen 10-day monitoring runs, with each run monitoring 4 subjects concurrently with the active personal DataRAM for particulate matter <2.5 μm ($PM_{2.5}$), elemental carbon (EC) and organic carbon (OC). One child per run had concurrent indoor/outdoor home monitoring. Measurements at central sites (24-hr $PM_{2.5}$, EC, OC) were taken daily and concentrations of PM_{25}, EC, and OC from traffic sources were calculated using the CALINE4 model for individual residences. We modelled outdoor concentrations of $PM_{2.5}$, EC and OC with multilinear regression including GIS and meteorological parameters and adjusted for auto-correlation between repeated measurements. **Results:** The model fit (R^2) for home outdoor estimates was 0.94, 0.74 and 0.80 for $PM_{2.5}$, EC and OC, respectively. Comparisons between these outdoor estimates and the personal measurements showed a good agreement for $PM_{2.5}$ (R^2=0.65-0.70) with a mean bias of –0.7±11.8μg for the smog receptor area, and 18.9±16.2μg for the traffic impacted area. However the outdoor estimates were not related to personal exposure for EC (R^2=0.01-0.29) and OC (R^2=0.03- 0.14). **Conclusions:** Predictions of outdoor concentrations can be used as approximations of personal exposure to $PM_{2.5}$. However, they are not appropriate for estimating personal exposure to traffic-related air pollutants including EC and OC in studies of acute exposure-response relationships.

Journal of Physics: Conference Series **151** (2009) 011002 doi:10.1088/1742-6596/151/1/011002

Impacts of Air Pollution Exposure on the Allergenic Properties of Arizona Cypress Pollens

Y Shahali[1,2], Pourpak Z[1] , Moein M[1], Zare A[1] and Majd A[2]

[1]Immunology, Asthma and Allergy Research Institute, Medical Sciences/ University of Tehran, I. R. Iran

[2]Department of Biology, Faculty of Sciences, Islamic Azad University, North Tehran Branch, I. R. Iran

E-mail: yshahali@yahoo.com

Abstract: Epidemiological studies have demonstrated that urbanization and high levels of vehicle emissions correlated with the increasing trend of pollen-induced respiratory allergies. Numerous works have investigated the role of pollutants in the pathogenesis of respiratory diseases but impacts of anthropogenic pollution on pollen allergenic properties are still poorly understood. The objective of this survey was to evaluate impacts of the traffic-related pollution on the structure and allergenic protein content of Arizona cypress (*Cupressus arizonica,* CA) pollens, recognized as a rising cause of seasonal allergy in various regions worldwide. According to our results, traffic-related air pollution by its direct effects on the elemental composition of pollens, considerably improved the fragility of the pollen exine, causing numerous cracks in its surface and facilitating pollen content liberation. Pollen grains were also covered by numerous submicronic orbicules which may act as effective vectors for pollen-released components into the lower regions of respiratory organs. On the other hand, this study provides us reliable explications about the low efficiency of standard commercial allergens in the diagnosis of the Arizona cypress allergy in Tehran. Although traffic related pollution clearly affects the allergenic components of CA pollens, the repercussions on the respiratory health of urban populations have yet to be clarified and need further investigations.

Journal of Physics: Conference Series **151** (2009) 011002 doi:10.1088/1742-6596/151/1/011002

Effects Of Urban Air Particulate Matter On Autonomic Control Of Heart Rhythm, Arrhythmia Predisposition And Arterial Blood Pressure In Adult Subjects With Different Health Conditions

Fossati, Serena[1]; Urso, Patrizia[1]; Ruggeri, Laura[1]; Fanetti, Anna Clara[1]; Metruccio, Francesca[2]; Peruzzo, Carlo[3]; Pini, Alessandro[4]; De Vito, Giovanni[5]; Cavallo, Domenico[3]; Carrer, Paolo[1]

1) University of Milan, Dept. of Occupational & Environmental Health, L. Sacco Hospital Unit, Milan IT
2) L. Sacco Hospital, International Centre for Pesticides and Health Risk Prevention, Milan IT
3) University of Insubria, Dept. of Chemical and Environmental Sciences, Como, IT
4) L. Sacco Hospital, Dept. of Cardiology, Milan IT
5) Milano-Bicocca University, Dept. of Clinical Medicine and Prevention, Monza IT

Corresponding author: serena.fossati@unimi.it

Abstract Exposure to particulate matter (PM) has been associated with increased cardiovascular morbidity and mortality in individuals with pre-existing cardiac and pulmonary disease. However the mechanisms behind this association are still unknown. *Aim* To assess, in the frame of the PM-CARE Study, the association between exposure to urban air particulate matter and changes in autonomic control of heart rhythm, arrhythmia predisposition and arterial blood pressure in adult subjects diagnosed with chronic ischemic heart disease, chronic asthma or chronic obstructive pulmonary disease and in subjects without diagnosis of the afore mentioned diseases. Materials and methods: 27 healthy individuals, 34 individuals with chronic ischemic heart disease, 20 with chronic asthma or COPD underwent a 24-hour exposure/clinical evaluation protocol during their habitual activities, both in the warm season (summer) and in the cold season (winter). Individual exposure to UFPs, $PM_{2.5}$ and PM_{10} was assessed for each subject, together with a 24hrs ambulatory ECG, for the assessment of heart rate variability, QT interval duration, and repeated arterial blood pressure measurement. The mean(±SD) age of the study population was 64(±10) years and 63% were male. The mean±SD exposure levels to monitored pollutants were 41.37±22.26 μg/m^3, 51.78±24.08μg/m^3 for $PM_{2.5}$ and PM_{10} respectively. UFPs mean was $23*10^3 \pm 11*10^3$ #/cm^3. Models that adjusted for health status, medication use, relative humidity and temperature did not demonstrate a significant association between an inter-quartile range increase in UFPs (15103 #/cm^3), $PM_{2.5}$ (26,17 μg/m^3) and PM_{10} (30,48 μg/m^3) and HRV indexes, QTec and BP, in the total sample and in the three groups. No significant association was observed between the investigated outcomes and PM exposure in healthy and in susceptible individuals, despite the high exposure levels monitored.

Journal of Physics: Conference Series **151** (2009) 011002 doi:10.1088/1742-6596/151/1/011002

Particulate air pollution and out-of-hospital cardiac arrests in Melbourne: a case crossover analysis

Martine Dennekamp, Andrew Forbes, Malcolm Sim, Michael Abramson

Department of Epidemiology and Preventive Medicine, Monash University, Melbourne, Australia

Abstract: Coronary artery disease accounts for about 20% of all Australian deaths. In most cases the coronary artery disease is clinically silent. This poses the question, what are the factors that determine whether on any given day, the disease remains silent or presents with an acute coronary syndrome or sudden arrhythmic death? One of the suggested causes has been exposure to air pollutants especially particulate air pollution. Worldwide, particles have consistently been shown to be associated with cardiovascular mortality and hospital admissions, however the mechanisms and causal factors driving these associations are still unclear. Current air quality standards in Australia are based on mass concentration of PM_{10}, however increasing attention has focused on $PM_{2.5}$ (measured by mass) and now the ultrafine particles (measured by number). The aim of this project is to examine associations between daily ambient air pollutants and the occurrence of out-of-hospital cardiac arrests in Melbourne and to determine which particle size metric is most important for this association. The study subjects include those who had an out-of-hospital cardiac arrest attended by a paramedic in the Melbourne metropolitan area, between January 2003 and December 2006. In Melbourne, there is a unique opportunity to obtaining information on out-of-hospital cardiac arrests through the Victorian Cardiac Arrest Registry. The Registry captures all cardiac arrests attended to by the Ambulance service. Information extracted from the registry includes date of birth, sex, date and time of event, address of the event, home address of the patient, pick-up location (e.g. house, street or public place), etc. Each year there are on average about 1,700 cardiac arrests with cardiac aetiology in the Melbourne metropolitan area which are attended by the Ambulance Service. Data for daily average PM_{10}, O_3, NO_2, CO and SO_2 were obtained from the Environmental Protection Agency Victoria and the Commonwealth Scientific and Industrial Research Organisation (CSIRO) for the study period from several monitoring stations in Melbourne. $PM_{2.5}$ data are available from 2003 onwards. Ultrafine particle data are available from one location from July 2005 onwards. Meteorological data (temperature, humidity, wind speed and direction) were collected from the Bureau of Meteorology in Melbourne and data is available from several locations in Melbourne. Daily concentrations of all pollution and meteorological variables were assigned to each postcode and statistical local area (SLA) by using the measurement of the monitors located closest to the postcode/SLA. The case-crossover method will be used to investigate whether a recent exposure to particulate air pollutants is related to the occurrence of a cardiac arrest. Only cases will be studied and their exposure at the time of the cardiac arrest will be compared with a matched "referent" period. A time-stratified referent period design will be employed to select control days associated with each index. In particular, time will be stratified into months and day of week, and the reference days associated with the date of each index case will be the same day of week within the same month. This will eliminate confounding by day of week and monthly trends.

Inhaled Particles X, (23–25 September 2008, Manchester) IOP Publishing
Journal of Physics: Conference Series **151** (2009) 011002 doi:10.1088/1742-6596/151/1/011002

Nanoparticle Risk to the Environment and Human Health

Birgit Gaiser[1], Philipp Rosenkranz[1], Jamie Lead[2], Charles Tyler[3], Mark Jepson[4], Teresa Fernandes[1], Vicki Stone[1]

1 Napier University, School of Life Sciences

2 University of Birmingham, Environmental Life Sciences

3 University of Exeter, School of Biosciences

4 Bristol University, Department of Biochemistry

b.gaiser@napier.ac.uk

Abstract: An important issue in the risk assessment of nanoparticles is their potential to cross barriers, such as the pulmonary epithelium, the blood-brain barrier, or the gastro-intestinal epithelium. The latter is important for nanoparticles contaminating water, entering the food chain and those used as food additives. Nanoparticles can enter the gastrointestinal tract through various points in the respiratory system, for example after removal through the mucociliary pathway or after deposition in the nasal region. Translocation into the bloodstream may also lead to exposure of cells such as hepatocytes. This project incorporates expertise in eco- and nano-toxicology, and aims at assessing the hazard of two particles, cerium dioxide (CeO_2) and silver (Ag), in aquatic invertebrates, fish, a human cell culture model for gastrointestinal translocation and in hepatocytes. The project compares interspecies differences between these models and how size influences the effects of the particle uptake and biological impact.

Development of a bio-mathematical model in rats to describe clearance, retention and translocation of inhaled nano particles throughout the body.

L MacCalman[1], C L Tran[1] and E Kuempel[2].

[1]Institute of Occupational Medicine, Edinburgh

[2]National Institute for Occupational Safety and Health, Cincinnati, Ohio, USA

laura.maccalman@iom-world.org

Abstract. Studies of the translocation of inhaled nanoparticles from rodent lungs to different target organs provide data to model the inhalation and translocation of nanoparticles. A compartmental model was developed based on biological information about the major organs and how they interrelate. This model, which is quantified by a set of differential equations 'describing' the passage of nanoparticles through the body, gives estimates of the particle mass present in each organ. Optimal parameter estimates were found by minimising the model mean squared error. Data from two different studies in rats (one endotracheal instillation and one inhalation exposure) were used to calibrate the model. Most of the nanoparticle mass remained in the lungs. The overall fit of the model to the total measured particle mass in the body was very good for both studies (R^2=0.98-0.99), although different parameter estimates were sometimes required to fit the study-specific data. The model fit to the measured particle mass by organ was very good for the lungs, brain, and spleen (R^2=0.81-0.98) in both studies, but not as good for the liver and kidney (R^2=0.32-0.53). While this model describes the retention and translocation of nanoparticles from the lungs reasonably well, further model evaluation and validation is needed using additional data.

Inhaled Particles X, (23–25 September 2008, Manchester) IOP Publishing
Journal of Physics: Conference Series **151** (2009) 011002 doi:10.1088/1742-6596/151/1/011002

Changes to the structure of blood clots formed in the presence of fine particulate matter.

Sofian Metassan[1,2], Robert A S Ariens[2], D Julian Scott[2] and Michael N Routledge[1]

[1]Molecular Epidemiology Unit and

[2]Cardiovascular and Diabetes Research Division, Leeds Institute for Health, Genetics and Therapeutics, The LIGHT Laboratories, University of Leeds, Leeds LS2 9JT

Abstract. Both long-term and short-term exposure (one to two hours) to particulate matter are associated with morbidity and mortality caused by cardiovascular diseases. The underlying mechanisms leading to cardiovascular events are unclear, however, changes to blood coagulability upon exposure to ultrafine particulate matter (UFPM, the smallest of which can enter the circulation) is a plausible mechanism. Objectives: This study aims to investigate the direct effects of particulate matter on fibrin polymerization, lateral aggregation and the formation of fibrin network structure. Methods: Standard Urban Particulate Matter (PM) was suspended in Tris buffer centrifuged and filtered with <200nm filter to obtain ultrafine PM or their water-soluble components. Purified normal fibrinogen was made to clot by adding thrombin and calcium chloride in the presence of varying concentrations of PM. Permeation properties (Darcy constant [Ks]) and turbidity of clots were measured to investigate the effects on flow-rate, pore size, and fibrin polymerization. In addition, confocal microscopy was performed to study detailed clot structure. Results: Total PM increased the Ks of clots in a dose dependant manner (Ks = 4.4, 6.9 and 13.2 x 10-9 cm2 for 0, 50 and 100 μg/ml total PM concentrations, respectively). Filtered PM also produced a significant increase in Ks at PM concentration of 17 μg/ml. Final turbidity measurements at 20min were obtained for varying concentrations of PM. Maximum optical density (OD) for 1 mg/ml fibrinogen at 0, 50, 100 and 200 μg/ml total PM concentrations were 0.39, 0.42, 0.45 and 0.46, respectively. The maximum OD for 0, 17, 34 and 68 μg/ml filtered PM concentrations were 0.39, 0.42 0.47 and 0.51, respectively, suggesting an increase in fibre diameter with increasing particulate concentration. The lag phase was significantly shorter and the rate of polymerisation was significantly faster in the presence of 68 μg/ml filtered PM. Confocal microscopy results showed decrease in fibre density without a significant increase in fibre diameter in the presence of total PM and filtered PM. Conclusion. The results indicate that total PM and filtered PM are capable of causing alterations to the fibrin polymerization and network structure as shown by the changes in permeation properties, the turbidity experiments as well as by confocal microscopy.

Journal of Physics: Conference Series **151** (2009) 011002 doi:10.1088/1742-6596/151/1/011002

Imaging carbon nanoparticles and related cytotoxicity

C Cheng[1], A E Porter[1], K Muller[2], K Koziol[3], J N Skepper[2], P Midgley[3] and M Welland[1]

[1]The Nanoscience Centre, University of Cambridge, 11 JJ Thompson Ave, Cambridge CB3 OFF, United Kingdom,

[2] Multi-imaging Centre, Dept of Physiology, Development and Neuroscience, Anatomy Building, University of Cambridge, Downing St, Cambridge, CB2 3DY, United Kingdom,

[3] Dept of Materials Science and Metallurgy, University of Cambridge, Pembroke St, Cambridge, CB2 3QZ, UK

Email: mew10@cam.ac.uk

Abstract. Carbon-based nanoparticles have attracted significant attention due to their unique physical, chemical, and electrical properties. Numerous studies have been published on carbon nanoparticle toxicity; however, the results remain contradictory. An ideal approach is to combine a cell viability assay with nanometer scale imaging to elucidate the detailed physiological and structural effects of cellular exposure to nanoparticles. We have developed and applied a combination of advanced microscopy techniques to image carbon nanoparticles within cells. Specifically, we have used EFTEM, HAADF-STEM, and tomography and confocal microscopy to generate 3-D images enabling determination of nanoparticle spatial distribution in a cell. With these techniques, we can differentiate between the carbon nanoparticles and the cell in both stained and unstained sections. We found carbon nanoparticles (C_{60}, single-walled carbon nanotubes (SWNT), and multi-walled carbon nanotubes (MWNT)) within the cytoplasm, lysosomes, and nucleus of human monocyte-derived macrophage cells (HMM). C_{60} aggregated along the plasma and nuclear membrane while MWNTs and SWNTs were seen penetrating the plasma and nuclear membranes. Both the Neutral Red (NR) assay and ultra-stuctural analysis showed an increase in cell death after exposure to MWNTs and SWNTs. SWNTs were more toxic than MWNTs. For both MWNTs and SWNTs, we correlated uptake of the nanoparticles with a significant increase in necrosis. In conclusion, high resolution imaging studies provide us with significant insight into the localised interactions between carbon nanoparticles and cells. Viability assays alone only provide a broad toxicological picture of nanoparticle effects on cells whereas the high resolution images associate the spatial distributions of the nanoparticles within the cell with increased incidence of necrosis. This combined approach will enable us to probe the mechanisms of particle uptake and subsequent chemical changes within the cell, essential for identifying the toxicological profiles of carbon nanoparticles.

Defining the biologically effective dose in particle toxicology, with special reference to nanoparticles

Ken Donaldson[1], Lang Tran[2], Paul Borm[3]

1 University of Edinburgh, Scotland;

2 Institute of Occupational Medicine, Scotland;

3 Zuyd University, Netherlands

Abstract: The internal dose is the quantity of a toxin that gains access to the body. The fraction of the deposited dose that remains in a long-lived compartment like the interstitium or in long-lived macrophage accumulations can take part in toxic reactions; it is less than the "total dose" that is breathed in and that deposits. This is then acted on by the milieu in the interstitium or in cells to cause dissolution of non-biopersistent particles or components. This may release harmful soluble components as well as harmless soluble components, which are cleared or metabolized. This can be understood in that all realistic particle exposures are particles that are multi-component and poly-dispersed and within this total dose, we can identify sub-fractions that are likely to drive an adverse effect- the biologically effective dose (BED). The BED is a useful concept, being that fraction of the total dose that is sufficiently biopersistent and also sufficiently active to cause an adverse effect such as oxidative stress, adduct formation, etc. For example, PM_{10} is measured by the mass metric, yet a variable and often substantial quantity of the mass of PM_{10} is sea salt, which is likely to be completely invisible to the lung following exposure at ambient levels, i.e. a harmless dose. Conversely, the transition metals can be seen to be driving the oxidative stress and inflammatory effects of many dusts, yet would contribute very little to mass. Particle surface reactivity and surface area also are key drivers of the adverse effect and contribute to the BED and this has special significance for nanoparticle toxicology. Shape also contributes to BED in the case of high aspect ratio particles (fibres). We will put forward an overall model for the particle BED and discuss its general application in particle toxicology. Nanoparticles in view of their propensity to translocate, may have more than one BED. The dose for translocation may be related to absolute size whilst their action at the target tissue may be related to surface area and reactivity as described above.

Journal of Physics: Conference Series **151** (2009) 011002 doi:10.1088/1742-6596/151/1/011002

In vivo Study with Quartz-Containing Ceramic Dusts: Inflammatory Effects of Two Factory Samples in Lungs after Intratracheal Instillation in a 28-Day Study with Rats

O Creutzenberg[1], C Ziemann[1], T Hansen[1], H Ernst[1], P Jackson[2], D Cartlidge[2], R Brown[3]

1 Fraunhofer Institute of Toxicology and Experimental Medicine, Hannover, Germany

2 CERAM Research Ltd., Stoke-on-Trent, UK

3 TOXSERVICES, Stretton, UK

E-mail: otto.creutzenberg@item.fraunhofer.de

Abstract. As various quartz polymorphs react differently in lungs, a differentiation of effects is needed while setting occupational exposure levels. The objective of this European Collective Research Project SILICERAM was to characterize differences in biological activity of four quartz species, i.) 2 quartz-containing materials collected at typical ceramic manufacturing sites (Tableware granulate, TG and Tableware cast, TC) versus ii.) a designed ceramic dust sample (Contrived Sample, CS) and iii.) ground quartz DQ12 (well-characterised standard quartz (Positive Control, PC) and TiO_2 **(negative control)**. TG and TC had been selected as the most promising two candidates based on a preceding *in vitro* screening of 5 factory samples. Total doses of 5 mg per rat of the TG and TC, 1.1 mg of the CS and 0.33 mg of the PC corresponding to 0.29, 0.16, 0.29 and 0.29 mg quartz per rat, respectively, were administered to rats by intratracheal instillation. After 3 days, bronchoalveolar lavagate (BAL) analysis resulted in polymorphonuclear neutrophil (PMN) levels of 15%, 25%, 0.6% and 25% in the TG, TC, CS and PC groups, respectively. At 28 days, the values were 29%, 20%, 7% and 45%. Histopathologically, the TG and TC groups showed very slight to slight effects, the PC group, however, stronger effects after the same period. In conclusion, the following ranking was found: PC > TG > TC > CS > TiO_2 > Vehicle Control. Thus, a clear differentiation of effects for TG and TC, CS and PC was found. From a regulatory point of view, the substance-specific toxic potentials of TG and TC may need to be considered when devising occupational exposure limits.

Inhaled Particles X, (23–25 September 2008, Manchester) IOP Publishing
Journal of Physics: Conference Series **151** (2009) 011002 doi:10.1088/1742-6596/151/1/011002

A 25 year experience of lung-retained fibre analysis in one institution: time trends and methodological findings

Bruce W. Case[1], Andre Dufresne[1] and Patrick Sebastien[2]

1 Department of Epidemiology, Biostatistics and Occupational Health;

2 McGill University, MONTREAL (Québec), CANADA.

Abstract: Since 1982, we have performed lung-retained fibre analysis in many contexts (cohort study, case-control study, medico-legal and compensation cases, background level establishment, etc.) using transmission electron microscopy (TEM) and energy-dispersive spectrometry of x-rays (EDS) to characterize fibres and to relate them to exposures, diseases, and demographics (e.g. place of residence). This paper is an attempt to look at some of this data across studies systematically to gain a new appreciation of methodological issues and trends over time. Particular questions to be answered include how much lung tissue from how many sites is needed or useful; whether counting short fragments (< 5 um length) versus regulatory fibres (> 5 um length) adds to complexity, reliability, or usefulness; what patterns of fibre type and concentration in mesothelioma cases have been over time, and whether inter-laboratory comparisons can be helpful. Types of study design are mentioned above; methodology relies on identification of lung tissue from individuals in study groups followed by chemical digestion and low-temperature ashing with TEM/ EDS examination using a standard carbon-replica technique at magnifications from 13,500 to 16,000X. Fibres of varying length intervals have been studied for different purposes; one focus of the current presentation is comparison of results for "short" structures (< 5 um) and "long" fibres (usually > 5 um but sometimes longer) counts in the same specimens and of comparative results using different numbers of samples in the same subjects. Where available, counts in our laboratory are compared to those in others. A systematic analysis of mesothelioma lung fibre content over time is also included. Overall, the database contains well over 1000 individual subjects. It includes asbestos textile workers (64) and miners (> 200); synthetic vitreous fibre workers (155), mesothelioma cases (209), workman's compensation cases (257), and many hundreds of population controls at all ages. Where counts for both fibres > 5 um and short structures < 5 um were available, correlations between the two were very high (> .9, p <.0001) without exception if group size exceeded 20. Where one to four samples per subject were available, there was no added utility beyond two samples. Correlations for total fibre counts between samples were in the range of .6 to .8 (p <.0001) regardless of sample numbers, again if group size exceeded 20. Fibre types were similar regardless of short or long counts and near-identical between samples, unlike some previously published studies. Overall, in mesothelioma cases, amosite and crocidolite remain by far the most common fibre types, but over time mean concentrations of both have decreased in keeping with recent observations for amphibole fibre clearance. With rare individual exceptions (which will be illustrated with case data), other fibre types are rarely above population background levels. Short tremolite structures are frequently seen in both cases and controls – including children – and may not represent "asbestos" at all; longer tremolite fibres are seen mainly in mesotheliomas among chrysotile miners and millers, in one case from the Libby, Montana vermiculite mining workforce, and in some talc-exposed individuals. For cohort and case-control studies having numbers greater than 20, a single sample per case or control is sufficient for analysis; for less, two samples are recommended if feasible. For individual cases (e.g. in litigation or workman's compensation), two samples per case should be used if available. Counts should be performed at a minimum for fibres longer than 5 um; counts of structures under 5 um do not add information but on the other hand if only such counts are available it is reasonable to conclude they represent long fibre distributions as well. Structures of length shorter than 0.5 – 1.0 um should not be counted as they invite contamination. Fibre concentrations remain the best available determination of past exposures in mesothelioma cases, are especially effective at finding unreported or unknown amphibole exposures, and are effective for chrysotile if time from last exposure is less than 15 years. Given the trend toward decreasing concentration, presumably due to clearance, levels considered within control range in the past are now considered elevated. Further decreases may diminish method usefulness in the future. Attempts are made using current clearance models to determine for how long lung-retained fibre counts will remain the gold standard for such exposure assessment. Immediate research needs include better determination of current population background levels in lung. Additional methodology is needed to determine whether short tremolite/ actinolite structures are asbestiform.

Inhaled Particles X, (23–25 September 2008, Manchester) IOP Publishing
Journal of Physics: Conference Series **151** (2009) 011002 doi:10.1088/1742-6596/151/1/011002

Biological effects of nano-nickel in rat lungs after administration by inhalation and by intratracheal instillation.

A Ogami[1] Y Morimoto[1] M Murakami[1] T Myojo[1] T Oyabu[2] and I Tanaka[2]

1 Department of Occupational Pneumology,

2 Department of Environmental Health Engineering,

Institute of Industrial Ecological Sciences, University of Occupational and Environmental Health, Japan, 1-1 Iseigaoka Yahatanishiku Kitakyushu, Japan 807-8555

E-mail: gamisan@med.uoeh-u.ac.jp

Abstract. We examined the biological effects of the nickel oxide (NiO) nanoparticles by inhalation and instillation study. Wistar male rats were exposed to NiO nanoparticles (nNiOs) for 4 weeks (6 hrs/day). The nNiOs was black-colored NiO (99.8%) and average particle size (APS) was 20 nm. The geometric mean diameter of the particles in the chamber and the daily average exposure concentration were 139 ± 12 nm and $1.0 \pm 0.5 \times 105$ particles/cc, respectively. The deposited amount of nNiOs in the rat lungs at 4 days after the inhalation exposure ended was 29 ± 4 µg. Although nNiOs exposure group showed temporal significant increase in the number of total cells in brochoalveolar lavage fluid (BALF) at 4 days after the exposure end, the difference were not seen at one month after an exposure end compared with control group. The histopathological change was not severe just after the inhalation nor throughout the observation time. Elemental mappings of nickel showed that nickel particles were located in agglomeration at the pulmonary macrophages.

Inhaled Particles X, (23–25 September 2008, Manchester) IOP Publishing
Journal of Physics: Conference Series **151** (2009) 011002 doi:10.1088/1742-6596/151/1/011002

Cell shape imaging analysis: a fast and reliable technique for the investigation of internalised carbon nanotubes in flat macrophages

F Tian[1], A Prina-Mello[2], G G Estrada,[3]A Beyerle,[1] W Kreyling[1] and T Stoeger[1]

[1] Institute for Inhalations Biology, German Research Centre for Environmental Health, Munich, Germany.

[2] Centre for Research on Adaptive Nanostructures and Nanosciences, Trinity College, Dublin, Ireland.

[3] Institute for Bioinformatics, German Research Centre for Environmental Health, Munich, Germany.

Email: furong.tian@helmholtz-muechen.de

Abstract: The aim of this work is to elucidate the mechanisms involved in the morphological adaptation and regulation of macrophages in the presence of internalised materials. This development will accelerate the toxicology assessment of novel nanomaterials and subsequently reduce their environmental and health exposure. For this purpose, we adapted our established in vitro culture system to investigate and measure cell shape changes with and without functionalized carbon nanotubes (CNTs). Two nanomaterials, such as fluorescent polystyrene (PS) beads and functionalized CNTs, were employed to track the material location under confocal microscopy, light microscopy and Transmission Electron Microscopy (TEM). It was found that particles equally spread throughout the entire cytoplasm in spherical macrophage; whereas when macrophages where forced to adhere to the substrate, via fibronectin coating, the accumulation of particles and tubes was limited to the vicinity of the nucleus due to the modified cellular micro architecture. TEM analysis also confirmed these findings and demonstrated that CNTs of about 5 μm laid at the bottom of adherent cells. Therefore, this cell shape analysis and manipulation may result very important for the quantification of internalised novel materials with high aspect ratio like nanotubes, nanorods and nanowires.

Journal of Physics: Conference Series **151** (2009) 011002 doi:10.1088/1742-6596/151/1/011002

Screening strategy to avoid toxicological hazards of inhaled nanoparticles for drug delivery: the use of α-quartz and nano zinc oxide particles as benchmark

Andrea Beyerle[1,2], Holger Schulz[1], Thomas Kissel[2], Tobias Stoeger[1]

[1] Institute of Inhalation Biology, Helmholtz Zentrum München, German Research Center for Environmental Health (GmbH), Ingolstädter Landstraße 1, D - 85764 Neuherberg, Germany; [2] Department of Pharmaceutical Technology and Biopharmacy, Philipps- University Marburg, Germany

e-mail: tobias.stoeger@helmholtz-muenchen.de

Abstract: Nanotechnology is a broad, revolutionary field with promising advantages for new medicine. In this context the rapid development and improvement of so called nanocarriers is of high pharmaceutical interest and some devices are already on the market. In our project we aim to develop well characterized nanoscaled drug delivery systems for an inhalative application. To this end, we focus on the most adverse side-effects within the lung, the cytotoxic and the proinflammatory responses to these nanoparticles (NPs). Before performing any animal experiments, we start with an in vitro screening for analyzing the cytotoxic and proinflammatory effects of the investigated particles on two murine lung target cell lines, the alveolar epithelial like typ II cell line (LA4) and the alveolar macrophage cell line (MH-S). Three different endpoints were estimated, (i) cellular metabolic activity, determined by the WST-1 assay, (ii) membrane integrity, by detection of LDH release and hemolytic activity, and (iii) secretion of inflammatory mediators. To analyze the relative particle toxicity we choose two reference particles as benchmarks, (i) fine α-quartz, and (ii) ultrafine ZnO particles. The investigation of dose-response and kinetics of proinflammatory and toxic effects caused to the named cell lines provide an insight to a close evaluation of our cell based screening strategy. α-quartz is well known for its inflammatory and toxic potential caused by inhalation, and nanosized ZnO particles - used in a broad field of nanotechnology like electronics, but also cosmetics and pharmaceuticals - is to a high degree cytotoxic and proinflammatory in vitro. Preliminary experiments indicated not only particle and cell specific inflammatory responses, but also different susceptibilities of the cell types being exposed to our benchmark particles regarding their size and surface activities. Exposure to the μm-sized α-quartz particles affected the viability of epithelia cells less than that of macrophages, pointing to the impact of particle uptake by phagocytosis. In contrast, the nanosized ZnO particles caused much stronger decrease in cell viability and higher levels of LDH in the macrophage cell line compared to epithelial cells, even though the hemolytic activity was much higher for the α-quartz particles than for the nanosized ZnO. For the proinflammatory effects, we observed a clear dose-dependent release of acute phase cytokines (TNF-α, IL-6, G-CSF> CXCL10>CCL2) for both alveolar cell lines after Min-U-Sil exposure. After ZnO treatment the cytokine responses were negligible compare to control cells. In conclusion, our data attach value to the use of different cell types to detect different pathways of toxicity generated by different particle properties. Therefore, we will establish both lung target cell lines for an in vitro screening to analyze proinflammatory and cytotoxicity effects of nanocarriers. The implementation of the two reference particles facilitate the validated classification of the cytotoxic responses caused by the NPs investigated.

Development of a Bolus Injection System for regional deposition studies of nanoparticles in the human respiratory system

V Koujalagi[1], S L Ramesh[2], G P P Gunarathne[2], S Semple[1] and J G Ayres[1]

1 Dept. of Environmental and Occupational Medicine, Liberty Safe Work Research Centre, University of Aberdeen, Foresterhill Road, Aberdeen AB25 2ZP, UK

2 School of Engineering, The Robert Gordon University, Schoolhill, Aberdeen AB10 1FR, UK

Email: g.gunarathne@rgu.ac.uk

Abstract. This study presents the work carried out in developing a precision bolus injection system in order to understand the regional deposition of nanoparticles (NP) in human lung. A real-time control system has been developed that is capable of storing graphite NP, assessing human breathing pattern and delivering a bolus of the stored NP at a pre-determined instance of the inhalation phase of breathing. This will form the basis for further development of a system to deliver radioactive nanoparticles to enable 3-dimensional lung imaging using techniques such as positron emission tomography (PET). The system may then be used to better understand the actual regional deposition in human lung, which could validate or challenge the current computational lung models such as that published by the International Commission for Radiation Protection (ICRP-1994). A dose related response to inhaled PM can possibly be shown, which can be used to review the current workplace exposure limits (WELs).

Journal of Physics: Conference Series **151** (2009) 011002 doi:10.1088/1742-6596/151/1/011002

Generation of Nanoparticle Agglomerates and their Dispersion in Lung Serum Simulant or Water

B A Wong[1], D G Nash[2], O R Moss[1]

[1]The Hamner Institutes for Health Sciences, PO Box 12137, Research Triangle Park, NC 27709

[2]Biostatistics, University of North Carolina, Chapel Hill, NC 27599

bwong@thehamner.org

Abstract. Nanoparticles released into the atmosphere, due to their high diffusivity, will likely begin to agglomerate. The state of agglomeration upon inhalation and the potential to disperse back into nanoparticles may affect the toxicity of the inhaled material. In order to investigate particle dispersion, a system was set up to generate aggregates from agglomerates. Primary particles, composed of zinc, were generated using zinc rods in a spark generator (Palas GFG-1000, Karlsrhue, Germany). These particles formed agglomerates which were passed through a room temperature aging chamber or through a tube furnace (Carbolite HST, Derbyshire, UK). Agglomerate size was measured with a scanning mobility particle sizer (SMPS model 3936, TSI Inc., Shoreview, MN). When furnace temperature was set near the zinc coalescence temperature, instead of decreasing in size, agglomerate size increased up to 30%; a percentage increase duplicated with the room temperature aging chamber. Starting with an aerosol of primary zinc particles, equal concentrations of agglomerate and aggregrate aerosol were produced. The extent of breakup and dispersion of agglomerates and aggregates to individual nanoparticles in lung serum simulant will be assessed using transmission electron microscopy.

Development of facilities for nanoparticle inhalation studies

M R Bailey, G Etherington, R Fielder, C B Howarth, A Hodgson and R L Maynard

Health Protection Agency, Centre for Radiation, Chemical, and Environmental Hazards, Chilton, Didcot, Oxon. OX11 0RQ.

Abstract: "Nanoparticles" are materials engineered to be smaller than 100 nanometres (nm) in at least one dimension because of the useful properties conferred by their small size. Their range and uses in industry, medicine and consumer products are expanding rapidly. Exposure to nanoparticles via inhalation, ingestion, or through the skin could occur during manufacture, during use, and as a result of product breakdown and disposal. Serious concerns about their possible health effects have been raised, for example, in the 2004 report by the Royal Society and the Royal Academy of Engineers, which called for major investment by the UK government into nano-toxicology. Work in the air pollution field gives concern about the potential toxicity of inhaled nanoparticles. Many experts believe that particles <100 nm (termed "ultrafine" in this context) account for much of the toxicity of the ambient aerosol, and are linked to cardiovascular disease. Concerns have also been raised because clearance from the lungs can be quite different from that of larger particles: those in the size-range 10–50 nm may be retained in the lungs much longer, while smaller particles may rapidly enter the blood, and transfer to other organs, including the unborn child. Hence the nano-toxicity of a material is not predictable from knowledge of its toxicity in other forms. Experts agree that carefully controlled inhalation studies are required for predicting effects on man. Recognising the need for such studies, and that no suitable inhalation facilities exist in the UK, the Health Protection Agency is developing a collaborative project between its Radiation Protection Division (RPD) and its Chemical Hazards and Poisons Division (CHaPD) to address these concerns. The new facility will have equipment and expertise to generate and characterise aerosols of nanoparticles and to administer them by inhalation to laboratory rodents under closely controlled conditions. The materials would be both "test particles", specifically fabricated with properties designed to investigate mechanisms, such as translocation of particles from the respiratory tract via the blood to other organs, and also subject well characterised nanoparticles to toxicological investigations. It is anticipated that once the facility was established, a wide range of research studies would be carried out in collaboration with UK research groups in academia and industry. RPD already has facilities for administering larger particles by inhalation to both laboratory rodents and human volunteers, and extensive experience of aerosol generation and characterisation. Plans are being developed to convert its small animal facility to administer aerosols of nanoparticles by inhalation (nose-only and intratracheal intubation). It is expected that by the time of the IPX Conference in September 2008 construction will be well advanced and some testing may have been carried out. Presentation of the project at the conference will be an ideal opportunity to bring this resource to the attention of potential collaborators, and also enable suggestions to be made that could be considered in its future development.

Journal of Physics: Conference Series **151** (2009) 011002 doi:10.1088/1742-6596/151/1/011002

Surface physico-chemistry dictates the potency and mechanism of platelet aggregation caused by similar-sized polystyrene latex nanoparticles

K Donaldson, C McGuinnes, L Tran, V Stone and R Duffin

Queens Medical Research Institute, University of Edinburgh , Edinburgh , Scotland

ken.donaldson@ed.ac.uk

Abstract. The translocation of combustion-derived nanoparticles from the lung surface into the bloodstream has been put forward as one possible mechanism for adverse cardiovascular effects described in the PM_{10} literature. Manufactured nanoparticles may pose the same potential risks. Limited studies have shown that NP in the blood may activate platelets causing increased adhesion and aggregation, so increasing the potential for, or severity of, thrombosis. There is a need to understand the effects of particle characteristics on the various endpoints of particle toxicity. In the present study we have examined the effect of a panel of 3 polystyrene latex nanoparticles (NP) of the same nominal size but possessing different surfaces as model NP, on whole blood and isolated platelets. The different surfaces were unmodified (umPLNP), aminated (aPLNP) and carboxylated (cPLNP). We show that the PLNP in whole blood associate with platelets and that the platelets then interact with monocytes to form monocyte platelet aggregates. Both cPLNP and the aPLNP cause platelet aggregation whilst the umPLNP do not. Whilst cPLNP cause aggregation by classical up-regulation of adhesion receptors aPLNP do not up-regulate adhesion receptors and appear to act by a more intense interaction and perturbation of the platelet membrane, directly revealing anionic phospholipids. Particle-derived oxidative stress is not the mechanism for these effects as shown by EPR analysis of the different PLNP.

Inhaled Particles X, (23–25 September 2008, Manchester) IOP Publishing
Journal of Physics: Conference Series **151** (2009) 011002 doi:10.1088/1742-6596/151/1/011002

A good practice guide for safe work with nanoparticles: the Quebec Approach

C Ostiguy[1], B Roberge[1], L Ménard[2] and C A Endo[3]

[1]: IRSST, 505 De Maisonneuve Blvd. West, Montréal, Québec, Canada, H3A 3C2

[2]: CSST, De Bleury, Montréal, Québec, Canada

[3]: NanoQuébec, Montréal, Québec, Canada

E-mail: claude.ostiguy@irsst.qc.ca

Abstract: new industrial revolution has already begun around nanotechnologies, letting us anticipate major scientific breakthroughs that will affect each economic activity sector and whose expected global economic impacts will exceed $1000 billion annually by 2012. Simultaneously, many studies reveal that nanoparticles represent different occupational health and safety (OHS) risks unique to them and that often differ from the risks related to the same chemical substances with larger dimensions. As the number of potentially exposed workers increases and much uncertainty persists about OHS risks, this extended abstract proposes a framework for occupational risk management with the objective of controlling exposure to NPs in a context of a major lack of specific data related to the hazards of these substances and to the level of occupational exposure. The framework takes into consideration the equal representation of both the employers and workers in the Québec legislation and accounts the potential routes of exposure and focuses on a structured approach dealing with hazard identification, exposure characterization, risk assessment and risk management through different control methodologies. These are included in a prevention program that must be followed up, once it has been implemented, and refined through an iterative approach as new data become available.

Inhaled Particles X, (23–25 September 2008, Manchester) IOP Publishing
Journal of Physics: Conference Series **151** (2009) 011002 doi:10.1088/1742-6596/151/1/011002

Measurement of number and mass concentration of particulate matter in rural and urban Pakistani households

I Colbeck and Z A Nasir

Department of Biological Sciences, University of Essex, Colchester, CO4 3SQ, UK

colbi@essex.ac.uk

Abstract. Most epidemiological studies have shown an association between particulate air pollution and acute and chronic respiratory morbidity both in adults and children. Particulate air pollution is a major emerging environmental concern in Pakistan, both outdoor and indoors, due to increased urbanization in cities and the use of biofuel in rural areas. In Pakistan, almost 70% of the population lives in rural areas and use wood, dung, crop residue or natural gas as a fuel for cooking and heating. Measurements of particulate mass and number concentration have been undertaken in rural and urban sites in Pakistan during October – November 2007. The values of the number concentration for Lahore were in the range 13,771 – 181,029 cm^{-3}. Smoking had a major share in the maximum number concentration values. Indoor number concentrations in Lahore were considerably greater than those observed outdoors in European cites. With reference to the mass concentration of particulate, at both the rural and urban sites, there was a large variation in mass of PM10, PM2.5 and PM1. At rural sites, cooking caused a remarkably high concentration of both mass and number concentration in the kitchen. In some cases a concentration of 300,000 cm^{-3} and 8000 $\mu g/m^3$ was recorded. These concentrations were due to incomplete combustion in traditional stoves in combination with poor ventilation. While in living rooms smoking and cleaning was the catalyst for the rise in concentration. In Lahore the indoor mass concentration fluctuated due to personal activities of the inhabitants. It was generally in range 131 $\mu g/m^3$ to 1200 $\mu g/m^3$. However during smoking and cleaning the concentration could go up to 7854 $\mu g/m^3$

Inhaled Particles X, (23–25 September 2008, Manchester) IOP Publishing
Journal of Physics: Conference Series **151** (2009) 011002 doi:10.1088/1742-6596/151/1/011002

Relationship between indoor particulate matter and carbon monoxide levels in Nepalese homes.

O P Kurmi[1,2], M Steiner[1], G D Henderson[1], S Semple[1], P P Simkhada[2], J G Ayres[1]

1 Department of Environmental and Occupational Medicine, University of Aberdeen, UK.

2 Department of Public Health, University of Aberdeen, UK

Abstract: In many developing countries the use of biomass fuels produces high indoor air concentrations of fine particulate matter and carbon monoxide. Some cross-sectional studies have examined the relationship between particulate matter (PM) and carbon monoxide (CO) from solid fuels. The present study compares data from a light-weight HOBO CO logger with that derived from a co-located PM sampler (TSI Sidepak and Dustrak). The CO device is a diffusive instrument and does not have the inherent disadvantages of noise and limited battery life that active pumped samplers suffer from. The study aimed to look at the relationship between indoor air concentrations of CO and PM in Nepalese homes using a range of fuels for cooking and lighting, Methods: We carried out a cross-sectional study in two rural areas and in urban Kathmandu Metropolitan City in Nepal. 24-hour indoor particulate matter ($PM_{2.5}$) concentrations (n=424) were measured with two types of photometric devices (TSI SidePak and DustTrak) and carbon monoxide levels (n=126) by HOBO CO logger. Results: Mean 24-hour indoor $PM_{2.5}$ was significantly higher in rural (n=210) compared to urban homes (n=221) (454μg/m^3 vs 133μg/m^3). Similarly, the mean 24-hour indoor CO was higher in rural (n=40) compared to urban homes (n=86) (17ppm vs 2.62ppm). The Pearson correlation between $PM_{2.5}$ (SidePak) and CO measurements in rural area is 0.49 which is significant at 0.05 levels. Similarly, the Pearson correlation between $PM_{2.5}$ (DustTrak) and CO measurements in rural area is 0.69 which significant at 0.01 level. $PM_{2.5}$ and CO measured in urban areas are not significant (p= 0.38). Conclusion: The study found that CO is a good proxy for $PM_{2.5}$ in rural household using biomass as the main fuel in traditional open stoves. CO cannot be used as a proxy for $PM_{2.5}$ in urban homes using LPG gas as their main fuels where there is near to complete combustion and vehicle traffic is the major source of CO.

Domestic work: exposure to Indoor Air Pollution in Nepalese homes

Om P Kurmi, Sean Semple, Markus Steiner, George D. Henderson, Jon G Ayres

Department of Environmental and Occupational Medicine, University of Aberdeen, Aberdeen, UK

Abstract. To measure particulate matter (PM) exposure of people involved in domestic work in urban and rural Nepal and to examine the performance of photometric devices in collecting this data. This paper details the results of these measurements and derives calibration factors for two photometric devices using gravimetric sampling as the gold standard. This work formed part of a cross sectional study to examine the relationship between indoor and outdoor air pollution in relation to respiratory and cardiac health among people in Nepal. Between April 2006 and February 2007 respirable PM levels were measured over a 24-hour period in 491 households in a range of urban and rural settings in the Kathmandu valley of Nepal. Sampling was carried out by photometric and gravimetric methods with the co-located gravimetric data used to derive a calibration factor for the photometric devices. The time weighted average (24 hour) respirable dust levels ranged from 280 to 8400 μg/m3 in the rural settings and 56 to 320 μg/m3 in the urban settings by photometric devices. Similarly, the time weighted average (24 hour) respirable dust levels ranged from 13 to 2600 μg/m3 in the rural settings and 2 to 110 μg/m3 in the urban settings by gravimetric methods. The co-located photometric and gravimetric devices indicate that the SidePak Monitor required a correction factor of 0.48 and 0.51 in rural and urban data respectively whereas the DustTrak device required a factor of 0.31 and 0.35 in rural and urban data respectively to correct for the particle size and density of the biomass smoke. Those involved in domestic work in rural Nepal are exposed to respirable dust concentrations that average approximately 1400 μg/m3 for a period of 24 hours. Converted to an 8-hour time-weighted average, this equates to more than the current UK Workplace Exposure Limit value for Nuisance dust (4 mg/m3). Homemakers, primarily women, spend a large proportion of their lives indoors in these high respirable dust concentrations and these exposures are likely to produce respiratory illness. Exposure can be controlled by the use of different fuel types and/or the use of flued stoves.

The UCB Particle Monitor: A tool for logging frequency of smoking and the intensity of second-hand smoke concentrations in the home

Sean Semple, Andrew Apsley, Gill Moir, George Henderson, Jon Ayres

Department of Environmental & Occupational Medicine, Liberty Safe Work Research Centre, University of Aberdeen, Foresterhill Road, Aberdeen AB25 2ZP, UK

E mail sean.semple@abdn.ac.uk

Abstract. Second-hand tobacco smoke (SHS) exposure generates a large public health burden. Recent legislation has moved to prohibit smoking in public places and there are concerns that this may lead to an increase in exposures in private homes. Measurement of SHS aerosol has tended to use active pumped samples or longer-term diffusive badges. Pumped methods are noisy and poorly tolerated in home settings while diffusive badges do not provide real-time data. The UCB particle monitor (UCB-PM) is a modified smoke-alarm device capable of logging changes in airborne particulate matter over extended periods and has been used successfully to measure biomass fuel smoke concentrations in developing world settings This study has examined the use of the UCB-PM to measure SHS aerosol in both controlled laboratory conditions and a pilot field trial over a 7 day period in a smoker's home. Comparisons with a pumped sampler (TSI Sidepak Personal Aerosol Monitor) indicate good agreement over a range of exposure concentrations but there is evidence of a threshold effect at approximately 0.5 mg/m^3 of fine particulate measured as $PM_{2.5}$. While this threshold effect undermines the ability of the device to provide useful data on the time-weighted average SHS concentration, the field trial indicates that that the UCB-PM has a sensitivity of about 71% and a specificity of 98%. The device has many advantages including zero noise operation, low cost and long battery life and may be a useful tool in quitting and smoke-free home intervention studies.

Personal exposure to airborne ultrafine particles in the urban area of Milan

A Cattaneo[1], G Garramone[2], M Taronna[2], C Peruzzo[2,3] and D M Cavallo[2]

1 Department of Occupational Health, University of Milan, via S.Barnaba 8, 20122 Milan, Italy

2 Department of Chemical and Environmental Sciences, University of Insubria, c/o Clinica del Lavoro "L.Devoto", via S.Barnaba 8, 20122 Milan, Italy

3 Occupational and Preventive Health Division, Macchi Foundation Hospital, Viale Borri 57, 21100 Varese, Italy

E-mail: domenico.cavallo@uninsubria.it

Abstract. The relevance of health effects related to ultrafine particles (UFPs; aerodynamic diameter < 100 nm) can be better evaluated using high-resolution strategies for measuring particle number concentrations. In this study, two different portable Condensation Particle Counters (CPCs) were used to measure personal exposure to UFPs in the central area of Milan for one week period during spring, with three sampling sessions per day. Experimental data were continuously collected along an established urban pathway, moving afoot or by different private and public means of transport. Correlation analysis between data measured by two CPCs was performed and general results showed a good agreement, especially at concentrations lower than 2×10^5 particles /cm^3. UFPs measures were divided on the basis of crossed environments or micro-environments, days of the week and day time (hours). The highest measured mean concentrations and data variability were observed during walking time and moving on motorized vehicles (bus and car), indicating that the highest exposure to UFPs can be reached near motorized traffic. The lowest exposures were observed in green areas and in office microenvironments. An appreciable difference between working and non-working days was observed. Concentration patterns and variation by days of the week and time periods appears related to time trends in traffic intensity.

Transport-Related Ultrafine Particle Exposure to Residents in the Suburbs (TRUERS)

A Leavey[1], I D Longley[1,3], P Harris[4], F de Vocht[2], S J Lindley[4], M W Gallagher[1]

[1]Centre for Atmospheric Science, the University of Manchester, Manchester, UK

[2]Centre for Occupational and Environmental Health, the University of Manchester

[3]National Institute of Water and Atmospheric Research Ltd, Auckland, New Zealand

[4]School of Environment and Development, the University of Manchester

anna.leavey@postgrad.manchester.ac.uk

Abstract: An increasing number of recent studies implicate Ultrafine Particles (UFP) s the more toxic fraction of PM10, responsible for cardiovascular complications and respiratory health effects. In the UK, road transport is considered the primary outdoor source of UFP, and as such it is within the vicinity of a main road that one expects the highest airborne concentrations. UFP concentrations are highly spatially variable and therefore traditional fixed-site monitoring stations may not be valid proxies for personal exposure. Given this variability, the assessment of the influence of distance from main roads on residential UFP exposure is important given the high proportion of time (typically 80-90%) that most people spend indoors. Whereas previous studies assessing UFP concentrations in association with distance to roads primarily focused on motorways or city centre roads, the aim of this project is to assess the spatial influence of roads in suburban locations on UFP concentrations, taking into account factors such as street geometry, traffic characteristics and meteorological conditions, to investigate whether residential locations near main roads experience elevated outdoor UFP concentrations compared to locations further away. Measurements were collected at 6 main roads during November and December 2007, in residential locations within the south Manchester area. Sites were selected based on the geometry of surrounding roads and distance from motorways and open park land. One portable condensation particle counter (TSI P-TRAK Model 8525) sampled on a curb by the road while a second similar counter sampled simultaneously at different distances (up to 200 meters) away from the road. Measurements were made at a 1 Hz frequency over a period of 5 minutes at 5 to 10 meter sampling intervals. Ambient background levels were recorded upwind from the measurement site, while meteorological data was collected using a hand held weather station. Finally, the frequency and type of traffic that passed during each sampling interval were also recorded. In total 6 roads were sampled an average of 6 times over 15 days. Average measured UFP concentrations at the different sites ranged from 15,000 to 98,000 (pt/cc) (curb side) with ambient ranging from 5,000 to 20,000 (pt/cc). Preliminary analyses suggest decreasing UFP concentrations with increasing distance from the road, characterised by a rapid exponential decline within the first 10 metres and stabilising at approximately 40 metres from the road. The variability of concentrations was also found to decline with increasing distance from the road. Additional analyses will assess this association as a function of street geometry, meteorology and traffic characteristics.

Inhaled Particles X, (23–25 September 2008, Manchester) IOP Publishing
Journal of Physics: Conference Series **151** (2009) 011002 doi:10.1088/1742-6596/151/1/011002

Comparing urban particle emission fluxes measured on the BT Tower (London) with measurements from Manchester, Edinburgh and Gothenburg

C L Martin[1], I D Longley[2], J R Dorsey[1], R M Thomas[1,3], M W . Gallagher[1], E Nemitz[3]

[1]Centre for Atmospheric Science, SEAES, University Of Manchester. UK

[2]NIWA Ltd, Newmarket, Auckland. New Zealand

[3]Centre for Ecology and Hydrology, Penicuik, Edinburgh. UK

claire.martin@postgrad.manchester.ac.uk

Abstract. Turbulent particle number emission fluxes have been measured directly using the eddy covariance technique in four European cities as part of the NERC SASUA project (Edinburgh, 1999-2001), the CEH Göte-2005 Project (Gothenburg, 2005), the NERC CityFlux project (Manchester, 2005-6) and the NERC/BOC Foundation REPARTEE project (central London, 2006). Urban areas have many emission sources of aerosol particulates including densely trafficked roads and dense industrial and commercial zones. As people both live and work in urban areas it is of particular importance to understand the atmospheric dispersion of particles in these areas. Urban particle concentrations are dependent upon the rate at which the urban canopy is ventilated, i.e. the vertical flux from the urban canopy. This paper aims to review the basic features of the flux data and compare the Edinburgh, Gothenburg, Manchester and London campaigns. In London, measurements were made on top of the BT Tower at a height of 189 m above ground level. These measurements occurred during October 2006 and ran concurrently with a particle characterisation measurement site in the centre of the Regent's Park approximately 1 km away. In Manchester, measurements were on top of one of the city's tallest buildings - the Portland Tower - at a height of 90 m above the city centre. This campaign consisted of three separate month-long periods in June 2005, February 2006 and May 2006. Gothenburg measurements were taken from atop the Skanska Lilla Bommen Tower during February 2005. CPC flux systems were deployed based upon a TSI 3010 or TSI 3025A, adapted from the system deployed above the city of Edinburgh through November 2000 using a TSI 3760. The basic features of the particle number flux data are compared for all the campaigns. Analysis indicates that in all campaigns the particle number fluxes exhibit a clear diurnal cycle with an early afternoon peak that is closely related to the diurnal sensible heat flux cycle. Dependencies on wind direction have been identified: there is an observable relationship between high particle number fluxes and areas of heavy urbanisation. The measured fluxes of all the campaigns have been related independently to traffic statistics where available. Further details on these campaigns and their analysis have been published by Martin et al. (Martin, C.L., Longley, I.D., Dorsey, J.R., Thomas, R.M., Gallagher, M.W., Nemitz, E., 2008. **Ultrafine particle fluxes above four major European cities.** *Atmospheric Environment,* Accepted for publication).

Novel aerosol sampling techniques for workplace health risk assessment

Richard Bateman [1], **Bob Muir** [2],

[1] CERAM Research Ltd, Queens Road, Penkhull, Stoke-on-Trent, ST4 7LQ, UK

[2] Naneum Ltd, Canterbury Enterprise Hub, University of Kent, Canterbury, Kent CT2 7NJ, UK

richard.bateman@ceram.com

Abstract. Health risks associated with respirable particles are known to be dependent on particle size which in turn directly influences exposure and dose. EU funded projects awarded to investigate the potential effects of respirable silica and lead particles on health, led to the development of two novel sampling instruments – a wide range aerosol sampler (WRAS) and a high volume sampler. The WRAS collects a series of size resolved samples across the whole aerosol in the range of particle sizes from 2nm- 30 microns and was first used to investigate exposures to lead containing aerosols arising from raw material handling, cutting and polishing processes at crystal glass factories. By analysing each collected size fraction using XRF, Lead concentration size distributions were then developed. Similarly, respirable crystalline silica (RCS) samples were collected at various ceramic factories across Europe using WRAS and the high volume sampler. The high volume sampler collected RCS particles in sufficient quantities for in vitro and in vivo toxicology studies. WRAS provided valuable data on PSD and thus what percentage of airborne material is likely to reach the alveolar region of the lungs with potential to cause damage. In the case of the lead particles, analysis showed no correlation between health risks and PM10, PM2.5 or PM1 but a strong correlation with PM0.2. For the respirable silica, in-vitro toxicology tests showed a wide variation in effects depending on the chemical nature of the respirable silica (which is dictated by processing history). The conclusion drawn from this work is that a "one size fits all" regulatory regime is unlikely to be appropriate.

Journal of Physics: Conference Series **151** (2009) 011002 doi:10.1088/1742-6596/151/1/011002

Occupational Exposure to Mineral Oil Metalworking Fluid (MWFs) Mist: Development of New Methodologies for Mist Sampling and Analysis. Results from an Inter-laboratory Comparison

C Khanh Huynh[1], H Herrera[1], J Parrat[2], R Wolf[3] and V Perret[4]

1. Institut Universitaire Romand de Santé au Travail (IST), CH-1005 Lausanne, Switzerland

2. Service des Arts et Métiers et du Travail (LIST), CH-2800 Delémont and Laboratoire intercantonal de santé au travail (LIST), CH-1034 Peseux, Switzerland

3. SUVA, CH-6002 Luzern, Switzerland

4. Service cantonal de toxicologie industrielle et de protection contre les pollutions intérieures (STIPI), CH-1211 Geneve, Switzerland

chhuynh@hospvd.ch

Abstract. Metalworking Fluids (MWFs) are largely used in the sector of undercutting, a large professional activity in Switzerland, in particular in the fine mechanic and watch making industry. France proposes a Permissible Exposure Limit (PEL) of 1 $mg.m^{-3}$ of aerosol. The American Conference of Governmental Industrial Hygienists (ACGIH) sets its value at 5 $mg.m^{-3}$ but a proposal to lower the standard ("intended changes") to 0.2 $mg.m^{-3}$ of aerosol is pending since 2001. However, it has not become a recognized threshold limit value for exposure. Since 2003, the new Swiss PEL (MAK) recommendations would be 0.2 $mg.m^{-3}$ of aerosol (oil with boiling point > 350°C without additives) and/or 20 $mg.m^{-3}$ of oil aerosol + vapour for medium or light oil. To evaluate evaporative losses of sampled oil, the German "Berufsgenossenschaftliches Institut für Arbeitssicherheit" (BGIA) recommends the use of a XAD-2 cartridge behind the filter. The method seems to work perfectly for MWFs in a clean occupational atmosphere free from interference of light vapour cleaning solvent such as White Spirit. But, in real situation, machine shop atmosphere contaminated with traces of White Spirit, the BGIA method failed to estimate the MWFs levels (over-estimation). In this paper, we propose a new approach meant to measure both oil vapours and aerosols. Five inter-laboratory comparisons are discussed, based on the production of oil mist in an experimental chamber under controlled conditions.

A mini-sampler for welding aerosol mounted in close vicinity of the mouth/nose

Göran Lidén[1] **and Jouni Surakka**[2]

[1] Department of Environmental Science, Stockholm University, Stockholm, Sweden
[2] Swedish Work Environment Authority, Solna, Sweden

eMail: Goran.Liden@itm.su.se

Abstract. A small personal aerosol mini-sampler to be used inside modern welding visors has been developed. The main object of the mini-sampler has been to sample manganese. The sampler is based on commercially available 13 mm filter holders but modified to incorporate an inlet nozzle made of aluminium. The nominal flow rate of the mini-sampler is 0,75 l/min. The sampler is to be worn mounted on a headset, modified from professional microphone headsets. The headset mounting arrangement was accepted by the welders. The sampling bias of the mini sampler versus the IOM sampler depends on the coarseness of the sampled aerosol. At the lowest concentration ratio of the open-face 25 mm filter holder to the IOM sampler equal to 0,65, the bias of the mini sampler is approximately -26% versus the IOM. The RMS sampling bias of the mini sampler versus the IOM sampler for manganese is -4,6%. The inhalable fraction of welding aerosol mass consists only of 25-55% of welding fume. The rest of the mass is made up of spatter particles and grinding particles. For manganese generally more than 65% is found in the fume.

Characterisation of fume from hyperbaric welding operations

John A S Ross[1,5], Sean Semple[1], Rodger Duffin[2], Frank Kelly[3], Joerg Feldmann[4] and Andrea Raab[4].

[1] Environmental and Occupational Medicine, University of Aberdeen.

[2] ELEGI Colt Laboratory, University of Edinburgh.

[3] Lung Biology Group, Kings College, University of London.

[4] Trace Element Speciation Laboratory, University of Aberdeen.

[5] Environmental and Occupational Medicine, University of Aberdeen, Aberdeen AB25 2ZP,

email j.a.ross@abdn.ac.uk, 01224 558197

Abstract. We report preliminary work characterising dust from hyperbaric welding trials carried out at increased pressure in a helium and oxygen atmosphere. Particle size and concentration were measured during welding. Samples for quartz and metal analysis and toxicity assessment were taken from a filter in the local fume extraction system. The residue of dust after metal extraction by nitric acid in hydrogen peroxide predominantly a non-metallic white powder assumed to be dust from welding rod coatings and thermal insulation material. Metallic analysis showed predominantly calcium, from the welding rod coating, and period 4 transition metals such as iron, manganese, magnesium and titanium (inductively coupled mass spectrometry, Agilent 7500c). The presence of zirconium indicated a contribution from grinding. The fume was nanoparticulate in nature with a mean particle diameter of 20-30 nm (MSI Inc WPS 1000XP). It showed an intermediate level of oxidative potential regarding the low-molecular weight respiratory tract lining fluid antioxidants ascorbate and glutathione and caused release of the inflammatory marker IL-8 in a human lung A 549 epithelial cell culture with no indication of cytotoxicity. The study findings have strong implications for the measurement techniques needed to assess fume exposure in hyperbaric welding and the provision of respiratory protection.

Investigation of methods for the sampling of airborne nanoparticles by electron microscopy

Delphine Bard[1], Andrew Thorpe[1], Derrick Wake[1], Garry Burdett[1] and Minnamari Vippola[2]

[1] Health and Safety Laboratory, Harpur Hill, Buxton SK17 9JN, UK

[2] Finnish Institute of Occupational Health, Topeliuksenkatu 41 a A, FI-00250 Helsinki, Finland

delphine.bard@hse.gov.uk

Abstract: Engineered nanomaterials include products manufactured at the nanoscale, with unique shapes and enhanced physical and chemical properties, compared with conventional materials of the same composition. Exposure to these materials during manufacturing and applications may occur through inhalation, dermal contact or ingestion. Occupational health risks and hazards associated with the manufacturing and use of nanoparticles are not yet clearly understood. Their development presents new challenges for the working environment, for the wider environment and also the public. There are a number of real time measurement instruments, which can assess airborne concentrations of nanoparticles in terms of mass, number and surface area. In addition to the concentration measurements, the physical and chemical characteristics of the engineered nanoparticles are important parameters for their discrimination from natural ultrafine particles or those produced from combustion (e.g. diesel particles). Physical and chemical characterisation techniques usually require the collection of a sample for off-line analysis. Analytical electron microscopy is the technique of choice but requires non-biased sampling onto an appropriate collection medium. This presentation discusses the on-going investigations into the efficiency of sampling airborne nanoparticles for electron microscopy analysis. This includes the collection of nanoparticles by point-to-plane electrostatic precipitation and filtration through polycarbonate filters and carbon films. The efficiency of these different sampling methods was assessed using a scanning mobility particle sizer and electron microscopy by generating nanoparticles of known concentration and size in a test chamber. Electron microscopy and quantitative analysis was used to quantify the particles collected. This study was partly funded by EU, NANOSH, NMP4-CT-2006-032777. The views and opinions expressed in this presentation do not necessarily reflect those of the European Commission.

Inhaled Particles X, (23–25 September 2008, Manchester) IOP Publishing
Journal of Physics: Conference Series **151** (2009) 011002 doi:10.1088/1742-6596/151/1/011002

Real-time measurement of dust in the workplace using Video Exposure Monitoring: farming to pharmaceuticals

P T Walsh, A R Forth, R D R Clark, K P Dowker and A Thorpe

Health & Safety Laboratory, Harpur Hill, Buxton, Derbyshire, SK17 9JN, UK.

peter.walsh@hsl.gov.uk

Abstract. Real-time, photometric, portable dust monitors have been employed for video exposure monitoring (VEM) to measure and highlight dust levels generated by work activities, illustrate dust control techniques, and demonstrate good practice. Two workplaces, presenting different challenges for measurement, were used to illustrate the capabilities of VEM: (a) poultry farming activities and (b) powder transfer operations in a pharmaceutical company. For the poultry farm work, the real-time monitors were calibrated with respect to the respirable and inhalable dust concentrations using cyclone and IOM reference samplers respectively. Different rankings of exposure for typical activities were found on the small farm studied here compared to previous exposure measurements at larger poultry farms: these were mainly attributed to the different scales of operation. Large variations in the ratios of respirable, inhalable and real-time monitor TWA concentrations of poultry farm dust for various activities were found. This has implications for the calibration of light-scattering dust monitors with respect to inhalable dust concentration. In the pharmaceutical application, the effectiveness of a curtain barrier for dust control when dispensing powder in a downflow booth was rapidly demonstrated.

Journal of Physics: Conference Series **151** (2009) 011002 doi:10.1088/1742-6596/151/1/011002

Multi-wall carbon nanotubes and the asbestos fibre pathogenicity paradigm

Craig A Poland[1], Rodger Duffin[1], Ian Kinloch[2]; Andrew Maynard[3], William AH Wallace[1], Anthony Seaton[4], Vicki Stone[5], Simon Brown[1], William MacNee[1] and Ken Donaldson[1*]

[1] MRC/University of Edinburgh Centre for Inflammation Research, Queen's Medical Research Institute, 47 Little France Crescent, Edinburgh EH16 4TJ, UK

[2] School of Materials, University of Manchester, Grosvenor Street, Manchester, M1 7HS, UK

[3] Woodrow Wilson International Center for Scholars, 1300 Pennsylvania Avenue, NW, Washington DC 20004-3027, USA

[4] Institute of Occupational Medicine, Research Avenue North, Riccarton, Edinburgh, EH14 4AP, UK

[5] School of Life Sciences, Napier University, Colinton Road, Edinburgh, EH10 5DT, UK

* Corresponding Author; ken.donaldson@ed.ac.uk

Abstract: Nanotechnologies promise a revolution in manufacturing, but there are accompanying risks in the potential exposure of workers and the general environment to new types of nanoparticles. Multi-wall carbon nanotubes (MWCNT) are a form of nanoparticle with increasing industrial potential with concomitant increases in production levels. However MWCNT have become of interest to toxicologists because of both their nano size and fibre-like shape, which may represent a hazard similar to that of asbestos. The structure/activity paradigm that explains the pathogenicity of asbestos is well established, defining the long rigid needle–like shape of the fibres and their ability to persist in the lungs, as requirements for pathogenicity. In order to understand whether samples of MWCNT comply with this paradigm a panel were studied with both long rctigid MWCNT and tangled/short MWCNT. Using carefully chosen controls, including long and short amphibole amosite asbestos, we used the mouse peritoneal cavity as an exposure model with several advantages for detecting potential asbestos-like properties. We demonstrated that MWCNT samples with long rigid fibres show asbestos-like behaviour in the mesothelial exposure model, but short and tangled MWCNT do not show such behaviour. Inflammation associated with long rigid MWCNT is persistent, leading to the formation of multi-nucleated giant cells and eventually diffuse fibrosis on the serosal surfaces of the peritoneal wall and viscera, similar to that seen with amphibole asbestos. These results confirm that caution is needed in introducing CNT products to market if a long-term legacy or harm is to be avoided, but that harmful asbestos-like behaviour can be avoided if nanotubes are short and/or curly.

Inhaled Particles X, (23–25 September 2008, Manchester) IOP Publishing
Journal of Physics: Conference Series **151** (2009) 011002 doi:10.1088/1742-6596/151/1/011002

In vitro models to predict toxicity and fibrogenicity of inhalable polymers and particles

Birgit Gaiser[1], Paul Daly[1], Leona Merolla[2], Philip Carthew[2], Rodger Duffin[1], Ken Donaldson[1]

[1] University of Edinburgh, Centre for Inflammation Research

[2] Safety and Environmental Assurance Centre, Unilever, Colworth

b.gaiser@napier.ac.uk

Abstract: This project is part of the Unilever R&D's Aerosol Safety Programme and aims at developing *in vitro* mono- and co-culture assays that can predict the fibrogenic potential of inhalable substances. The organic polymers S2218600, S2429901 and S2219200 (hereafter referred to as P1, P2 and P3, respectively) were used as model substances. The particles TiO_2 and Min-U-Sil (quartz) were used as negative and positive controls for fibrogenesis. The polymers caused inflammation (P3>P1>P2) and a fibrotic response (P1>P3>P2) in male Wistar rats following intra-tracheal instillations. Animals treated with P1 and P3 developed fibrosing granuloma, and all polymers caused thickening of alveolar walls. Polymers, particularly P2, increased the size of alveolar macrophages. Min-U-Sil caused fibrosis and a small increase in macrophage size. The *in vitro* cytotoxicity of the polymers as determined by LDH assay was ranked P3>>P1>P2. Collagen in HFL-1 cells was increased compared to the untreated control after treatment with P1 in fibroblast mono-cultures, and in the order P1>P3>P2 for fibroblasts in co-culture with AE2 cells. Alpha-smooth muscle actin, a protein up-regulated in fibroblasts in fibrosis and in epithelial cells undergoing epithelial-mesenchymal transition, was up-regulated in RFL-6 and RLE-6TN cells. These changes were not observed for Min-U-Sil and TiO_2.Genomic analysis of whole rat lung after instillation of the polymers revealed a list of genes, including IGFBP-2 and -3, CCL-7, □-2-macroglobulin and transcriptional regulators ID1-3, which were changed in cell cultures as well as in rat lung, and are potential candidates for *in vitro* screening assays. *In vivo*, the polymers and Min-U-Sil influenced expression of genes involved in cytokine signalling, extra-cellular matrix remodelling, oxidative stress and inflammation. In summary, *in vitro* assays can be used to predict some chronic effects of polymers. The ranking of the polymers established for inflammation *in vivo* was reflected in the cytotoxicity, and the fibrogenic responses *in vivo* by the increases in collagen and □-sma *in vitro*. Transcriptomic analysis lacked sufficient correlation between *in vivo* and *in vitro* gene expression to warrant the use of whole genomic analysis as an alternative for animal testing. Polymers and particles showed very different profiles, and appear to work by different mechanisms of action.

Alveolar macrophage accumulation rates, for 28 nm and 250 nm PSL, are mediated by separate mechanisms.

O R Moss and V A Wong

The Hamner Institutes for Health Sciences, Research Triangle Park, NC 27509-2137, USA;

moss@thehamner.org

Abstract. When macrophages accumulate 28 nm and 250 nm diameter polystyrene latex (PSL) beads, the accumulation rates should reflect differences in molecular and cellular function. We used a confocal microscope to measure the accumulation rates of nanoparticles by F344-rat-alveolar macrophages (~25,000 cells adhered to a 0.7 cm^2 surface). Over the cells were layered 0.1 ml of media, and 0.1 ml of media-with-beads. Fresh cells were introduced for each exposure scenario. The maximum possible individual macrophage exposures were as follows: $8x10^6$, $8x10^5$, and $8x10^4$ 28 nm beads per macrophage; and $8x10^4$ and $1.12x10^4$ 250 nm beads per macrophage. Accumulation rates were estimated over 23 minutes. The increase in bead accumulation-rate matched changes in bead-availability: 7x increase for 250 nm beads; 100x increase for 28 nm beads; and 700x increase for all bead availabilities. The maximum sustained 28 nm bead accumulation rate was > 30,000 /min (for 5 min). Increases in bead accumulation could be explained by two mechanisms: bead-diffusion; and, for the macrophage, macropinocytosis. Also for the highest concentrations of 28 nm beads, we saw a colligative threshold -- possibly due to beads masking the cell surface or obstructing cellular mechanisms.

Dissolution and reactive oxygen species generation of inhaled cemented tungsten carbide particles in artificial human lung fluids

A B Stefaniak, S S Leonard, M D Hoover, M A Virji and G A Day

National Institute for Occupational Safety and Health, Morgantown, WV, USA, 26505

E-mail: AStefaniak@cdc.gov

Abstract. Inhalation of both cobalt (Co) and tungsten carbide (WC) particles is associated with development of hard metal lung disease (HMD) via generation of reactive oxygen species (ROS), whereas Co alone is sufficient to cause asthma via solubilization and hapten formation. We characterized bulk and aerodynamically size-separated W, WC, Co, spray dryer (pre-sintered), and chamfer grinder (post-sintered) powders. ROS generation was measured in the murine RAW 264.7 cell line using electron spin resonance. When dose was normalized to surface area, hydroxyl radical generation was independent of particle size, which suggests that particle surface chemistry may be an important exposure factor. Chamfer grinder particles generated the highest levels of ROS, consistent with the hypothesis that intimate contact of metals is important for ROS generation. In artificial extracellular lung fluid, alkylbenzyldimethylammonium chloride (ABDC), added to prevent mold growth during experiments, did not influence dissolution of Co (44.0±5.2 vs. 48.3±6.4%); however, dissolution was higher ($p<0.05$) in the absence of phosphate (62.0±5.4 vs. 48.3±6.4%). In artificial macrophage phagolysosomal fluid, dissolution of Co (36.2±10.4%) does not appear to be influenced ($p=0.30$) by the absence of glycine (29.8±2.1%), phosphate (39.6±8.6%), or ABDC (44.0±10.5%). These results aid in assessing and understanding Co and W inhalation dosimetry.

The role of macrophage mediators in respirable quartz-elicited inflammation

D van Berlo[1], C Albrecht[1], AM Knaapen[2]*, FJ van Schooten[2], and RPF Schins[1]

[1]Institut für Umweltmedizinische Forschung (IUF) at the Heinrich Heine University Duesseldorf, Germany.

[2]Nutrition and Toxicology Research Institute Maastricht (NUTRIM), Department of Health Risk Analysis and Toxicology, University of Maastricht, The Netherlands. * current address: Department of Toxicology and Drug Disposition, Schering-Plough, 5340 BH Oss, The Netherlands

roel.schins@uni-duesseldorf.de

Abstract. The instigation and persistence of an inflammatory response is widely considered to be critically important in quartz-induced lung cancer and fibrosis. Macrophages have been long recognised as a crucial player in pulmonary inflammation, but evidence for the role of type II epithelial cells is accumulating. Investigations were performed in the rat lung type II cell line RLE and the rat alveolar macrophage cell line NR8383 using Western blotting, NF-κB immunohistochemistry and qRT-PCR of the pro-inflammatory genes iNOS and COX-2, as well as the cellular stress gene HO-1. The direct effect of quartz on pro-inflammatory signalling cascades and gene expression in RLE cells was compared to the effect of conditioned media derived from quartz-treated NR8383 cells. Conditioned media activated the NF-κB signalling pathway and induced a far stronger upregulation of iNOS mRNA than quartz itself. Quartz elicited a stronger, progressive induction of COX-2 and HO-1 mRNA. Our results suggest a differentially mediated inflammatory response, in which reactive particles themselves induce oxidative stress and activation of COX-2, while mediators released from particle-activated macrophages trigger NF-κB activation and iNOS expression in type II cells.

Journal of Physics: Conference Series **151** (2009) 011002 doi:10.1088/1742-6596/151/1/011002

Building an industry-wide occupational exposure database for respirable mineral dust – experiences from the IMA Dust Monitoring Programme

Remko Houba[1], Jelle Vlaanderen[2], Richard Jongen[1] and Hans Kromhout[2]

[1] Arbo Unie, Expert Centre for Chemical Risk Management, Nijmegen, The Netherlands

[2] Institute for Risk Assessment Sciences, Utrecht University, The Netherlands

Corresponding author: h.kromhout@uu.nl

Abstract: Building an industry-wide database with exposure measurements of respirable mineral dust is a challenging operation. The Industrial Minerals Association (IMA-Europe) took the initiative to create an exposure database filled with data from a prospective and ongoing dust monitoring programme that was launched in 2000. More than 20 industrial mineral companies have been collecting exposure data following a common protocol since then. Recently in 2007 ArboUnie and IRAS evaluated the quality of the collected exposure data for data collected up to winter 2005/2006. The data evaluated was collected in 11 sampling campaigns by 24 companies at 84 different worksites and considered about 8,500 respirable dust measurements and 7,500 respirable crystalline silica. In the quality assurance exercise four criteria were used to evaluate the existing measurement data: personal exposure measurements, unique worker identity, sampling duration not longer than one shift and availability of a limit of detection. Review of existing exposure data in the IMA dust monitoring programme database showed that 58% of collected respirable dust measurements and 62% of collected respirable quartz could be regarded as 'good quality data' meeting the four criteria mentioned above. Only one third of the measurement data included repeated measurements (within a sampling campaign) that would allow advanced statistical analysis incorporating estimates of within- and between-worker variability in exposure to respirable mineral dust. This data came from 7 companies comprising measurements from 23 sites. Problematic data was collected in some specific countries and to a large extent this was due to local practices and legislation (e.g. allowing 40-h time weighted averages). It was concluded that the potential of this unique industry-wide exposure database is very high, but that considerable improvements can be made. At the end of 2006 relatively small but essential changes were made in the dust monitoring protocol and the data collection sheet. In addition a system of quality control was set up and each new set of data is thoroughly investigated before inclusion into the database. Recently, it became apparent that more than 80% of the measurement data collected since winter 2005/2006 is of high quality. The IMA Dust Monitoring Programme Database contains personal measurements of more than 2,000 monitored workers who are representative of in total 5,000 workers in the industrial minerals production. This unique prospective exposure database will prove to be very valuable when health effects due to exposure to respirable mineral dust among these workers will be evaluated in the future.

Inhaled Particles X, (23–25 September 2008, Manchester) IOP Publishing
Journal of Physics: Conference Series **151** (2009) 011002 doi:10.1088/1742-6596/151/1/011002

Sanding dust from nanoparticle-containing paints: physical characterisation.

I K Koponen, K A Jensen, T Schneider

National Reseach Centre for Working Environment (NRCWE), Lersø Park Alle 105, DK-2100 Copenhagen, Denmark

E-mail : ikk@nrcwe.dk

Abstract. Increasing use of nanoparticles in different industrial applications has raised a new potential health risk to the workers as well as to the consumers. This study investigates the particle size distributions of sanding dust released from paints produced with and without engineered nanoparticles. Dust emissions from sanding painted plates were found to consist of five size modes; three modes under 1 μm and two modes around 1 and 2 μm. We observed that the sander was the only source of particles smaller than 50 nm and they dominated the number concentration spectra. Mass and surface area spectra were dominated by the 1 and 2 μm modes. Addition of nanoparticles caused only minor changes in the geometric mean diameters of the particle modes generated during sanding of two paints doped with 17 nm TiO2 and 95 nm Carbon Black nanoparticles as compared to the size modes generated during sanding a conventional reference paint. However, the number concentrations in the different size modes varied considerably in between the two NP-doped paints and the reference paint. Therefore, from a physical point of view, there may be a difference in the exposure risk during sanding surfaces covered with nanoparticle-based paints as compared to sanding conventional paints.

Journal of Physics: Conference Series **151** (2009) 011002 doi:10.1088/1742-6596/151/1/011002

Toxicity of Inhaled Traffic Related Particulate Matter

Miriam E Gerlofs-Nijland[1]; Arezoo Campbell[2]; Mark R Miller[3]; David E Newby[3]; Flemming R Cassee[1]

1 National Institute for Public Health and the Environment, Bilthoven, Netherlands.

2 Western University of Health Sciences, Pomona, CA, USA.University of Edinburgh,

3 Edinburgh, United Kingdom.

Email: miriam.gerlofs@rivm.nl

Abstract. Traffic generated ultrafine particulates may play a major role in the development of adverse health effects. However, little is known about harmful effects caused by recurring exposure. We hypothesized that repeated exposure to particulate matter results in adverse pulmonary and systemic toxic effects. Exposure to diesel engine exhaust resulted in signs of oxidative stress in the lung, impaired coagulation, and changes in the immune system. Pro-inflammatory cytokine levels were decreased in some regions of the brain but increased in the striatum implying that exposure to diesel engine exhaust may selectively aggravate neurological impairment. Data from these three studies suggest that exposure to traffic related PM can mediate changes in the vasculature and brain of healthy rats. To what extent these changes may contribute to chronic neurodegenerative or vascular diseases is at present unclear.

Mortality in an extended follow-up of British coal workers.

MacCalman, L and Miller, B G

Institute of Occupational Medicine, Edinburgh.

laura.maccalman@iom-world.org

Abstract. The Pneumoconiosis Field Research (PFR) programme was established in the 1950s, to evaluate effects of coal mining exposures on the health and mortality of British coal workers. Surveys of working miners were carried out at 5-yearly intervals, initially in 24 collieries but later concentrating on 10, collecting detailed work histories and health information for each recruit. Here we report on cause-specific mortality in a cohort of almost 18,000 men from 10 British collieries, followed up for periods up to 47 years, yielding over 516,000 life-years of follow-up. External analyses compared cause-specific death rates in the cohort to those of the population of the regions in which the collieries were situated, using Standardised Mortality Ratios (SMRs). The causes investigated included lung cancer, stomach cancer, non-malignant respiratory disorders and cardiovascular disorders. SMRs showed evidence of an initial healthy worker effect diminishing over time. Several causes, including non-malignant respiratory disease and lung cancer, showed a significant deficit of mortality at the start of the study period with an excess in the latter part of the follow-up period. In these results, effects of working conditions are likely to be confounded with smoking habits. Overall, we believe our results may be generalised to the British coal industry since nationalisation.

Journal of Physics: Conference Series **151** (2009) 011002 doi:10.1088/1742-6596/151/1/011002

Trends in pneumoconiosis mortality and morbidity for the United States, 1968–2005, and relationship with indicators of extent of exposure

M D Attfield, K M Bang, E L Petsonk, P L Schleiff, J M Mazurek

Division of Respiratory Disease Studies, National Institute for Occupational Safety and Health, Centers for Disease Control, 1095 Willowdale Road, Morgantown, WV, 26505, USA

E-mail: mda1@cdc.gov

Abstract. This surveillance report examines trends in selected pneumoconioses in the U.S. for 1968–2005 and their relationship with past indicators of extent of exposure. Numbers of deaths with asbestosis, silicosis, and coal workers' pneumoconiosis (CWP) were tabulated by time and age at death. Worker monitoring CWP prevalence data were tabulated by tenure group. Information on indicators of extent and intensity of exposure were obtained from various sources. Asbestosis deaths from 1968–2005 closely followed the historical trend in asbestos consumption, and appear to be declining in most age groups. Given appropriate exposure control, asbestosis could be eliminated by 2050. Silicosis deaths decreased substantially from 1968–2005, but levelled off after 1998 in all age groups, indicating a continuing occupational risk. In the anthracite coal region, CWP mortality has been declining rapidly. If there is no resurgence in the industry, CWP could disappear in that region by 2030. In the much larger bituminous region, deaths have declined over time but may be increasing among younger individuals. In addition, although CWP prevalence in working coal miners declined substantially from 1970 to 1994, it increased from 1995 to 2006. This indicates the need for increased vigilance in dust control in underground coal mining.

Inhaled Particles X, (23–25 September 2008, Manchester) IOP Publishing
Journal of Physics: Conference Series **151** (2009) 011002 doi:10.1088/1742-6596/151/1/011002

European and International Standards on health and safety in welding

A Howe

Health and Safety Laboratory, Harpur Hill, Buxton, Derbyshire, SK17 9JN, UK

E-mail: alan.howe@hsl.gov.uk

Abstract. A number of European and International Standards on health and safety in welding have been published in recent years and work on several more is nearing completion. These standards have been prepared jointly by the International Standards Organization (ISO) and the European Committee for Standardization (CEN). The standards development work has mostly been led by CEN/TC 121/SC 9, with excellent technical input from experts within Europe; but work on the revision of published standards, which has recently gathered pace, is now being carried out by ISO/TC 44/SC 9, with greater international involvement. This paper gives an overview of the various standards that have been published, are being revised or are under development in this field of health and safety in welding, seeking to (i) increase international awareness of published standards, (ii) encourage wider participation in health and safety in welding standards work and (iii) obtain feedback and solicit comments on standards that are currently under development or revision. Such an initiative is particularly timely because work is currently in progress on the revision of one of the more important standards in this field, namely EN ISO 10882:2001 *Health and safety in welding and allied processes — Sampling of airborne particles and gases in the operator's breathing zone — Part 1: Sampling of airborne particles*

Historical exposure levels of inhalable dust in the Polish rubber industry compared to levels in Western Europe.

F de Vocht[1], H Kromhout[2], W Sobala[3], and B Peplonska[3]

[1] Occupational & Environmental Health Research Group, School of Translational Medicine, Faculty of Medical and Human Sciences, The University of Manchester, Manchester, UK

[2] Division of Environmental Epidemiology, Institute for Risk Assessment Sciences, Utrecht University, Utrecht, The Netherlands

[3] Department of Occupational and Environmental Epidemiology, NOFER Institute of Occupational Medicine, Lodz, Poland

Frank.devocht@manchester.ac.uk

Abstract. Although studies have been carried out to assess inhalable dust exposure levels in the rubber manufacturing industry, the levels of exposure in factories in Eastern Europe are less well documented. Routine stationary sampling for compliance testing of inhalable aerosols has however been conducted in a large factory producing tires and tubes in Poland between 1981 and 1996 (N=6,152). This study was conducted to assess historical inhalable aerosol levels in different departments in this rubber plant and to compare the results with estimates based on European data from the United Kingdom, Sweden, the Netherlands and Germany, and also Poland (EXASRUB project). Geometric mean (GM) concentrations in the factory ranged from 2.41 mg/m^3 to 5.82 mg/m^3 and were to a large extent associated to the actual production capacity of the plant and flow of the production process. Whereas 3-4 fold differences between departments existed prior to about 1985, stronger reduction of exposure in the raw materials and finishing departments (-12%/year) compared to other departments (range -5%/yr to -3%/yr), resulted in comparable levels in the 1990s. However, in the pre-treating departments, average concentrations were still about a factor 2-3 higher than in other departments, which could presumably be attributed to the use of anti-tacking agents. GM concentrations have been modelled using (1) stationary measurements collected in the Polish factory only, or (2) all European data collected in the EXASRUB project. Comparison of the estimates showed that these were fairly similar for both datasets. This analysis showed that the levels of inhalable aerosols in the Polish rubber industry have been at least a factor three to four higher than in Western European countries in the 1980s and 1990s, depending on the department, but that these differences were getting smaller in the 1990s. Furthermore, the estimates based on all European data from EXASRUB provides valid estimates compared to factory-specific data.

Inhaled Particles X, (23–25 September 2008, Manchester) IOP Publishing
Journal of Physics: Conference Series **151** (2009) 011002 doi:10.1088/1742-6596/151/1/011002

Personal Exposure to Inhalable Cement Dust among Construction Workers

Susan Peters[1], Yngvar Thomassen[2], Edeltraud Fechter-Rink[3] and Hans Kromhout[1]

1. Institute for Risk Assessment Sciences, Environmental Epidemiology Division, Utrecht University, Utrecht, The Netherlands

2. National Institute of Occupational Health, Oslo, Norway

3. TÜV SÜD Industrie Service GmbH, Mannheim, Germany

Corresponding author: s.peters@uu.nl

Abstract. A case study was carried out in 2006-2007 to assess the actual cement dust exposure among construction workers involved in a full-scale construction project and as a comparison among workers involved in various stages of cement and concrete production. Full-shift personal exposure measurements were performed for several job types. Inhalable dust and cement dust (based on analysis of elemental calcium) concentrations were determined. Inhalable dust exposures at the construction site ranged from 0.05 to 34 mg/m3, with a mean concentration of 1.0 mg/m3. For inhalable cement dust mean exposure was 0.3 mg/m3 (range 0.02-17 mg/m3). Reinforcement and pouring workers had the lowest average concentrations. Inhalable dust levels in the ready-mix and pre-cast concrete plants were, on average, below 0.5 mg/m3 for inhalable dust and below 0.2 mg/m3 for inhalable cement dust. Highest dust concentrations were measured in cement production, particularly during cleaning tasks (inhalable dust GM=55 mg/m3; inhalable cement dust GM=33 mg/m3) at which point the workers wore personal protective equipment. Elemental measurements showed highest but very variable cement percentages in the cement plant and very low percentages of cement during reinforcement work and pouring.

Silica dust control in small-scale building/structure demolition operations using good work practice guidance

C V Muianga [1,2,3], C H Rice[1] and P Succop[1]

[1]Department of Environmental Health, College of Medicine University of Cincinnati. 3223 Eden Ave., Kettering Laboratory, Cincinnati, OH 45267-0056, USA

[2]Center for Industrial Studies, Safety and Environment, Eduardo Mondlane University, P.o.Box # 257, Maputo, Mozambique

[3]Email: muiangcv@email.uc.edu

Abstract. Work practices can influence exposure, especially in small-scale operations conducted by mobile work crews. This study evaluated the use of information on good work practice in control guidance sheets adapted from UK Silica Essentials guidance sheets by trained workers and supervisors employed in small-scale concrete and masonry demolition operations. A one-page employee silica task-based control guidance sheet for each of four demolition tasks and multiple-page silica control guidance for supervisors were developed. Interactive, hands-on worker training on these task-based good work practice controls was developed. Training was presented to 26 participants from two demolition crews. Feedback on the training and task-based good work practice control guidance sheets was elicited. Observations of work practices were made before and after training. Participants indicated gains in knowledge and checklists were used to document skill attainment. The quality of the training and usefulness of the material/skills was rated high by trainees. Increased use of water to suppress dust and wet cleaning methods on the job were documented following the training. Additional follow-up after training is required to determine long-term impact on sustained changes in work practices, and to evaluate the need for refresher training.

Journal of Physics: Conference Series **151** (2009) 011002 doi:10.1088/1742-6596/151/1/011002

Temporal changes in the variability of respirable mineral dust exposure concentrations

Hans Kromhout[1], Jelle Vlaanderen[1], Richard Jongen[2], Remko Houba[3]

1 Institute for Risk Assessment Sciences, Utrecht University, The Netherlands

2 ArboUnie Occupational Health Service, The Netherlands

3 ArboUnie Expert Centre for Chemical Risk Management, Nijmegen, The Netherlands

Corresponding author: h.kromhout@uu.nl

Abstract: In the last decade a lot of evidence with regard to temporal trends in exposure concentrations in predominantly Western industrial countries has become available. In a recent literature review (Creely et al. 2007) overall percentage of yearly declines up to 32% were presented. To what extent these temporal declines also affect the variability in exposure concentrations is unknown. The main reason is lack of longitudinal data including repeated measurements that would allow evaluating trends in personal and temporal components of exposure variability. The recently elaborated exposure database from the Industrial Minerals Association Dust Monitoring Programme provided an opportunity to study these trends for exposure to respirable mineral dust. This database currently contains more than 11,000 measurements from more than 20 companies and 80 sites throughout Europe. About one-third of the measurement data comprised repeated measurements within a specific site-job-survey combination. Linear mixed models were used to estimate variance components. Variance components were consequently plotted against year of measurements. For 377 groups of workers (with number of workers>2, total number of observations >5 and repeats >1.25) the fold range of the total variability (${}_{t}R_{.95}$) appeared to go down significantly with 3% per campaign (half year): from 34 in summer 2002 to 20 in winter 2005/2006. When the variability was teased apart the fold range for the temporal variability (${}_{ww}R_{.95}$) appeared to decrease significantly as well with 3% per campaign (half year): from 17 in summer 2002 to 10 in winter 2005/2006. The between-worker variability did not show a temporal trend and stayed constant with on average a fold-range (${}_{bw}R_{.95}$) of approximately 4. Downward temporal trends in exposure level of respirable (crystalline silica) dust seem to coincide with downward trends in the size of temporal variability. Fold-ranges of average exposure of individual workers within a job at a particular site appear to be stationary. Implications are that attenuation of exposure-response associations in epidemiogical studies will decline and that fewer measurements will have to be collected to arrive at accurate estimates of long-term exposure to respirable mineral dust in this industry.

Inhaled Particles X, (23–25 September 2008, Manchester) IOP Publishing
Journal of Physics: Conference Series **151** (2009) 011002 doi:10.1088/1742-6596/151/1/011002

Dust particles during routine callus reduction in Podiatry

G Scott[1], N McLarnon[1], J G Burrow[1], M Hepher[2]

1 Division of Podiatric Medicine and Surgery, School of Health and Social Care, Glasgow Caledonian University, Cowcaddens Road, Glasgow, Scotland, UK, G4 0BA

2 School of the Built and Natural Environment, Glasgow Caledonian University, Cowcaddens Road, Glasgow, Scotland, UK, G4 0BA

N.A.McLarnon@gcal.ac.uk

Abstract. This unique study was conducted to ascertain the presence of both respirable and inhalable dust particles, following the routine reduction of callosities using a callous burr. All previous literature related to dust inhalation within the Podiatric Profession has focused upon the inhalation of nail dust particles, identifying a gap in literature and a need to highlight the health implications and risks from inhaling dust released during callous reduction. Samples, using the SKC sample pump, were obtained from fifteen participants prior to and following the reduction of callosities. Samples were weighed pre and post callous reduction and the results indicate that both respirable dust (0.001) and inhalable dust (0.235) volumes increase following treatment. Computer enhanced images of dust particles were also gained using Scanning Electron Microscopy and the particles were highlighted as being irregular and pleomorphic in shape. Overall, the results illustrate that callous burring techniques produce an occupational hazard to practicing podiatrists due to significantly increased quantities of respirable dust, which have the potential to irritate the respiratory tract when inhaled.

Journal of Physics: Conference Series **151** (2009) 011002 doi:10.1088/1742-6596/151/1/011002

A study of the effect of particulate deposit upon fibrous filter efficiency.

Sarah J Dunnett[1], Charles F Clement[2]

[1]Department of Aeronautical and Automotive Engineering, Loughborough University, Loughborough, Leics. LE11 3TU, U.K.

[2]15 Witan Way, Wantage, Oxon, OX12 9EU, U.K.

s.j.dunnett@lboro.ac.uk

Abstract. The aim of this work is to develop numerical methods to model the effects that particle deposit collected by fibrous filters has upon the flow field within the filter and hence upon further deposition. A single fibre model has been developed with the deposit modelled as a porous layer on the fibre surface. Using mathematical techniques the flow field outside and within the porous layer are determined. Once the flow field for a particular deposit has been obtained the equations of motion of the particles are solved and the feedback effects of the deposit upon further deposition investigated. The mechanisms of interception and diffusion are considered. It is found that for the smaller particles the porosity of the deposit has an insignificant effect upon the flow field. However the porosity becomes increasingly important as the relative size of the particle to the fibre increases.

Size selective isocyanate aerosols personal air sampling using porous plastic foams

Cong Khanh Huynh, Trinh Vu Duc

Institut Universitaire Romand de Santé au Travail (IST), 21 rue du Bugnon - CH-1011 Lausanne, Switzerland

chuynh@hospvd.ch

Abstract. As part of a European project (SMT4-CT96-2137), various European institutions specialized in occupational hygiene (BGIA, HSL, IOM, INRS, IST, Ambiente e Lavoro) have established a program of scientific collaboration to develop one or more prototypes of European personal samplers for the collection of simultaneous three dust fractions: inhalable, thoracic and respirable. These samplers based on existing sampling heads (IOM, GSP and cassettes) use Polyurethane Plastic Foam (PUF) according to their porosity to support sampling and separator size of the particles. In this study, the authors present an original application of size selective personal air sampling using chemical impregnated PUF to perform isocyanate aerosols capturing and derivatizing in industrial spray-painting shops.

Journal of Physics: Conference Series **151** (2009) 011002 doi:10.1088/1742-6596/151/1/011002

Characteristics of Welding Fume Aerosol Investigated in Three Swedish Workshops

C Isaxon[1], J Pagels[1], A Gudmundsson[1], C Asbach[2], A C John[2], T A J Kuhlbusch[2], J E Karlsson[3], R Kammer[3] , H Tinnerberg[3], J Nielsen[3] and M Bohgard[1]

1. Division of Ergonomics and Aerosol Technology (EAT), Lund University, Faculty of Engineering, PO Box 118, SE-221 00, Lund, Sweden
2. Institute of Energy and Environmental Technology, (IUTA), Duisburg, Germany
3. Occupational and Environmental Health, Lund University, University Hospital, Lund, Sweden

Abstract: Potentially high human exposures to nanometer sized airborne particles occur due to welding and other thermal processes in industrial environments. Detailed field measurements of physical and chemical particle characteristics were performed in three work-shops in Sweden. Measurements were performed both in the plume 5-20 cm above the welding point and in the background air (more than 5 m away from the nearest known particle source). Particle number and mass concentrations were measured on-line. A low pressure impactor was used for size-resolved chemical particle composition. The in-plume measurements generated the chemical signatures for different welding processes. These signatures were then used to identify contributions from various processes to the particle concentrations in different size classes. The background number and mass concentrations increased by more than an order of magnitude during intense activities in the work-shops compared to low activities during breaks.

Journal of Physics: Conference Series **151** (2009) 011002 doi:10.1088/1742-6596/151/1/011002

Parallel particle impactor – novel size-selective particle sampler for accurate fractioning of inhalable particles

S Trakumas[1] and E Salter[2]

[1] SKC Inc., 863 Valley View Road, Eighty Four, PA 15330
[2] SKC Ltd., 11 Sunrise Park, Higher Shaftesbury Road, Blandford Forum, Dorset, DT11 8ST

E-mail: saulius@skcinc.com

Abstract. Adverse health effects due to exposure to airborne particles are associated with particle deposition within the human respiratory tract. Particle size, shape, chemical composition, and the individual physiological characteristics of each person determine to what depth inhaled particles may penetrate and deposit within the respiratory tract. Various particle inertial classification devices are available to fractionate airborne particles according to their aerodynamic size to approximate particle penetration through the human respiratory tract. Cyclones are most often used to sample thoracic or respirable fractions of inhaled particles. Extensive studies of different cyclonic samplers have shown, however, that the sampling characteristics of cyclones do not follow the entire selected convention accurately. In the search for a more accurate way to assess worker exposure to different fractions of inhaled dust, a novel sampler comprising several inertial impactors arranged in parallel was designed and tested. The new design includes a number of separated impactors arranged in parallel. Prototypes of respirable and thoracic samplers each comprising four impactors arranged in parallel were manufactured and tested. Results indicated that the prototype samplers followed closely the penetration characteristics for which they were designed. The new samplers were found to perform similarly for liquid and solid test particles; penetration characteristics remained unchanged even after prolonged exposure to coal mine dust at high concentration. The new parallel impactor design can be applied to approximate any monotonically decreasing penetration curve at a selected flow rate. Personal-size samplers that operate at a few L/min as well as area samplers that operate at higher flow rates can be made based on the suggested design. Performance of such samplers can be predicted with high accuracy employing well-established impaction theory.

High efficiency CIP 10-I personal inhalable aerosol sampler

P Görner, R Wrobel, X Simon

INRS, Rue du Morvan, CS 60027, 54519 Vandoeuvre les Nancy, France

E-mail: peter.gorner@inrs.fr

Abstract. The CIP 10 personal aerosol sampler was first developed by Courbon for sampling the respirable fraction of mining dust. This respirable aerosol sampler was further improved by Fabries, then selectors for sampling thoracic and inhalable aerosols were designed. Kenny *et al.* evaluated the particle-size dependent sampling efficiency of the inhalable version in a large-scale wind tunnel using a life-size dummy. The authors found that the overall sampling efficiency decreases more rapidly than the CEN-ISO-ACGIH target efficiency curve. Görner and Witschger measured the aspiration efficiency of the CIP 10 omni-directional inlet. They found that the aspiration efficiency was high enough for inhalable aerosol sampling. This result led to the conclusion that the low sampling efficiency is due to some internal losses of the aspirated particles before they reach the final sampling stage, namely the CIP 10 rotating filter. Based on the assumption that the inhalable particles are selected at selector aspiration level, an experimental research project was conducted to improve particle transmission to the collection stage of the sampler. Two different inhalable selectors were designed by Görner and tested in a laboratory wind tunnel. The transmission efficiency of both models was measured by Roger following an experimental protocol described by Witschger. The T-shaped air flow circuit was finally adopted to draw the aspirated particles into the final collection stage of the CIP 10. Actually, in this selector, the almost horizontally aspirated particles should be conducted vertically to the rotating cup. In two previous prototypes, particles could be deposited in certain places by inertia (where the aerosol was forced to deviate drastically) or by sedimentation (where the aerosol decelerated). The aerodynamic behaviour of the adopted solution causes the particles to accelerate radially between two horizontal plates before they enter a vertical tube. This acceleration avoids the particles being deposited on the lower horizontal plate. At the beginning of the vertical tubing, the mutually opposing particle trajectories limit particle wall deposition by virtual impaction effect. The inner selector walls are polished to avoid particles being stopped by eventual surface asperities. Particle size-dependent sampling efficiency was measured in the laboratory wind tunnel. The experimental aerosol was composed of polydisperse glass micro-spheres. The size analysis of the particles collected was done by the Coulter Counter technique. The transmission efficiency (reciprocal to wall losses) was found to be close to 100 % for the entire range of particle sizes, and indicated no particle loss. The overall sampling efficiency was measured using a rotating bluff body at an external wind speed of 1 m/s. The rotating bluff body represents a scaled torso of an operator. The "high efficiency" CIP 10-I (I for inhalable) responds fairly well to the conventional CEN-ISO-ACGIH criteria for sampling the inhalable health-related aerosol fraction.

Inhalability for aerosols at ultra-low windspeeds

Darrah K. Sleeth and James H. Vincent

Department of Environmental Health Sciences, School of Public Health, University of Michigan, Ann Arbor, MI, 48109, U.S.A.

jhv@umich.edu

Abstract. Most previous experimental studies of aerosol inhalability were conducted in wind tunnels for windspeeds greater than 0.5 ms^{-1}. While that body of work was used to establish a convention for the inhalable fraction, results from studies in calm air chambers (for essentially zero windspeed) are being discussed as the basis of a modified criterion. However, information is lacking for windspeeds in the intermediate range, which – it so happens – pertain to most actual workplaces. With this in mind, we have developed a new experimental system to assess inhalability – and, ultimately, personal sampler performance – for aerosols with particle aerodynamic diameter within the range from about 9 to 90 μm for ultra-low windspeed environments from about 0.1 to 0.5 ms^{-1}. This new system contains an aerosol test facility, fully described elsewhere, that combines the physical attributes and performance characteristics of moving air wind tunnels and calm air chambers, both of which have featured individually in previous research. It also contains a specially-designed breathing, heated, life-sized mannequin that allows for accurate recovery of test particulate material that has been inhaled. Procedures have been developed that employ test aerosols of well-defined particle size distribution generated mechanically from narrowly-graded powders of fused alumina. Using this new system, we have conducted an extensive set of new experiments to measure the inhalability of a human subject (as represented by the mannequin), aimed at filling the current knowledge gap for conditions that are more realistic than those embodied in most previous research. These data reveal that inhalability throughout the range of interest is significantly different based on windspeed, indicating a rise in aspiration efficiency as windspeed decreases. Breathing flowrate and mode of breathing (i.e. nose versus mouth breathing) did not show significant differences for the inhalability of aerosols. On the whole however, the data obtained here are within the range of inhalability data that exist from the large body of the previous experimental work performed at the higher windspeeds. These latest findings are an important contribution to the ongoing discussion in international standards-setting bodies about the possible adjustment of the quantitative definition of what constitutes the inhalable fraction.

Inhaled Particles X, (23–25 September 2008, Manchester) IOP Publishing
Journal of Physics: Conference Series **151** (2009) 011002 doi:10.1088/1742-6596/151/1/011002

Modelling exposure to pharmaceutical agents

J W Cherrie [1], A T Gillies [2], A Sleeuwenhoek [1], M van Tongeren [1], P McDonnell [3], M Coggins [3], S R Bailey [4]

1 Institute of Occupational Medicine, Edinburgh, EH14 4AP, UK.

2 Gillies Associates Ltd, 34 Holman Road, Aylsham, Norfolk NR11 6BZ, UK.

3 National University of Ireland, Department of Physics, Galway, Ireland.

4 GlaxoSmithKline, Southdownview Way, Worthing, BN14 8NQ, UK.

Email: john.cherrie@iom-world.org

Abstract. Aerosol exposure to the active ingredients of pharmaceuticals arises in research, development and manufacturing. In most instances it is only possible to make small numbers of measurements of exposure and given the inter- and intra-individual variability it is often difficult to obtain sufficient objective data to make reliable decisions about the appropriateness of control measures. This paper describes the development and validation of an exposure model with the potential to predict airborne exposure from both new and existing operations. The model could be used to more efficiently target exposure measurement resources.

Inhaled Particles X, (23–25 September 2008, Manchester) IOP Publishing
Journal of Physics: Conference Series **151** (2009) 011002 doi:10.1088/1742-6596/151/1/011002

An analysis of employee exposure to organic dust at large-scale composting facilities.

P Sykes, J A Allen, J D Wildsmith and K P Jones

University of Wales Institute Cardiff (UWIC), Cardiff School of Health Sciences, Western Avenue, Cardiff, CF5 2YB.

psykes@uwic.ac.uk

Abstract. The occupational health implications from exposure to dust, endotoxin and 1-3 β Glucan at commercial composting sites are uncertain. This study aims to establish employee exposure levels to inhalable and respirable dust, endotoxin and 1-3 β Glucan during various operational practices in the composting process. Personal samples were collected and the inhalable and respirable dust fractions were determined by gravimetric analysis. Endotoxin concentrations were determined using a Limulus Amebocyte Lysate assay (LAL). 1-3 β Glucan levels were estimated using a specific blocking agent to establish the contribution that these compounds gave to the original endotoxin assay. Employees' exposure to dust was found to be generally lower than the levels stipulated in the Control of Substances Hazardous to Health Regulations (COSHH) 2002 (as amended), (median inhalable fraction 1.08 mg/m^3, min 0.25 mg/m^3 max 10.80 mg/m^3, median respirable fraction 0.05 mg/m^3, min 0.02 mg/m^3, max 1.49 mg/m^3). Determination of the biological component of the dust showed that employees' exposures to endotoxin were elevated (median 31.5 EU/m^3, min 2.00 EU/m^3, max 1741.78 EU/m^3), particularly when waste was agitated (median 175.0 EU/m^3, min 2.03 EU/m^3, max 1741.78 EU/m^3). Eight out of 32 (25%) of the personal exposure data for endotoxin exceeded the 200 EU/m^3 temporary legal limit adopted in the Netherlands and thirteen out of 32 (40.6%) exceeded the suggested 50 EU/m^3 guidance level suggested to protect workers from respiratory health effects. A significant correlation was observed between employee inhalable dust exposure and personal endotoxin concentration ($r = 0.728$, $p<0.05$) and also personal endotoxin exposure and 1-3 β Glucan concentration ($r = 0.817$, $p<0.05$). Further work is needed to explore the possibility of using inhalable dust concentration as a predictor for personal endotoxin exposure. The general dust levels stipulated in the COSHH Regulations 2002 (as amended) are inadequate for managing the potential health risks associated with endotoxin exposure at composting sites. Employee exposure levels and dose-response disease mechanisms are not well understood at this present time. Consequently, in light of this uncertainty, it is recommended that a precautionary approach be adopted in managing the potential health risks associated with inhalation of organic dusts at composting sites.

Occupational rhinitis and occupational asthma; one airway two diseases?

M J Seed, M Gittins, F De Vocht and R M Agius

Occupational and Environmental Health Research Group, University of Manchester.

Correspondence to Dr Martin Seed, Occupational & Environmental Health Research Group, School of Translational Medicine, Faculty of Medical & Human Sciences, University of Manchester, Ellen Wilkinson Building, Oxford Road, Manchester M13 9PL, UK. Martin.seed@manchester.ac.uk

Abstract. The concept of 'one airway, one disease' refers to the frequent comorbidity of asthma and rhinitis. However, only limited research has been done on this association for the diverse range of occupational respiratory sensitisers. The relative frequency of rhinitis was determined for the 15 respiratory sensitisers reported to cause at least 10 cases of rhinitis or asthma to The Health and Occupation Reporting (THOR) network between 1997 and 2006. Of 1408 cases, 1190 were sole diagnoses of asthma, 138 sole diagnoses of rhinitis and in 80 cases asthma coexisted with rhinitis. The six sensitisers for which rhinitis featured in over 15% of cases were all particulates and known to cause release of mast cell mediators, either directly or through IgE antibodies. Four of the other nine sensitisers often exist as vapours and only two have been consistently associated with IgE-mediated disease mechanisms. Particle size did not appear to correlate with the relative frequency of rhinitis. Despite its limitations this study would support the hypothesis that there are at least two mechanistic categories of respiratory sensitisation with rhinitis being relatively more common where the mechanism is IgE-mediated. Particulate nature may be another important factor to consider in future studies.

Journal of Physics: Conference Series **151** (2009) 011002 doi:10.1088/1742-6596/151/1/011002

Respiratory Tract Deposition of Particles from Biomass Combustion

Jakob Löndahl[1], Erik Swietlicki[1], Joakim Pagels[2], Andreas Massling[1], Christoffer Boman[3], Jenny Rissler[1], Anders Blomberg[4] and Thomas Sandström[4]

1. Department of Physics, Lund University, PO Box 118, SE-221 00, Lund, Sweden

2. Division of Aerosol Technology (EAT), Lund Institute of Technology, PO Box 118, SE-221 00, Lund, Sweden

3. Energy Technology and Thermal Process Chemistry, Umeå University, SE-901 87 Umeå, Sweden

4. Department of Respiratory Medicine and Allergy, University Hospital, SE-901 85 Umeå, Sweden

E-mail: jakob.londahl@nuclear.lu.se

Abstract. The respiratory tract deposition of particles from two types of biomass combustion was measured for 10 healthy subjects. The aerosol was extensively characterised. Model calculations of the deposition were made based on the particle properties. The results show that particle water absorption has substantial impact on deposition. The particles from biomass combustion obtained a size in the respiratory tract at which the deposition probability is close to its minimum.

Inhaled Particles X, (23–25 September 2008, Manchester) IOP Publishing
Journal of Physics: Conference Series **151** (2009) 011002 doi:10.1088/1742-6596/151/1/011002

Rat inhalation test with particles from biomass combustion and biomass co-firing exhaust

B Bellmann, O Creutzenberg, H Ernst, and H Muhle

Fraunhofer Institute of Toxicology and Experimental Medicine, Nikolai-Fuchs-Str.1, 30625 Hannover, Germany

E-mail: bernd.bellmann@item.fraunhofer.de

Abstract. The health effects of 6 different fly ash samples from biomass combustion plants (bark, wood chips, waste wood, and straw), and co-firing plants (coal, co-firing of coal and sawdust) were investigated in a 28-day nose-only inhalation study with Wistar WU rats. Respirable fractions of carbon black (Printex 90) and of titanium dioxide (Bayertitan T) were used as reference materials for positive and negative controls. The exposure was done 6 hours per day, 5 days per week at an aerosol concentration of 16 mg/m^3. The MMAD of all fly ash samples and reference materials in the inhalation unit were in the range from 1.5 to 3 µm. The investigations focused predominantly on the analysis of inflammatory effects in the lungs of rats using bronchoalveolar lavage (BAL) and histopathology. Different parameters (percentage of polymorphonuclear neutrophils (PMN), interleukin-8 and interstitial inflammatory cell infiltration in the lung tissue) indicating inflammatory effects in the lung, showed a statistically significant increase in the groups exposed to carbon black (positive control), C1 (coal) and C1+BM4 (co-firing of coal and sawdust) fly ashes. Additionally, for the same groups a statistically significant increase of cell proliferation in the lung epithelium was detected. No significant effects were detected in the animal groups exposed to BM1 (bark), BM2 (wood chips), BM3 (waste wood), BM6 (straw) or titanium dioxide.

Inhaled Particles X, (23–25 September 2008, Manchester) IOP Publishing
Journal of Physics: Conference Series **151** (2009) 011002 doi:10.1088/1742-6596/151/1/011002

Characterization of Particle Emissions from the Combustion of Different Australian Vegetation

Hansen D[1], Porter N[1], Elms T[1], Reisen F[2] and Meyer C[2]

1. RMIT University, Melbourne, VIC 3001

2. CSIRO Marine & Atmospheric Research, Aspendale VIC 3195

Abstract: It is well established that a fireground contains a range of hazards, both physical and chemical, that make firefighting an extremely risky occupation. In particular, bushfires generate a range of air toxics that are potentially harmful to human health and safety. Of these air toxics, particulate matter (PM) is a major health interest. It is well known that Australian bushfires generate high levels of PM over a broad size range; however, it is not known what types, and levels, of organic compounds and heavy metals are being transmitted to the firefighters, via these particulates. Studies have been conducted in Europe and the US that have evaluated the different toxics contained in smoke, and of the exposure of firefighters. However, as Australia has different fire fighting techniques, a different climate, and different vegetation, these studies cannot be directly applied to Australian firegrounds. Therefore, a specific study is required to evaluate the PM material produced by Australian vegetation. The level of particle emissions, composition, and concentrations, all need to be evaluated for different Australian vegetation types under the different conditions likely to be found on the fireground. This paper presents work undertaken to characterise the volatile organic components and heavy metals adsorbed to particulates generated in bushfires. In order to achieve this characterization, comprehensive chemical analysis has been conducted, including analysis for Poly Aromatic Hydrocarbons [PAHs], a range of water soluble ions, Total Carbon [TC], Elemental Carbon [EC], Organic Carbon [OC], levoglucosan and heavy metals. Investigations have also evaluated firefighter particle exposure, the determination of emission factors for typical vegetation, the investigation of noted "high emission" vegetation, and the investigation of the effect of fuel conditions, fuel load, and fuel size on emissions. The first stage of this process involved work monitoring firefighter exposure during Australian bushfires in 2006/2007/2008 and at prescribed burning activities conducted throughout Australia. Atmospheric sampling focused on two main areas: 1. personal monitoring, with samples collected in the breathing zone of active firefighters, and 2. local environment samples, collected with vehicle mounted instrumentation. The second stage has involved small scale simulated burns, with sample collection using a specially designed high volume sampler. By establishing a carbon balance, emission factors for specific chemical species have been determined for different fuel types under different conditions. The results of these trials and field work will generate better risk management procedure, based upon the nature and intensity of individual fires and Australian firefighting techniques, which will be used to mimimize firefighter exposure to hazardous emission materials. A secondary outcome will be more prescriptive and efficient assessment of dangers, both to the firefighters and the general community, presented by individual bush fire scenarios.

Inhaled Particles X, (23–25 September 2008, Manchester) IOP Publishing
Journal of Physics: Conference Series **151** (2009) 011002 doi:10.1088/1742-6596/151/1/011002

Links between urban ambient particulate matter and health – A time series analysis of particle metrics

R W Atkinson[1], G W Fuller[2], H R Anderson[1], R Harrison[3], B Armstrong[4]

1 – Department of Community Health Sciences, St. George's, University of London, Cranmer Terrace, London. SW17 0RE

2 – Environmental Research Group, King's College London, 4th Floor, Franklin Wilkins Building, 150 Stamford Street, London. SE1 9NH

3 – Division of Environmental Health and Risk Management, School of Geography, Earth & Environmental Sciences, University of Birmingham, Edgbaston, Birmingham. B15 2TT

4 – Public and Environmental Health Research Unit, London School of Hygiene and Tropical Medicine, Keppel Street, London. WC1E 7HT

Abstract: There is convincing evidence that inhaled particulate matter in the outdoor air is associated with adverse health effects. Particulate matter is a mixture of particles of varying size, number and composition and the nature of this mixture varies according to emission sources, secondary chemical reactions in the atmosphere, weather conditions and other factors. For the protection of public health it would be desirable to know which component of the particulate mixture to target with regulation. However, knowledge about this is limited. The pollutants of most concern are those that are derived from combustion sources, but it is not clear to what extent health effects are caused by those directly emitted from combustion sources (primary particles) or those formed by complex chemical processes in the atmosphere which may have occurred many miles away (secondary particles). In size terms, particles less than 10 microns in diameter are of most concern since these can penetrate the deep airways (PM_{10}). However, this metric encompasses two fractions which differ in source and size: 1) the fine fraction usually measured as $PM_{2.5}$ and derived from combustion sources and gas to particle conversion processes (i.e. formation of sulphate, nitrate and secondary organic carbon) and 2) the coarse fraction usually measured as $PM_{10-2.5}$ which includes particles from non-combustion sources (such as wind-blown dust) which differs in composition and, probably, health effects. Ultrafine particles ($PM_{0.1}$) comprise the greatest number but the least mass and it has been postulated that this fraction may be important for health effects. The heath effects of short term exposure to air pollutants have been investigated using an epidemiological study design known as a time series study. In these studies, daily counts of a health outcome (e.g. numbers of deaths) in a population, usually an urban one, are related to daily measurements of outdoor air pollution concentrations. It is known from previous time-series studies in London that particles, measured as PM_{10}, and black smoke show short-term associations with increases in daily counts of mortality and hospital admissions. In recent years a number of newer and potentially more informative metrics have been introduced, including components of PM_{10} such as carbon, sulphate or nitrate, size fractions such as $PM_{2.5}$, $PM_{10-2.5}$ and particle number concentration. This provides an opportunity to carry out new time-series analyses to identify which component is most important for health effects. These results would then inform policy for the protection of public health. This project, funded by DEFRA, aimed to analyse, using time series methods, the health effects of various particle metrics within London. This analysis aimed to identify which of the particle metrics are most important for health impact considerations. The particle metrics studied were: carbon (total, organic and elemental), anions (nitrate, sulphate and chloride), gravimetric PM_{10}, $PM_{2.5}$ and $PM_{10-2.5}$, particle number concentrations (PNC), modelled particle source apportionment, TEOM measurements for PM_{10} and $PM_{2.5}$, black smoke and for a limited period particle measurements made using FDMS (Filter Dynamic Measurement System) samplers. Most of these measurements were made at North Kensington - a background monitoring station in central London. Daily numbers of hospital admissions and deaths in London for respiratory and cardiovascular causes were studied. Data were assembled for the period January 2000 to December 2005 inclusive. The availability of the particle metric data over the six year study period was variable. Some pollutants had measurements available for most days whilst others were missing for a large number of days and the patterns of missing data varied from metric to metric. As only preliminary findings were available at the time of the meeting a small number of initial results were presented and discussed.

Application of a dynamic population-based model for evaluation of exposure reduction strategies in the baking industry

Tim Meijster[1,2], Nick Warren[3], Dick Heederik[2], Erik Tielemans[1]

[1] TNO Quality of Life, Business unit Quality and Safety, Zeist, The Netherlands

[2] Utrecht University, Institute of Risk Assessment Sciences, Division of Environmental Epidemiology, Utrecht, The Netherlands

[3] Health and Safety Laboratory, Harpur Hill, Buxton, Derbyshire, United Kingdom

Tim.meijster@tno.nl

Abstract. Recently a dynamic population model was developed that simulates a population of bakery workers longitudinally through time and tracks the development of work-related sensitisation and respiratory symptoms in each worker. Input for this model comes from cross-sectional and longitudinal epidemiological studies which allowed estimation of exposure response relationships and disease transition probabilities This model allows us to study the development of diseases and transitions between disease states over time in relation to determinants of disease including flour dust and/or allergen exposure. Furthermore it enables more realistic modelling of the health impact of different intervention strategies at the workplace (e.g. changes in exposure may take several years to impact on ill-health and often occur as a gradual trend). A large dataset of individual full-shift exposure measurements and real-time exposure measurements were used to obtain detailed insight into the effectiveness of control measures and other determinants of exposure. Given this information a population wide reduction of the median exposure with 50% was evaluated in this paper.

1. Introduction

Exposure to flour dust and related enzymes is one of the most common occupational exposures associated with the development of occupational respiratory symptoms especially occupational asthma [1, 2]. Recently several studies reported no or only a small decreasing time trend in occupational exposure to flour dust and fungal α-amylase [3, 4]. These observations indicate the need for an evidence base for effective interventions to decrease occupational exposure to flour dust and as a result the prevalence of respiratory diseases.

In the Netherlands a health surveillance system was recently installed in all major flour processing industries as part of a covenant to monitor the disease status of the population [5]. In addition, several large exposure measurement studies were conducted that provide detailed information on exposure levels and determinants of exposure including control measures [4, 6, 7]. The information from the health surveillance system in combination with the exposure data was used to develop a dynamic population-based model. This model can be used to simulate a population of bakery workers longitudinally through time and tracks the development of work related sensitization and respiratory symptoms in each worker related to their exposure. This model enables (prospective) health impact assessment of different intervention scenarios and can provide insight in the impact of exposure reductions (or other interventions) on the population burden of disease. This can provide researchers and policy makers with information on trends in disease development in the population resulting from exposure reductions and help decide on the most cost-effective and best achievable intervention strategy.

2. Methodology

Figure 1 gives a graphical representation of the health impact assessment approach as in our study; the cost-benefit analysis is a separate addition that is beyond the scope of this abstract but will be discussed briefly in the presentation.

2.1. The dynamic population-based model

Basically the core of our health impact assessment is the developed dynamic population-based model that integrates the available epidemiological and exposure information, in combination with general population information relevant for the modeled exposure-disease relationship (e.g. atopy, background rates of sensitization, distribution of career length etc.). This model enables the evaluation of the impact of changes in exposure on the disease burden of the population. This might then provide input for a cost-benefit analysis.

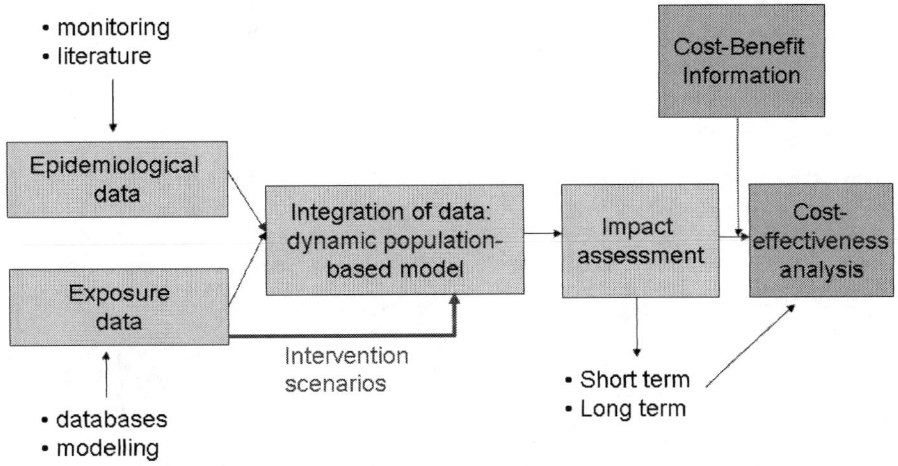

Figure 1. The health impact assessment approach

The development of our dynamic population-based model is described in a paper by Warren *et al.* [8]. In short, the model simulates a population of bakery workers longitudinally through time and tracks the development of work related sensitization and respiratory symptoms in each worker related to their exposure. The main part is a multi stage disease model that simulates the transition of workers between different disease states. The main states incorporated are work related sensitization to wheat allergens and fungal α-amylase, lower- and upper respiratory symptoms and finally work disability. Disease progression is modelled with a yearly probability that depends upon the workers' simulated exposure to flour dust and fungal α-amylase. The dose-response slopes for these transition probabilities were estimated using a substantial dataset of bakery workers in the Netherlands [9]. In our model, work related sensitization and atopy are key determinants for the development of respiratory symptoms and upper respiratory symptoms are assumed to generally precede the development of lower respiratory symptoms. Separate transition probabilities have been estimated for atopic- and non-atopic workers.

2.2. Intervention scenario

We used the model to evaluate the effect of exposure reductions on the development and occurrence of disease in this population. The scenario discussed in this paper involves a population wide reduction in median exposure to flour dust of 50%. A 50% reduction is assumed to be feasible taking into account the state-of-the-art knowledge of efficacy of control measures [7]. For fungal α-amylase, direct evidence of efficacy of control measures is lacking, although a very recent time trend analysis does show reductions in amylase exposure in the Netherlands over the past decade [4]. However, no information is available on the associated determinants of this decrease. For this worked example we assume the decrease in fungal α-amylase is equal to that of flour dust with a 50% reduction in median exposure at the start of the simulation period. Variability in both exposures is assumed to remain the same over time.

2.3. Simulation

The intervention scenario was simulated by reducing the population median exposure at the time of the intervention (nominally the year 2000) and running the model for 20 years in order to observe changes in disease prevalence. The results from the simulation were used to create trend plots, and to observe the trend in relative prevalence for the different disease states as well as for the healthy population. Confidence intervals for the relative prevalence were obtained through an uncertainty analysis that repeated the procedure within a Monte Carlo algorithm, using random model parameters generated from distributions to represent parameter uncertainty. In the presented results we have aggregated the different disease states into three main disease states; sensitized, work-related upper respiratory symptoms and work-related lower respiratory symptoms.

3. Results

In our scenario the population median exposure decreased by 50%, prior to exposure reduction a geometric mean exposure level of 1.8 mg/m^3 for flour dust and 1.0 ng/m^3 for fungal α-amylase was observed. Figure 2 shows the relative trend in prevalence of our three main disease states, the width of the bands reflects the uncertainty in our estimates of the relative prevalence of the different disease states for a given year.

As can be seen from Figure 2 a downward trend is predicted for all three groupings of disease states with strongest trend for lower respiratory symptoms. The plot also shows that lower respiratory symptoms have the largest associated uncertainty. Another important observation is the fact that although the exposure reductions occurred at the start of the simulation period, the effect of this reduction is still ongoing several years after the introduction. This is most likely caused because complete turnover of a working population takes several decades. Overall we can say that a 50% reduction in exposure will eventually lead to a 35 to 65% reduction in asthma cases after more then a decade, a 20 to 35% reduction of workers with upper respiratory symptoms and 10 to 20% less sensitized workers.

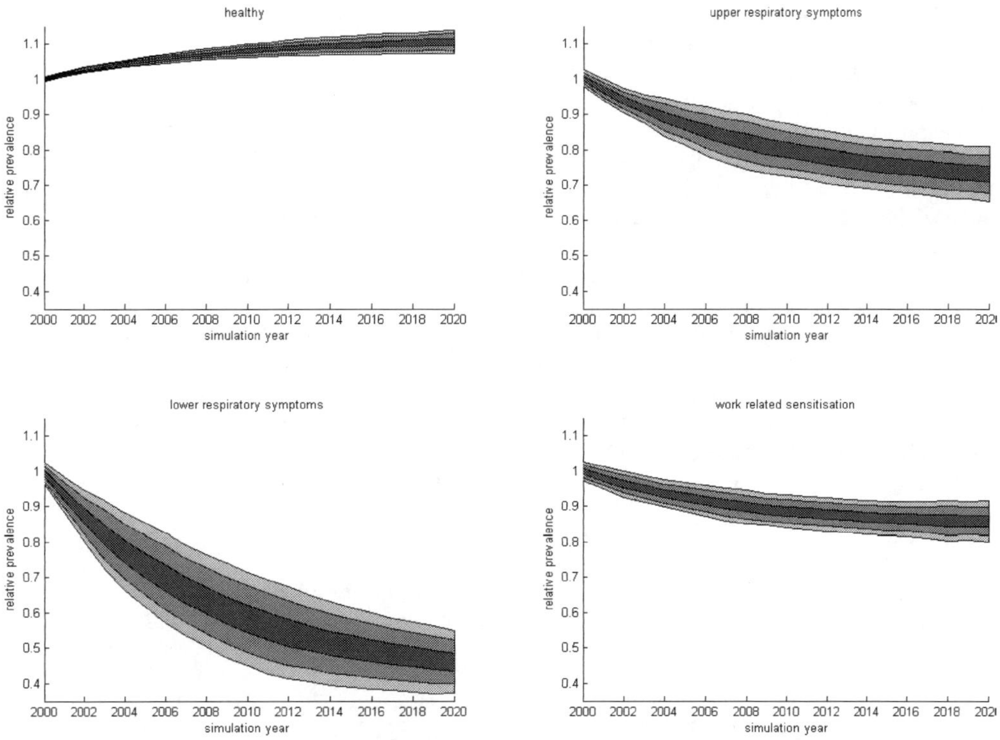

Figure 2. Disease prevalence for our three major disease states with time

4. Conclusions

The example presented here shows how health impact assessment can use available exposure and epidemiological data to obtain more insight in the relation between exposure reductions (e.g. interventions) and trends in the population burden of disease. The results indicate that the impact on health might develop relatively slow and take several years to take full effect, this is probably associated with the worker population turn over rate. This is important information for both policy makers (who might enforce and monitor these changes) as well as employers and employees who initiate changes in exposure to reduce disease prevalence. Nevertheless it is possible that other more specific intervention scenarios will lead to quicker results. For example exposure reduction in combination with health surveillance might be more effective. This, and other possibilities are explored in another study, the preliminary results of which will be available soon.

The example presented here has some flaws, we decrease the median of the population exposure distribution but assume that variability remains the same. It is much more likely that a hygiene intervention will have a different impact on high and low exposed workers, as was also shown in an earlier paper [7]. This would result in a more drastic change of the population exposure distribution, it will not only shift as is the case in our example, but the variability (e.g. right tail of the distribution) will also decrease. This might change the impact on our disease outcomes. Another simplistic assumption is that exposure decreases all at once at the start of the simulation when, in general, these reductions are much more likely occur in a stepwise manner [3, 10, 11].

As can be seen from our results the uncertainty in our prevalence estimation for lower respiratory symptoms is larger then for the other states. The main reason for this is that the number of cases in our dataset was small. As a result the estimated dose response curve for this disease state had larger confidence intervals resulting in bigger uncertainty in our simulation model. This could be decreased if a larger dataset or good longitudinal information becomes available.

In general the input parameters for the disease model, especially the transition probabilities were estimated using information from a cross-sectional study. Although the use of sophisticated statistical techniques allowed us to do this, longitudinal data with which to check the validity of these parameters are scarce. The general comparisons that we made and also the comparisons of cross-sections of data from our model (data not presented here) with observational studies suggest that the model performs extremely well. Nevertheless it would be preferable to obtain better longitudinal data to perform a more rigid validation of the model.

Overall, we have demonstrated the added value of using health impact assessment especially when planning strategies to prevent occupational diseases. The methodology enables decision makers to weigh different options and also to discuss the implications of the observed trends. In the near future a more elaborate analysis will be presented comparing several different realistic intervention strategies and evaluating the validity of our model in a broader sense.

Reference List

[1] Baur X, Degens PO and Sander I. Baker's asthma: still among the most frequent occupational respiratory disorders. *J Allergy Clin Immunol* 1998 Dec;**102**(6 Pt 1):984-97.

[2] Bernstein IL, Chan-Yeung M, Malo JL, Bernstein DI 2006 *Asthma in the workplace*. 3 ed. (New York, Taylor & Francis Group)

[3] Creely K, Tongeren van M, While D, Soutar A, Tickner J, Bolton A, et al. Trends in inhalation exposure: mid 1980s till present. Suffolk: HSE Books; 2006. Report No.: RR 460.

[4] Meijster T, Tielemans E and Heederick D. The impact on exposure to flour dust and fungal á-amylase among Dutch workers exposed to flour dust: A time trend analysis. *submitted.*

[5] Meijer E, Suarthana E, Monchy de J, Rooy van F, Rooijakkers J, Meijster T, et al. Diagnostic Research in Occupational Allergy: Results of a nationwide surveillance among bakery workers. In: Suarthana E. Predicting Occupational Lung Disease. *Thesis* Utrecht: Utrecht University; 2008. p. 89-101.

[6] Meijster T, Tielemans E, de Pater N and Heederik D. Modelling Exposure in Flour Processing Sectors in The Netherlands: a Baseline Measurement in the Context of an Intervention Program. *Ann Occup Hyg* 2007 Mar 17;**51**(3):293-304.

[7] Meijster T, Tielemans E, Schinkel J and Heederik D. Evaluation of peak exposures in the Dutch flour processing industry: implications for intervention strategies. *Ann Occup Hyg* Accepted

[8] Warren N, Meijster T, Heederik D and Tielemans E. A dynamic population-based model for the development of work related respiratory health effects amongst bakery workers. *submitted.*

[9] Jacobs J, Meijster T, Meijer E, Suarthana E and Heederik DJJ. Wheat allergen exposure and the prevalence of work-related sensitization and allergy. *Allergy* 2008; Epub ahead of print.

[10] Kromhout H, Vermeulen R. Long-term trends in occupational exposure: Are they real? What causes them? What shall we do with them? *Ann Occup Hyg* 2000;**44**(5):325-7.

[11] Miller BG, Cherrie JW, Groat S and Kauffer E. Changes in workplace concentrations of airborne respirable fibres in the European ceramic fibre industry 1987 1996. *Ann Occup Hyg* 2007 Aug;**51**(6):501-7.

Journal of Physics: Conference Series **151** (2009) 012002 doi:10.1088/1742-6596/151/1/012002

GIS-based models for ambient PM exposure and health impact assessment for the UK

John R Stedman, Susannah Grice, Andrew Kent and Sally Cooke

AEA Energy & Environment, The Gemini Building, Fermi Avenue, Harwell

Didcot, Oxfordshire, OX11 0QR

John.stedman@aeat.co.uk

Abstract. GIS-based models have been developed to map ambient PM_{10} and $PM_{2.5}$ mass concentrations across the UK. The resulting maps are used for the assessments of air quality required by the EU ambient air quality directives, health impact assessment and the development of UK air quality policy. Maps are presented for 2006 along with projections to 2020. The largest single contribution to the UK population-weighted mean annual mean background concentrations of PM_{10} in 2006 is estimated to be from secondary PM (43%), followed by the contribution from primary PM (24%). Concentrations are predicted to decline by 15% for PM_{10} and 13% for $PM_{2.5}$ over the period from 2006 to 2020. The extent of exceedence of the 24-hour limit value is predicted to decline from 1.9% to 0.1% of urban major roads over the same period. The potential health benefits of reductions in ambient PM are large. A reduction in concentration of 0.93 μg m^{-3} as a result of a possible package of measures has been estimated within the UK Air Quality Strategy to result in a reduction in life years lost of approximately 2 – 4 million over a period of 100 years.

1. Introduction

There remains considerable concern about the health effects of airborne particulate matter [1]. This has lead to the adoption of objectives and limit values for PM_{10} for the protection of human health at a UK national scale via the Air Quality Strategy [2] and a European scale via the recent Ambient Air Quality Directive [3]. Objectives and exposure reduction targets have also recently been adopted for $PM_{2.5}$ [2,3].

A model of ambient particulate matter concentration across the UK has been developed within a Geographical Information System (GIS). This model is known as the Pollution Climate Mapping (PCM) model. The development of this model has been described by Stedman et al [4], who presented maps of both PM_{10} and $PM_{2.5}$ for 2004. This paper describes the further development of the PCM model to provide maps of concentration for 2006. The resulting maps are used to supplement ambient monitoring data in the compilation of the annual air quality assessments required by the EU ambient air quality directives [3]. Maps of projections of future air quality are also compiled and are used to underpin the development air UK air quality policy. The calculation of maps for a baseline and various different policy scenarios enables the difference between the baseline and different options to be assessed in terms of compliance with air quality objectives and limit values. A key summary statistics derived from the mapping studies is the population-weighted annual mean PM concentration, which can be used to estimate short and long term health impacts. A value can then be assigned to these health effects and compared with the costs of different policy options within a cost benefit analysis [2,5].

Maps of background PM concentrations are calculated by summing contributions from a wide range of different sources. The maps therefore provide a useful resource for the assessment of the relative contributions to total PM mass from the different chemical components and sources. Maps for 2006 are presented along with population-weighted mean and exceedence statistics for these maps and projections to 2020. Examples of health impact assessments included in the Air Quality Strategy published in 2007 [2] are also presented.

2. Method

2.1. Introduction

Maps of background PM concentrations are calculated within the PCM model by summing contributions from the following sources within the GIS:

- Secondary inorganic aerosol
- Secondary organic aerosol
- Large point sources of primary particles
- Small point sources of primary particles
- Area sources of primary particles
- Regional primary particles
- Iron and calcium rich dusts
- Sea salt

Maps have been calculated on a 1 km x 1km grid for the whole of the UK land area. The listed components have been found to provide approximate mass closure and no additional residual is required in order to provide good agreement with measured concentrations. A roadside increment model has also been used to calculate estimates of concentrations for 10,027 individual urban major road links.

2.2. Secondary aerosol

Secondary inorganic aerosol concentrations have been derived by interpolation and scaling of measurements of sulphate, nitrate and ammonium at 28 rural sites [6]. The scaling factors are required to take account of the contribution from bound water, counter ions and to apportion the mass between the fine ($PM_{2.5}$) and coarse (PM_{10} – $PM_{2.5}$) fractions [7]. Secondary organic aerosol (semi-volatile organic compounds formed by the oxidation of non-methane volatile organic compounds) concentrations have been estimated from results from the HARM/ELMO model [8].

2.3. Primary particulate matter

Estimates of the emission of primary particulate matter (PM_{10} or $PM_{2.5}$) are available from the UK National Atmospheric Emission Inventory [9] (NAEI). The contribution to ambient PM of emissions from identified large point sources with emissions of greater than 500 tonnes per year have been modelled explicitly using the air dispersion model ADMS [10]. The contribution from point sources of primary particles with annual emission of less than 500 tonnes have been estimated using a generic small points model [7] derived using ADMS.

The emissions from area sources of primary particles are represented within the NAEI as estimates for 1 km x 1 km grid squares across the UK. The contribution from local area sources (within 15 km) to ambient PM concentrations has been estimated using a dispersion kernel based model derived by running the air dispersion model ADMS for unit emissions. The contribution from more distant sources of primary particles can be described as a regional background of primary PM. This was estimated using the TRACK [11] receptor oriented, Lagrangian statistical model and emissions estimates from the NAEI and EMEP.

2.4. Iron and calcium rich dusts

The contribution to ambient PM concentrations from iron and calcium rich dusts was estimated from a combination of measurements made in Birmingham [12] and surrogate variables (vehicle movements and population, respectively) for the spatial distribution of the emission associated with these dusts. Rural concentrations across the UK were set to the values measured in the Birmingham area, while the additional urban increment (urban background – rural) was set to vary with the surrogate emission variables such that the modelled concentration in Birmingham matched the measurements.

2.5. Sea salt

The contribution from sea salt was derived by interpolation of measurements of chloride at the 28 rural sites [6] and scaling to take account of the sodium counter ion.

2.6. Roadside increment

The concentration of PM at the roadside (within approximately 10 m) can be considered as the sum of the background concentration described above and a roadside increment to take account of the very local emission from the vehicles on the adjacent road. We have used an empirical model [7] to estimate the roadside increment from the road link specific emissions estimates provided by the NAEI.

2.7. Monitoring data

There are a wide variety of methods for measuring PM mass concentrations and the mass obtained is dependent on the instrument used and this is an area of active research in itself [13]. A very brief summary of the comparability of UK measurements with our mapping studies is included here. The reference method for comparison with the EU directive limit values is a gravimetric instrument. Tapered Element Oscillating Microbalance (TEOM) instruments fitted with a Filter Dynamics Measurement System (FDMS) have been shown to be equivalent to the reference method, as have a number of β-attenuation instruments if a suitable scaling factor is applied [14]. The standard TEOM instrument, widely used in the UK, has been shown not to be equivalent even if the commonly applied scaling factor of 1.3 is used. A method utilizing the results from TEOM FDMS instruments to correct standard TEOM data, the Volatile Correction Model (VCM) has, however, been described [15].

Monitoring data from the UK national monitoring networks is available from the UK Air Quality Archive [16].

2.8. Model calibration and verification

Ambient monitoring data is required to calibrate the area source dispersion model component of the PCM model and the roadside increment model, while the models for the remaining components are not calibrated. Measurements of PM_{10} and $PM_{2.5}$ from Partisol 2025 gravimetric instruments have been used to calibrate the models for 2006. There is a known high blank filter artefact with these measurements, thought to be related to filter conditioning and humidity problem with the quartz filters used, although the exact causes are not fully understood. We have therefore subtracted a total correction of 6.59 $\mu g\ m^{-3}$ from the reported annual means for 2006. This is made up of two components, 4.09 $\mu g\ m^{-3}$ for laboratory blank issues and 2.5 $\mu g\ m^{-3}$ for the blank due to the use of quartz filters, as opposed to telflon coated filters. The choice of these corrections has been informed by consideration of PM_{10} and $PM_{2.5}$ monitoring data and PCM models for 2005 and 2006. FDMS measurements are available for 24 national network monitoring sites for 2007 and corrected Partisol 2025 for 2007 are consistent with a provisional model for 2007 calibrated using FDMS data only.

The PCM maps of PM_{10} and $PM_{2.5}$ for 2006 have been calibrated using the corrected Partisol 2025 monitoring data and the results verified by comparison with TEOM data for London and the South East as corrected using the VCM model. Further indications of model performance have been obtained by comparison with TEOM instruments multiplied by 1.3 and data suitably scaled data from β-attenuation instruments.

2.9. Projections of concentration to 2020.
Maps of PM concentration in 2010, 2015 and 2020 have also been calculated for a baseline scenario incorporating our best estimate of the likely impact of currently agreed national and international policies. The emissions of primary PM for future years have been provided by the NAEI. Secondary inorganic aerosol concentrations have been assumed to decline in line with the impact of changes in precursor emissions on concentration predicted by the EMEP model [17]. The concentrations of secondary organic aerosol, iron and calcium rich dusts and sea salt have been assumed to remain unchanged from the values in 2006.

2.10. Population-weighted means
Population-weighted mean concentrations have been calculated by multiplying the 1 km x 1 km background maps by 1 km x 1 km population statistics from the 2001 census. The values for all of the grid squares are summed and then divided by the total population to calculate the population-weighted mean. This is a useful summary statistic, which is related to human health impacts if the dose response function is assumed to be linear with no threshold.

3. Results

3.1. Mapped PM_{10} and $PM_{2.5}$ concentrations for 2006
Mapped PM_{10} and $PM_{2.5}$ concentrations for 2006 are shown in Figure 1. PM_{10} concentration are higher than $PM_{2.5}$, the main contributors to the different include coarse sea salt, iron and calcium rich dusts and nitrate. The south east to north west gradient is driven by the gradient in secondary inorganic aerosol and the higher concentrations in urban areas and close to major road corridors are due to emissions of primary PM and iron and calcium rich dusts.

Figure 2 shows verification plots for the background and roadside maps of PM_{10} concentration for 2006. As expected the agreement is good for the national network sites used to calibrate the models. The agreement with the TEOM data corrected using the VCM model is also good. This provides some additional support for our choice of correction applied to the Partisol 2025 data prior to model calibration. Comparison with TEOM data x 1.3 indicates that these measurements are systematically higher than predicted by the model but that the ranking of the sites is quite reasonable. This suggests that TEOM x 1.3 is likely to overestimate PM_{10} concentrations at many locations.

3.2. Maps of PM_{10} components for 2006
Figure 3 shows the individual contributions to ambient PM_{10} from primary PM, secondary PM, iron and calcium rich dusts and sea salt. The population-weighted mean contribution from each of these sources are listed in Table 1. The largest single contribution is from secondary PM (43%), followed by the contribution from primary PM (24%).

Table 1. UK Population weighted mean annual mean PM_{10} concentrations in 2006 ($\mu g\ m^{-3}$).

Component	Concentration
Primary PM	3.91
Secondary PM	7.06
Iron and calcium rich dusts	3.22
Sea salt	2.30
Total	16.49

a) PM_{10} b) $PM_{2.5}$

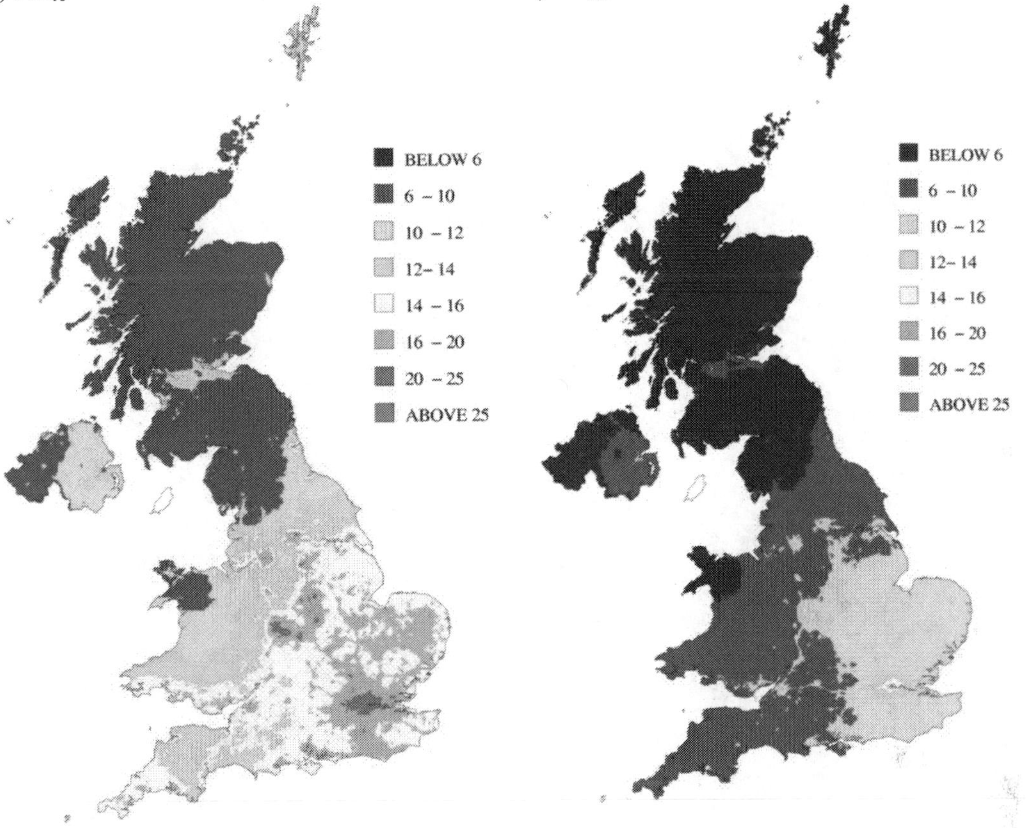

Figure 1. Mapped annual mean background PM concentrations in 2006 ($\mu g\ m^{-3}$)

a) Background b) Roadside

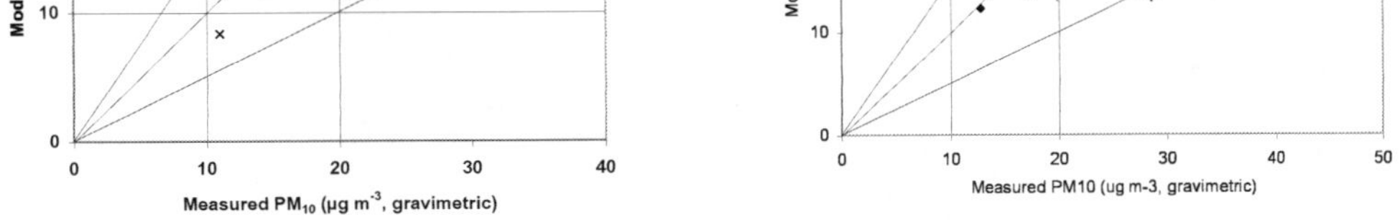

Figure 2. Verification plots for mapped PM_{10} in 2006.

IOP Publishing
doi:10.1088/1742-6596/151/1/012002

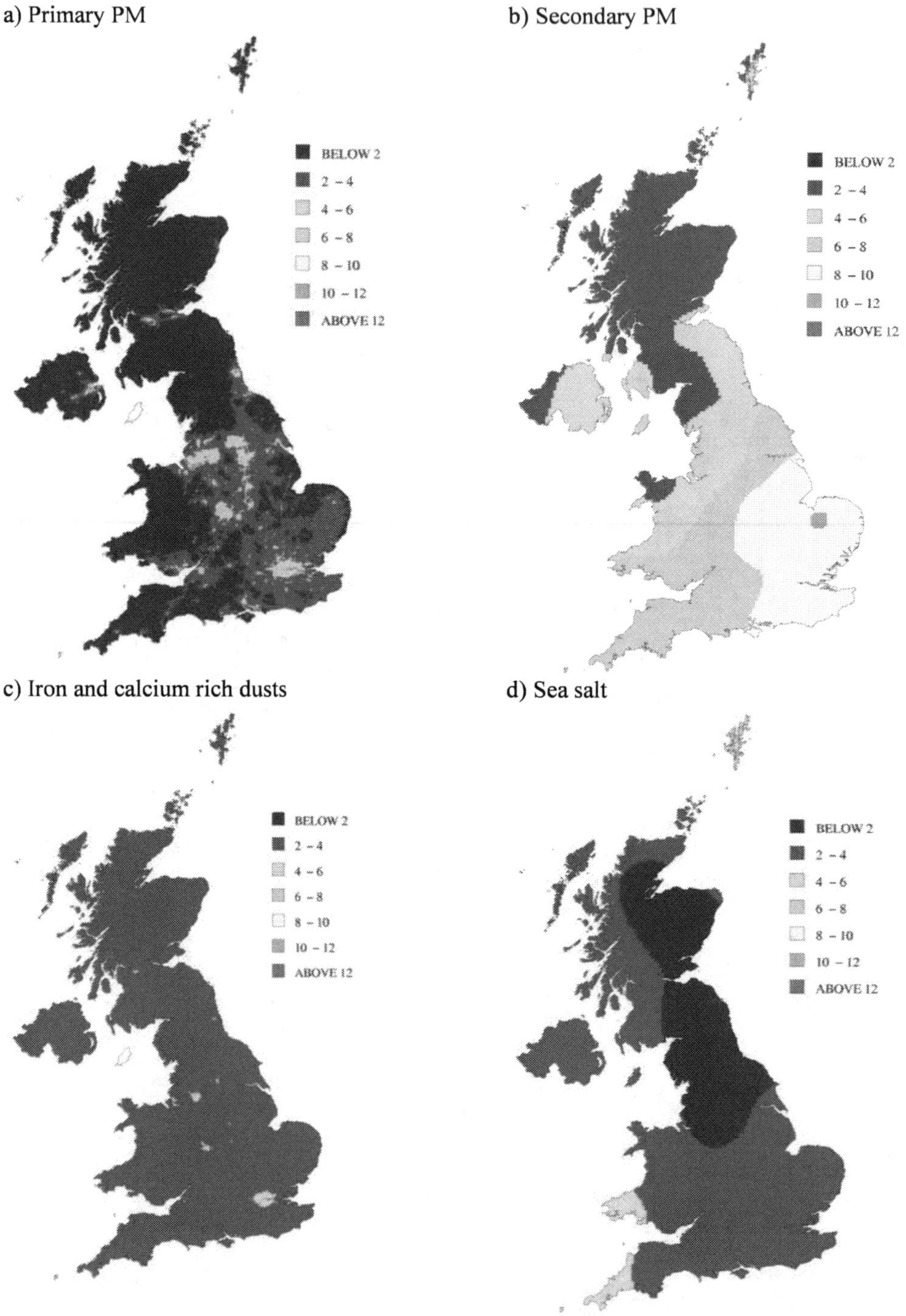

Figure 3. The contributions to mapped annual mean background PM_{10} concentrations in 2006 from different components ($\mu g\ m^{-3}$)

3.3. Projections to 2020

Figure 4 shows population-weighted mean PM_{10} and $PM_{2.5}$ concentration projections for our baseline scenario to 2020. Concentrations are expected to decline but not dramatically, by 15% for PM_{10} and 13% for $PM_{2.5}$ over this period. The decline is significantly less steep than the decline in primary PM emissions or the precursor emissions of secondary inorganic PM. This is because the concentrations of secondary PM are expected to decline more slowly than precursor emissions and there are a number of components, such as iron and calcium rich dusts and sea salt, which are not expected to decline. The modelled PM_{10} value in 2006 is close to the provisional FDMS-calibrated concentration for 2007, which provides additional confidence that an appropriate correction has been applied to the 2006 Partisol 2025 measurement data.

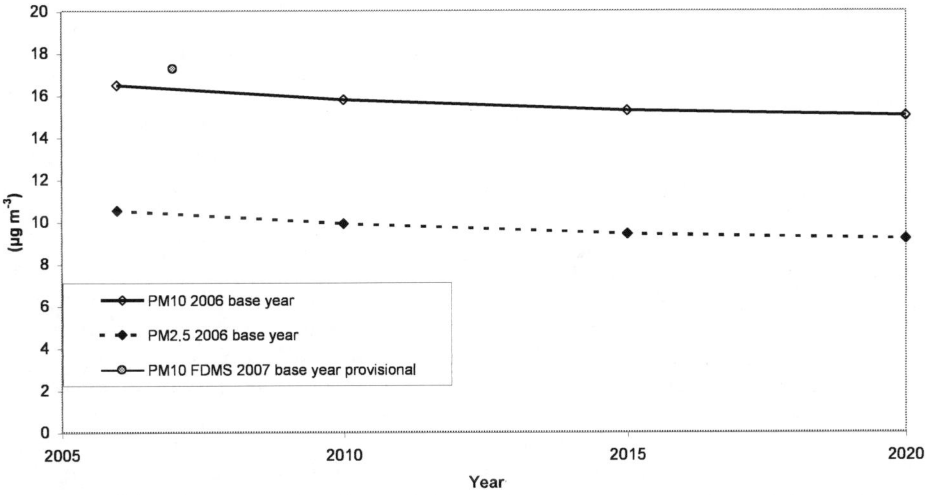

Figure 4. UK Population weighted mean annual mean PM_{10} and $PM_{2.5}$ concentrations 2006 – 2020 (μg m^{-3}).

Table 2 shows the expected change in the extent of exceedence of the 24-hour EU limit value for PM_{10} at the roadside in urban areas between 2006 and 2020. Concentration are typically higher at the roadside than in background locations, so it is roadside locations that are most at risk of exceeding the limit value. The 24-hour limit value came into force in 2005 and specifies that concentrations should not exceed 50 μg m^{-3} more on than 35 days per year. We have modelled annual mean concentrations and have taken an annual mean concentration of more than 31.5 μg m^{-3} to be equivalent to an exceedence of the 24-hour limit value. This relationship has been derived from an examination of ambient monitoring data [7].

The UK has been divided into 43 zones for air quality assessment. These are 28 large cities and 15 geographical areas. Eight zones are estimated to have exceeded the limit value in 2006, reducing to just 10 km of road in one zone (London) by 2020. The new air quality directive specifies that the contribution from natural sources can be excluded prior to the air quality assessment. Table 2 shows that subtracting the contribution from sea salt significantly reduces the extent of exceedence and exceedences are then confined to the busiest roads in the largest cities in 2006.

Table 2. Predicted extent of exceedence of the 24-hour PM_{10} limit value at the roadside (values in brackets with sea salt contribution subtracted).

	Total assessed	2006	2010	2015	2020
Length of road (km)	13,880	266(138)	182(66)	45(9)	10(2)
Percentage of road length (%)	100	1.9(1.0)	1.3(0.5)	0.3(0.1)	0.1(0.0)
Number of zones	43	8(3)	2(2)	1(1)	1(1)

4. Discussion

4.1. Possible uses of the maps within epidemiology studies

The type of maps presented here can be used as an alternative to using individual monitoring site data for epidemiology studies of the health impact of air pollutants [18]. We believe that this method of estimating individual exposure to outdoor air pollution has some advantages over others. Whichever technique is used, there will always be some exposure misclassification. Most commonly, researchers have used single monitoring station data to assign exposure to all individuals living in large areas, which are likely to reflect individual exposure poorly because of the distance from the monitoring station and individual mobility [19]. Other researchers have used population density [20] or estimated traffic exposure [21] in an attempt to reduce exposure misclassification. The disadvantage of this type of approach (based on land use) is that the variables employed to improve upon fixed site monitoring may not always be the best markers for air pollutant emission densities.

We consider that the use of a GIS based PCM model of ambient air quality concentrations has considerable advantages over regression-based models. This is because our model incorporates detailed information from the emission inventory and meteorological data is used to model the dispersion of these emission in that atmosphere. A range of different tools have been used to estimate of the contributions from various other PM sources to ambient PM including the use of measurement data for different chemical species as appropriate. An additional advantage of this type of approach is that there is scope to investigate the relative contributions to the total PM mass from the different chemical components and sources.

4.2. Health impact assessment

The type of maps presented here have be extensively used to support the development of UK air quality policy through the review and revision of the Air Quality Strategy [2]. We have adopted the methods for health impact quantification recommended by the Interdepartmental Group on Costs and Benefits (IGCB) [5]. Table 3 shows examples of the estimated health benefits of a package of measures for 2020 illustrated in the Air Quality Strategy [2]. The package of measures included early uptake of Euro 5/6 measures to reduce emissions from cars and Euro VI measures for heavy duty vehicles, increased uptake of low emission vehicles and measures to reduce emission from international shipping. The base year for this analysis was 2003. The life table calculations carried out for the Air Quality Strategy [2] used changes in concentration for 100 years the assess the change in life expectancy due to the long-term effect of PM. The change in annual mean PM_{10} concentration in 2020 has been assumed to apply for 100 years. A coefficient of 6.0% per 10 $\mu g\ m^{-3}$ change in PM has been used in these calculations. The results of the health impact assessment for the package of measures are shown in Table 3. The long-term effects (life years saved) dominate the health impact assessment. The range for this statistic represents the impact of different assumptions concerning the lag between the change in concentration and the health impact (0 – 40 years) [5]. These health benefits have been calculated from a reduction in UK population-weighted mean annual mean PM_{10} concentration of 0.93 μg m^{-3}. The value of the total health benefits of this measure have been calculated by the IGCB to be £886 – 2,039 million (annual net present value). This shows the large potential value of reductions in PM concentrations.

Table 3. Results of the health impact assessment for PM10 for the package of measures 'R' in 2020 [5]

Life years saved 2010-2109	Respiratory hospital admissions (2020 per year)	Cardiovascular hospital admissions (2020 per year)
2,202,000 – 3,805,000	325	326

Acknowledgements

This work was funded by the UK Department for Environment, Food and Rural Affairs, Welsh Assembly Government, the Scottish Executive and the Department of the Environment in Northern Ireland under contract CPEA 15.

References

[1] Pope, C. A. and Dockery, D. W. (2006) Health Effects of Fine Particulate Air Pollution: Lines that Connect. *J. Air & Waste Mange. Assoc.* **56**: 709-742

[2] Defra et al (2007) Department for Environment, Food and Rural Affairs, The Scottish Executive, Welsh Assembly Government and The Department of the Environment for Northern Ireland.The Air Quality Strategy for England, Scotland, Wales and Northern Ireland: Volume 1 and 2. (Cmnd 7169). The Stationery Office 2007 http://www.defra.gov.uk/environment/airquality/strategy/index.htm

[3] Directive on ambient air quality and cleaner air for Europe (2008) http://register.consilium.europa.eu/pdf/en/07/st03/st03696.en07.pdf

[4] Stedman, J. R., Kent , A. J., Grice, S., Bush, T. J. and Derwent, R. G. (2007). A consistent method for modelling PM_{10} and $PM_{2.5}$ concentrations across the United Kingdom in 2004 for air quality assessment. *Atmospheric Environment*, **41**, 161-172.

[5] DEFRA et al (2007). Department for Environment, Food and Rural Affairs, The Scottish Executive, Welsh Assembly Government and The Department of the Environment for Northern Ireland. An Economic Analysis to Inform the Air Quality Strategy: Updated Third Report of the Interdepartmental Group on Costs and Benefits. http://www.defra.gov.uk/environment/airquality/strategy/index.htm

[6] Personal communication, Sim Tang, CEH (2004). CEH Edinburgh, Centre for Ecology and Hydrology, Natural Environment Research Council.http://www.nbu.ac.uk/cara/networks.htm

[7] Kent, A J, Grice S, Stedman, J R, Cooke, S, Bush, T J, Vincent K J, Abbott J, (2008). UK air quality modelling for annual reporting 2006 on ambient air quality assessment under Council Directives 96/62/EC, 1999/30/EC and 2000/69/EC. AEA Energy & Environment. Report AEAT/ENV/R/2502.

[8] Whyatt, J.D., Metcalfe, S.E., Nicholson, J., Derwent, R.G., Page T. and Stedman, J. (2007) Regional scale modelling of particulate matter in the UK: source attribution and an assessment of uncertainties. *Atmospheric Environment*, **41**, 3315-3327.

[9] Dore, C. J., Watterson, J. D., Murrells, T. P., Passant, N. R. et al., (2007). UK Emissions of Air Pollutants 1970 to 2005. National Atmospheric Emissions Inventory, AEA Technology, Report AEAT/ENV/R/2526. ISBN: 0-9554823-3-X http://www.airquality.co.uk/archive/reports/cat07/0801140937_2005_Report_FINAL.pdf

[10] CERC (2005) Cambridge Environmental Research Consultants

http://www.cerc.co.uk/software/adms3.htm

[11] Lee, D. S., Kingdon, R. D., Jenkin, M. E., and Garland, J. A. (2000) Modelling the atmospheric oxidised and reduced nitrogen budgets for the UK with a Langrangian multi-layer long-range transport model. *Environmental Modelling and Assessment, 5, 83-104*

[12] Yin, J. and Harrison, R. M. (2008). Paragmatic mass closure study for $PM_{1.0}$, $PM_{2.5}$ and PM_{10} at roadsdie, urban background and rural sites. *Atmospheric Environment*, **42**, 980-988.

[13] Air Quality Expert Group (AQEG, 2005). Particulate Matter in the United Kingdom. http://www.defra.gov.uk/environment/airquality/aqeg/particulate-matter/index.htm

[14] Harrison, D., Maggs, R and Booker, J. (2006). UK Equivalence Programme for Monitoring of Particulate Matter. BV/AQ/AD202209/DH/2396 http://www.airquality.co.uk/archive/reports/cat05/0606130952_UKPMEquivalence.pdf

[15] Green, D., Baker, T. and Fuller, G. (2007). King's College London Volatile Correction Model for PM_{10}. KCLERG\MT\FDMS\EQ. http://www.airquality.co.uk/archive/reports/cat13/0711261345_KCL_Volatile_Correction_Model_for_PM10.pdf

[16] UK Air Quality Archive (2008). http:/www.airquality.co.uk.

[17] EMEP (2008) http://www.emep.int/mscw/mscw_publications.html#2007

[18] Chronic exposure to outdoor air pollution and markers of systemic inflammation. Forbes, L. J. L., Patel, M. D., Rudnicka, A. R., Cook, D. G., Bush, T. Stedman, J. R. Whincup, P. H., Strachan, D. P., Anderson, H. R. Submitted to Epidemiology.

[19] Pope CA, III, Burnett RT, Thun MJ, Calle EE, Krewski D, Ito K et al (2002). Lung cancer, cardiopulmonary mortality, and long-term exposure to fine particulate air pollution. *JAMA* **287**(9):1132-1141

[20] Hoek G, Brunekreef B, Goldbohm S, Fischer P, van den Brandt PA (2002). Association between mortality and indicators of traffic-related air pollution in the Netherlands: a cohort study. *Lancet* 360(9341):1203-1209.

[21] Jerrett M, Burnett RT, Ma R, Pope CA, III, Krewski D, Newbold KB et al (2005). Spatial analysis of air pollution and mortality in Los Angeles. *Epidemiology* 16(6):727-736.

Inhaled Particles X, (23–25 September 2008, Manchester) IOP Publishing
Journal of Physics: Conference Series **151** (2009) 012003 doi:10.1088/1742-6596/151/1/012003

Aerosol exposure: concepts, criteria, standards and applications

James H. Vincent

Department of Environmental Health Sciences, School of Public Health, University of Michigan, Ann Arbor, MI 48109, U.S.A.

jhv@umich.edu

Abstract. This paper places *Inhaled Particles X* in the context of the whole sequence of such symposia, going back to the first one in 1961. It draws together some of the essential principles that have been learned since that earlier meeting about the nature of exposure and exposure assessment and thus provides a framework by which to integrate the new knowledge presented at this latest one. In the process, the importance of understanding the formal definition of aerosol exposure is stressed, including the distinction between exposure intensity and exposure history, and how that relates to some measure of cumulative dose which, in turn, may be linked with knowledge about intrinsic toxicity, etc. This then leads to a definition of exposure standards, and the important ingredients of criteria, sampling and limit values. A summary is provided of the current set of particle size-selective criteria that have been widely agreed in the international occupational and environmental health community. Some ideas are presented about how this set might be expanded for certain applications, the important case of ultrafine aerosols being one of them.

1. Introduction

The first BOHS-sponsored "Symposium on the physical, chemical and physiological factors governing the entry of harmful substances into the body via the respiratory system" took place in Oxford in March 1961, and the proceedings – Inhaled Particles and Vapours – appeared the next year, edited by C. Norman Davies, one of the pioneers of what we now know as 'aerosol science'. Over the years since, a succession of nine such symposia has appeared, each reflecting the latest (and best) work in what has continued to emerge as a field important to occupational and environmental health (and to public health more generally). In the 1997 proceedings of the eighth symposium in the series, W. Henry Walton – who himself was editor of many of the later volumes – published a short history of what had been achieved over the preceding years, noting that the original statement of intent made no mention of clearance, biological responses or epidemiology arising from aerosol exposures, but that these became important components of the later meetings.[1]

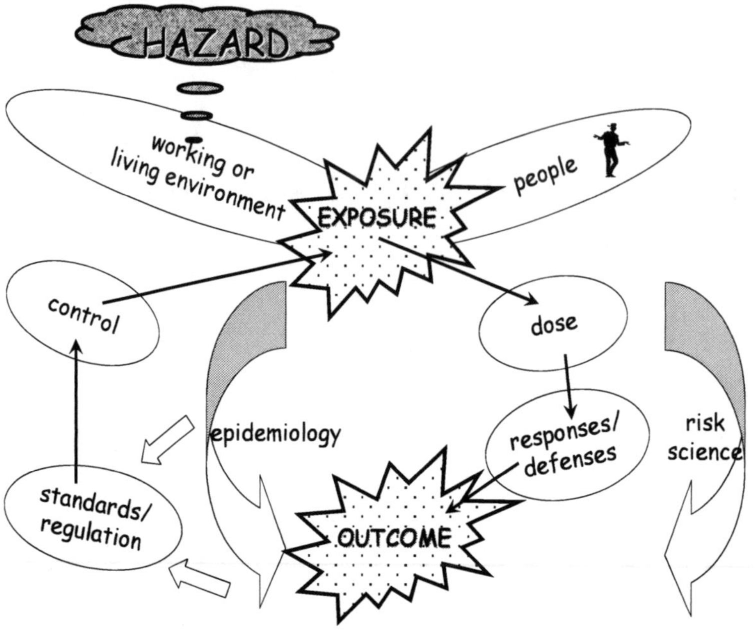

Figure 1: Framework on which to base discussion of aerosol exposures, effects and outcomes, and options for regulation and control.

Now, at the Tenth Inhaled Particles Symposium (IPX), marking an important milestone in this unique series, it is appropriate to re-visit the original aims, and to relate them to how the symposia has evolved and expanded over the years. As a starting point, Figure 1 draws together schematically the various environmental health considerations as they apply to aerosols, and argues that exposure is the central issue. The aerosol hazard may be present in the environment, but exposure occurs only when people interact with that environment. Once exposure occurs, however, it may be followed by a succession of pharmacokinetic and toxicologic processes inside the exposed organism that may or may not lead to an adverse outcome in the form of a clinically or pathologically identifiable disease. The relationship between the exposure and some such outcome may be studied by enquiring into the qualitative and quantitative nature of those intermediate processes, by what we may term as the '*risk sciences*'. Or it may be studied by exploring the quantitative associations by means of the tools of epidemiology. Or by some combination of the two. Here, it should be noted that difficult questions about the detailed roles of toxicity and biochemistry are particularly important – indeed crucial – for exposures where the aerosols comprise complex mixtures, even more so for environmental than for most occupational exposures.

Once the exposure-outcome relationship is determined, by whatever means, there emerges a basis for setting a meaningful standard. The purpose of the standard then becomes the instrument by which to determine the extent to which people may be at risk and to indicate where remedial measures are needed. For the latter there are options for technical control directly of the hazard itself (e.g., by elimination), or of the environment (e.g., by ventilation), or of the way in which people interact with their environment (e.g., by job rotation, or respiratory protection). Beyond these options, there may be some situations (e.g., exposure to certain biological organisms) where there might also be the possibility of intervention in the biological processes inside the exposed individual (e.g., by immunization). The picture described in Figure 1, or something like it, becomes a comprehensive framework by which we may think of the problem, putting it all together, and identifying the diverse roles of the participating individuals and their respective disciplines.

2. Background

For aerosol exposures, the question of the nature of exposure was addressed closely by the occupational health community in the years leading up to the first Symposium. It had long been recognized that certain types of disease among mineworkers were strongly associated with the exposures to airborne dust. In the early 1900s clear evidence emerged – from examination of lung tissue from miners recovered at autopsy – that pneumoconiosis was a disease of the alveolar region of the lung; and further that only a fine particle fraction of inhaled aerosol was able to penetrate so deeply into the respiratory tract.[2] This knowledge provided the basis of the first criteria for aerosol exposure assessment in which only the fine particles considered to be relevant to the disease were to be included. Not long afterwards, a number of novel particle size-selective sampling instruments appeared. A famous example was the konimeter, widely used in the mining industries in several countries, in which sampled aerosol was deposited by impaction onto a glass slide and particles with width (when viewed under an optical microscope) less than 10 μm were subsequently viewed by microscope, sized and counted.[3] Much later, during the 1950s, the role of aerosol physics in the penetration of particles into – and regional deposition in – the human respiratory tract began to be clarified by the efforts of some of the leading aerosol scientists that had been drawn into this field, including most notably the aforementioned Norman Davies and Henry Walton. Their insights led to the definition of the *respirable fraction* by the British Medical Research Council.[4] Based on the earliest experimental inhalation studies with human volunteer subjects, the respirable fraction set off in the first instance to reflect the aerodynamic particle size dependent-penetration of particles down to the alveolar region. The convention that was ultimately adopted was chosen on the basis of how such penetration might be mimicked in a sampling instrument, specifically the horizontal elutriator.[5] In making this choice, it was tacitly assumed that the proportion of inhaled fine particles that remained airborne for long enough to be exhaled could be neglected. The convention took the form of a curve describing the probability of penetration as a function of particle aerodynamic diameter (d_{ae}) - defined as diameter of an equivalent spherical particle of density 10^3 kg/m^3 having the same falling speed in air as the particle in question - and immediately became known as the '*BMRC curve*'. This was later recognized by the wider occupational health community during the Johannesburg Pneumoconiosis Conference in 1959, and subsequently became known to many as the '*Johannesburg curve*'.[6] Soon afterwards, other versions of the respirable fraction were proposed, including one by the American Conference of Governmental Industrial Hygienists (ACGIH) that was capable of being followed by a small personal sampler based on a miniature cyclone.[7]

3. Exposure concepts

The thinking behind these early definitions of a fine aerosol subfraction of what is inhaled provides a basis for a general definition of aerosol exposure. One version is that exposure may be defined as *the intensity of the agent of interest at the relevant interface between the environment and the biological system representing the exposed subject*. Here, therefore, in relation to the preceding, the relevant interface is the alveolar surface of the deep lung, and the intensity is the aerosol concentration there reflecting the flux of particulate material that may pass to the subject. This definition, however pedantic, is useful because it can be generalized to other aerosol fractions (e.g., that deposited in the airway region). Indeed, it may also be applied in environmental health more widely to other potentially harmful agents.

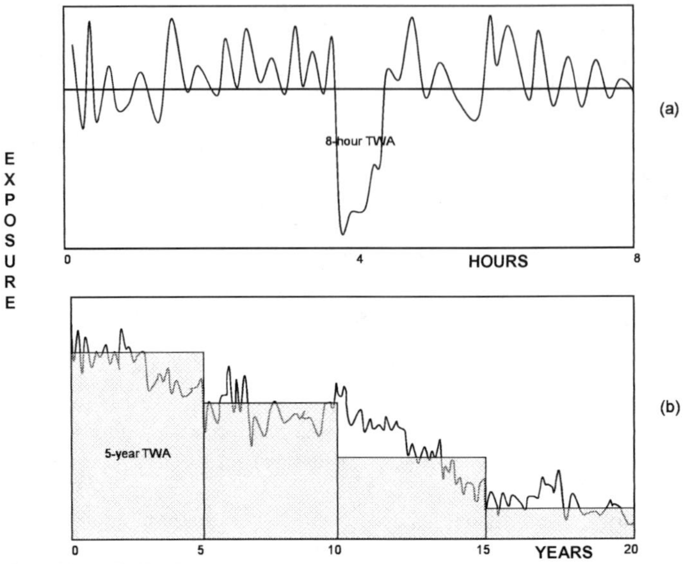

Figure 2: A hypothetical (yet typical) aerosol exposure scenario of an individual worker: (a) instantaneous exposure intensity over a single 8-hour working shift; and (b) trend of 8-hour time-weighted average exposure intensity over the individual's working life, also showing extended 5-year time-weighted averages of the type characterized in some occupational exposure scenarios.

Some qualifications are required. Exposure intensity is a scalar quantity that takes instantaneous values. Or it may be averaged over time, which may be short (e.g., from a few minutes up to many hours, a working shift perhaps) or long (e.g., a subject's lifetime). A hypothetical yet typical aerosol exposure scenario of an individual worker is given in Figure 2. Figure 2a illustrates how the exposure might vary during a single working day, including a period of low exposure during the individual's lunch break. The continuously-varying line indicates the changing instantaneous levels of exposure, and the flat, straight line represents the corresponding 8-hour time-weighted average (TWA). Figure 2b extends this idea to a much longer period, which might be as long as the individual's working life. Now the continuous line represents the changing 8-hour TWA levels, and the shaded boxes represent averages over much longer intervals that might represent the results of periodic sampling campaigns (e.g., as obtained during the large, long-running project conducted in the British coal mining industry known as the *Pneumoconiosis Field Research*, [8]). In both the short and the long terms, it is well known that exposure is highly variable, driven by factors relating to both the nature of the hazard (and how it is generated), the environment itself, and the individual human subject. In addition, Figure 2b shows hypothetically how, over the years, the exposure levels may fall steadily as standards become stricter and more stringent controls are implemented. Figure 2b may therefore be regarded as the worker's *exposure history*.

Exposure history is one ingredient in the risk science that is applied within the framework outlined in Figure 1. Another is the dose of the particulate material at relevant sites in the respiratory tract and beyond, reflecting how much is deposited and where, and its subsequent fate. How much is cleared, and how fast? How much remains, or is transferred to other tissue? Some of the physiological mechanisms governing dose are quite fast, with time constants of the order of hours or days, others quite slow, of the order of months and years. Then there is the toxicity of the material, referring to its action at given locations and biological responses that are induced, along with the biochemistry that governs changes in chemical form or persistence of the material in its ability to be harmful. All these remain outside the scope of this paper, but are the subjects of many contributions elsewhere at this Symposium. Suffice to say for the present that all events and processes leading to adverse health outcomes are driven in the first instance by the exposure.

A significant point within this discussion is contained in the question: to what extent might the influence of recent exposures influence the adverse outcome in relation to exposures that happened in the distant past? It is unlikely that their impact will be the same. With this in mind it might be argued that, in occupational epidemiology, the invocation of a cumulative exposure obtained by simply adding together all the exposures over the lifetime of the exposed individual is not quite ideal. In fairness, many epidemiologists have recognized this notion and the weakness it brings to their analyses, and – indeed – some have attempted to account for it by introducing concepts such as *lag time*.[9]

4. Exposure standards

Occupational and environmental exposure standards may take two forms: (a) regulatory standards, applied and enforced by national governments, built within overarching occupational health policies; and (b) guidance standards, usually offered by private organizations (e.g., ACGIH for occupational exposures) aimed at providing benchmarks to guide the efforts of health professionals.

An occupational or environmental health *policy* may be described as the outcome of discussion between individuals (or groups of individuals), often representing parties with competing interests, who strive to influence an end point which takes the form of occupational health-related legislation and regulation. The content of such discussion is usually based on existing public health data records describing the incidence of disease and its associations with environmental factors. The end point is usually based on some sort of consensus about the ways in which occupational health is valued in a responsible society, and about what is and what is not acceptable. A *standard* for a given agent derives from the specific value that society gives to a particular aspect of adverse health outcome, and to the factors which influence it or are influenced by it. If the policy justifies and defines the problem and provides a regulatory framework for addressing it, incorporating all diverse factors which society at large needs to take into account, a standard provides a measurable reference point consisting of specific guidelines by which the desired objective can be quantified and achieved.

As long ago as 1993,[10] this author defined an exposure standard in terms of a set of ingredients: (a) a set of one or more criteria providing a scientific basis by which to define quite specifically the agent of interest, including which chemical species and – for aerosols – which particle size fraction is the most appropriate to the health effect in question; (b) measurement instrumentation, accompanied by appropriate sampling strategies and analytical procedures, that allow quantitation of the agent of interest in a way matching the criteria defined at the outset; and finally (c) a limit value that provides a quantitative yardstick for what is, and what is not, permissible exposure. Specification of the limit value must match the initial criteria and must relate to the identified measurement methods. So (c) cannot be assigned without prior reference to (a) and (b).

5. Exposure criteria for aerosols

Much of the discussion about exposure criteria for aerosols relates to the physical properties of the particles governing how much is deposited in the human respiratory tract, and where. It is well known that the entry of particles into the body through the nose and mouth during inspiration, and deposition at defined regions within the respiratory tract, and governed by physical processes, including for the larger particles (generally larger than 1 μm in diameter) the aerodynamics of inertial movement and gravitational settling, and for smaller particles (generally smaller than 1 μm) the thermodynamics of diffusion. The early recognition of the role of finer particles in the dose of particulate material to the alveolar region of the lung has already been described. Over the years since those pioneering efforts, a more comprehensive set of particle size-selective criteria has emerged, based on the large amount of experimental data from wind tunnel experiments with inert-but-breathing mannequins and studies of particle inhalation under highly-controlled conditions using human volunteer subjects. Now there is wide international acceptance of a framework that includes, for occupational exposures, the *inhalable fraction* (the particle size-dependent fraction of particles that enter through the nose and/or mouth during breathing), the *thoracic fraction* (the fraction of particles that penetrate below the larynx and into the lung) and – again – the *respirable fraction* (the fraction of particles that penetrate down to the alveolar region). The latter part is consistent with the earlier definitions of the respirable fraction, but the detailed form of the definition has been updated in the light of later information. The set of criteria that has gained widespread acceptance is the harminized version promulgated by the Comité Europëen Normalisation (CEN), the International Standards Organization (ISO) and the American Conference of Governmental Hygienists (ACGIH), respectively.[11, 12, 13] The set is often referred to as the '*CEC/ISO/ACGIH criteria*'. Each of the individual criteria listed is expressed in terms of a probability (i.e., of inhalation or penetration) as a function of particle aerodynamic diameter. Thus each may be described by a curve, as already mentioned, and the agreed versions for the three fractions listed above are shown in Figure 3. Also shown there are the aerosol (or *particulate matter*) fractions defined by the United States Environmental Protection Agency (EPA); namely, PM_{10} (which is a definition for the thoracic fraction, but where the fraction is defined by a curve that is slightly different especially at the larger-particle end) and $PM_{2.5}$ (a more recent addition, aimed at separating out combustion-related aerosols in ambient atmospheric air).[14, 15] Following the rationale outlined above, such curves become the yardsticks against which the measured sampling efficiency performance data for candidate sampling instruments should be compared. That is, for a given fraction, the performance curve for the ideal sampling instrument should match the curve defining the criterion. Here the definition of what is and what is not a 'match', takes into account both the proximity of the instrument's performance curve to the curve representing the actual original criterion, along with considerations of how well the sampler collects the ideal mass.

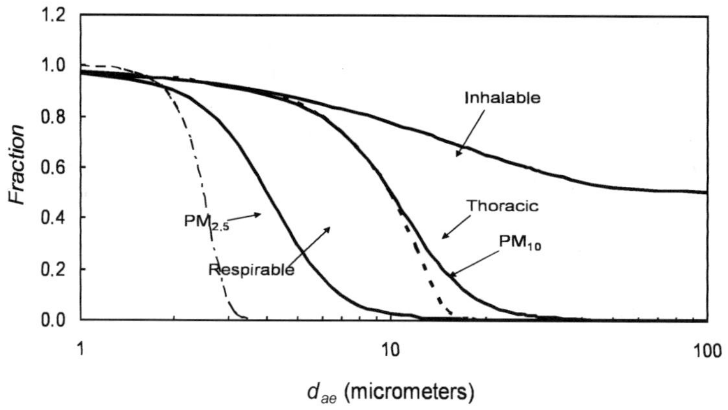

Figure 3: Summary of the latest, currently accepted, particle size-selective criteria for health-related aerosol exposure assessment.[17]

The fractions described above would appear to be sufficient for the needs of most standards. However it is important to note that none of them actually defines a fraction that relates directly to dose. That is, they refer to penetration to a given region of the respiratory tract, and hence not necessarily to what is actually deposited. For the purpose of some epidemiology, it has been argued that it would be more appropriate to define a deposition fraction. For example, McCawley has advocated an alveolar deposition fraction that is more directly related to dose than the penetration-based respirable fraction.[16] The same rationale may be extended to other fractions, including ones for very fine and ultrafine particles that are currently not embodied in the CEN/ISO/ACGIH criteria.

As discussed in a recent book by this author,[17] a rich source of information on which to base any extension of the currently-accepted set of particle size-selective criteria may be found in the comprehensive 1994 report of the International Commission on Radiological Protection.[18] This report describes how all human lung deposition data available at the time were modeled to provide a large set of empirical equations for the specific deposition efficiencies for each region of the respiratory tract, and for particles both in the aerodynamic and thermodynamic regimes respectively. These equations may be used in turn to determine the deposition efficiency (as a proportion of what was inhaled in the first place) for particles of all sizes in any region of the human respiratory tract, taking into account particle deposition occurring during both the inspiration and the expiration phases of the breathing cycle. Once the deposition efficiency is calculated in this way for representative breathing conditions (e.g., corresponding to nose or mouth breathing, corresponding to level of work activity undertaken, etc.), a curve representing a criterion for specific applications may be created, and – in turn – an appropriate sampling instrument may be sought, either an existing instrument, modified accordingly, or one designed from scratch.

There are at present no health-related particle size-selective criteria for ultrafine particles in the range of diameter below 0.1 μm. But, because of the interest stimulated by concerns about health effects in populations exposed to ultrafine particles in working and ambient atmospheric environments, much current research is striving to provide the information needed in order to develop a standard. Some of the more recent developments will be described elsewhere at this Symposium. In order to establish a firm scientific basis for criteria for ultrafine aerosols, and subsequently exposure assessment, there are several issues that need to be addressed and integrated. In the first instance they must include the particle size-dependent deposition of ultrafine particles in relevant regions of the respiratory tract. It is generally believed that there are two significant target regions of the human respiratory tract for particles in this size range, the nasal passages (for nose breathing) and the alveolar region (for mouth breathing). There are thought to be distinctly different, potentially-serious, health effects thought to arise from deposition of ultrafine particles in both of these regions. This awareness provides a useful starting point, suggesting the need for *two* criteria for ultrafine particles. For ultrafine particles, however, it is not possible to separate the criteria from the methods by which the sampled particles are assessed. Although, as described earlier, there are data for the particle size-dependent deposition efficiency of very small particles in the nose and alveolar region, we are not yet sure of the manner in which particle size influences the nature and level of the toxic response. In turn, therefore, we are uncertain about the most appropriate metric for aerosol concentration. From inhalation toxicology, *surface area concentration* would appear to be more appropriate than the *mass concentration* which is customarily applied nowadays for most aerosols.[19] Whichever is eventually deemed to be the most appropriate index of exposure concentration, however, it will remain necessary to integrate that knowledge with the deposition curve for either of the regions indicated in order that dose may be estimated. It is worth summarizing how this might be achieved.

Based on the equations for nasal and alveolar deposition for very fine particles presented in the ICRP report, we may develop the deposition criteria, say, $UN(d)$ and $UA(d)$, respectively (where d is now the thermodynamic particle diameter. The particle equivalent volume diameter is usually considered appropriate for this). Here it is noted that special care should be taken in choosing the most appropriate representative breathing flowrate, since inspiration through the nose usually takes place only at lower volumes corresponding to relatively light work. Popular current instrumentation for evaluating ultrafine aerosols is the differential mobility analyzer (DMA), which provides the particle size distribution – say $f(d)$ – of the aerosol in the size range of interest. It is therefore a simple matter to apply such information directly in one or other of the following

$$_{nasal}Dose(d_V) = f(d_V)UN(d_V) \quad \text{and} \quad _{alveolar}Dose(d_V) = f(d_V)UA(d_V)$$

That is, the measured particle size distribution is combined mathematically with the fraction of interest, leading to an accurate estimate of what is actually deposited. Scientifically, it is not sufficient to simply define and measure ultrafine particles as those below a certain size (e.g., 0.1 μm, as has been suggested as one possible threshold).

6. Conclusions

This paper has argued that due attention to the correct definition of exposure, coupled with a full understanding of the basis of exposure assessment as it might be applied in the context of either standards or epidemiology, can provide the most representative measure of aerosol exposure relevant to specific health effects. A framework exists by which to draw together the contributions of the individual disciplines that are brought to bear in dealing with the behaviour and effects of inhaled particles.

References

[1] Walton W H 1997 Inhaled Particles VIII: History of the Symposia Inhaled Particles VIII, Proceedings of the Eight International Symposium on Inhaled Particles, Annals of Occupational Hygiene **41**, Supplement 1 pp. xvii-xix

[2] McCrea J 1913 The ash of silicotic lungs, Publication of the South African Institute of Medical Research (Johannesburg, South Africa)

[3] Kotzé Sir R 1916 Final Report of the Miners' Phthisis Prevention Committee, Union of South Africa (Johannesburg, January 10th 1916)

[4] British Medical Research Council (BMRC) 1952 Recommendations of the BMRC panels relating to selective sampling, from the minutes of a joint meeting of Panels 1, 2 and 3 held on March 4th 1952

[5] Walton W H 1954 Theory of size classification of airborne dust clouds by elutriation British Journal of Applied Physics **5** (Supplement 3) S29-S40

[6] Orenstein A J 1960 Recommendations adopted by the Pneumoconiosis Conference Proceedings of the Johannesburg Pneumoconiosis Conference ed A J Orenstein (London: Churchill) pp619-621

[7] American Conference of Governmental Industrial Hygienists (ACGIH) 1968 Threshold limit values of airborne contaminants (Cincinnati: American Conference of Governmental Industrial Hygienists)

[8] Attfield M D and Kuempel E D 2003 Commentary: pneumoconiosis, coalmine dust and the PFR Annals of Occupational Hygiene **47** 525-529

[9] Salvan A, Stayner L, Steenland K and Smith R 1995 Selecting an exposure lag period, Epidemiology **6**, 387-390

[10] Vincent J H 1993 Perspectives on international standards for health-related sampling of airborne contaminants Applied Occupational and Environmental Hygiene **8** 233-238

[11] Comité Européen Normalisation (CEN) 2002 Workplace atmospheres – assessment of performance instruments for measurement of airborne particle concentrations, European Standard EN 13205 (Brussels, Belgium)

[12] International Standards Organization (ISO) 1995 Air quality - particle size fraction definitions for health-related sampling, ISO Standard ISO 7708 (Geneva, Switzerland)

[13] American Conference of Governmental Hygienists (ACGIH) 2007 Threshold Limit Values for Chemical Substances and Physical Agents and Biological Exposure Indices (Cincinnati, American Conference of Governmental Industrial Hygienists)
[14] Environmental Protection Agency (EPA) 1996 Air Quality Criteria for Particulate Matter, EPA/600/P-95/001 (Washington, DC)
[15] Peters T M, Vanderpool R W and Wiener R W 2001 Design and calibration of the EPA $PM_{2.5}$ Well Impactor Ninety-Six (WINS) Aerosol Science and Technology **34** 389-397
[16] McCawley M A 1999 Particle size-selective criteria for deposited submicrometer particles, Particle Size-Selective Sampling for Particulate Air Contaminants ed J H Vincent (Cincinnati: American Conference of Governmental Industrial Hygienists)
[17] Vincent J H 2007 Aerosol Sampling: Science, Standards, Instrumentation and Applications (Chichester: Wiley & Sons)
[18] International Commission on Radiological Protection (ICRP) 1994 Human respiratory tract model for radiological protection ICRP Publication 66, Annals of the ICRP **24** 1-482
[19] Oberdörster G 2000 Toxicology of ultrafine particles: in vivo studies Philosophical Transactions of the Royal Society of London, Series A **358** 2719-2740

Journal of Physics: Conference Series **151** (2009) 012004
doi:10.1088/1742-6596/151/1/012004

Occupational Exposure to Polycyclic Aromatic Hydrocarbons in Wood Dust

C K Huynh, P Schüpfer and P Boiteux

Institute for Work and Health, rue du Bugnon 21, CH-1005 Lausanne, Switzerland

chuynh@hospvd.ch

Abstract. Sino-nasal cancer (SNC) represents approximately 3% of Oto-Rhino-Laryngology (ORL) cancers. Adenocarcinoma SNC is an acknowledged occupational disease affecting certain specialized workers such as joiners and cabinetmakers. The high proportion of woodworkers contracting a SNC, subjected to an estimated risk 50 to 100 times higher than that affecting the general population, has suggested various study paths to possible causes such as tannin in hardwood, formaldehyde in plywood and benzo(a)pyrene produced by wood when overheated by cutting tools. It is acknowledged that tannin does not cause cancer to workers exposed to tea dust. Apart from being an irritant, formaldehyde is also classified as carcinogenic. The path involving carcinogenic Polycyclic Aromatic Hydrocarbons (PAHs) emitted by overheated wood is attractive. In this study, we measured the particle size and PAHs content in dust emitted by the processing of wood in an experimental chamber, and in field situation. Quantification of 16 PAHs is carried out by capillary GC-ion trap Mass Spectrometric analysis (GC-MS). The materials tested are rough fir tree, oak, impregnated polyurethane (PU) oak. The wood dust contains carcinogenic PAHs at the level of $\mu g.g^{-1}$ or ppm. During sanding operations, the PU varnish-impregnated wood produces 100 times more PAHs in dust than the unfinished wood.

1. Introduction

Since the 1960s, a large excess risk for adenocarcinomas SNC has been observed among furniture makers and other workers exposed to wood dust [1-3]. Among the occupational causes of SNC, the International Agency for Research on Cancer (IARC) has classified wood dust as definitively carcinogenic to humans (Group 1) on the basis of epidemiological evidence [4], although the exact carcinogenic compounds have not been identified [5-6]. The Occupational Exposure Limits (OEL) of the European Union (EU) regulation for wood dust is 1 $mg.m^{-3}$ and the Threshold Limit Value Time-Weighted Average (TLV-TWA, USA) for Western red cedar Wood dust is 0.5 $mg.m^{-3}$ and 1 $mg.m^{-3}$ for all other species. The Scientific Committee for Occupational Exposure Limits (SCOEL) of the European Union (EU) has stated that exposure to wood dust above 0.5 $mg.m^{-3}$ induces pulmonary effects and should be avoided [7]. Based on possible cancer risk, occupational exposure to wood dust was assessed in a European project WOODEX [8], in 2000-2003, concluding that about 3.6 million workers (2.0% of the employed population in the 25 member states of the EU) were occupationally exposed to wood dust. Swiss Occupational exposure limit, designated as "Valeur limite Moyenne d'Exposition" (VME), regulated of 2 $mg.m^{-3}$ of inhaled hardwood dust and 3 $mg.m^{-3}$ of softwood dust would appears to not effectively protect workers, and should be lower.

The high proportion of wood workers contracting a SNC, has suggested various studies paths to possible causes such as tannin in hardwood, formaldehyde in plywood [5] and benzo(a)pyrene produced by wood overheated by cutting tools. It is acknowledged that chronic tea dust exposure can cause increased prevalence of respiratory symptoms. However, tannin does not cause cancer under exposure to tea dust [9]. Formaldehyde is an irritant, and is classed carcinogenic (Group 1), but it is only present in plywood, particleboard or fiberboard. Specific tasks, such as sanding, cause the emission of fine dust, which can remain in suspension in air for many hours in badly ventilated locations.

The pathway involving carcinogenic Polycyclic Aromatic Hydrocarbons (PAHs) emitted by wood when overheated by cutting tools is attractive. In this study, we have examined the content of PAHs in relation to operations performed on wood; first in an experimental chamber, and second in a field situation. We have measured the particle size of the generated dust, the suspended dust level and amount of PAHs on collected dust by personal sampling.

2. Methodology

The collecting of suspended dust at flow rate of 28.3 $L.min^{-1}$ (1 CFM), according to the size of particles, is performed with an Andersen impactor (Ambient Particle Sizing Sampler, model 2000, Andersen Inc., USA) equipped with glass fiber filters. To study the size distribution of the wood particles in relation to operations such as sawing, planning or sanding, we simulated these operations in our experimental chamber of 10 m^3.

The materials tested are rough fir tree, oak and impregnated polyurethane (PU) oak. Suspended dust was generated inside the experimental chamber by wood working tools such as sander, circular saw and plane. The aerodynamic median diameters of suspended particles in the air were measured by weighing the collected dusts on 7 different Andersen impactor stages and backup filter, and computerized using individual calibrated penetration curves of each stage. Continuous monitoring of the inhaled dust level during floor sanding operations were also recorded using a real-time dust monitor (MIE personal DataRAM, model pDR-1200AN, Thermo Andersen, USA), operated with data logging at 1 s interval time.

2.1. Sampling of wood dust

Closed face Polystyrene cassette (SKC Inc., Eighty Four, PA, USA), placed upstream of the pumping system, containing a Ø 37 mm GF/B glass fiber filter (Gelman Sciences Inc., MI, USA) between two spacer rings, was used to collect the wood dust at a flow rate of 2 $L.min^{-1}$ over 3 to 5 hours.

2.2. Solvents

Toluene, dichloromethane, n-hexane, cyclohexane, dimethylformamide and methanol (analytical grade quality, Ammann-Technik AG, Switzerland) were redistilled before use. SiO_2, cartridges, vials, glassware and other hardware devices were rinsed with distillate methanol to ensure general low blank control.

2.3. Standards

PAHs were purchased from the Bureau of Certified Reference (BCR, Brussels). Their purity was higher than 99%. Standard and sample solutions were stored in dark glass bottles (to avoid photodecomposition) in a freezer at –20°C until analysis. Indeno[1,2,3-cd]fluoranthene was used as the internal standard. Other certified perdeuterated PAHs compounds, such as Naphthalene-d_8 (51.2 $ng.\mu L^{-1}$), Acenaphthene-d_{10} (5.06 $ng.\mu L^{-1}$), Pyrene-d_{10} (49.5 $ng.\mu L^{-1}$), Benzo[a]pyrene-d_{12} (24.8 $ng.\mu L^{-1}$), Perylene-d_{12} (19.86 $ng.\mu L^{-1}$) and Benzo[ghi]perylene-d_{12} (23.49 $ng.\mu L^{-1}$) in hexane/toluene solution (SRM 2270 PAH-II solution, National Institute of Standards and Technology, NIST, USA) were also used as internal standards.

Journal of Physics: Conference Series **151** (2009) 012004 doi:10.1088/1742-6596/151/1/012004

2.4. Purification and enrichment scheme

The general scheme for sample treatment and clean-up steps aimed at the determination of PAHs was previously reported with details [10]. The procedure comprises an ultrasonic extraction with boiling toluene, which has been proved to be efficient for high molecular weight PAHs. The various steps include liquid-liquid partition, SiO_2 micro-column purification and HPLC fractionation prior to injection. The purified and enriched final extract material appears as a clear yellow liquid in toluene ready for analysis by GC-MS.

2.5. Ion-trap GC-MS determinations

Experimental conditions were fully described elsewhere [10-12]. Briefly, we report here a few details. GC-MS analysis of 1 µL of the final clean extract was carried out with a Varian Saturn 2000 MS (Varian, Inc., PaloAlto, Calif) equipped with an "on-column" injector (model 1078, Varian), a DB-17ms fused silica column of 30 m length, 0.25 mm inner diameter (ID) and 0.25 µm film thickness (SGE, Infochroma, Switzerland) connected at the inlet to a 2 m x 0.53 mm ID retention gap. Helium was used as the carrier gas and the head pressure was programmed as follows: 10 psi followed by ramping at 20 psi/min to 30 psi. The initial oven temperature was 80°C for 2 min, followed by a first ramping at 20°C/min to 120°C and a second ramping at 5°C/min to the final temperature at 320°C maintained for 20 min. Autosampler : CombiPAL, Varian, injection speed : 0.2 µl/s. Temperature programming of the injector: 40°C – 200°C/min – 300°C (58 min).

The MS detection was operated in the electron ionization mode with an electron energy of 50 eV, an electron multiplier voltage of 2000 V, and an emission current of 10 µA. The mass range scanned was from 55 to 350 m/z at 2 scan/second. The axial modulation of the ion trap was set at 4.0 V. Mass calibration was performed with perfluorotributylamine (FC-43).

The detection and identification of PAHs in samples were made by comparing the spectrum of a corresponding standard at the same retention time. Quantifications of PAHs were performed on the basis of ratio of the peak areas for the PAH of interest relative to the internal standard (indeno[1,2,3,cd-fluoranthene) in their specific mass fragmentograms.

An aliquot of clean extract was also analyzed by High Performance Liquid Chromatography coupled with a Fluorescent Detector (HPLC-FL). However, for quantification the capillary GC-ion trap mass spectrometric technique was preferred because of the insufficient resolution of the HPLC column and the possible quenching or co-elution effect of matrix compounds.

3. Results and discussions

3.1. Size distribution of the wood particles

The materials tested were rough fir tree (soft wood), oak (hard wood) and impregnated PU oak. The picture 1 shows the dimensions of oak wood particles collected on a glass fibre filter during sanding operation. The picture was taken by optical microscopy instrument equipped with digital camera (Leica, Model DM2500P, Optical magnitude 20x). A particle of wood at dimensions of 352.9 x 29.7 µm (length x diameter) was observed suspended in air and collected on the filter. This observation demonstrated that very coarse particles of wood dust, more than 100 µm, were collected on the personal sampler, and potentially inhaled by the workers. In the actual situation, the authors pointed out the difficulty of monitoring with accuracy the personal exposure to wood dust, using actual sampling heads with low performance against coarse particles [13-14]. Figure 1 shows the mass-related size distribution of the suspended particles in the air of the experimental chamber according to the operations of planning, sanding or sawing: the aerodynamic median diameters are very similar, about 11 µm, with a standard deviation of ± 2 µm.

3.2. Monitoring the dust level in a parquet sanding operation

In figure 2, we present a plot of the inhaled wood dust level monitored during a parquet floor sanding operation in a real situation (picture 2). The mean average level of recorded dust exceeded the Swiss

Journal of Physics: Conference Series **151** (2009) 012004 doi:10.1088/1742-6596/151/1/012004

VME of 2 $mg.m^{-3}$, especially when the worker was in sitting position, near the radiator, when the observed level was 40 $mg.m^{-3}$. As seen in earlier studies [15-16], sanding separates out as a wood working process of high exposure.

3.3. PAHs in wood dust

Table 1 presents geometric means (N=3) for the levels of 16 PAHs in dusts of different tree species (fir tree, oak and PU impregnated oak), together with the result from personal sampling and fixed station during parquet floor sanding operation. PAHs are produced by the sanding or cutting tools excessively heating at the surface layer of the wood, which induces a partial combustion of the organic matter in wood or the PU varnish. There is no notable difference in the content of PAHs in softwood (fir tree) and hardwood (oak). In contrast, the PU varnish-impregnated wood produces 100 times more PAHs than the unfinished wood during sanding operations. Focusing on amounts of benzo(a)pyrene, a particularly high carcinogenic potential compound, we observed high productions in PU impregnated wood: 5.90 $\mu g.g^{-1}$ comparing to 0.008 $\mu g.g^{-1}$ in both soft and hard wood dust. If we focus only on the PAHs contents, the difference between PAHs levels in soft dust compared to those in hard dust is not significant: 1.775 $\mu g.g^{-1}$ against 2.296 $\mu g.g^{-1}$, which does not explain the different expressed in Swiss regulation: Swiss VME regulated of 2 $mg.m^{-3}$ of inhaled hardwood dust and 3 $mg.m^{-3}$ of softwood dust. But when we compare with PU impregnated wood, the total amount of PAHs is 100 fold higher: 191.45 $\mu g.g^{-1}$. Personal exposure to PAHs was significant: for example Chrysene at 0.499 $\mu g.m^{-3}$, Benzo(a)pyrene at 0.057 $\mu g.m^{-3}$ and Indeno(1,2,3-cd)pyrene at 0.041 $\mu g.m^{-3}$. This is on the same order of magnitude as PAHs levels in asphalt occupational exposure [11-12].

4. Concluding remarks

In this paper, we have studied the PAHs content in wood dust in relation to operations performed on different wood species in an experimental chamber and in a real situation. As already mentioned, there are very few previous experiences in the measurement of PAHs concentration in wood dust. The perspective involving carcinogenic PAHs emitted by wood overheated by cutting tools is attractive and is confirmed in this study. In fact, the amount of PAHs in wood dust is more important than the nature of the wood, insofar as carcinogenic compounds are concerned.

One possible hypothesis to explain a large excess risk for sino-nasal cancer, particularly adenocarcinomas, observed among furniture and other workers exposed to wood dust is the size distribution of wood dust. In field situation, we observed that very coarse particles of wood dust, more than 100 µm, were collected on the personal sampler, and such particles were potentially inhaled by the workers. These particles, which have a sedimentation velocity of more than 25 $cm.s^{-1}$ [17, 18], probably settled in the nose of exposed workers, not penetrating so far in the lungs. Indeed, the occupational exposure to wood dust was associated predominantly with adenocarcinomas of the nose and sinuses, contrary to tobacco smoke and other occupational exposure where lung cancer is the target [3]. In the present situation, no sampling heads were available for coarse particles higher than 100 µm, and published data in term of mass concentration of wood dust based on conventional sampling devices could under-estimate the real occupational exposure level in specific cases.

It is crucial that the affected workers be aware of the danger of wood dust exposure and that they should adopt adequate protection equipments, such as dust protection mask or improving workshop ventilation. Tools fitted with dust collection bags produce less airborne dust and contribute to better protection. Could the presence of PAHs in wood dust be a cause of adenocarcinoma in joiners and cabinetmakers? Additional investigations are necessary to better determine the woodworker's exposure to the PAHs and its health effects.

Acknowledgements
The authors wish to thank the SUVA Research Found and Syngenta Crop Protection AG for their contribution to the financial support of the work, and F Perret and M Meagher for their comments and remarks.

References

[1] Acheson E D 1968 *Br. Med. J.* **2** 587-596
[2] Demers P A et al. 1995 *Amer. J. Ind. Med.* **28** 151-166
[3] Mannetje A. et al. 1999 *Amer. J. Ind. Med.* **36** 101-107
[4] International Agency for Reasearch on Cancer (IARC) 1995 *IARC monographs on the evaluation of carcinogenic risks to humans* **62** Wood dust and formaldehyde (Lyon: WHO)
[5] Luce D, Gerin M, Leclerc A, Morcet J F, Brugere J and Golberg M 1993 *Int. J. Cancer* **21** 224-231
[6] Leclerc A, Martinez Cortes M, Gerin M, Luce D and Brugere J 1994 *Am. J. Epidemiol.* **140** 340-349
[7] Scientific Committee for Occupational Exposure Limits (SCOEL) **2003** SCOEL/SUM/102 final
[8] Kauppinen T et al. 2006 *Amer. J. Ind. Med.* **50** 549-561
[9] Jayawardana P L and Udupihille M 1997 *Occup. Med.* **47** 105-109
[10] Vu Duc T, Huynh C K and Boiteux P 1995 *Micro Chimica Acta* **120** 271-280
[11] Huynh C K, Vu Duc T, Le Coutaller P, Surmont F and Deygout F 2007 *Polycyclic Aromatic Compounds* **27** 107–121
[12] Vu-Duc T, Huynh C K and Binet S 2007 *J. Occup Environ Hyg* **4** (S1):245-248
[13] Görner P et al **2001** *Ann OccupHyg* **45** 43-54
[14] Kennedy N K and Hinds W C **2002** *J Aerosol Sci* **33** 237-255
[15] Jones P A and Smith L C **1986** *AnnOccup Hyg* **30** 171-84
[16] Alwis U, Mandryk J, Hocking A D et al. **1999** *Am Ind Hyg Assoc J* **60** 641-6
[17] INRS Guide for ventilating practice, 6. Oil mist extraction and processing **1990** INRS
[18] Su W C and Vincent J H **2002** *J Aerosol Sci* **33** 103-118

Picture 1. Optical microscopy picture of oak wood dust generated during planing operation and collected on the filter. Instrument: Optical microscope (Leica, model DM2500P, optical magnitude: 20x). Particle of wood at dimensions of 352.9 x 29.7 μm (length x diameter) was observed suspended in air.

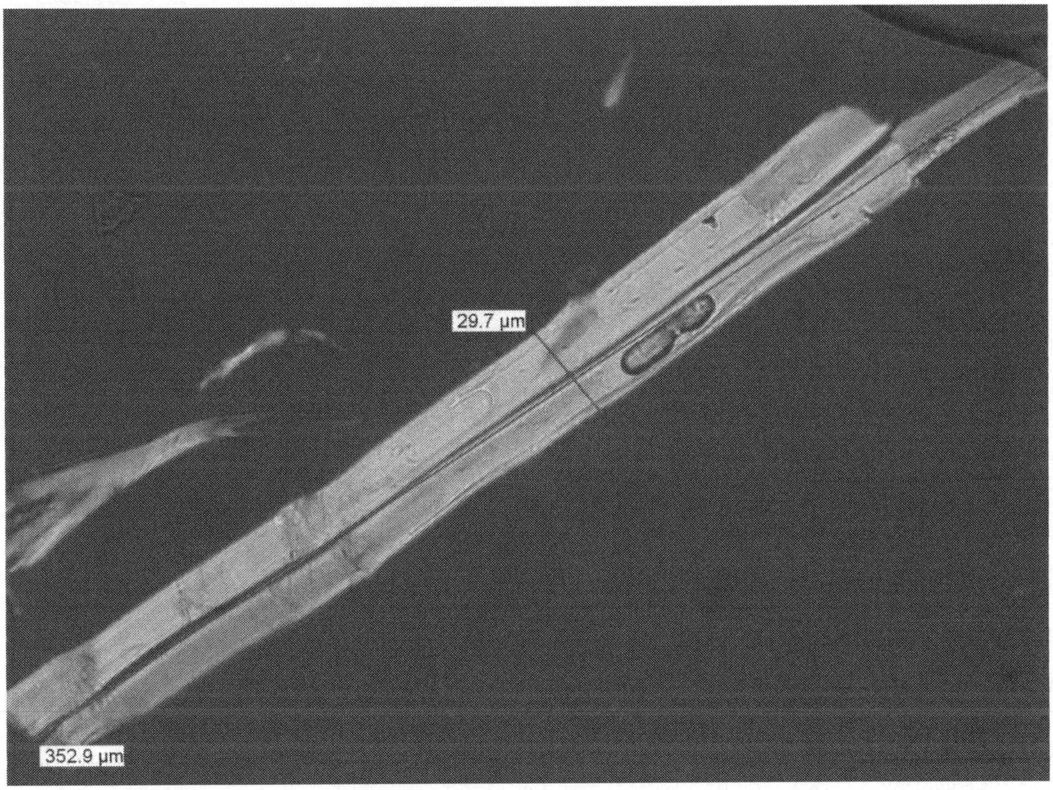

Figure 1. Size distribution and median diameter of the suspended particles in the air of the experimental chamber according to the operations of planing, sanding or sawing, determined by Andersen impactor and gravimetric measurements.

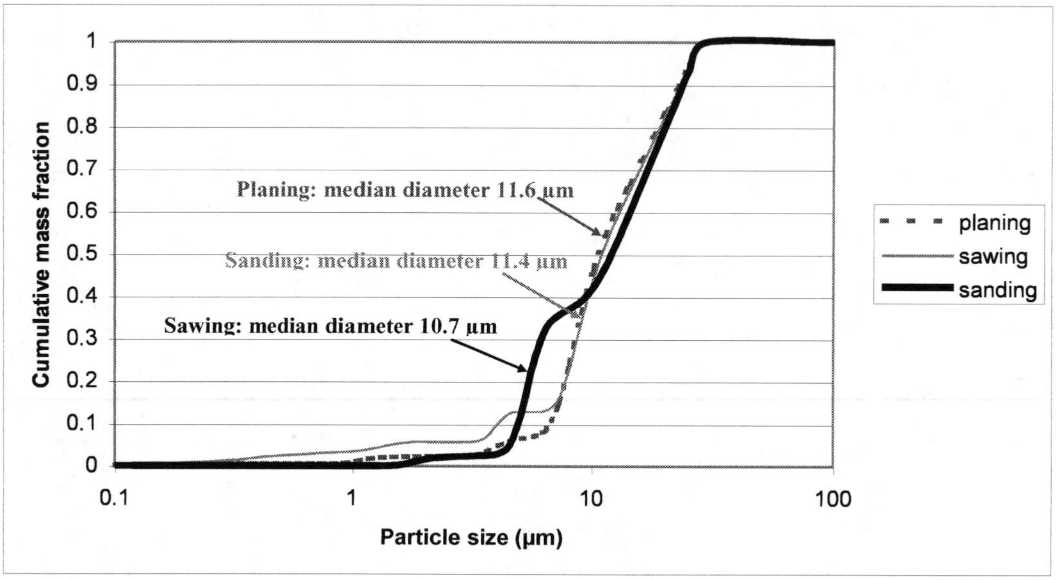

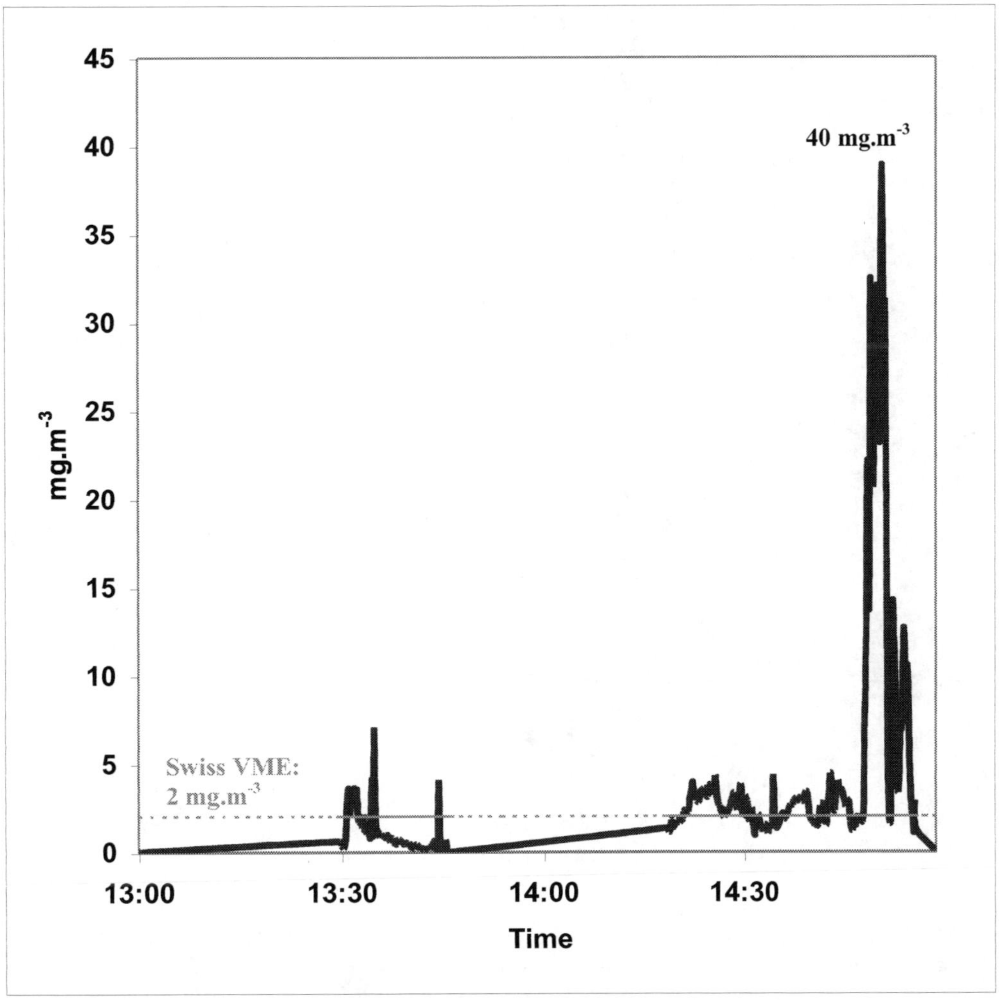

Figure 2. A personal wood dust exposure level was recorded using dust real time monitor (DataRAM) during a sanding operation (standing position) of the parquet floor. A high level of inhaled particles (40 $mg.m^{-3}$) was observed during a manual sanding operation, in sitting position, near the radiator.

Picture 2: A personal wood dust exposure level was recorded using dust real time monitor (DataRAM) during a sanding operation (standing position) of the parquet floor. Dust collected by personal sampling device (pump with cassette filter) was also done.

Table 1. PAHs contents in dust ($\mu g.g^{-1}$ or ppm) and in air ($\mu g.m^{-3}$) in relation with dust categories produced by sanding operation.

Products	Concentration in dust ($\mu g.g^{-1}$ or ppm)			Air concentration ($\mu g.m^{-3}$) Impregnated PU oak	
	Fir tree	Oak	Impregnated polyurethane (PU) oak	Fixed station	Personal
Sampling time (min)	-	-	-	127	127
Volume of air (L)	-	-	-	292.1	698.1
Dust concentration ($mg.m^{-3}$)	-	-	-	29.63	32.37
Naphthalene	0.153	0.401	0.13	-	-
Acenaphthene	-	-	-	-	-
Acenaphthylene	-	-	1.65	-	-
Fluorene	0.038	0.034	0.07	-	-
Phenanthrene	0.544	0.887	34.10	0.037	0.132
Anthracene	0.137	0.188	5.80	0.011	0.035
Fluoranthene	0.290	0.240	28.50	0.056	0.105
Parene	0.204	0.185	20.0	0.043	0.150
Benzo(a)anthracene	0.016	0.063	5.90	0.021	0.099
Chrysene	0.232	0.184	57.90	0.112	0.499
Benzo(b)fluoranthene	0.081	0.051	20.80	0.029	0.166
Benzo(k)fluoranthene	0.020	0.011	5.30	0.004	0.061
Benzo(a)pyrene[a]	0.008	0.008	5.90	0.025	0.057
Dibenzo(a,h)anthracene	0.003	0.005	2.80	-	0.036
Benzo(g,h,i)perylene	0.022	0.012	1.80	-	-
Indeno(1,2,3-cd)pyrene	0.027	0.027	0.80	0.003	0.041
Total	**1.775**	**2.296**	**191.45**	**0.341**	**1.381**

Remarks:
- not detected or low than the limit of detection of the method (< 0.001 ppm or <0.001 $\mu g.m^{-3}$)
[a]: Swiss regulation VME for Benzo(a)pyrene: 2 $\mu g.m^{-3}$

Exposure to respirable crystalline silica in south african farm workers

Andrew Swanepoel[1,2], David Rees[1,2], Kevin Renton[2], Hans Kromhout[3]

[1] University of the Witwatersrand, School of Public Health, Johannesburg, South Africa.

[2] National Institute for Occupational Health, Johannesburg, South Africa.

[3] Environmental Epidemiology Division, Institute for Risk Assessment Sciences, University of Utrecht.

andrew.swanepoel@wits.ac.za

Abstract Although listed in some publications as an activity associated with silica (quartz) exposure, agriculture is not widely recognized as an industry with a potential for silica associated diseases. Because so many people work in agriculture; and because silica exposure and silicosis are associated with serious diseases such as tuberculosis (TB), particular in those immunological compromised by the Human immunodeficiency virus (HIV), silica exposure in agriculture is potentially very important. But in South Africa (SA) very little is known about silica exposure in this industry. The objectives of this project are: (a) to measure inhalable and respirable dust and its quartz content on two typical sandy soil farms in the Free State province of SA for all major tasks done on the farms; and (b) to characterise the mineralogy soil type of these farms. Two typical farms in the sandy soil region of the Free State province were studied. The potential health effects faced by these farm workers from exposure to respirable crystalline silica are discussed.

1. Introduction

To date, there have been no studies published in South African on personal quartz exposure (commonly known as respirable crystalline silica) and agriculture and therefore it is unknown to what extent quartz exposure is a risk in the SA agriculture industry. South Africa has a large agricultural sector and is a net exporter of farming products such as maize, wheat and sunflower. Out of a population of 47.9 million people, the agricultural industry contributes around 8.8% (1.164 million people) of formal employment (13.234 million people) in South Africa.

Although listed in some publications as an activity associated with silica (quartz) exposure, agriculture is not widely recognized as an industry with a potential for silica associated diseases. Several good international published studies on silica exposure in agriculture indicated variable exposure levels between various operations. Studies have reported that farming activities in sandy soils can produce quartz levels well above generally accepted occupational exposure limits (OELs) and that sandy soils are commonplace in drier regions of many countries.

Because so many people work in agriculture; and because silica exposure and silicosis are associated with serious diseases such as tuberculosis (TB), particularly in those immunological compromised by the Human immunodeficiency virus (HIV), silica exposure in agriculture is potentially very important.

2. Objectives

The objectives of this study are: (a) to measure respirable dust and its quartz content on three typical sandy soil and sandy loam soil farms in the Free State and North West provinces of central South Africa for all major tasks done on the farms; and (b) to characterise the mineralogy soil type of these farms.

3. Methods

Two typical farms in the sandy soil region of the Free State province (Farm 1 and 2) and a typical farm in the sandy-loam region of the North West province (Farm 3) were identified by a pedologist. The actual farms were selected by convenience sampling because access to the farms was available. Personal breathing zone samples were measured in all workers on the farms involved in the following operations: maize planting and harvesting, watermelon and pumpkin handling, sunflower planting and ground preparation.

Respirable dust was measured using the standard widely used Health and Safety Executive (HSE) laboratory Method for the Determination of Hazardous Substances (MDHS 14/3 – General methods for sampling and gravimetric analysis of respirable and inhalable dust) and a Higgens-Dewell cyclone was used operating at a flow rate of 2.2 ℓ/min. Respirable quartz was measured and analysed using X-ray diffraction (XRD) at the National Institute for Occupational Health (NIOH) as specified in the HSE analytical method MDHS 101 (Crystalline silica in respirable airborne dust – Direct-on-filter analysis by infrared spectroscopy and X-ray diffraction).

A hundred and forty nine breathing zone measurements of approximately eight hours in length (mean 467 min) were collected over a complete cycle of a year on the three identified farms. Data was analysed using the S-Plus 6.2 statistical software package. Written informed consent was obtained from all participants in the study and the study was approved by the University of the Witwatersrand Human Research Ethics Committee. Representative soil samples were collected at a depth of 3– 4 centimeters (cm) for analysis by XRD to describe the mineralogy of the soils.

4. Results

A hundred and forty nine (n=149) breathing zone measurements of approximately eight hours in length (mean 467 min) were collected over a complete cycle of a year on the three identified farms. Values below the detection limit were assigned half the detection limit for statistical analysis.

Table 1. Respirable quartz exposures during farming operations (mg/ m^3) on three central South African farms.

Farms	n	Mean ±SD	Median	Min	Max	Inter Quartile Range	%≥ 0.1[a] mg/m^3	%≥ 0.05[b] mg/m^3	%≥ 0.025[c] mg/m^3
Farm 1	13	0.0.1 ±0.05	0.03	0.020	0.630	1.872	30.7%	30.8%	53.9%
Farm 2	77	0.06 ±0.12	0.03	0.021	0.314	0.029	14.3%	26%	50.7%
Farm 3	59	0.04 ±0.01	0.02	0.021	0.166	0.010	6.8%	15.3%	37.3%
Total: Farm 1,2,3	149	0.05 ±0.01	0.02	0.020	0.630	0.020	12.8%	22.1%	45.6%

[a] 0.1 mg/m^3 is the South African occupational exposure limit (OEL) for quartz
[b] 0.05 mg/m^3 is the NIOSH recommended exposure limit (REL)
[c] 0.025 mg/m^3 is the ACGIH Threshold Limit Value (TLV)

Journal of Physics: Conference Series **151** (2009) 012005 doi:10.1088/1742-6596/151/1/012005

The overall mean Time Weighted Average (TWA) respirable quartz concentration was 0.05 mg/m^3 with a maximum of 0.63 mg/m^3 measured on farm 1 during sunflower planting operations of the study (Table 1). Thirteen percent of the measurements exceeded the South African Department of Labor (DoL) OEL of 0.1 mg/m^3, 22% exceeded the National Institute for Occupational Health and Safety (NIOSH) recommended exposure limit (REL) of 0.05 mg/m^3 and 46% of the measurements exceeded the widely used reference value of the American Conference of Governmental Industrial Hygienists (ACGIH) of 0.025 mg/m^3 (TLV-TWA).

The overall mean TWA respirable dust concentration was 0.57 mg/m^3 with a maximum of 6.06 mg/m^3 measured on farm 1 during sunflower planting operations of the study (Table 2). Only 7% of the measurements exceeded the South African Department of Labor (DoL) OEL of 2 mg/m^3.

Table 2 Respirable dust exposures during farming operations (mg/ m^3) on three central South African farms

Farms	n	Mean ±SD	Median	Min	Max	Inter Quartile Range	%≥2[a] mg/m^3
Farm 1	13	1.472 ±0.47	0.101	0.017	6.062	1.872	23.1%
Farm 2	77	0.6106 ±0.12	0.246	0.009	5.689	0.363	7.8%
Farm 3	59	0.3111 ±0.58	0.170	0.019	2.803	0.309	1.7%
Total: Farm 1,2,3	149	0.57 ±0.01	0.227	0.009	6.062	0.473	6.7%

[a] 2 mg/m^3 is the South African occupational exposure limit (OEL) for respirable dust.

5. Discussion

Many studies indicated that quartz exposure in agriculture might be a risk to farm workers, but generally no systematic data are available on the magnitude of the risk. Although exposure to quartz during agricultural activities is variable, exposure can extend throughout the growing season. The duration of specific activities varies due to many factors, and lifetime exposure to quartz may be considerable.

A number of possible explanations exist for the paucity of cases of silicosis. Exposure assessment in agriculture is difficult because of the varied and cyclic nature of the farmers work and the diverse locations of the farms. Exposures may vary with local farming practice and specific tasks performed on the farm, commodities grown or raised, geography, environmental factors (i.e. temperatures, season changes, relative humidity, rainfall, wind speeds, wind directions) and other factors such as soil characteristics (i.e. soil temperatures, mineralogy of soil, soil moisture) and tractor characteristics (tractor speed and enclosed cabins). Variations in soil components (e.g., silicates, iron oxides) can affect the pathogenicity of silica. Silica potency is reduced by other soil components such as iron oxides and clay silicates.

In addition, silica in soils is likely to be aged, with most farming activities producing little freshly fractured material. Fresh-fractured silica dust is known to be more pathogenic by contact with the air (e.g., most silica in soil), however little is known about the potential for fresh-fractured silica in the soil that might be caused by contact with metal during activities such as plowing and discing. Only a small portion of the soil is in contact with the machinery, while the majority of dust is from the disturbed soil. Thus, it seems likely that fresh fractured dust could not comprise more than a small fraction of repirable soil dust. Under-diagnosis is also a possible explanation. Cases of silicosis have been reported, however, and silica exposure in farming deserves further research.

Notably, these widely used "safe" limits for quartz of 0.05–0.1 mg/m3 have been shown not to be protective. The American Conference of Governmental Industrial Hygienists (ACGIH) has recommended a threshold limit value (TLV) for quartz of 0.025 mg/m3.

6. Conclusion

These data suggest that quartz exposure is a possible risk in the farming industry. If it is, this is a serious public health concern in South Africa given the country's high TB rates and the number of people involved in agriculture. Therefore it needs to be further investigated

Keywords: silica, quartz exposure, respirable, agriculture, farm workers.

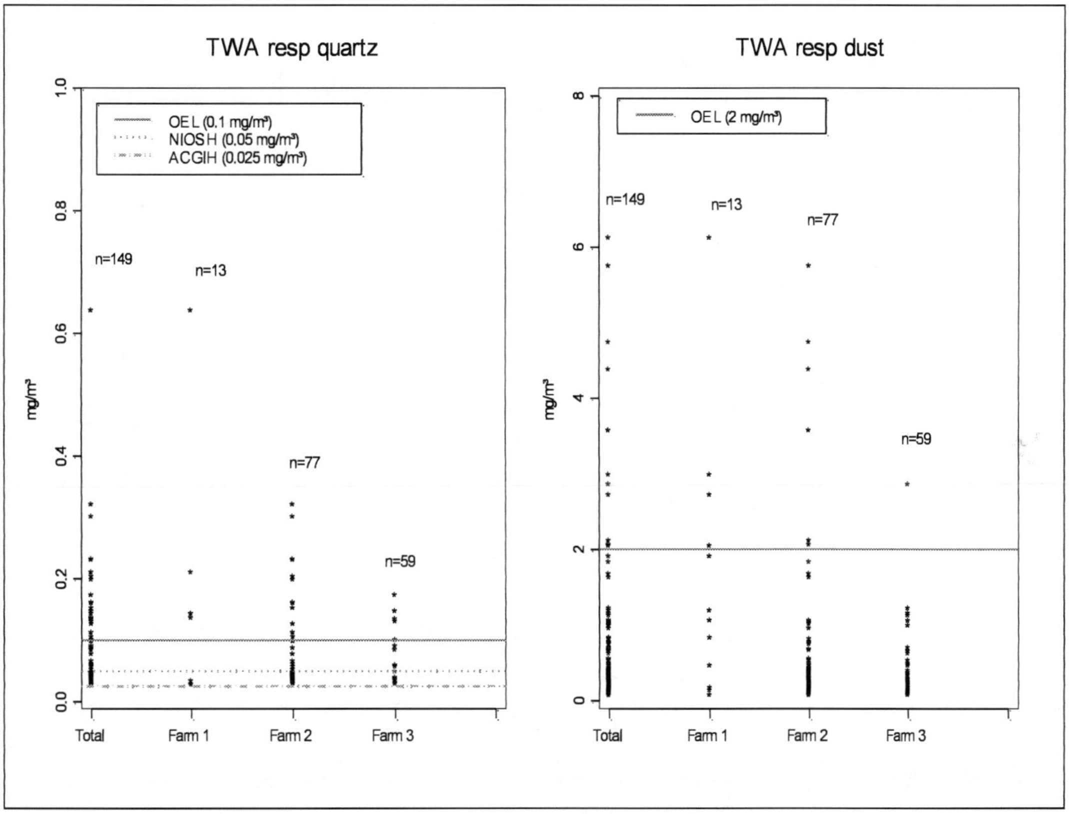

TWA resp quartz = the time weighted average concentration of respirable quartz.
TWA resp dust = the time weighted average concentration of respirable dust.

Figure 1. Respirable quartz and respirable dust concentrations (mg/m^3) during farming operations (mg/ m^3) on three central South African farms (*n*=149).

Inhaled Particles X, (23–25 September 2008, Manchester) IOP Publishing
Journal of Physics: Conference Series **151** (2009) 012006 doi:10.1088/1742-6596/151/1/012006

Indoor concentrations of fine particles and particle-bound PAHs in Gothenburg, Sweden

S Johannesson[1], K Bergemalm-Rynell[1], B Strandberg[1], G Sällsten[1]

[1] Department of Occupational and Environmental Medicine, Sahlgrenska University Hospital and Academy, Gothenburg, Sweden

E-mail: sandra.johannesson@amm.gu.se

Abstract. Fine particles are formed in a variety of processes, both natural and anthropogenic. Epidemiological studies have shown an association between exposure to particulate matter and adverse health effects. Airborne particles contain a variety of compounds, including polycyclic aromatic hydrocarbons (PAHs), and several of the PAHs are known or suspected carcinogens. In this study, stationary measurements of $PM_{2.5}$ were performed in the residences of 20 study participants along with simultaneous monitoring at an urban background site. The collected particle mass was then analyzed for its content of some particle-bound PAHs using GC-MS. The median level of $PM_{2.5}$ indoors was 7.3 μg/m^3 and in urban background 5.3 μg/m^3. For benzo(a)pyrene (B(a)P) the corresponding results were 10 pg/m^3 and 35 pg/m^3, respectively. There were significant correlations between indoor and ambient levels for both $PM_{2.5}$ (r_s=0.58, p=0.02) and B(a)P (r_s=0.67, p=0.007). No significant correlation was, however, found between the concentration of $PM_{2.5}$ and the associated levels of the investigated PAH compounds. This finding implies that exposure to B(a)P or other particle-bound PAH components needs to be separately assessed.

1. Introduction

Fine particles are formed in a variety of processes, both natural and anthropogenic. Traffic, burning of biomass and other fuels for heating purposes and industry are important sources of fine particles in urban areas. Epidemiological studies have shown an association between exposure to particulate matter and adverse health effects such as cancer, cardiovascular diseases and respiratory illnesses [1, 2]. In southern Sweden, long-range transported particles give a substantial contribution to ambient levels.

Airborne particulate matter contains a variety of compounds, including polycyclic aromatic hydrocarbons (PAHs), which are formed during incomplete combustion processes. Domestic wood-burning and traffic are the two main contributors to the release of PAHs in Sweden [3]. Several known or suspected carcinogens are found among the PAHs, and many of these are 5- and 6-ring compounds that are associated with particles. Benzo(a)pyrene (B(a)P) is the most studied among the carcinogenic PAHs. According to the unit risk estimate from the WHO [4], the exposure to B(a)P that will cause an excess lifetime cancer risk of 1/100 000, is 0.1 ng/m^3. This level is used as a Swedish health based guideline value [3].

Within the frame work of the Swedish environmental protection agency, a project concerning the general population's exposure to carcinogenic compounds was started in the year 2000. The project includes five Swedish cities, one of them being Gothenburg. The health related survey includes

Journal of Physics: Conference Series **151** (2009) 012006 doi:10.1088/1742-6596/151/1/012006

measurements of benzene, 1,3-butadiene, formaldehyde and nitrogen dioxide. Measurements of fine particles were added to the project in 2005. This study presents results from the measurements of $PM_{2.5}$ and some particle-associated PAHs in Gothenburg during the fall 2006.

2. Materials and methods

2.1. Study group

Adult citizens (age 20-50 years) living in Gothenburg were randomly selected from the population register. Among 40 study participants, 10 individuals were randomly selected for home indoor measurements of fine particles. In addition to these 10 individuals, another 10 volunteers were recruited among the employees at the Department of Occupational and Environmental Medicine in Gothenburg.

2.2. Monitoring

Stationary measurements of fine particles ($PM_{2.5}$) were performed indoors (living room) in the homes of the 20 study participants during the fall of 2006. The sampling was performed for 48 hours using GK2.05 (KTL) cyclones connected to BGI400S sampling pumps (BGI Inc., Waltham, MA, USA) with a flow rate of 4 L/min. The flow rate was adjusted prior to monitoring and controlled at the end of the sampling period. The average flow rate was then used to calculate the total volume of air drawn through the filter. Simultaneously with the indoor sampling (for 17 of the 20 subjects), urban background monitoring was conducted at a central site within the city using a PQ100 Basel $PM_{2.5}$ sampler with an impactor cutoff system (EPA WINS, BGI) at 16.7 L/min. On four occasions, indoor sampling was performed simultaneously in the homes of two study subjects. Consequently, on these occasions two indoor samples were accompanied by a single urban background sample (one sample missing due to instrument failure).

2.3. Analyses

Teflon filters, 2 µm pore size (Pall Teflo) were used for all measurements. All filters were weighed before and after exposure. The filters were conditioned for 24 hours prior to weighing in a climate chamber controlled for temperature and humidity (temperature: 23 ±0.5°C, relative humidity (RH): 50±5 %). The mass concentration was determined using a Mettler Toledo MX5, and the limit of detection was 1 $\mu g/m^3$.

The particle mass was then analyzed for its content of particulate PAHs. Analysis and detection was performed using high-resolution gas chromatography (GC) connected to a low-resolution mass spectrometer (MS) in selective ion recording (SIR) mode. Field blanks were processed in parallel with the samples. An internal standard (16 PAHs) was used in the analysis. A reference material (SRM 1649a urban dust) certified for 12 PAH components was used as control samples. The results for these control samples showed acceptable agreement with the certified results for all components. The detection limit was 5 pg/m^3 for the indoor samples, and 1 pg/m^3 for the urban background samples.

2.4. Statistics

Statistical calculations were performed with the SAS System for Windows, version 9.1 (2003). Correlations were assessed using the Spearman rank correlation coefficient (r_s) and differences between pairs of indoor and ambient levels were calculated using the Wilcoxon signed rank test. Statistically significant refers to 5 % significance level in two-tailed tests ($p<0.05$). If the particle mass was below the LOD, the LOD divided by the square root of 2 was used [5].

Journal of Physics: Conference Series **151** (2009) 012006 doi:10.1088/1742-6596/151/1/012006

3. Results

In total, the particle mass was analyzed for nine different PAHs. For three of the components, benzo(a)anthracene, perylene and dibenzo(a,h)anthracene, more than 50 % of the indoor samples were below the detection limit. The median concentration, mean and range for the indoor and urban background measurements of $PM_{2.5}$ and the six detectable PAH components are listed in table 1.

Table 1. Indoor and urban background concentrations of $PM_{2.5}$ and PAHs.

Compound	Indoors				Urban background			
	N	Median (pg/m^3)	Mean (pg/m^3)	Range (pg/m^3)	N	Median (pg/m^3)	Mean (pg/m^3)	Range (pg/m^3)
$PM_{2.5}$ (µg/m^3)	20	7.3	9.7	3.2-45	11	5.3	7.3	2.9-18
Chrysene	20	6	23	<5-290	11	30	39	30-110
Benzo(b)fluoranthene	20	15	40	<5-370	11	72	83	26-280
Benzo(k)fluoranthene	20	7	27	<5-340	11	67	85	10-230
Benzo(a)pyrene	20	10	32	<5-340	11	35	46	11-140
Indeno(1,2,3-c,d)pyrene	20	41	69	<5-390	11	83	91	28-280
Benzo(g,h,i)perylene	20	37	73	<5-390	11	85	95	28-260

3.1. Particle mass

The median concentrations from the parallel measurements of $PM_{2.5}$ indoors in living rooms and in urban background are presented in figure 1. The median concentration of $PM_{2.5}$ indoors was 7.3 µg/m^3 (95 % CI 5.9-9.3 µg/m^3) and in urban background 5.3 µg/m^3 (95 % CI 3.3-17 µg/m^3). The highest level of $PM_{2.5}$ was found in the home of the one smoker (subject 9) who also smoked indoors.

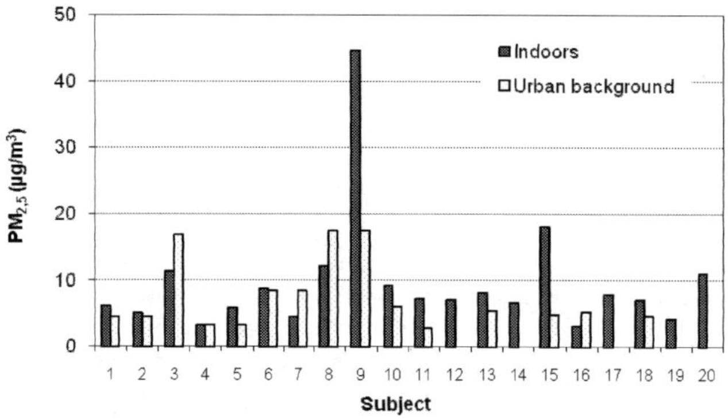

Figure 1. Median levels of $PM_{2.5}$ indoors and in urban background, respectively. Subject 1, 7 and 9 were smokers, but only subject 9 smoked indoors.

3.2. Benzo(a)pyrene

Results from the parallel indoor and urban background measurements of B(a)P are shown in figure 2. The median B(a)P concentrations indoors and in urban background were 10 pg/m^3 (95 % CI 5-23 pg/m^3 and 35 pg/m^3 (95 % CI 13-110 pg/m^3), respectively. Also for B(a)P, the highest concentration was found in the home of the smoker who also smoked indoors.

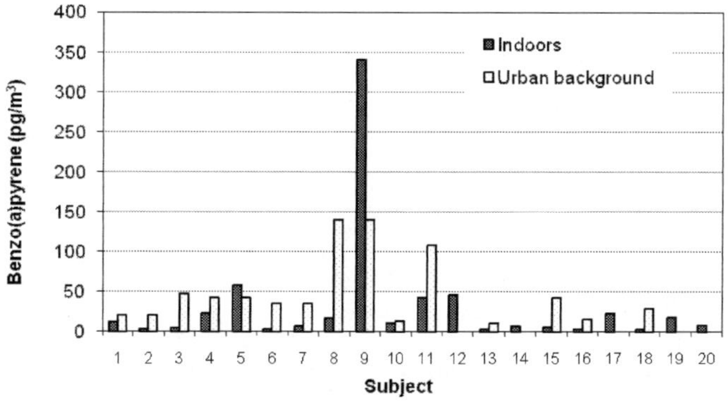

Figure 2. Median levels of B(a)P indoors and in urban background, respectively. Subject 1, 7 and 9 were smokers, but only subject 9 smoked indoors.

3.3. Relations between concentrations in different microenvironments

There was a statistically significant correlation between indoor and ambient concentrations for both $PM_{2.5}$ ($N=15$; $r_s=0.58$, $p=0.02$) and B(a)P, ($N=15$; $r_s=0.67$, $p=0.007$). Omitting the one smoker who also smoked indoors (subject 9) from the dataset resulted in a nearly significant correlation for $PM_{2.5}$ and a slightly reduced, but still significant correlation coefficient for B(a)P, see figure 3 (a) and (b).

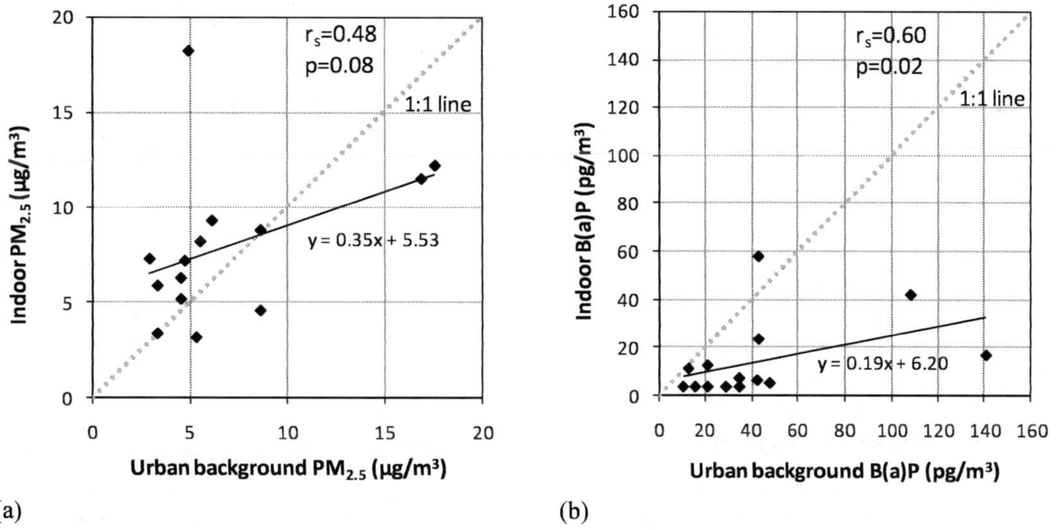

Figure 3. Indoor versus urban background concentrations of (a) $PM_{2.5}$ and (b) B(a)P. Included in each graph is also the best fit regression line for the 14 subjects.

A significant correlation between indoor and urban background levels was not only shown for B(a)P, but also for the other compounds (r_s=0.66-0.84). Urban background concentrations were, however, significantly higher than indoor levels for all the investigated PAHs. In general, the PAH components correlated well with each other, which holds for both the indoor and outdoor measurements. There was, however, no correlation between particle mass concentrations and the different PAH components for neither the indoor measurements, nor at the urban background site.

4. Discussion

Indoor concentrations of $PM_{2.5}$ were low compared to levels found in studies conducted in central and southern Europe [6], but comparable to levels found in Helsinki [7] and a previous study in Gothenburg [8]. Also the measured levels of B(a)P must be considered low. However, Clayton et al. [9] found comparable indoor B(a)P levels in a study conducted in Minnesota, while most of the concentrations measured in the study from United States by Naumova et al. [10] were higher. Also the study from Berlin [11] reported higher median levels of B(a)P compared with Gothenburg.

The method for analysis of the PAHs in this study is the same as in the study by Pleil et al.[13], where $PM_{2.5}$ filters were analyzed for their content of PAHs. In that study, a mean concentration of B(a)P of 87 pg/m^3 was reported from outdoor measurements in North Carolina. In another Swedish study, in an area where residential wood-burning is common, elevated indoor levels of several PAHs were found in the residences of study subjects who used wood-burning appliances for space heating [12]. In that study, median indoor B(a)P concentrations of 520 pg/m^3 was found in the wood-burning homes and 120 pg/m^3 in the reference homes. In the present study from Gothenburg, however, none of the study participants used wood for space heating.

PAHs with 3-4 rings are classified as semi-volatile and can have a significant fraction in both the gaseous and particulate phases [3, 12]. This might have contributed to the quite low levels of chrysene detected in the collected particle mass. However, the remaining five detected PAH components are 5-6 ring compounds that are all found in the particulate phase [3, 13].

No correlation was found between $PM_{2.5}$ and levels of B(a)P or the other individual PAH components. This is in consistence with the findings by Ohura et al. [14]. In the study by Fromme et al. [11], no correlation was found between particle mass and a majority of the investigated PAHs, apart for measurements conducted in smoker households. The finding of no association between particle mass and PAHs implies that exposure to B(a)P or other particle-associated PAHs needs to be characterized separately. Finally, in the present study we found the median levels of $PM_{2.5}$, both indoors and at the urban background site, to be below the WHO guideline value of 10 μg/m^3 [15]. Furthermore, the median concentration of B(a)P was well below the Swedish health based guideline value.

Acknowledgements

The study was funded by the Swedish Environmental Protection Agency.

References

[1] Pope CA and Dockery DW. Health effects of fine particulate air pollution: lines that connect. *J Air Waste Manag Assoc.* 2006 Jun;56(6):709-42.

[2] Schlesinger RB, Kunzli N, Hidy GM, Gotschi T and Jerrett M. The health relevance of ambient particulate matter characteristics: coherence of toxicological and epidemiological inferences. *Inhal Toxicol.* 2006 Feb;18(2):95-125.

[3] Bostrom CE, Gerde P, Hanberg A, Jernstrom B, Johansson C, Kyrklund T, Rannung A, Törnqvist M, Victorin K and Westerholm R. Cancer risk assessment, indicators, and guidelines for polycyclic aromatic hydrocarbons in the ambient air. *Environ Health Perspect.* 2002 Jun;110 Suppl 3:451-88.

[4] WHO. Air Quality Guidelines for Europe. 2nd Edition ed: WHO 2000.

[5] Hornung RW and Reed LD. Estimation of average concentration in the presence of nondetectable values. *Applied Occupational and Environmental Hygiene.* 1990;5(1):46-51.

[6] Gotschi T, Oglesby L, Mathys P, Monn C, Manalis N, Koistinen K, Jantunen M, Hanninen O, Polanska L and Künzli N. Comparison of black smoke and PM2.5 levels in indoor and outdoor environments of four European cities. *Environ Sci Technol.* 2002 Mar 15;36(6):1191-7.

[7] Koistinen KJ, Hanninen O, Rotko T, Edwards RD, Moschandreas D and Jantunen MJ. Behavioral and environmental determinants of personal exposures to PM2.5 in EXPOLIS - Helsinki, Finland. *Atmospheric Environment.* 2001;35(14):2473-81.

[8] Johannesson S, Gustafson P, Molnar P, Barregard L and Sallsten G. Exposure to fine particles (PM2.5 and PM1) and black smoke in the general population: personal, indoor, and outdoor levels. *J Expo Sci Environ Epidemiol.* 2007 Nov;17(7):613-24.

[9] Andrew Clayton C, Pellizzari ED, Whitmore RW, Quackenboss JJ, Adgate J and Sefton K. Distributions, associations, and partial aggregate exposure of pesticides and polynuclear aromatic hydrocarbons in the Minnesota Children's Pesticide Exposure Study (MNCPES). *J Expo Anal Environ Epidemiol.* 2003 Mar;13(2):100-11.

[10] Naumova YY, Eisenreich SJ, Turpin BJ, Weisel CP, Morandi MT, Colome SD, Totten L, Stock T, Winer A, Alimokhtari S et al. Polycyclic aromatic hydrocarbons in the indoor and outdoor air of three cities in the U.S. *Environ Sci Technol.* 2002 Jun 15;36(12):2552-9.

[11] Fromme H, Lahrz T, Piloty M, Gebhardt H, Oddoy A and Ruden H. Polycyclic aromatic hydrocarbons inside and outside of apartments in an urban area. *Sci Total Environ.* 2004 Jun 29;326(1-3):143-9.

[12] Gustafson P, Östman C and Sällsten G. Indoor levels of polycyclic aromatic hydrocarbons in homes with or without wood burning for heating. *Environ Sci Technol.* 2008(DOI: 10.1021/es800304y).

[13] Pleil JD, Vette AF and Rappaport SM. Assaying particle-bound polycyclic aromatic hydrocarbons from archived PM2.5 filters. *J Chromatogr A.* 2004 Apr 9;1033(1):9-17.

[14] Ohura T, Amagai T, Sugiyama T, Fusaya M and Matsushita H. Characteristics of particle matter and associated polycyclic aromatic hydrocarbons in indoor and outdoor air in two cities in Shizuoka, Japan. *Atmospheric Environment.* 2004;38(14):2045-54.

[15] WHO. *WHO Air Quality Guidelines Global Update 2005.* Copenhagen, Denmark: World Health Organization, Regional Office for Europe, Bonn, Germany.; 2005.

Mesothelioma in Qingdao, PRC (2000 – 2007)

Arthur L. Frank [1], Pang Zengchang [2], Zhang Huaqiang [3], Zhang Yun [2]

[1] Drexel University School of Public Health, Philadelphia, PA 19102

[2] Qingdao Center for Disease Control and Prevention, Qingdao, PRC

[3] Qingdao Institute of Women's and Children's Health Care, Qingdao, PRC

E-mail: alf13@drexel.edu

Abstract. The city of Qingdao, PRC has been the site of two asbestos product facilities that operated for almost fifty years, as well as a shipyard. Because of a new computerized data collection system for death certificates, almost all 48,000 yearly deaths from a population base of 7.5 million are now recorded with cause of death and "usual occupation". All mesothelioma deaths from 2000 through 2007 are reviewed and the unusual finding is that of a predominance of cases in females. The issues of competing causes of death and potential underreporting are discussed.

1. Introduction

The city of Qingdao, on the coast of the Yellow Sea about 500 miles south of Beijing, has been the home to two of China's largest asbestos production manufacturing facilities, and to a large shipyard known for its extensive repair work on ships from dozens of nations. Now closed, at the two factories a variety of products have been made. When operating, the factories were staffed by about 1,300 workers, and about 3,000 former workers had retired due to age or illness and were being supported by active workers. The factories operated from the mid-1950s until the end of the 20th century. At one, using about 98% chrysotile, much of it Chinese with some Canadian, building products were made including asbestos corrugated roofing materials and asbestos pipe. A small amount of crocidolite was the amphibole occasionally used. At the other factory, using primarily Chinese chrysotile, with some Canadian chrysotile, the products included textiles, brake shoes, and rubberized asbestos mats.

The shipyard employed about 4,000 workers and was also opened in the 1950s and closed and moved in 2004 to another site elsewhere on the Yellow Sea.

The Qingdao Center For Disease Control and Prevention (QCDC) has for many years been active in a variety of international research projects including a multi-center twin study, work in the area of non-communicable diseases including cardiovascular disease, and the authors have worked together for more than a decade on asbestos-related questions [1].

In the past few years certain leading research-oriented health centers the computerized operation began in 2000 in Qingdao. China, such as the QCDC, have been computerizing health data, including regional deaths certificates. The city of Qingdao itself has a bit over two million persons, but the QCDC records cover an area of some 7.5 million persons both in the city proper and regionally. Progress has been made in computerizing death certificates, and to attempt to have as complete reporting as possible from the region. Qingdao has been one of the fastest growing cities in China.

At the present time from the 7.5 million persons in the region some 48,000 death certificates are recorded yearly, representing mostly hospital records as well as some deaths recorded elsewhere. This number represents about 95% of all deaths in the area, a higher percentage than even just a few years ago when a file card system was used and many deaths went unrecorded, and cause of death could not be easily studied.

The leading cause of cancer deaths in some Western countries in both males and females is lung cancer. In China, at present, about 66% of all males are cigarette smokers while only about three percent of women smoke, though there is a rising trend among young women. Although smoking rates among women have been low, lung cancer rates have been surprisingly high. This has been attributed to the burning indoors, often in confined spaces, of waste vegetable or animal materials for cooking and heating. Coal is sometimes used as well. Considerable combustion products are found in homes and regularly expose women. Such exposures would not lead to mesotheliomas, however.

Mesotheliomas are recorded as part of the developing computerized data base and the findings from recent years are described here. Death certificates list the usual occupational, but there is no comprehensive list of former occupational exposures, nor any way to assess environmental exposures. Smoking histories are not currently computerized, but are recorded on a separate, non-computerized, file card system.

Data about past causes of death has been difficulty in the past to obtain in China for a variety of reasons [2], and with regard to the diagnosis of mesotheliomas the difficulties of correctly making this diagnosis has been documented [3]. Although unable to prove easily, the likelihood is that the diagnosis of mesothelioma is underreported in the data of the QCDC, as has been the case elsewhere [4].

A recent paper by Nishikawa and colleagues [5] reviews pleural mesothelioma mortality on a global basis, and plots such data along with increasing or decreasing use of asbestos. China is not covered, but data is given for more than thirty countries, and the pleural mesothelioma deaths range from 31 per million per year in the U.K. to 0.5 in Brazil and Ecuador. The author is personally aware of the gross underdiagnosis and underreporting in Brazil, and the Chinese data given here is reflective of a setting in which minimal diagnostic testing is done, special staining is virtually unknown, and underreporting clearly the case.

2. Results

Data from the eight years covering 2000 – 2007 is now available for causes of death including for mesothelioma. Mindful of the area population base of some 7.5 million, and the number of total deaths each year being about 48,000, it is clear that the number of mesothelioma deaths recorded by the QCDC appears to be significant.

In the Qingdao area, for the year 2000 there were 19 mesothelioma cases, all pleural, with 14 in females, 2 in males, and the gender of three cases was unknown. Five of the cases listed as the usual occupation "farmer".

In 2001 there were again 19 mesothelioma cases, 17 in females, 1 in a male, and the additional case had no gender recorded. Again, all were pleural mesotheliomas.

For 2002 some 15 cases of mesothelioma were recorded on death certificates, representing two males and 13 females. One case was in a child, aged twelve. No data was recorded that year for the location of the lesion. For 2003 there were six cases, four recorded as pleural and two as peritoneal. The three cases among males were all pleural, while the three female cases contained the two peritoneal cases.

In 2004 six mesothelioma cases, all pleural, were recorded, four in males, and two in females. For 2005 there were also six cases, one a peritoneal case in a female, with four male pleural and one female pleural cases recorded.

For 2006 there were 17 mesotheliomas, 12 in males and 5 in females. "Farmers" was again the leading occupation listed. Eleven were listed as pleural mesotheliomas, with one peritoneal mesothelioma and five as "other". The youngest death was at 19 years of age.

For 2007 there was 6 mesotheliomas, equally split between men and women. There was one pleural mesothelioma in a woman, three peritoneal cases (2 women), and two in men listed as "other". One individual was 90. The age range between 2000-2007 was 12 to 93, with a predominance among women, as noted in Table 1.

The recorded job categories vary greatly, and as noted above, are not record as a lifetime work history, but only as "usual occupation". Short periods of factory work would not be recorded nor would cases reflecting household or other environmental exposure. Many cases were recorded among farmers, with a variety of factory workers noted. These reports included transportation facility workers, a health worker, a housewife, and even a lawyer. No Chinese asbestos mines are in this area.

Taken together, there were 94 mesotheliomas recorded in the Qingdao area between 2000 and 2007. The majority, quite different from data elsewhere in the world, were found among woman. As elsewhere, most were pleural lesions. The age range was as noted elsewhere [6].

3. Discussion

With the advent of electronic data bases it will now be possible to better record cases of mesothelioma in the Qingdao area. A continuing problem is one of potential under-diagnosis since the level of pathologic sophistication may not yet rival the experiences in North America or Europe. Special staining techniques, often useful to establish the diagnosis, may be lacking for some cases. Also, to a greater extent than in other venues, diagnoses are more often made on clinical grounds. The extent of underreporting of mesothelioma cases cannot be accurately assessed, but may be significant.

Striking in Qingdao is the overwhelming proportion of females among these cases, with roughly twice as many cases seen in women compared to men. This may not be as striking as it seems at first because many asbestos exposed workers in the two Qingdao factories were women. Open since the 1950s, and closed in 2000, the two asbestos factories in Qingdao operated in the city making a wide variety of asbestos containing products including textiles, asbestos/rubber matting, brake shoes, and asbestos cement building products, with one-half the workforce being female. With few exceptions, such as gas mask workers during World War II, the majority of workers in Europe and North America, as well as at mine sites throughout the world, have been males. There was also large shipyard that operated in Qingdao until recently, doing major ocean-going vessel repairs.

Also, with a much higher smoking rate among men, there may be a considerably higher death rate from lung cancer in men compared to women, and as a competing cause of death men may not live long enough to develop mesotheliomas, the disease with the longest average latency after exposure to asbestos.

In addition to traditional occupational exposures there are several sources of exposure in the general population, leading to potential significant environmental exposure. This comes from the use of asbestos cement water pipes, as well as home construction materials built of asbestos cement. Some water collection from run-off from asbestos corrugated sheets might further add to the overall exposure.

A deficit in our analysis is not having anything like a complete exposure history for any of these individuals, and when available only having "usual occupation" recorded. Farmer was a frequently noted job category but also found were such occupations as factory manager, health care worker, chemical worker, smelter worker, transport facility workers, lawyer, and others.

As noted earlier, one Qingdao factory used only chrysotile, the other only about 2% crocidolite, and mined Chinese asbestos appears to be of the chrysotile variety, and it has been reported to lack tremolite contamination [7].

In future efforts more detailed exposures may be undertaken to better assess asbestos exposure. It may also be possible to review mesothelioma rates in other Chinese locations with similar computerized data systems such as in Harbin, Shanghai and Wuhan. Unfortunately, Shenyang, a major industrial city with a major asbestos facility does not yet have as good a data-recording system.

As noted earlier, data in China has been difficult to collect and analyze, but with new systems in place this should soon improve in some respects. Both the percentage of deaths recorded, and the recording in an electronic system will facilitate further work. The potential to study other aspects of asbestos-related disease, such as the unsettled matter of excess cervical and uterine cancer rates, may be possible in the future.

Acknowledgement

The authors wish to acknowledge the contribution of the staff of the Department of Non-Communicable Disease Prevention at the Qingdao Center for Disease Control and Prevention which has set up and maintains the data collection system.

References

[1] Pang Z, Zhang Z, Wang Y, and Zhang H 1997 *Am. J. Indst. Med.* **32** 442
[2] Frank A 1995 *Med. Lav.* **86** 457
[3] Sterman D and Albelda S 2005 Respirology **10** 266
[4] Camridge D, Stockton D and Bain M 2006 *Br. J. Ca.* **95** 649
[5] Nishikawa K, Takahashi K, Karjalainen A et al. 2008 Env. Health Persp. Online 14 August 2008.
[6] Selikoff I and Lee D 1978 *Asbestos and Disease* (New York, NY: Academic Press)
[7] Yano E, Zhi-Ming W, Xiao-Rong W, et al 2001 *Am J. Epi* **154** 538

Table 1
Mesothelioma Cases Recorded in Qingdao, PRC (2000 – 2007)

Year	Total #	# Male	# Female	# Gender Unknown	Pleural	Peritoneal	"Other"
2000	19	2	14	3	19	0	0
2001	19	1	17	1	19	0	0
2002	15	2	13	0	NR	NR	NR
2003	6	3	3	0	4	2	0
2004	6	4	2	0	6	0	0
2005	6	4	2	0	5	1	0
2006	17	12	5	0	11	1	5
2007	6	3	3	0	1	3	2

NR – Not reported

Journal of Physics: Conference Series **151** (2009) 012008 doi:10.1088/1742-6596/151/1/012008

Heterogeneity of exposure and attribution of mesothelioma: Trends and strategies in two American counties.

B W Case[1,2] and J L Abraham[3]

[1] Department of Pathology and School of Environment, McGill University, Montréal, Québec, Canada. Email: bruce.case@mcgill.ca

[2] Professeur invité, Epidemiology & Biostatistics Unit, INRS-Institut Armand-Frappier, Université du Québec, Laval, Québec, Canada.

[3] Department of Pathology, SUNY Upstate Medical University, Syracuse, NY 13210, USA.

bruce.case@mcgill.ca

Abstract. As mesothelioma risk has begun to decline in the United States, two trends are gaining relative importance. "Legacy" exposures causing this disease are most important in locales having past asbestos industry, shipyards, and/or local distribution of asbestos amphibole-containing material as a result. "Future" exposures are of particular concern in relation to so-called "naturally occurring asbestos" (NOA) areas which include unequivocally asbestiform amphibole. In this paper, Jefferson Parish, Louisiana is used as an example of the first trend, and El Dorado County, California as an example of the second. Available tumor registry, epidemiology, historical and mineralogical data, and lung-retained fibre content are used as indicators of disease and exposure. Jefferson Parish, LA was chosen as the prototype of "legacy" exposures on the basis of historical evidence of asbestos plants with known mesotheliomas in the workforce, known shipyards in the same area, EPA records of distribution of crocidolite-containing scrap to and remediation of over 1400 properties, NIOSH published data on mesothelioma by county, and exposure data including lung-retained fibre analyses in victims, where available. El Dorado, CA was chosen as the prototype of NOA amphibole exposures on the basis of tumor registry data, activity-based EPA sampling data in one area, and lung-retained fibre analyses in area pets, and future risk assessment based on tremolite-specific modelling in Libby, Montana and elsewhere. As expected, the legacy exposure area was high in mesothelioma incidence and mortality. Lung-retained fibre content confirms crocidolite exposures in exposed plant-workers and those exposed to crocidolite-containing scrap, and amosite in shipyard workers. In contrast, to date, cancer registry data in the NOA-amphibole ("future") county does not show a clear increase in incidence or mortality, but grouped county data from the area show a shift in higher incidence rates to the NOA areas and away from California "legacy" (e.g., shipyard) areas from 1988-2005. EPA active sampling has confirmed excess tremolite/ actinolite fibre(s) in air, although there is debate about its nature and the appropriateness of the area sampled. Lung-retained fibre in local pets shows unequivocally elevated asbestiform tremolite/ actinolite in areas thought to be most affected, but numbers are small. Future risk is expected to rise due to a vastly increased population base coupled with exposures potentially created by related construction activities. Although legacy exposures are producing smaller numbers of cases with time, they continue to occur at high rates, and new sources of legacy exposure are being discovered in highly localized "hotspots". Differential exposure sources remain a problem in attribution, but continued remediation seems the best strategy for prevention. In the "future" risk county and surrounding areas, incidence trends are less clear, but again highly localized exposures as opposed to broad areas seem important. Activity-based air sampling; targeted soil samples, and lung-retained fibre analyses may be useful in defining areas of highest future risk and potential prevention.

1. Introduction

Patterns of geographical distribution of mesothelioma cases in the United States were last studied in depth over 20 years ago [1,2]. Authors noted that, with some exceptions, highest concentrations of cases were likely to be noted in counties having either asbestos-using industrial plants, shipyards, or both. Most studies of mesothelioma incidence in the United States have relied on the Surveillance, Epidemiology, and End Results Program (SEER) data, a subset which is incomplete but which also supplies cases for nine jurisdictions [3]. According to this source of information, mesothelioma incidence among U.S. men peaked before the turn of the century and has begun a slow decline [3,4], possibly due to decline of use of commercial amphiboles since the late 1960's [3].

While the decline is welcome, it brings new challenges for mesothelioma ascertainment. These can be divided into three parts:

- a deficit in the SEER data itself
- a persistence of the "legacy" exposures both in their traditionally known forms (as in continuing risk from past direct and indirect exposures from transformation plants and shipyards) and newly discovered offshoots (as in the distribution of asbestos-containing (AC) scrap from such sites), and
- exposure from mineral deposits not commercially exploited, either directly via disturbance of the material *in situ* or indirectly via transport by wind or water; this is sometimes called *naturally occurring asbestos* (NOA), although the distinction is in lack of present commercial use, not geological occurrence. Past sources of commercial exploitation can become current sources of NOA exposure, or they can occur through weathering or development not specifically aimed at extraction of the minerals, such as construction activities or real-estate development siting.

1.1. Deficits in the SEER data itself.

SEER data is by definition limited to those jurisdictions from which it is derived. Both overestimates and underestimates of mesothelioma incidence nationwide are theoretically possible, depending on the representativeness of the data. It has been suggested that, for example, studies of attributable risks of varying causes of mesothelioma based on the SEER data may have overestimated shipyard, ship-building and ship repair, since the data-set includes the Los Angeles County area (as well as cases and controls from the Veterans Administration). Los Angeles County had a disproportionate share of such activities and 46% of incident cases were attributable to exposures from those sources there, not "directly representative of the whole population of the United States" [5]. The accuracy of diagnosis data within the SEER data has also been questioned, but this incidence data matches national mortality data quite well for 1998-1999 if compared with ICD-10 mortality data for mesothelioma for the same areas [6].

1.2. Persistence of legacy exposures.
Commercial amphibole exposures occurred long after the late 1960's in the United States. For example, crocidolite consumption in the manufacture of products constituted nine per cent of U.S. total asbestos consumption in 1981, 95% of which was in the production of asbestos cement pipe [7]. Amosite consumption in insulation production continued to a lesser degree after 1972, when the U.S. Navy formally discontinued asbestos use (amosite had constituted up to 85% by weight, by specification, of naval ship insulation during World War II [8]). However, exposure to amosite in installed insulation and other products in secondary trades continued (and continues), as amosite by one account was identifiable in the lungs of 82% of all mesothelioma victims analyzed over the last 25 years [9].

An emerging source of exposures to commercial fibre is domestic use of industrial scrap, which unless painstakingly remediated may remain in a neighborhood virtually forever. This has occurred for example in Italy near a crocidolite-using asbestos cement plant, where close examination of exposures of mesothelioma cases identified "the use of AC residuals such as mixing them into the soil to create hard pavement and improve water absorption or applying layers of finely ground AC—rich in asbestos fibres—in the lofts for thermal insulation" [10]. Distribution of crocidolite-containing scrap throughout an entire county by an American cement pipe plant is an example offered in this paper, for Jefferson Parish, Louisiana; a county which included other asbestos product plants and shipyards as well. Exposures from all three "legacy" sources have contributed to the burden of mesothelioma, making it one of the highest-incidence counties in the nation (see "Results").

1.3. Exposure from mineral deposits not commercially exploited (NOA).
Of the 50 American states, 35 have been identified by the United States Geological Survey (USGS) as containing asbestos occurrences, most of which have not been exploited. These are largely in two areas. The first begins in Eastern Canada and runs down the eastern coastal and inland states; USGS counts 60 former asbestos mines and 331 "Natural asbestos occurrences" in this area, including not only chrysotile but former anthophyllite asbestos mines in Georgia, North Carolina, Virginia, Maryland, Connecticut, and Massachusetts [11]. The second most important belt of occurrences runs from Alaska through British Columbia, Washington State, Oregon and California; as yet, it has not been as well characterized although California was the site of the largest chrysotile mining operations in the western USA and chrysotile was also mined in northern British Columbia. Other occurrences have been identified in the central and "Rocky Mountain" states [12,13]. Most recently, USGS has published tables and maps of NOA in the southwest including Arizona, Utah and Nevada, including information on 113 occurrences, most in Arizona where there are 46 former chrysotile mines, the last of which closed in 1982 [14]. Tremolite, apart from being possibly associated with occurrences of other minerals, was also specifically mined in Utah; tremolite and actinolite are commonly encountered but not well characterized as to whether they are "asbestiform" or "non-asbestiform but fibrous", a distinction which may lack biological significance [15].

Except for the unusual Zonolite Mountain deposit of mixed asbestiform amphibole at Libby, Montana, which had a very high proportional mortality for mesothelioma among the workforce [16-18] and residents [18] in and near the vermiculite mine located there, there have been no real investigations of NOA mesothelioma causation – or even of the workforces of the over 100 previous asbestos mines. A single exception looked at residential proximity to California asbestos deposits as the independent variable for mesothelioma incidence 1988-97 in the California Cancer Registry, and found a significant relationship [19]. Like all ecologic studies this one was potentially hampered by inadequate occupational and other individual exposure data, including residential proximity itself other than that at time of diagnosis, but the positive results do suggest caution in residential siting. El Dorado County, on the western slope of the Sierras and known to contain deposits of asbestiform tremolite/ actinolite, has been an area of particular concern in this regard and is the county used as the NOA example in this paper.

Journal of Physics: Conference Series **151** (2009) 012008 doi:10.1088/1742-6596/151/1/012008

2. Methods

2.1. Selection of counties and sources of information.
The choices of Jefferson Parish, Louisiana as representative of a "legacy exposure" county and of El Dorado, California, as representative of an "NOA exposure" county were made *a priori* on the basis of familiarity of the authors with these areas. As a first step in seeking out information, state cancer registries in the two areas were examined (although as noted below this proved unsatisfactory in Louisiana). Historical exposure data, including published papers, geological maps, and documentation of exposure in the two areas by regulatory agencies was sought out in general terms, including identification by bibliographic search (e.g. Pubmed). Individual mesothelioma cases were culled from the files of one of us (BWC) for Jefferson Parish; neither of us had a file on a case in El Dorado county, which we regard as an area primarily of future risk rather than past exposure, although certainly exposures may have occurred in the past. Information on potential exposure(s) was obtained from deposition testimony, social security records, medical records, and records of employment where available. Where possible, lung tissue was obtained and analysed as outlined in the following section.

2.2. Lung-retained fibre analysis.
Mesothelioma cases from Jefferson Parish for which lung-retained fibre analysis were performed by one of us (BWC) were analysed according to the methods published for the McGill University laboratory [20-25]. Briefly, lung samples were obtained from paraffin blocks after identification from corresponding stained slides, deparaffinized with filtered xylene, and prepared for analytical electron microscopy following digestion with filtered household bleach followed by oxidation in a low-temperature asher with grid preparation using a carbon replica technique. Fibre identification and counting were performed at a magnification of 13,500X in a Phillips 420™ electron microscope equipped with an energy dispersive spectrometer permitting the capture of X-Ray energy dispersive spectra (EDS) of individual particles. Fibres were defined and counted as having length greater than 5 um and aspect ratio greater than 3:1. EDS spectra were observed and evaluated online; reference standards for comparison were spectra obtained from similarly prepared grids holding UICC samples of amosite, crocidolite, and chrysotile and a Research Triangle Park sample of tremolite, among others. Fibres were counted in two samples from different parts of lung to a theoretical maximum of sixty fields per sample (detection limit 34.37 fibres / mg dry lung).

For El Dorado county, no mesothelioma cases were available: for exposure assessment cat and dog lungs obtained from local veterinarians were obtained and analysed using both the McGill methods as above and the SUNY lab methods. The latter were similar with respect to sample preparation and fibre identification, but scanning electron microscopy was used at a magnifications of 4000X and 8000X rather than transmission electron microscopy as in previous work from that laboratory [26-28].

3. Results

3.1 Legacy exposures in Jefferson Parish, Louisiana.
The Louisiana Tumor Registry proved not to be a useful source for county-by-county comparisons of mesothelioma rates. This is because that registry groups counties together into regions for some purposes, and therefore Jefferson Parish – the chosen county for this paper – is grouped with Orleans Parish (which includes the city of New Orleans proper) as well as St.-Bernard Parish. The latest figures for the three counties combined nonetheless show the *region* to have the highest age-adjusted (to the 2000 U.S.

population) annual incidence rate in the state at 2.1 per hundred thousand for the five years from 2000-2004, based on 104 recorded cases (69 in white males; 19 in white females), as compared to 1.3 for the state as a whole [29]. The county was not listed as one of the high mesothelioma rate counties in 1987 [1,2], and is not included in SEER data (nor is any county in Louisiana), but the New Orleans "state economic area", which includes Jefferson Parish, had a "significantly high" SMR for "pleural cancer" (ICD8 code 163.0) among white females for U.S. death certificate data 1968-78, although not for males [2]. Overall, 14 of 174 such economic areas in the US had "significantly high" female pleural cancer mortality, and of these five, including the New Orleans area, had both shipyard(s) and asbestos manufacturing plant(s). At least two of the latter were present in Jefferson Parish; workforces for these were studied by Hughes and Weill [30]. One of the two was the Johns Manville plant, which included a pipe-manufacturing facility which used crocidolite "steadily" in manufacture of "transite" cement pipe. Hughes et al. identified eight mesothelioma cases in workers from this plant, and found further in an internal case-control analysis that mesothelioma was related to proportion of time spent in the pipe area, as well as duration of employment. Although the plant existed from 1928, the pipe plant opened in 1947. According to US EPA and Louisiana Department of Environmental Quality (LDEQ) documents obtained through freedom of information requests, through the 1950's and 1960's, in addition to atmospheric pollution from the plants, plant personnel produced an asbestos-containing aggregate which was distributed throughout the communities of the surrounding area (the "Westbank" of the Mississippi River). This contained both chrysotile and crocidolite asbestos, in a proportion varying from 1:3 respectively to 1:1, and determined by testing to be 35% to 45% total asbestos by weight. It was created in a hammer mill and mixed with a filler, usually composed of gypsum, dolomite, or calcite. The asbestos aggregate and filler formed "a concrete-like material when mixed with water and therefore was considered by many local residents to be a concrete substitute for construction purposes" [31]. It was distributed to local residents free of charge, often by pickup or dump trucks. It was first identified as a potential hazard in 1990; then again in 1992 and 1994, but by 1996 it had become friable and was considered an "imminent threat" to public health, with mesothelioma mentioned specifically as a possible outcome from exposures [31]. The material was "located in residential yards and driveways, school playgrounds, around day care centers, and in other areas easily accessed by the public" and "Children were seen playing on driveways composed of friable ACM with toys and basketballs...vehicles were observed creating dust clouds when passing over areas that contained the ACM. All of these routine activities (were) expected to increase ACM friability and dramatically increase human exposures" [31]. In 1996 remediation activities began, and currently over 1400 properties have been remediated; however, a walk-through of the residential areas near the plant in 2008 by one of us (BWC) clearly identified areas without mitigation, including some in the drainage paths from the plant and from dump sites. In 1999, according to NIOSH data, Jefferson Parish had the eleventh-highest age-adjusted mortality rate at 39.9 per million population for U.S. residents age 15 and over, based on six deaths in women and eight in men (the corresponding overall U.S. rate ranged between 11.32 and 11.64 in the four years from 1999-2002) [32]. Recently NIOSH published an up-to-date supplement of this data for the five years from 2000 to 2004, finding 75 deaths from mesothelioma in that time period (age-adjusted rate 42.1 deaths per million, of which 34.7% were female) in Jefferson Parish [33]. This is by far the largest current number of mesothelioma victims in high-rate counties in the USA, exceeding by fifteen-fold for example the five deaths seen in the same time period in Lincoln County, Montana, where the Libby "epidemic" is situated [33].

3.1.1. Total cases identified. Twenty-two Jefferson Parish medicolegal mesothelioma cases (ten female and twelve male) were identified by one of us (BWC) from our files. At least six of the 22 were related; two sets of siblings and one parent and child. All were resident at some time in their lives (most frequently for most of their lives) in Jefferson Parish. All had pathologically verified mesothelioma, confirmed with immunohistochemistry. All but one were pleural. The peritoneal case was a 39 year old female with no definite history of asbestos exposure. All other cases (12 male and 9 female) were pleural and had a definite history of asbestos exposure including one or more of direct occupational exposure in asbestos manufacturing plant(s) or shipyard(s) in the area; indirect exposures (household or domestic exposure, usually via the laundering of clothes from such workplaces by family members); or environmental exposures through living near asbestos manufacturing plants or at home addresses where asbestos-containing materials had been directly deposited. Some had also attended schools where the ACM scrap was placed in the yards during childhood. Many had suggested possible exposures from multiple sources including two or more of the above, making attribution difficult in some cases.

3.1.2. Abbreviated exposure histories for pleural mesothelioma cases for which lung tissue analysis could not be performed. For 12 of the 21 pleural cases (six men and six women), no lung tissue was available for assessment of retained fibres to aid in exposure assessment. Of these, four were men who worked in shipyards; two for at least twenty years, one in his "college years", and one before becoming a fireman. The latter recalled exposure to amosite-containing insulation and many other asbestos-containing products and also had a crocidolite scrap driveway as a child. A fifth man had crocidolite-containing scrap material deposited at his residence in childhood; a sixth had such material in the neighborhood but could not recall his childhood address. Both of the latter also claimed chrysotile exposure in automotive brake work. Of the six women with no lung fibre analysis five had crocidolite-containing scrap deposited on their property as children, and one of the five lived approximately three hundred feet from the pipe plant for twenty years. Four laundered clothes of fathers, husbands and/or sons who worked at one time either in the asbestos manufacturing plants in the area, in shipyards, or in both.

3.1.3. Cases identified with previous fibre analysis or with lung tissue available. Lung tissue had been analysed previously in other laboratories in two cases. One was a 66 year old woman who had a crocidolite-containing scrap driveway as a child and one crocidolite-cored asbestos body and two tremolite fibres were identified from lung tissue; the other was a shipyard electrician who had an unspecified number of amosite fibres identified in his lung tissue with relatively low (4000 amosite fibres per gram wet lung) concentrations. Close examination of the methods in the latter case however suggest this result may not be reliable, as the analyst had no past experience with the technique, no description of methods and no published or unpublished control values. Samples were available for our own analysis and provided in seven cases. Results for these cases are shown in Table 1.

Table 1. Jefferson Parish mesothelioma cases with lung-retained fibre analysis[a,b].

Age, Sex Latency	Primary exposure	Other exposures	Crocidolite	Amosite	Other[c]
49 M; 49 yrs	Residential[d]	Barges (no insulation)	344, 344	n.d.[e], n.d.	n.d., 34 (Ch) 2640 (AB)
64 M; 42 yrs	Shipyard electrician	Residential	n.d.	3093, 2690	Fibres n.d. by TEM; 15,360 (AB)
52 F; 52 yrs	Residential	none	172, 206	n.d.	Fibres n.d. by TEM; 1480 (AB)
57 M; 52 yrs	Residential	Sales, home improvement	n.d.	n.d.	Fibres n.d. by TEM; AB n.d. by PCOM
60 F; 54 yrs	Residential	Domestic[f]	309, 109	n.d.	n.d., 69 (Ch) 8000 (AB)
75 M; 58 yrs	Occupational-Pipe plant	Occupational-Chem. Plant	722, 1134	69, 137	34, n.d. (Ch); n.d., 69 (T) 24,000 (AB)
57 M; 52 yrs	Residential	Shipyard, < 1 yr.	1134, 825	n.d.	No other fibres by TEM; 2933 (AB)

[a] Fibres > 5 micrometers per mg dry lung; aspect ratio > 3:1; detection limit 34 f/ mg (i.e. 1 fibre = 34 f/mg).
[b] Where two values are given, two separate analyses were performed from two parts of lung.
[c] (Ch) = chrysotile (background = 60-80 [15, 20-22]); (T) = tremolite (background = 40-60[15,20-22]); (AB) = asbestos bodies by phase-contrast optical microscopy per gram dry lung (background < 1000).
[d] Residential = crocidolite-containing scrap at residence in childhood.
[e] n.d. = none detected
[f] Alleged household exposure via laundering of clothes of relative who worked in shipyard < 2 yrs

3.2. Assessment of NOA exposures in El Dorado County, California

3.2.1. Previous work by others. The possibility of tremolite/ actinolite asbestos exposures in the then thinly-populated county of El Dorado, east of California's capital Sacramento, first came on the national scene with a series of newspaper articles by Chris Bowman of the Sacramento Bee. Mr. Bowman was following up on the dilemma of a local citizen who had felt forced to leave his home in the county because of the newly-discovered risk. It has long been known that such fibre was present; it is clearly indicated for example on the 1970 state geological maps used by Pan et al. [19] in their assessment of mesothelioma risk and proximity to naturally occurring asbestos deposits. In an intraperitoneal animal experiment designed to highlight differences in risk between "asbestiform" and "nonasbestiform" tremolite/ actinolite (the latter has not been regulated as asbestos since 1992 and there is debate about health effects of the very wide range of fibres and particles covered by the definition[15]), the asbestiform fibre found to produce mesothelioma most rapidly and consistently was one from a source near this area of California [34]. Citizen concern in the area sparked regulatory agency attention, with an activity-based sampling regimen applied by US EPA in a study performed in 2005 [35]. During the study, EPA technical personnel simulated recreational activities at schools and a park, including cycling, team sports and running. Personal air samplers were worn during the activities and dust was collected from breathing zones to represent potential exposures to both children and adults. Several protocols were used in electron microscopic fibre counts, including modified ISO 10312; "Phase Contrast Microscopy Equivalent" (PCME) counts of fibers longer than 5 micrometers, with aspect ratios greater than 3:1 (to conform to measures used in exposure assessment in historical epidemiology studies using phase contrast optical microscopy). Results were summarized by EPA: "U.S. EPA found asbestos fibers in almost all El Dorado Hills air samples, including those collected nearby, but outside the area of activity. In general, personal asbestos exposures from simulated sports and play activities were significantly elevated over levels observed in the nearby asbestos air samples taken outside the area of activity. The dominant asbestos fiber type detected was amphibole, which is considered to be more toxic than chrysotile" [35]. There was some debate between EPA, an industry lobbying group, and some elements in the El Dorado community governance about the appropriateness of counting methods, and the degree to which identified amphiboles were asbestiform (as opposed to "cleavage fragments"); leading to the US Geological Survey conducting their own investigation in the area [36]. They concluded that both types of "fiber" were present at different sites, and that "Based on aspect ratio data of the less than 3 um diameter particles, the morphologies of the amphiboles in the El Dorado Hills area are intermediate between what might generally be considered a population of commercial-grade asbestos particles and a population of cleavage fragments, produced by milling of massive amphibole" [35]. They stressed further that measurement, not health effects evaluation of any population of fibres, is the province of mineralogists and allied scientists: "It is the obligation of the analytical and mineralogical communities to provide accurate, unbiased, and scientifically sound information to the health and regulatory communities so that appropriate and informed, health-related policy and regulatory decisions can be made" [36].

This series of events left many in the community confused, divided and angry. Some saw a playing down of risk based on obscure mineralogical criteria with no proven health implications. Conversely, others saw unfair singling out of their community (California being well-known to contain much serpentine rock – indeed, it is the "state rock" – often home to asbestos deposits which are largely chrysotile), even though the instigation and emphasis of the investigations had always been on the basis of the presence of amphiboles, and any arguments between scientists had had no relationship to chrysotile asbestos). Although further studies had been planned, including government-sponsored "expert" conferences and risk assessments, none were performed, leaving the situation in limbo.

3.2.2. California Cancer Registry Data. We looked at figures in the California Cancer Registry using their online tool, which allows county-by-county comparisons for any time period demarcated by years between 1988 and 2005. Given the small population at risk, long latency for mesothelioma and time window we expected this might not be sufficient for detecting trends. Mesothelioma (ICD-10) was not coded but pleural cancer was. In fact, there were 33 cases registered for "Pleural Cancer" from 1988-2004 for an age-adjusted rate per 100,000 of 1.36; seventh highest in the state. Also of interest, two of the six higher-incidence counties (sometimes grouped) were immediately adjacent and also on the Western Slope of the Sierra, where tremolite/ actinolite deposits are scattered: Mariposa – Toulomne (30 cases; rate 1.93) and Alpine - Amador – Calaveras (23 cases; rate 1.86). For comparison purposes, Los Angeles County, included as noted above in SEER data and having in the past 46% of cases attributable to exposure from shipyards [2] had a rate of only 0.85. However, this was based on 1044 cases 1988-2004, with a "population at risk" two orders of magnitude higher. Nevertheless, 95% CI on rates did not overlap, showing a significantly increased mesothelioma rate for El Dorado County vs. Los Angeles County (and for the other Western Slope counties) over this time period (difficulties with interpretation of this type of comparison are evident and are discussed later).

3.2.3. Lung-retained fibre analysis. It has not been possible to obtain lung samples from mesothelioma victims – or from hospitals, coroners' offices, or any other source – in El Dorado County, despite four years of efforts in that direction. In the interim, we elected to begin to collect as possible pets (cats and, preferably, dogs) who died of natural causes in county. These were provided by concerned citizens and lungs were obtained *in toto* at post-mortem by local veterinarians and supplied to our laboratories for analysis. To date, we have received lung samples from two cats and four dogs; five have been analysed in both laboratories, as reported here; analysis is not complete in the McGill laboratory for the final dog. Results for the from both McGill and SUNY laboratories to date for tremolite/ actinolite concentrations at two fibre lengths are shown in Table 2.

Table 2: Lung fibre concentrations in El Dorado County pets from two laboratories[a]: Tremolite/ actinolite fibres/ gram dry lung, aspect ratio > 3:1

ANIMAL	Exposure	SUNY Lab > 5 um length	McGill Lab > 5 um length	SUNY Lab > 10 um length	McGill Lab > 10 um length
Cat 1	none	0	0	0	0
Cat 2	9 years, indoors	86,000	157,000	86,000	157,000
Dog 1	> 2 years	1,250,000	566,500	192,000	412,000
Dog 2	> 8 years	9,162,000	928,000	2,030,600	928,000
Dog 3	unknown	186,000	204,000	72,000	47,500
Dog 4	unknown	961,000	-	448,000	-

[a]McGill Laboratory: each value is the mean of two samples; analysis by Transmission Electron Microscopy (TEM) ; SUNY (State University of New York) Laboratory: each value is the mean of two to three samples; analysis by Scanning Electron Microscopy (SEM). Both laboratories used similar sample preparation and fibre identification for tremolite/ actinolite (by EDS) methodology.

Of the six animals shown, the first (Cat 1) did not live in an area known to contain tremolite/ actinolite. The second cat did, but (like most cats) spent most of its life indoors. Dogs 1 and 2 both lived in "exposed" areas but as shown there was greater lung concentration of tremolite/ actinolite shown in both laboratories for the dog with longer residence time. For the third and fourth dogs we have not yet obtained "residential histories", and analysis is incomplete. Other fibers were also found but in small numbers and concentrations except for Dog 3 in the McGill Laboratory; this dog had 204,000 fibres (mean of two counts from two portions of lung based on eight fibres seen) tremolite/ actinolite with mean length 10.4 micrometers but chrysotile concentration was, unlike in the other cases, very high at mean over 2.5 million fibres per gram dry lung and very long with mean fibre length over 13 micrometres. However, this was not replicated in the SUNY laboratory in which no chrysotile was seen at a detection limit varying between 7000 and 20,000 fibres per mg dry lung.

4. Discussion

Results in these two counties show two very different patterns relating to exposure and mesothelioma risk. Each presents scientific and medical dilemmas; each can be aided by obtaining accurate statistics on mesothelioma incidence and mortality, and ideally, lung-retained fibre analysis (where possible). In this section we discuss the central dilemmas for legacy exposures and for future risk from NOA exposures and suggest ways in which addressing them may be improved, as well as current deficiencies both in the available data from these counties and in our own analysis. Finally, addressing the different problems for these two situations has some common – and common-sense – solutions.

4.1. Legacy exposures

Although asbestos use (consumption) in the USA peaked in 1973 [37] and mesothelioma incidence is said based on SEER data to have peaked in the 1990's [3,4] U.S. mesothelioma *mortality* remains approximately flat at 2500 cases per year for the most recently available (1999-2002) National Center for Health Statistics multiple cause-of-death data (ICD-10 data). As may be seen in this report, latency remains very long, although there is some indication that the linear increase with time from first exposure after approximately 20 years flattens out after 50 years from first exposure [38]. It is therefore not surprising that most cases continue to be due to legacy exposures; indeed on the basis of consumption figures alone incidence would be expected to keep rising for another decade. New trends have appeared in addition to the known exposures, which in the USA were maximal in counties that had asbestos manufacturing plants and shipyards. Some legacy exposures are due to secondary exposure from the latter, both through past neighborhood and domestic exposures and past and continuing exposure from material distributed from manufacturing plants, such as that described here for Jefferson County, Louisiana. The exposure profile overall in such a neighborhood can therefore historically be quite complex, leading to concomitant complexity in attribution. The latter is relatively easy to estimate in a population by, for example, type of exposure or occupation. Attribution of the *individual case*, however, sometimes called "specific causation", remains a problem. One approach has been to apply what we know about *population* risks – namely, that they are related to differential fibre potency [39] and to the third to the fourth power of time from first exposure [40]. One model has been constructed to apply dose reconstruction for these two powerful factors in *general causation* to specific causation [41]. While this has some appeal, it is controversial. For one of us (BWC) it lacks biological rationale for the individual case, since it does not seem possible given the accepted tenets of carcinogenesis that in the individual case a large number of exposures could simultaneously, or even in sequence, have been causal. On the other hand, experts disagree on this matter, and indeed that includes the two authors of this paper. One of us (JLA) feels it is, conversely, quite biologically plausible that ALL exposures contribute to the resultant mesothelioma, and that it is not currently possible to exclude any one documented exposure completely, *if* had that one lesser exposure been the only one, there would be no argument that it was sufficient and the only cause.

Whichever view is taken, this leaves those concerned with compensation for the individual case with a dilemma – when more than one exposure has *occurred*, how does one decide biologically which exposure(s) is (or are) causal? In many jurisdictions such a decision must be made, and some US Appeal Courts do not allow "expert" opinions with wordings such as "each and every exposure contributed to…" or "each and every breath of asbestos-containing air was a cause of…" the disease.

While it appears sensible, where possible, to assign attribution to the most important exposure (in terms of dose or some surrogate for dose, with the above-noted corrections for fibre type and for time from first exposure), it is not always possible. Evidently, if science cannot determine which exposure is the one which *was* causal in the individual case, and multiple possibilities exist, there needs to be a balance between fairness to the individual with disease and fairness to those, if any, deemed "responsible" for the exposures that constitute those possibilities. Where available, lung-retained fibre analysis can provide one component of the solution to this difficulty. This is evident in the Jefferson Parish cases having unequivocal lung-retained fibre results in which it is clear either that a professional (for example, amosite from shipyard insulation or crocidolite from cement pipe manufacture) or residential (for example, from crocidolite-containing scrap on a person's property to which that person was exposed in childhood) exposure was most significant. Arguments may be raised that rapid differential clearance may cause an exposure to (for example) chrysotile to "disappear"; conversely however it is just this rapid clearance that has often been raised as the explanation for the wide differences in risk between fibre types

[3, 39, 41]. Of course, lung tissue is in any case not always available from medical procedures such as diagnostic lung biopsy, therapeutic pleuropneumonectomy, or autopsy requested by relatives.

Better cancer registries – or preferably, a dedicated mesothelioma registry of the type available in parts of France [42] and in the past (1980-85) in Australia [43] – could contribute as well, although their main contribution would be to a better ascertainment, diagnosis, and treatment of the disease. Unfortunately the Australian example with the demise of the quality of the registry with defunding after 1985 [43] indicates that obtaining adequate funding for this endeavour will not be a simple matter. Pending legislation in the US Congress has suggested funding a Registry there, but the details remain vague.

No matter what solution is suggested, there will inevitably be controversy associated with it, as fixing blame on a maximum number of potential parties increases potential rewards in tort litigation both for patients and lawyers, while also spreading financial risk more thinly among defendants.

4.2. Future risk from NOA exposures.

As with legacy exposures, exposures such as NOA, which create *future* risk, may be aided in improved assessment via cancer registries and by better exposure assessment techniques which adequately address both current and past exposures, including lung-retained fibre analysis where possible, as long as performance of the latter is not made an excuse for inaction. The latter is self-evident, and can easily be achieved through acquisition of lung samples from previously taken therapeutic lung biopsies and resections, routine autopsies from hospitals, or from coroners' departments. In addition, there are many already-analyzed series of residents of given communities which may be found in the literature and used for this purpose. Two examples of autopsy series used for exposure assessment via lung-retained fibre analysis are our analysis of the lungs of children for US EPA [44] and our assessment of Canadian background lung asbestos values by obtaining right upper lobe samples from coroners in nine of the ten provinces and both Territories [45]. Examples from a wide range of patients from many states were provided in both study subjects and controls for exposure assessment related to the man-made mineral fibre industry [46]. Direct assessment in relation to proximity to asbestos sources has shown that lung content varies inversely with distance from chrysotile mines and directly with duration lived within ten km of such mines [15,20,21]. All of the latter studies were funded through peer review by public agencies in Canada and the USA, so there is every reason to believe the same could be achieved for studies relating to proximity to NOA.

4.2.1. Translating NOA exposure assessment to risk assessment for mesothelioma. While better cancer registries will provide better characterization of where disease occurs, and lung-retained fibre analysis (as well as the type of activity-based air sampling practiced by US EPA in El Dorado Hills, California [35]) will help in exposure assessment, linking the two in risk assessment is no simple matter. Cancer registries are by definition ecological in nature, "counting cases" in given areas and comparing them to national or other rates. They cannot produce individual risk data, and the same deficiency is evident for studies such as that of Pan et al. which ultimately rely on cancer registries for their disease data [19]. The confounding effect of legacy exposures, particularly those related to occupation which still cause the highest proportion of cases nation- and world-wide, are a particular problem. Only dedicated studies which provide individual measures of both disease and exposure are likely to provide definitive answers.

4.3. "Current" legacy exposures and NOA exposures: A common problem.

As past occupational exposures become an increasingly less important source of mesothelioma risk with increasing time, current NOA exposures and legacy exposures which are due to unremediated waste or scrap left over from the past will become the source of an increasing proportion of mesothelioma risk. Although their origins are quite different, their effects are likely to be similar, depending on exposure patterns. The most important task we face is to *identify* them, as has been the case in Jefferson Parish, Louisiana and El Dorado County (and the rest of the Western Slope of the Sierras). Once identified however further tasks await. Once found, sources must be characterized; some will be serious and require remediation or other corrective measures. Those which provide disease risk are of particular concern, but we should not wait until disease has been established to begin to prevent exposures. Epidemiology is a useful tool for identifying past "epidemics", whether of infectious diseases or cancers caused by toxic materials. But we cannot wait for the comfort of epidemiological proof where exposures we know to be analogous to those which have produced disease in the past exist. Once identified, both legacy exposures and NOA exposures must be eliminated as possible causes of future disease.

Acknowledgments
Funding of the lung-retained fibre analyses reported in Table 1 for the Jefferson Parish cases was provided by the law firms who referred the cases to one of us (BWC). Funding of the lung-retained fibre analyses for the El Dorado county pets reported in Table 2 was provided directly by the authors with assistance with resources from their university departments, and additional assistance from Mr. Bryan Burnett of San Diego and Mr. Terry Trent, also of California. We also thank Mr. Trent and Dr. Marci R. Culley of Georgia State University for invaluable assistance in understanding the dynamics and time course of the events in El Dorado County which precipitated the interest and public health concerns there from 1998 to the present. Mr. Burnett and Mr. Trent are also co-authors of a previous abstract of preliminary results in four of the six pets reported in Table 2 [47].

Competing Interests
Drs. Case and Abraham have acted as expert witnesses for law firms representing defendants and/ or plaintiffs in asbestos litigation and/or compensation board proceedings in the United States and Canada, and have been paid and unpaid consultants to regulatory agencies and compensation boards in North America. Both have also provided unpaid assistance to community groups and residents having concerns with the types of asbestos exposure discussed in this paper, including residents of El Dorado County, California.

References

[1] Enterline P E and Henderson V L 1987 Geographic patterns for pleural mesothelioma deaths in the United States, 1968-81 J Natl Cancer Inst 79 31-7
[2] Connelly R R, Spirtas R, Myers M H, Percy C L and Fraumeni J F, Jr. 1987 Demographic patterns for mesothelioma in the United States J Natl Cancer Inst 78 1053-60
[3] Weill H, Hughes J M and Churg A M 2004 Changing trends in US mesothelioma incidence Occup Environ Med 61 438-41
[4] Price B and Ware A 2004 Mesothelioma trends in the United States: an update based on Surveillance, Epidemiology, and End Results Program data for 1973 through 2003 Am J Epidemiol 159 107-12.
[5] Spirtas R, Heineman E F, Bernstein L, Beebe G W, Keehn R J, Stark A, Harlow B L and Benichou J 1994 Malignant mesothelioma: attributable risk of asbestos exposure Occup Environ Med 51 804-11
[6] Pinheiro G A, Antao V C, Bang K M and Attfield M D 2004 Malignant mesothelioma surveillance: a comparison of ICD 10 mortality data with SEER incidence data in nine areas of the United States Int J Occup Environ Health 10 251-5
[7] Clifton R A. 1982 Asbestos *Minerals Yearbook Centennial Edition 1981* Washington: United

States Bureau of Mines pp 103-12.
[8] Rushworth D H 2005 The Navy and Asbestos Thermal Insulation Naval Engineers Journal **Spring** 35-48
[9] Roggli V L and Vollmer R T 2008 Twenty-five years of fiber analysis: what have we learned? *Hum Pathol* **39** 307-15
[10] Maule M M, Magnani C, Dalmasso P, Mirabelli D, Merletti F and Biggeri A 2007 Modeling mesothelioma risk associated with environmental asbestos exposure *Environ Health Perspect* **115** 1066-71
[11] Van Gosen B S 2005 Reported Historic Asbestos Mines, Historic Asbestos Prospects, and Natural Asbestos Occurrences in the Eastern United States: U. S. Geological Survey Open-File Report 2005-1189, 1 plate.
[12] Van Gosen B S 2006 Reported Historic Asbestos Prospects and Natural Asbestos Occurrences in the Central United States: U. S. Geological Survey Open-File Report 2006-1211, 1 plate.
[13] Van Gosen B S 2007 Reported Historic Asbestos Mines, Historic Asbestos Prospects, and Natural Asbestos Occurrences in the Rocky Mountain States of the United States (Colorado, Idaho, Montana, New Mexico, and Wyoming): U. S. Geological Survey Open-File Report 2007-1182, 1 plate.
[14] Van Gosen B S 2008 Reported Historic Asbestos Mines, Historic Asbestos Prospects, and Natural Asbestos Occurrences in the Southwestern United States (Arizona, Nevada, and Utah): U. S. Geological Survey Open-File Report 2008-1095, 1 plate.
[15] Case B W 1991 Health effects of tremolite. Now and in the future *Ann N Y Acad Sci* **643** 491-504
[16] Sullivan P A 2007 Vermiculite, respiratory disease, and asbestos exposure in Libby, Montana: update of a cohort mortality study Environ Health Perspect **115** 579-85
[17] McDonald J C, Harris J and Armstrong B 2004 Mortality in a cohort of vermiculite miners exposed to fibrous amphibole in Libby, Montana *Occup Environ Med* **61** 363-6
[18] Case B W 2006 Libby Mesothelioma Update: Occupational and Non-Occupational. *Lung Cancer* **54** S10
[19] Pan X L, Day H W, Wang W, Beckett L A and Schenker M B 2005 Residential proximity to naturally occurring asbestos and mesothelioma risk in california *Am J Respir Crit Care Med* **172** 1019-25
[20] Case B W and Sebastien P 1987 Environmental and occupational exposures to chrysotile asbestos: a comparative microanalytic study *Arch Environ Health* **42** 185-91
[21] Case B W and Sebastien P 1989 Fibre levels in lung and correlation with air samples *IARC Sci Publ* **90** 207-18
[22] McDonald J C, Armstrong B, Case B, Doell D, McCaughey W T, McDonald A D and Sebastien P 1989 Mesothelioma and asbestos fiber type. Evidence from lung tissue analyses *Cancer* **63** 1544-7
[23] Case B W 1994 Biological indicators of chrysotile exposure *Ann Occup Hyg* **38** 503-18, 410-1
[24] Case B W and Dufresne A 1997 Asbestos, asbestosis, and lung cancer: observations in Quebec chrysotile workers *Environ Health Perspect* **105 Suppl 5** 1113-9
[25] Nayebzadeh A, Case B W, Masse J and Dufresne A 2006 Mineralogical and exposure determinants of pulmonary fibrosis among Quebec chrysotile miners and millers *Int Arch Occup Environ Health* **79** 227-36
[26] Wright R S, Abraham J L, Harber P, Burnett B R, Morris P and West P 2002 Fatal asbestosis 50 years after brief high intensity exposure in a vermiculite expansion plant *Am J Respir Crit Care Med* **165** 1145-9
[27] Abraham J L, Burnett B R and Hunt A 2002 Quantification of non-fibrous and fibrous particulates

in human lungs: Twenty-year update on peumoconiosis database. *Ann Occup Hyg* **46(S1)** 397-401

[28] Abraham J L. Analysis of fibrous and non-fibrous particles. 2006 In: Rom W and Markowitz S, eds. *Environmental and Occupational Medicine*. 4th ed. Philadelphia: Lippincott Williams and Wilkins 2006 pp 277-97.

[29] Peters E S, Schmidt B A, Andrews P A, Hsieh M, Mumphrey B, Wu X and Chen V W 2007 *Cancer in Louisiana, 2000-2004* New Orleans: Louisiana Tumor Registry 91 pages

[30] Hughes J M, Weill H and Hammad Y Y 1987 Mortality of workers employed in two asbestos cement manufacturing plants *Br J Ind Med* **44** 161-74

[31] Martin J J 1996 Memorandum: Request for a time-critical removal action and an exemption from the $2 million statutory limit at the Westbank asbestos site Marrero, Jefferson Parish, Louisiana Dallas, Texas: US Environmental Protection Agency Region 6 14 pages.

[32] NIOSH 2003 Malignant Mesothelioma. *Work-Related Lung Disease Surveillance Report 2002.* Cincinnati Ohio Department of Health and Human Services pp157-68.

[33] NIOSH 2008. Malignant mesothelioma: Counties with highest age-adjusted death rates (per million population), U.S. residents age 15 and over, 2000–2004. Ref. No. 2007T07-10, Work-Related Lung Disease (WoRLD) Surveillance Report (update), Surveillance Branch, Division of Respiratory Disease Studies, NIOSH, Morgantown, WV 26505. June 23, 2008. Available online at http://www2.cdc.gov/drds/WorldReportData/pdf/2007T07-10.pdf (Accessed October 5, 2008).

[34] Davis J M, Addison J, McIntosh C, Miller B G and Niven K 1991 Variations in the carcinogenicity of tremolite dust samples of differing morphology *Ann N Y Acad Sci* **643** 473-90

[35] US Environmental Protection Agency Region 9 2005 *U.S. EPA Asbestos Assessment for El Dorado Hills.* (San Francisco, CA: US Environmental Protection Agency Region 9). 6 pages.

[36] Meeker G P, Lowers H A, Swayze G A, Van Gosen B S, Sutley S J and Brownfield I K. 2006 *Mineralogy and Morphology of Amphiboles Observed in Soils and Rocks in El Dorado* Hills, *California.* (Reston, VA: U.S. Geological Survey) Open-File Report 2006-1362 47 pages

[37] Buckingham D.A. and Virta R.L. U.S. Geological Survey 2007 Asbestos Statistics in: Kelly, T.D., and Matos, G.R., comps., *Historical statistics for mineral and material commodities in the United States: U.S. Geological Survey Data Series 140.* Available online at http://minerals.usgs.gov/ds/2005/140/asbestos.pdf (Accessed June 26, 2008)

[38] McDonald J C, Harris J M and Berry G 2006 Sixty years on: the price of assembling military gas masks in 1940 *Occup Environ Med* **63** 852-5.

[39] Hodgson J T and Darnton A 2000 The quantitative risks of mesothelioma and lung cancer in relation to asbestos exposure *Ann Occup Hyg* **44** 565-601

[40] Peto J, Seidman H and Selikoff I J 1982 Mesothelioma mortality in asbestos workers: implications for models of carcinogenesis and risk assessment *Br J Cancer* **45** 124-35

[41] Price B and Ware A 2005 Mesothelioma: risk apportionment among asbestos exposure sources *Risk Anal* **25** 937-43

[42] Goldberg M, Imbernon E, Rolland P, Gilg Soit Ilg A, Saves M, de Quillacq A, Frenay C, Chamming's S, Arveux P, Boutin C, et al. 2006 The French National Mesothelioma Surveillance Program *Occup Environ Med* **63** 390-5

[43] Yeung P and Rogers A 2001 An occupation-industry matrix analysis of mesothelioma cases in Australia 1980-1985 *Appl Occup Environ Hyg* **16** 40-4

[44] Case BW K M, Harrigan M, Dufresne A. 1994 Lung fibre content of American children aged 8-15 years. *Ann Occup Hyg* **38** 639-45

[45] Case B W, Sebastien P and McDonald J C 1988 Lung fiber analysis in accident victims: a

biological assessment of general environmental exposures *Arch Environ Health* **43** 178-9.

[46] McDonald J C, Case B W, Enterline P E, Henderson V, McDonald A D, Plourde M and Sebastien P 1990 Lung dust analysis in the assessment of past exposure of man-made mineral fibre workers *Ann Occup Hyg* **34** 427-41

[47] Abraham J, Case BW, Burnett B, Trent T. 2005 Lung-Retained Fiber in Pets Confirms Environmental Exposure to Naturally-Occurring Asbestos in Western El Dorado County, California. *Am J Resp Crit Care Med* **171(S1)** A818.

Inhaled Particles X, (23–25 September 2008, Manchester) IOP Publishing
Journal of Physics: Conference Series **151** (2009) 012009 doi:10.1088/1742-6596/151/1/012009

Trends in pneumoconiosis and other lung diseases, as reported to a UK-based surveillance scheme for work-related ill-health

S Turner[1], R McNamee[2], M Carder[1], R Agius[1]

[1]Occupational and Environmental Health Research Group, University of Manchester, UK

[2]Biostatistics, Health Methodology Research Group, University of Manchester, UK

E-mail: susan.m.turner@manchester.ac.uk

Abstract: The changing nature of industries associated with exposure to hazardous dusts in manufacture or in use, as well as better control methods, might be expected to be associated with a reduction in incidence of pneumoconiosis and other lung diseases. Data collected by the University of Manchester's ODIN/THOR network on work-related ill-health in the UK (as diagnosed by specialist physicians) can be used to estimate time trends in the lung diseases reported to the surveillance schemes. Reporters of work-related lung diseases in THOR (previously ODIN) mainly comprise two groups, namely clinical specialists in respiratory medicine and occupational physicians. These reporters return information on work-related cases of ill-health using postal reporting cards or an on-line web form. 'Report cards' are returned even if no new cases are seen, with responses recorded each month i.e. whether a card is returned and number of cases returned. Probabilities of a non response and, for returned cards, of a 'zero' return were modelled as a function of calendar time and/or membership time using 2-level logistic models. Annual change in disease incidence (all work-related respiratory disease and specific diagnoses) was estimated using 2-level Poisson models controlling for reporter characteristics, season, and whether or not a first report. The impact of membership time on reporting was also investigated. Case reports include information on patient demographics, diagnoses, industry, occupation, and suspected agents/exposures. These case details are coded and analysed using SPSS. Annual change in incidence of all work-related respiratory disease reported by specialist chest physicians (1999-2006) was -1.7% (95% CI: -3.1%, -0.2%). Specific diagnoses reported by chest physicians showed that the annual change in incidence for asthma was -3.1% (95% CI: -5.8%, -0.4%), for mesothelioma was -4.1% (95% CI: -6.7%, -1.5%), for benign pleural plaques was +1.1 (95% CI; -1.0%, +3.2%), and for pneumoconiosis was -2.6 (95% CI: -6.6, +1.5) over the same time period. Occupational physicians' reporting showed a change in incidence of -6.1% (95% CI: -11.6%, -0.4%) for all respiratory disease, and -8.4% (95% CI: -15.3%, -0.9%) for asthma. Given variation between reporter groups, and according to model assumptions, time trends from surveillance data need to be interpreted with caution, but may have some place in planning interventions aimed at improving the health of a workforce. Further work to investigate case details (such as suspected agent/exposures) should also add to this knowledge base.

1. Background

The changing nature of industries associated with exposure to hazardous dusts in manufacture or in use, as well as better control methods, might be expected to be associated with a reduction in incidence of pneumoconiosis and other lung diseases. The 'Revitalising Health and Safety' (RHS) initiative announced in 1999 [2] set out a headline target to reduce the incidence rates of cases of work-related ill-health by 20% by 2010, and to achieve half the improvement by 2004. This was re-affirmed in 'Securing Health Together' [3]. National disease incidence data include sources such as the Self-reported Work-related Illness [10] survey, which reports information that is subjective (as it is dependent on the perceptions and beliefs of the sufferer), and voluntary surveillance schemes for occupationally related ill-health, which are based on medical reports.

In the UK, a pioneering scheme for work-related respiratory disease reporting (SWORD) was set up in 1989 [6]. Following this, sister schemes to cover other work-related diagnostic areas were developed (such as skin disease, psychological ill-health and musculoskeletal disorders), and collectively formed the Occupational Disease Intelligence Network (ODIN) which existed until 2002, when the schemes were relaunched as The Health and Occupation Reporting (THOR) network [11].

In THOR, occupational respiratory disease is reported by clinical specialists in respiratory medicine, by occupational physicians, (on a smaller scale) by consultants in communicable disease control, and by general practitioners (since 2005). Data collected on work-related respiratory disease in the UK by the University of Manchester's ODIN/THOR networks have been studied in detail in order to estimate true rates of change in incidence of work-related respiratory diseases from 1996-2005 [5]. In this study, reporting to SWORD by clinical specialists in respiratory disease showed a fall in incidence for pneumoconiosis of -5.7% (95% CI: -10.5, -0.5) per year. More recent data are now available from THOR which can be used to extend the trends in incidence study period. In addition, information on specific diagnoses, for example pneumoconiosis and non-malignant pleural disease (NMPD), and analysis of causal agents can be undertaken.

2. Aims

To investigate time trends for lung diseases reported to a UK wide surveillance scheme for work-related ill-health.

3. Methods

Reporters of lung diseases in SWORD comprise clinical specialists in respiratory medicine who return information on work-related cases of ill-health using postal reporting cards or an on-line web form. Reporting cards or web forms are returned even if no new cases are seen, with responses recorded each month i.e. whether a card is returned and the number of cases returned. There are two groups of reporters within SWORD; 'core' reporters who are asked to return a reporting card (or web form) every month, and 'sample' reporters who return a card (or web form) for one randomly allocated month each calendar year. The 'core' reporting group mainly comprises respiratory specialist physicians with a specific interest in work-related lung disease, while the 'sample' group also consists of respiratory specialist physicians but most of these do not have such a specific interest in occupational diagnoses.

Probabilities of a non response and, for returned cards, of a 'zero' return were modelled as a function of calendar time using 2-level logistic models. Annual change in disease incidence (all work-related respiratory disease and specific diagnoses) was estimated using a 2-level (hierarchical) Poisson regression model controlling for reporter characteristics (e.g. 'core' or 'sample' reporting), season, and whether or not a first report. The impact of membership time on reporting was also investigated. The STATA software command xtnbreg was used to fit longitudinal, negative binomial (i.e. over-dispersed) Poisson models with random effects.

Case reports include information on age, gender, geographical location (first half of postcode) job title, industry, and suspected agent (up to six agents can be recorded) for each case. The occupational information within the case reports is coded using the Standard Occupational Classification (SOC) for the job title, and the Standard Industrial Classification (SIC) for data on industry [9],[7],[1]. The postcode information allows grouping of information according to the geographical classification in the Labour Force Survey [8]. All coding is undertaken independently by two researchers, and any discrepancies are reconciled by a third person.

Case details were analysed for all respiratory disease, reports of NMPD, and diagnoses of pneumoconiosis, using SPSS v15.0.

4. Results

Annual change in incidence of all work-related respiratory disease reported by specialist chest physicians (1999-2006) was -1.7% (95% CI: -3.1%, -0.2%). Specific diagnoses reported by chest physicians showed that the annual change in incidence for pneumoconiosis was -0.8 (95% CI: -4.8, +3.3), and for NMPD was +2.2 (95% CI; -1.0%, +3.2%), over the same time period (Figures 1 and 2).

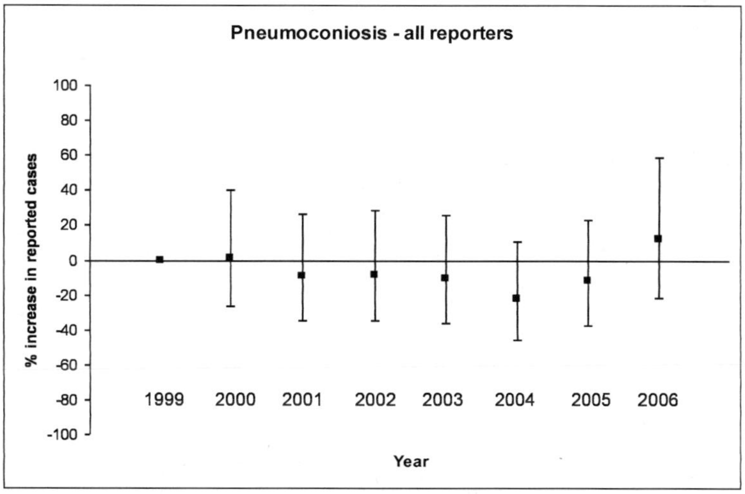

Percentage change = - 0.8 (95% CI: - 4.8, + 3.3)

Figure 1. Percentage change in incidence of pneumoconiosis in SWORD (1999-2006)

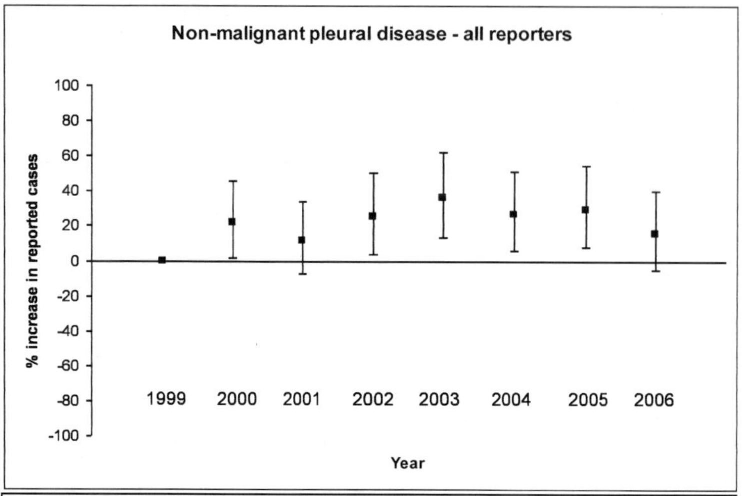

Percentage change = +2.2 (95% CI; -1.0%, +3.2%)

Figure 2. Percentage change in incidence of non-malignant pleural disease in SWORD (1999-2006)

The annual change in incidence for pneumoconiosis and NMPD was also calculated for the different reporter groups (i.e 'core' and 'sample' reporters), as shown in Table 1.

Table 1. Annual change in incidence for pneumoconiosis and NMPD by reporter group

Diagnosis	Reporter group	Percentage change (95% CI)
Pneumoconiosis	'core'	+1.2 (-3.4, +6.1)
	'sample'	-9.9 (-17.7, -100.0)
NMPD	'core'	+1.0 (-1.4, +3.4)
	'sample'	+5.6 (+0.6, +10.9)

Between 1999 and 2006, 3697 cases of non-malignant pleural disease (NMPD) were reported to SWORD. Nearly all (97.7%) were males, and the mean age was 66.3 years (most cases [55.6%] being in the 60-75 year age range). The most frequently reported employment sectors were construction (37.0%); 'manufacture of other transport equipment' (20.3%) i.e. ship building; and 'electricity, gas, steam and hot water supply' (8.2%) i.e. employment in power stations. Nearly all cases (98.8%) cited asbestos as a causal agent.

There were fewer reports of pneumoconiosis than NMPD over this 7 year time period, with 637 cases reported to SWORD. Again, the vast majority (99.1%) were males, and the age demography was similar to that seen for NMPD reporting; a mean age of 69.1 years and most cases [54.9%] being in the 60-75 year age range. Construction (27.0% cases) and 'manufacture of other transport equipment' (17.4%) were the two most frequently reported employment sectors, but 'mining of coal and lignite; extraction of peat' (which was solely coal mining as reported to SWORD) featured as the third most frequently reported industrial sector, contributing 12.6% cases. Asbestos was reported as a causal agent in three quarters (75.2%) of the cases, coal dust in 10.2%, and silica in 8.9%.

5. Conclusions

We cannot conclude with confidence whether the incidence of pneumoconiosis is changing using these data, but there is a suggestion of increasing incidence for NMPD. We also found variations in trends between 'core' and 'sample' reporter groups for the two diagnoses.

For pneumoconiosis, it is likely that potential cases of work-related lung disease are referred to (and therefore reported by) tertiary respiratory centres who make up the 'core' reporting group within SWORD, thereby accounting for the differences in incidence between 'core' and 'sample' reporting. It is also possible that, if pneumoconiosis incidence is falling (or perceived to be falling), the tertiary centres with specialist interest in occupational disease see proportionately more of the cases than chest physicians without special interest or expertise in this area.

In comparison, the greater increase in trend for 'sample' (rather than 'core') reporting for NMPD might be explained by changes in medical practice, which may artefactually affect trends. For example the increasing availability and sensitivity of modern imaging techniques (such as computerised tomography) might produce incidental findings including pleural plaques, and result in apparent upward trends, especially for the 'sample' reporting group who may have less expertise in, and experience of, work-related lung disease.

When addressing work-related disease incidence, validity is also vulnerable to systematic changes in, or awareness of, the epidemiological evidence of causal links, and in government policy (such as compensation), and we cannot rule out reporting bias for such reasons. Brief guidance on attribution is available to SWORD reporters, but imposition of strict criteria for disease or its attribution is deliberately avoided so as to encourage and maintain participation. Such compromises are typical of surveillance systems – characterised by "practicality, uniformity and rapidity, rather than by accuracy or completeness" [4] - but may also be inevitable in any long-term, voluntary data collection system [5].

In summary, although the monitoring of trends as described here need to be interpreted with caution, they may have a place in planning or evaluating interventions aimed at improving the health of a workforce. For pneumoconiosis and NMPD such interventions may have limited scope in view of historical aspects relating to exposure, but these methodologies may be important for ongoing monitoring of newer agents, such as 'asbestos substitutes'. Further work to investigate case details (e.g. suspected agents/exposures, industrial sectors and occupations) may also add to our knowledge base of work-related lung diseases.

Acknowledgements:

We are grateful to Professors Corbett McDonald and Nicola Cherry. Corbett McDonald initiated the SWORD scheme, and he and Nicola Cherry were co-directors of ODIN (THOR's predecessor) from 1996 to 2002. We are also grateful to all physicians in the UK who participate in the reporting schemes, for their invaluable contribution and co-operation. Thanks are also due to Catherine Roberts for her research and administrative assistance in SWORD. Physicians who wish to join THOR and participate in reporting can find further details at http://www.medicine.manchester.ac.uk/coeh/thor

Funding acknowledgement:

THOR is partly funded by the UK Health and Safety Executive (HSE), but the views presented here are those of the authors and not necessarily of the HSE.

References

[1] Central Statistical Office. Indexes to the Standard Industrial Classification of Economic Activities 1992. London: HMSO, 1993.

[2] HSE Strategy Statement (2000) Available from: URL: http://www.hse.gov.uk/revitalising/strategy.htm

[3] HSE SH2, 2000 Available from: URL: http://www.hse.gov.uk/sh2/

[4] Last J. Dictionary of Epidemiology. 4th Edition. New York: Oxford University Press, 2001. ISBN13: 978-0-19-514169-6 ISBN10: 0-19-514169-5

[5] McNamee R, Carder M, Chen Y, Agius R. Measurement of trends in incidence of work-related skin and respiratory diseases, UK 1996-2005. Occup Environ Med (Published Online First: 16 April 2008. doi:10.1136/oem.2007.036731)

[6] Meredith SK, Taylor VM, McDonald JC. Occupational respiratory disease in the United Kingdom 1989: a report to the British Thoracic Society and the Society of Occupational Medicine by the SWORD project group. Br J Ind Med 1991;48:292-298.

[7] Office for National Statistics. Standard Occupational Classification. Norwich: The Stationery Office, 2000.

[8] Office for National Statistics. Labour Force Survey 2002-2005.

[9] Office of Population Censuses and Surveys. Standard Occupational Classification. London: HMSO, 1990.

[10] SWI, 2006/07 Labour Force Survey. Self-reported work-related ill-health (SWI) 2006/07. Available at http://www.hse.gov.uk/statistics/swi/0607.pdf

[11] THOR - The Health and Occupation Reporting network. Available from: URL: http://www.medicine.manchester.ac.uk/coeh/thor

Respirable crystalline silica – a failure to control exposure!

J R Cain MSc MBA MFOH CBiol FIBiol Dip Occ Hyg

HM Regional Specialist Inspector (Occupational Hygiene), Health and Safety Executive, Marshalls Mill, Marshall Street, Leeds LS11 9YJ

john.cain@hse.gsi.gov.uk

Abstract: Several sites were visited to monitor stonemason exposure to respirable crystalline silica (RCS), inhalable dust and respirable dust. At all sites, exposure to RCS exceeded the Workplace Exposure Limit of 0.1 mg/m^3 8-hour TWA. There was therefore a continuing high risk of workers developing silicosis unless the appropriate measures were instigated to prevent or control exposure. Exposure control was ineffective at all sites e.g. water wall extraction systems were not well designed. There was evidence that foreign workers were at a greater exposure risk. But even with appropriate controls to mitigate exposure to RCS it may not be possible to sustain exposure to below 0.1 mg/m^3 8-hour TWA without on-going HSE intervention.

1. Introduction

There has been an increase in the number of companies in the UK carrying out stonemasonry work. This has lead to a subsequent increase in the risks of developing silicosis in particular progressive silicosis.

As part of a Health and Safety Executive (HSE) enforcement initiative, in 2007 several site visits were made to companies in North East England to measure stonemason exposure to respirable crystalline silica (RCS), inhalable dust and respirable dust and to assess the measures used to prevent or mitigate exposure to stone dust.

There was evidence that during stonemasonry work exposure to RCS could not be controlled to below the Workplace Exposure Limit (WEL) of 0.1 mg/m^3 8-hour Time Weighted Average (TWA) or exposure to inhalable dust and respirable dust to below 10 mg/m^3 and 4 mg/m^3 8-hour TWAs respectively (occupational exposure limits for the dusts). There was also evidence that foreign workers mainly Eastern European were at a greater exposure risk.

2. Methods

Seven sites were visited. Most of the sites employed over five workers but not all workers were involved with stonemasonry work. The stonemasonry work at each site varied but included mainly ornamental /architectural work and the cutting of fire surrounds. Almost all the stonemasons worked for a full 8-hour shift. Eastern European stonemasons (mainly from Poland) tended to work excess overtime, worked unsupervised and were therefore at a greater risk of exposure.

Stonemasons and other relevant workers were sampled to measure exposure to RCS, inhalable dust and respirable dust using the methods as outlined in MDHS 14/3 " *General methods for sampling and analysis of respirable and inhalable dust*". The RCS analysis was carried out using x-ray diffraction and as outlined in MDHS 101 "*Crystalline silica in respirable airborne dusts*".

The measures taken by each company to prevent or control exposure to dust were assessed in particular the use of water wall extraction systems.

3. Results

Table 1 lists the exposure controls at each site. In general the water wall extraction systems were not effective at capturing dust because of poor design e.g. insufficient enclosure and poor maintenance. There was widespread evidence that stonemasons who wore RPE (mainly disposable half-face dust masks) had not been face fit tested and that most of them had not been trained on their use.

Most companies did not provide adequate health surveillance and training. Because of the lack of effective measures to prevent or control exposure to RCS during stonemasonry work all workers were at a high risk of developing respiratory disease in particular the Eastern European workers who tended to have communication problems and were not familiar with working conditions in the UK.

Table 2 shows the distribution of the 8-hour TWAs for all the personal samples collected i.e. RCS, inhalable dust and respirable dust. Interestingly, 50% of all the personal samples collected to measure exposure to RCS were above 0.1 mg/m^3 8-hour TWA and therefore exceeded the WEL.

The ranges and the geometric means of the personal inhalation exposures to RCS, inhalable dust and respirable dust at each site i.e. 8-hour TWAs are given in Table 3. At six sites the geometric means for RCS were above 0.1 mg/m3 8 hour TWA (for all sites it was 0.14 mg/m3 8-hour TWA). The two companies with the highest geometric mean personal exposures to RCS i.e. A and G had evidence of silicosis. The maximum exposures to inhalable dust at these two sites were very high i.e. at A – 134.7 mg/m3 8-hour TWA and at G - 120.2 mg/m3 8-hour TWA whereas exposure to respirable dust was high at only one of these sites i.e. A- 12.8 mg/m3 8-hour TWA.

Enforcement action has been taken against all companies to mitigate exposure to RCS. Two of the companies have been prosecuted.

4. Discussion

The monitoring survey indicated that during stonemasonry work the WEL for RCS was exceeded at all sites thereby increasing the worker risks of developing lung disease (including lung cancer?). Recent evidence suggests that the higher the exposure to RCS the greater the risk of silicosis. The risk rises faster the higher the exposure i.e. it is multiplicative. At the two sites with the highest exposures to RCS and inhalable dust, stonemasons had developed silicosis.

The situation in the North East of England is likely to be repeated elsewhere. There are therefore thousands of workers who may be exposed to significantly high levels of RCS. The number of cases of silicosis in the UK is likely to increase. This is especially important for foreign stonemasons who are at a particular high risk of dust exposure because they tend to work long hours, have poor communication with their employers and lack the necessary training that leads to the use of inappropriate controls.

There was a lack of knowledge by employers on the risks associated with stonemasonry work. This was reflected in the poor training given to employees, on the implementation of ineffective controls (in particular the widespread use of inappropriate water wall extraction systems) and on the failure to provide adequate health surveillance and suitable and sufficient risk assessments.

HSE will need to continue to develop its corporate strategies e.g. enforcement to ensure that the risks to stonemasons of developing silicosis are significantly reduced nationally. Such an approach will be critical if the WEL for RCS is reduced to 0.05 mg/m^3 8-hour TWA at some stage in the future. Furthermore, there will need to be employer/employee behavioural changes to ensure that the measures used to mitigate exposure to RCS are sustained in the long-term.

5. Conclusions

Many stonemasons are exposed to high levels of RCS and other dusts during work activities. Unless the appropriate exposure control measures are implemented for example, the use of effectively designed and well maintained LEV systems, the use of water suppression, the use of on-tool extraction for powered hand tools and the wearing of appropriate RPE/PPE, it is highly likely that the WEL for RCS will continue to be exceeded. But even with appropriate controls it may not be possible to sustain exposure to below 0.1 mg/m^3 8-hour TWA without on-going HSE intervention.

Table 1: Details of Exposure Controls in use at each site

Site		(RPE)	Equipment (PPE)		
A.)	❑ LEV (for Polish workers). ❑ Two inside areas with ineffective water wall extraction systems. ❑ Enclosed booth for smoothing/polishing of stonework together with some carving and shaping. It was not effective. The LEV had not undergone thorough examination and testing.	- respirator with FFP3 classification. No face fit testing had been carried out. Storage boxes had recently been provided for the RPE to be stored in.	The masons worked in day-to-day clothing.	Water cooled saws.	surveillance and training.
B.)	stem was installed but it was not in use. No other LEV used	-face mask with FFP3 classification. No face fit testing had been carried	polishers.	s	surveillance and assessments.
	This had undergone thorough examination and testing i.e. COSHH Reg 9.	Disposable half-face masks with FFP3 classification. No face fit testing had been carried out	Workers wore their own clothes and overalls that were laundered at the site.		Health surveillance carried out. No effective training.

Table 1: Details of Exposure Controls in use at each site

Site		Respiratory protective equipment (RPE)	Personal Protective Equipment (PPE)		
D.)	capturing hood used. It was ineffective. This had not undergone thorough examination and testing as per regulation 9 of COSHH.	disposable half-face dust mask FFP3 was worn by all the workers. No face fit testing had been carried out. In the hand masonry area, powered (fan assisted) RPE was worn by all the workers. These were examined regularly on a monthly basis.	Safety boots, eye hi-visibility vests and hearing defenders. In the brick making area the workers wore their own clothing. In the hand masonry area disposable coveralls were used.		carried out. Unsuitable and insufficient risk assessments.
E.)	They were not effective.	-face dust mask FFP2 was worn by all the workers in the brick making area. Face fit testing had been carried out.	Safety shoes/boots, -visibility vests and hearing defenders. The overalls were supplied by the company who also Latex gloves were issued	Water cooled saws and wet polishing.	surveillance carried out.

Table 1: Details of Exposure Controls in use at each site

Site		tective equipment (RPE)	Personal Protective Equipment (PPE)	Other Controls	Comments
F.)	ally enclosed extraction booth had been recently installed. Some powered tools had extraction.	disposable half-face dust masks FFP3 worn. Face fit testing had been carried out where appropriate.	d rubber palmed fabric gloves. The company had recently introduced arrangements for the laundering of work	and manual when polishing.	assessment.
G.)	It was not very effective. It was used by the masons using powered tools only; there was no LEV for the masons using non-powered hand tools	-face dust mask FFP3. The masons had not been face fit tested.	Low dust release eralls, safety shoes and ear defenders.		No health surveillance carried out.

Table 2: Distribution of the 8hrTWA exposures for all the samples collected

Sample Type	**Exposure levels 8hr TWA mg/m^3**	**Percentage of samples (%)**
Inhalable Dust	<=10	62.4
	> 10	37.6
Respirable Dust	<= 4	89.6
	> 4	10.4
RCS	<= 0.05	29.2
	> 0.05 to <= 0.1	20.8
	> 0.1 to <= 0.3	19.7
	> 0.3	30.3
	Check Sum	100
RCS	<= 0.05	29.2
	<= 0.1	50
	<= 0.3	69.7

Table 3: The ranges and the geometric means of personal exposure to RCS, inhalable dust and respirable dust

Site	**Number of samples and sample type**	Inhalable Dust 8hr TWA mg/m³			Respirable Dust 8hr TWA mg/m³			Respirable Crystalline Silica 8hr TWA mg/m³		
		Parameter			Parameter			Parameter		
		Min	**Max**	**Geo Mean**	**Min**	**Max**	**Geo Mean**	**Min**	**Max**	**Geo Mean**
All Sites	38(Respirable Dust) 36(RCS) 36(Inhalable Dust)	0.26	134.700	6.84	0.015	12.800	0.420	0.010	4.120	0.14
C.)	7 (Resp Dust+RCS) 5 (Inhalable Dust)	1.9	16.	4.65	0.06	0.54	0.21	0.04	0.23	0.09
F.)	1 (Resp Dust, RCS and Inhalable Dust)	6.3	6.3	6.3	0.18	0.18	0.18	0.12	0.12	0.12
E.)	9 (Resp Dust, RCS and Inhalable Dust)	0.26	24.90	1.51	0.02	2.51	0.10	0.01	1.10	0.03
B.)	4 (Respirable + inhalable Dust) 3 (RCS)	2.11	36.65	6.75	0.42	8.34	1.40	0.05	0.26	0.11
A.)	9 (Resp Dust, RCS and Inhalable Dust)	2.70	134.7	26.39	0.25	12.80	1.1.99	0.09	4.12	0.89
D.)	6 (Resp Dust, RCS and Inhalable Dust)	1.700	46.80	7.61	0.23	0.99	0.42	0.04	0.79	0.14
G.)	2 (Resp Dust, RCS and Inhalable Dust)	6.8	120.2	28.59	0.18	0.66	0.34	0.06	0.49	0.17

Journal of Physics: Conference Series **151** (2009) 012011 doi:10.1088/1742-6596/151/1/012011

Rat- and Human-based Risk Estimates of Lung Cancer from Occupational Exposure to Poorly-Soluble Particles: A Quantitative Evaluation

E D Kuempel, [1] **R J Smith,** [1] **D A Dankovic,** [1] **LT Stayner**[2]

[1]National Institute for Occupational Safety and Health, Education and Information Division, Risk Evaluation Branch, Cincinnati, Ohio, USA

[2] University of Chicago School of Public Health, Division of Epidemiology and Biostatistics, Chicago, Illinois, USA

ekuempel@cdc.gov

Abstract. In risk assessment there is a need for quantitative evaluation of the capability of animal models to predict disease risks in humans. In this paper, we compare the rat- and human-based excess risk estimates for lung cancer from working lifetime exposures to inhaled poorly-soluble particles. The particles evaluated include those for which long-term dose-response data are available in both species, i.e., coal dust, carbon black, titanium dioxide, silica, and diesel exhaust particulate. The excess risk estimates derived from the rat data were generally lower than those derived from the human studies, and none of the rat- and human-based risk estimates were significantly different (all p-values$\geq$0.05). Residual uncertainty in whether the rat-based risk estimates would over- or under-predict the true excess risks of lung cancer from inhaled poorly-soluble particles in humans is due in part to the low power of the available human studies, limited particle size exposure data for humans, and ambiguity about the best animal models and extrapolation methods.

1. Introduction

Quantitative risk assessment (QRA) requires dose-response data to evaluate the relationship between exposure to a toxicant and the probability of developing an adverse health effect. When epidemiological exposure-response data are available (e.g., from an occupational cohort), risk estimates can be obtained directly from humans. However, human exposure-response data can be inadequate for QRA (e.g., insufficient exposure data or follow-up time). The goal of risk assessment is to determine the probability of disease or injury so that risk management intervention measures can be taken to prevent adverse effects in humans. QRA often relies on data from laboratory animals to estimate disease risks in humans.

Chronic inhalation studies in rats exposed to poorly-soluble particles have shown significantly elevated lung tumor response, while studies in mice and hamsters have not [1-3]. Occupational exposures to some respirable, poorly-soluble particles have been associated with elevated lung cancer in some studies, including carbon black [4,5]; diesel exhaust particulate [6-8]; and titanium dioxide [9], although generally without evidence of an exposure-response relationship; no association with particle exposure and lung cancer was found in other studies of titanium dioxide [10], carbon black

[11], or coal dust [12-14]. Significant exposure-response relationships have been reported for crystalline silica in several studies [15, 16] including a pooled analysis of six cohorts [17], but not in a recent U.K. study with lower exposures [18]. Because epidemiology studies typically do not have sufficient power to detect disease risk (e.g., lung cancer) at levels of concern to regulatory agencies (e.g., 1/1000), negative studies generally cannot be used to rule out such risks [19]. Thus, experimental studies in animals are often used to determine dose-response relationships and estimate risk in humans.

There is some debate about the extent to which rodent data may be useful for predicting human lung cancer risk of inhaled poorly soluble particles, particularly in regards to the role of particle overload of lung clearance and inflammation and cell proliferation processes in the rat lung tumor response [20-23]. These evaluations have largely been qualitative, and quantitative dose-response data in both animals and humans are needed.

In this study, we quantitatively compare working lifetime excess risk estimates of lung cancer derived from both rat and human dose-response data of respirable poorly-soluble particles including coal dust, carbon black, titanium dioxide, silica, and diesel exhaust particulate. These comparisons provide a quantitative evaluation of the relevance of the rat model of particle-induced lung cancer for risk assessment in humans.

2. Methods

Comparison of the animal and human-based excess risk estimates included several steps, including: estimate working lifetime excess risk of lung cancer in humans; predict the retained particle lung burden in humans; estimate the rat-equivalent lung burden; compute the excess risk in rats at that lung burden; and finally compare the human- and rat-based excess risks (Figure 1).

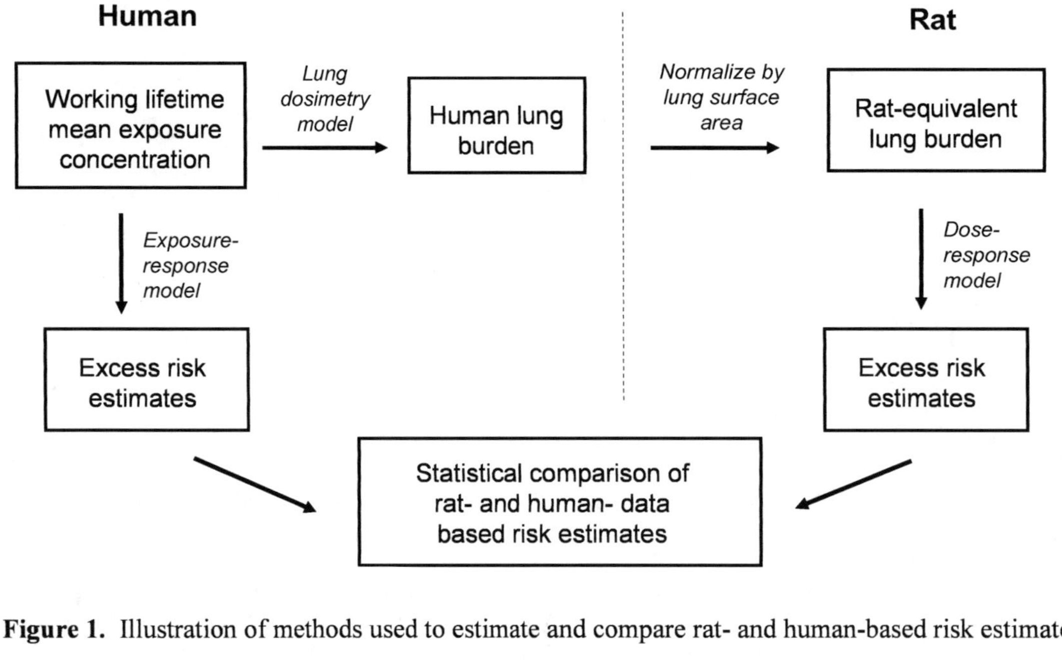

Figure 1. Illustration of methods used to estimate and compare rat- and human-based risk estimates.

Journal of Physics: Conference Series **151** (2009) 012011 doi:10.1088/1742-6596/151/1/012011

Excess risk (ER), or exposure-attributable risk, is defined here as the increased probability of an adverse response at a given dose compared to that in unexposed individuals:

$$ER(dose) = P(dose) - P(0) \quad (1)$$

where P(dose) and P(0) represent the probability of tumor response at a given dose or zero dose, respectively. Dose was expressed as either cumulative exposure to airborne particles (human studies) or the retained lung burden as particle mass or particle surface area at the end of exposure (rat studies).

The methods shown in Figure 1 were performed for each of the particle types investigated in this study (i.e., coal dust, carbon black, titanium dioxide, and crystalline silica). The models used at each step are described in Sections 2.1 and 2.2.

2.1. Human data
Epidemiological studies were selected which had quantitative estimates of cumulative exposure to respirable particles and lung cancer mortality rates. Studies with clear evidence of confounding exposures were excluded, but no additional criteria for study quality were imposed. These studies include: crystalline silica [15, 16, 18]; coal dust [14]; carbon black [4] and titanium dioxide [9, 10]. For silica, the pooled analysis [Steenland et al. 2001] was not used here due to more complicated methods required to estimate excess risks; and the three most recent individual studies were selected, one of which [15] was included in the pooled analysis. In addition, studies of diesel exhaust particulate studies evaluated in Stayner et al. [6] are illustrated separately.

The lifetime excess risks of lung cancer were estimated using Poisson regression models and lifetable analyses that account for competing causes of death, in U.S. males up to age 85 after a 45-year working lifetime [24]. A linear relative rate model was used to fit the data for crystalline silica [15, 16, 18], coal dust [14], carbon black [4], and one of the titanium dioxide studies [9]. In fitting this model to the Brown and Rushton study [18], which reported results as categorical relative risks, correlations were not accounted for in the categories. All models were fit by iteratively re-weighted least squares. Goodness of fit ($p>0.05$) was evaluated for each model, except in Fryzek et al. [10], since the coefficient for that model was provided by personal communication (Fryzek to NIOSH in 2004). The linear relative rate model was of the form:

$$RR(X) = 1 + \beta * X \quad (2)$$

A log-linear relative rate model was used to fit the data in the other titanium dioxide study [10], since that study reported relative risks based on this model form:

$$RR(X) = \exp(\beta * X) \quad (3)$$

where RR(X) is the relative rate in the exposed population; β is the coefficient for cumulative exposure; and X is the cumulative exposure to airborne particles (mg/m^3 x years) (respirable size fraction, except Fryzek et al. [10], which used total particulate).

Excess risks were estimated at the airborne mean concentrations (8-hour time-weighted average) that are equivalent to the cohort mean or median cumulative exposures reported in each epidemiology study, averaged over a 45-year working lifetime. For example, a cohort mean cumulative exposure of 64 mg/m^3 x yr [14] averaged over 45 years is equivalent to an airborne mean concentration of 1.4 mg/m^3. Similarly, a mean concentration of 0.83 mg/m^3 is estimated from a 45-year exposure at 37.5 mg/m^3 x yr (the midpoint of the middle cumulative exposure group for all-plants) [4]. In Boffetta et al. [9], a mean concentration of 0.044 mg/m^3 is estimated from a 45-year exposure at 1.98 mg/m^3 x yr (the cohort median cumulative exposure). Insufficient information was available in Fryzek et al. [10] to estimate a mean concentration, and the excess risk was estimated at the mean concentration from Boffetta et al. [9].

The 15-year lag model for cumulative exposure was used in Fryzek et al. [10]. Cumulative exposure group midpoints were used to fit the linear relative rate models in the other studies, and for Boffetta et al. [9], a value of 78.1 mg/m^3 x yr was assumed for the highest category.

A human lung dosimetry model (multiple-path particle deposition model, or MPPD) [25] was used to estimate the retained particle lung burdens after 45-years of exposure at these mean concentrations (assuming 8 hrs/day, 5 days/wk, 50 wks/yr). The Yeh-Schum deposition model within MPPD was selected, along with the default model parameters, except for the following, which represent occupational exposures: breathing frequency (17.5 breaths/min) [26] and tidal volume (1143 ml, to yield a minute volume of 20 L/min) [26]. In addition, breathing scenario was assumed to be oronasal-normal augmenter, and the inhalability adjustment was selected. Particle sizes and input parameter values are provided in Appendix A.

2.2. Rat data

Chronic inhalation studies in rats were selected on the same types of particles as those in the epidemiological studies. The rat studies included: coal dust [27, 28]; carbon black [3, 29], titanium dioxide [3, 30, 31]; and crystalline silica [31]. The dose-response data analyzed include the retained lung burden after two years of inhalation exposure and the lung tumor proportion, excluding squamous cell keratinizing cystic tumors (which have been considered a noncancerous lesion in rats [32]). The particle mass lung dose was used for coal dust, carbon black, and silica. For titanium dioxide, data were available in both fine and ultrafine particle sizes, so the mass lung doses of titanium dioxide were converted to particle surface area doses (using specific surface area) in order to utilize the data available for both particle sizes.

The rat dose-response data were fit with a multistage model, using the benchmark dose software (BMDS) [33]:

$$P(D) = 1 - \exp(-Q0 - Q1*D - Q2*D^2) \tag{4}$$

where P(D) is the probability of lung tumors at a given dose; Q0 is the coefficient for the background tumor rate; Q1 and Q2 are the coefficients for dose; and D is dose (particle mass or surface area lung burden). Multistage models have been used for many years in cancer risk assessment. A 2nd degree polynomial model form was used to allow for nonlinearity in the dose-response curve while not over-fitting the data, and the coefficients were restricted to be nonnegative. Male and female rat data were generally combined, except for carbon black, where only the female rat data were used [3, 29] due to the heterogeneity in the data from the lack of an exposure-related increase in lung tumors in male rats [29]. Model fits to the dose-response data were considered adequate at p-values > 0.05 (goodness of fit test).

The rat-based excess risks associated with the doses of interest from the human studies were estimated by linear extrapolation below the 10% BMD (estimated by the multistage model; target BMD was derived from the slope of the line from 10% BMD to zero). Dose was normalized between humans and rats based on the ratio of alveolar lung surface area [34] – using the values of 102.2 and 0.4 m^2, respectively [35]. An assumption of the interspecies dose extrapolation is that equal responses will occur at an equivalent lung dose [36, 37], either the total mg or m^2 of particles/g lung. A two-year inhalation exposure in the chronic studies in rats and a 45-year working lifetime in humans were assumed to be equivalent durations of exposure relative to an average lifetime in each species.

2.3. Comparison of rat and human excess risk estimates

The rat- and human-based maximum likelihood estimate (MLE) of excess risk and the 95% upper confidence limit (95% UCL) were computed using the models for cumulative exposure reported in the epidemiology studies and in rats at the equivalent lung burdens (Sections 2.1 and 2.2). The rat- and human-based excess risk estimates were quantitatively compared using a statistical test of the hypothesis that the animal and human-based excess risks are equal. The test statistic is the difference in these risk estimates divided by the standard error of the difference. P-values are based on the standard normal distribution.

Excess risk estimates for diesel exhaust particulate, based on the rat and human studies reported in Stayner et al. [6], are provided here in graphical form. These estimates are expressed as unit cancer risk, which is defined as the excess risk of lung cancer after 45-yr occupational exposure to 1 $\mu g/m^3$ of diesel exhaust particulate.

3. Results

All multistage models provided adequate fit to the rat dose-response data (p-values 0.3 – 1.0; goodness of fit test). For carbon black, the initial models including male and female rat data from Nikula et al. [29] did not provide adequate fit (p-values 0.01 – 0.02); and models applied to the male rat data alone did not converge due to the lack of a positive dose-response relationship. Models applied to the female rat data from Nikula et al. [29] and Heinrich et al. [3] (which used female rats only) provided adequate fit (p-values 0.3-0.8). All human models evaluated provided adequate fit (p-values >0.05).

The rat- and human-based excess risk estimates are provided in Table 1 for the poorly-soluble, low toxicity particles (coal mine dust, carbon black, and titanium dioxide) and in Table 2 for respirable crystalline silica. Statistical comparison of these rat- and human-based excess risk estimates show that none of the excess risk estimates based on rat or human data is significantly different (all p-values $\geq$ 0.05).

Regarding the magnitude of the excess risk estimates, the rat-based MLEs were clearly higher than the human-based estimate for coal dust (which was negative); however, the rat-based estimates (MLEs and 95% UCLs) did not exceed the 95% UCL from the human study (Table 1). For carbon black, the rat-based excess risk estimates exceeded those from the human study, but the differences were not statistically significant. For titanium dioxide, the rat-based excess risk estimates (MLE and 95% UCL) were lower than the 95% UCL of the human studies, although the MLE from Fryzek et al. [10] was negative. For crystalline silica, the rat-based MLEs and 95% UCLs were very similar in magnitude to the human-based estimates MLEs and 95% UCLs from all three epidemiological studies, except for the negative MLE from Brown and Rushton [18] (Table 2).

For diesel exhaust particulate, the rat- and human-based excess risk estimates for lung cancer are shown in Figure 2. Unit cancer risks are shown (i.e., mean concentration of 1 $\mu g/m^3$ diesel exhaust particulate over a 45-yr occupational exposure). The rat 95% UCL are not shown because they are not sufficiently different from the MLEs to be discernable on this log scale. Figure 2 shows that, in general, the rat studies underpredict the excess risks based on the human studies.

Table 1. Comparison of human- and rat-based excess risk estimates of lung cancer following 45-year working lifetime exposure by particle type and mean concentration.

Particle type and human study	Mean concentration (mg/m^3) [a]	MLE (95% UCL) Human	MLE (95% UCL) Rat [b]	Comparison of human & rat excess risks (p-value)
Coal mine dust Attfield and Kuempel [14]	1.4	-2.5×10^{-3} (6.6×10^{-3})	4.1×10^{-3} (1.0×10^{-2})	0.3
Carbon black Sorahan et al. [4]	0.83	5.6×10^{-3} (1.1×10^{-2})	1.4×10^{-2} (2.0×10^{-2})	0.09
Titanium dioxide Fryzek et al. [10]	0.044	-9.4×10^{-4} (6.8×10^{-4})	7.0×10^{-5} (8.2×10^{-5})	0.2
Titanium dioxide Boffetta et al. [9]	0.044	4.4×10^{-4} (7.1×10^{-4})	7.0×10^{-5} (8.2×10^{-5})	0.1

Abbreviations: Maximum likelihood estimate (MLE); 95% upper confidence limit (95% UCL).
[a] Estimated from the cohort mean or median cumulative exposure averaged over 45 years of exposure.
[b] Rat data from: coal dust [27, 28]; carbon black [3, 29] (female rats); titanium dioxide [3, 30, 31].

Table 2. Comparison of human- and rat-based excess risk estimates of lung cancer following 45-year working lifetime exposure to respirable crystalline silica.

Human study	Mean concentration (mg/m^3) [a]	MLE (95% UCL) Human	MLE (95% UCL) Rat [b]	Comparison of human & rat excess risks (p-value)
Rice et al. [15]	0.162	3.8×10^{-2} (8.0×10^{-2})	3.7×10^{-2} (6.5×10^{-2})	1.0
Attfield and Costello [16]	0.047	1.3×10^{-2} (2.0×10^{-2})	1.1×10^{-2} (1.9×10^{-2})	0.8
Brown and Rushton [18]	0.025	-7.8×10^{-4} (1.0×10^{-2})	5.7×10^{-3} (1.0×10^{-2})	0.3

Abbreviations: Maximum likelihood estimate (MLE); 95% upper confidence limit (95% UCL).
[a] Estimated from the cohort mean cumulative exposure averaged over 45 years of exposure: 7.3 $mg/m^3 \times yr$ [15]; 2.2 $mg/m^3 \times yr$ [16]; 1.12 $mg/m^3 \times yr$ [18].
[b] Rat data from Muhle et al. [31].

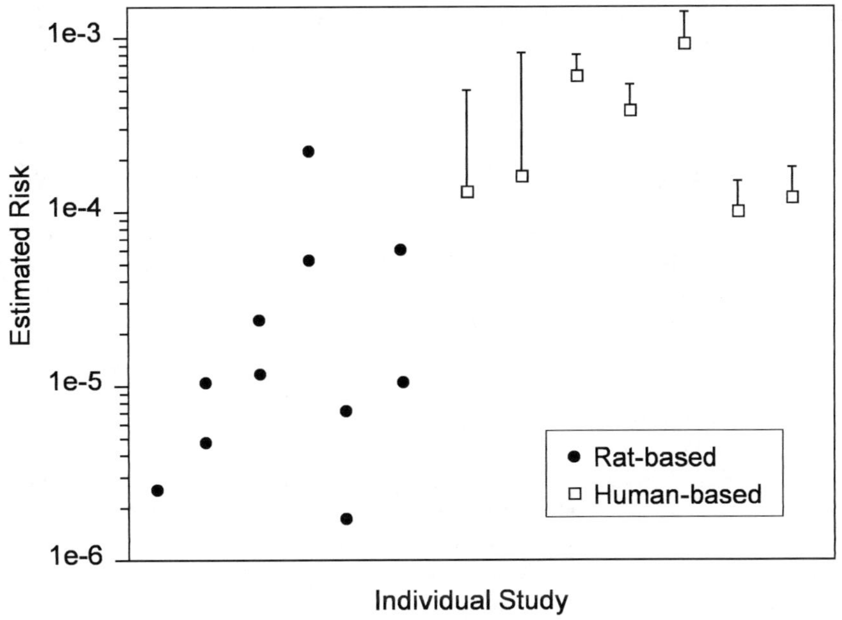

Figure 2. Diesel exhaust particulate and lung cancer: comparison of rat- and human-based excess risk estimates across studies (estimates from Stayner et al. [6]). Maximum likelihood estimates are shown for both rat and human studies; 95% upper confidence limits are shown for the human studies.

4. Discussion

The findings of the epidemiological studies have been mixed with regard to elevated lung cancer and occupational exposure to poorly-soluble particles. All of the studies with cumulative exposure data have been included in this study, whether or not a relationship between particle exposure and lung cancer was observed. In contrast, each of the rat studies found statistically-significant elevations in lung tumors after chronic inhalation exposure to coal dust, carbon black, titanium dioxide, diesel exhaust particulate, or silica.

It may be hypothesized that if the rat is not a reasonable model of particle-elicited lung cancer, e.g., due to a rat-specific oversensitive lung response to particle overload, then the rat-based risk estimates may exceed those based on the human data. If so, then risk estimates based on the rat dose-response data may over-predict the unknown, true excess risks of particle-elicited lung cancer in humans.

The purpose of this study was to quantitatively and statistically compare the rat- and human-based excess risk estimates of lung cancer for working lifetime exposures to each of these particle types. This analysis provides comparisons to assess whether the rat-based risk estimates over- or under-predict the human lung cancer risk from exposure to poorly-soluble particles of various types. These findings may be relevant to other poorly-soluble particles for which dose-response data are available in rats but not humans.

4.1. Quantitative evaluation

This analysis showed statistically consistent excess risk estimates based on the rat and human studies. That is, there were no statistically significant differences in the rat- and human based estimates of lung cancer excess risk from long-term exposure to poorly-soluble respirable particles (all p values > 0.05).

Among the poorly-soluble low toxicity particles, the rat-based MLE estimates for coal dust and titanium dioxide were lower than the human 95% UCL, even when the human study did not show an association between particle exposure and lung cancer. For carbon black, the rat-based MLE estimate was higher than the human 95% UCL, although not significantly. In the coal miner study [14], a potential exists for bias in the exposures associated with the SMRs including for lung cancer. Since the cumulative exposure data were available only until the start of follow-up, the cumulative exposure assigned to each miner is biased downward, and for the younger miners with shorter tenures at start of follow-up, the relative bias could be substantial. For this reason, possible bias in the excess risk estimates from the human study (Table 1) cannot be dismissed, which could also influence the comparison with the rat data.

For crystalline silica, the rat- and human-based excess risk estimates (MLEs and 95% UCLs) were remarkably similar. Although only one chronic inhalation study in rats exposed to respirable crystalline silica was available (and only a linear model was possible due to the limited data), the epidemiological studies were based on three different occupational cohorts. Unlike other poorly-soluble particles with low inherent toxicity, crystalline silica is cytotoxic, and significant exposure-response relationships have been observed in the rat [31] and human studies [16, 17]. Brown and Rushton [18] did not observe a positive exposure-response relationship, which may be due to the low exposure concentrations and durations.

For diesel exhaust particulate, the rat-based excess risk estimates generally under-predicted those based on the human studies. While the rat dose-response relationships to diesel exhaust particulate and carbon black appear to be similar [3, 29], it may be that the human excess risk estimates include additional risk due to the carcinogenic activity of the adsorbed poly-aromatic hydrocarbons on the diesel exhaust particulate, or that other unmeasured factors in the humans studies increases their apparent exposure-attributable risk.

4.2. Study limitations

These rat- and human- based excess risk comparisons are limited to exposures that are consistent with the human studies. Because the models used to describe human excess risk based on the human data differ from the models based on the rat data, it is possible that some comparisons could yield nonsignificant differences over one range of exposures and significant differences over another range for a given set of data. However, the practical implication of this limitation may be modest for comparisons of risk estimates at the exposures reported in the human studies.

Another limitation of this analysis is the relatively low power of the epidemiological studies. Human studies with large variability in the excess risk estimates may be consistent with a wide range of possible risk estimates, and even a non-significant exposure coefficient could be consistent with a positive exposure-response relationship in studies with large imprecision [19]. Thus, although this analysis showed that the rat- and human-based excess risk estimates are statistically consistent, the large variability in the human excess risk estimates may result in low power to detect a true difference in the animal- and human- based risk estimates.

The animal data were also relatively sparse due to relatively few dose groups (which were nonetheless typical of chronic bioassays); the data for silica and coal dust were limited to one and two dose groups, respectively, in addition to controls. Thus, while the standard errors of the human-based excess risk estimates are typically larger than those from the animal studies (by approximately an order of magnitude), the standard errors are similar for the rat- and human-based risk estimates for silica and coal dust. More dose-response data in both animals and humans would be needed to reduce the uncertainty in these excess risk comparisons.

There is uncertainty in what model would best describe the rat dose-response relationship for particle lung burden and tumor response. The rat-based excess risk estimates were based on multistage stage model with linear extrapolation below the 10% BMD. The dose-response relationships in the range of the data are clearly nonlinear for the poorly-soluble particles with low inherent toxicity (e.g., titanium dioxide and carbon black). Models and methods that account for low-dose curvature in the rat data (e.g., other BMD models or Bayesian model averaging of those models [38]) may provide different rat-based excess risk estimates, which may be lower. However, since the rat-based estimates from the multistage model were generally below the human-based risk estimates (and none was significantly different), it is unlikely that these other models would alter the overall findings.

In addition, the risk estimates may vary depending on the choice of lung dosimetry model and parameter values. For example, predicted human mean lung burdens associated with occupational exposures varied by several-fold [39] between the ICRP [26] clearance model (used in the MPPD model [25]) and those developed by Kuempel et al. [40] and Tran and Buchanan [41] based on coal miner data. The values selected to normalize lung burden between rat and human can also influence the risk estimates (e.g., lung mass or lung surface area), including whether the mean or distribution of these values is considered.

4.3. Qualitative evaluation

Although the mechanisms of particle-induced lung disease have not been fully elucidated, the rat and human lung responses to some poorly-soluble particles (e.g., coal dust and silica) appear to be qualitatively similar in many of the key steps for which there are data [23, 42, 43]. In humans, workers in dusty jobs had elevated pulmonary inflammation markers [44, 45], and in coal miners a relationship between dust exposure, pulmonary inflammation, and fibrosis was observed [46]. Although not linked to particle exposure, individuals with chronic pulmonary inflammation and epithelial cell proliferation due to idiopathic pulmonary fibrosis have an increased risk of lung cancer [47].

Semi-quantitative comparisons of the lung responses in rats and humans to several types of particles (diesel exhaust, coal dust, silica, or talc) indicate both similarities and differences – including similar regions of particle retention in the lungs but at different proportions [48], and greater fibrotic response in humans but greater inflammation and epithelial hyperplasia responses in rats [49]. Mice and hamsters have been shown to have lower lung tumor response and to give false negatives in bioassays for some particulates that have been classified by IARC as human carcinogens (limited or sufficient evidence), including crystalline and diesel exhaust [2].

Lung clearance of particles is slower in humans than in rats, by approximately an order of magnitude [26, 50], and some humans in dusty jobs (e.g., coal miners) have exposures and lung burdens that would be considered overloaded in rats [51]. Thus, the doses that cause overloading in rats may be relevant to estimating disease risk in workers with high dust exposures. However, the lack of exposure-related elevation in lung cancer rates in coal miners raises questions about whether the rat model would overpredict disease risk in humans.

Comparison of the lung doses of rats and humans shows that although the mean lung burdens have been relatively high in coal miners historically (approximately 10-20 mg/g lung) [40, 41], these lung burdens are several times lower than the mean lung burdens associated with the excess lung tumors in rats. For example, in rats chronically exposed to coal mine dust, a lung burden of 57 mg/g was associated with 11% lung tumors (vs. 0% in unexposed controls) [27]. In rats exposed to fine TiO_2, lung burdens up to 55 mg/g were not associated with lung tumors, and lung tumors were elevated only in rats with lung burdens greater than 200 mg/g (approximately 16% in male and female rats, excluding keratinizing cystic tumors) [30]. Thus, as seen in rat studies, doses that cause overloading may not be high enough to cause lung cancer. However, particle size and surface reactivity are also important, and smaller mass doses of ultrafine particles (e.g., TiO_2) or cytotoxic particles (e.g., quartz) are associated with elevated lung tumor response in rats [3, 31].

5. Conclusions

The rat- and human-based excess risk estimates of lung cancer from long-term exposure to poorly-soluble respirable particles are statistically consistent based on available data. Residual uncertainty in whether the rat-based risk estimates would over- or under-predict the true excess risks of lung cancer in humans is due in part to the low power of the available human studies, limited particle size exposure data for humans, and ambiguity about the best animal models and extrapolation methods.

Disclaimer: The findings and conclusions in this paper are those of the authors and do not necessarily represent the views of the National Institute for Occupational Safety and Health.

References

[1] Driscoll KE 1996 *Particle overload in the rat lung and lung cancer, implications for human risk assessment* eds JL Mauderly and RJ McCunney (Washington, DC: Taylor and Francis) pp 139-53
[2] Mauderly JL 1997 *Environ. Health. Perspect.* **105**(Suppl 5) 1337–46
[3] Heinrich U, Fuhst R, Rittinghausen S, Creutzenberg O, Bellmann B, Koch W and Levsen K 1995 *Inhal. Toxicol.* 7(4) 533–556
[4] Sorahan T and Harrington JM 2007 *Am. J. Indust. Med.* **50**(8) 555-64
[5] Wellmann J, Weiland SK, Neiteler G, Klein G and Straif K 2006 *Occup. Environ. Med.* **63**(8) 513-21
[6] Stayner L, Dankovic D, Smith R and Steenland K 1998 *Am. J. Indust. Med.* **34**(3) 207-19
[7] Steenland K, Deddens J and Stayner L 1998 *Am. J. Ind. Med.* **34**(3) 220–28
[8] Garshick E, Laden F, Hart JE, Rosner B, Smith TJ, Dockery DW and Speizer FE 2004 *Environ. Health. Perspect.* **112**(15):1539–43
[9] Boffetta P, Soutar A, Cherrie JW, Granath F, Andersen A, Anttila A, Blettner M, Gaborieau V, Klug SJ, Langard S, *et al.* 2004 *Cancer Causes Control* **15** 697–706
[10] Fryzek JP, Chadda B, Marano D, White K, Schweitzer S, McLaughlin JK and Blot WJ 2003 *J. Occup. Environ. Med.* **45** 400–9
[11] Dell LD, Mundt KA, Luippold RS, Nunes AP, Cohen L, Burch MT, Heidenreich MJ and Bachand AM 2006 *J. Occup. Environ. Med.* **48**(12) 1219-29
[12] Miller BG and Jacobsen M 1985 *Br. J. Indust. Med.* **42**(11) 723-33
[13] Morfeld P, Lampert K, Ziegler H, Stegmaier C, Dhom G and Piekarski C 1997 *Appl. Occup. Environ. Hyg.* **12**(12) 909-14
[14] Attfield MD and Kuempel ED 2008 *Am. J. Indust. Med.* **51**(4) 231-45
[15] Rice FL, Park R, Stayner L, Smith R, Gilbert S and Checkoway H 2001 *Occup. Environ. Med.* **58**(1) 38-45
[16] Attfield MD and Costello J 2004 *Am. J. Indust. Med.* **45**(2) 129-38
[17] Steenland K, Mannetje A, Boffetta P, Stayner L, Attfield M, Chen J, Dosemeci M, DeKlerk N, Hnizdo E, Koskela R, *et al.* 2001 *Cancer Causes Control* **12**(9) 773-84
[18] Brown TP and Rushton L 2005 *Occup. Environ. Med.* **62**(7) 446-52
[19] Stayner LT and Smith RJ 1993 *Epidemiol. Prev.* **14**(53) 32-9
[20] Watson AY and Valberg PA 1996 *Particle overload in the rat lung and lung cancer, implications for human risk assessment* eds JL Mauderly and RJ McCunney (Washington, DC: Taylor and Francis) pp 227–57
[21] ILSI 2000 *Inhal. Toxicol.* **12**(1-2) 1-17
[22] Hext PM, Tomenson JA and Thompson P 2005 *Ann. Occup. Hyg.* **49**(6) 461-72
[23] Baan RA 2007 *Inhal. Toxicol.* **19**(Suppl. 1) 213-28
[24] BEIR IV 1998 *Health risks of radon and other internally deposited alpha-emitters* (Washington, DC: National Academy Press) pp 131-36
[25] CIIT and RIVM 2006 *Multiple-path particle dosimetry model (MPPD V 2.0, 2002-2006)* (Research Triangle Park, NC: Centers for Health Research and The Netherlands: National Institute for Public Health and the Environment).
[26] ICRP 1994 *Human respiratory tract model for radiological protection (Annals of the ICRP* Publication No. 66*)* ed H Smith (Tarrytown, New York: International Commission on Radiological Protection)
[27] Martin JC, Dániel H and LeBouffant L 1977 *Inhaled particles* part 1 (Oxford, UK: Pergamon Press) 361–70
[28] Lewis TR, Morrow PE, McClellan RO, Raabe OG, Kennedy GL, Schwetz BA, Goehl TJ,

Roycroft JH and Chhabra RS 1989 *Toxicol. Appl. Pharmacol.* **99**(3) 377–83
[29] Nikula KJ, Snipes MB, Barr EB, Griffith WC, Henderson RF and Mauderly JL 1995 *Fundam. Appl. Toxicol.* **25** 80–94
[30] Lee KP, Trochimowicz HJ and Reinhardt CF 1985 *Toxicol. Appl. Pharmacol.* **79** 179–92
[31] Muhle H, Bellmann B, Creutzenberg O, Dasenbrock C, Ernst H, Kilpper R, MacKenzie JC, Morrow P, Mohr U, Takenaka S, Mermelstein R 1991 *Fund. Appl. Toxicol.* **17** 280–99
[32] Boorman GA, Brockman M, Carlton WW, Davis JMG, Dungworth DL, Hahn FF, Mohr U, Reichhelm H-BR, Turusov VS, Wagner BM 1996 *Toxicol. Pathol.* **24** 564–72
[33] US EPA 2003 *Benchmark dose software*, version 1.3.2 (Washington, DC: US Environmental Protection Agency, National Center for Environmental Assessment)
[34] US EPA 2005 *Guidelines for carcinogen risk assessment* (Washington, DC: US Environmental Protection Agency, Risk Assessment Forum) EPA/630/P-03/001F
[35] Mercer RR, Russell ML, Roggli VL and Crapo JD 1994 *Cell. Mol. Biol.* **10** 613-24
[36] Jarabek AM, Asgharian B and Miller FJ 2005 *Inhal. Toxicol.* **17**(7-8) 317-34
[37] Kuempel ED, Tran CL, Castranova V and Bailer AJ 2006 *Inhal. Toxicol.* **18**(10) 717-24
[38] Wheeler MW and Bailer AJ 2007 *Risk Anal.* **27**(3) 659-70
[39] Kuempel ED and Tran CL 2002 *Ann. Occup. Hyg.* **46**(Suppl 1) 337–41
[40] Kuempel ED, O'Flaherty EJ, Stayner LT, Smith RJ, Green FHY and Vallyathan V 2001 *Reg. Toxicol. Pharmacol.* **34**(1) 69-88
[41] Tran CL and Buchanan D 2000 *Development of a biomathematical lung model to describe the exposure-dose relationship for inhaled dust among U.K. coal miners* (Edinburgh, UK: Institute of Occupational Medicine) IOM Research Report TM/00/02
[42] Vallyathan V, Shi X and Castranova V 1998 *Environ. Health Perspect.* **106**(suppl 5) 1151-55
[43] Castranova V 2000 *Inhal. Toxicol.* **3** 7–14
[44] Rom WL 1991 *Am. J. Indust. Med.* **19** 15–27
[45] Lapp NL and Castranova V 1993 *Occup. Med.* **8**(1) 35–56
[46] Kuempel ED, Attfield MD, Vallyathan V, Lapp NL, Hale JM, Smith RJ and Castranova V 2003 *J. Biosci.* **28**(1) 61-9
[47] Katabami M, Dosaka-Akita H, Honma K, Saitoh Y, Kimura K, Uchida Y, Mikami H, Ohsaki Y, Kawakami Y and Kikuchi K 2000 *Am. J. Respir. Crit. Care Med.* **162** 295–300
[48] Nikula KJ, Vallyathan V, Green FH and Hahn FF 2001 *Environ. Health Perspect.* **109**(4) 311-8.
[49] Green FH, Vallyathan V and Hahn FF 2007 *Toxicol. Pathol.* **35**(1) 136-47
[50] Snipes MB 1996 *Particle overload in the rat lung and lung cancer, implications for human risk assessment* eds JL Mauderly and RJ McCunney (Washington, DC: Taylor and Francis) pp 91-109
[51] Freedman AP and Robinson SE 1988 *Respirable dust in the mineral industries: health effects, characterization, and control* eds RL Frantz and RV Ramani (University Park, PA: The Pennsylvania State University) pp 181-6
[52] Roller M and Pott F 2006 *Ann. N.Y. Acad. Sci.* **1076** 266-80
[53] Vallyathan V, Schwegler D, Reasor M, Stettler L, Clere J and Green FHY 1988 *Ann. Occup. Hyg.* **32**(suppl 1) 279-89

Journal of Physics: Conference Series **151** (2009) 012011 doi:10.1088/1742-6596/151/1/012011

Appendix A.

The particle parameters used in the lung dosimetry modeling, and the lung burden predictions in rats and humans are provided in Table A-1.

Table A-1. Particle characteristics used in lung burden estimation.

Particle Type	Density[a]	MMAD (GSD)[b]	Mean concentration (mg/m^3)[c]	Human lung burden (mg/lung)	Rat-equivalent lung burden (mg/lung)[d]
Coal dust	1.5	3.4 (2)	1.4	1,159	4.5
Carbon black	1.85	0.64 (2.06)	0.83	884	3.4
TiO_2 - F	4.25	1.6 (2)	0.044	52	0.20
SiO_2	2.5	3.5 (2)	0.16	135	0.53
			0.047	39	0.15
			0.025	21	0.082

Abbreviations: TiO_2-F: titanium dioxide (fine); SiO_2: crystalline silica; MMAD: mass median aerodynamic diameter; GSD: geometric standard deviation.

[a] Reported by various internet sources and consistent with values in Roller and Pott [52].

[b] Reported in: Vallyathan et al. [53]: coal dust and SiO_2, MMAD; Heinrich et al. [3]: carbon black MMAD and GSD; Lee et al. [30]: TiO_2-F, MMAD. Assumed GSD of 2 for coal dust, TiO_2-F, and SiO_2, which is similar to values reported for the other particles.

[c] Mean concentration equivalent to the cohort mean or median cumulative exposure averaged over 45-year working lifetime (8 hr/d, 5 d/wk, 50 wk/yr).

[d] Normalized from human to rat by the ratio of lung surface areas (0.4 rat/102.2 human) [35].

Journal of Physics: Conference Series **151** (2009) 012012 doi:10.1088/1742-6596/151/1/012012

Occupational characteristics of respiratory cancer patients exposed to asbestos in Lithuania

R Petrauskaitė Everatt[1,3], G Smolianskienė[1,5], A Tossavainen[2], S Cicėnas[3,4] and R Jankauskas[1]

[1] Institute of Hygiene, Etmonu 3/6, LT-01129 Vilnius, Lithuania

[2] Finnish Institute of Occupational Health, Topeliuksenkatu 41a A, FI-00250 Helsinki, Finland

[3] Institute of Oncology, Vilnius University, Santariškių 1, LT-08660 Vilnius, Lithuania

[4] Institute of Rehabilitation, Sport Medicine and Nursing, Antakalnio 57, LT-10207, Vilnius, Lithuania

[5] Author to whom correspondence should be addressed. E-mail: *grazina.smolianskiene@dmc.lt*

Key words: asbestos; exposure assessment.

Abstract. Objective: To assess characteristics of asbestos exposure in respiratory cancer patients in Lithuania. **Methods.** Information on occupational exposure to asbestos was collected by personal interviews and occupational characteristics were evaluated among 183 lung cancer and mesothelioma patients with cumulative asbestos exposure ≥0.01 fibre years hospitalized at the Institute of Oncology, Vilnius. Additionally, some results of workplace air measurements were reviewed. **Results.** Cases with estimated cumulative exposure ≥5 fibre years had worked mainly in the construction industry (49%), installation and maintenance (13%), foundry and metal products manufacturing (6%), heating trades and boilerhouses (6%) as fitters/maintenance technicians, construction workers, welders, electricians or foremen. Typical asbestos materials used by the patients were asbestos powder, asbestos cement sheets and pipes, asbestos cord, brake and clutch linings. Patients were exposed to asbestos when insulating boilers, furnaces, pipes in power stations, industrial facilities, ships, locomotives, buildings, while covering and repairing roofs, at the asbestos cement plant or unloading asbestos products. Most patients with estimated cumulative exposure of ≥0.01-4.9 fibre years worked as lorry, bus or tractor drivers and motor vehicle mechanics. In 2002-2007 workplace air asbestos concentrations exceeded the limit value of 0.1 f/cm^3 in 11 samples out of 208 measurements. **Conclusion.** The results of this study indicate that since the 1960s occupational exposure to chrysotile asbestos was extensive in Lithuania.

1. Introduction

Since the year 1961 in Lithuania about 700,000 tons of raw chrysotile asbestos were used for the production of asbestos cement sheets and pipes, for the insulation of boiler stations and pipes and as friction materials [1,2]. All asbestos was imported from Russia. There were two factories manufacturing asbestos cement products as well as four main shipyards situated in western Lithuania. The production of corrugated asbestos cement sheets started in 1956. The annual output rose to over 100 million asbestos cement sheets annually in the 1970s, remaining high until 1991 and then declining (figure 1) [3].

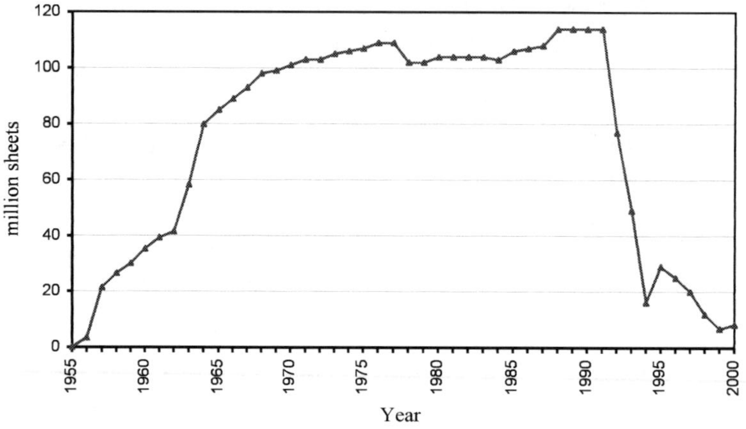

Figure 1. Production of corrugated asbestos cement sheets in Lithuania.

The first regulations on the production and use of asbestos were passed in 1998. Asbestos import and use were banned completely in 2004. The current limit for the concentration of asbestos fibres in the workplace air is 0.1 f/cm^3 for all types of asbestos [4].

In Lithuania no cancer case has been diagnosed or compensated for as an asbestos-related occupational disease so far, although about 1500 new lung cancer cases and 10 mesotheliomas are registered annually [5,6]. Few explanations for this situation were suggested, including insufficient recognition among workers and physicians of the occupational etiology of asbestos-related cancers and unclear criteria for occupational disease certification [5]. There is little information about the past occupational exposure to asbestos in Lithuania in relation to duration, latency period and distribution of asbestos-exposed workers over industries and about the sources of definite, probable and possible exposure to asbestos. Such knowledge could be useful for informing workers on their individual risk or its assessment, planning preventive measures, medical follow-up and for occupational disease diagnosis.

The aim of this study is to assess characteristics of asbestos exposure in respiratory cancer patients in Lithuania.

2. Methods

Information on occupational exposure to asbestos, as well as demographic characteristics and smoking habits, was collected by personal interviews of 302 lung cancer and mesothelioma patients who were hospitalized at the Institute of Oncology, Vilnius, between November 1, 2003 and May 15, 2005. Here we provide a description of occupational data on 183 persons exposed to asbestos. From the occupational history of each patient, the main source of asbestos exposure was assessed. The duration of exposure and time since the first exposure was evaluated based on dates of employment if not indicated otherwise by the patient. Exposure assessments were based on job descriptions and the German BK-report "Faserjahre" [7]. For diagnostic and compensation purposes this documentation is used to evaluate the cumulative lifetime exposure of a patient with suspected occupational disease. A dosis exceeding 25 fibre years/cm^3 has been estimated to cause a two-fold risk of lung cancer [7]. The cumulative exposures were classified as 0.01 – 4.9 fibre years, 5 – 24.9 fibre years, and ≥25 fibre years. Nonexposed persons or persons with an estimated exposure to asbestos of <0.01 fibre years (119 persons) were not included into the present study.

Measurements of asbestos fibre concentrations in the air for the period 2002-2007 were performed at the Laboratory of Chemical Hazards Investigation of Hygiene Institute using phase–contrast optical microscopy.

3. Results

Details of 302 study participants and results of the assessment of their cumulative exposure to asbestos as well as findings of fibre burden analysis on 23 lung tissue samples were reported previously [8]. In this paper we focused on 183 patients for whom the estimated cumulative exposure to asbestos was ≥ 0.01 fibre years.

Among the group of 67 patients (65 men and 2 women) with cumulative exposure ≥ 5 fibre years, most cases had worked in the construction industry (49%, 33 cases), the installation, maintenance and metal constructions industry (13%, 9 cases), heating trades and boilerhouses (6%, 4 cases), and foundry and metal products manufacturing (6%, 4 cases). The rest had worked in asbestos cement production, shipbuilding, glass factories, power stations or railways. Fitters/maintenance technicians (19%), construction workers (13%), welders (9%), electricians (8%) and foremen (6%) were the most frequent occupations in this group (table 1). The majority of patients (94%) had started their work with asbestos exposure before the year 1990. The mean time between the first asbestos exposure and the disease diagnosis was 34 years (min 9, max 53). A majority of patients (76%) had worked with asbestos for more than 10 years. Typical asbestos materials used by the patients were asbestos powder, corrugated asbestos cement sheets, asbestos cement pipes, asbestos cord and brake/clutch linings. The patients were exposed to asbestos when insulating boilers, furnaces, pipes in power stations, industrial facilities, ships, locomotives and buildings.

Table 1. Industries, occupations and materials reported by 67 patients with a cumulative exposure of ≥5 fibre years.

Industry	Occupation	Asbestos materials	No of patients
Asbestos cement or building materials	Fitter/maintenance technician[a]	Asbestos powder, asbestos cement	1
	Workman	Pipes, corrugated asbestos cement sheets; bystander (asbestos dust)	2
Construction	Warehouse worker, loader	Corrugated asbestos cement sheets, asbestos powder in sacks, asbestos cement products	2
	Bricklayer	Corrugated asbestos cement sheets, pipes, asbestos powder, asbestos cord, sheets, other asbestos cement products	6
	Plasterer	Asbestos sheets, asbestos powder, corrugated asbestos cement sheets	1
	Welder	Asbestos sheets, cord, powder, asbestos cement products, asbestos insulation removal	4
	Joiner-woodworker	Corrugated asbestos cement sheets	1
	Tractor driver, crane operator	Asbestos cord, asbestos cement products, brakes, clutch, bystander	2
	Builder; laborer	Corrugated asbestos cement sheets, seals, cord, cardboard, asbestos powder, asbestos insulation removal	9

	Foreman	Asbestos cord, powder, removal of asbestos insulation, asbestos cement pipes, corrugated asbestos cement sheets	4
	Insulator[a]	Asbestos powder, asbestos insulation removal	1
	Construction smith	Asbestos powder, asbestos cardboard, sheets	1
	Electrician, pipe fitter	Asbestos cord, sheets, cardboard, pipes, removal of insulation, laminate	2
Installation, maintenance	Fitter/maintenance technician, assembler	Asbestos cord, cardboard, sheets, textile, removal asbestos insulation	3
	Plumber	Asbestos cord, pipes, powder	2
	Foreman[a]	Asbestos cord, sheets, cardboard, paronit, asbestos powder	1
	Concrete worker	Asbestos cord, asbestos pipes, corrugated asbestos cement sheets	1
Boilerhouse, heating company	Foreman[b], boiler operator	Asbestos sheets, cardboard, cord, bystander	4
Foundry, metal products or metal construction manufacturing	Fitter/maintenance technician[a]	Asbestos powder, cord, sheets, cardboard, asbestos. textiles, asbestos insulation removal	3
	Metal worker	Bystander	1
	Welder[a]	Asbestos powder, asbestos insulation materials	2
Power station	Electrician	Asbestos insulation on wires, asbestos powder	1
	Fitter/maintenance technician[a]	Asbestos powder, cord, sheets, patterns, cardboard, asbestos insulation removal	1
Railway	Fitter/maintenance technician[a]	Asbestos powder, sheets	1
Glass factory	Fitter/maintenance technician	Asbestos cord, cardboard, laminate	1
	Sheet metal worker[a]	Asbestos sheets, cords, seals	1
Shipyard	Smith, electrician[a]	Asbestos cords, cardboard, powder	1
Fishing ship	Ship mechanic	Asbestos gloves, cord, board	1
Motor vehicle repair	Fitter/maintenance technician	Brake linings, asbestos cord	1
Rubber factory	Electrician[a]	Asbestos powder	1
Port	Clearing and forwarding agent	Bulk asbestos in sacks	1
Other companies and industrial plants	Electrician	Asbestos cord, sheets, cardboard, textiles, pipes, insulation removal, laminate	2
	Fitter/maintenance Technician	Asbestos sheets, bystander	1
	Joiner	Corrugated asbestos cement sheets	1

[a] – heavily exposed patient (cumulative exposure ≥25 fibre years)
[b] – mesothelioma case

Heavily exposed patients (≥ 25 fibre years) had worked as fitter/maintenance technicians, welders, foremen, insulators or electricians in foundries, construction, installation, shipyard, asbestos-cement products manufacture, glass industry, power plant, maintenance of railway locomotives and in the chemical industry. The occupations and industries, as well as asbestos materials used by heavily exposed patients, are presented in Table 2. Their first occupational exposure to asbestos occurred during the period 1958-1978 (table 2).

Most of the 116 patients with estimated cumulative exposure of 0.01-4.9 fibre years worked as lorry, bus or tractor drivers (40%) and motor vehicle mechanics (13%). Welders, fitters/maintenance technicians or metal workers accounted for 12% of all patients in this group. The most common industries were construction, agriculture, transport and land reclamation, heating companies and power stations. The main sources of asbestos exposure were brake/clutch linings and corrugated asbestos cement sheets.

All patients who worked at shipyards or in the asbestos cement industry reported some exposure to asbestos or were bystanders. No interviewed patient mentioned spraying asbestos. One patient had removed sprayed asbestos while working abroad.

Table 2. Industry, occupation and asbestos materials reported by 10 heavily exposed patients with a cumulative exposure of ≥25 fibre years.

No	Industry	Occupation	Years of exposure	Asbestos materials used
1	Asbestos cement plant	Fitter/maintenance technician	1970-1976	Asbestos powder, asbestos cement
2	Installation, maintenance	Foreman, plumber	1978-1994	Asbestos powder, cord, sheets, cardboard, laminate
3	Construction	Foreman/insulator	1978-1992	Asbestos powder
4	Metal products manufacturing	Welder	1960-1971	Asbestos powder
	Construction	Plumber	1981-1983	Asbestos cord, asbestos powder
5	Power station	Fitter/maintenance technician	1958-2004	Asbestos powder, rope, sheets, forms, cardboard
6	Railway	Fitter/maintenance technician	1958-2000	Asbestos powder, board
7	Glass factory	Sheet metal worker	1967-1993	Asbestos sheets, cord, seals
8	Shipbuilding	Fitter/maintenance technician, welder, electrician	1965-1991	Asbestos cord, cardboard, powder
9	Foundry	Fitter/maintenance technician	1970-1991	Asbestos powder, cardboard, cord, sheets, textiles, insulation removal
10	Chemical industry	Electrician	1964-1974	Asbestos powder

Most of the 208 measurements of asbestos fibres in the workplace air during the period 2002-2007 were performed at the plants in the energy sector (64%) and construction (11%) (table 3). In 11 samples workplace air asbestos concentrations exceeded the limit value of 0.1 f/cm^3. High levels of asbestos fibres were found in the workplaces of insulators and plumbers. The highest concentration was measured in the construction industry during the demolition of structures containing asbestos (1.8 f/cm^3).

Table 3. Concentrations of asbestos fibres in workplace air in 2002-2007.

Economic activity of the plant	Occupation	Number of measurements exceeding 0.1 f/cm^3/total	Concentration of asbestos fibres in the workplace air, f/cm^3		
			Average	Min.	Max.
Energy	Insulator	3/133	0.03	< 0.01	0.63
Construction	Plumber	1/23	0.09	< 0.01	1.8
Machinery production	Smith-repairer, machine tool setter, pressman	3/16	0.04	< 0.01	0.19
Public service	Officer	0/13	< 0.01	< 0.01	< 0.01
Transport	Motor vehicle mechanic, pressman	2/12	0.05	< 0.01	0.13
Chemistry	Smith-repairer	2/7	0.06	< 0.01	0.15
Food production	Operator	0/4	< 0.01	< 0.01	< 0.01
All		11/208	0.04	< 0.01	1.8

4. Discussion

In our previous publication it was estimated that 3.4% of Lithuanian lung cancer cases could be attributed to heavy asbestos exposure at work (≥ 25 fibre years) [8].

The results of the present study confirm that chrysotile asbestos was widely used in Lithuania. Construction, maintenance, heating, boilerhouse and foundry workers constitute a big group among patients heavily/moderately exposed to asbestos (≥ 5 fibre years). Most patients with relatively low estimated exposure worked as lorry, bus or tractor drivers, motor vehicle mechanics, welders, fitters/maintenance technicians or metal workers. No patient reported spraying asbestos, and one patient had removed sprayed asbestos abroad. Our interview data suggests that the most common method of insulation was mixing asbestos powder with water and cement and its application on pipes, furnaces and other machinery. Workplace air of insulators and plumbers had the highest concentrations of asbestos fibres.

Our results are similar to those from a survey in Estonia where it was found that mechanics-locksmiths, machine operators, construction workers and thermal insulators were most exposed to asbestos. Thermal insulation was particularly dangerous, because asbestos fibres are readily released, producing airborne concentrations from 0.30-0.49 f/cm^3 [9].

As was discussed in our earlier publication, the strength of the present study is a detailed questionnaire, including 33 specific questions about job tasks, asbestos materials used and location, the duration and frequency of every job activity involving asbestos exposure presumably increased the accuracy of exposure assessment. However, some misclassification of exposure could occur, because the exact estimation of cumulative asbestos exposure was complicated by the high number of jobs in the past, and because the workers were not able to know their past exposures. Another limitation is the selection of the patients for the study. As few patients from the north or the west of Lithuania come to the Institute of Oncology for cancer treatment, the majority of the study patients (73%) were residents of the eastern part of Lithuania. The underestimation of patients from asbestos cement and shipbuilding industries cannot be ruled out. It is also difficult to evaluate the representativeness of measurement data of asbestos fibre concentrations in workplace air. Measurements were performed with various purposes, but it was not a specifically designed study with a carefully selected sample representative of all workplaces in the country. The number of asbestos fibre concentration measurements in the workplace air is rather small. Therefore, the results should be interpreted with caution. Despite this limitation, the results of asbestos measurements indicate that in some workplaces workers are still exposed to high asbestos concentrations that exceed permissible levels.

The results of this study have shown that occupational risks are important, in addition to the risks from smoking. Inquiring into them should be an important part in the medico-legal compensation of occupational diseases. This study may help to define more clearly diagnostic criteria of occupational lung cancer, to plan preventive measures and to inform workers about their individual risks.

Among most patients with cumulative asbestos exposure ≥ 5 fibre years, the first exposure to asbestos had occurred before the 1990s, when the import of raw asbestos and the manufacture of asbestos cement products was high. The decrease in the consumption of asbestos in Lithuania started about 20 years later than in Finland or the UK, and about 10 years later than in the Netherlands or Germany [10-12]. Therefore, it is likely that the incidence of asbestos-related diseases will still increase during the next years in Lithuania.

Acknowledgments

Partially the study was funded by Lithuanian State Science and Studies Foundation.

References

[1] Smailytė G, Filipauskienė J, Cicėnas S and Kurtinaitis S 2003 Pleural mesothelioma incidence in Lithuania *Public Health* **2**(21):61-5 (in Lithuanian)
[2] State Labour Inspection 2001 *Careful, Asbestos!* (Vilnius: Garnelis) p 32 (in Lithuanian)
[3] Department of Statistics to the Government of the Republic of Lithuania (Statistics Lithuania)

http://www.std.lt/web/main.php
[4] The Standard of Hygiene of Lithuania HN 23:2007 2007 The limit values of occupational exposure of chemical substances. The general requirements of measuring and risk assessment (exposure) *State news* **108**:4434
[5] Everatt R, Jankauskas R and Kučiauskaitė A 2006 Status of occupational cancer in Lithuania *Int. J. Occup. Environ. Health* **12** 24-7
[6] Lithuanian Cancer Registry 2005 http://www.is.lt/cancer-reg/
[7] BK-Report 1/97 1997 *Hauptverband der gewerblichen Berufsgenossenschaften. Faserjahre: Berufsgenossenschaftliche Hinweise zur Ermittlung der kumulativen Asbestfaserstaub-Dosis am Arbeitsplatz (Faserjahre) und Bearbeitungshinweise zur Berufskrankheit Nr 4104 Lungenkrebs-Kehlkopfkrebs* [Calculation of cumulative exposure to asbestos in the recognition of lung cancer as occupational disease]. Sankt Augustin, Germany
[8] Petrauskaitė Everatt R, Smolianskienė G, Tossavainen A, Cicėnas S and Jankauskas R 2007 Occupational asbestos exposure among respiratory cancer patients in Lithuania *Am. J. Ind. Med.* **50** 455-63
[9] Kangur M, Krooni P 2003 Asbestos-related health risks in Estonia *Int. J. Occup. Saf. Ergon.* **9**(1) 97-101
[10] Hagemeyer O, Otten H and Kraus T 2006 Asbestos consumption, asbestos exposure and asbestos-related occupational diseases in Germany *Int. Arch. Occup. Environ. Health* **79**(8):613-20
[11] Burdorf A, Dahhan M and Swuste P 2003 Occupational characteristics of cases with asbestos-related diseases in The Netherlands *Ann. Occup. Hyg.* **47**(6) 485-92
[12] Hodgson JT, McElvenny DM, Darnton AJ, Price MJ and Peto J 2005 The expected burden of mesothelioma mortality in Great Britain from 2002 to 2050 *Br. J. Cancer* **92** 587-93

Journal of Physics: Conference Series **151** (2009) 012013 doi:10.1088/1742-6596/151/1/012013

Tungsten Oxide Fiber Dissolution and Persistence in Artificial Human Lung Fluids

A B Stefaniak and M Chirila

National Institute for Occupational Safety and Health, Morgantown, WV, USA, 26505

E-mail: AStefaniak@cdc.gov

Abstract. Tungsten is a dense metal that is used in a range of industrial applications, including non-sag wire for light bulb filaments. During the conversion of tungsten oxide powder into tungsten metal powder for use as filaments, aerosols may be generated which contain tungsten sub-oxide particles having fiber morphology. To evaluate whether these fibers pose a yet unrecognized inhalation hazard due in part to their biodurability, we characterized the physicochemical properties and measured relative dissolution of fiber-containing ($WO_{2.81}$, $WO_{2.66}$, $WO_{2.51}$) and isometric-shaped ($WO_{3.00}$, $WO_{2.98}$) powders in artificial lung fluids. Raman spectroscopy results present a shift in the main frequencies for tungsten oxide samples that were sonicated in surfactant, confirming a decrease in the size of the crystalline domains by de-agglomeration. Geometric mean fiber aspect ratios were 8.3 ($WO_{2.81}$), 7.9 ($WO_{2.66}$), and 6.9 ($WO_{2.51}$). In artificial extracellular lung fluid, alkylbenzyldimethylammonium chloride (ABDC), added to prevent mold growth during experiments, inhibited ($p < 0.05$) dissolution of $WO_{2.98}$, $WO_{2.81}$, and $WO_{2.66}$. Less ($p < 0.05$) of the fibrous $WO_{2.66}$ and $WO_{2.51}$ dissolved relative to W metal; however, biodurability was only modestly greater than W metal. These data are useful for understanding the inhalation dosimetry of fibrous and non-fibrous forms of tungsten oxide materials.

1. Introduction

Tungsten is a dense metal that is used in a wide range of applications, including non-sag wire for filaments in light bulbs [1-3]. The manufacture of tungsten for non-sag wire begins with the calcining of ammonium paratungstate to yield tungsten oxide powder, which in turn, is reduced under controlled conditions to yield tungsten metal powder. During the production of tungsten metal powder, aerosols may be generated which contain tungsten sub-oxide particles having fibrous morphology [4-8]. The potential toxicity of tungsten sub-oxide fibers is poorly understood; however, in one preliminary study, these materials generated free radicals [7]. The capacity, if any, of tungsten sub-oxide fibers to contribute to development of adverse pulmonary health effects is currently not well understood.

Results of limited air sampling in occupational environments suggests that tungsten oxide fibers are respirable, with fibers generally having diameter <0.3 μm and very few fibers having length >5 μm [4-8]. Both single fibers and aggregates of fibers have been identified in air samples [4-6]. Fiber dimensions determine the extent to which fibers are inhaled, retained, and cleared by the lung. In humans engaged in mouth breathing, deposition in the alveolar region of the lung is maximal for fibers having aerodynamic diameter of 2 μm and aspect ratios (length/width) in the range 3 to 20 [9]. Biopersistence of inhaled fibers is determined by the rate at which they are physically cleared from the lung by mechanical action and the rate at which they dissolve in the lung (biodurability) [9]. Fibers with length greater than 20 μm probably cannot be engulfed by a single macrophage and will remain in the lung if durable. Fibers with length between 5 and 20 μm may inhibit lung macrophage mobility (i.e., mechanical clearance) in a graded fashion. Mechanical clearance of fibers <5 μm by lung macrophages is probably not influenced by fiber length [10]. The chemistry of fibers, in part, determines their biodurability via dissolution, which occurs in the extracellular lung fluid and/or within lung macrophage phagolysosomes [9].

We characterized the physicochemical properties of five tungsten oxide materials and began to evaluate the hypothesis that tungsten oxide fibers are more biodurable than isometric-shaped tungsten oxide particles and tungsten metal in lung fluids. Particle solubility was measured in variations of artificial extracellular lung fluid and lung alveolar macrophage phagolysosomal fluid to evaluate solvent effects on observed dissolution rates.

2. Materials and Methods

2.1. Study powders

Five bulk tungsten oxide powders were obtained from an industrial producer for study: $WO_{3.00}$, $WO_{2.98}$, $WO_{2.81}$, $WO_{2.66}$, and $WO_{2.51}$. With the exception of the $WO_{3.00}$ material, the oxidative states of these powders are average values. The $WO_{3.00}$ and $WO_{2.98}$ powders consisted of isometric-shaped particles and accounted for approximately 80% of this manufacturer's production. The $WO_{2.81}$, $WO_{2.66}$, and $WO_{2.51}$ powders contained fiber-shaped particles and accounted for the remaining 20% of industrial production. Physicochemical properties of these five tungsten oxide materials, and for comparison a pure tungsten metal powder, were characterized prior to estimation of biodurability in artificial lung fluids. A static dissolution technique was chosen for the biodurability studies because it has been shown to provide results for beryllium oxide that were nearly identical to in vivo models [11].

2.2. Characterization of study powders

Physical characterization of study materials included scanning electron microscopy (SEM) to evaluate particle morphology and size, helium pycnometry to determine density, and nitrogen gas adsorption to determine specific surface area (SSA). Raman spectroscopy was used to investigate the morphology and aggregation state of tungsten oxide fibers. For these experiments, we compared dry tungsten oxide powders (as received) with samples prepared by mixing the powder with artificial extracellular lung fluid (SUF) at a concentration of 4 mg mL^{-1}. The liquid samples were sonicated for 30 min. For collection of spectra, a small amount of powder was set to a glass fiber filter on a microscope slide. In the case of liquid samples, two silicon wells were firmly attached to a microscope slide and 100 μL of suspension was added to the slide. Raman spectra were collected using a with an excitation wavelength of 514.5 nm. The laser beam was focused through a microscope down to an approximate diameter of 10 μm and the power measured before the microscope was 1 mW. A CCD was used to detect the Raman signal in the region (80 – 1000) cm^{-1} where tungsten oxide exhibits main vibrations frequencies characteristic to both bulk and microcrystalline domains [12,13].

Chemical characterization of study materials included O content and 35 impurities: Ag, Al, As, Ba, Be, C, Ca, Cd, Co, Cr, Cu, Fe, H, K, La, Li, Mg, Mn, Mo, N, Ni, P, Pb, S, Sb, Se, Sn, Sr, Te, Ti, Tl, V, Y, Zn, and Zr. X-ray photoelectron spectroscopy was used to determine surface elemental chemistry (to a depth of approximately 20 Å).

2.3. Artificial lung fluid solubility studies

To estimate solubility, a known mass of each powder was weighed on a 0.025-μm pore size 47-mm diameter mixed cellulose ester filter (Millipore, Bedford, MA) then covered with a second filter. This filter "sandwich" was secured tightly in a static dissolution chamber (Intox Products, Moriarity, NM) using four nylon screws to provide a particle-tight seal around the edge of the sandwich. Each chamber was immersed in 80 mL of SUF or artificial macrophage phagolysosomal simulant fluid

(PSF) in a polypropylene plastic beaker and maintained at 37 °C for the duration of the study. For each dissolution experiment, triplicate samples were prepared for each study powder. At each time point, the chamber was removed from its beaker, the fluid transferred to a borosilicate glass jar, and fresh equilibrated fluid added to the sample beaker. Samples were stored frozen at -5°C and thawed prior to quantification of dissolved W in accordance with U.S. OSHA Method 213 (inductively coupled plasma-atomic emission spectroscopy).

2.3.1. *Artificial extracellular lung fluid.* SUF was prepared as previously described [14]. The pH of SUF was maintained at 7.3±0.1 for the duration of the dissolution experiments by blanketing the headspace in each sample beaker with a 5% CO_2/95% air mixture. The recipe for preparation of SUF includes 50 ppm of ABDC, which is added to prevent mold growth during experiments. ABDC is not a constituent of human extracellular lung fluid. To determine whether the presence of this exogenous constituent artificially influenced dissolution, dissolution of each study powder was measured using SUF prepared with and without ABDC; solvent was changed at 1, 4, 24, and 72 hours.

2.3.2. *Artificial macrophage phagolysosomal fluid.* To simulate the acidic conditions encountered by particles that are phagocytised by lung alveolar macrophages, PSF was prepared as described by Stefaniak et al. [15]. The pH of the PSF solution was maintained at 4.55±0.10 for the duration of the dissolution experiments. As with SUF, a preliminary experiment was performed using each study powder to evaluate whether ABDC artificially influenced dissolution. Our *a priori* assumption was that material dissolution would be faster in the more acidic PSF solvent relative to the neutral pH of the SUF solvent. As such, solvent in each sample beaker was changed at 1, 4, 8, 24, 48, and 72 hours to evaluate whether dissolution behaviour was linear or multiphasic under acidic conditions.

3. Results and Discussion

3.1. Characteristics of study powders

Table 1 summarizes the physical properties of the five study powders. As expected, density generally increased as powder oxygen content decreased. SSA varied by a factor of 15 among powders. Figure 1 illustrates the morphology of the dry tungsten oxide study powders. $WO_{3.00}$ and $WO_{2.98}$ were large (*ca.* 30 μm) compact cube-shaped particles, some with rounded edges, whereas $WO_{2.81}$, $WO_{2.66}$, and $WO_{2.51}$ were mixtures of fibers and compact cube-shaped particles. Tungsten metal was aggregate clusters of 1.2 μm primary particles. Among all fiber-containing materials, geometric mean (GM) fiber lengths were ≤ 5 μm and diameters ranged from 0.5 to 0.6 μm. GM fiber aspect ratios (length/width) determined from SEM images decreased as powder oxygen content decreased. The GM aerodynamic fiber diameter, calculated from the equation of Cox [16] assuming motion perpendicular to the fiber's axis, was 2.3, 3.1, and 2.1 μm for $WO_{2.81}$, $WO_{2.66}$, and $WO_{2.51}$, respectively. Fibers with these aerodynamic sizes and aspect ratios have high probability of depositing in the non-ciliated alveolar region of the lung [9]. The dimensions of fibers in these bulk tungsten oxide powders are consistent with previous reports that airborne tungsten oxide fibers were generally respirable, with most fibers having diameter <0.3 μm and very few fibers having length >5 μm [4-8]. Additionally, the SEM images illustrate both single and aggregated fiber, consistent with morphologies observed for airborne tungsten oxide fibers in workplace air.

Figure 2 presents the Raman spectra of the tungsten oxide study materials. No major differences were observed between spectra when tungsten oxide powder was suspended in SUF with or without ABDC (data not shown). The WO_3, $WO_{2.98}$, and $WO_{2.81}$ powders displayed intense bands around 718 and 807 cm^{-1} which are characteristic to stretching mode of ν(O – W – O) [12]. As the concentration of oxygen decreased these two bands decreased as shown for the $WO_{2.66}$ and $WO_{2.51}$ powders [17]. Other important Raman bands observed from our samples include the in-plane bending modes δ(O – W – O) with frequencies at 275 and 325 cm^{-1} and lattice modes at 94 cm^{-1} and 136 cm^{-1}, respectively.

Relative to the main Raman frequencies observed for the dry tungsten oxide powders, when these materials were suspended in SUF most of the bands were observed to shift to lower frequencies. Additionally, a new lattice mode at 94 cm^{-1} becomes noticeable. These results can be interpreted as a decrease in size of the microcrystalline domain from about 100 nm to 35 nm [13] after the tungsten oxide was sonicated in SUF. Note that when we mixed the tungsten oxide powders with phosphate buffered saline solution, no shift or new bands were observed. The time and power of the ultrasonic treatment was not believed to be sufficient to break or damage the particles. As such, the decrease in the microcrystalline domain was interpreted to be the result of de-aggregation of larger particles.

Table 1. Physical properties of tungsten oxide and metal powder study materials.

				GM (GSD)		
Study Material	Formula	Density ($g\ cm^{-3}$)	SSA ($m^2\ g^{-1}$)	L (μm)	W (μm)	AR
"Yellow" tungsten oxide	WO_3	7.2±0.01	0.75±0.03	NA[a]	31.1 (1.8)	NA
TB "Blue" tungsten oxide	$WO_{2.98}$	7.0±0.02	11.10±0.76	NA	29.8 (2.0)	NA
"Blue" high tungsten oxide	$WO_{2.81}$	7.1±0.00	2.85±0.02	4.0 (1.9)	0.5 (1.5)	8.3 (1.8)
"Blue" tungsten oxide	$WO_{2.66}$	8.2±0.03	1.50±0.07	4.9 (1.5)	0.6 (1.4)	7.9 (1.7)
"Blue" low tungsten oxide	$WO_{2.51}$	7.7±0.05	2.10±0.02	3.1 (1.7)	0.5 (1.4)	6.9 (1.8)
Tungsten metal	W	19.6±0.30	0.68±0.09	NA	1.2 (1.4)[b]	NA

[a] NA = Not applicable for isometric-shaped particles
[b] Primary particle size

Figure 1. Scanning electron micrographs of tungsten oxide materials illustrating particle morphologies.

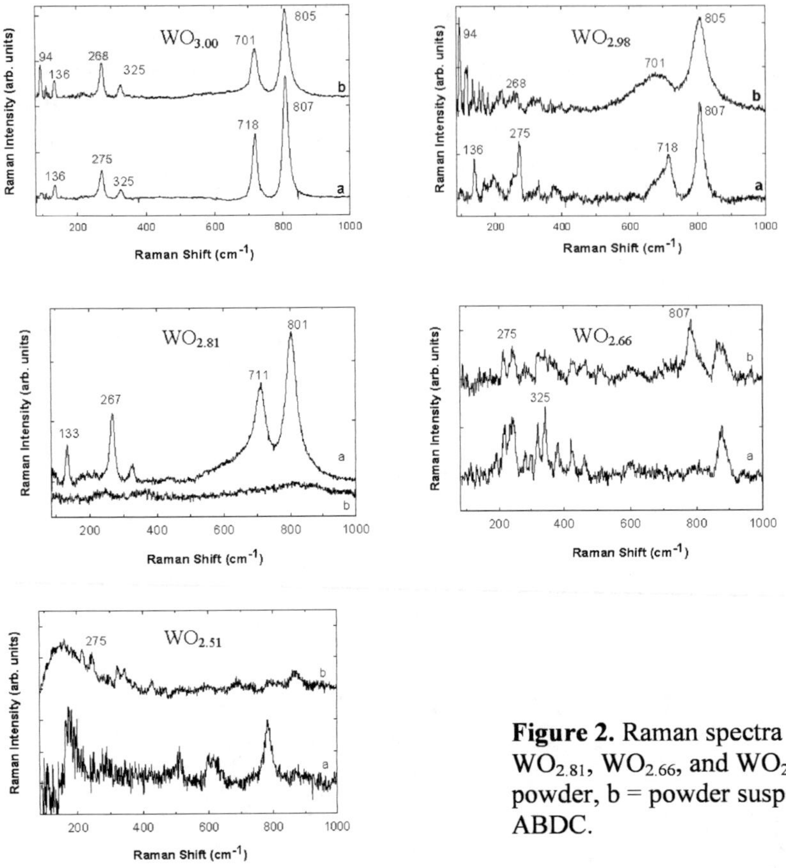

Figure 2. Raman spectra from $WO_{3.00}$, $WO_{2.98}$, $WO_{2.81}$, $WO_{2.66}$, and $WO_{2.51}$ powders. a = dry powder, b = powder suspended in SUF with ABDC.

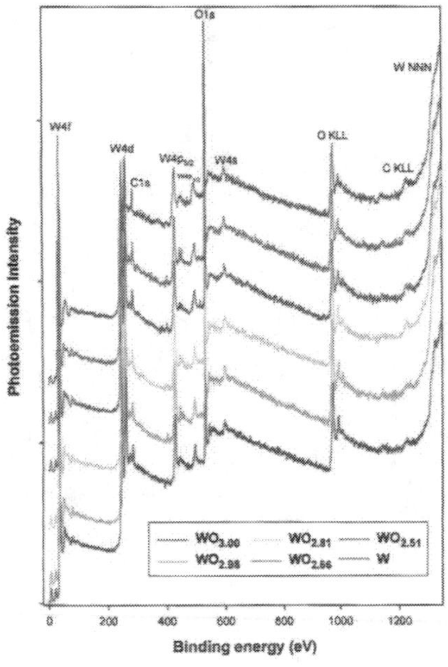

Figure 3. X-ray photoelectron spectroscopy survey spectra of tungsten oxide materials and tungsten metal for comparison. Surfaces of the tungsten oxide and tungsten metal powders were W, O, and minor amounts of C.

As illustrated in figure 3, tungsten oxide powder surfaces were composed of W, O, and minor amounts of C which was consistent with tungsten oxide. Levels of impurities were all $\leq 0.001\%$ by weight in all powders. Thus, the study powders were high purity tungsten oxide materials.

3.2. Artificial lung fluid solubility studies

Figure 4 is a plot of the mass fraction of tungsten dissolved from each of the five bulk tungsten oxide materials in SUF. At the time of this writing, quantification of dissolved tungsten from bulk tungsten oxide materials exposed to PSF was in progress; therefore, only results from studies with SUF are reported herein. A t-test of sample means indicated that, in SUF, ABDC inhibited ($p < 0.05$) dissolution of $WO_{2.98}$, $WO_{2.81}$, and $WO_{2.66}$ but not $WO_{3.00}$ or $WO_{2.51}$. The exact mechanism by which ABDC inhibited dissolution of certain tungsten oxide materials is unknown; however, this anti-fungal agent is a cationic surfactant which may chemically interact with powder surfaces to inhibit dissolution. There was no statistical difference in dissolution between the isometric-shaped $WO_{3.00}$ and tungsten metal particles. The fibrous $WO_{2.66}$ and $WO_{2.51}$ materials dissolved more slowly ($p < 0.05$) than the isometric tungsten metal particles, indicating potential for prolonged biodurability of these fiber materials in human extracellular lung fluid due to the unique chemistry. Note that the biological relevance of this observed difference in dissolution rates has yet to be elucidated. Biopersistence of inhaled fibers is determined by the rate at which they are mechanically cleared from the lung and their biodurability, which is influenced by fiber chemistry [9] and the biological fluid to which the fiber is exposed. As such, fiber biodurability via dissolution may differ between the neutral pH extracellular lung fluid and the acidic fluid encountered upon engulfment by lung macrophage cells and sequestration in phagolysosomes.

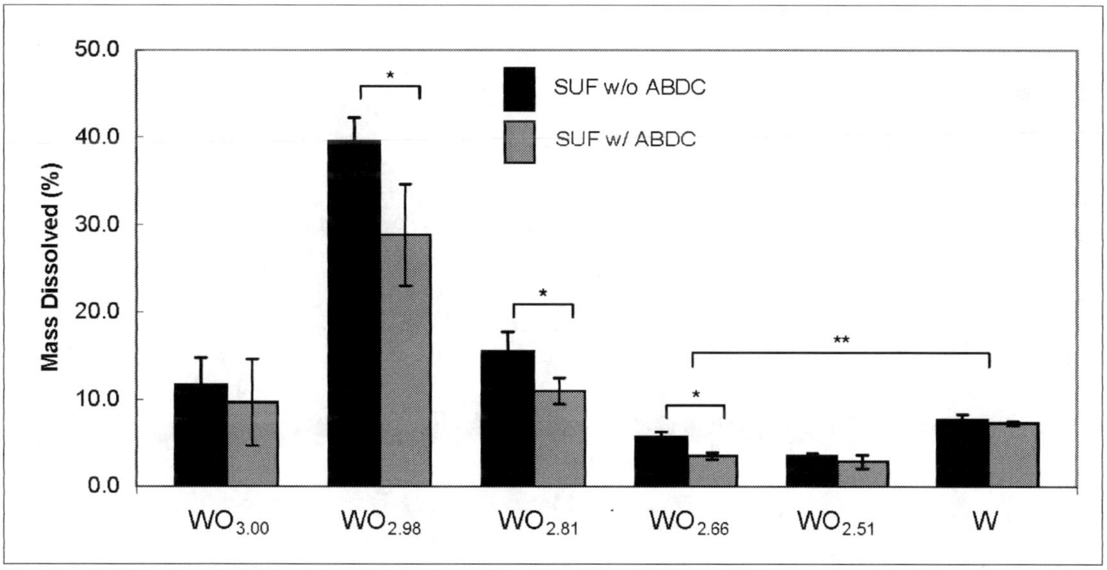

Figure 4. Total mass fraction of tungsten dissolved in 72 hours from various bulk tungsten oxide powders using serum ultrafiltrate (SUF) without and with alkylbenzyldimethylammonium chloride (ABDC), which is added to prevent mold growth during experiments. For comparison, the mass fraction of tungsten dissolved from bulk tungsten metal (W) powder in SUF without and with ABDC is included in the plot. * = difference in dissolution for SUF without and with ABDC ($p < 0.05$ using two-sided t-test assuming equal variances). ** = dissolution in SUF without ABDC lower than tungsten metal ($p < 0.05$ using one-sided t-test assuming equal variances).

4. Summary

Five tungsten oxide powder materials were characterized and their dissolution behavior evaluated in SUF relative to pure W metal. Powder physicochemical properties were unique among study materials. Additionally, Raman spectroscopy was shown to be useful for determination of the morphology and crystalline domain of tungsten oxides. On a mass basis, the rank order of tungsten dissolution (from greatest to least) was: $WO_{2.98} > WO_{2.81} > WO_{3.00} > W > WO_{2.66} > WO_{2.51}$. Fibrous $WO_{2.66}$ and $WO_{2.51}$ materials are potentially more biodurable than isometric shaped tungsten oxide and tungsten metal powders; however, all three materials are poorly soluble. Knowledge gained from these dissolution studies will aid in determining whether the biodurability of these tungsten oxide fiber materials contribute a potential hazard due to prolonged biopersistence in the lung.

Acknowledgements

This research was supported in part by the National Toxicology Program through Interagency Agreement number Y1-ES-9045-10 with the National Institute for Occupational Safety and Health. Mention of a specific product or company does not constitute endorsement by the Centers for Disease Control and Prevention. The findings and conclusions in this report are those of the authors and do not necessarily represent the views of the National Institute for Occupational Safety and Health.

References

[1] Lassner E and Schubert W-D 1995 *Int. J. Refract. Metals Hard Mat.* **13** 111-17
[2] Lunk H-J, Ziemer B, Salmen M, and Heidemann D 1993-1994 *Refract. Metals Hard Mat.* **12** 17-26
[3] Bartha L, Kiss AB, and Szalay T 1995 *Int. J. Refract. Metals Hard Mat.* **13** 77-91
[4] Sahle W 1992 *Chest* **102** 1310
[5] Sahle W, Laszlo I, Krantz S, and Christensson B 1994 *Ann. Occup. Hyg.* **38** 37-44
[6] Sahle W, Krantz S, Christensson B, and Laszlo I 1996 *Sci. Total Environ.* **191** 153-67
[7] Leanderson P and Sahle W 1995 *Toxicol. In Vitro.* **9** 175-83
[8] McKernan JL, Toraason MA, Fernback JE 2008 *J Occup Environ Hyg.* **5** 463-474
[9] Oberdörster G 2002 *Inhal. Toxicol.* **14** 29-56
[10] Davis JMG 1994 *Environ. Health Perspect.* **102(Suppl 5)** 113-17
[11] Stefaniak AB, Day GA, Hoover MD, Breysse PN, Scripsick RC 2006 *Toxicol. In Vitro.* **20** 82-95
[12] Ingham B, Chong SV, and Tallon JL 2005 *J. Phys. Chem. B* **109** 4936-40
[13] Boulova M, Lucazeau G 2002 *J. Solid State Chem.* **167** 425-34
[14] Finch GL, Mewhinney JA, Eidson AF, Hoover MD, and Rothenberg SJ 1988 *J. Aerosol Sci.* **19** 333-42
[15] Stefaniak AB, Guilmette RA, Day GA, Hoover MD, Breysse PN, and Scripsick RC 2005 *Toxicol. In Vitro.* **19** 123-34
[16] Cox RG 1970 *J. Fluid Mech.* **44** 791-810
[17] Wang XP, Yang BQ, Zhang HX, Feng PX 2007 *Nanoscale Res. Lett.* **2** 405-09

Journal of Physics: Conference Series **151** (2009) 012014 doi:10.1088/1742-6596/151/1/012014

Genomic instability in quartz dust exposed rat lungs: Is inflammation responsible?

C Albrecht[1], AM Knaapen[2]*, G Cakmak Demircigil[3], Erdem Coskun[3], FJ van Schooten[2], PJA Borm[4] and RPF Schins[1]

1 Institut für Umweltmedizinische Forschung (IUF) at the Heinrich Heine University Duesseldorf, Germany.

2 Nutrition and Toxicology Research Institute Maastricht (NUTRIM), Department of Health Risk Analysis and Toxicology, University of Maastricht, The Netherlands

3 Gazi University, Faculty of Pharmacy, Dept. of Toxicology, Ankara, Turkey

4 Centre of Expertise in Life Sciences (Cel), Hogeschool Zuyd, Heerlen, Netherlands

* current address: Department of Toxicology and Drug Disposition, Schering-Plough, 5340 BH Oss, The Netherlands

e-mail: catrin.albrecht@uni-duesseldorf.de

Abstract. Exposure to quartz dusts has been associated with lung cancer and fibrosis. Although the responsible mechanisms are not completely understood, progressive inflammation with associated induction of persistent oxidative stress has been discussed as a key event for these diseases. Previously we have evaluated the kinetics of pulmonary inflammation in the rat model following a single intratracheal instillation of 2mg DQ12 quartz, either in its native form or upon its surface modification with polyvinylpyridine-N-oxide or aluminium lactate. This model has been applied now to evaluate the role of inflammation in the kinetics of induction of DNA damage and response at 3, 7, 28, and 90 days after treatment. Bronchoalveolar lavage (BAL) cell counts and differentials as well as BAL fluid myeloperoxidase activity were used as markers of inflammation. Whole lung homogenate was investigated to determine the induction of the oxidative and pre-mutagenic DNA lesion 8-hydroxy-2-deoxy-guanosine (8-OHdG) by HPLC/ECD, while mRNA and protein expression of oxidative stress and DNA damage response genes including hemeoxygenase-1 (HO-1) and apurinic/apyrimidinic endonuclease (APE/Ref-1) were evaluated using Western blotting and real time PCR. Isolated lung epithelial cells from the treated rats were used for DNA strand breakage analysis using the alkaline comet assay as well as for micronucleus scoring in May-Gruenwald-Giemsa stained cytospin preparations. In the rats that were treated with quartz, no increased 8-OHdG levels were observed, despite the presence of a marked and persistent inflammation. However, DNA strand breakage in the lung epithelial cells of the quartz treated rats was significantly enhanced at 3 days, but not at 28 days. Moreover, significantly enhanced micronucleus frequencies were observed for all four time points investigated. In the animals that were treated with the PVNO modified quartz, micronuclei scores did not differ from controls, while in those treated with the aluminium coated quartz intermediate effects were found. These findings were in line with the kinetics of inflammation and epithelial proliferation in the rat lungs for the different treatments. Notably, a highly significant correlation was observed between neutrophil numbers and micronucleus frequencies, indicative for a role of inflammation in eliciting genomic instability in target cells of quartz-induced carcinogenesis. Our ongoing investigations focus on the evaluation of the causality between both in relation to quartz exposure.

1. Introduction

Inflammation has been recognised as an important factor in cancer development and progression [1, 2, 3]. Chronic inflammation is also considered to be crucial in driving lung cancer formation after inhalation of quartz as well as poorly-soluble low toxicity dusts. Herein, specifically a mechanism is considered to be involved whereby inflammatory cells release excessive amounts of reactive oxygen species (ROS) which are known to be able to cause oxidative DNA damage and mutagenesis [reviewed in 4]. This mechanism can be referred to a secondary genotoxicity, in contrast to primary genotoxicity, which is considered to involve a direct genotoxic action of a particle [5]. Previous studies have demonstrated the carcinogenicity of quartz as a variably entity [6]. We and others have shown that especially the surface reactivity of the particles contributes to their biological response, since surface modifications by polyvinylpyridine-N-oxide (PVNO) or aluminium lactate (AL) are able to reduce the inflammatory and genotoxic potential of DQ12 quartz particles [7, 8, 9, 10]. The present study was undertaken to evaluate the relation between the kinetics of quartz-induced lung inflammation and genotoxicity *in vivo* at 3, 7, 28 and 90 days after single intratracheal (i.t.) instillation. Inflammation was characterised by inflammatory cell counts of the bronchoalveolar lavage (BAL) and differentials. For the evaluation of genotoxicity three independent methods were used. In whole lung tissue homogenate we determined the formation of the oxidative and pre-mutagenic lesion 8-hydroxydeoxyguanosine [11]. Two further measurements of genotoxicity were determined specifically in the lung epithelial cells, isolated from the treated animals. These cells were specifically analysed because of their relevance as target cells in pulmonary carcinogenesis [8]. For the determination of DNA strand breakage the alkaline comet assay was used. This assay is a sensitive genotoxicity indicator assay that allows for quantification of DNA-damage in individual cells [12]. The isolated lung epithelial cells were also evaluated for micronuclei (MN), which are small extranuclear bodies, arising in dividing cells from chromosome fragments or a whole lagging chromosome. MN analysis in isolated lung epithelial cells from rats has been successfully applied for the evaluation of *in vivo* genotoxicity of hard metal particles and carbon nanotubes after i.t. instillation [13, 14].

2. Materials and methods

2.1. Particle characteristics and coating procedure

Dörentruper quartz (DQ12) was used as native quartz as well as for the preparation of surface-modified quartz samples. The procedure for particle coating with aluminium or PVNO and quality control evaluation (i.e. determination of coating efficiency and size distribution) have been described elsewhere [8, 9, 10].

2.2. Animals and treatment

Female Wistar rats (Janvier, Le Genest St. Isle, France) were maintained in an accredited on-site testing facility, responding to the guidelines of the Society for Laboratory Animals Science (GV-SOLAS). At a age of eight weeks (220±1g) animals were i.t. instilled following anesthetization (Isofluran, Essex Pharma GmbH, Munich, Germany) using 400µl volume containing phosphate buffered saline (PBS) as vehicle control, 2mg DQ12 quartz or DQ12 (2mg) coated with either PVNO or with AL. At days 3, 7, 28 and 90 post-instillation, animals were sacrificed by deep anesthetization with pentobarbital (50 mg/kg body weight) followed by exsanguination via the A. abdominalis.

2.3. Broncho-alveolar lavage and analysis
Lungs of 5 animals per treatment group and time point were lavaged as described previously (Albrecht et al., 2004). The BAL fluid was spun (500g, 10min, 4°C), and cells were collected for cell counting and May-Grunwald-Giemsa (MGG)-stained cytospin preparations. Inflammatory and cytotoxic data are published elsewhere [9].

2.4. DNA isolation and analysis of 8-hydroxy-2'-deoxyguanosine by HPLC/ECD
Lung tissue was removed, chopped, aliquots were snap frozen in liquid nitrogen and stored at –80°C until later measurement of 8-OHdG using high performance liquid chromatography with electrochemical detection (HPLC-ECD) [15]. Values are expressed as the total amount of 8-OHdG per 106 nucleotides.

2.5. Lung epithelial cell isolation
In order to investigate the specific target cell for particle-induced lung tumours, lung epithelial cells were isolated according to the method of Richard and colleagues [16] as modified by Knaapen and co-worker [8]. A part of the obtained epithelial cell fraction was used to prepare MGG-stained cytospin preparations in order investigate the appearance of MN. Another part was used for the evaluation of DNA strand breakage by the alkaline comet assay.

The purity of the isolated cell fraction, as determined by microscopy evaluation of the MGG-stained slides was found to be 63 – 75 % epithelial cells. Other cell types were neutrophils (4 – 18 %) and macrophages (13 – 20 %).

2.6. Comet assay
DNA strand breakage was investigated immediately after cell isolation using the alkaline comet assay as described previously [8] according to the guidelines published by Tice and Colleagues [17]. For each animal three independent slides were prepared, and on each single slide 50 cells were randomly evaluated. Cells were categorised according to presence or absence of a tail. The comet assay analysis was performed for 2 time points, i.e. at 3 days and 28 days.

2.7. Micronucleus evaluation in lung epithelial cells
In order to determine the proportion of the micronucleated cells, in each MGG-stained slide 1000 lung cells were counted at x1000 magnification with immersion oil using a Zeiss Axioscope 2 microscope. The identification and scoring of the MN was done according to Tolbert and colleagues [18]. Characteristics of each evaluated cell are an intact cytoplasm and an intact nucleus with a smooth and distinct nuclear perimeter. A cellular structure was then identified as a MN if its size was less than a third of the diameter of the associated nucleus, but large enough to discern the round shape and a comparable staining intensity to the nucleus.

2.8. Statistical analysis
Data are expressed as mean ± SD. Statistical analysis was performed using SPSS version 15.0 for Windows, post hoc Tukey-HSD. Correlation between MN formation and percentage of PMN was tested using Person's r-test. Differences were considered as statistically significant when $p < 0.05$.

3. Results
Investigation of whole lung tissue homogenate showed no enhanced 8-OHdG/dG ratios in the animals after treatment with native quartz at all investigated time points (Figure 1). Surprisingly, at all time points 8-OHdG levels tended to be higher in the lungs from rats that were exposed to the coated quartz preparations than from those that were treated with the original quartz sample. At the 7 day time point, 8-OHdG levels were significantly enhanced in animals that were treated with the PVNO-coated quartz sample ($p<0.05$, Fig. 1).

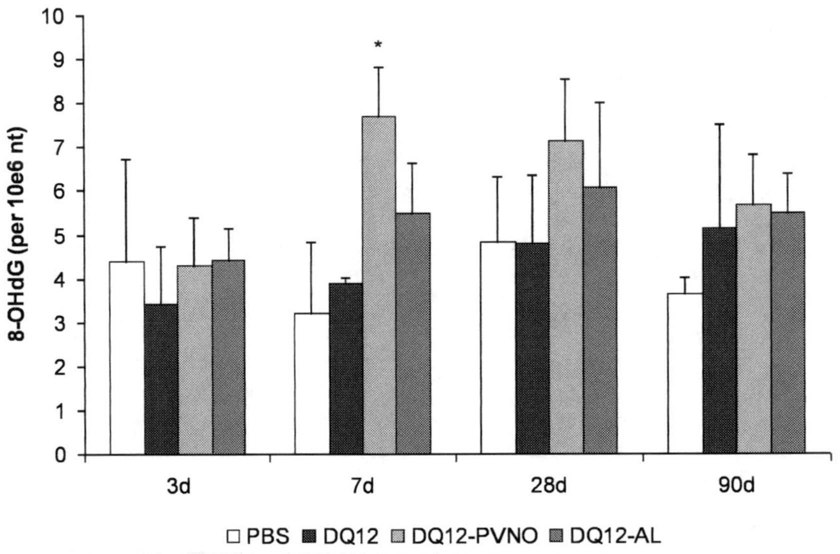

Figure 1: 8-OHdG analysis by HPLC/ECD in lung tissue, obtained from rats exposed to a single dose of 2 mg non-coated DQ12 or DQ12 coated with PVNO or AL at 3, 7, 28 and 90 days after i.t. instillation. Data are shown as mean ± SD (n=5). *p<0.05 vs. PBS

The comet assay was used for the determination of DNA strand breakage in freshly isolated epithelial cells. The measurements were performed at the 3 day and the 28 day time points (Table 1).

Table 1: Percent of isolated lung epithelial cells showing DNA damage as determined by the alkaline comet assay. Cells were isolated from the lungs of rats that were exposed to a single dose of 2 mg non-coated DQ12 or DQ12 coated with PVNO or AL (* p< 0.05).

Treatment	3 days (% of cells with tail)	28 days (% of cells with tail)
PBS	32.4 ± 6.6	56,4 ± 13.8
DQ12	41.1 ± 11.3 *	51,2 ± 14.6
DQ12 – PVNO	28.6 ± 12.9	62,6 ± 14.7
DQ12 – AL	31.3 ± 10.2	60,0 ± 10.4

Three days after quartz instillation a significant higher number of cells with DNA damage were observed, when compared to the control animals. Particle surface modification with PVNO and AL inhibited the ability of quartz to induce these DNA damage. After 28 days, these differences were not present. In fact, the DQ12 treated animals tended to show the lowest DNA damage at this time point.

As a further marker for the genotoxic potential of the quartz particles, the appearance of MN were evaluated at MGG-stained cytospin preparations of lung epithelial cells. In quartz-treated animals significantly higher MN % were observed at all times points (Figure 2). Interestingly, for animals that had received AL-coated quartz also enhanced MN frequencies were found at the 28 and 90 day time points. In contrast, animals that were treated with the PVNO-coated quartz MN % in the lung epithelial cells isolates were not increased.

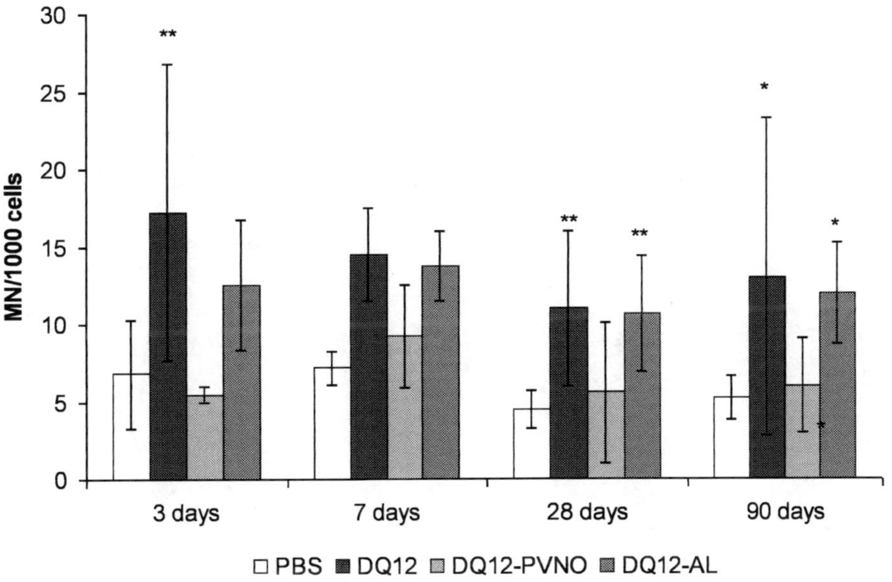

Figure 2: Data of MN formation calculated as absolute number per 1000 cells evaluated from MGG - stained cytospin preparations. Cells were obtained from rats exposed to a single dose of 2 mg non-coated DQ12 or DQ12 coated with PVNO or AL at 3, 7, 28 and 90 days after single i.t. instillation. Data are shown as mean ± SD (n=5). *p < 0.05 vs. PBS; ** p< 0.01

In order to evaluate the relation between the inflammation as induced by the different quartz preparations and genotoxicity, correlation analyses were performed between inflammatory cell counts from bronchoalveolar lavage and the MN frequencies as determined in the lung epithelial cells. For each of the four investigated time points there was a significant linear correlation between the % of neutrophils and the MN frequencies when data were considered on the single animal level (3, 7, 28 and 90 days, all: $p<0.01$). The relation between neutrophilic inflammation and epithelial lung cell genotoxicity was also observed on the group level for the entire study (Figure 3). This effect was also clearly exhibited on the single animal level ($n = 76$; $R^2 = 0.46$; $p < 0.0001$).

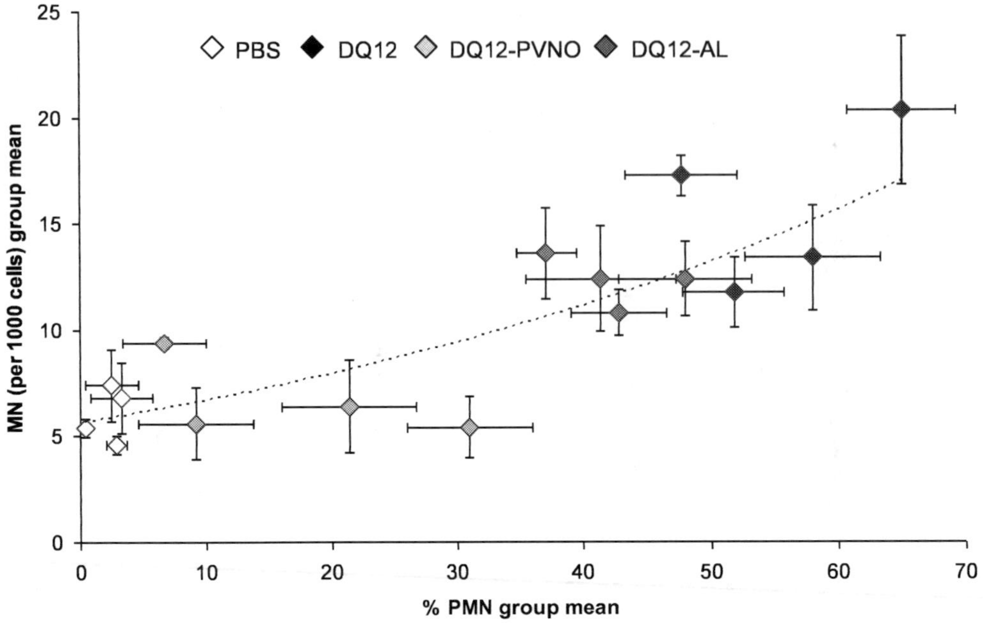

Figure 3: Correlation analysis on the group level between percentage of PMN in the BAL and formation of MN in epithelial cells (n = 16, R2 = 0.70 Pearson). The line depicts an exponential fit. Each point in the graph represents mean ± SD of the percentages of PMN and MN of four to five animals per treatment and time point.

4. Discussion and conclusion

The major goal of our study was to evaluate the relation between inflammation and genotoxicity, considered to be a crucial mechanism of particle-induced carcinogenesis. Therefore, we evaluated the kinetics of lung inflammation after single i.t. instillation of different preparations of quartz in rat lungs in relation to various genotoxicity readouts. The application of two different surface-modifications of DQ12 quartz (i.e. PVNO and aluminium) along with the original DQ12 sample resulted in levels of inflammation which largely differed in extent as well as over time [9]. In this study we have now examined over the time different markers of particle-induced genotoxicity. Being the most-well investigated oxidative DNA lesion, the induction of 8-OHdG was measured in whole lung tissue homogenates by HPLC/ECD [11, 15]. Increased 8-OHdG immunoreactivity in lung alveolar cells has been previously shown by Seiler and colleagues [19] three weeks after a single i.t. instillation of 1.5 mg quartz in rats, while 3 days after instillation such an increase was absent. Surprisingly, no significant 8-OHdG induction was observed by the original quartz sample in the current study despite a persistent inflammation, neither with HPLC/ECD nor with supportive investigations by immunohistochemistry (data not shown). The divergence between the different *in vivo* quartz studies may have resulted from use of different doses, exposure times or quartz samples, i.e. as an effect of a "batch to batch variation" [6]. Also the possibility of an artificial induction of 8-OHdG during DNA extraction and processing as required for HPLC/ECD analysis, leading to high background levels would neither explain positive findings by others [e.g. 11] as well as our unexpected effects in the animals which received PVNO-coated quartz samples.

In contrast to the lack of detectable increases in 8-OHdG upon DQ12 treatment, a significant increase in DNA strand breakage of lung epithelial cells from quartz-treated animals could be measured three days after quartz instillation. This difference might be explained by the specific isolation of the lung epithelial cells and the high sensitivity of the comet assay method. Importantly, DNA-damage was not observed in the animals that were treated with the surface modified quartz samples. As these data were related to the extent of neutrophilic inflammation as induced by the different sample treatments for this time point, they are in support of the causal role of inflammation in particle-induced genotoxicity [8]. However, 28 days after the instillation, no difference in DNA strand breakage could be detected in the lung epithelial cells from the different treatment groups. These findings may be explained by an induction of DNA repair, as indicated from our previous investigations were we observed enhanced expression of the oxidative DNA damage repair protein apurinic/apyrimidinic endonuclease/redox factor 1 (APE/Ref1) [10]. However, the role of epithelial cell proliferation and differentiation known to occur after quartz treatment should also be considered. In this regard, the third genotoxicity assay evaluated in our current study, i.e. MN in the lung epithelial cells, are considered to be the most relevant: While the comet assay is a measure of the bare presence of a potentially broad spectrum of DNA lesions which may - or may not - lead to a mutation, a MN represents the product of a clastogenic or aneugenic effect and cell proliferation. The relevance of MN test has been further supported by recent observations that demonstrated the predictive value of their increases in peripheral blood lymphocytes for lung cancer risk in humans [20].

In the present study, MN frequencies were found to be enhanced in lung epithelial cells up to the period of 90 days after quartz instillation. In the knowledge that the quartz treatment in our study has been shown to lead to a persistent increasing inflammation over this time period [9], it may be proposed that the observed *in vivo* genotoxicity results from (a) yet to be identified mechanism(s) operating in inflamed lung tissues. This is further supported by observed gradients in inflammation and MN in the lungs of the rats that received AL- and PVNO-modified DQ12 respectively, visualised by a highly significant correlation between the percentages of neutrophils and MN in the over all study. Although previous *in vitro* studies by us and others have demonstrated that both neutrophils and quartz particles can cause genotoxic insult to epithelial cells [reviewed in 5] our present data are in further support of the secondary mechanism of genotoxicity. Noteworthy in this regard, the MN frequency in epithelial cells of the control rats as well as those that were treated with PVNO-coated DQ12 was about 0.5 – 1 % indicative of the absence of an increased genotoxicity up to an approximate 30 % of neutrophils in the lung as observed with the DQ12- PVNO group.

In conclusion, our study demonstrates that modification of the reactive surface of quartz particles can have significant impact on pulmonary inflammation and associated genotoxicity and therefore provides further clues for the observed variable cancer risk of quartz exposure [6]. Moreover, our data are the first to show a marked *in vivo* association between neutrophilic inflammation in the lung and the formation of MN in lung target cells for (particle-induced) lung carcinogenesis. Additional experiments are needed to establish the causality of this relation.

Acknowledgements.
The study was financially supported by the Ministerium of Wirtschaft, Mittelstand, Technologie und Verkehr Nordrhein-Westfalen, the Silikosegesellschaft Nordrhein-Westfalen and the Federal Ministry of Environment. A.M. Knaapen was supported by a postdoctoral fellowship from the Netherlands Organisation for Scientific Research (NWO, grant 916.46.092). The authors wish to thank Dr. K. Unfried for his help with the animal instillation. We want to acknowledge Mrs. A. Winzer, K. Ledermann, C. Weishaupt and V. Suri for their technical support.

References

[1] Bartsch H and Nair J 2006 Langenbecks Arch. Surg. **391** 499
[2] Engels EA 2008 Expert. Rev. Anticancer Ther. **8** 605
[3] Mantovani A, Allavena P, Sica A and Balkwill F 2008 Nature **454** 436
[4] Knaapen AM, Borm PJ, Albrecht C and Schins RP 2004 Int. J. Cancer **109** 799
[5] Schins RP and Knaapen AM 2007 Inhal. Toxicol. **19** 189

[6] Donaldson K and Borm PJA 1998 Ann. Occup. Hyg. **42** 287
[7] Duffin R, Gilmour PS, Schins RPF, Clouter A, Guy K, Brown DM, MacNee W, Borm PJA, Donaldson K and Stone V 2001 Toxicol. Appl. Pharmacol. **176** 10
[8] Knaapen A M, Albrecht C, Becker A, Höhr D, Winzer A, Haenen G R, Borm PJA and Schins RPF 2002 Carcinogenisis **23** 1111
[9] Albrecht C, Schins RPF, Höhr D, Becker A, Shi T, Knaapen AM and Borm PJA 2004 Am. J. Respir. Cell Mol. Biol. **31** 292
[10] Albrecht C, Knaapen AM, Becker A, Höhr D, Haberzettl P, van Schooten FJ, Borm PJA and Schins RPF 2005 Resp. Res. **6** 129
[11] Yamano Y, Kagawa J, Hanaoka T, Takahashi T, Kasai H, Tsugane S and Watanabe S 1995 Environ. Res. **69** 102
[12] Singh NP, McCoy MT, Tice RR and Schneider EL 1988 Exp. Cell. Res. **175** 184
[13] De Boeck M, Hoet P, Lombaert N, Nemery B, Kirsch-Volders M and Lison D 2003 Carcinogenesis **24** 1793
[14] Muller J, Decordier I, Hoet PH, Lombaert N, Thomassen L, Huaux F, Lison D and Kirsch-Volders M 2008 Carcinogenesis **29** 427
[15] Knaapen AM, Seiler F, Schilderman PA, Nehls P, Bruch J, Schins RPF and Borm PJA 1999 Free Radic. Biol. Med. **27** 234
[16] Richards RJ, Davies N, Atkins J and Oreffo VI 1987 Lung **165** 143
[17] Tice RR, Agurell E, Anderson D, Burlinson B, Hartmann A, Kobayashi H, Miyamae Y, Rojas E, Ryu JC and Sasaki YF 2000 Environ. Mol. Mutagen. **35** 206
[18] Tolbert PE, Shy CM and Allen JW 1992 Mutat. Res. **271** 69
[19] Seiler F, Rehn B, Rehn S, Hermann M and Bruch J 2001 Am. J. Respir. Cell. Mol. Biol. **24** 492
[20] El-Zein RA, Schabath MB, Etzel CJ, Lopez MS, Franklin JD and Spitz MR 2006 Cancer Res. **66** 6449

Pyrogenic effect of respirable road dust particles

Umesh Jayawardena, Linda Tollemark, Christer Tagesson and Per Leanderson

Occupational and Environmental Medicine, University Hospital, S-581 85 Linköping, Sweden

E-mail: per.leanderson@lio.se

Abstract. Because pyrogenic (fever-inducing) compounds on ambient particles may play an important role for particle toxicity, simple methods to measure pyrogens on particles are needed. Here we have used a modified *in vitro* pyrogen test (IPT) to study the release of interleukin 1β (IL-1β) in whole human blood exposed to respirable road-dust particles (RRDP). Road dusts were collected from the roadside at six different streets in three Swedish cities and particles with a diameter less than 10 μm (RRDP) were prepared by a water sedimentation procedure followed by lyophilisation. RRDP (200 μl of 1 – 10^6 ng/ml) were mixed with 50 μl whole blood and incubated at 37 °C overnight before IL-1β was analysed with chemiluminescence ELISA in 384-well plates. Endotoxin (lipopolysaccharide from Salmonella minnesota), zymosan B and Curdlan (β-1,3-glucan) were used as positive controls. All RRDP samples had a pyrogenic effect and the most active sample produced 1.6 times more IL-1β than the least active. This formation was of the same magnitude as in samples with 10 ng LPS/ml and was larger than that evoked by zymosan B and Curdlan (by mass basis). The method was sensitive enough to determine formation of IL-1β in mixtures with 10 ng RRDP/ml or 0.01 ng LPS/ml. The endotoxin inhibitor, polymyxin B (10 μg/ml), strongly reduced the RRDP-induced formation of IL-1β at 1μg RRDP/ml (around 80 % inhibition), but had only marginal or no effects at higher RRDP-concentrations (10 and 100 μg /ml). In summary, all RRDP tested had a clear pyrogen effect in this *in vitro* model. Endotoxin on the particles but also other factors contributed to the pyrogenic effect. As opposed to the *limulus* amebocyte lysate (LAL) assay (which measures endotoxin alone), IPT measures a broad range of pyrogens that may be present on particulate matter. The IPT method thus affords a simple, sensitive and quantitative determination of the total pyrogenic potential of ambient particles.

1. Introduction

It is now widely acknowledged that exposure to particulate air pollution is associated with respiratory and cardiovascular morbidity and mortality [1, 2]. For example, (epidemiological studies have shown that) exposure to fine particulate air pollution is associated with the incidence of cardiovascular disease and death among postmenopausal women [3]. Out of various environmental factors, ultra-fine particles, PM_{10} and carbon dioxide were significantly associated with out-of-hospital coronary deaths in Rome, Italy [4]. One important type of particulate air pollution is respirable road dust particles (RRDP). These are heterogeneous by nature and built up by materials that have their origin in mechanical wear processes in the tire-road interface or from vehicle break, combustion-generated material or from organic materials from plant and soil [5]. The frequent use of studded tires during wintertime in the Nordic countries results in an enhanced formation of wear-induced road particles but due to the wet climate there are generally low concentrations of airborne respirable particulate matter during winter. However, when the roads become dryer in March and April the level of RRDP in the air rises considerably and may often exceed 50 μg/m^3 [6].

After inhalation and deposition in the lung, ambient particles can interact with a variety of lung cells and may also have systemic effects [7, 8, 9, 10]. These interactions eventually lead to a release of various cytokines with ensuing inflammatory reaction and tissue injury. There is presently much interest in the factors that may cause such a cytokine release. One possibility is the presence of surface-bound pyrogenic (fever-inducing) substances on the particles, such as bacterial endotoxin (lipopolysaccharide, LPS, from Gram-negative bacteria), lipoteichoic acid (from Gram-positive bacteria), peptidoglycan (from bacterial cell walls), fungal spores, and viral pathogens. Because the presence of pyrogens may determine particle toxicity, simple methods to measure pyrogens on particles are needed. The most commonly used method for measuring endotoxin is the *limulus* amebocyte lysate (LAL) assay [11]. This is a sensitive assay for endotoxin but does not detect other pyrogenic agents. A broader spectrum of pyrogens can be detected with the *in vitro* pyrogen test (IPT) which is a model where the sample is mixed with a small volume of human whole blood, incubated overnight, and the levels of the cytokine interleukin-1 beta (IL-1β) are determined with ELISA-technique [12]. This method has previously been successfully used to study air-borne pyrogens from samples collected on membrane filters [13] and immune-stimulating compounds on surfaces [14]. IPT has been proposed as an alternative to the animal-based LAL-test and the rabbit pyrogen test [15], which both are used to test pyrogens in pharmaceutical products but have several shortcomings. The IPT (which uses human cells) has also been validated against these two animal-based techniques and found to be sensitive, accurate and cost efficient [16].

In the present work, we have used the IPT to measure pyrogenic compounds on RRDP collected at six different locations in three Swedish cities during wintertime. The pyrogenic effect of the RRDP samples was also compared with that of lipopolysaccharide (LPS), zymosan B and β-glucan.

2. Materials and Methods

2.1. Preparation of respirable road dust particles (RRDP)

Road surface-samples were collected from six different streets in three different Swedish cities during February 2007. All sites had similar surroundings with residential areas and greenery but with varying traffic intensity and mean annual concentrations of PM_{10} (table 1).

There were no factories, animal facilities or other similar potential sources of endotoxin in the vicinity of the sites. The crude samples were transferred into plastic bags and kept at 4 °C and used within one week. To preparer a fraction of respirable particles the material was mixed with Milli-Q water to a slurry in a polypropylene tube (1 kg road course-sample per litre of pyrogen free water). After a short agitation, 800 ml slurry was poured into a 1 l sterile glass flask (Schott) that was left in room temperature for 18 hours. The upper phase was then carefully decanted into a 2 l sterile round-bottom flask where it was frozen and lyophilized. The dry fraction of RRDP was then transferred to a pyrogen-free glass vial. A scanning electron microscope micrograph of a typical sample is shown in figure 1.

Journal of Physics: Conference Series **151** (2009) 012015 doi:10.1088/1742-6596/151/1/012015

Table 1. Road dust samples used in the study.

Sample	Street (city)[a]	Coordinates[b]	PM_{10} concentration[c, d] ($\mu g/m^3$)	Daily traffic intensity[d] (x 10^3 cars per day)
1	N. Tanneforsvägen (Li)	58 24′ 08.86 N, 15 38′ 54.54 E	23.2	15.7
2	Drottninggatan (Li)	58 24′ 29.81 N, 15 37′ 15.86 E	25.2	12.2
3	Östra promenaden (No)	58 35′ 28.78 N, 16 11′ 51.08 E	28.4	24.7
4	G. Tanneforsvägen (Li)	58 24′ 29.31 N, 15 38′ 48.47 E	23.7	15.3
5	Ålidbacken (Um)	63 48′ 50.10 N, 20 18′ 18.07 E	No data	7.7
6	Söderleden (Li)	58 23′ 41.48 N, 15 38′ 51.22 E	24.4	10.5

[a] (Li) = Linköping, (No) = Norrköping, (Um)= Umeå
[b] Coordinates from Google Earth
[c] Mean annual
[d] Information given by respective municipality.

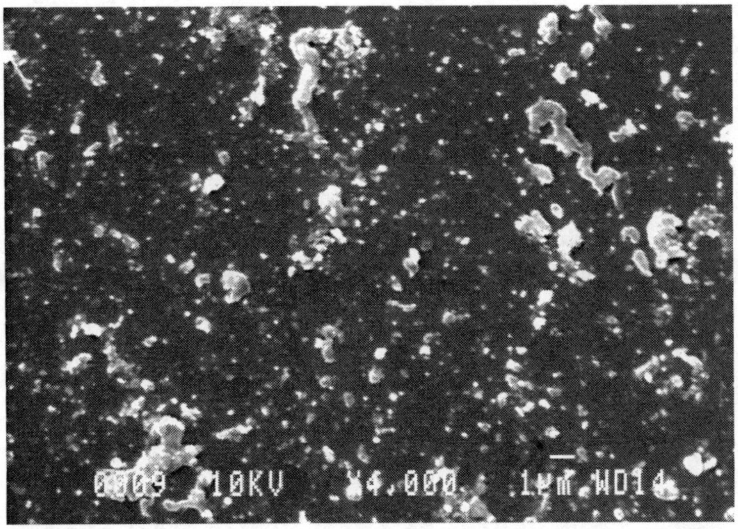

Figure 1. Scanning electron micrograph of one of the RRDP samples examined for pyrogenic effect (no. 6).

2.2. In vitro Pyrogen Test (IPT)

Blood was drawn from the forearm of healthy volunteers (personal in the laboratory) into heparinised tubes (BD Vacutainer, Plymouth, UK) and kept at room temperature until use within one hour. Heparin tubes could sometimes be contaminated by endotoxins [17]. Although this was not a problem in the present investigation, it is necessary that a batch of endotoxin-free tubes is tried out and used. Samples of RRDP (1 – 10^6 ng/ml), LPS (10^{-4} - 10^2 ng/ml from *Salmonella minnesota*, Sigma L-6261), zymosan B (10^{-1} to 10^5 ng/ml, Sigma Z-4250), or Curdlan (10^{-1} to 10^5 ng/ml, β-1.3-glucan from Wako Pure Chemical Industries ltd) were diluted in 0.9 % NaCl solution (Ecotainer® 0.9 % NaCl, B. Braun Melsungen, Germany). To examine if particles *per se* caused formation of IL-1β, experiments were performed in which well-characterised polystyrene particles (1 µg/ml in NaCl-solution) were incubated with whole blood as described above (uniform microspheres, PA02N, polystyrene particles with a mean diameter of 0.15 µm from Bang Laboratories Inc., Fishers, IN, USA). In some experiments the endotoxin inhibitor, polymyxin B, was used (1-10 µg/ml, Sigma P0972). RRDP (1, 10 or 100 µg/ml or LPS (1 ng/ml) were then treated with polymyxin B for 1 hour and incubated together with whole blood as described below. Each sample (200 µl) was mixed with 50 µl whole blood in wells of a 96-well plate (Polypropylene, Round bottom plate, Nunc A/S, Roskilde, Denmark, no; 267245) and incubated covered with a lid for 18 hours in a cell culture chamber (37 °C and 5% CO_2). Samples were then transferred to 1.5 ml Eppendorf tubes and centrifuged at 10000 x g for 3 minutes before 40 µl of the supernatant was pipetted in duplicate into anti IL-1β-precoated wells on a 384-well plate (Maxisorp, Nalge Nunc International, Rochester, USA, no; 460372).

The ELISA was performed with antibodies and standards from Diaclone (Human IL-1β Eli-pair, Diaclone, Besancon, France, nr; 851-610-010). The protocol from the manufacturer was followed with the exception that 384-well polystyrene plates were used instead of clear 96-well plates and chemiluminescence instead of absorbance to quantify the signal. These modifications made it possible to use less sample and reagents and also resulted in an improved sensitivity and a larger span in which the standard curve was linear. In brief, 40 µl coating antibody in 0.05M carbonate buffer pH 9.6 was added to each well and the plate was covered with adhesive tape and incubated in a refrigerator over night. After two washings with 100 µl washing solution (phosphate buffered saline with 0.05 % Tween 20), 100 µl of 5 % bovine serum albumin (BSA) in phosphate buffered saline was added and the plate was incubated for 2 hour at room temperature (RT) under agitation (200 r/min). After two additional washings with 100 µl washing solution, 40 µl of sample, standard, or blank and 40 µl of biotinylated detection antibody (diluted in PBS with 1 % BSA) were added to wells in duplicate and the plate were incubated for 3 h at RT under agitation (200 r/min). In each washing step the plate was tapped firmly on sheets of paper. After three washings with 100 µl washing solution/well, 40 µl Streptavidin-HRP (in PBS with 0.1 % Tween 20 and 1 % BSA) was added to each well. The plate was incubated for 20 min in RT and under agitation (200 r/min) before each well was washed three times with 100 µl washing solution and two times with 100 µl water. Finally 40 µl ECL plus (diluted 1+3 with MilliQ-water) was added to each well (Western Blotting Detection System, no; RPN 2132, GE Healthcare, Little Chalfont, UK). The plate was then incubated for 15 min in darkness before the chemiluminescence was read in a plate reader (Lumistar, BMG labtechnologies, Offenburg, Germany). The coefficient of variation (CV) when blood from one person was exposed for 1 ng LPS /ml and incubated in nine wells and then analysed as separate samples in duplicate was 7 % (n=9). When one blood sample was incubated with 1 ng LPS /ml in 12 wells and then pooled and analysed as 12 different samples in duplicate the CV was 6 % (n=12).

3. Results

The pyrogenic effect of RRDP (sample no. 6), LPS, zymosan B and Curdlan is shown in figure 2. The lowest concentration of RRDP that caused a detectable IL-1β formation was 10 ng RRDP/ml. At higher RRDP concentrations, the IL-1β formation increased in a dose-dependent way up to 1 mg RRDP/ml. On a mass basis, the RRDP sample was more pyrogenic than zymosan B and Curdlan but less pyrogenic than LPS. An elevated IL-1β formation was determined at LPS concentrations as low as 0.01 ng/ml, and the levels were increased in mixtures up to 1 ng LPS/ml where after the dose-response curve was flattened out. The dose-response curve for LPS did therefore appeared different from that of RRDP, zymosan B and Curdlan (figure 2).

To examine if pure particulate matter *per se* caused formation of IL-1β, whole blood was stimulated with 0.15 μm polystyrene particles. The formation of IL-1β in samples with polystyrene particles (1 μg/ml) was found to be around 40 pg IL-1β/ml which was just over the background and only 3 % of that caused by the same concentration of RRDP.

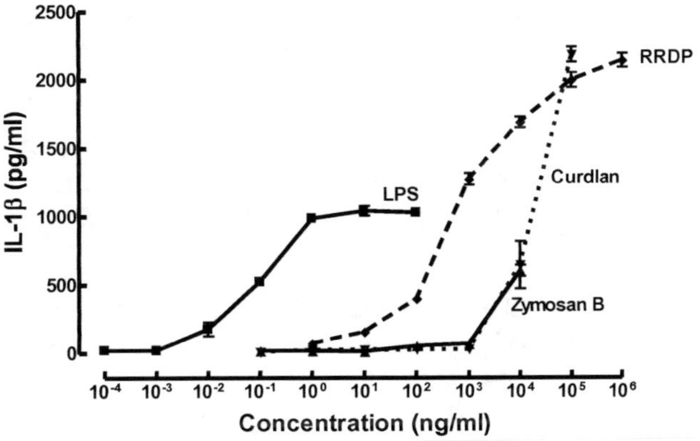

Figure 2. Formation of IL- 1β in diluted whole blood exposed for 18 h to different concentrations of RRDP (sample no. 6), Zymosan B, Curdlan or LPS. Results are presented as mean ± SEM. Further details are given in Materials & Methods.

As shown in figure 3, the LPS inhibitor, polymyxin B, largely abolished the pyrogenic effect of RRDP at low concentrations (1 μg/ml). Here, the IL-1β level in samples with 10 μg polymyxin B/ml was reduced by 82 % compared to that in samples without polymyxin B. Higher concentrations of polymyxin B did not reduce the IL-1β formation any further (not shown in the figure). On the other hand, at higher concentrations of RRDP (10 μg/ml), polymyxin B (10 μg/ml) reduced the IL-1β formation by only 37% and in samples containing 100 μg RRDP/ml it had no influence on the IL-1β formation (figure 3). The formation of IL-1β by LPS (1 ng/ml) was completely abolished by all tested concentrations of polymyxin B.

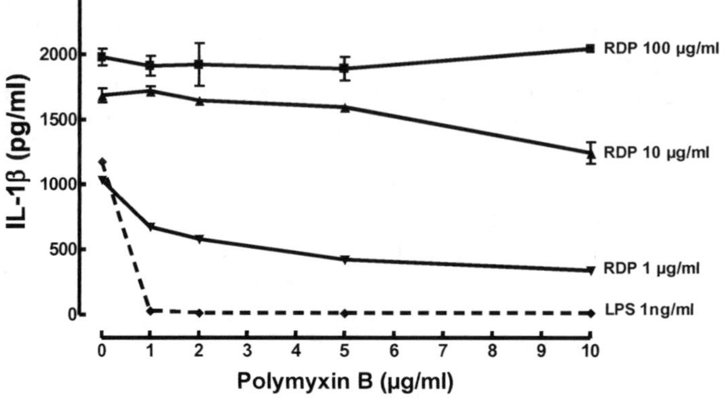

Figure 3. The influence of the endotoxin inhibitor, polymyxin B, on the formation of IL-1β in whole blood exposed to different concentrations of RRDP (sample no. 6). Results are presented as mean ± SEM. Further details are given in Materials & Methods.

Figure 4 shows the IL-1β formation from the six RRDP samples collected at different locations. All the samples had a pyrogenic effect and the most active (no. 6) produced 1.6 times more IL-1β than the least active. Samples 3, 4, 5 and 6 caused significantly more IL-1β than samples 1 and 2. The presence of polymyxin B (10 µg/ml) reduced the IL-1β formation substantially (81-90 %) in all the RRDP samples. No correlation was found between the formation of IL-1β and the mean annual PM_{10} or daily traffic intensity at the sample sites (table). The formation of IL-1β by LPS (10 ng/ml) was reduced by 94 % in the presence of polymyxin B.

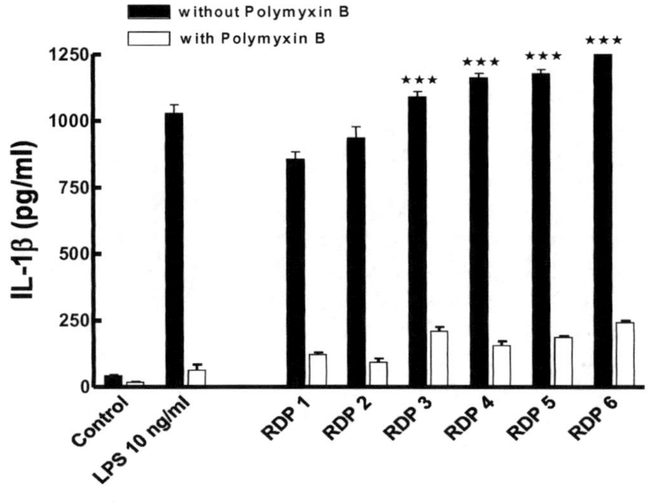

Figure 4. Formation of IL-1β in diluted whole blood incubated for 18 h with six different RRDP samples. RRDP (1 µg/ml) or LPS were incubated without (filled bars) or with (open bars) the endotoxin inhibitor, polymyxin B (10 µg/ml). Results are presented as mean ± SEM. Further details are given in Materials & Methods. *** = $p < 0.001$ compared with RRDP sample number 1.

4. Discussion

In this modified IPT, we used only 50 µl whole blood per sample and measured IL-1β with chemiluminescence ELISA in 384-well plates. This enabled a sensitive analysis of a large number of samples simultaneously. Increased levels of IL-1β were detected at RRDP concentrations as low as 10 µg/ml, which was tree orders of magnitude higher than that of LPS but three orders of magnitude lower than that of zymosan B and Curdlan, respectively (figure 2). The dose-response curve for RRDP was different from that of LPS: with increasing LPS concentrations, the IL-1β-formation increased up to a concentration of 1 ng LPS/ml and was then flattened out while no such effect could be seen with increasing concentrations of RRDP (up to 1 mg/ml). The presence of the endotoxin inhibitor, polymyxin B, resulted in a dose-related reduction in IL-1β formation in mixtures that contained 1µg RRDP/ml (figure 3). It is likely, therefore, that endotoxin played a significant role for the *in vitro* pyrogenic effect of the RRDP.

Endotoxin can bind to particle surfaces [18, 19] and the levels in urban ambient $PM_{2.5}$ particulate samples have been found to be higher in spring and summer than in winter [20]. A rather big week to week fluctuation in the endotoxin concentration has been found in particulate air samples collected from the same place, while a good correlation has been found between endotoxin levels on particles collected at the same time in two cities that lay 80 km apart [21] or at 40 different sites in and around Munich [22]. Previous studies have indicated an important role of endotoxin in the formation of cytokines (IL-6 and TNF-α) by cultured macrophages exposed to ambient particles [6]. Moreover, particles collected in the spring was found to cause more cytokine formation than particles collected in the winter, and it was suggested that this difference was mainly due to differences in endotoxin content [6]. Aside from endotoxin, there are a number of other cytokine-inducing compounds in bacteria [23], and β-glucans [24], fungal spores [25] and viral pathogens [26] can also act as pyrogens. It cannot be precluded that such non-bacterial pyrogens could contribute to the pyrogenic effects of RRDP as seen in the present study.

A number of epidemiological studies have shown that individuals exposed to high levels of particulate air pollution have an increased risk of acute cardiovascular events [27]. It is possible that this increased risk is the result of a particle-evoked inflammation in the lung with systemic cardiovascular involvement, or due to an extrapulmonary translocation of particles into the circulation with subsequent pro-atherosclerotic effects [28]. For the moment, it is not clear to what extent ambient ultrafine particles can be translocated into the bloodstream in humans. In rats, radiolabeled polystyrene particles with a diameter of 54.4 nm that had been instilled in the lung were found to enter into the circulation. The translocation was rather small, however, but increased significantly in animals that had been pre-treated with LPS, implicating that pulmonary inflammation enhances the extrapulmonary translocation of nanoparticles [29]. Nemmar and co-workers have found that carbon nanoparticles labelled with technetium-99 translocate from the lung into the circulation in humans [30] but this finding has not been confirmed by other investigators [31, 9, 32]. Thus, it is presently under debate if a systemic translocation of ultrafine carbon particles takes place in healthy human subjects and more research has to be conducted in order to elucidate this [32]. Yet, as mentioned above, animal studies have shown that pretreatment with LPS enhances the translocation of polystyrene nanoparticles [29] and it is possible that the presence of LPS (or other pyrogens) on RRDP might enhance their transfer into the circulation in humans. This is yet another reason why simple methods to measure pyrogens on RRDP might be useful. It should also be pointed out that endotoxin *per se* has great potential as a proinflammatory mediator of atherosclerosis [34] and that RRDP might act as a vector for the transfer of endotoxin into the circulation.

In summary, our results show that IPT can be used to measure the pyrogenic potential of ambient particulate matter. All RRDP tested had a clear pyrogen effect in this *in vitro* model. Endotoxin on the particles but also other factors contributed to the pyrogenic effect. As opposed to the *limulus* amebocyte lysate (LAL) assay (which measures endotoxin alone), IPT measures a broad range of pyrogens that may be present on particulate matter. The IPT method thus affords a simple, sensitive and quantitative determination of the total pyrogenic potential of ambient particles.

References

[1] Peters A, Wichmann H E, Tuch T, Heinrich J and Heyder 1997 Respiratory effects are associated with the number of ultrafine particles *Am. J. Resp. Crit. Care Med.* **155** 1376-83

[2] Schwartz J and Morris R 1995 Air pollution and hospital admissions for cardiovascular disease in Detroit, Michigan *Am. J. Epid.* **142** 23-35

[3] Miller K A, Siscovick D S, Sheppard L, Shepherd K, Sullivan J H, Anderson G L and Kaufman J D. 2007 Long-term exposure to air pollution and incidence of cardiovascular events in women *N. Engl. J. Med.* **356** 447-58

[4] Forastiere F, Stafoggia M, Picciotto S, Bellander T, D'Ippoliti D, Lanki T, von Klot S, Nyberg F, Paatero P, Peters A, Pekkanen J, Sunyer J and Perucci C A 2005 A case-crossover analysis of out-of-hospital coronary deaths and air pollution in Rome, Italy *Am. J. Resp. Crit. Care Med.* **172** 1549-55

[5] Veranth J M, Moss T A, Chow J C, Labban R, Nichols W K, Walton J C, Watson J G and Yost G S 2006 Correlation of *in vitro* cytokine responses with the chemical composition of soil-derived particulate matter *Envir. Health Persp.* **114** 341-9

[6] Salonen R O, Hälinen A I. Pennanen A S, Hirvonen M R, Sillanpää M, Hillamo R, Tingming S, Borm P, Sandell E, Koskentalo T and Aarnio P 2004 Chemical and *in vitro* toxicological characterization of wintertime and springtime urban-air particles with an aerodynamic diameter below 10 µm in Helsinki. *Scand. J. Work Envir. Health* **30** 80-90

[7] Sandhu R S, Petroni D H and George W J 2005 Ambient particulate matter, C-reactive protein, and coronary artery disease *Inhal. Tox.* **17** 409-13

[8] Rundell K W, Hoffman J R, Caviston R, Bulbulian R, Hollenbach A M 2007 Inhalation of ultrafine and fine particulate matter disrupts systemic vascular function *Inhal. Toxicol.* **19** 133-40

[9] Mills N L, Amin N, Robinson S D, Anand A, Davies J, Patel D, de la Fuente J M, Cassee F R, Boon N A, Macnee W, Millar A M, Donaldson K and Newby D E 2006 Do inhaled carbon nanoparticles translocate directly into the circulation in humans? *Am. J. Resp. Crit. Care Med.* **173** 426-31

[10] Inoue K, Takano H, Yanagisawa R, Hirano S, Kobayashi T, Fujitani Y, Shimada A and Yoshikawa T 2007 Effects of inhaled nanoparticles on acute lung injury induced by lipopolysaccharide in mice *Toxicology* **238** 99-110

[11] Devleeschouwer M J, Cornil M F and Dony J 1985 Studies on the sensitivity and specificity of the *Limulus* amebocyte lysate test and rabbit pyrogen assays *Appl. Envir. Microbiol.* **50** 1509-11

[12] Hartung T *et al* 2001 Novel pyrogen tests based on the human fever reaction. The report and recommendations of ECVAM Workshop 43. European Centre for the Validation of Alternative Methods. European Centre for the Validation of Alternative Methods *Altern. Lab. Anim.* **29** 99–123

[13] Kindinger I, Daneshian M, Baur H, Gabrio T, Hofmann A, Fennrich S, von Aulock S and Hartung T 2005 A new method to measure air-borne pyrogens based on human whole blood cytokine response *J. Immun. Meth.* **298** 143– 53

[14] Hasiwa M, Kullmann K, von Aulock K, Klein C and Hartung T 2007 An *in vitro* pyrogen safety test for immune-stimulating components on surfaces *Biomaterials* **28** 1367–75

[15] Weary M E and Wallin R F 1973 The rabbit pyrogen test *Lab. Anim. Sci.* **23** 677-81

[16] Hoffmann S *et al.* 2005 International validation of novel pyrogen tests based on human monocytoid cells *J. Immun. Meth.* **298** 161-73

[17] Redl H, Bahrami S, Leichtfried G and Schlag G. 1992 Special collection and storage tubes for blood endotoxin and cytokine measurements *Clin. Chem.* **38** 764-5

[18] Brooks R A, Wimhurs J A and Rushton N 2002 Endotoxin contamination of particles produces misleading inflammatory cytokine responses from macrophages *in vitro* *J. Bone Joint Surg.* **84** 291-5

[19] Cho D R, Shanbhag A S, Hong C Y, Baran G R and Goldring A R 2002 The role of adsorbed endotoxin in particle-induced stimulation of cytokine release *J. Orthop. Res.* **20** 704-13

[20] Carty C L, Gehring U, Cyrys J, Bischof W and Heinrich J 2003 Seasonal variability of endotoxin in ambient fine particulate matter *J. Environ. Monit.* **5** 953-8

[21] Heinrich J, Pitz M, Bischof W, Krug N, Borm P J A 2003 Endotoxin in fine ($PM_{2.5}$) and coarse ($PM_{2.5-10}$) particle mass of ambient aerosols. A temporo-spatial analysis *Atm. Envir.* **37** 3659-67

[22] Morgenstern V, Carty C L, Gehring U, Cyrys J, Bischof W and Heinrich J 2005 Lack of spatial variation of ambient particulate matter across a German metropolitan area *Atmo. Envir.* **39** 6931-41

[23] Henderson B and Wilson M 1996 Cytokine induction by bacteria: beyond lipopolysaccharide *Cytokine* **8** 269-82

[24] Brown G D and Gordon S 2001 Immune recognition. A new receptor for beta-glucans *Nature* **413** 36-7

[25] Tran P, Ahmad R, Xu J, Ahmad A and Menezes J 2003 Host's innate immune response to fungal and bacterial agents *in vitro*: up-regulation of interleukin-15 gene expression resulting in enhanced natural killer cell activity *Immunology* **109** 263-70

[26] Bowie A G and Haga I R 2005 The role of Toll-like receptors in the host response to viruses *Mol. Immunol.* **42** 859-67

[27] Mills N L, Törnqvist H, Robinson S D, Gonzalez M C, Söderberg S, Sandström T, Blomberg A, Newby D E and Donaldson K 2007 Air pollution and atherothrombosis *Inhal. Tox.* **19 Suppl 1** 81-9

[28] Duffin R, Mills N L and Donaldson K 2007 Nanoparticles- a thoracic toxicology perspective *Yonsei Med. J.* **48** 561-72

[29] Chen J, Tan M, Nemmar A, Song W, Dong M, Zhang G and Li Y 2006 Quantification of extrapulmonary translocation of intratracheal-instilled particles in vivo in rats: effect of lipopolysaccharide *Toxicology* **222** 195-201

[30] Nemmar A, Hoet PH, Vanquickenborne B, Dinsdale D, Thomeer M, Hoylaerts MF, Vanbilloen H, Mortelmans L and Nemery B 2002 Passage of inhaled particles into the blood circulation in humans *Circulation* **105** 411-4

[31] Brown J S, Zeman K L, and Bennett W D 2002 Ultrafine particle deposition and clearance in the healthy and obstructed lung *Am. J. Resp. Crit. Care Med.* **166** 1240–7

[32] Möller W, Felten K, Sommerer K, Scheuch G, Meyer G, Meyer P, Häussinger K and Kreyling W G 2008 Deposition, retention, and translocation of ultrafine particles from the central airways and lung periphery *Am. J. Resp. Crit. Care Med.* **177** 426-32

[33] Kreyling W G, Semmler-Behnke M and Möller W J 2006 Ultrafine particle-lung interactions: does size matter? *Aerosol Med.* **19** 74-83

[34] Stoll L L, Dennig G M and Weintraub N L 2004 Potential role of endotoxin as a proinflammatory mediator of atherosclerosis *Arterios. Thromb. Vasc. Biol.* **24** 2227-36

Effective biological dose from occupational exposure during nanoparticle synthesis

Evangelia Demou[1], Lang Tran[2] and Christos Housiadas[3]

[1] ETH Zurich, Institute of Environmental Engineering, Zurich, Switzerland
[2] Institute of Occupational Medicine, Edinburgh, UK
[3] "Demokritos" National Centre for Scientific Research, Athens, Greece

Email: evangelia.demou@ifu.baug.ethz.ch

Abstract. Nanomaterial and nanotechnology safety require the characterization of occupational exposure levels for completing a risk assessment. However, equally important is the estimation of the effective internal dose via lung deposition, transport and clearance mechanisms. An integrated source-to-biological dose assessment study is presented using real monitoring data collected during nanoparticle synthesis. Experimental monitoring data of airborne exposure levels during nanoparticle synthesis of $CaSO_4$ and $BiPO_4$ nanoparticles in a research laboratory is coupled with a human lung transport and deposition model, which solves in an Eulerian framework the general dynamic equation for polydisperse aerosols using particle specific physical-chemical properties. Subsequently, the lung deposition model is coupled with a mathematical particle clearance model providing the effective biological dose as well as the time course of the biological dose build-up after exposure. The results for the example of $BiPO_4$ demonstrate that even short exposures throughout the day can lead to particle doses of 1.10·E+08 #/(kg-bw·8h-shift), with the majority accumulating in the pulmonary region. Clearance of particles is slow and is not completed within a working shift following a 1 hour exposure. It mostly occurs via macrophage activity in the alveolar region, with small amounts transported to the interstitium and less to the lymph nodes.

1. Introduction

Ultrafine particles due to their increased specific surface area may present an increased hazard and risk to the health of humans and workers. This is especially true for the case of nanoparticles. These particles are engineered to have novel, enhanced and specific properties, which may not be toxicologically tested.

Assessment of nanoparticle health impacts requires the characterization of the exposure and equally important the estimation of intake through lung deposition, translocation from the deposition sites and the clearance of the deposited particles. Retention in the respiratory system of ultrafine carbon particles, for instance, is significant and up to 75% retention has been observed 24-hours from exposure [1]. Additionally, studies suggest that the role of macrophages in ultrafine and nanoparticle clearance from the lungs may not be sufficient [2] and occurs at a lesser extent than their micro-sized counterparts [3].

A number of models have been developed examining particle uptake in the respiratory system. The International Commission on Radiological Protection [4] developed the ICRP66 model to calculate particle deposition and internal doses with size differentiated uptake parameters [4]. More models have been developed since then, examining particle uptake as well as the distribution in the different organs of the body and clearance processes [5-7].

This paper presents an integrated source-to-biological dose study in which all above points, i.e., exposure, lung deposition and clearance of inhaled particles following exposure during nanoparticle production, are combined based on realistic data collected in an operating nanoparticle synthesis setting. The analysis of exposures incurred in real production settings of different particles allows for the examination of the differences and the importance of the aerosol dynamics in the internally delivered biological dose.

2. Methods

2.1. Methodology

Data from experimental monitoring of airborne exposure levels during the synthesis of nanoparticles in research laboratories is coupled with a transport and deposition model. This model considers aerosol dynamics to translate the exposure concentration to lung deposition based on the physical-chemical properties of the investigated particles. In a subsequent step, the lung deposition mathematical model is coupled with a mathematical model of particle clearance to provide the effective biological dose in target organs by inhalation for the exposed workers. The physiological parameters used were those of Caucasian adults undergoing light exercise.

2.1.1. Experimental monitoring. The experimental part involved detailed, size-resolved measurements of $CaSO_4$ and $BiPO_4$ airborne material during their synthesis. Number and mass particle concentrations as well as the size distributions of the airborne particles were monitored. A Scanning Mobility Particle Sizer (SMPS model 3936, TSI Inc) scanned in the 20 – 673 nm size range. Condensation Particle Counters (CPC 3022A and 3007, TSI Inc.) quantified number concentrations, while the DustTrak™ (model 8520, TSI Inc.) and SidePak™ (model AM510, TSI Inc.) monitored PM1 and PM10 mass concentrations, respectively. Besides monitoring results, the time-activity patterns of the workers present in the setting were also recorded during working hours. A survey on production, workers' behavior and employment duration was completed by all researchers providing estimates of total exposure duration in such settings. The resulting doses estimated from the monitored exposure conditions represent a worst case scenario, as particle retention by personal protective equipment is not considered.

2.1.2. *Lung deposition modelling.* The lung deposition mathematical model used in this study is a mechanistic model that solves in an Eulerian framework the general dynamic equation for polydisperse aerosols, taking into account the processes of growth and coagulation when significant [8]. Deposition is determined in all generations of the respiratory tract. Particle deposition is assumed to be the result of the mechanisms of gravitational settling, Brownian diffusion, inertial impaction, and flow turbulence, acting simultaneously. The employed modeling approach is shown to be specifically appropriate for the study of the lung deposition of nanoparticles [8]. This model has been extensively validated against a large body of experimental and numerical respiratory data, for both inert and hygroscopic aerosols [4, 9-15].

2.1.3. Retention and clearance modelling. The particle retention and clearance mathematical model describes the distribution of the internalized dose in different target systems (interstitium and lymph nodes) beyond the portal of entry organ, which is in this case the lung [7]. Additionally the model describes the time course of the build-up of the dose after exposure. The model has been calibrated with data from various experimental studies [7].

3. Results

3.1. Monitoring results

During the synthesis of these compounds using the flame-spray pyrolysis technique, the production of $CaSO_4$ resulted in elevated airborne concentrations of 5 to 13 times the background concentration and size distributions peaked between 115 and 170 nm (Figure 1) [16]. The latter compound displayed similar airborne size distributions with the peak ranging between 135 and 165 nm. However, the airborne exposure levels were 37 to 42 times the background level, mainly due to a production system change.

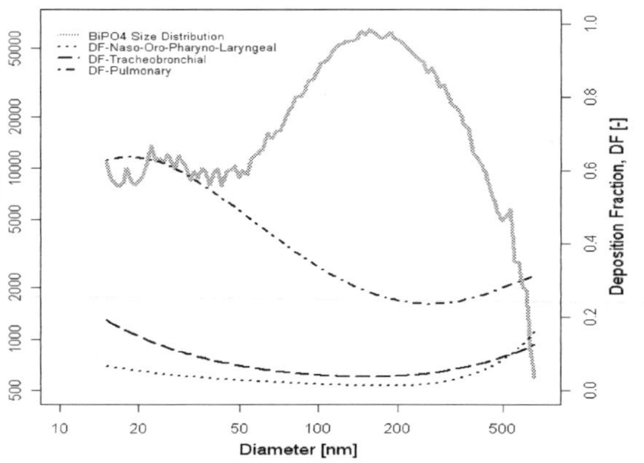

Figure 1. $BiPO_4$ number concentration during a single production. The estimated deposition fractions in the extrathoracic, tracheobronchial and pulmonary regions of the respiratory system are also plotted.

Production runs lasted between 8 to 9 min on average, whereas total production time for $CaSO_4$ and $BiPO_4$ was 162 and 63 min, respectively. During this time workers were within 1 meter from the production unit. Exposure conditions relapsed back to background with 1-2 min after production ends.

3.2. Particle lung deposition

The lung deposition mathematical model was used to estimate the particle specific deposition fractions (DF_i) of the emitted particles in the extrathoracic (NOPL), tracheobronchial (TB), and pulmonary (PL) regions of the human respiratory system (Figure 1) [8]. The deposited particle dose in terms of number concentration of particles in these regions was then estimated (Equation 1) by using the total exposure time during production and physiological parameters of Caucasian adults undergoing light activity [4].

$$D_i = \frac{C \cdot IR \cdot T}{BW} \cdot DF_i \qquad \textit{Equation}\ 1$$

where D_i is the particle dose in region i of the respiratory tract (# /(kg-bw · working shift)), C is the particle number concentration (#/cm^{-3}), IR the inhalation rate of a Caucasian adult (cm^3/h), T the duration of exposure per working shift (h/working shift), BW the body-weight of a Caucasian adult (kg-bw), and DF_i (-) the size differentiated particle deposition fraction in region i of the respiratory tract.

Figure 2a displays the variation in deposited particle dose for $BiPO_4$ in the different regions of the respiratory system per production run, as well as the total daily dose based on the exposure duration of 63 min. The total working shift deposited dose of $BiPO_4$ during synthesis amounts to 1.10·E+08 #/(kg-bw·8h shift). The maximum average deposited dose occurs in the pulmonary region and is 1.10·E+07 #/(kg-bw·8h shift).

3.3. Clearance Model

The particle specific pulmonary dose in terms of number concentration estimated with the lung deposition model was then fed to the retention and clearance model, assuming the same exposure conditions. Particle retention in the alveolar surface and clearance via macrophage activity and to the interstitium and subsequently to the lymph nodes was estimated. In the example of $BiPO_4$ exposure, the model shows that during exposure there is a steady build-up of particles in all four investigated regions. Once exposure ceases, clearance of inhaled particles from the alveolar free surface is very slow, whereas build-up continues throughout the remaining working shift in the other regions. Figure 2b shows the integrated particle dose over an 8h working shift following 63 min exposure at the beginning of the working day. Most particles within this time frame are retained in the alveolar region, whereas only a small fraction is transported to the interstitium and lymph nodes.

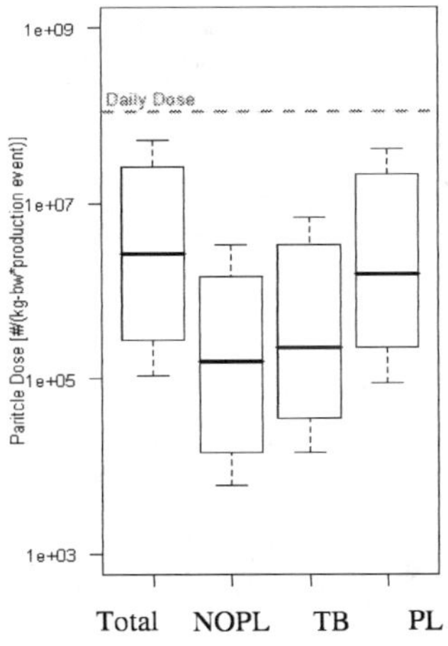

Figure 2a. $BiPO_4$ dose variation per production runs and daily dose

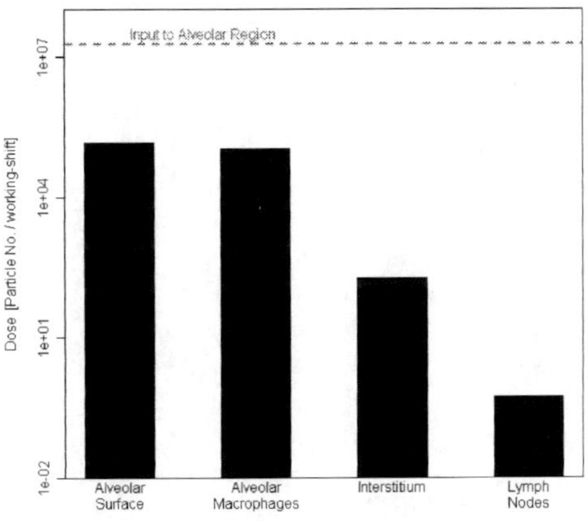

Figure 2b. $BiPO_4$ effective dose in alveolar, interstitium and lymph nodes, regions of the human body

4. Discussion and conclusion

The employed modeling approach is shown to be specifically appropriate for the study of the lung deposition of nanoparticles. On one hand, the Eulerian approach, contrary to the commonly employed Lagrangian approach, permits to properly account for the axial dispersion of the particles, which is of importance for very small particle sizes. On the other hand, the use of the general dynamic equation permits to fully account for aerosol processes, in particular hygroscopic growth and coagulation of very fine particles. Also, it allows expressing deposition in terms of a number of relevant exposure metrics, namely number or surface, and not only in terms of mass. This is of value because surface or number, as well as properties such as solubility are considered as more appropriate surrogates for describing the biological effects of nanoparticles.

The combination of the deposition model with the clearance model completes the description of the pathway from a worst-case external exposure to internal dose. Further research is required on the extension of this model to account for the distribution of the inhaled particles to other targets organs in the human body and to construct a mathematical description of the dose-response relationship. Further experimental data are needed to complete this step.

References

[1] Moller W, Felten K, Sommerer K, Scheuch G, Meyer G, Meyer P, et al. 2008 Deposition, retention, and translocation of ultrafine particles from the central airways and lung periphery. *Am J Respir Crit Care Med.***177**:426-32.

[2] Geiser M, Casaulta M, Kupferschmid B, Schulz H, Semirriler-Behinke M, Kreyling W. 2008 The role of macrophages in the clearance of inhaled ultrafine titanium dioxide particles. *American Journal of Respiratory Cell and Molecular Biology.***38**(3):371-6.

[3] Semmler-Behnke M, Takenaka S, Fertsch S, Wenk A, Seitz J, Mayer P, et al. 2007 Efficient elimination of inhaled nanoparticles from the alveolar region: Evidence for interstitial uptake and subsequent reentrainment onto airway epithelium. *Environ Health Perspect.***115**(5):728-33.

[4] ICRP. 1994 Human Respiratory Tract Model for Radiological Protection. *Annals of the ICRP.***Publication 66, 24**((1-3)).

[5] Hofmann W, Sturm R. 2004 Stochastic model of particle clearance in human bronchial airways. *J Aerosol Med-Depos Clear Eff Lung.***17**(1):73-89.

[6] Sturm R. 2007 A computer model for the clearance of insoluble particles from the tracheobronchial tree of the human lung. *Computers in Biology and Medicine.***37**(5):680-90.

[7] Tran CL, Jones AD, Cullen RT, Donaldson K. 1999 Mathematical modeling of the retention and clearance of low-toxicity particles in the lung. *Inhal Toxicol.***11**(12):1059-76.

[8] Mitsakou C, Helmis C, Housiadas C. 2005 Eulerian Modelling of lung deposition with sectional representation of aerosol dynamics. *Journal of Aerosol Science.***36**(1):75-94.

[9] Asgharian B, Hofmann W, Bergmann R. 2001 Particle deposition in a multi-path model of the human lung. *Aerosol Sci Technol.***34**:332-9.

[10] Blanchard JD, Willeke K. 1984 Total deposition of ultrafine sodium chloride particles in human lungs. *J Appl Physiol.***57**:1850-6.

[11] [Chan TL, Lippmann M. 1980 Experimental measurements and empirical modelling of the regional deposition of inhaled particles in humans. *AIHA J.***41**:399-409.

[12] Lippmann M. Regional depositionof particles in the human respiratory tract in: Handbook of Physiology, Section 9:Reactions to Environmental Agents. In: Society AP, editor. Bethesda, MD; 1977. p. 213-32.

[13] Londahl J, Massling A, Pagels J, Swietlicki E, Vaclavik E, Loft S. 2007 Size-resolved respiratory-tract deposition of fine and ultrafine hydrophobic and hygroscopic aerosol particles during rest and exercise. *Inhal Toxicol.***19**:109-16.

[14] Stahlhofen W, Gebhart J, Heyder J. 1980 Experimental determination of the regional deposition of aerosol particles in the human respiratory tract. *AIHA J.***41**:385-98.

[15] Tu KW, Knutson EO. 1984 Total deposition of ultrafine hydrophobic and hygroscopic aerosols in the human respiratory system. *Aerosol Sci Technol.***3**:453-65.

[16] Demou E, Hellweg S, Stark WJ, Pratsinis SE. Researchers exposure to particles in nanomaterial production laboratories. European Aerosol Conference 2007; Salzburg; 2007.

Simultaneous on-line size and chemical analysis of gas phase and particulate phase of mainstream tobacco smoke

J McAughey[1], T Adam[2], C McGrath[1], C Mocker[2,3] and R Zimmermann[2,3,4]

1 British American Tobacco, Group R&D Centre, Southampton SO15 8TL, UK

2 Institute of Ecological Chemistry, Helmholtz Zentrum München, Germany

3 Research Center for Environmental Health, D-85764 Neuherberg, Germany3Institute of Chemistry, University of Rostock, D-18051 Rostock, Germany

4 Bavarian Institute of Applied Environmental Research & Technology GmbH, D-86167, Augsburg, Germany

E-mail: john_mcaughey@bat.com

Abstract. Tobacco smoke is a complex and dynamic physical and chemical matrix in which about 4800 components have been identified. It is known that deposition efficiencies of smoke particles in the lung in the lung (60-80%) are greater than expected for smoke particles of 150–250 nm count median diameter (CMD). Various mechanisms have been put forward to explain this enhanced deposition pattern, including coagulation, hygroscopic growth, condensation and evaporation, changes in composition, or changes in inhalation behaviour. This paper represents one of three studies seeking to better quantify smoke chemistry, inhalation behaviour and cumulative particle growth. This information will improve dosimetry estimates in quantitative risk assessment tools as part of a harm reduction process. In this study smoke particle size and chemistry were measured simultaneously in real-time using electrical mobility spectrometry and soft-ionisation, time-of-flight mass spectrometry respectively. Qualitative puff-by-puff resolved yields of three selected compounds (acetaldehyde, phenol, and styrene) are shown and compared with particle number and count median diameter from different smoking intensities and filter ventilation. Yields of chemical analysis, particle diameter and concentration are in good agreement with the intensity of the smoking regime and the dilution of smoke by filter ventilation.

1. Introduction

Tobacco smoke is a complex and dynamic matrix consisting of more than 4800 compounds. It is composed of a gas phase and a particulate phase and many semi-volatile substances are partitioned between these two phases. The majority of substances can be found in the particulate phase; the gas phase consists of 400 to 500 individual compounds of which about 300 can be classified as semi-volatiles. Most of these smoke constituents are at trace levels. The chemical composition and partition between both phases can change continuously and is strongly influenced by time, temperature, chemistry and dilution of smoke [1].

Tobacco smoke analysis is usually carried out by off-line techniques with phases separate. Moreover, many analytical methods involve such techniques as trapping, storage, separation and derivatisation, all of which may alter smoke composition and phase. In addition, the results obtained are mostly for total yields in smoke from whole cigarettes. Therefore, short-time fluctuations and variations in concentration by puff are not resolved. This may be of particular interest for smoke particles which are known to be changing rapidly through coagulation and condensation [2]

In order to gain dose-based predictions of smoke components relevant for human smokers, it would be helpful to investigate particle diameter and the chemistry of both phases simultaneously during the smoking process in real time, in fresh rather than aged smoke.

Gas phase measurements by puff have been characterised for example by Fourier transform infrared spectroscopy [e.g.3] and gas chromatography-mass spectrometry [e.g. 4]. However, limiting factors for many techniques are either low time resolution, ageing of smoke, risk of artefact formation due to usage of chemicals or the fact that only a few substances can be analysed at the same time within complex spectra of primary compounds and their fragments..

In this context, two photon ionisation techniques (Single Photon Ionisation: SPI, Resonance Enhanced Multiphoton Ionisation: REMPI) coupled to time-of-flight mass spectrometry (TOFMS) have recently proven to be well suited for on-line analysis of gaseous and semi-volatile compounds of the cigarette smoking process [5].

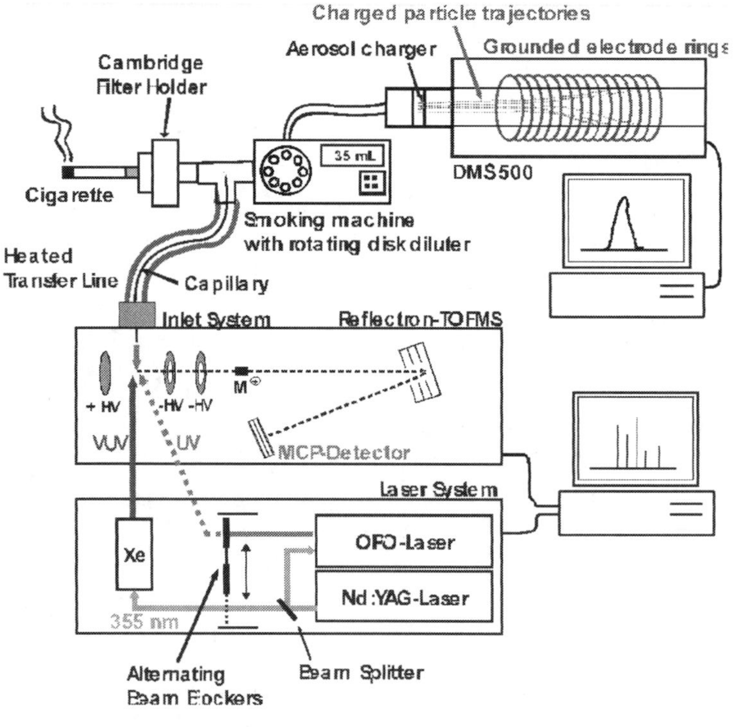

Figure 1. Schematic of experimental set-up of SPI/REMPI-TOFMS for on-line gas phase analysis, Cambustion DMS-500 electrical mobility spectrometer for on-line size particle analysis coupled to a Matter MD-19E combined cigarette smoking machine and diluter.

The REMPI technique uses at least two photons for photo-ionisation, which takes place via an optical resonance absorption step. Only those molecules that exhibit a suitable electronic transition with a respective excitation wavelength together with an ionisation energy (IE) lower than the combined energy of the photons, may be ionised by two-photon ionisation. The REMPI technique, in particular, is well suited for the on-line analysis of aromatic compounds [e.g.6]

The SPI technique using vacuum ultraviolet (VUV) photons for ionisation can be applied to detect additional compounds, e.g. aliphatic hydrocarbons or carbonyl compounds [e.g. 7]. As with REMPI, SPI generally causes no fragmentation of molecules. However, the selectivity of the SPI process is less compared to REMPI, because all compounds with an IP lower than the photon energy may be ionised. Despite this, most background gases such as nitrogen, oxygen, carbon dioxide and water are not subject to ionization, because their IEs are higher than the energy of the most commonly used VUV sources (<11 eV) and thus are not detected in the mass spectrometer. In this context, puff-by-puff resolved quantification of several potential toxicants in mainstream smoke by SPI-TOFMS has been reported recently [8,9].

For real-time particle sizing, a differential mobility particle spectrometer was used (Model DMS-500, Cambustion, Cambridge, UK). This technique enables real-time particle diameter and concentration measurements at 38 channels at 10 Hz resolution in the range 5 to 1000 nm [10].

A range of cigarettes of different tobacco type (Virginia, Burley and Oriental) and filter ventilation (0, 35 and 70%) were analysed for gas phase and whole smoke chemistry under two smoking regimes with measurements of particle diameter and concentration. Whilst multiple compounds were quantified, this paper will focus on chosen examples of vapour phase, particle phase and semi-volatile behaviour .

2. Methods

The smoke was provided by a smoking engine and diluter (Model MD-19E, Matter Engineering, Wohlen, Switzerland), set to a square-wave smoking profile controlling puff volume, puff duration, and puff frequency. Two puffing regimes were used; either a 35 ml (ISO) or 70 ml puff (INT) of 2 seconds duration, once every sixty seconds. A rotating disk diluter allows a 50:1 real-time dilution of the whole smoke, prior to particle diameter and concentration measurement by an electrical mobility spectrometer at 10 Hz resolution in the range 5–1000 nm (Model DMS-500, Cambustion, Cambridge, UK). A sub-sample from the excess non-diluted smoke was sampled directly by capilliary at 10 $ml.min^{-1}$, in real-time, for mass spectrometry. Two soft photo-ionisation techniques were applied; SPI with VUV ionisation at 118 nm and REMPI with laser pulses at 240 nm and 270 nm. For vapour phase only measurements, particulate material was intercepted with a 44mm diameter Cambridge filter pad located directly behind the cigarette filter. Cigarettes with Virginia tobacco were pre-conditioned at 22°C and 60% relative humidity for at least 48 hours prior to measurement.

A schematic of the combined sampling system is shown in Figure 1.

Measurements were carried out in triplicate. In order to qualitatively account for the total amount per cigarette all on-line recorded mass spectra were summed up resulting in a single summed mass spectrum and mean as well as standard deviation were calculated for the three measurements. The same procedure was carried out for all individual puffs of all measurement series leading to puff-by-puff resolved results for the SPI/REMPI-TOFMS. In principle, quantification of several compounds is possible by applying external gas standards as demonstrated previously [8,9]. In this study, the focus was on gaining a qualitative overview on a large number of species where there were known differences in the smoke particle size.

Regarding the DMS-500 results count median diameter, particle number, total surface area, and mass load were calculated for each individual puff as well as the whole cigarette by usage of the provided software package. Mean and standard deviation were calculated for the nine corresponding measurements (3 × SPI, 3 × REMPI at 240 nm, 3 × REMPI at 270 nm) of each cigarette type.

3. Results

The use of different levels of filter ventilation in the cigarettes and different flow rates of 1.05 $l.min^{-1}$ (ISO) and 2.1 $l.min^{-1}$ (INT) were chosen to give different transit times of smoke through the cigarette. As coagulation is the driving force for particle growth in this phase, it was anticipated that differences in particle diameter would be observed, against which any differences in chemical composition might be related.

Count median diameter (CMD) averaged over the cigarette varied from 182–260 nm (Table 1) and increased with increasing filter ventilation and lower puff flow rates, a consequence of increasing smoke residence time and coagulation within the rod. Puff-by-puff data showed increasing particle concentration and decreasing diameter as the tobacco was consumed and the coagulation period decreased (Figure 2a).

Table 1. Average count median diameter for 7 puffs per cigarette at 1.05 (ISO) and 2.1 l.min-1 (INT) (n=9)

Filter ventilation	0 %	35 %	70 %
ISO	182.8 + 2.6	194.0 + 2.7	219.8 + 4.5
INT	226.4 + 5.7	241.9 + 5.5	258.1 + 4.3

Particle concentration also changes with ventilation, smoking intensity and puff number as shown in Figure 2b; generally increasing coagulation time tends to give fewer, larger particles.

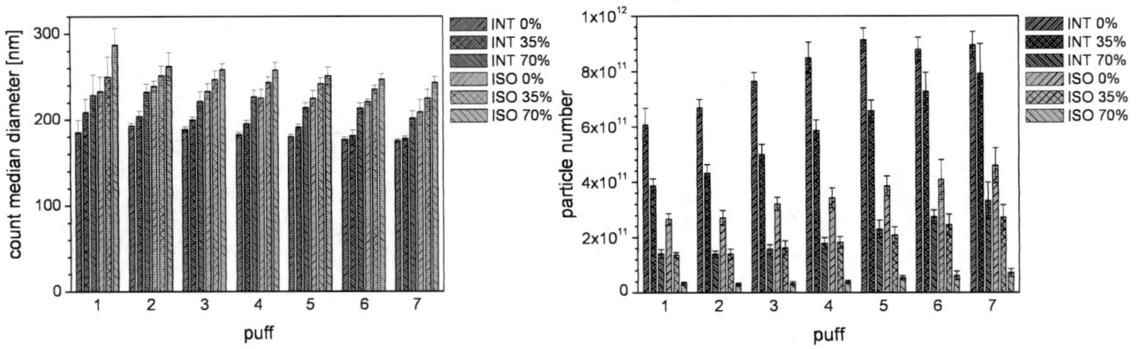

Figure 2 : Puff resolved count median diameter and particle number for Virginia cigarettes of 0, 35 and 70% filter ventilation under two puffing regimes

Initial mass spectrometry data show that most smoke constituents feature a continuous increase from the first to the last puff, as filtration by the tobacco rod, air dilution and outward gaseous diffusion decrease. However, there are some substances, in particular unsaturated hydrocarbons e.g. butadiene and isoprene (not shown here), which show different behaviour by having the highest amounts in the first puff. This is likely to be related to the different combustion and pyrolysis conditions when the cigarette is lit [8,9].

Figure 3 shows typical summed mass spectra demonstrating the wide variety of compounds accessible by the used photo-ionisation techniques, SPI at 118 nm (top), REMPI at 240 nm (middle), and REMPI at 270 nm (bottom). In all measurements a pure Virginia tobacco cigarette having a filter ventilation of 0 % was smoked under the 35 ml puffing smoking conditions from the first to the last puff. The, all on-line recorded mass spectra were summed with the three illustrations on the left side showing whole smoke, with vapour phase post-filter shown on the right. It can be seen that SPI is suited to detect lower molecular weight organic species, with a list of all accessible tobacco smoke constituents published elsewhere [11]. It can also be noted that the Cambridge filter does not greatly affect the compounds in the lower mass range below about 100 *m/z*. With the REMPI technique aromatic compounds and their homologues can be determined with great sensitivity.

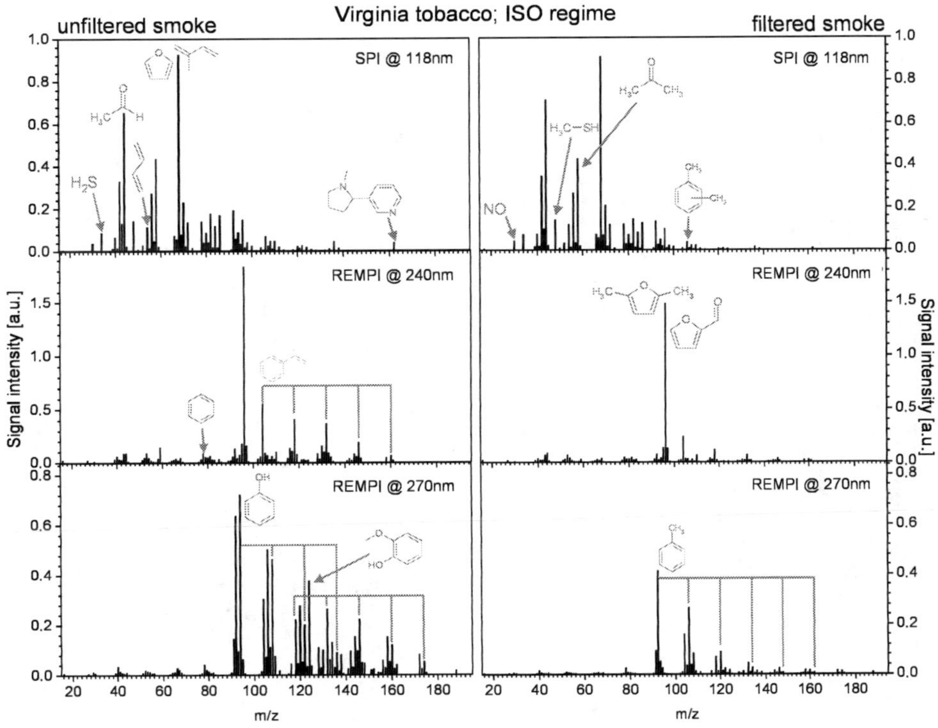

Figure 3: Summed mass spectra of filtered (gas phase) and unfiltered mainstream (whole smoke) smoke of a complete Virginia cigarette (0 % ventilation; ISO smoking regime) when ionised by SPI, REMPI at 240 nm, and REMPI at 270 nm.

To illustrate the physical behaviour of smoke constituents in different phases, puff-by-puff resolved yields for acetaldehyde (44 *m/z*), styrene (104 m/z) and phenol (94 m/z) were determined and plotted in Figure 4. For all measurements the yields increase with increasing puff number. In addition, it can be noted that the yields for the ventilated cigarettes decrease as the ventilation rate increases and increase when smoking under the more intense flow conditions [INT].

Comparing whole smoke (WS) and gas phase (GP) spectra highlights differing behaviour. For acetaldehyde, there are few observable differences for the puff-resolved values demonstrating that acetaldehyde is completely present in the gaseous phase. The same holds for various other species most of them having rather low molecular weights.

In contrast, styrene (104 *m/z*) features a different behaviour whereby the yields are lowered when a particle filter is applied to the smoke stream, supporting the hypothesis that styrene exhibits semi-volatile behaviour, partitioning between the vapour and particle phases.

A different behaviour was observed for phenol (94 *m/z*) which was almost completely trapped by the Cambridge filter, implying it is almost wholly associated with the particulate phase.

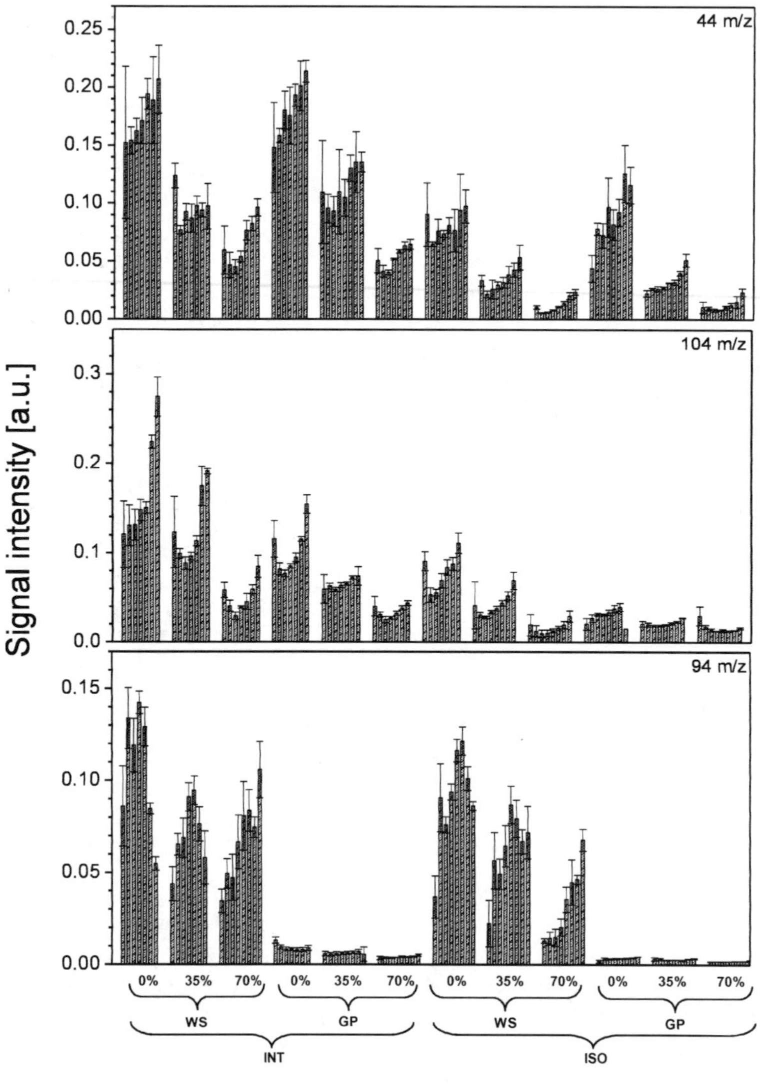

Figures 4 A to C: Puff-resolved signals of acetaldehyde (m/z=44), styrene (m/z=104), and phenol (m/z=94) for Virginia tobacco research cigarettes having 0 %, 35 % or 70 % filter ventilation, smoked

under the ISO or the intense smoking regime with the Cambridge filter pad incorporated (gas phase) or not (whole smoke)

In conclusion, the combined measurement system for aerosol size, concentration and composition in real-time using combined electrical mobility and soft-ionisation mass spectrometry has proven successful. The techniques have illustrated the influence of smoking intensity and filter ventilation on yields with clear discrimination between volatile semi-volatile, and involatile particulate-bound species. Further preliminary data analysis suggests that when normalised for puff by puff mass yield and excluding the lighting puff, many species remain at a relatively consistent proportion of the smoke composition, regardless of particle diameter. However, significant first puff elevations are observed for a number of chemical species. Further work will focus on improving real-time quantitative measurement by MS particularly for key toxicants and nicotine and sampling exhaled smoke. This will allow improved estimates of the efficiency and regional deposition of toxicant chemical species and particles to better improve dosimetry and quantitative risk assessment.

Acknowledgements
This work was partly funded by British American Tobacco.

References
[1] Baker, R. R. (1999) *in* Tobacco: Production, Chemistry, and Technology (Davis, L. D., and Nielsen, M. T., Eds.), pp. 398 - 439, Blackwell Science, Oxford, U. K.
[2] Phalen, RF, Oldham, MJ and Mannix, RC. (1994) *Aerosol Sci. Technol.* 20: 215-226.
[3] Parrish, M. E., Lyons-Hart, J. L., and Shafer, K. H. (2001) *Vibrational Spectroscopy* 27, 29-42.
[4] Thomas, C. E., and Koller, K. B. (2001) *Beiträge zur Tabakforschung International* 19, 345-351.
[5] Mitschke, S., Adam, T., Streibel, T., Baker, R. R., and Zimmermann, R. (2005) *Analytical Chemistry* 77, 2288-2296.
[6] Zimmermann, R., Heger, H. J., Kettrup, A., and Boesl, U. (1997) *Rapid Communications in Mass Spectrometry* 11, 1095-1102.
[7] Mühlberger, F., Zimmermann, R., and Kettrup, A. (2001) *Analytical Chemistry* 73, 3590-3604.
[8] Adam, T., Mitschke, S., Streibel, T., Baker, R. R., and Zimmermann , R. (2006) *Chemical Research in Toxicology* 19, 511 - 520.
[9] Adam, T., Mitschke, S., Streibel, T., Baker, R. R., and Zimmermann , R. (2006) *Analytica Chimica Acta* 572, 219 - 229.
[10] Symonds, J. P. R., Reavell, K. S. J., Olfert, J. S., Campbell, B. W., and Swift, S. J. (2007) *Journal of Aerosol Science* 38, 52 - 68.
[11] Mitschke, S. (2007) *PhD thesis,* Technische Universität München, pp. 221.

Real-time measurement of inhaled and exhaled cigarette smoke: Implications for dose

Conor McGrath[1], Nigel Warren[1], Philip Biggs[1] and John McAughey[1]

[1] British American Tobacco, Group R&D Centre, Southampton, SO15 8TL, UK

conor_mcgrath@bat.com

Abstract. Inhalation of tobacco smoke aerosol is a two-step process involving puffing followed by inhalation. Measured smoke deposition efficiencies in the lung (20-70%) are greater than expected for smoke particles of 150 – 250 nm count median diameter (CMD). Various mechanisms have been put forward to explain this enhanced deposition pattern, including coagulation, hygroscopic growth, condensation and evaporation, changes in composition, or changes in inhalation behaviour. This paper represents one of a series of studies seeking to better quantify smoke chemistry, inhalation behaviour and cumulative particle growth. The studies have been conducted to better understand smoke dosimetry and links to disease as part of a wider programme defining risk and potential harm reduction. In this study, the average CMD of inhaled smoke was 160 nm while the average CMD of exhaled smoke was 239 nm with an average growth factor of 1.5.

1. Introduction

Tobacco smoke is a complex and dynamic matrix consisting of gaseous compounds and particulate material, in which over 4800 constituents have been identified [1]. The subsequent deposition of smoke constituents in the respiratory tract of smokers is an equally complex process both in terms of quantification of dose and location. Indeed an improved understanding of the dose and deposition location of tobacco smoke constituents in the human airways may well elude further information regarding the mechanisms of tobacco related diseases. The paper of Baker and Dixon [2] describes the work which has been carried out in the last century. The over-riding message from this paper is that there still remains a great deal of uncertainty as to the principal driving mechanisms governing tobacco smoke deposition in the human airways. A number of studies have also investigated the links between deposition and the physical properties of the aerosol, most notably the particle size of the inhaled aerosol. Richardson [3] noted that the retention of smoke particles in the human lung was higher than would be expected on the basis of their size when measured as they enter the mouth during smoking. Richardson suggested that the aerosol must grow within the humid environment of the human airways to alter the particle size distribution and therefore result in enhanced deposition [3].

The ambiguity which exists between experimental data and predicted data from deposition models [4, 5] has still not been fully addressed in the literature. Tobacco smoke particles are typically 150 – 250 nm CMD [6] and their high concentration (10^{12} particles per cigarette) and the hygroscopic nature of the smoke droplets means that this particle diameter may change rapidly via coagulation and condensation [7,8]. Despite the small diameter of the smoke particles, smoke deposition efficiencies of 40-97% in the lung have been reported in the review of Baker and Dixon [2].

This paper describes a novel method to measure the puff by puff count median diameter (CMD) of inhaled smoke, post smoking by replicating the measured smoking profile, and also the puff by puff diameter of exhaled smoke in real time, therefore allowing a puff by puff growth factor to be calculated. This will define a boundary for the cumulative effect of growth and shrinkage of the smoke particles in the body. Particle deposition efficiencies were also measured, in separate exhale capture experiments using solanesol as a tobacco-specific chemical marker for smoke particles.

2. Methods

The inhaled and exhaled smoke diameters and solanesol deposition efficiencies were measured for seven volunteer smokers (5 males; 2 females) from whom informed consent had been obtained. These measurements were made with a research filter-tipped cigarette of conventional construction, using only Virginia tobacco, with an ISO tar yield of 8.8 mg per cigarette, measured as nicotine-free dry particulate matter (NFDPM). During smoking a pre-calibrated portable smoking analyser was used to record puff profiles (puff number, volume, interval and duration) for every experimental run.

Particle size and concentration measurements for smoke used a fast, electrical, differential mobility spectrometer (DMS-500, Cambustion, UK) at 10 Hz resolution [9, 10]. After each puff and inhalation, volunteers exhaled through a tube, from which a sub-sample of the total exhalate was taken into the DMS-500 particle spectrometer. The spectrometer sampling line requires a constant flow rate of 8.5 $l.min^{-1}$ and therefore variable dilution of the exhaled smoke occurred. Thus, exhaled smoke particle size can be accurately measured at this time, but not exhaled particle number or mass concentration. Sample losses in transport to the spectrometer were calculated to be <0.1% for 150 - 270 nm CMD particles in an 8 mm diameter tube at 8.5 $l.min^{-1}$.

Each volunteer smoked three replicates and inhaled to a depth of their choice, as per their normal smoking behaviour. This is referred to as 'free smoking', or VC_{free}. Each volunteer took a puff and inhalation once a minute, and then immediately exhaled into the exhalate sampling line.

The inhaled particle size distribution was subsequently measured by re-smoking cigarettes on a smoking simulator (Smoking Cycle Simulator (SCS), Cambustion, UK) using human puffing profiles which had previously been recorded by the portable smoking analyser. The SCS has been described previously [10]. In short, the simulator continually delivers smoke and diluting air to the spectrometer via a critical orifice. At any time-point, flow through the cigarette, and primary dilution is regulated by altering the diluting air flow at 12.5 Hz resolution. Secondary dilution within the spectrometer is also controlled; thus mass is conserved in the sampling system and particle number and mass concentration can be quantified. The SCS was connected to the DMS500 spectrometer allowing the inhaled particle size distribution for each volunteer's smoking profile to be measured. Measurements on each smoking profile were recorded in triplicate, giving nine inhaled profiles per volunteer.

For calibration purposes, the puff volumes generated by the SCS were compared against the portable smoking analyser quality control (QC) file, which consists of Bell, Triangle, Square and Near Triangle puffs at volumes of 25, 50, 75 and 100 ml. The QC file was played by the SCS with the smoking analyser used to record the actual puff volumes achieved. Figure 1 shows a snapshot of the agreement between the SCS volumes and the actual volumes from the QC file, while Figure 2 shows the good agreement between a complex human smoking profile recorded from a lighting puff, as recorded by the smoking analyser and as re-created on the SCS.

Particle deposition efficiency in the lung was estimated by Liquid Chromatography-Mass Spectrometry measurement of solanesol, a tobacco-specific high boiling point alcohol ($C_{45}H_{74}O$: MW = 630) which remains associated with the particulate phase of smoke [11] and represents approximately 3% of the particulate matter mass. For any cigarette type, the solanesol as a proportion of the particulate matter (by mass or UV measurement) remains constant at smoking flow rates from 1.05 to 4.5 $l.min^{-1}$, covering measured human smoking flows. Inhaled particulate phase solanesol was calculated from solanesol measurement in the residual cigarette filter tip using filtration efficiency measurements from prior calibrations. Exhaled solanesol was measured from an exhalate capture filter pad. These two values were used to calculate particle deposition efficiency in the respiratory system for each test.

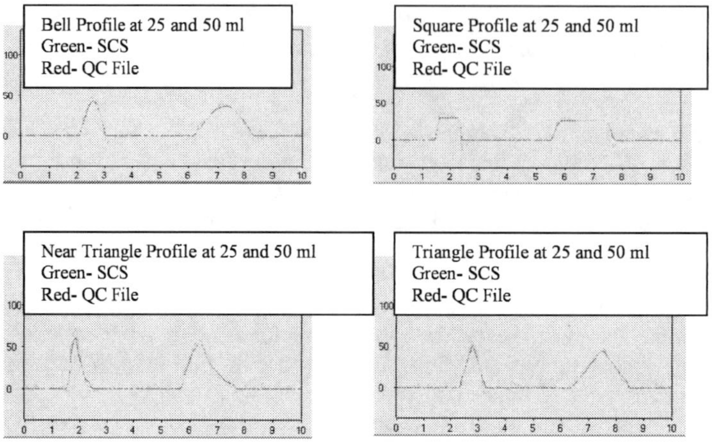

Figure 1 SCS Profile shape and volumes

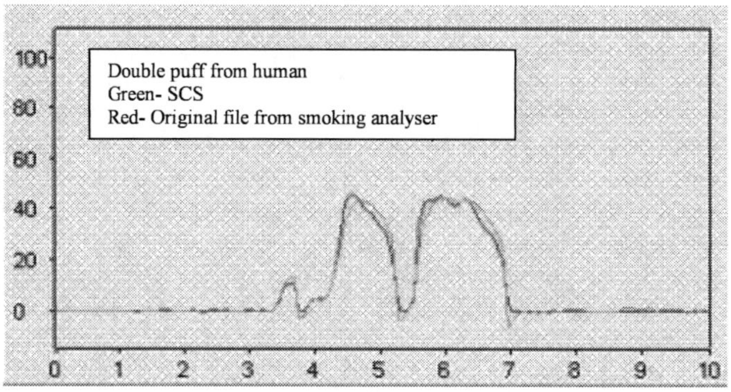

Figure 2: Human Double Puff Profile on the SCS

3. Results

Inhaled smoke particle size was calculated from a weighted average of the diameter of individual puffs and was found to be in the range of 150–175 nm CMD across the seven subjects. Equivalent exhaled smoke particle size in the range of 220–270 nm CMD with relatively consistent growth factors across volunteers, but with differences in absolute growth values between volunteers (1.3–1.8), consistent with measured smoking behaviour. The resultant mean particulate (solanesol) deposition estimate was found to be 49±17%. Table 1 shows the mean inhaled and exhaled CMD, the mean solanesol deposition and the mean growth factor from all volunteer measurements. There was also good consistency in terms of the inhaled CMD between the three replicates of each replayed smoking profile using the SCS. The average coefficient of variation for the inhaled CMD for nine measurements per subject (triplicate measures of each of three profiles) was found to be 3%.

Table 1: Summary data from all volunteer measurements

Subject	Inhaled CMD (nm)	Exhaled CMD (nm)	Growth Factor	Solanesol Retention (%)
1	152.2 ± 4.1	267.3 ± 5.8	1.76	52
3	157.5 ± 4.0	238.0 ± 3.1	1.51	62
9	158.9 ± 5.9	227.4 ± 0.4	1.43	49
12	161.8 ± 2.6	220.8 ± 2.2	1.37	21
102	156.8 ± 5.1	260.3 ± 5.1	1.66	57
105	174.4 ± 4.7	230.5 ± 3.9	1.32	67
112	159.4 ± 3.9	227.9 ± 3.2	1.43	22
Mean	160.1 ± 6.4	238.9 ± 16.5	1.50 ± 0.15	47 ± 17

An example of inhaled versus exhaled size on a cigarette-weighted average basis for each volunteer inhaling to a VC_{free} inhalation depth is shown in Figure 3. In general, normalized for all subjects, inhaled puffs became progressively smaller puff-by-puff as the residence time in the cigarette decreased, thus decreasing the time available for coagulation. This trend was significant and is shown in Figure 4. At the high concentrations in fresh inhaled smoke, coagulation is the principal mechanism for growth. In contrast, the particle size for all exhaled puffs normalised across all subjects was reasonably consistent from puffs 1-6, suggesting subsequent coagulation, hygroscopic, condensation and evaporation mechanisms in the lung were complete by the time of exhalation. In practice, this pattern suggested a progressive increase in growth factor puff-by-puff, observed in all subjects.

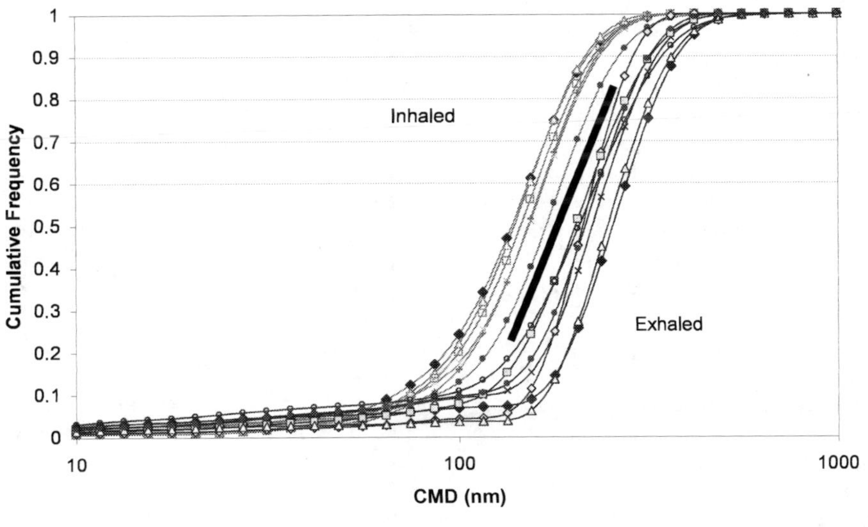

Figure 3: Cumulative frequency plots for inhaled (red) and exhaled (blue) particle size (CMD) for each volunteer

The data from Figure 3 represent the largest particles observed, but it is notable that > 90% of the exhaled particles are smaller than 500 nm CMD.

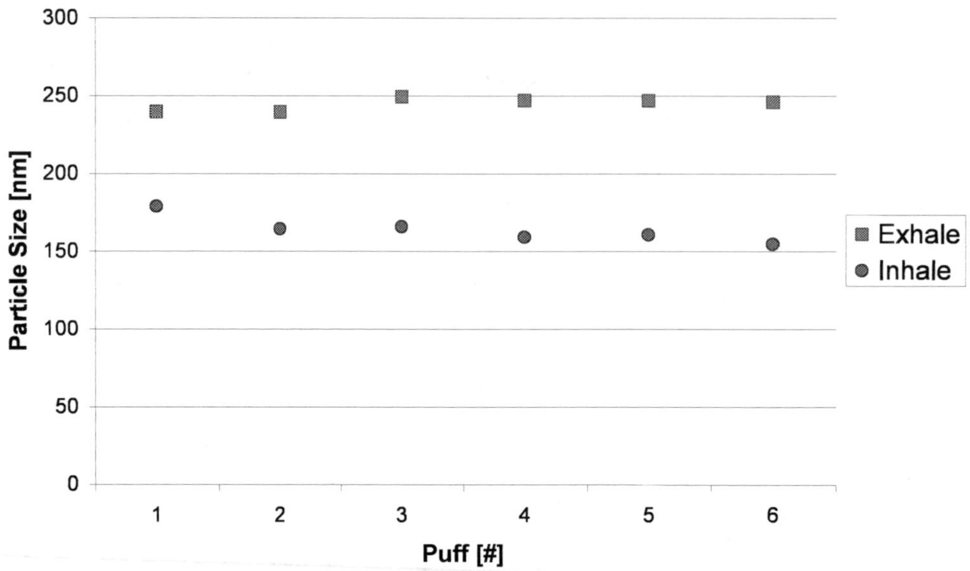

Figure 4 - ANOVA general linear model fitted means for inhaled and exhaled particle size (CMD) against puff number

4. Discussion

The work carried out here was a study designed to test and validate new measurement tools to help understand particle size dynamics in the smoking process. This is important as observed particle deposition efficiencies (e.g. 47 ± 17% in this study) are significantly greater than those predicted by deposition models [4,5] of 10-20% for particles of the diameter of smoke. The models reflect deposition by a combination of impaction sedimentation and Brownian Motion, with a lung deposition minimum typically observed at approximately 500 nm. Various authors [12-14] have subsequently modelled particle growth through a combination of coagulation, hygroscopic growth, charge, breathing patterns, colligative (dense cloud) behaviour and hydrodynamic interactions with the surrounding air. These are addressed more fully in a recent review by Longest and Xi [15]. In their study, subsequent modelling of condensational growth suggested that smoke may grow to several microns in diameter, although this is not observed experimentally.

The cumulative growth factors of 1.3–1.7 measured here are consistent with previous published data [16] but do not achieve the cumulative growth calculated by modelling approaches. An alternative model suggests an alternative competing mechanism of particle shrinkage by evaporation and selective deposition of semi-volatile species, e.g. nicotine [17}, and carbonyl species [18] which forms a significant proportion of the undiluted particle mass. If particle diameters become smaller, this would also enhance deposition relative to modelled data.

Our data reported here show clear evidence of particle growth of cigarette smoke from inhalation to exhalation. However, it is not yet possible to fully deconvolute these data to give a full understanding of the competing mechanisms involved at various parts of the smoking and inhalation cycle. Two areas in particular need further work. Further data is required to measure hold times in the mouth and lung and to measure inhaled and exhaled flow. It is also possible that selective deposition of the smaller particles may increase the exhaled d_{50} and enhance the measured growth factor.

In conclusion, the use of time-resolved electrical mobility analytical techniques is offering fresh insight into the aerosol behaviour of tobacco smoke. In turn, this further improves our understanding of the mechanisms driving regional deposition and local dose at each generation in the lung. A combination of these physical data, along with data on chemical composition and puffing and inhalation behaviour will allow a better understanding of effective dose and risk. Characterisation of these dosimetric processes offers scope for better understanding of sites and causes of disease, with appropriate dose application to *in vitro* models of disease.

Ethical Considerations

This study was conducted in line with the principles of the Declaration of Helsinki. Volunteers were all current smokers recruited from the workforce at BAT GR&D Centre. Each volunteer was interviewed and provided with study information prior to enrolment. Informed consent was obtained from each participant and each was made aware they were free to leave the study at any time.

References

[1] Baker, RR 1999. Smoke Chemistry. In *Tobacco: Production, Chemistry, and Technology*; Blackwell: Oxford, U.K.; 398
[2] Baker RR. and Dixon M 2006 Inhalation Toxicology **18,** 255–294
[3] Richardson RB 1975 *BAT R&D Report* RD.1185-R
[4] ICRP International Commission on Radiological Protection 1994. Human Respiratory Tract Model for Radiological Protection. **24** (1-3). Pergamon, Oxford
[5] National Council on Radiation Protection and Measurements 1997 NCRP Report No. 125 Bethseda, Maryland
[6] Bernstein, DM 2004 Inhalation Toxicology. **16**: 675-689
[7] Ingebrethsen, B.J. 1986 Rec. Adv. Tob. Sci. **12**: 54-142
[8] McRae, D.D. 1990 Rec. Adv. Tob. Sci. **16**: 233-323
[9] Reavell, K. *et al*. 2002. SAE Technical Paper, 2002-01-2714
[10] McAughey *et al* 2007. American Association for Aerosol Research 26th Annual Conference, Reno NV.
[11] Ogden M.W. and Maiolo K.C. 1989. Environ. Sci. Technol. **23**: 1148-1154.
[12] Phalen RF *et al* 1994. Aerosol Science & Technology **20:** 215-226
[13] Robinson RJ and Yu CP 2001. Aerosol Science & Technology **34:** 202-215.
[14] Broday DM and Robinson R 2003. Aerosol Science & Technology **37:** 510-527.
[15] Longest PW and Xi J 2008. Aerosol Science & Technology **42:**579-602.
[16] Hicks JF *et al* 1986. In: *Aerosols, Formation and Reactivity* Pergamon, Oxford, 244-247.
[17] Ingebrethsen BJ 2006. Inhalation Toxicology **18**:1071-1076.
[18] Moldoveanu S *et al* 2007. Beitr. Tabakforschung Int. **22:** 346-357.

Journal of Physics: Conference Series **151** (2009) 012019 doi:10.1088/1742-6596/151/1/012019

Puffing and inhalation behaviour in cigarette smoking: Implications for particle diameter and dose

Colin Dickens, Conor McGrath, Nigel Warren, Philip Biggs and John McAughey

British American Tobacco, Group R&D Centre, Southampton, SO15 8TL, UK

colin_dickens@bat.com

Abstract. Inhalation of tobacco smoke aerosol is a two-step process involving puffing followed by inhalation. Measured smoke deposition efficiencies in the lung (20-70%) are greater than expected for smoke particles of diameter 150 – 250 nm CMD. Various mechanisms have been put forward to explain this enhanced deposition pattern, including coagulation, hygroscopic growth, condensation and evaporation, changes in composition, or changes in inhalation behaviour. This paper represents one of a series of studies seeking to better quantify smoke chemistry, inhalation behaviour and cumulative particle growth. The studies have been conducted to better understand smoke dosimetry and links to disease as part of a wider programme defining risk and potential harm reduction. In this study, it was noted that particle deposition increased with increasing inhalation depth, and that smoke inhalation volumes were generally greater than normal tidal breathing volumes. A weak association was observed between particle diameter and puff flow, but no strong association between particle diameter and retention efficiency.

1. Introduction

Tobacco smoke is a complex and dynamic matrix consisting of gaseous compounds and particulate material, in which over 4800 constituents have been identified [1]. The subsequent deposition of smoke constituents in the respiratory tract of smokers is an equally complex process both in terms of quantification of dose and location. Indeed an improved understanding of the dose and deposition location of tobacco smoke constituents in the human airways may well elude further information regarding the mechanisms of tobacco related diseases. The paper of Baker and Dixon [2] describes the work which has been carried out in the last century. The overriding message from this paper is that there still remains a great deal of uncertainty as to the principal driving mechanisms governing tobacco smoke deposition in the human airways. A number of studies have also investigated the links between deposition and the physical properties of the aerosol, most notably the particle size of the inhaled aerosol.

Smoking behaviour involves drawing a puff into the mouth cavity with the soft palate closing the throat. Puff volume may be up to 100-120ml. The particle size of the smoke aerosol is principally dependent on its residence time in the cigarette and the time available for coagulation. But it will also be influenced by how long the puff is held in the mouth. Therefore the higher the flow intensity during puffing, the smaller the initial particle size. In contrast, the higher the ventilation level of the cigarette filter, the slower the smoke flows through the rod - resulting in a larger initial particle size. The particle size will continue to grow through coagulation as it is held in the mouth.

The ambiguity which exists between experimental data and predicted data from deposition models [3] has still not been fully addressed in the literature. The published review by Bernstein [4] supports the hypothesis that smokers inhale more deeply than normal tidal breathing.

2. Methods

Informed consent was obtained from seven volunteer smokers (5 males; 2 females) and each given a unique identifier code from a database used for all volunteer or assessment panel studies within the laboratory. The puffing behaviour, inhaled and exhaled particle diameter, and solanesol deposition efficiencies were measured for these seven volunteer smokers at two controlled and one free inhalation depth. These measurements were made using a research cigarette of conventional construction, made only from Virginia tobacco, with an ISO tar yield of 8.8 mg per cigarette as nicotine-free dry particulate matter (NFDPM).

Smoking behaviour was measured using a Smoking Analyser (SA7), developed in-house. Flows leave the cigarette and are measured inside the holder as they move to the smoker's mouth. The SA7 DAT unit uses a pressure transducer to measure pressure differences at 25Hz and converts them to flow and volume values against known flows and pressures. Data on puff number, duration, interval, flow and volume are also analysed. The human profile can be replicated off-line to measure total smoke or individual constituent deliveries, or to measure particle diameter. The screenshot in Figure 1 from the SA7 shows puff flow, puff volume, puff duration, elapsed time, pressure drop and (pre-calibrated) optical tar.

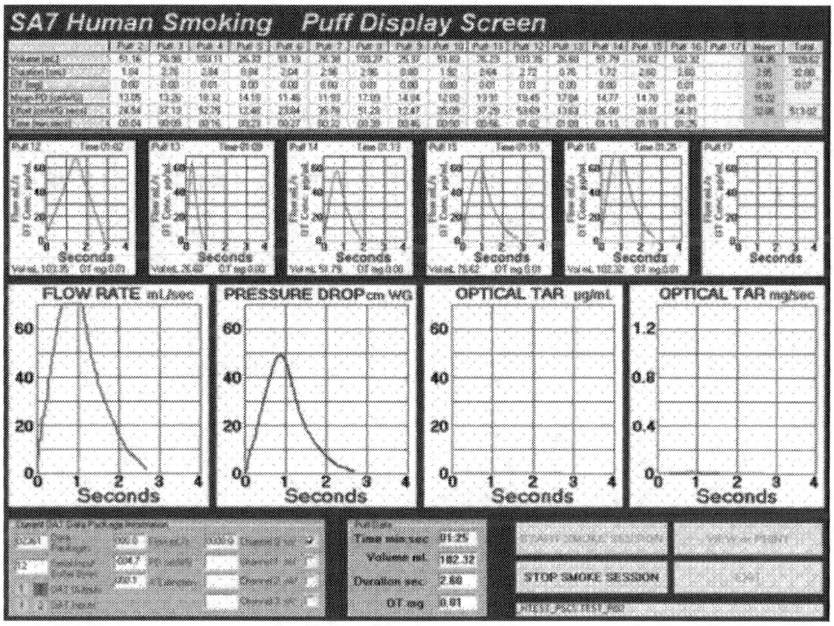

Figure 1 : Smoking profile analyser

Inhalation behaviour was controlled by pre-screening and measuring the lung vital capacity (VC) of each subject. Pre-prepared inhalation volumes of 15% and 25% of their vital capacity (VC_{15} and VC_{25}) were used as well as a free inhalation regime, under the free will of the subject. Three measurements were conducted per inhalation profile, with a further three free inhalation replicates used in a parallel experiment to measure exhaled particle diameter

Particle deposition efficiency in the lung was estimated by Liquid Chromatography-Mass Spectrometry measurement of solanesol, a tobacco-specific high boiling point alcohol ($C_{45}H_{74}O$: MW = 630) which remains associated with the particulate phase of smoke [5] and represents approximately 3% of the particulate matter mass. For any cigarette type, the solanesol as a proportion of the particulate matter (by mass or UV measurement) remains constant at smoking flow rates from 1.05 to 4.5 $l.min^{-1}$, covering measured human smoking flows. Inhaled particulate phase solanesol was calculated from solanesol measurement in the residual cigarette filter tip using filtration efficiency measurements from prior calibrations. Exhaled solanesol was measured from an exhalate capture filter pad. These two values were used to calculate particle deposition efficiency in the respiratory system for each test. Solanesol losses in the exhale capture system were assessed by wiping the internal surfaces of the pad holder, with no measureable losses detected.

Particle size and concentration measurements for smoke used a fast, electrical, differential mobility spectrometer (DMS-500, Cambustion, UK) at 10 Hz resolution [6,7] with the inhaled particle size distribution subsequently measured by re-smoking cigarettes on a smoking simulator (Smoking Cycle Simulator (SCS), Cambustion, UK) using human puffing profiles which had previously been recorded by the SA7 portable smoking analyser. The SCS has been described previously [7]. Sample losses in transport to the spectrometer were calculated to be <0.1% for 150 nm CMD particles in an 8 mm diameter tube at 8.5 $l.min^{-1}$ (8).

3. Results

3.1. Particle size and puffing behaviour

Table 1 shows average puffing behavior for each volunteer, showing significant differences in the total puff volume and average puff durations between volunteers, but less variation in the average puff flow.

The data were analysed to look for correlation between the puffing behavior and the particle size of the inhaled smoke. Puffing behavior was recorded in terms of total puff volume, total puff duration and average puff flow. Statistical analysis showed inhaled particle size was not correlated to total puff volume or total puff duration. However, there was a weak correlation between inhaled particle size and average puff flow, with particle size decreasing with increasing puff flow (Figure 2). Linear regression between inhaled particle size and average puff flow gave a coefficient of determination of 17.1%.

Table 1 - Puffing behaviour and inhaled particle size per cigarette for each volunteer averaged over triplicate measurements and quoted with ± standard deviation

Subject	Total Puff Volume (ml)	Mean Puff Volume (ml)	Average Puff Duration (s)	Average Puff Flow ($ml.s^{-1}$)	CMD (nm)
1	572	95 ± 2	3.3 ± 0.2	29.5 ± 2.4	154 ± 5
3	576	72 ± 1	1.7 ± 0.3	40.7 ± 1.7	156 ± 4
9	376	63 ± 13	1.9 ± 0.3	35.0 ± 2.9	159 ± 9
12	400	57 ± 1	1.7 ± 0.1	34.5 ± 1.1	159 ± 1
102	717	119 ± 7	2.6 ± 0.5	27.5 ± 3.0	160 ± 8
105	440	73 ± 4	1.5 ± 0.2	35.0 ± 0.9	172 ± 7
112	441	55 ± 5	3.3 ± 0.2	29.5 ± 2.4	156 ± 1

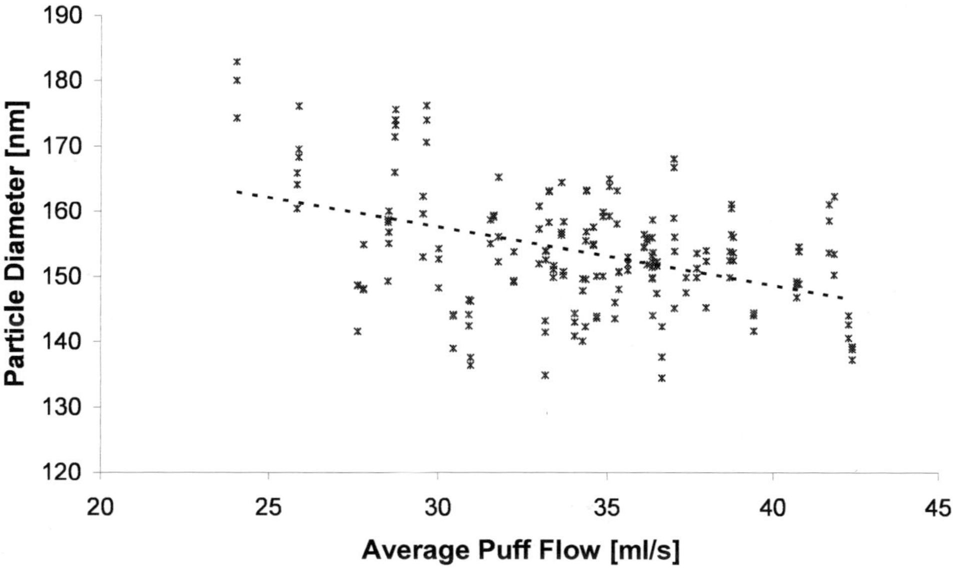

Figure 2 - Particle diameter (CMD) versus average puff flow

3.2. Particle size and retention

Analysis by ANOVA showed that there were statistically significant differences between only a few of the volunteers in terms of the particle size of inhaled cigarette smoke, but in general there were not significant differences between all subjects. In addition, the particle size measurements and the solanesol retention measurements were performed during separate experiments, so it is not possible to draw any direct conclusions in regards to the effect of particle size on retention from these data.

3.3. Retention and inhalation regime

Analysis by ANOVA showed that retention was statistically different for the VC_{15} regime compared to VC_{25} and Free, but not statistically different between VC_{25} and VC_{Free}. The average retention values for VC_{15}, VC_{25} and Free were 40.2%, 51.2% and 47.8% respectively. This implies that when allowed to inhale freely, the volunteers inhaled closer to VC_{25} than VC_{15}.

Figure 3 plots average retention for each subject by inhalation regime, with the subjects arranged by increasing retention for the Free regime. As shown by the ANOVA analysis, in general the retention when using the VC_{25} regime is higher than for VC_{15}, with the Free regime giving retention figures much closer to VC_{25} than VC_{15}. The two volunteers with the lowest average VC_{15} retentions (volunteer 12 and 112) were also the only volunteers where retention during the Free regime was closer to VC_{15} than VC_{25}. Thus in any study comparing different cigarette types, it may be important to identify or control for the two different behaviour types.

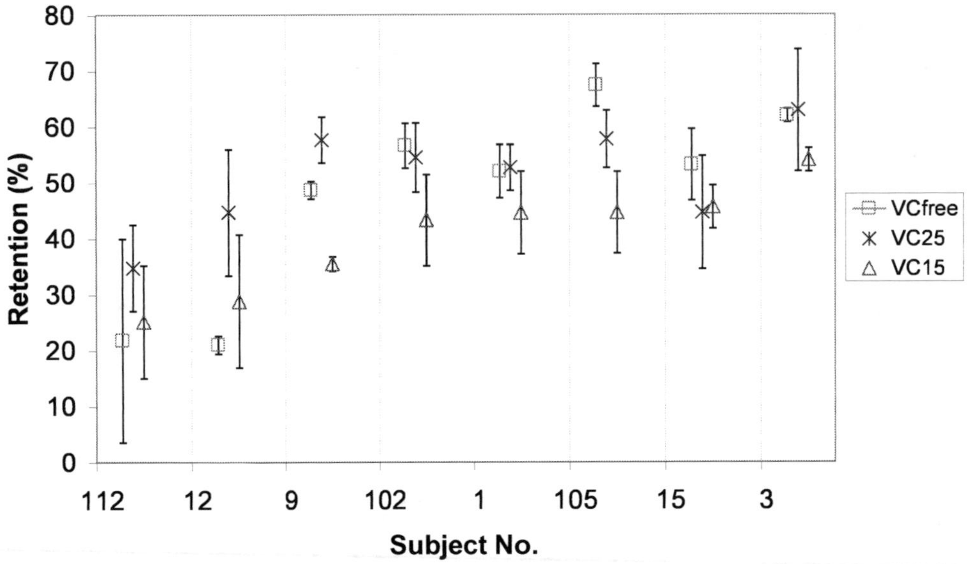

Figure 3 – Average deposition (Mean ± SD for triplicate measures) for each volunteer by inhalation regime, with the volunteers arranged by increasing deposition at VC15.

4. Discussion

These measurements show the importance of inhalation regime on the retention of the particulate phase of cigarette smoke. There is a clear trend of increasing retention when moving from inhalation at VC_{15} to VC_{25}. Additionally, when volunteers were allowed to inhale freely, they tended to achieve retention closer to those measured for VC_{25} than VC_{15}. In retention studies it is common to control smoke inhalation, by forcing the volunteer to inhale to either VC_{15} or VC_{25}. The results from this study imply that in general using VC_{25} will give a retention measurement closer to a volunteer's retention when no inhalation regime is imposed. However, for some volunteers this will give misleading results.

The study also showed that smoking profile had a small effect on the particle size of the inhaled cigarette smoke, with a decrease in particle size with increasing average puff flow. However, the correlation was not strong. The data were not sufficient to show any effect of particle size on retention.

Future work will provide better data to examine the effect of particle size on retention, by providing quantitative measurement of the particle size distribution and number concentration for both inhaled and exhaled smoke.

Ethical Considerations

This study was conducted in line with the principles of the Declaration of Helsinki. Volunteers were all current smokers recruited from the workforce at BAT GR&D Centre. Each volunteer was interviewed and provided with study information prior to enrolment. Informed consent was obtained from each participant and each was made aware they were free to leave the study at any time.

References

[1] Baker, R.R. (1999). Smoke Chemistry. In Tobacco: Production, Chemistry, and Technology; Blackwell: Oxford, U.K.; 398

[2] Baker RR, Dixon M 2006. The Retention of Tobacco Smoke Constituents in the Human Respiratory Tract. Inhalation Toxicology **18**: 255 - 294

[3] ICRP Publication 66 1995. Human Respiratory Tract Model for Radiological Protection, 66. International Commission on Radiological Protection.

[4] Bernstein DM 2004. Inhalation Toxicology. **16**: 675-689.

[5] Ogden M.W. and Maiolo K.C. (1989) *Environ. Sci. Technol.* **23**: 1148-1154

[6] Reavell, K. *et al*. 2002. SAE Technical Paper, 2002-01-2714

[7] McAughey J *et al* 2007. American Association for Aerosol Research 26th Annual Conference, Reno NV.

[8] Baron P and Willeke K, 2001 *Aerosol Measurement*, 2nd Edition, J Wiley and Sons.

Regional lung deposition of aged and diluted sidestream tobacco smoke

W Hofmann[1], R Winkler-Heil[1] and J McAughey[2]

[1]Division of Physics and Biophysics, Department of Materials Engineering and Physics, University of Salzburg, Hellbrunner Str. 34, 5020 Salzburg, Austria
[2]British American Tobacco, Group R&D Centre, Southampton SO15 8TL, UK

E-mail: Werner.Hofmann@sbg.ac.at

Abstract. Since aged and diluted smoke particles are in general smaller and more stable than mainstream tobacco smoke, it should be possible to model their deposition on the basis of their measured particle diameters. However in practice, measured deposition values are consistently greater than those predicted by deposition models. Thus the primary objective of this study was to compare theoretical predictions obtained by the Monte Carlo code IDEAL with two human deposition studies to attempt to reconcile these differences. In the first study, male and female volunteers inhaled aged and diluted sidestream tobacco smoke at two steady-state concentrations under normal tidal breathing conditions. In the second study, male volunteers inhaled aged and diluted sidestream smoke labelled with ^{212}Pb to fixed inhalation patterns. Median particle diameters in the two studies were 125 nm (CMD) and 210 nm (AMD), respectively. Experimental data on total deposition were consistently higher than the corresponding theoretical predictions, exhibiting significant inter-subject variations. However, measured and calculated regional deposition data are quite similar to each other, except for the extra-thoracic region. This discrepancy suggests that either the initial particle diameter decreases upon inspiration and/or additional deposition mechanisms are operating in the case of tobacco smoke particles.

1. Introduction

Despite widespread prevalence of tobacco smoking in the population, very few studies of the total and regional deposition of smoke particles in the lung have been reported, either for mainstream (active) or environmental (passive) tobacco smoke. A recent review [1] has summarised experimental retention efficiency data and shown efficiencies of 60-80% for mainstream smoke, but 10-40% for environmental tobacco smoke (or aged and diluted sidestream smoke). These differences may arise from differences in particle diameter and chemistry, but may also be influenced by normal inhalation versus smoking behaviour [2].

Aged and diluted sidestream smoke particles are in general smaller and more stable than mainstream tobacco smoke. On dilution, coagulation is no longer significant, a significant mass fraction of the particles evaporates, and little hygroscopic growth potential is observed. Thus it should be possible to model deposition behaviour of these smoke particles on the basis of their measured diameter. However in practice, measured deposition values are greater than those predicted by the model for the controlled breathing conditions used experimentally [3].

2. Model description

Deposition of inhaled cigarette smoke was computed by an updated version of the stochastic Lagrangian transport and deposition model IDEAL, which was originally developed by Koblinger and Hofmann [4] and Hofmann and Koblinger [5]. In this computational model, geometric airway parameters, such as diameters, lengths, branching and gravity angles are randomly selected from probability density distributions and correlations among some of these parameters, derived from published morphometric data. The random walk of single particles, inhaled at randomly selected times during inhalation, through the stochastic, asymmetric lung structure on inspiration and expiration is simulated by Monte Carlo methods. Particle deposition in single airways is calculated by using analytical equations in straight and bent tubes for the physical deposition mechanisms affecting submicron particle deposition, such as Brownian motion or inertial impaction. Hence deposition of an individual particle is based on the average deposition behaviour of many particles.

To demonstrate the validity of the stochastic deposition model for ultrafine particles, predicted total deposition values were compared in Figure 1 with the experimental data of Heyder et al. [6] and Schiller et al. [7] for monodisperse spherical particles under oral breathing conditions. Considering the effect of inter-subject variability on lung deposition, excellent agreement between experimental data and theoretical predictions was obtained.

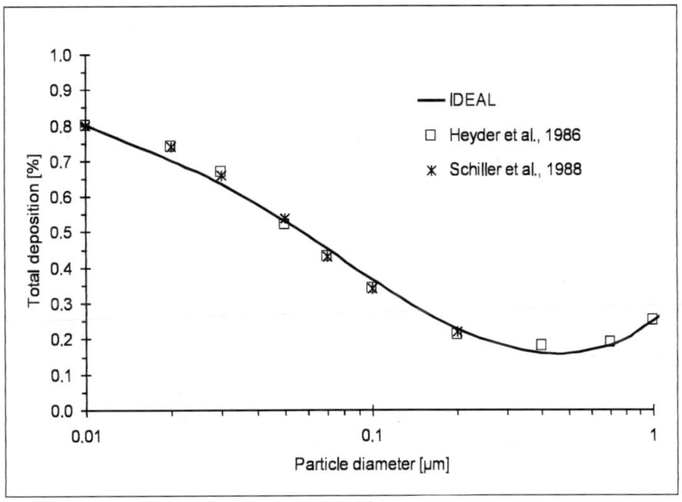

Figure 1. Comparison of theoretical predictions of total deposition with experimental data in human test subjects for a flow rate of 250 ml.s^{-1}, a tidal volume of 1000 ml, and a breathing frequency of 7.5 min^{-1} [6,7] under oral breathing conditions.

3. Results

In the current paper, data from two earlier human deposition studies with additional regional and inhalation information have been re-analysed to attempt to reconcile the observed differences between experimental data and modelling predictions.

3.1. Exhale capture study

In the first study [7], ten non-smoking male and ten female subjects inhaled aged and diluted sidestream tobacco smoke via an oro-nasal mask at two steady-state concentrations (termed high and low, on average, 980±140 and 150±60 µg.m^{-3} respectively) over a period of 60 minutes with exhale capture to a vacuum-assisted filter pack. Particle diameter was measured using a quartz crystal microbalance (MMD = 140 nm, GSD = 2.5) and a Las-X spectrometer (CMD = 125 nm, GSD = 1.5). Retention efficiency for UVPM (UV absorbance at 325 nm [9]), solanesol [10] and nicotine was carried out with retention approximately 40 ± 20% for the two particulate markers and 80 ± 20% for the vapour phase nicotine. Inhalation patterns for both genders showed average depths of 17 ± 6% of FVC, implying normal tidal breathing was approached during the exposure. Average breathing patterns measured were 7.5 x 0.94 $l.min^{-1}$ (male), and 9.4 x 0.63 $l.min^{-1}$ (female).

Table 1. Comparison of measured total deposition (retention) data with modelling predictions.

Condition	Male retention (%)		Female retention (%)	
	exp	mod	exp	mod
UVPM high	41 ± 14	28	17 ± 10	26
UVPM low	36 ± 20	30	-	-
Solanesol high	40 ± 20	28	27 ± 14	22

While theoretical predictions were consistently lower than the corresponding experimental data, as shown in Table 1, they still fall within the range of the reported standard deviations.

Since high inter-subject variations were observed in the experiments for both men and women for all three markers, the individual exposure data were used to compute individual deposition data for both male (Figure 2) and female (Figure 3) volunteers. While measured total deposition in the male subjects is consistently higher than the corresponding computed values, agreement is obtained for the average of the female subjects. A number of low UVPM values were unavailable for the females as the exhaled amount was below the limit of quantification.

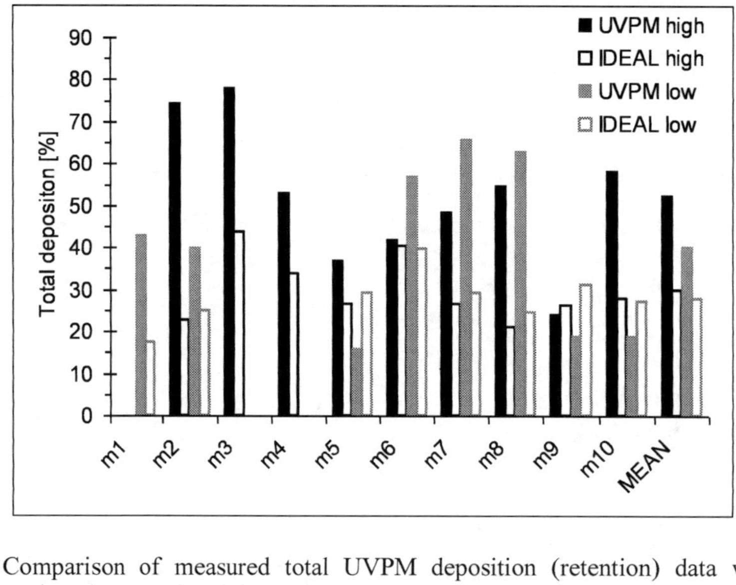

Figure 2. Comparison of measured total UVPM deposition (retention) data with modelling predictions for male volunteers for high and low concentrations.

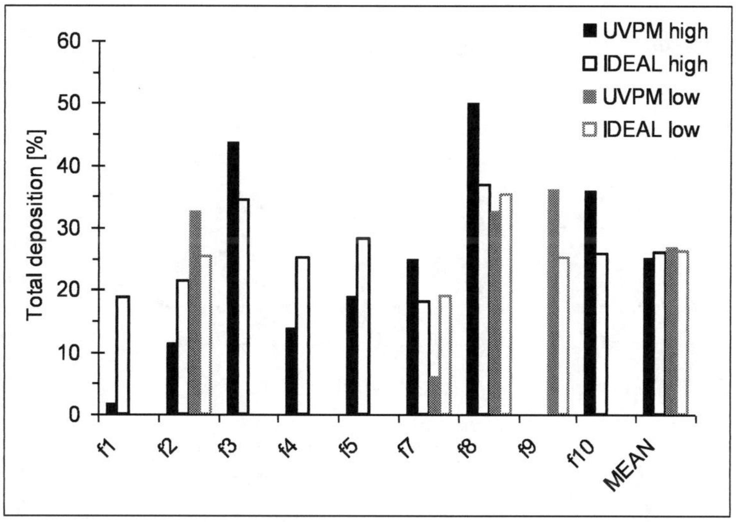

Figure 3. Comparison of measured total UVPM deposition (retention) data with modelling predictions for female volunteers for high and low concentrations.

3.2. Radiotracer study

In the second study [11], nine non-smoking male volunteers inhaled aged and diluted sidestream smoke labeled with ^{212}Pb [12] to fixed inhalation patterns. This smoke was aged for 15 minutes to meet radiological protection needs. Particle diameters were measured using a quartz crystal microbalance (MMD = 180 nm, GSD = 1.5) and a Delron low pressure impactor (AMD = 210 nm, GSD = 1.4). The measured half-time of bronchial clearance ranged on average from 8 – 12 hours. Total deposition is shown in Table 2 and regional deposition patterns are listed in Table 3.

Table 2. Average measured and modelled total deposition for three different breathing patterns.

Breathing pattern	Measured retention (%)	Modelled retention (%)
Nose: 6 x 1.0 $l.min^{-1}$	54 ± 10	27
Mouth: 6 x 1.0 $l.min^{-1}$	43 ± 17	25
Mouth: 12 x 0.5 $l.min^{-1}$	22 ± 8	14

Experimental data on total deposition for the different breathing patterns are consistently higher than the corresponding theoretical predictions. However, measured and calculated regional deposition data are quite similar proportionally to each other, except for the extra-thoracic region. This suggests that the same factors, such as size reduction upon inhalation or additional deposition mechanisms affect deposition in all regions of the respiratory tract.

Table 3. Average measured and modelled regional deposition for three different breathing patterns.

Regional deposition	Measured rentention (%)	Modelled retention (%)
Extrathoracic	7 – 11	1 – 9
Bronchial	18 – 22	21 – 30
Alveolar	67 – 73	68 – 76

4. Conclusions

In both studies analysed in this paper, experimental data on total deposition are generally higher than the corresponding theoretical predictions. These findings are consistent with the experimental results of Morawska et al. [3], who observed similar differences. Since model predictions exhibit excellent agreement with experimental data for inert spherical particles [6,7], these differences may be attributed to either changes in particle diameter when entering the warm and humid atmosphere of the human lung, e.g. evaporation of volatile compounds or condensation-induced restructuring, and/or to additional deposition mechanisms which operate specifically for tobacco smoke particles, e.g. electrical charge effects, chain aggregation or cloud behaviour [13]. For ultrafine particles, Brownian motion is the dominating physical deposition mechanism, increasing deposition with decreasing particle diameter. Thus the experimental data suggest aerodynamic behaviour more consistent with smaller particles. In contrast, measured and calculated regional deposition data are quite similar to each other, indicating that the above mechanisms affect deposition in all regions of the respiratory tract.

Further work will seek to visualise sidestream smoke under similar and ageing conditions with electron microscopy and mobility measurements to determine the plausibility of shape factors, aggregation and charge.

Ethical Considerations

This work involved a re-analysis of data from two volunteer studies conducted by one of the current authors whilst at AEA Technology. Details of the procedures for ethical oversight for the original studies are contained in references [8,9].

References

[1] Baker RR and Dixon M 2006 *Inhal Toxicol* **18**, 255-294
[2] Bernstein DM 2004 *Inhal Toxicol* **16**, 675-689
[3] Morawska L, Hofmann W, Hitchins-Loveday J, Swanson C and Mengersen K 2005 *J Aerosol Sci* **36**, 939-957
[4] Koblinger L and Hofmann W 1990 *J Aerosol Sci* **21**, 661-674
[5] Hofmann W and Koblinger L 1990 *J Aerosol Sci* **21**, 675-688
[6] Heyder J, Gebhart J, Rudolf G, Schiller CF and Stahlhofen W 1986 *J Aerosol Sci* **17**, 811-825
[7] Schiller CF, Gebhart J, Heyder J, Rudolf G and Stahlhofen W 1988 *Ann Occup Hyg* **32**, Suppl. 1, 41-49
[8] McAughey JJ, Knight DA, Black A and Dickens CJ 1995 *Inhal Toxicol* **6**, 615-633

[9] Conner JM, Oldaker GB III, and Murphy JJ 1990 *Environ. Technol.* **11,** 189-196
[10] Ogden MW and Maiolo KC 1989 *Environ. Sci. Technol.* **23,** 1148-1154
[11] Strong JC, Black A, Knight DA, Dickens, CJ and McAughey JJ 1994 J *Radiat Prot Dosim* **54,** 47-56
[12] Strong JC, Knight DA and Black A 1994 *J Aerosol Sci* **25**, 99-107
[13] Hofmann W, Morawska L and Bergmann R 2001 *J Aerosol Med* **14**, 317-326

Journal of Physics: Conference Series **151** (2009) 012021 doi:10.1088/1742-6596/151/1/012021

Diesel exhaust particulate material expression of in vitro genotoxic activities when dispersed into a phospholipid component of lung surfactant

Shi X C, Keane M J, Ong T M, Harrison, J C, Slaven J E, Bugarski A D, Gautam M, Wallace W E.

US Centers for Disease Control and Prevention, National Institute for Occupational Safety and Health, Health Effects Laboratory Division, Morgantown, WV, USA

mjk3@cdc.gov

Abstract. Bacterial mutagenicity and mammalian cell chromosomal and DNA damage in vitro assays were performed on a diesel exhaust particulate material (DPM) standard in two preparations: as an organic solvent extract, and as an aqueous dispersion in a simulated pulmonary surfactant. U.S. National Institute for Standards and Technology DPM SRM 2975 expressed mutagenic activity in the *Salmonella* reversion assay, and for in vitro genotoxicity to mammalian cells as micronucleus induction and as DNA damage in both preparations: as an acetone extract of the DPM mixed into dimethylsulfoxide, and as a mixture of whole DPM in a dispersion of dipalmitoyl phosphatidyl choline. Dispersion in surfactant was used to model the conditioning of DPM depositing on the deep respiratory airways of the lung. DPM solid residue after acetone extraction was inactive when assayed as a surfactant dispersion in the micronucleus induction assay, as was surfactant dispersion of a respirable particulate carbon black. In general, a given mass of the DPM in surfactant dispersion expressed greater activity than the solvent extract of an equal mass of DPM.

1. Introduction

Some organic solvent extracts of filter-collected diesel exhaust particulate material (DPM) express in vitro genotoxic activities to bactetria or mammalian cells [1,2]. These activities vary with engine design and operating conditions, fuels, and other engineering factors [3]. We have reported that some filter-collected DPM also can express bacterial mutagenicity and in vitro mammalian cell chromosomal or DNA damage activities when the material is mixed into an aqueous dispersion of simulated lung surfactant, e.g., aqueous dispersion of dipalmitoyl phosphatidyl choline (DPPC) [4,5]. Genotoxic activity is not expressed by the DPPC extract [6]; instead it resides with the non-dissolved but surfactant-dispersed solid ultrafine particles.

In the study reported here, a U.S. National Institute for Standards and Technology standard DPM reference material, SRM 2975, was assayed as both a solvent extract and as a surfactant dispersion in three in vitro assays: bacterial mutagenicity as Salmonella reversion to histidine independence, using the YG1024 and YG1029 tester strains with and without microsomal enzyme S9 activation; mammalian cell clastogenic activity as micronucleus induction in Chinese hamster fibroblast-derived V79 cells; and DNA damage as single- or double-strand breaks in V79 cells. SRM 2975 was prepared as an acetone extract mixed into dimethylsulfoxide; or as whole DPM mixed into an aqueous dispersion of DPPC. To investigate the role of particles *per se*, that is, particles bearing no solvent-extractable genotoxicants, micronucleus assays also were performed on DPPC surfactant dispersion of the residue of previously acetone-extracted SRM 2975, and on a respirable particulate carbon black.

2. Materials and methods

2.1. Diesel particle material (DPM)

Standard Reference Material, SRM 2975, a forklift-generated diesel exhaust particulate material from the U.S. National Institute of Standards and Technology, was used for the study.

2.2. Treatment of samples

2.2.1. *Solvent extraction* For preparation of DPM- solvent extract, 500 mg of the SRM 2975 DPM and 100ml of acetone were sonicated together in water bath for 2 h. Then the liquid was evaporated under N2 to reduce the volume to <40ml, and then centrifuged at 3000xg for 30 min. The supernatant was removed and syringe-filtered. The filtrate was evaporated to dryness under N2. This acetone extraction procedure yielded extract material equal to 4% of the mass of the original SRM2975 DPM when performed for samples for mutagenicity testing; 5.6% for samples for micronucleus and DNA damage assays, and 8 % for samples for micronucleus comparisons of DPPC-dispersed carbon black with extracted or dispersed DPM. This evaporated solvent extract residue was mixed into dimethylsulfoxide (DMSO) at 36 mg DPM-extract per ml DMSO solution as a stock solution. Samples were diluted serially with complete Eagle's minimal essential medium (MEM, Gibco) to desired concentrations for in vitro assay.

2.2.2. *Surfactant Dispersion* For the preparation of surfactant dispersion, 2.5mg DPPC (Calbiochem, La Jolla, CA) per ml was mixed with 0.85% physiological sterile saline (PSS) and sonicated under N2 at 40 W (Heat system-ultrasonic, Plainview, NY) for 10 min. The probe was wiped with 95% ethanol and was covered with parafilm during sonication. Two mg DPM per ml DPPC stock dispersion was made after agitating the sample in a water bath for 60 min at 37°C. The sample was diluted serially with complete Eagle's MEM medium to desired concentrations.

For surfactant dispersion of acetone-extracted SRM 2975 residue or of carbon black, 2mg per ml DPPC dispersions were prepared and diluted serially with complete medium to required concentrations.

2.2.3. Ames Salmonella Assay Mutagenic activity was analyzed by the pre-incubation variant of the Ames Salmonella microsuspension assay system [7]. Two bacterial test strains, YG1024 and YG1029, were used to identify frame-shift and base- pair substitution mutations. A 10% concentration of rat liver microsomal fraction S9 was used for metabolic activation for half of the samples. Co-incubation of bacteria, test sample, and S9 was performed by a rotary shaker at 37°C for 30 min before plating. Duplicate plates were made for each concentration of test sample and the control. Three readings of revertant colonies per plate were scored by an automatic colony counter (Accu Count 1000, Biologics Inc., Gainville, VA) at 48 h for YG1029 and at 72 h for YG1024 after incubation. The average number of revertant colonies and standard deviation per concentration were calculated. A sample was considered to be positive for mutagenic activity if the number of revertant colonies per plate was twice as high as that of the DMSO solvent control or DPPC surfactant control in at least one concentration and if there was a clear concentration-response relationship.

2.1.4 Micronucleus (MN) Assay Micronucleated cells (MNC) were scored according to specific criteria [8,9]. Micronuclei located within the cytoplasm but separated from the main nuclei and with diameter between one-twentieth and one-fifth that of the nucleus were scored as a MN. The frequency of micronucleated cell (MNC) expressed as the number of MNC per 1000 cells was based on 6000 cells scored for each treatment. Chinese hamster lung fibroblasts (V79 cells) at a concentration of 1

×106 were cultured in a 25ml flask containing 5 ml of Eagle's minimal essential medium (MEM, Gibco) supplemented with 10% inactivated fetal bovine serum and 2% penicillin-streptomycin in a humidified incubator at 37°C in an atmosphere of 5% CO2 for 24 h. Cells were treated for 24 h in 5ml of fresh complete MEM medium containing the test sample which had been prepared as a solution in 0.5% DMSO or dispersion in DPPC. Cytotoxicity of V79 cells after treatment was determined by cell density measurement and viability was determined by using the trypan blue exclusion assay. Cells were rinsed with PBS and re-incubated in 5ml fresh complete MEM medium for an additional 24 h. Cells were harvested by trypsinization, collected by centrifugation, and re-suspended in medium at about 4-5× 105/ml. Slides were prepared by a cytocentrifuge (Shandon Cytospin 3, Shandon Inc., Pittsburgh, PA). An aliquot of 75µl of the cell suspension was added into a chamber and the cells were pelleted onto the slide at 500 rpm for 5 min. Slides were dried on a slide warmer at 37°C for 10 min, fixed with absolute methanol, and stained with Diff-Quik stain (Dage Diagnostics, Aguada, Puerto Rico).

2.1.5 Single Cell Gel Electrophoresis (SCGE) The procedure for the SCGE assay followed that reported by Tice et al. [10] with minor modifications. V79 cell culture and sample treatment for SCGE assay was the same as that for micronucleus assay. After challenge, cells were rinsed with phosphate-buffed saline (PBS, Gibco) and harvested by trypsinization. The cell suspensions were centrifuged at 600 x g at 4°C for 10 min and the cell pellets were resuspended in 0.5 ml of complete medium (approximately 1.5 - 2 x 106/ml). Tubes containing the resuspended cells were kept on ice for making sandwich gel slides.

Electrophoresis was carried out at 300mA and approximately 25 volts (0.7 v $cm^{-1^{-1}}$) for 30 min at room temperature. Slides were transferred to Coplin Jars, soaked in Tris buffer for 5 min three times to neutralize the gel. The slide was then stained with 50 µl of 2 µg/ml ethidium bromide (Sigma-Aldrich) and covered with a cover slip for 5 min, followed by rinsing in distilled water. The finished slides were covered with a cover slip and stored in a humidified airtight container at 4°C for analysis. Image analysis was performed at 200x magnification using a fluorescence microscope (Zeiss, 7082 Oberkochen, Germany) equipped with a 100 W Hg vapor lamp, a 515 – 560 nm excitation filter and a 590 nm emission filter. The tail length of DNA migration was determined with an eyepiece micrometer by measuring the distance between the edge of the nucleus to the end of the tail. For each treatment, 25 cells per slide were analyzed for 4 replicate slides.

3. Results

3.1. Mutagenicity

Results of the salmonella gene mutation test are shown in Table 1, and data for the assay using the YG1024 tester strain without S9 activation are graphed in Figure 1, showing response as revertants per plate, while concentration values in the Table and Figure are in terms of the mass of SRM 2975 DPM dispersed for the DPM-DPPC dispersed samples and in terms of mass of SRM 2975 DPM subjected to extraction for the DPM-Solvent Extract samples. Regression equations and corresponding p-values are given in Table 2 for linear fits to the concentration-response data. Both DPM acetone solvent extract and DPM dispersion in DPPC showed statistically significant positive mutagenic activity to both YG1024 and YG1029 strains both with and without S9 activation; and YG1024 gave a stronger mutagenic response than YG1029 to both DPM solvent extract and DPM dispersion in DPPC samples The mutagenic activity in general, was stronger without S9 metabolic activation. Overall, activity for DPM dispersion in DPPC was stronger than that of DPM solvent extract samples.

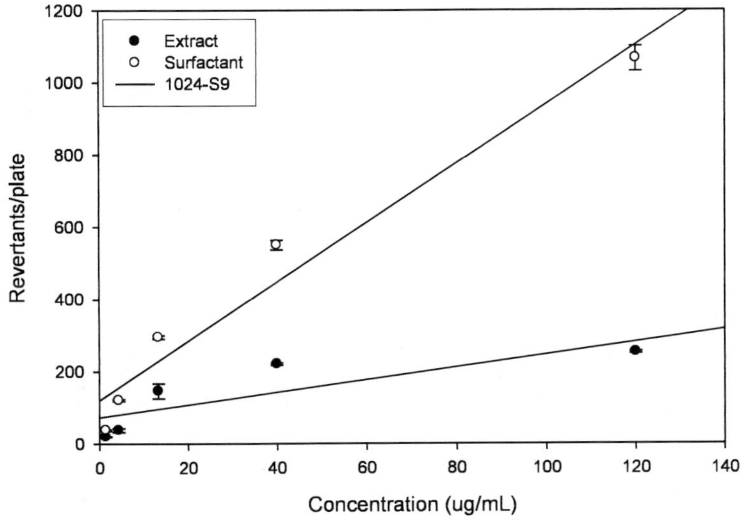

Figure 1: Mutagenic Activity of SRM 2975 Diesel Particulate Material as solvent extract or as surfactant dspersion using Salmonella YG1024 reversion to histidine independence, without sample activation by S9 microsomal enzyme.

Table 1: Revertant Colonies Induced by Diesel Particle Material (DPM) SRM 2975 in Salmonella Typhimurium YG1024 and YG1029

Test Sample	Concentration[a] (μg/plate)	Average number of revertant colonies/plate[b] YG1024		YG1029	
		w/out S9	with S9	w/out S9	with S9
DMSO	10[c]	18 ±2	29±5	27± 4	26±6
DPM Solvent Extract	13.3	146 ±51	24±3	63± 1	51±5
DPM Solvent Extract	40	221± 8	77±4	152± 23	79±7
DPM Solvent Extract	120	252± 7	435±32	351± 11	192±2
DPPC	10[c]	23 ±1	41±5	62± 3	79±2
DPM-DPPC	13.3	295 ±13	167±3	173± 3	99±13
DPM-DPPC	40	550± 33	456±65	357± 25	188±16
DPM-DPPC	120	1065± 84	1007±81	750± 60	560±9
2AA[d]	0.05				794±16
1-NP[e]	0.05	403±10		428±2	

[a] Concentration based on original DPM subjected to extraction or dispersion.

[b] Average of 6 counts. [c] μl/plate. [d] 2-aminoanthracene. [e] 1-nitropyrine.

Journal of Physics: Conference Series **151** (2009) 012021 doi:10.1088/1742-6596/151/1/012021

Table 2: Regression Equations for Salmonella mutagenic activity in Revertants/plate versus sample concentration in terms of original SRM2975 extracted or dispersed

	Extract	Surfactant	p-value
1024+S9	y = -11.84 + 3.57x	y = 64.14 + 8.06x	< 0.0001
1024-S9	y = 76.37 + 1.70x	y = 119.86 + 8.20x	< 0.0001
1029+S9	y = 34.81 + 1.29x	y = 53.94 + 4.13x	< 0.0001
1029-S9	y = 27.07 + 2.74x	y = 89.91 + 5.62x	< 0.0001

3.2. Micronucleus Induction

Results of the comparative study of micronucleus induction in V79 cells, based on equal concentration of SRM 2975 DPM subjected to acetone extraction or DPPC dispersion, are given in Table 3 and plotted in Figure 2. DPM solvent extract induced significant micronucleus formation at concentrations of 180µg/ml ($P<0.05$) and 240 µg/ml ($P<0.01$). DPM dispersion in DPPC induced significant micronucleus formation with positive concentration-response behavior for all of the concentrations tested. Micronucleus formation activity in V79 cells was stronger for surfactant-dispersed whole soot particles than for the organics extracted from an equal mass of soot. Tables 3, 4 and 5 and Figure 2 also show the results of micronucleus assay of DPPC-dispersion of the solid residue of acetone solvent-extracted SRM 2975 DPM, and of DPPC-dispersed ultrafine carbon black. The residue of acetone-extracted SRM 2975 was inactive for micronucleus induction when assayed as a dispersion in DPPC. Carbon black, assayed as an ultrafine particulate material bearing no extractable genotoxicants, also was inactive for micronucleus induction as a dispersion in DPPC.

Table 3: Comparative Studies of Micronucleus Induction by Diesel Particle Material (DPM) SRM 2975 in V79 Cells

Test sample	Concentration[a] (µg/ml)	MNC/1000 cells[b] (Mean±SD)
DMSO	5[c]	5.6± 2.3
DPM Solvent Extract	40	6.3± 1.6
DPM Solvent Extract	180	7.6± 2.3*
DPM Solvent Extract	240	9.9± 2.2**
DPPC		5.8 ±1.9
DPM-DPPC	40	7.2± 2.3*
DPM-DPPC	180	11.1 ±2.8**
DPM-DPPC	240	16.1 ±2.2**
DPM Solvent Residue-DPPC	40	6.5± 1.7
DPM Solvent Residue-DPPC	180	6± 3
DPM Solvent Residue-DPPC	240	5.2± 2
Carbon Black-DPPC	40	5.6± 2.6
Carbon Black-DPPC	180	6.6± 3.3
Carbon Black-DPPC	240	5.6± 1.8
MNNG[d]	0.25	23.4± 2.8**

[a] Based on original DPM extracted or dispersed. [b] Micronucleated cell counts were based on 12000 cells scored. [c] µl/ml. [d] N-methy-N-nitro-nitrisoquanine. * $P\leq 0.05$ compared with negative control. ** $P \leq 0.01$

Additional micronucleus induction assays were performed comparing the activity of a mass of SRM 2975 DPM dispersed in DPPC with an equal mass of extract obtained by acetone extraction of a (larger) mass of SRM 2975 DPM. Shown in Table 4 and Figure 3, both preparations caused significant micronucleated cells at all such concentrations ranging from 13.3 to 180 μg/ml, and both exhibited a concentration-dependent increase in activity.

Table 4: Regression Equations for number of micronucleated cells versus concentration in terms of original SRM2075 DPM extracted or dispersed.

Extract:	$y = 5.44 + 0.016x$
Surfactant:	$y = 5.25 + 0.041x$
Carbon Black:	$y = 5.72 + 0.001x$
Residue:	$y = 6.83 - 0.006x$

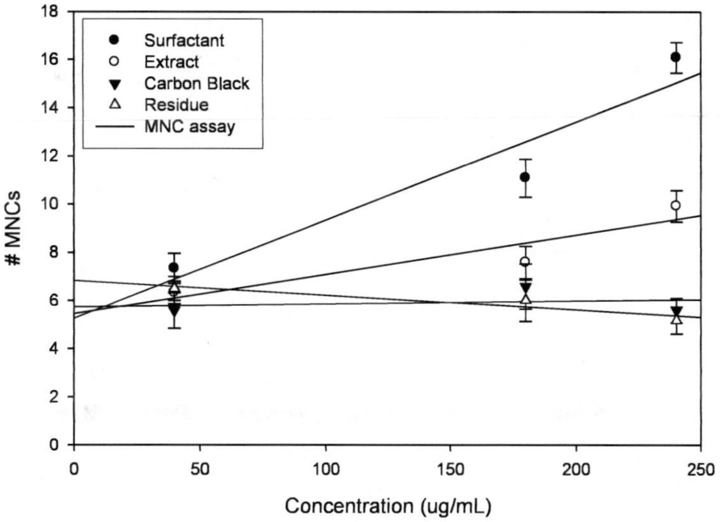

Figure 2. Micronuclei Induced by SRM2979 DPM dispersed in DPPC surfactant or extracted by acetone; by carbon black dispersed in DPPC; and by DPPC dispersion of the residue of acetone extraction of DPM.

Table 5: Induction of Micronucleus by Diesel Particulate Material SRM 2975 in V79 Cells

Test sample	Concentration[a] (μ g/ml)	MNC/1000 cells[b] (Mean±SD)	Multiple Nucleus cell/1000 cells (Mean±SD)
DMSO	5[c]	5.5 ±2.3	11.8± 3.4
DPM Solvent Extract	13.3	15.7± 2.2**	22.8± 3.1**
DPM Solvent Extract	40	17.5 ±2.1**	26.7 ±3.8**
DPM Solvent Extract	120	21.7± 6.3**	25.0 ±7.1**
DPM Solvent Extract	180	32.5 ±2.8**	39.8 ±5.7**
DPPC		7.8± 2.5	18.6± 3.0
DPM-DPPC	13.3	9.8 ±3.5	26.2± 8.2*
DPM-DPPC	40	12.2 ±1.6**	39±.3 4.6**
DPM-DPPC	120	15.3± 4.1**	34.2± 12.6**
DPM-DPPC	180	19.3 ±3.6**	57.0± 13.7**
NC		7.5± 2.6	15.5± 1.9
MNNG[d]	0.25	29.7± 4.2**	11.5 ±2.5

[a] For solvent extraction the concentration was based on extracted material, for DPPC dispersion the concentration was based on original DPM dispersed. [b] Micronucleated cell count (MNC) was based on 1000 cells. A total of 6000 cells were scored. [c] μl/ml. [d] N-methy-N-nitro-nitrisoquanine * P< 0.05 compared with negative control. ** P < 0.01 compared with negative control.

Table 6: DNA damage Induced by Diesel Particle Material (DPM) SRM in V79 Cells with SCGE

Test Sample	Concentration[a] (μg/ml)	Damaged Cells/ 100 Cells	Tail Length (μ m)[b] (Mean±SE)
DMSO	5[c]	14	1.1± 0.28
DPM Solvent Extract	13.3	20	1.1± 0.25
DPM Solvent Extract	40	22	1.5 ±0.29
DPM Solvent Extract	120	33	2.4± 0.37**
DPM Solvent Extract	180	54	2.6± 0.27**
DPPC		13	1.2 ±0.32
DPM-DPPC	13.3	10	0.8± 0.25
DPM-DPPC	40	29	2.1 ±0.34*
DPM-DPPC	120	38	2.6 ±0.36**
DPM-DPPC	180	41	2.9± 0.37**
Complete Medium	0	9	0.7± 0.21
MNNG[d]	0.25	89	8.1± 0.35**

[a] For solvent extraction the concentration was based on mass of extracted material, for DPPC dispersion the concentration was based on mass of original DPM dispersed. [d] N-methy-N-nitro-nitrisoquanine. [b] Based on 100 cells. [c] μl/ml * P< 0.05 ** p< 0.01 vs. control.

3.3. DNA damage

Results from the SCGE assay of DNA damage in V79 cells were scored as tail length (TL) in µm. Assays were performed comparing the activity of a mass of SRM 2975 DPM dispersed in DPPC with an equal mass of extract obtained by acetone extraction of a (larger) mass of SRM 2975 DPM. Any cell with a nonzero tail was classified as damaged cell. Both exposure of V79 cells to DPM solvent extract and to DPM dispersion in DPPC caused significant DNA migration at concentrations ranging from 120 to 180 and 40 to 180 µg/ml, respectively and displayed a concentration-dependent increase in tail length in Table 6 and Figure 2. The percentage of damaged cells increased with concentration-response manner in both DPM solvent and dispersion in DPPC.

Table 7: Logistic Regression Equations for number of DNA-damaged cells /100cells versus concentration in terms of original SRM2075 DPM extracted or dispersed.

Extract:	$y = 1.6248 - 0.0092x$
Surfactant:	$y = 1.6217 - 0.0079x$

4. Discussion

Numerous in vitro genotoxicity studies have been performed upon diesel exhaust particulate materials (DPM) collected by filtration and assayed as organic solvent extracts. In many but not all cases such extracts have been found to express positive genotoxic activities. However, the physiologically anomalous use of organic solvent extracts of DPM questions the biological availability of DPM-borne genotoxicants under conditions of deposition in the lung. And such conventional in vitro genotoxicity assays applied to DPM organic solvent extracts cannot identify the possible effects of exhaust particulate size or structure on genotoxic activity.

We have reported that some DPM can express in vitro genotoxic activities as mixtures into aqueous dispersion of some principal phospholipid pulmonary surfactants, e.g., dipalmitoyl phosphatidyl choline, DPPC. Extraction of DPM by DPPC does not result in significant genotoxic activity by the extract; however, the non-dissolved particulate phase of direct mixture of DPM into DPPC dispersion can express in vitro genotoxic activities. Results suggest that particles adsorb surfactant onto their surfaces from aqueous dispersion, making the so-conditioned particle surfaces hydrophilic and permitting their dispersion in aqueous media. This retains particle structure and its effects on the expression of toxicities, partially modeling particle surface conditioning upon inhalation and deposition in the deep lung. This permits the application of in vitro bioassays in a manner which avoids anomalous test preparation destruction of particle size and structure, factors which may be important in the expression of toxicities by ultrafine or nano-particles. Further, it partially models the effects on DPM toxicity of the conditioning of respirable particle surfaces which occurs promptly upon respirable particle deposition in the deep lung onto the surfactant-rich aqueous hypophase coating of the respiratory bronchioles and alveoli. Surfactant dispersion concentration requirements for this have been determined in terms of particulate surface area [11]

The current study objective was to compare some in vitro genotoxic activities between solvent-extraction and surfactant- dispersion preparations for a generally available diesel exhaust reference material, U.S. National Institute for Standards and Technology DPM standard reference material SRM 2975. Assays were chosen to represent bacterial mutagenicity and mammalian cell chromosomal or DNA damage.

In Salmonella mutagenicity assays using either of two tester strains, both solvent extract and surfactant dispersions of the DPM were active, expressing for at least one sample concentration more than twice the solvent or surfactant control activity, and demonstrating a positive concentration – mutagenic activity response. Comparing mutagenic activities versus concentrations measured on the basis of mass of original DPM dispersed into DPPC or mass of original DPM subjected to solvent extraction, it was found that the surfactant dispersion preparation was significantly more active for a given concentration than the solvent extract.

Those, similarly, were the results for micronucleus induction in mammalian-derived V79 cells: both surfactant dispersion and solvent extract preparations of SRM 2975 DPM were active, with surfactant dispersion being significantly more active than the solvent extract for concentration based upon mass of DPM extracted or dispersed. Comparison also was made using the micronucleus assay to identify a possible contribution to the activity from the presence of non-dissolved particles per se in the surfactant dispersion preparations. This was approached by assaying a carbon black particulate material in surfactant dispersion, and by assaying in surfactant dispersion the non-dissolved particulate residue surviving organic solvent extraction of SRM2975 DPM. Surfactant dispersions of both types of insoluble particles carrying no extractable genotoxicants were inactive. Micronucleus induction also was compared between the two preparations also on the basis of mass of original SRM 2975 DPM dispersed in surfactant versus equal mass of solvent extracted DPM material (from a larger amount of original SRM 2975 DPM). In that comparison, a mass of extracted material is significantly more active than an equal mass of dispersed DPM. It is noted that a given mass of extract requires the extraction of a much larger mass of DPM for this SRM2975 DPM with solvent extractables of about 10% the mass of the DPM extracted.

DNA damage was compared between the two preparations only on the basis of concentrations of surfactant dispersed SRM 2975 DPM versus equal concentrations of material extracted from a (larger) amount of SRM 2975 DPM. In this SCGE assay under this manner of concentration comparisons, there was no significant difference in activity between the two preparations. Again, with the low solvent extractable fraction of this SRM 2975 DPM, this suggests surfactant dispersion of this DPM is at least as active and reasonably more active than the solvent extract of an equal mass of original SRM 2975 DPM.

The implication of these results is that for this SRM 2975 DPM, a given mass of extractable DPM-associated genotoxic materials express greater in vitro genotoxic activities if they are associated with surfactant-dispersed particles than is expressed by the genotoxicants when extracted from the particles. Lack of in vitro genotoxic activity of surfactant-dispersed carbon black and of surfactant-dispersed SRM2975 which had been stripped of genotoxic compounds by acetone extraction, suggest the greater activities seen for this DPM in surfactant dispersion may be a synergistic effect of particle-genotoxicant association rather than an additive effect of activity of the carried genotoxicants plus activity of carrier genotoxicant-barren particles.

That not all DPM solvent extracts express genotoxic activity; and that diesel engine design and operational factors affect the amount of emissions and specific genotoxic activities of DPM solvent extracts, has suggested that in vitro assays of DPM might be a source of convenient first-tier guidance for diesel design or exhaust control technology development with the objective of minimizing hazardous diesel exhaust emissions. The results of in vitro assays of surfactant-dispersed DPM suggest that such efforts can be extended to include a more physiologically-relevant preparation of DPM for in vitro testing which is not blind to considerations of the size and structure of DPM and its conditioning upon deposition in the lung. Our current research includes extension of the findings of in vitro activities of surfactant-dispersed DPM to the concept of collection of exhaust DPM directly into surfactant dispersion for bioassay.

References

[1] International Agency for Research on Cancer, 1989. Diesel and gasoline engine exhausts and some nitroarenes. Vol. 46. Monographs on the Evaluation of Carcinogenic Risks to Humans. World Health Organization, IARC, Lyon, France

[2] US Environmental Protection Agency, 2002. Health Assessment Document for Diesel Engine Exhaust. USEPA EPA/600/8-90/057F. 01 May 2002. U.S. EPA, Office of Research and

Development, National Center for Environmental Assessment, Washington, DC.

[3] McMillian MH., Cui M, Gautam M, Keane M, Ong T, Wallace W, Robey E [2002]. Mutagenic Potential of Particulate Matter from Diesel Engine Operation on Fischer-Tropsch Fuel as a Function of Engine Operating Conditions and Particle Size. Soc.Auto.Engineers Technical Paper 2002-01-1699, pp. 1-18.

[4] Wallace WE, Keane MJ, Hill CA, Xu J, Ong TM [1987]. "Mutagenicity of Diesel Exhaust Particles and Oil Shale Particles Dispersed In Lecithin Surfactant". Journal of Toxicology and Environmental Health, 21 163-171.

[5] Wallace W, Keane M, Murray M, Chisholm W, Maynard A, Ong T . "Phospholipid Lung Surfactant and Nano-Particle Surface Toxicity: Lessons from Diesel Soots and Silicate Dusts. Journal of Nanoparticle Research 9(1): 23-38.

[6] King, LC., Kohan MJ, Austin AC, Claxton LD, Hunsingh JL [1981]. Evaluation of the release of mutagens from diesel particles in the presence of physiological fluids. Environ. Mutag. 3:109-129.

[7] Ames, B., McCann J, Yamasaki E [1975]. Methods for detecting carcinogens and mutagens with the Salmonella/mammalian microsome mutagenicity test. Mutat. Res. 31: 347-364.

[8] Countryman PI and Heddle JA [1976]. The production of micronuclei from chromosome aberrations in irradiated cultures of human lymphocytes. Mutat Res. 41(2-3):321-32.

[9] Roberts CJ, Morgan GR, Holt PD [1986]. A critical comparison of the micronucleus yield from high and low LET irradiation of plateau-phase cell populations. Mutat Res. 160(3):237- 242.

[10] Tice RR, Strauss GSH,Peters WP [1992]. High-dose combination alkylating agents with autologous bone-marrow support in patients with breast cancer: preliminary assessment of DNA damage in individual peripheral blood lymphocytes using the single cell gel electrophoresis assay. Mutat. Res. 271:101-113.

[11] Wallace WE, Keane MJ, Gautam M, Shi X-C, Murray D, Ong T "Dispersion of Nanoparticles in Pulmonary Surfactants for In Vitro Toxicity Studies: Lessons from Ultrafine Diesel Exhaust Particles and Fine Mineral Dusts". In; Nanotoxicology: Characterization, Dosing, and Health Effects on Target Organs, Edited by N. Monteiro-Riviere & L. Tran; Informa Healthcare USA, Inc., New York. ISBN-10: 1-4200-4514-8. Chapter 11; pp. 153-171.

Quartz-Containing Ceramic Dusts: *In Vitro* Screening of the Cytotoxic, Genotoxic and Pro-Inflammatory Potential of 5 Factory Samples

C Ziemann[1], P Jackson[2], R Brown[3], G Attik[4], B H Rihn[4], O Creutzenberg[1]

[1]Fraunhofer Institute of Toxicology and Experimental Medicine, Hannover, Germany
[2]CERAM Research Ltd., Stoke-on-Trent, UK
[3]TOXSERVICES, Stretton, UK
[4]Nancy-University, Faculté de Pharmacie, Nancy, France

E-mail: christina.ziemann@item.fraunhofer.de

Abstract. Inhalation of some respirable crystalline silica (MMAD ≤ approx. 4 μm) leads to inflammatory and malignant diseases. Comprehensive physicochemical/biological data and suitable *in vitro/in vivo* methods may distinguish between more or less harmful quartz-varieties. Within the European Collective Research Project SILICERAM an *in vitro* screening battery was established to evaluate cytotoxicity (LDH-release, MTT-assay), genotoxicity (Comet-assay) and pro-inflammatory potential (PGE_2-liberation, TNF-α mRNA expression) of 5 respirable quartz-containing dusts from ceramic plants: brickwork (BR: 7.8% quartz), tableware granulate/cast (TG/TC: 5.8%/3.1%), tiles (TI: 8.1%), refractory (RF: 3.7%). DQ12 (87% α-quartz) and Al_2O_3 were used as particulate positive and negative controls, respectively. Primary rat alveolar macrophages and the macrophage cell line NR8383 served as model systems. Aluminium lactate was used as inhibitor of biologically active silica, enabling differentiation of silica- and non-specific toxicity. At 200μg/cm^2 (2h) the dusts did not alter significantly LDH-release (except TC), whereas the MTT-assay demonstrated the mainly quartz-independent rank order: DQ12>RF>TG>TI≥BR>TC>Al_2O_3. DNA-damage was maximal for BR and TI followed by DQ12>TG>TC>RF>Al_2O_3. All dusts induced PGE_2-liberation (DQ12>BR>TC>TG>TI>RF>Al_2O_3) at 50μg/cm^2 (4h), but TNF-α mRNA (10μg/cm^2, 24h) was only increased by DQ12, TG (quartz-dependently), and TC. In conclusion, these *in vitro* tests were an adequate approach to screen the toxic potential of quartz-containing ceramic dusts, but the quartz-content was too low to differentiate the various quartz-varieties.

1. Introduction

Due to its low solubility and its reactive surface, respirable crystalline silica (RCS) may exert harmful effects on biological targets such as DNA, membranes, and proteins. Inhalation of RCS may thus lead to adverse health effects in the lung, e.g. inflammation, silicosis, and cancer. The debate on adverse effects of RCS and subsequent air limits at workplaces was reinforced by the decision of the International Agency on Research in Cancer (IARC) in 1997 to classify some exposures to RCS as category 1 carcinogen. At present, the different EU states have their own occupational limits for RCS and setting of common RCS exposure limits on a mg/m3 level is difficult, because lung-reactivity of RCS-containing dusts may differ substantially depending on quartz polymorph and industrial sector.

Considering the ubiquitous occurrence, the diverse use of quartz and the complex characteristics of its toxicity, comprehensive physicochemical/biological data sets and suitable *in vitro/in vivo* methods are strongly needed to allow differentiation of various quartz varieties and different quartz processing routes. For that reason, an *in vitro* screening battery was established and used in the European Collective Research Project SILICERAM to evaluate the cytotoxic, genotoxic, and pro-inflammatory potential of RCS-containing ceramic dusts from traditional subsectors of the ceramic industry: brickwork, tableware, tiles, and refractory ware.

2. Materials and Methods

2.1. Test and reference materials

For incubations, test and reference particles were dosed on a mass basis ($\mu g/cm^2$) and not on RCS basis due to theire unexpectedly low RCS content. Particles were suspended in culture medium without fetal calf serum (FCS) and sonicated for 10 min in a water bath.

2.1.1. *Ceramic dusts.* With 2 high-volume samplers, developed within the SILICERAM project, 5 RCS-containing ceramic dusts were collected in different countries from various industrial sectors or manufacturing plants: brickwork (BR: 7.8% quartz, weight-% by XRD), tableware granulate (TG: 5.8%), tiles (TI: 8.1%), tableware cast (TC: 3.1%), and refractory ware (RF: 3.7%) with geometric mean diameters (weighted by mass) of 5.17, 5.97, 2.14, 4.36, and 7.70 μm, respectively.

2.1.2. *Reference samples.* Quartz DQ12 (mid-size, Dörentrup, Bergbauforschung, Essen, Germany, 87% α-quartz/13% amorphous silica) was used as positive, Al_2O_3 (active, neutral type 507C, Aldrich, Taufkirchen Germany) as particulate negative control with geometric means weighted by mass of 3.01 and 4.23 μm, respectively.

2.2. Cells Systems

Primary rat alveolar macrophages (PAM) isolated from adult Wistar rats and the rat alveolar macrophage cell line NR8383 were used as test systems. PAM were pre-cultured for 24 h in DMEM with GlutaMAX^{TM}, 4.5 g/l D-glucose, 110 mg/l sodium pyruvate, 10% FCS and antibiotics. They were plated at a density of 1.05×10^5 cells/cm^2 on 24-well plates with a hydrophobic culture surface (Comet-assay, LDH-liberation) or on conventional 4-well plates (PGE_2-release). NR8383 cells were cultured in DMEM supplemented with 200 mM L-glutamine, 15% FCS and antibiotics and were split every 3-4 days. They exhibited 50% adherent and 50% floating morphology. Both cell types were cultured and incubated at 37°C in a humidified atmosphere of 5% CO_2 / 95% air using an incubator.

2.3. Lactate dehydrogenase (LDH)-release

To determine cell viability/membrane integrity of dust-treated PAM, lactate dehydrogenase (LDH)-activity was measured in the incubation supernatant of the Comet-assay. LDH-activity was evaluated according to the manufacturer's protocol using the "Cytotoxicity Detection Kit (LDH)" from Roche Diagnostics (Mannheim, Germany) after 2 h of incubation with 200 $\mu g/cm^2$ of the different dusts ± 10 μM aluminium lactate (AL, to quench quartz-specific biological effects).

2.4. MTT-assay

The metabolic activity/viability of dust-treated NR8383 cells was investigated using the MTT-assay adapted from [1]. In brief, NR8383 cells were seeded in 96-well plates (1×10^5 cells per well) and pre-cultured overnight. After incubation for 2 h with 50 - 200 $\mu g/cm^2$ of the different particles (± 10 μM AL) cells were washed and 3-(4,5-dimethylthiazol-2-yl)-2,5-diphenyltetrazolium bromide (MTT)-solution was added. After incubation for 3 h at 37°C the MTT-solution was discarded and the blue MTT-reduction product formazan was dissolved in DMSO. Formazan was determined spectrophotometrically by measuring absorbance at 540 nm, using a microplate reader. The results were expressed as percentage of viability, namely the ratio of absorbance of exposed to unexposed cells x 100.

2.5. Comet-assay

To determine the DNA-damaging potential of the ceramic dust samples, the alkaline *in vitro* Comet-assay (pH>13) according to [2] and [3] with minor modifications was used. Pre-cultured PAM were incubated for 2 h with 200 μg/cm^2 of the dust samples (± 10 μM AL). Cells were then detached by placing on ice for 10 min. All of the following steps were performed under red light to avoid unspecific DNA-damage. After centrifugation cells were resuspended in pre-heated low-melting agarose (LMA) and applied to pre-coated slides. After gelation, a second layer of LMA was applied and slides were finally immersed in lysis solution (2.5 M NaCl, 100 mM Na_2EDTA, 10 mM Tris-HCl, 8 g/l NaOH, 1 % Triton-X100, 10% DMSO) and stored overnight at 4°C. They were then placed on a cooled electrophoresis platform and covered with pre-cooled electrophoresis buffer (300 mM NaOH, 1 mM Na_2EDTA, pH 13). After DNA-unwinding (20 min) and electrophoresis (20 min at 24V/300 mA) slides were neutralized and stained with ethidium bromide. The tail moment of the nuclei was analyzed using a fluorescence microscope and the Comet Assay III software from Perceptive Instruments (Haverhill, UK). One hundred nuclei were scored per slide.

2.6. Prostaglandin E_2 (PGE_2)-liberation

To assess the pro-inflammatory potential/phagocytic stress induced by the ceramic dusts, PGE_2 was measured in the incubation supernatant of PAM, incubated for 4 h with 50 μg/cm^2 of the ceramic dust or the reference samples. PGE_2 was determined by a highly sensitive and specific competitive enzyme immunoassay using a monoclonal antibody from clone E2R1 [4]. The detection limit was 3 pg/ml of a 96-well plate. Results were calculated from the standard curve by cubic spline interpolation.

2.7. TNF-α mRNA expression

To assess TNF-α mRNA expression, NR8383 cells were cultured for 24 h (10^6 cells/ml) with 10 μg/cm^2 of the ceramic dusts or the reference particles (± 10 μM AL). RNA was extracted, using the RNeasy Mini Kit (QIAGEN). Reverse-transcription was performed using TNF-α- and glyceraldehyde-3-phosphate (GAPDH)-specific primer [5] and *Moloney* murine leukaemia virus reverse-transcriptase, leading to amplicon lengths of 175 and 168 bp, respectively. Resulting cDNA was amplified with Taq polymerase and 10 x Long Expand^{TM} Buffer from Roche (Meylan, France). Amplification reaction consisted of melting (3 min, 94°C), followed by 30 cycles of melting (30 sec, 94°C), annealing (30 sec, 58°C), and extension (2 min, 72°C), and a final extension step (7 min, 72°C). Amplification products were run on 1% agarose gels with Tris·borate/EDTA-buffer (pH 8.0), stained with ethidium bromide, and visualized by UV-transillumination. Relative TNF-α mRNA expression was expressed as the ratio of fluorescence intensities of TNF-α andGAPDH (house-keeping gene) bands.

3. Results

The *in vitro* screening battery was initially optimized concerning suitability, sensitivity, concentrations, and incubation times. Optimized protocols were finally used for the main experiments, because of limited amounts of the 5 factory dusts. For the MTT-assay, however, 3 concentrations were used to evaluate concentration-dependency (see table 1). DQ12 and Al_2O_3 were defined as appropriate particulate positive and negative controls (table 1 and figures 1-4). As expected, AL efficiently inhibited the DQ12-induced biological effects (table 1 and figures 2-4). By testing the factory samples (200 μg/cm^2, 2 h) in PAM, all dusts, except DQ12 and TC, mediated only moderate, mostly not quartz-specific, increases in LDH-levels, as compared to Al_2O_3 (figure 1). At 200 μg/cm^2 (2 h), the MTT-assay with NR8383 cells demonstrated the mainly quartz-independent rank order of toxicity: DQ12 > RF > TG > TI ≥ BR > TC (table 1). For DQ12, BR, TG, and TC but not for TI and RF decrease in cell viability was concentration-dependent. The marked, concentration-independent effect of RF was most likely of artificial nature due to the dark color of the dusts, thus interfering with photometric formazan

Journal of Physics: Conference Series **151** (2009) 012022 doi:10.1088/1742-6596/151/1/012022

determination. In the Comet-assay with primary rat alveolar macrophages (200 μg/cm², 2 h), BR and TI exhibited the highest DNA-damaging potential, followed by DQ12, TG, TC, RF, and Al_2O_3 (figure 2). About 50 % of the TG-, BR- and TI-mediated effects were quartz-related (as determined by AL-treatment), whereas TC and RF showed values of 20 % and 40 %, respectively. In pre-experiments with BR and TG clear concentration-dependent increase in DNA-damage could be detected (data not shown). All factory dusts induced higher PGE_2-liberation in primary rat alveolar macrophages than Al_2O_3 as particulate negative control (DQ12 > BR > TC > TG > TI > RF; figure 3). Relative TNF-α mRNA expression was significantly increased in NR8383 cells only by DQ12 and TG, both quartz-related, and by TC, not quartz-related, as compared to the vehicle control (figure 4).

Table 1. Results of the MTT-assay with NR8383 cells. Data represent relative means ± standard deviations of 3 independent experiments with 5 replicates per experiment. Significantly different from the corresponding vehicle control: $^{\#}P \leq 0.05$, $^{\#\#}P \leq 0.001$, Student's *t*-test for unpaired values.

	50 μg/cm²		100 μg/cm²		200 μg/cm²	
Treatment	without AL	with AL	without AL	with AL	without AL	with AL
Vehicle control	100 ± 5	100 ± 7	100 ± 5	100 ± 9	100 ± 6	100 ± 3
Al_2O_3	98 ± 4	112 ± 4	95 ± 8	105 ± 7	99 ± 2	99 ± 3
DQ12	86 ± 6	113 ± 7	58 ± 5#	100 ± 7	29 ± 5##	95 ± 9
Brickwork (BR)	94 ± 7	97 ± 5	70 ± 9#	91 ± 5	71 ± 6#	75 ± 9
Tableware (TG)	81 ± 9	88 ± 4	64 ± 1#	73 ± 3	57 ± 7#	52 ± 8
Tiles (TI)	71 ± 9#	72 ± 8	66 ± 8#	68 ± 6	73 ± 7#	74 ± 9
Tableware (TC)	93 ± 1	95 ± 11	89 ± 1	80 ± 1	79 ± 8#	71 ± 7
Refractory (RF)	54 ± 12#	57 ± 11	50 ± 11#	53 ± 7	49 ± 5#	61 ± 8

AL = 10 μM aluminium lactate

4. Discussion

All 5 ceramic dust samples mediated more or less cytotoxic, genotoxic and pro-inflammatory effects, with different dusts in each case to exhibit the highest potential. As judged by the experiments with AL there seemed to be no clear correlation between the different biological effects and the quartz content (see 2.1.1 and 2.1.2) of the dust samples. For example, the tile sample demonstrated an even higher clastogenic potential than the positive control DQ12, although the tile sample contained only 8.1 % versus 87.0 % quartz, respectively. Interestingly, in all other assays the toxic potential of the tile sample was rather low, indicating that there were special dust components within the sample with high DNA-damaging potential. In addition, except for the positive control DQ12, cytotoxicity (metabolic capacity and membrane integrity) of the factory samples was nearly RCS-independent, because there was no significant effect of parallel AL treatment. This lack in clear correlation in the *in vitro* tests indicated that other dusts components e.g. special metal compounds, like iron (substantial FeO_2 content in some of the samples), may also account for adverse effects and that dust components like clay were perhaps able to quench the biologic activity of RCS in the ceramic dust samples.

In conclusion, the established *in vitro* battery has been found to be an adequate approach to screen and differentiate the multifactorial toxic potential of quartz-containing ceramic dusts. However, in the present study, the quartz-content of the ceramic dusts was too low to clearly identify and differentiate quartz activities in the various dust samples. To really approach this goal dusts with higher RCS content or the pure quartz varieties used for production in ceramic plants should be compared.

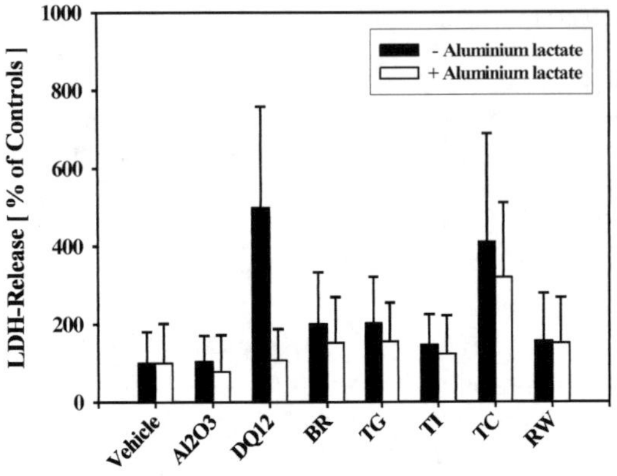

Figure 1. LDH-release. PAM were incubated for 2h with or without 200µg/cm^2 of the dusts ± 10µM AL and LDH-activity was measured in the culture supernatants. Data represent means ± SD of 3 independent experiments.

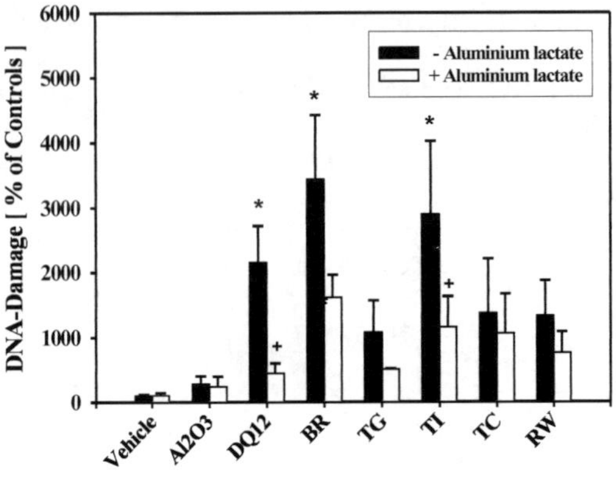

Figure 2. Comet-assay. PAM were incubated for 2h ± 200µg/cm^2 of the dusts ± 10µM AL and DNA-damage was measured. Data represent means ± SD of 3 independent experiments. Significantly different from the vehicle control: **P*≤0.05; from the respective sample without AL: $^+P \leq 0.05$, Student's *t*-test for paired values.

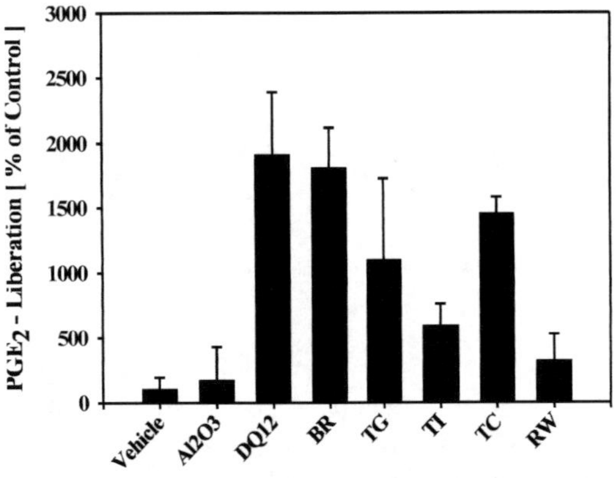

Figure 3. PGE_2-liberation. PAM were incubated for 4h with or without 50µg/cm^2 of the test and reference dusts and PGE_2-liberation was evaluated by a competitive enzyme immunoassay. Data represent means ± SD of 2 independent experiments each performed in duplicate.

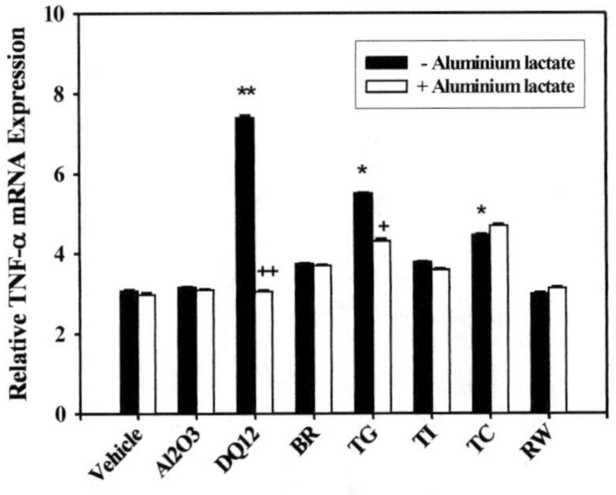

Figure 4. TNF-α mRNA expression. NR8383 were incubated for 24h ± 10µg/cm^2 of the dusts ± 10µM AL and TNF-α mRNA expression was determined by PCR. Data represent means ± SD of 3 independent experiments each performed in triplicate. Significantly different from the vehicle control: **P* ≤0.05, ***P*≤0.01; from the respective sample without AL: $^+P \leq 0.05$, $^{++}P \leq 0.05$; Student's *t*-test for unpaired values.

Acknowledgment
The work was supported by the EC through the COLL-CT-2003-500 896 project SILICERAM.

References

[1] Mosman T 1983 *J. Immunol. Methods* **65** 55
[2] Singh NP, McCoy MT, Tice RR, Schneider EL 1988 *Exp. Cell Res.* **175** 184
[3] Tice RR, Agurell E, Anderson D, Burlison, B, Hartmann A, Kobayashi H, Miyamae Y, Rojas E,

Ryu J-C, Sasaki YF 2000 *Environ. Mol. Mutagen.* **35** 206
[4] Schäfer D, Lindenthal U, Wagner M, Bolcskei PL, Bänkler HW 1996 *Thorax* **51** 919
[5] Giulietti A, Overbergh L, Valckx D, Decallonne B, Bouillon R and Mathieu C 2001 *Methods.* **25** 386

Airborne Fibre and Asbestos Concentrations in System Built Schools

Garry Burdett*, Steve Cottrell and Catherine Taylor

Health and Safety Laboratory, Harpur Hill, Buxton, Derbyshire, UK, SK17 9JN

e-mail Garry.Burdett@hsl.gov.uk

Abstract: This paper summarises the airborne fibre concentration data measured in system built schools that contained asbestos insulation board (AIB) enclosed in the support columns by a protective steel casing. The particular focus of this work was the CLASP (Consortium of Local Authorities Special Programme) system buildings. A variety of air monitoring tests were carried out to assess the potential for fibres to be released into the classroom. A peak release testing protocol was adopted that involved static sampling, while simulating direct impact disturbances to selected columns. This was carried out before remediation, after sealing gaps and holes in and around the casing visible in the room (i.e. below ceiling level) and additionally round the tops of the columns, which extended into the suspended ceiling void. Simulated and actual measurements of worker exposures were also undertaken, while sealing columns, carrying out cleaning and maintenance work in the ceiling voids. Routine analysis of these air samples was carried out by phase contrast microscopy (PCM) with a limited amount of analytical transmission electron microscopy (TEM) analysis to confirm whether the fibres visible by PCM were asbestos or non-asbestos. The PCM fibre concentrations data from the peak release tests showed that while direct releases of fibres to the room air can occur from gaps and holes in and around the column casings, sealing is an effective way of minimising releases to below the limit of quantification (0.01 f/ml) of the PCM method for some 95% of the tests carried out. Sealing with silicone filler and taping any gaps and seams visible on the column casing in the room, also gave concentrations below the limit of quantification (LOQ) of the PCM method for 95% of the tests carried out. The data available did not show any significant difference between the PCM fibre concentrations in the room air for columns that had or had not been sealed in the ceiling void, as well as in the room. Occupant exposures during normal classroom teaching activities in areas that had undergone remediation were also monitored with much greater use of TEM analysis to determine the asbestos fibre concentration. No asbestos fibres were found in the TEM analysis of samples of classroom air from seven schools, after remediation had taken place. Occupant exposures in one school and one office building have been monitored before remediation and only one asbestos fibre was found in each set of samples analysed. The average monitored occupant exposures to asbestos in CLASP buildings was <0.00005 f/ml, an order of magnitude lower than previous measurements in UK buildings.

Key Words: asbestos, buildings, schools, airborne, concentrations, PCM, TEM, CLASP, sampling.

1. Introduction

In the summer of 2006, after asbestos removal contractors had carried out asbestos removal work at a system built (CLASP) school building, the laboratory carrying out clearance monitoring obtained concentrations of airborne fibres above the 'clearance level', when, as part of deliberate disturbance they struck the support columns in the rooms. This finding was reported to the Health and Safety Executive (HSE) in September and prompted a rapid assessment of the causes of the fibre release by HSE inspectors, with support from the Health and Safety Laboratory (HSL).

The initial HSE/HSL investigation confirmed the clearance laboratories own findings that the airborne fibre release came from the columns and consisted mainly of amosite fibres generated from the asbestos insulating board (AIB) that had been installed to provide fire protection to the structural steel columns supporting the building structure. The AIB was enclosed by a metal casing around the column but leaked fibres into the room when the column was struck. The main source of fibres appeared to come from gaps where the two halves of the steel casing joined together. Also some of the casings had been damaged due to drilling through them to attach various items or from improper window replacement. Some debris from the original installation also appeared to have been swept into and around the base of the column in the gap behind the skirting board. The failure of the enclosure to prevent airborne fibre release into the classroom when the column casings were subjected to impacts was a source of concern, as it would increase the asbestos exposure and risk to the occupants and maintenance personnel. Also as enclosing asbestos is one of the remediation methods recommended in HSE guidance, an evaluation of the causes of the airborne release and assessing ways to improve the effectiveness of the enclosures was seen as a high priority.

After further field sampling work and data gathering had confirmed the primary cause for the release of fibres into the room (damaged and /or poor sealing column casing) and that sealing of gaps in the casing was an effective ways to minimise the release into the room, an initial advisory note was prepared and circulated by HSE in mid-October, before the half-term holiday. The note was sent to the Directors of Education/Children's Services in England, Scotland and Wales and Governing Bodies of Foundation, Voluntary Aided and Independent schools, informing them of the potential for fibre release and requiring them to identify Mark 4 and 4B CLASP buildings; and to take action to seal any visible gaps in and around the column casings through which asbestos could escape directly into the room. Following this initial advice further work was carried out to assess airborne fibre levels using both routine phase contrast microscopy (PCM) with the use on selected samples of analytical transmission electron microscopy (TEM) to confirm the actual concentration of asbestos fibres present. This information was fed back to a CLASP advisory group [1, 2] (Burdett, 2007 & Burdett, 2008), which included representatives from HSE policy, sector and operational groups, HSL, Department for children, schools and families (DCFS), local government employers, unions and the building manufacturer. Its terms of reference were to ensure duty holder awareness of the issues, providing guidance, evaluating compliance and effectiveness of remedial action and agreeing a communication strategy for the education sector.

This paper summarises the airborne fibre concentration data collected in CLASP system built schools that had asbestos insulating board (AIB) in the support columns.

1.1. Description of the CLASP construction and occurrence of asbestos

The school where the escape was first noticed was a particular type of system built construction known as CLASP (Consortium of Local Authorities Special Programme). The CLASP concept was started in 1957 for the purposes of improving the construction and delivery of schools and used a systematic form of construction based largely on pre-fabricated elements. The system has a wide usage with some 3500 buildings in the UK including: primary and secondary schools, hospitals, universities, offices fire and ambulance stations. Several marks / redesigns of the building took place from 1957 –1980 when asbestos containing products were specified for use in the buildings. These were the mark 2, 3, 4 and some mark 5 structures. It should be noted while some asbestos products may always have been included, the majority of buildings were locally adapted by private and public architects, who may have added additional asbestos products, or even have substituted some of the commonly specified asbestos products. The wide involvement of local architects means that there is considerable variety in the design between different buildings and where the asbestos may or may not be present. Details on where and which types of asbestos would normally be present were first circulated by CLASP to all Local Authorities in 1974 and this information has since been updated.

The use of AIB insulation in the columns was a common design feature and must be assumed to be present, in CLASP schools built before 1980, unless the building survey shows otherwise. It should be noted that in some mark 4 buildings the column casings were moulded from asbestos wood containing about 25% chrysotile, however no examples of these were found in the buildings examined in this survey. Figure 1 shows the external appearance of the column with a gap in the columns casing. Figures 2 - 4 shows examples of the use and appearance of AIB in the columns. A variety of configurations were found with AIB attached to the column rather than the casing and residual layers of material where the AIB had been removed. Off-cuts and debris could sometimes be present at the top and bottom of the columns and was probably placed there during the installation.

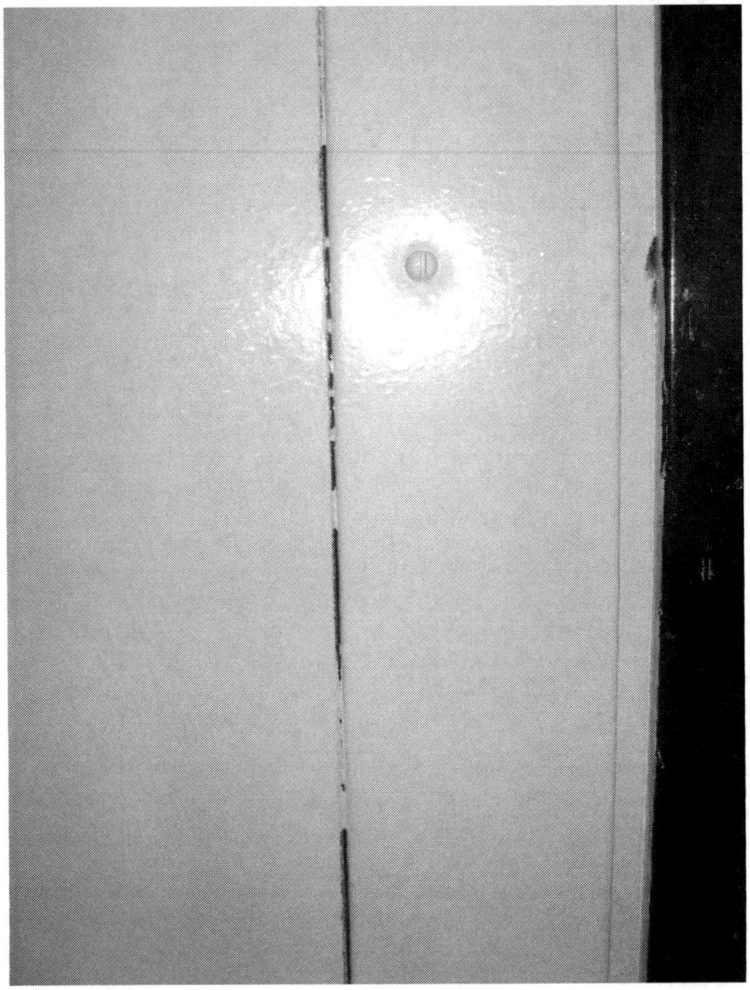

Figure 1: External appearance of the column showing an example of a gap in the casing.

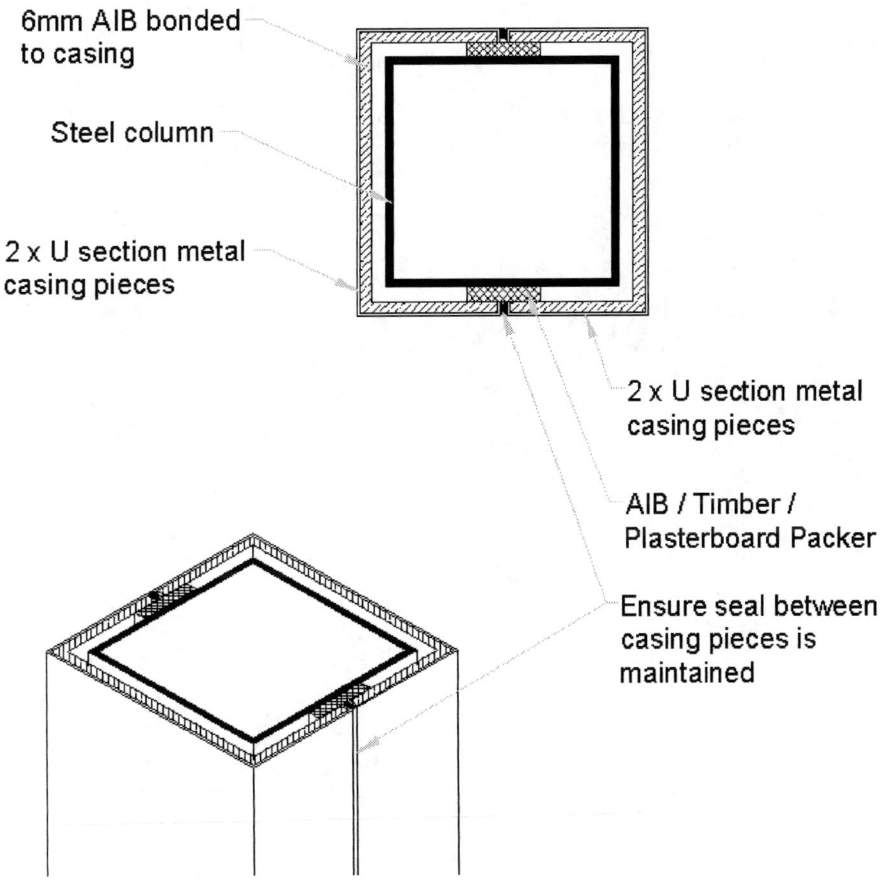

Figure 2: Schematic diagram of use of AIB in the CLASP columns

Figure 3: Example of a column casing with AIB attached to one side of the casing. The torch beam shows where the AIB has fractured due to a screw being placed into the casing. A hole is visible in the side of the casing and visible debris shows that friable material was present.

Figure 4: View of base of column from which the casing was removed.

AIB was also commonly used in fire critical evacuation routes, fire risk areas or areas where there may be damp e.g. in suspended ceilings in entrances, in porches or over recessed areas and kitchens or in stud walls in stairwells, at high levels in halls and kitchens. Most AIB contained about 20% asbestos with Amosite being the predominant asbestos type present. Other forms of asbestos products are also commonly found in pre-1980 CLASP buildings, the most common being vinyl floor tile, asbestos cement sheet and moulded fittings, stair fittings, cloth fire breaks in ceilings voids. The vast majority of these would contain chrysotile asbestos. Some of the pre-cast concrete panels on the outside of the buildings also contained actinolite asbestos.

2. Method

2.1. Sampling strategy

As peak releases of airborne asbestos fibres will occur only infrequently it is much more convenient to simulate a peak release episode, using a standardised protocol. Simulation also allows pre-planning and the use of appropriate controls to prevent the spread of asbestos and the exclusion of other building occupants. The extent of force applied and the amount of damage inflicted on the ACM during the simulation will have a direct influence on the amount of airborne fibres released. The monitored fibre concentration is also influenced by a number of other sampling and environmental factors, e.g. the period of sampling and of disturbance, the closeness to the source of the fibre emissions, the use of personal sampling, the use of containment, the amount of ventilation / mixing of air.

Personal sampling using small lower flow rate battery pumps to sample air through a filter placed close to the breathing zone is the preferred method of collecting workplace occupational samples. While this is appropriate for maintenance workers who may carry out work in less accessible areas (e.g. ceiling voids) and directly disturb the asbestos, its applicability for monitoring pupils and teachers for background levels is less clear cut, besides the practical difficulties involved. Even for simulated short-term peak releases from impacts, the sampling rates achievable from personal sampling pumps will allow only concentrations above the control limit to be measured (e.g. a 10 minute personal sample has a limit of quantification (LOQ) of 0.24 f/ml ; see HSG 248 [3]). Therefore higher flow rate static sampling was used for occupancy sampling and most of the peak release simulation sampling to allow an improved LOQ.

Longer-term sampling to measure the average exposure during normal occupation is more demanding particularly in schools where there is a greater need to supervise the sampling and where pump noise will be more of an issue. In general, the shorter school day and the need for increased supervision, favours the use of higher flow rate (e.g. 8 –16 l/minute), heavier and more noisy sampling pumps, to take static samples over one school day. The analytical sensitivity of a clearance / background sample (based on 480 litres sampled onto a 22 mm diameter exposed filter area with 1.5 mm^2 examined by PCM) is 0.0003 f/ml based on a half fibre counted. However, in practice as blank filters have background fibre counts, the limit of detection is some ten times higher 0.003 f/ml. For the purpose of clearance indicator sampling, when it is important to determine whether the measured fibre concentrations can be considered to have a high probability to be above background, the acceptable concentration is set at <0.010 f/ml.

Fibre counting has poor statistical precision. This means that individual results must be compared carefully to assess whether any differences are actually present. Particles sampled onto a filter at best have a random distribution. This means that the precision of the count is limited by the underlying 'Poisson' statistics. The precision is usually expressed in terms of the confidence interval, which defines the upper and lower limits expected for a defined percentage of repeat counts. For example 95% confidence limits mean that on average 19 of the 20 values from repeat counts would be within the upper and lower limits. For low counts the lower confidence limit is 0, so a one-sided upper 95% confidence interval is used. For a count of 0 it is 95% probable that the count is < 3 fibres. This is used as the limit of quantification for TEM analysis where there is normally no background of asbestos fibres on the blank membrane filter.

For any sampling in schools extensive communications with the local authorities, head teachers, governors, teachers, teaching unions and parents is necessary particularly if occupant sampling is to be carried out. Gaining access and sampling is therefore very time consuming and it was realised early on that working with local authorities and a number of local United Kingdom Accreditation Service (UKAS) accredited local sampling and PCM analysis laboratories was the only possible way to gather significant amounts of data in the short time period available. Due to several laboratories being involved, it was important that familiar routine sampling protocols were used and that the results of the PCM analysis was fed back to the interested parties and passed on to HSL.

The difficulties of access and communications, the use of local resources and the short time available to collect the data meant that it was not possible to implement a statistical sampling design to randomise the schools, classrooms and columns tested. The concern was to focus on whether we could detect columns that were likely to release airborne fibres and effectively remediate the release. This inevitably biased the sampling to the schools, which were thought to have the greatest amount of potential damage and the testing of columns that had physical damage or visible gaps. HSL in a later sampling exercise did set out to test columns that were judged to be in a reasonable to good condition.

2.2. Sampling and analysis methods

To assess peak releases from simulated disturbance of the columns a disturbance protocol was adopted, which was closely based on what is routinely used by PCM laboratories for the clearance air testing, in stage three of the four-stage clearance procedure detailed in HSG 248. At the school where the problem was first found the sampling laboratories had already developed a protocol which involved hitting of the column casing, slamming doors and sitting/jumping on the window shelf a number of times. The exact disturbance activity depended on the columns under test with the disturbances being repeated between 3-5 times at the start of the sampling period. This procedure along with the requirements for static air sampling as described in HSG 248, was adopted for the peak release testing of columns for both before and after remediation.

Static air samples were collected by sampling air through a 0.8 or 1.2 μm pore size mixed cellulose ester membrane filter at calibrated sampling rates of 8 l/minute or greater. This was used to collect a minimum sampling volume of 480 l over a period of around 60 minutes (or less at higher flow rates). The standard PCM fibre counting method was used (see HSG 248 annex 1) at X500 magnification to assess the concentration of visible >5 μm long fibres with aspect ratios > 3:1 and widths <3 μm. It should be noted for the purposes of this report that all data collected were recalculated based on the number of fibres and fields of view counted and the air volume sampled, assuming a 380 mm^2 exposed filter area and a graticule area of 0.0075 mm^2.

For sampling occupant exposure to asbestos during normal occupation, higher volumes of air need to be collected, preferably by extending the sampling period to get a more representative sample of the average exposure. A target sample volume of 3000 l over a minimum period of one school day was set with the filter being monitored for signs of overloading of particulates (colour changes from white to light-mid grey). After sampling was completed a half filter was cut using a scalpel blade and prepared and counted at x 500 by PCM analysis in the normal way (see HSG 248). A portion of the remaining filter was prepared and analysed by analytical electron microscopy. The gold standard for this type of analysis is the International Standard Organisation ISO 10312:95 [4] method based on transmission electron microscopy (TEM) which uses a combination of both energy dispersive x –ray analysis (EDXA) and electron diffraction (ED) to identify the asbestos fibres and magnifications of 10,000 to 20,000 to count and measure the asbestos fibres. The results can be expressed in terms of PCM equivalent asbestos fibres (approximately the same size of fibres counted by PCM) for the purposes of assessing exposure and risk.

One day of sampling represents only a snapshot of the occupant's average exposure and several days of sampling were preferred. In practise, it was found that a sequence of daily samples spread over several weeks was the more acceptable method for the occupants for collecting samples. This gave several individual results that can be averaged or "pooled" to give a better estimate of the cumulative average exposure to the occupants, as well as the range of individual daily exposures.

Personal sampling for fibre concentrations with PCM analysis of the fibre concentrations, as described in HSG 248 was used to monitor actual and simulated maintenance and cleaning activities.

The PCM and TEM results represent data collected from a number of laboratories, all of which were UKAS accredited and hence meet the requirements of ISO 17025[5].

3. Results

Table 1 summarises the PCM data from schools for various activities, some later PCM data from specific maintenance activities are given separately.

3.1. PCM results from disturbance of unsealed columns.

After HSE's initial investigation of the reported findings and it had been confirmed that amosite asbestos fibres were being released to air, the emphasis of the monitoring was on the effectiveness of the remedial action. The PCM data available from peak disturbance testing from some 31 air samples are summarised in figure 5. Some 60% of the samples exceeded 0.01 f/ml with the highest value in a school building being 0.44 f/ml.

The columns selected for peak release testing were generally chosen to represent a worst-case situation where there was visible damage to the casing (e.g. from installation of fire alarms) and / or obvious gaps were present. Columns adjacent to improperly or poorly replaced windows and with attached doors were also preferentially chosen to test, as the AIB would have been subject to greater damage. Some columns have been tested more than once or had more than one air sample collected. Generally, it was reported that in the same building, the more damage to the column casing and the greater area of visible gaps, the higher the fibre concentration measured.

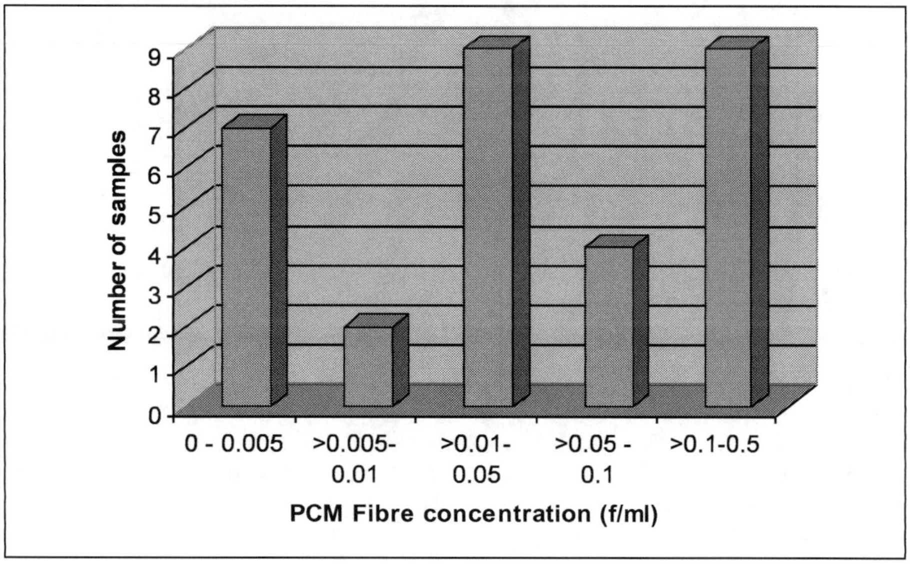

Figure 5: Distribution of PCM fibre concentrations from disturbing hitting the columns.

3.2. PCM results from samples taken after sealing columns and cleaning the area.
The key issue for HSE and the duty holders was to determine whether the recommended advice to inspect the column casings for obvious holes and gaps and to seal them to increase the integrity of the enclosure was successful in minimising peak releases. Figure 6 summarises the results from 95 samples collected during the peak release testing of columns after fully sealing all gaps visible in the room as well the top of the column in the ceiling void and cleaning the area. The distribution of the fibre counts is grouped around the limit of detection with a slightly elongated tail, suggesting that a few columns were still giving peak fibre release above the LOQ, even after complete sealing has taken place (including above ceiling). The overall level of the peak release was relatively low with a mean of 0.004 f/ml with about 5% of the samples exceeding the LOQ (0.01 f/ml), the highest result being 0.058 f/ml.

The results from peak release testing of 14 columns where all the gaps visible in the room has been sealed and taped but the top of the column left unsealed and the area not cleaned, gave mean and median concentrations of 0.006 and 0.002 f/ml respectively, with the highest single result of 0.032 f/ml.

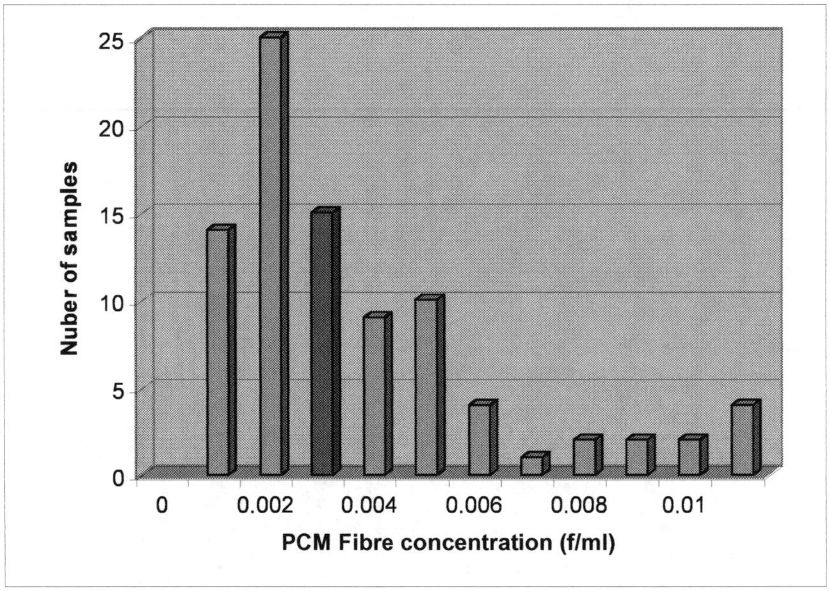

Figure 6: PCM results from disturbance testing of fully sealed columns after cleaning the room (the different coloured column at 0.003 f/ml represents the limit of detection).

3.3. PCM results measured during normal use after remediation.
Figure 7 summarises 96 measurements after the classrooms were reoccupied and in normal use. This has a similar distribution to that found in the peak release testing in figure 6. However in an occupied room there are several sources of non-asbestos fibres (e.g. paper fibres and fine synthetic fibres) from the presence and activities of the occupants and is likely to substantially overestimate the asbestos fibre concentration. A number of samples (44) were taken after the remediation but before reoccupation and are also summarised in figure 7. Although some limited activity may have been present, this represents the best assessment estimate of the background level of fibre counts without any known disturbance.

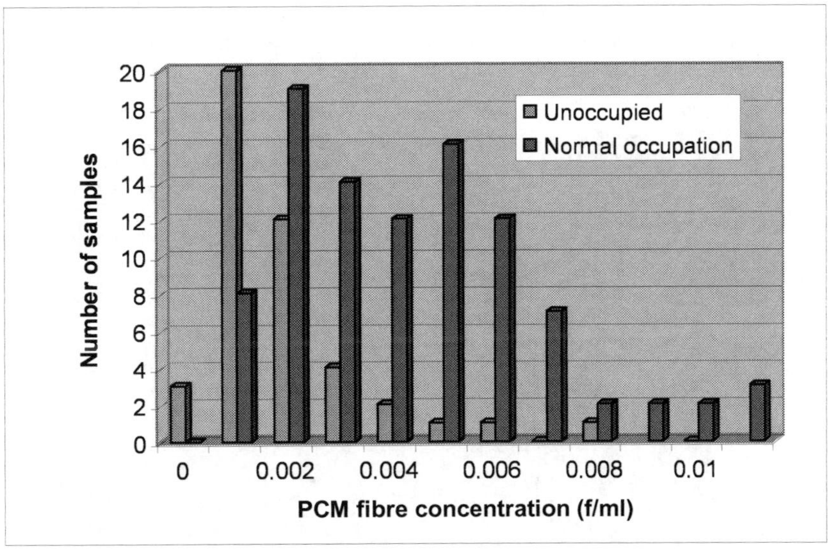

Figure 7: Comparison of PCM fibre concentrations measured in unoccupied and normally occupied classrooms after remediation.

3.4. Further PCM results from investigations of the effectiveness of remediation.

The effectiveness of sealing to the ceiling height was studied in the original school with columns that gave high releases. Table 2 gives an example of the PCM results found when testing a column where several of the ceiling tiles around the column were damaged / displaced and shows that there were a significant reduction in PCM fibre concentrations even when the ceiling void was poorly isolated from the room. The TEM analysis of samples after sealing gaps below the ceiling detected 1 PCME amosite fibre, equivalent to an analytical sensitivity of 0.0016 f/ml.

At other columns in the same school releases from columns during peak disturbance testing were monitored adjacent to the columns at both ground level and inside the ceiling void (with ceiling tiles replaced). The results prior to remediation in table 3, showed that peak fibre emissions were relatively low at room level but increased significantly at one position after remediation. A similar low value followed by an increase was seen in all three ceiling void samples. The most likely explanation for these increases was that the maintenance personnel working in the ceiling void released fibres into the air so that when sampling after remediation commenced (<1 hour later) increased airborne fibre concentrations were still present. However as the ceiling voids were interconnected it is not impossible that some brush testing in the ceiling void elsewhere may have also contributed to the levels. Either way this data showed that disturbing and working in the ceiling void (in this case using experienced asbestos removal contractors who had sealed many tens of such columns) there is a greater possibility for exposures to occur.

3.5. PCM results while carrying out remedial work on columns.

Only a limited amount of data (12 samples) was available during the sealing of the column casings and the tops of the columns in the ceiling void, which gave a mean and median of 0.004 f/ml with >95% of the results being below the LOQ. However, as these were static samples at room level they may have underestimated the exposure from work in the ceiling voids. A larger number of static samples were collected when filling gaps and taping of the seams of the column casings in the room areas. As the tape was being firmly pressed against column casing up to the ceiling, this flexing of the casing would cause a more vigorous disturbance than someone leaning or pushing against the column. The distribution of the individual results are given in figure 8, based on 188 individual PCM samples.

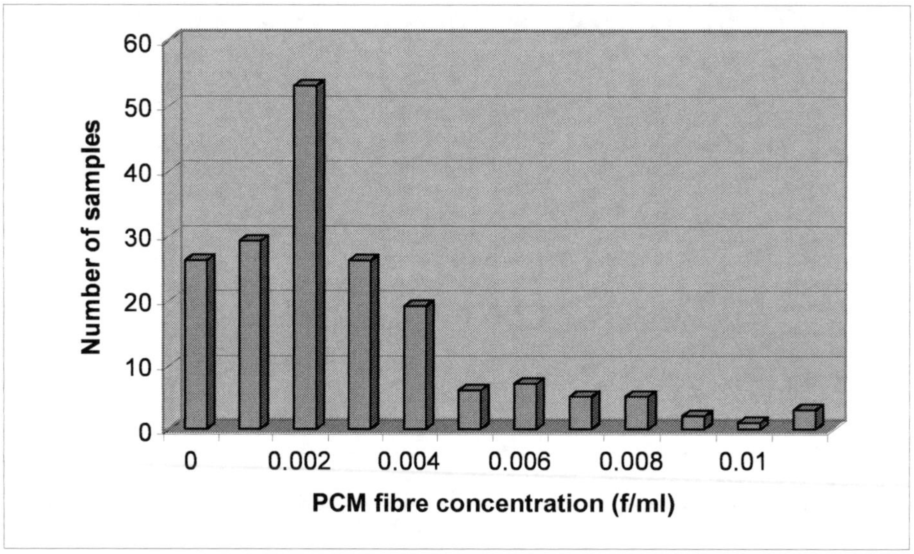

Figure 8: PCM air monitoring results from 188 samples taken during the taping of the vertical seams of the columns and surrounding joints.

3.6. Cleaning and maintenance worker exposures

Twenty PCM analyses of samples collected by persons inspecting columns and ceiling areas gave mean and median concentrations of 0.004 and 0.002 f/ml respectively. The majority of samples were static samples by the inspection areas and may underestimate personal exposures when looking into the ceiling voids.

Three school cleaners who wore personal samplers during normal cleaning activities in a school that was considered to have a potential problem had measured PCM concentrations of 0.003, 0.011 and 0.014 f/ml giving an average of <0.01 f/ml.

Some simulated maintenance work was carried out at the most contaminated school and one other. The PCM results are summarised in table 4. The highest results came from an ceiling area where AIB ceiling tiles were present, so the settled dust on joists etc. was brushed rather than the surface of the AIB tile.

3.7. TEM results for asbestos fibre concentrations in schools before remediation

Since advice was issued so rapidly it was not possible to carry out a programme of monitoring asbestos fibre concentrations in schools by analytical TEM before remediation had taken place. The one school where data was obtained was reported to have a chrysotile asbestos board in the column and cannot be seen as representative of the amosite containing AIB. The calculated pooled result from two samples taken in different areas gave an analytical sensitivity (based on 1 asbestos fibre seen) of 0.00005 f/ml with a limit of detection of <0.0003 f/ml. Only one PCME asbestos fibre (chrysotile) was detected showing that the airborne exposures were low.

It was however possible to carry out prolonged sampling in an ex-school CLASP 4 building used as offices. No remediation of the columns had been carried out, which contained amosite AIB attached to the columns casings. The sampling strategy was to collect week long samples during the day (8 am to 6 pm) to capture normal occupation and cleaning activity. Two samplers were set up one at a flow rate of 3.2 l/minute and one at 1.6 l/minute in two office rooms. Both sampling heads were positioned above radiators where there would be a good convective mixing of room air and close to sets of double doors. One sampler was particularly close to a set of doors, which vibrated the column when opened and closed and visible gaps were present in the columns around the windows. Even though extended counting was carried out on the samples, only 1 asbestos fibre (amosite) was found in the TEM analyses of the 10 samples collected over 5 weeks. As ~100 grid openings were counted in each sample, the average values for the building can be calculated from the pooled individual samples (see Table 1). This gave a calculated concentration for PCME asbestos fibres equal to the analytical sensitivity of 0.000005 f/ml and an upper 95% confidence value of <0.000024 f/ml for continuous daytime monitoring over the five-week period. TEM airborne asbestos concentrations in occupied schools after remediation

In total, 31 individual samples taken in eight occupied schools were analysed by TEM. Seven samples from two schools, which had both the tops and below ceiling sections of the columns sealed, had no asbestos fibres detected in the classroom air samples taken during normal use and occupation. A further 21 samples from five schools where the below ceiling column casings only were sealed, also had no asbestos fibres detected in the classroom air samples taken during normal use and occupation. However, TEM analysis of three samples taken in a school corridor between two sets of double swing doors did find one PCME amosite fibre. As the doors were mounted on the columns and were frequently opened and banged shut, causing movement and vibration to the building structure, this was a particularly active area. The pooled result from 3 samples gave a calculated concentration equivalent to the analytical sensitivity of 0.0001 f/ml and was below the limit of detection (0.0005 f/ml).

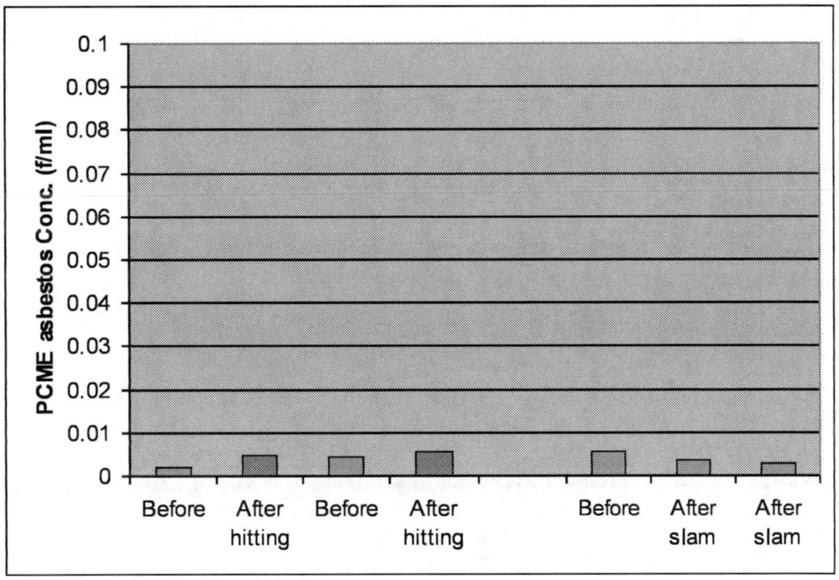

Figure 9: TEM results of the PCME asbestos fibre concentrations in air before and after striking columns in reasonably good condition and slamming doors mounted on the column. All air concentrations given in terms of the calculated limit of detection (i.e. 95% confidence that concentrations are below this value). Asbestos fibres were only detected after hitting the column casings).

3.8. TEM results for releases from columns in good condition

A CLASP school where columns were judged to be in reasonable / good condition was sampled before and then during and after peak release testing. The results in terms of PCME amosite fibres are summarised in figure 9. These results further add to the evidence based observation that airborne fibre releases from the columns to the room air, occur much more readily when there is substantial direct disturbance to the column casing and the casings have damage and/or gaps along the seams or edges in the room (i.e. below the ceiling void).

4. Discussion

4.1. Mechanisms for release of airborne fibres

Although asbestos is a hazardous carcinogen, a risk will occur only when the asbestos fibres in the material are made airborne, inhaled and deposited in the lung. Generally for fibres to be released from an asbestos containing material (ACM) to air, some form of disturbance is necessary. Disturbance due to maintenance activity is a recognised as a way of producing short-term "peak" exposure episodes and a specific duty to manage asbestos is present in the control of asbestos regulations (CAR, 2006) [6] to minimise the possibility of uncontrolled maintenance releases. Direct disturbance and impact damage to ACMs by the building occupants will also result in short-term "peak" releases of fibres to air. Hence, the duty to manage also requires regular inspections for signs of damage or deterioration to the ACMs in the building, with appropriate management action being taken to minimise further damage and release, as far as reasonably practicable. If damage to ACMs occurs and the debris is not cleaned-up appropriately, there is the possibility that this may increase the background airborne fibre concentrations in the building, due to the dust and debris being disturbed and re-suspended by occupants and cleaning personnel.

Background concentrations due to environmental-related factors are usually several orders of magnitude lower than "peak" disturbance releases. Expansion and contraction of the building structure due to thermal changes, vibration, wind-pressure and air erosion and normal occupancy, have all been suggested as possible disturbances, which can cause fibre release and give raised background concentrations. Although a combination of these are usually present, the airborne fibre concentrations produced are rarely above the limit of quantification that a single sample measurement can realistically achieve. Conversely, short-term peak releases are measurable but occur relatively infrequently. This is why in this report extensive use of simulations and peak release testing were carried out to assess whether the remediation measures were effective in minimising airborne releases and could be applied with minimal disturbance and fibre release. It is important to distinguish that peak release testing does not represent the average exposure but may contribute to it.

The magnitude of any release of airborne asbestos fibres will depend on a range of release factors (e.g. type of ACM; the condition of the ACM, the type of disturbance taking place; the magnitude of the disturbance and force applied; the length of time the disturbance takes place; the surface area or volume of the ACM being disturbed) and on other mitigating factors (e.g. wet, enclosed, encapsulated); ventilation rates, distance the source of the release. etc). In particular, for the CLASP columns the amount of airborne fibre release depends on the integrity of the casing and the: presence of gaps and holes, visual damage and denting, looseness / amount of movement when pushed; maintenance work carried out after installation (e.g. fixtures attached to the casing or holes drilled into the casing) or improper installation (e.g. when new windows installed) etc. As most of these release factors can be assessed visually from inside the room inspection of the column casing and sealed by the use of silicone fillers and cover tapes this provides a simple strategy to detect and prevent airborne releases to the occupants.

4.2. Monitored releases from columns without remediation

The highest concentrations found from peak release testing in schools under the simulation conditions described and in a sealed area was 0.44 f/ml, next to the column. Higher airborne concentrations can be released if the column is more damaged and hence less efficient in protecting the AIB and containing the release. AIB can be quite friable and can be damaged relatively easily giving rise to peak airborne fibre concentrations of a few f/ml. During poor dry removal where AIB ceiling tiles are broken and dropped to the ground airborne concentrations of well over 10 f/ml can be generated. Therefore there is a potential for higher fibre concentrations to be released if the casings are removed or detached. The highest value recorded to date (2.37 f/ml) during peak release testing was in a non-school situation, where the casing had become detached and was left dangling by a single screw at the top of the column. The higher levels of release possible once the casing is detached or removed, underlines the current requirement (see Reg 4 of CAR, 2006) for the management and the routine checking of all ACMs and any control measures (such as enclosures) for signs of damage.

A draft report (personal communication, 2008) [7] of some testing undertaken for publicity purposes in a school was also sent to HSE. As described the report appeared to have carried out far greater numbers of violent disturbances to the column casing and surrounds than used by other laboratories. The column was also described as having a gap down the front of the column. In addition, a small enclosure surrounded the column (<10 m^3) and the ceiling tiles had been taken down and disturbance was carried out around the top of the column. The comparative static samples collected over a period of one hour gave results of 0.72 and 0.49 f/ml, which given the intensity of the disturbance was consistent with the results summarised above.

4.3. Effectiveness of remediation

The focus of the work was not to carry out a survey to find the highest release from an unremediated column but to engineer and assess a suitable way to minimise the release of airborne fibres into the room where occupants are present. This assessment was carried out using the same peak release testing procedure. The results showed that by sealing any gaps in and around the casing that were visible in the room, was an effective way of reducing occupant exposure. The mean and median PCM fibre concentrations after sealing were 0.006 and 0.002 f/ml respectively, with the highest single result of 0.032 f/ml. When the top of the column was also sealed and the whole area cleaned before re-testing this gave a mean and median of 0.004 and 0.003 f/ml, respectively, with the highest result being 0.058 f/ml.

The requirement under regulation 11 of CAR, 2006 is to prevent exposure as far as reasonably practicable, or to reduce exposure to the lowest level reasonably practicable. Given that the PCM data from peak release tests had mean values no more than twice the limit of detection and ~95% of measurements were below the limit of quantification it is clear that exposures are as to close to as low as reasonably measurable by the PCM method. It is important to note that the PCM count does not discriminate between asbestos and non-asbestos fibres so the PCM count represents the maximum value for the asbestos fibre component. Peak release testing is not representative of the average asbestos exposures to the occupants as it is clearly meant to replicate an infrequent high release event to assess the efficacy of the controls.

To derive an estimate of the average asbestos exposure, longer sampling during normal occupation with analytical TEM analysis is necessary. When the TEM analysis was used to analyse 28 samples from seven schools, which had been re-occupied after remediation, no asbestos fibres were detected in the analysis of high volume samples collected inside classrooms. Most individual samples had analytical sensitivities of ~0.0002 f/ml and a limit of detection of ~ 0.0006 f/ml. Taken as a group representing occupied CLASP schools which had under gone remediation (two had complete remediation), an overall analytical sensitivity of 0.000016 f/ml was achieved and the average level in schools following remediation was below the limit of detection <0.000048 f/ml. This represents the upper 95% value for PCME asbestos fibres and is some ten times lower than the average background previously found in UK asbestos containing buildings (Burdett and Jaffrey, 1986). A single asbestos fibre found during the analysis of three samples taken between two sets of swing doors in a school corridor are included the calculated value for the asbestos concentration (and analytical sensitivity) was 0.000014 f/ml.

Although, prior to remediation, no measurements were made in schools containing AIB materials in the columns, five weeks of monitoring in an ex-school building now used as a busy office was carried out. This gave a calculated concentration for PCME asbestos fibres equal to the analytical sensitivity of 0.000005 f/ml and an upper 95% confidence value of <0.000024 f/ml for continuous daytime monitoring over the five-week period. It is notable that during one of the sampling periods a severe gale took place causing widespread damage across the country but no asbestos fibres were detected in the air samples taken over the period.

Often the casing may not extend directly to the roof space or the floor above and may finish above the suspended ceiling, leaving a few centimetres of the AIB projecting above the casing. Although the projecting part of the AIB is unlikely to be directly disturbed, the top of the column is unsealed and provides a possible route for airborne fibres to be released into the roof space / ceiling void. In many schools the level of interconnectivity and air change between the classroom and void areas is likely to be limited unless the ceiling tiles have been damaged and removed. The measurements carried out in occupied classrooms, where the tops had or had not been sealed, did not detect any significant difference in the fibre concentrations and the results of peak release tests at room level also showed no significant differences in the data available. However, monitoring in the ceiling void at some sites showed that exposures to workers who have to enter the ceiling voids to carry out this sealing may occur, while it is far from clear that there is any significant additional reduction to occupants, from sealing the tops of the columns. Also, the possibility for further spread of dust and debris into the classroom from work above the sealing remains as well as other risks such as working at height.

4.4. Comparison to other work

The average air concentrations measured in occupied classrooms after remediation (<0.00005) f/ml) and in the occupied office (0.00002 f/ml) before remediation were an order of magnitude lower than the average background found in other asbestos containing buildings [8,9] (i.e. 0.0005 f/ml Burdett and Jaffrey, 1986; 0.0004 f/ml HEI, 1991). A recent update of US asbestos containing buildings (Lee and Van Orden, 2008) reported the average PCME asbestos fibre concentrations in 317 occupied schools as 0.0001 f/ml and an average of 752 buildings as 0.00008 f/ml. The outdoor environment close to the buildings had an average concentration of 0.00002 f/ml. These averages were based on pooling relatively large numbers of samples e.g. 3979 samples for inside all buildings and 1678 outdoor samples). Although the types of asbestos and construction will differ and the results in this paper are based on far fewer samples, there is no significant difference between the US TEM results and those from the CLASP schools.

4.5. Exposure limits and measurement issues

Currently, the EU wide exposure limit for workers is based on an 8-hour time weighted exposure of 0.1 f/ml as counted by the WHO method [11] for light microscopy (WHO, 1997). The UK operates a tighter standard based on 0.1 f/ml monitored over a 4-hour period but also limits peak exposures to a maximum of 0.6 f/ml over a 10 minute sampling period. These are regarded as the upper limit and in general there is an expectation to reduce levels to as low as reasonably practicable through the use of appropriate controls. There are no EU or UK limits for environmental exposure but the WHO has recommended a limit of 0.001 f/ml based on the annual cumulative exposure. The UK clearance indicator, which is used to determine if it is appropriate to remove an enclosure after work on asbestos has been completed, is 0.01 f/ml. This is not a health based limit but a peak disturbance / clearance indicator to show that this level would unlikely to be exceed by any subsequent routine activity that is likely to occur.

The vast majority of static samples collected for PCM analysis, sampled 480 l of air on the filter and counted 200 fields of view. This gives a calculated limit of detection (LOD) of 0.003 f/ml based on there being a 95% confidence that the filter background counts do not exceed 5 fibres and any additional fibres were likely to be collected from the air. However due to the poor precision of counting low numbers of fibres, HSG 248 recommends that a LOQ for reporting individual samples of

0.010 f/ml (based on 20 fibre counts). As the purpose of this paper was to summarise many such counts, using the same sampling and analysis protocol, a more meaningful average could be obtained by reporting all individual measurements as their calculated values and taking the average of the individual results. If substantially different volumes of air or areas were analysed the samples were pooled together and the mean calculated by dividing the total numbers of fibres counted by the total volume of air analysed during the PCM count. Therefore although data were often initially reported as <0.01 f/ml, all values were recalculated and these individual values were used to calculate the mean. The range and distribution of the individual sample concentrations have also been given.

5. Conclusions

Direct impacts and flexing of the column casings which disturbed the AIB (especially when the AIB was directly attached to the casing) appeared to be the main mechanism for fibres to be disturbed and released inside the column. These fibres can then disperse directly into the room, through gaps or damage in the column casing or where it interfaces to other surfaces inside the room.

Peak release testing by banging the columns, slamming doors attached to or adjacent to columns as well as similar disturbance of window areas, was shown to give rise to airborne concentrations of PCM fibres, many of which were shown to be amosite fibres by TEM analysis of the PCM equivalent fibres.

Visual assessment of the column casing and its immediate surroundings below the ceiling level for gaps and damage is an important way of assessing whether peak releases may take place. The duty to manage in the control of asbestos regulations (CAR, 2006) already requires the duty holder in schools to make an assessment and carry out regular re-inspections for signs of damage or deterioration to the ACMs in the building. The duty holder also has to take appropriate management action to minimise and control damage and fibre release, as far as reasonably practicable.

The PCM fibre concentrations data from the peak release tests showed that while direct releases of fibres to the room air can occur from gaps and holes in and around the column casings, sealing is an effective way of minimising releases to below the limit of quantification (0.01 f/ml) of the PCM method for some 95% of the tests carried out. Sealing with silicone filler and taping gaps and seams on the column casing below ceiling also gave concentrations below the LOQ of the PCM method for 95% of the tests carried out. The data available did not show any significant difference between the PCM fibre concentrations in the room air for columns that had or had not been sealed in the ceiling void, as well as in the room.

Occupant exposures during normal teaching activities in areas that had undergone remediation were also monitored, with much greater use of TEM analysis to determine the asbestos fibre concentration. No asbestos fibres were found in the TEM analysis of samples of classroom air from eight schools, after remediation had taken place. Occupant exposures in only one school and one office building have been monitored before remediation, and only one asbestos fibre was found in each case. The average monitored occupant exposures to asbestos in CLASP buildings was <0.00005 f/ml) an order of magnitude lower than previous measurements in UK buildings.

If sealing of the tops of the columns is to be undertaken, the additional disturbance and potential exposures of maintenance personnel, along with the risks from work at height, should be considered in the risk assessment.

Acknowledgements

To ensure all data available could be collected all participating local authorities, schools and laboratories were assured that the data would be presented in a non-attributable form. Therefore there are many people who have contributed to this data set and summary, whose help is gratefully acknowledged. The HSL results relied heavily on the support and efforts of HSL staff.

References

[1] Burdett, G.J. (2007) Summary of fibre concentrations in CLASP construction schools containing asbestos. http://www.hse.gov.uk/research/hsl_pdf/2007/hsl0722.pdf.

[2] Burdett, G. (2008) Further measurements of fibre concentrations in CLASP construction buildings. http://www.hse.gov.uk/research/rrhtm/rr624.htm

[3] HSG248, Asbestos: The analyst's guide for sampling, analysis and clearance procedure, HSE 01/05; ISBN 0-7176-2875-2, 2005. Available from HSE Books.

[4] ISO 10312:1995. Ambient Air - Determination of asbestos fibres - direct transfer transmission electron microscopy procedure. International Standards Organisation. Geneva. 1995. Available from British Standards Institute, 389 Chiswick High Road, London W4 4AL

[5] BS EN ISO/IEC 17025:2000 General requirements for the competence of testing and calibration laboratories. Available from British Standards Institute, 389 Chiswick High Road, London W4 4AL

[6] CAR, Control of Asbestos Regulations, SI 2006/2739 The Stationary Office, ISBN 0 11 075191 4, 2006.

[7] Personal communication (from Johnathon Gibson, ITN Ltd. Draft report by G & L Consultancy, Taunton, Somerset, TA3 5PX)

[8] Burdett, G.J.; Jaffrey, S.A.M.T. (1986) Airborne asbestos concentrations in buildings, Annals of Occupational Hygiene, 30, 2, 185-199.

[9] Health Effects Institute - Asbestos Research, Asbestos in public and commercial buildings: A literature review and synthesis of current knowledge. 1991.

[10] Lee, R.J. and Van Orden, D.R. (2008) Airborne asbestos in buildings. Reg. Tox. and Pharm., 50, 218-225.

[11] WHO, Determination of airborne fibre number concentration. A recommended method by phase-contrast optical microscopy (membrane filter method). World Health Organisation, Geneva. 1997.

Table 1: Summary of the PCM air concentration data for schools.

Statistic	Peak release testing of columns before remediation)	Peak release testing of columns after remediation	Peak release testing of columns after sealing columns in room only.	After remedial work before occupation	Normal occupation after remedial work	Exposure during sealing and taping columns in the room and ceiling	Exposure during sealing and taping columns in the room	Inspection of ceiling area
Number of samples	31	95	14	44	96	12	188	20
Mean (f/ml)	0.094	0.004	0.006	0.002	0.005	0.004	0.002	0.004
Median (f/ml)	0.033	0.003	0.003	0.001	0.004	0.004	0.002	0.002
Largest value (f/ml)	0.441	0.058	0.032	0.008	0.022	0.006	0.039	0.019
Smallest value (f/ml)	0.001	<0.001	0.001	<0.001	0.001	0.001	0	0

Table 2: PCM results from peak release testing before and after remediation and while partially sealing the column with minor gaps.

Room number and sample position	PCM fibre concentration (f/ml)
Class 43 adjacent to undamaged column during and following striking of the column.	0.07
Class 43 teachers position (corridor end) during and following striking of the column.	0.03
Adjacent fixed point sample while sealing up column in room 43.partial remediation	0.006
Adjacent fixed point sample, during and following the striking of column after partial remediation	0.010

Table 3: Results from peak release testing of column casings in good condition before and after full remediation at room levels and in the ceiling void.

Sample position	PCM fibre concentration (f/ml)			
	Before remediation in room	After remediation in room	Before remediation in ceiling void	After remediation in ceiling void
Corridor outside HM office	0.005	0.008	0.005	0.013
Staff room adjacent to windows	0.007	0.025, 0.031	0.022	0.347
Room opposite HM office	0.012	0.006	0.007	0.024

Table 4: Results from simulated maintenance activity using personal sampling from schools considered to have potential high amount of AIB contamination in ceiling areas.

Simulated maintenance activity	PCM fibre concentration (f/ml)
Ceiling void inspection 14 tiles lifted in staff room.	0.012, 0.022
In ceiling void of staff room shaking and moving cables	0.023
Static samples in room below disturbed area	0.007, 0.002
Brushing in ceiling void of staff room (5 minutes brushing) 10 minutes sampling	0.18
Brushing in ceiling void of lobby (1.5 minutes brushing) 10 minutes sampling	0.18
Lifting tiles in kitchen area and disturbing cables (different school) Note: air extraction on	0.011, 0.029

Journal of Physics: Conference Series **151** (2009) 012024 doi:10.1088/1742-6596/151/1/012024

Emphysema and Pulmonary Impairment in Coal Miners: Quantitative Relationship with Dust Exposure and Cigarette Smoking

E D Kuempel, Ph.D.[1]**; V Vallyathan, Ph.D.**[2]**; F H Y Green, M.D.**[3]

1 National Institute for Occupational Safety and Health, Education and Information Division, Risk Evaluation Branch, Cincinnati, Ohio, USA

2 National Institute for Occupational Safety and Health, Health Effects Laboratory Division, Pathology and Physiology Research Branch, Morgantown, West Virginia, USA

3 Department of Pathology, Faculty of Medicine, University of Calgary, Calgary, Alberta, Canada

E-mail: ekuempel@cdc.gov

Abstract. Coal miners have been shown to be at increased risk of developing chronic obstructive pulmonary diseases including emphysema. The objective of this study was to determine whether lifetime cumulative exposure to respirable coal mine dust is a significant predictor of developing emphysema at a clinically-relevant level of severity by the end of life, after controlling for cigarette smoking and other covariates. Clinically-relevant emphysema severity was determined from the association between individuals' lung function during life (forced expiratory volume in one second, FEV_1, as a percentage of predicted normal values) and emphysema severity at autopsy (as the proportion of lung tissue affected). In a logistic regression model, cumulative exposure to respirable coal mine dust was a statistically significant predictor of developing clinically-relevant emphysema severity, among both ever-smokers and never-smokers. The odds ratio for developing emphysema associated with FEV_1 <80% at the cohort mean cumulative coal dust exposure (87 mg/m^3 x yr) was 2.30 (1.46-3.64, 95% confidence limits), and at the cohort mean cigarette smoking (among smokers: 42 pack-years) was 1.95 (1.39-2.79).

1. Introduction

Emphysema is characterized by abnormal, permanent enlargement of the airspaces distal to the terminal bronchiole, with destruction of their walls, and it is assessed morphologically at autopsy [1] or by computerized tomography. Emphysema is classified by type and severity (as percentage of the lung tissue affected) using standardized schema [2-4]. Previous studies have shown that emphysema severity is associated with cumulative dust exposure in U.K. coal miners [5], and with coal dust lung burden in Australian and U.S. coal miners [6, 7]. Emphysema severity has also been associated with lung function deficits during life in Australian coal miners [6] and in African gold miners [8]. An inverse relationship was observed between the emphysema severity at autopsy and FEV_1 in those studies [6, 8].

In this study, we quantitatively examine the relationship between emphysema severity, cumulative dust exposure, and other parameters in an autopsied population of U.S. coal miners and non-miners [7]. The effects of cumulative dust exposure and cigarette smoking on the probability of developing clinically-relevant emphysema severity were estimated in this cohort.

2. Methods

The study group consists of 722 autopsied individuals, including 616 coal miners from southern West Virginia and 106 non-miners from West Virginia and Vermont who died during 1957-1978. Characteristics of the full study population are described elsewhere [7, 9]. Briefly, data on work history, smoking, race, and age at death were obtained from medical records and questionnaire completed by next-of-kin. Cumulative dust exposures for each miner were estimated using individuals' work history data and job-specific measurements of the airborne mean concentration of respirable coal mine dust, based on a method by Attfield and Morring [10]. Whole lung sections were prepared from sagittal slices of left lung and classified and graded by two pathologists (FHYG and VV) using standardized schema [2, 3]. Each of the 10 lung zones was graded for emphysema severity up to a maximum score of 100 for each zone, for a possible maximum score or index of 1000 for an individual's lungs (the total emphysema severity index divided by 10 is equivalent to the percentage of the total lung tissue affected). In a subset of 116 individuals, data were available on FEV_1 from spirometry tests during life. A smaller subset of 65 miners had FEV_1 tests performed within 5 years of death, which is a criterion used in previous studies [6, 8]. The spirometry methodology in these U.S. coal miners was reported previously [11].

Clinically-relevant emphysema severity was defined in this study based on the relationship between emphysema severity at autopsy and lung function during life and on the standardized criteria for clinically-relevant pulmonary impairment. These standardized criteria include FEV_1 <80% of predicted normal values – approximately equal to the 5th percentile lower limit of normal [12, 13] – and FEV_1 <65% of predicted normal values – which has been associated with severe exertional dyspnea in U.K. coal miners [14, 15]. The emphysema severity index associated with an FEV_1 of either 80% or 65% of the predicted normal values was calculated from a linear regression model of the relationship between FEV_1 and emphysema severity index in the subset of individuals with FEV_1 data (n=116). Nonlinearity in this relationship was tested by the addition of a quadratic term for FEV_1. The emphysema severity indexes associated with an FEV_1 of either 80% or 65% (of predicted normal values) were used as cutoff-points for clinically-relevant emphysema, and treated as dichotomous response variables in logistic regression models. An individual was coded as a responder if his emphysema severity index was greater than these defined cut-off values.

Logistic regression models [16] of the expected proportion of responders in this cohort were used to estimate the probability of disease (i.e., emphysema severity index associated with FEV_1 <80% or <65%). The predictor variables in these models, which were identified in the full analysis of this cohort [9], include: cumulative exposure to respirable coal mine dust, age at death, race (white/non-white), and pack-years of smoking. The probability of developing clinically-relevant emphysema was estimated at various values of these predictor variables. Statistical tests (Wald) were performed to compare the effects of cumulative dust exposure and cigarette smoking.

3. Results

3.1. Study characteristics

The characteristics of this study population include: 85% miners; 78% ever-smokers; 72% whites; 64.6 years average age at death; 42.4 average pack-years of smoking (among smokers). For the subset of individuals for whom FEV_1 data were available (n=116), 98% were miners; 87% were ever-smokers; 71% were whites; average age at death was 64 years; and average pack-years of smoking (among smokers) was 41.4 years.

3.2. Emphysema severity and FEV_1

Figure 1 shows the relationship between the emphysema severity index and FEV_1 ($p<0.0001$; R^2= 0.17) among the 116 individuals with FEV_1 data. A quadratic term for FEV_1 was not significant (p=0.064). The mean emphysema severity index associated with 80% FEV_1 was 285, and the mean emphysema severity index associated with 65% FEV_1 was 392. These values were used as the cutoff-points to define clinically-relevant emphysema severity in the logistic regression models. Similar values were obtained among the 65 individuals with FEV_1 data within five years of death; i.e., the mean emphysema severity index associated with 80% or 65% FEV_1 was 277 or 383, respectively.

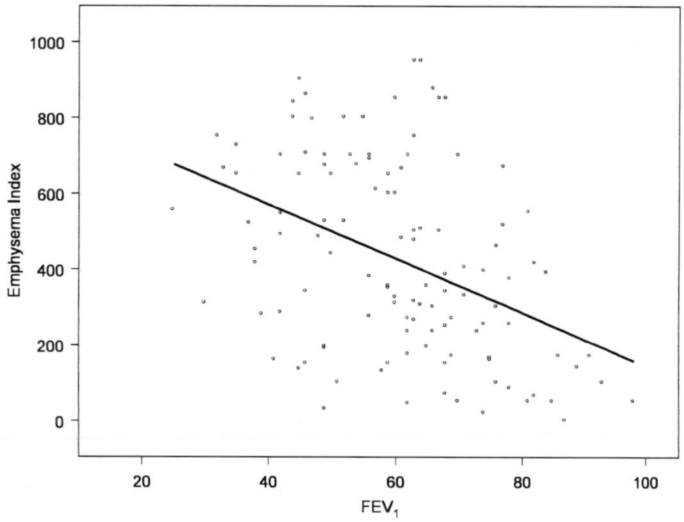

Figure 1. Relationship between emphysema severity index and forced expiratory volume in one second (FEV_1) (as percentage of predicted normal values); data for 116 individuals with FEV_1 fit with a simple linear regression model.

3.3. Logistic regression models

Tables 1 and 2 show the logistic regression modeling results for the expected proportion of individuals developing an emphysema severity index associated with FEV_1 <80% or <65%, respectively, by the end of life. Cumulative exposure to respirable coal mine dust, age at death, and race are statistically significant predictors of developing either of these clinically-relevant degrees of emphysema. No significant interactions were observed, although a pack-year x age interaction was observed in the main analysis of this cohort in a weighted least squares linear regression model [9].

Table 1. Logistic Regression Model: Predictors of Emphysema Severity Associated with FEV_1 <80% (of predicted normal values)[a]

Parameter[†]	Estimated Coefficient	Standard Error	P-value
Intercept	-5.0	0.83	
Cumulative exposure (mg/m^3 x years)	0.0095	0.0027	0.0004
Cigarette smoking (packs/day x years)	0.016	0.0042	0.0002
Age at death (years)	0.048	0.012	0.0001
Race (non-white)	0.88	0.28	0.002

[a] N=342 miners and non-miners. Likelihood ratio (4 d.f.)=75.1, p<0.0001.

Table 2. Logistic Regression Model: Predictors of Emphysema Severity Associated with FEV_1 <65% (of predicted normal values)[a]

Parameter	Estimated Coefficient	Standard Error	P-value
Intercept	-4.5	0.82	
Cumulative exposure (mg/m^3 x years)	0.010	0.0027	0.0003
Cigarette smoking (packs/day x years)	0.0099	0.040	0.01
Age at death (years)	0.036	0.012	0.003
Race (non-white)	0.80	0.28	0.004

[a] N=342 miners and non-miners. Likelihood ratio (4 d.f.)=56.3, p<0.0001.

Based on the model in Table 1, the odds ratios of developing emphysema associated with <80 % FEV_1, for one unit of increase in either cumulative coal dust exposure or pack-years of cigarette smoking, were: 1.010 (1.004-1.015) and 1.016 (1.008-1.024), respectively (as point estimate and 95% confidence interval, CI). For emphysema associated with <65 % FEV_1 (Table 2), these odds ratios were: 1.010 (1.005-1.015) and 1.010 (1.002-1.018), respectively, for one unit of increase in either cumulative coal dust exposure or pack-years of cigarette smoking. The effects of cumulative dust exposure and cigarette smoking were not statistically significantly different at either disease response (p-values: 0.2 and 1.0, respectively).

In separate models by smoking group (ever, n=405; never, n=115), the effect of cumulative coal dust exposure (mg/m^3 x yr) on clinically-relevant emphysema remained statistically significant in each group ($p<0.05$ among ever smokers and $p \leq 0.01$ among never smokers), although the coefficients for cumulative exposure were higher among never smokers. For example, in the model for emphysema associated with <65% FEV_1, the coefficients (and standard errors) were: 0.0077 ($\pm$0.0031) among ever smokers, and 0.016 ($\pm$0.0057) among never smokers.

Figure 2 shows the probability of developing clinically-relevant emphysema (associated with FEV_1 <80% or <65% of predicted normal values) with increasing cumulative exposure to respirable coal mine dust, based on the models shown in Tables 1 and 2. These relationships are illustrated for either ever-smokers or never-smokers, among whites, at the cohort-average age-of-death and pack-years of smoking (among smokers). A background probability (i.e., among individuals without dust exposure) of developing these degrees of emphysema severity is predicted to be approximately 10-20%, with the higher levels predicted among smokers. The probability of developing either level of emphysema severity is predicted to increase with increasing cumulative exposure to respirable coal mine dust. The probability of developing emphysema associated with <80% FEV_1 is greater than that for developing emphysema associated with <65% FEV_1, which is expected since emphysema severity of lower degree is more prevalent than emphysema of a more severe degree.

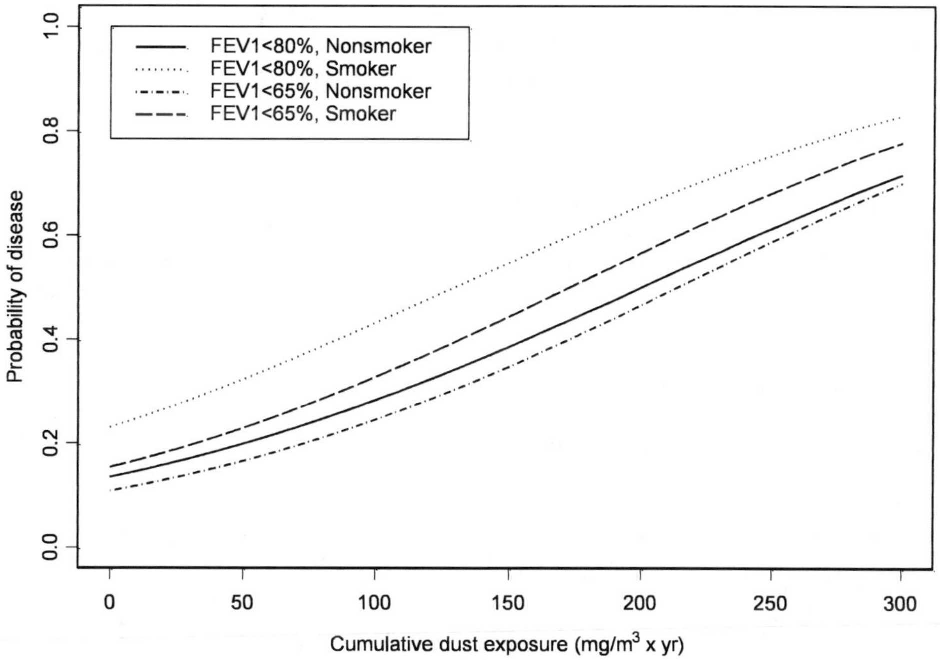

Figure 2. Relationship between the probability of developing clinically-relevant emphysema (i.e., associated with FEV_1 of <80% or <65% of predicted normal values) and the cumulative exposure to respirable coal mine dust – among white never-smokers at the cohort-average age of death.

At the cohort average cumulative dust exposure (87 mg/m^3 x yr) and age of death (65 yr), the maximum likelihood estimate (MLE) of the probability of developing emphysema associated with FEV_1 <80% was 0.25 (95% CI: 0.18-0.34) among white never-smokers, and 0.39 (0.33-0.47) among white ever-smokers. For emphysema associated with FEV_1 <65%, the estimates for the same conditions were 0.22 (0.16-0.30) among never smokers, and 0.30 (0.24-0.36) among smokers. Among non-whites, the comparable estimates were nearly two-fold higher within each smoking group.

The effects of cumulative coal dust exposure and pack-years of smoking were additive (on the log odds scale) in this logistic regression model. Thus, the probability of developing clinically-relevant emphysema was higher among smokers, for a given age of death and race. The relative effects of cumulative dust exposure and cigarette smoking in this cohort can be evaluated by examining their odds ratios. The odds ratio of developing emphysema associated with FEV_1 <80% was 2.30 (1.46-3.64) at the cohort mean cumulative coal dust exposure (87 mg/m^3 x yr), and 1.95 (1.39-2.79) at the cohort mean cigarette smoking (among smokers: 42 pack-years). For emphysema associated with FEV_1 <65%, the odds ratio was 2.39 (1.51-3.83) at the cohort mean cumulative coal dust exposure, and 1.52 (1.10-2.13) at the cohort mean cigarette smoking (all values as point estimate and 95% CI).

Thus, the contribution of cumulative dust exposure was greater than that of cigarette smoking at the cohort mean values, although not significantly so (p-values: 0.6 and 0.1, respectively, for emphysema associated with FEV_1 <80% or <65%).

4. Discussion

This study provides a clinical perspective to our earlier findings [9] based on emphysema severity as a continuous variable. The current analyses describe the quantitative relationship between FEV_1, emphysema severity, and cumulative dust exposure in an autopsied population of U.S. coal miners and non-miners. Results show that the probability of developing clinically-relevant emphysema severity (i.e., associated with FEV_1 of <80% or <65% of predicted normal values) is related to the cumulative exposure to respirable coal mine dust, cigarette smoking pack-years, age at death, and race. The higher probabilities predicted for non-whites might be attributed to unmeasured factors related to race, such as higher dust exposures than those reported in the work histories or non-occupational exposures related to cooking or heating methods used in the home. Cumulative dust exposure and cigarette smoking had similar and additive estimated effects on the probability of developing clinically-relevant emphysema.

Associations between dust exposure, emphysema severity, and pulmonary function have been observed in previous studies of miners in several countries. Ruckley et al. [5] reported that centriacinar emphysema was associated with the amount of dust retained in the lungs of U.K. coal miners who had dust-related fibrosis. Studies in living miners in the U.K. [14, 15, 17] and the U.S. [18] observed independent effects of coal dust exposure and cigarette smoking on lung function deficits, including miners with negative chest x-rays for pneumoconiosis. Leigh et al. [6] found that emphysema severity in Australian coal miners was predicted by the coal dust content in the lungs, age, and smoking; and that emphysema severity was highly negatively correlated with FEV_1 obtained within five years of death.

Similar relationships have been observed for other mineral dusts. In a study of South African gold miners, Hnizdo et al. [8] reported that the average FEV_1 was 52.0% (S.D. 19.4%) of predicted normal values among the 170 individuals with emphysema severity of 35-65% of lung tissue affected. In this analysis of U.S. coal miners, the mean FEV_1 was 58.1% (S.D. 15.1%) of predicted normal values among the 34 individuals with an equivalent emphysema severity of 350-650 (corresponding to 35-65% of lung tissue affected),

Differences in the spirometry and emphysema grading criteria apparently did not have a large effect. Hnizdo et al. [8] used the 1979 ATS spirometry standards and a National Centre for Occupational Health method for grading emphysema severity using whole lung sections and 20 radiating zones (summing to 100%), while the data in this study are based on FEV_1 prediction formulas from Miller (as reported in Rasmussen et al. [11]) and the Thurlbeck method to determine emphysema severity at autopsy [2, 3].

These consistent findings across studies suggest: (1) the emphysema severity-FEV_1 relationship may be broadly applicable to other dust-exposed cohorts, perhaps through a common mechanism for emphysema disease development regardless of the causative agent, and (2) the emphysema severity index is associated with lung functional deficits during life. However, it should be noted that FEV_1 is a relatively nonspecific test for measuring airflow limitation, as FEV_1 deficit can be caused by diseases such as chronic bronchitis and asthma in addition to emphysema [19]. These factors likely account for some of the variability observed in the relationship between FEV_1 and emphysema severity. Also, individuals' responses to inhaled dust can vary, such that the average decrements in the emphysema index or FEV_1 observed in a population may not accurately portray the effects of dust exposure on the pulmonary impairment and disability in an individual miner.

In this study, cumulative coal dust exposure and cigarette smoking were each statistically significant predictors of developing clinically-relevant emphysema. These effects were additive (on the log odds scale) – indicating the need to reduce exposures to both agents in order to reduce the probability of emphysema. The odds ratios for developing clinically-relevant emphysema (i.e., associated with FEV_1 <80% or <65%) for cumulative coal dust exposure (2.30 or 2.39, respectively) were higher (although not significantly) than those for cigarette smoking (1.95 or 1.52, respectively) at the cohort mean values. It is important to note that never-smoking coal miners also had a significantly elevated probability of developing clinically-relevant emphysema severity. Reducing exposures to respirable coal mine dust, as recommended by NIOSH [20] would reduce the probability of dust-related disease including emphysema.

5. Conclusions

Cumulative exposure to respirable coal mine dust was a statistically significant predictor of the probability of clinically-relevant emphysema among both ever- and never-smokers. The observed odds ratios for developing clinically-relevant emphysema were somewhat higher for cumulative dust exposure than for cigarette smoking in this cohort.

References

[1] Kleinerman J, Green F, Harley RA, Lapp NL, Laqueur W, Naeye RL, Pratt P, Taylor G, Wiot J and Wyatt J 1979 Pathology standards for coal workers pneumoconiosis: Report of the Pneumoconiosis Committee of the College of American Pathologists to the National Institute for Occupational Safety and Health. *Arch. Pathol. Lab. Med.* **103** 375-432

[2] Thurlbeck WM, Dunnill MS, Hartung W, Heard BE, Heppleston AG and Ryder RC 1970 A comparison of three methods of measuring emphysema *Hum. Pathol.* **1**(2) 215-26

[3] Saito K and Thurlbeck WM 1995 Measurement of emphysema in autopsy lungs, with emphasis on interlobar differences *Am. J. Respir. Crit. Care Med.* **151**(5) 1373-76

[4] Green FHY, Brower PL, Vallyathan V and Attfield M 1998 Coal mine dust exposure and type of pulmonary emphysema in coal workers *Advances in the prevention of occupational respiratory diseases* eds Chiyotani K, Hosoda Y and Aizawa Y (Amsterdam, The Netherlands: Elsevier Science BV) pp 948-53

[5] Ruckley VA, Gauld SJ, Chapman JS, Davis JMG, Douglas AN, Furney JM, Jacobsen M and Lamb D 1984 Emphysema and dust exposure in a group of coal workers *Am. Rev. Respir. Disease* **125** 528-32

[6] Leigh J, Driscoll TR, Cole BD, Beck RW, Hull BP and Yang J 1994 Quantitative relation between emphysema and lung mineral content in coal workers *Occup. Environ. Med.* **51** 400-7

[7] Vallyathan V, Green FHY, Brower P and Attfield MD 1997 The role of coal mine dust exposure in the development of pulmonary emphysema *Ann. Occup. Hyg.* **41**(suppl 1) 352-7

[8] Hnizdo E, Murray J and Davison A 2000 Correlation between autopsy findings for chronic obstructive airways disease and in-life disability in South African gold miners *Int. Arch. Occup. Environ. Health.* **73**(4) 235-44

[9] Kuempel ED, Wheeler MW, Smith RJ, Vallyathan V and Green FHY Contributions of dust exposure and cigarette smoking to emphysema severity in U.S. coal miners [Submitted]

[10] Attfield MD and Morring K 1992 Derivation of estimated dust exposures for U.S. coal miners before 1970 *Am. Indust. Hyg. Assoc. J.* **53** 248-55

[11] Rasmussen DL, Laqeur WA, Futterman P, Warren HD and Nelson CW 1968 Pulmonary impairment in southern West Virginia coal miners *Am. Rev. Respir. Dis.* **98**(4) 658-67

[12] ATS 1991 Lung function testing: selection of reference values and interpretive strategies *Am. Rev. Respir. Dis.* **144** 1202-18

[13] Boehlecke B 1986 Laboratory assessment of respiratory impairment for disability evaluation *Occupational respiratory diseases* ed Merchant JA (Cincinnati, OH: U.S. Department of Health and Human Services, Public Health Service, Centers for Disease Control, National Institute for Occupational Safety and Health) DHHS (NIOSH) Publication No. 86-102, pp 181-216

[14] Marine WM, Gurr D and Jacobsen M 1988 Clinically important respiratory effects of dust

exposure and smoking in British coal miners *Am. Rev. Respir. Dis.* **137**(1)106-12

[15] Soutar C, Campbell S, Gurr D, Lloyd M, Love R, Cowie H, Cowie A and Seaton A 1993 Important deficits of lung function in three modern colliery populations: relations with dust exposure *Am. Rev. Respir. Dis.* **147**(4) 797-803

[16] SAS 2003 Release 9.1 for Windows (Cary, NC: SAS Institute)

[17] Lewis S, Bennett J, Richards K and Britton J 1996 A cross sectional study of the independent effect of occupation on lung function in British coal miners *Occup. Environ. Med.* **53**(2):125-8

[18] Attfield MD and Hodous TK 1992 Pulmonary function of U.S. coal miners related to dust exposure estimates *Am. Rev. Respir. Dis.* **145**(3):605-9

[19] Wright JL 2001 Diseases of the small airways *Lung* **179**(6):375-96

[20] NIOSH 1995 *Criteria for a Recommended Standard: Occupational Exposure to Respirable Coal Mine Dust* (Cincinnati, Ohio: U.S. Department of Health and Human Services, Public Health Service, Centers for Disease Control and Prevention, National Institute for Occupational Safety and Health) DHHS (NIOSH) Publication No. 95-106

Acknowledgments

The authors are grateful to Dr. Donald Rasmussen for providing information regarding the spirometry tests and data collection in this cohort. We would like to thank Mr. Randall Smith for providing input on the statistical methods; Dr. Michael Attfield and Mr. Mathew Wheeler for providing data and analyses to estimate individuals' cumulative exposures to respirable coal mine dust; and Ms. Patricia Schleiff for providing an earlier version of the dataset. We respectfully acknowledge the late Dr. Werner Laqueur of the Beckley Appalachian Regional Hospital, West Virginia, who initiated the data collection and pathological evaluation for this cohort.

Disclaimer

The findings and conclusions in this paper are those of the authors and do not necessarily represent the views of the National Institute for Occupational Safety and Health.

Journal of Physics: Conference Series **151** (2009) 012025 doi:10.1088/1742-6596/151/1/012025

High Temporal Resolution Measurements of Roadside Particle Size Distributions and Their Implications for Exposure

A S Tomlin, D T Young, J J N Lingard and E L Agius

Energy and Resources Research Institute, University of Leeds, Leeds, LS2 9JT, UK

A.S.Tomlin@leeds.ac.uk

Abstract High time resolution, size segregated studies of ultrafine particle number concentrations (PNC) were performed at two UK roadside locations using a DMS500 particle spectrometer including sampling with and without a thermodenuder. Intercomparisons between the DMS500, an SMPS and a CPC, indicate that the SMPS can underestimate PNC due to the transient nature of excursions in concentrations and the longer scan time of the instrument. Short duration transient periods of high number concentrations made considerable contributions to hourly average PNC. The transient nature of changes in PNC potentially affects the type of exposure experienced in the roadside environment, as well as indoor outdoor exchange. The transient shifts were dominated by Aitken mode particles during high traffic periods, with the Aitken mode comprised of internally mixed particles with a low volatility core (thought to be primarily soot) and a surface of condensed and absorbed material. Transient periods of high nucleation mode PNC were also found, but were relatively more important during night-time and afternoon periods, due to secondary formation processes and lower concentrations of pre-existing particles. The thermodenuder studies indicated the presence of highly volatile liquid droplets in the nucleation mode which were lost on heating. Transient events and complexities in particle mixing state within the ultrafine fraction, coupled with discrepancies between instruments, make the potential adoption of a total number based air quality standard more complicated than first appears. Nucleation mode particle numbers are highly dependant on atmospheric and traffic conditions as well as pre-existing particle surface area and may be toxicologically less significant if they are predominantly liquid droplets. Aitken mode particles have been shown in previous studies to have higher toxicological significance due to their chemical composition. Their hourly average PNC correlated well with $PM_{2.5}$ and estimated PM_1 with lower significance of equivalent correlation of nucleation mode particles.

1. Introduction

Air quality standards for particulate matter in the UK are currently based on 24 hour and annual means for PM_{10} concentrations, where PM_{10} represents the mass of particles that can pass through a size selective inlet with a collection efficiency of 50% for particles of aerodynamic diameter 10μm. In order to comply with standards, control efforts are therefore aimed at reducing the mass of particles below 10μm, and current European engine emission standards are PM_{10} based. A number of studies have shown that PM_{10} within UK urban areas can be dominated by the coarser fraction, and have indicated a lack of correlation between the numbers of ultrafines (Uf) and PM_{10}, particularly at background sites. Whilst Uf $\leq$100 nm in diameter (Dp) only contribute 1 to 20 % of the particle mass, they can represent >90 % of the total PNC in a diesel exhaust [1] or at an urban roadside [2]. It follows that achieving the current standards will not necessarily lead to a reduction in the numbers of Uf particles, although this would depend on the sources of the emissions under control. Since engine emissions are dominated by smaller particles, control technologies aimed at reducing overall mass would have a positive impact reducing parts of the Uf fraction. On the other hand, some studies indicate that as particle mass concentrations in ambient air decrease, increased levels of particle numbers are detected [3]. Proposals for a $PM_{2.5}$ based standard are under discussion and would also help to focus controls to the finer fractions. However, toxicological studies have suggested that high numbers of Uf particles in the atmosphere can contribute to inflammatory responses and this raises the question as to whether mass based standards are the optimal way to reduce the impact of particulate pollution on health [4]. The possible additional use of a number concentration based standard for emissions has been outlined in proposals for Euro VI legislation for light duty vehicles (EC regulation No 715/2007).

There are a growing number of studies which aim to compare the use of different measures of particle concentration in terms of their association with end health effects such as daily mortality, hospital admissions [5], cardiorespiratory symptoms [6], stroke mortaility [7] and asthma medication use [8]. Many studies suggest stronger associations of fine particles ($PM_{2.5}$) with effects such as hospital admission by age [5], respiratory symptoms [6,9] and peak expiratory flow, than coarse particles [9], with particular emphasis on fine sulphate particles in [9]. There are conflicting results on the influence of ultrafines with some studies showing weaker associations of ultrafines than fines with certain health outcomes [6,10] and others showing comparable effects [7]. The chemical complexity and daily, seasonal and spatial variability in the composition within the particle size distribution is a possible reason for the lack of consensus between different studies. Perhaps one area of concern is the presence of semi-volatile material within ambient particles which may not be well understood from a toxicological point of view, and in addition can strongly influence measured particle concentrations. The TEOM (Tapered Element Oscillating Microbalance) has received significant attention in this area since it uses a heated sample of 50°C in order to reduce the effects of humidity. This however also causes potential loss of semi-volatile material which means that comparisons between TEOM and other gravimetric measurements show an underestimation of mass by the TEOM [11,12]. An average correction factor of 1.3 has been suggested [13] in order to correct for total PM mass but the percentage of semi-volatile material within a PM10, PM2.5 or PM1 sample may vary with source location, time of day (see later discussion), meteorological conditions or time of year. This makes the use of an average correction factor problematic. In addition, since much of the volatile material is present in the finer fractions, the shift towards $PM_{2.5}$ or PM_1 would undoubtably change the relationship between TEOM and other mass based measurements.

The potential adoption of a number based standard for air quality however, also raises further considerations with regard to the presence of semi-volatile material. High PNC can be dominated by nucleation mode particles. These can be characterised as predominantly droplets of up to 20nm formed in part from semi-volatile gases such as unburnt fuel and lube oil, as well as secondary organics [2,14]. Their formation is highly dependant on atmospheric and dilution conditions as well as pre-existing coarser particle concentrations [15-18]. Another main contributer to high number concentrations in urban areas is the so called Aitken mode (approximately 30 – 100 nm, [2,18]). These are generally internally mixed particles, often with a soot based core and a surface coating of semi-volatiles, potentially including polyaromatic hydrocarbons (PAHs) and other toxic organic and inorganic compounds which may be carried to sites of blood air exchange [19]. The main sources of Aitken mode particles within urban areas are primary exhaust emissions as they are usually associated with incomplete combustion. They are potentially more biologically persistent than nucleation mode particles due to their internally mixed structure. The issue of adopting appropriate metrics for the control of particulate pollutants in order to optimally reduce their impact on the populations' health is not a straightforward one. This has been acknowledged in Euro VI emissions limit proposals by the suggestion that emissions based measurements should be based on techniques that exclude the volatile nucleation mode fraction. Air quality measurements of number concentration do not usually consider such exclusions however. In order to make appropriate decisions regarding particle metrics for air quality, it is first critical that we understand the nature of the particles driving population exposure in terms of particle size, mixing state and composition. This study contributes to improving our understanding by carrying out detailed, high time resolution measurements of size segregated PNCs, as well as particle volatility, at two UK roadside environments. The results serve to highlight the transient nature of exposure to elevated particle concentrations at the roadside, as well as illustrating the complex mixing state of particles in the Uf region.

2. Methodology

2.1. Number concentration study

Short time-scale measurements (1Hz) of PNC and particle size distributions from 4.87<Dp<1000nm were performed using a Cambustion DMS500 (differential mobility sizer) at a road side location in Leicester, UK. Sampling took place over four weeks during February and March 2005 along a portion of Narborough Road (grid ref. SK 573030) which is an arterial road (A5460) running between Narborough and Leicester with an average bi-directional traffic flow of 1100 vehicles hour^{-1}. From a study by Harris et al. [20] this can be expected to be a daytime fleet composition of 88.3% cars, 7.6% LGVs, 1.7% HGVs, 1.7% buses and 0.7% motorcycles, changing to a night fleet composition of 90.7% cars, 7.6% LGVs and 1.7% HGVs. The sampling inlet of the DMS500 was placed at a height of 1.5 m from the ground and 0.3 m from the road. The DMS500 classifies particles based on their mobility in an applied electrical field. A series of grounded electrometer rings situated on the outside of an annular classifier allow simultaneous measurements of distinctly charged particles across multiple size classes, making the DMS500 a fast particulate spectrometer, with a response time of approximately 300 ms allowing near real time measurements [17,21]. Measurements of wind speed (10Hz), solar radiation, traffic detector occupancy (1/4 second) and concentrations of a range of gas phase pollutants (NO, NO_2, CO) were also obtained within the study [17]. In addition, in order to investigate the types of vehicles responsible for large peaks in PNC on Narborough Road, video footage was captured for certain periods during the sample campaign so that specific vehicle events responsible for transient increases in particle count could be identified. Such data presented here is from night-time studies since the identification of individual vehicle events is easier due to the more isolated nature of both the transient increases in particle counts and also the isolated vehicles.

2.2. Volatility study

In a subsequent study, 24 hour particle sampling took place on weekdays between July and September, 2007, at a roadside site situated on Kirkstall Rd in Leeds, UK (grid ref. SE 271347). Average bi-directional traffic flow on Kirkstall Rd was found to be 1250 vehicles hour^{-1}. Vehicle classification gave an average fleet composition of 89.0% cars, 6.7% LGVs, 1.1% HGVs, 1.6% buses and 1.6% motorcycles. To provide insight into urban particle volatility and mixing state characteristics, measurements of particle size distributions and PNC were performed using a Dekati thermodenuder (TD) in series with the DMS500. In addition, a condensation particle counter (CPC, TSI Model 3025A) and a TSI scanning mobility particle sizer (SMPS, Models 3080, 3085, 3024A) were used to provide reference measurements of PNC and size distributions during the experiments. The SMPS scanned for 4.7<Dp<160 nm over 210 s, with a down-scan of 20 s leading to a full cycle time of 4 mins. The sample inlet was situated on top of an Air Monitoring Enclosure (AME) approximately 2.8m from the ground and 4.5 m from the road. Full details of the experimental set up can be found in [22]. Eight weeks of sampling took place, including four weeks at ambient temperature and one week each at TD temperatures of 50, 75, 100 and 125°C.

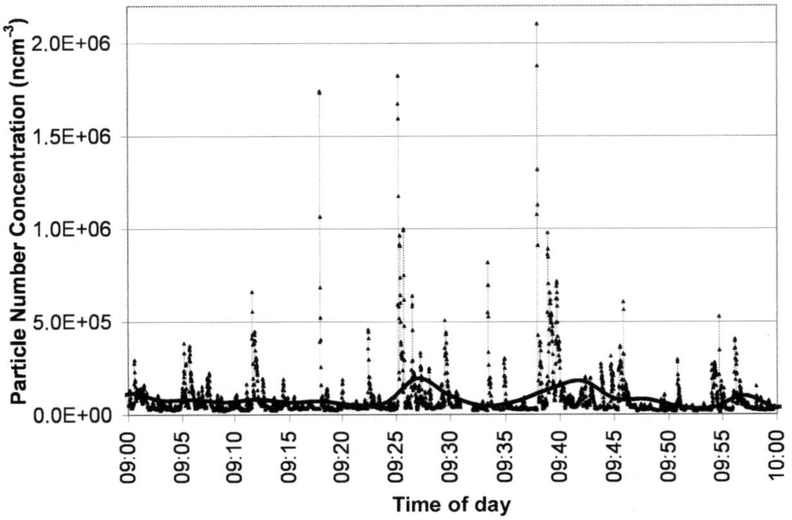

Figure 1. Transient nature of roadside particle counts shown by comparison between 1 second (grey line with black symbols) and 3 minute averages (black line) on Narborough Road, 2nd March 2005.

3. Results

3.1. Calibration study and instrument inter-comparison

Total PNC validation was performed for the DMS500 by the authors using laboratory measurements of urban background air compared with CPC measurements. An SMPS was also tested against the CPC for comparison. The linear relationship between total PNC measured by the DMS (y) and the CPC (x) was y = 0.97x (R^2 = 0.82), and by the SMPS (y) was y = 0.62x (R^2 = 0.52), based on a 24 hour diurnal average. This indicates that the DMS500 gives a better representation of the total PNC than the SMPS, perhaps because the latter has longer scanning times causing it to miss potentially important transient increases in concentration. This result alone has implications for the use of number concentrations as a particle standard, since their transient nature means that measured concentrations can be highly dependant on instrument scan times. The high temporal resolution measurements facilitated by the DMS500 help to illustrate this point as discussed below.

3.2. Transient roadside measurements

At the roadside on Narborough Road, particle counts on a second by second basis were often an order of magnitude higher than the 3 minute mean PNCs usually monitored at urban monitoring stations using SMPS instruments as illustrated in figure 1. During certain periods of the day these excursions were found to contribute up to one fifth of total PNC and yet only lasted for a few minutes in total during each hour, as shown in figure 2. Figure 2 also shows diurnal variability in the contribution of these transient peaks to hourly averaged PNC. As expected, peaks in the contribution of transient increases in PNC were seen during rush hour traffic periods (08.00-10.00, 18.00-20.00) but there are also smaller peaks during the night-time (03.00) and in the early afternoon (14.00-15.00) where traffic activity is lower. The exposure of individuals in the roadside environment is regularly dominated by these short term transient peaks, which is often not reflected in toxicological studies [23]. An important question arises as to whether the effects of short term exposure to elevated transient particle concentrations lasting for periods of seconds, are different from those resulting from exposure to the same integrated PNC, but over a much longer period of time. Michaels and Kleinman [24] explored the effects in rats of exposure to aerosols with differing degrees of short term excursions from the mean concentration, and found differences in the development of lesions between the two exposure conditions. The excursions in the study were however, 5 minutes long. So far there seem to be no published studies that have examined the effect of shorter transient particle bursts, as have been identified during the current roadside study. The potential importance of such peaks merits further investigation into their sources, and in particular the nature of the size distribution and mixing state of the contributing particles.

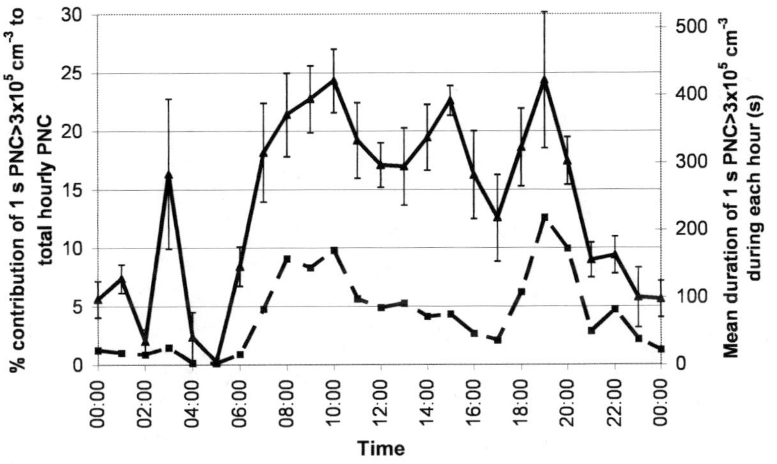

Figure 2. Average contribution of 1 s particle counts > 3 x 10^5 cm^{-3} (solid line) to total hourly PNC with mean duration per hour of these exceedences (dotted line) over all data for Narborough Road.

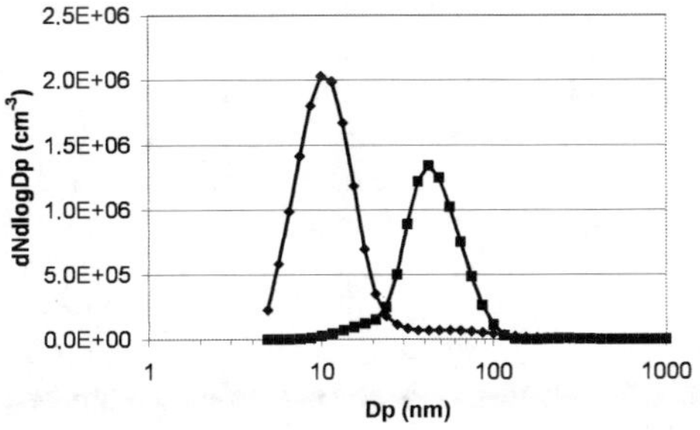

Figure 3. Two particle size distributions taken from 4th March 2005, each one a 1 second mean measured at the peak of an elevated number concentration event. Diamonds – nucleation dominated event. Squares – Aitken dominated event.

Figure 3 shows two examples of particle size distributions measured during a peak event. Clearly both have sources of different natures since one is dominated by a burst of nucleation mode particles with a peak around 10nm in diameter, and the other by a burst of Aitken mode particles with a peak diameter of around 42nm. The nature of these particles is potentially different, with the nucleation mode expected to comprise mainly liquid aerosol and the Aitken mode an internally mixed aerosol with a solid core and potentially semi-volatile components absorbed or condensed onto its surface.

This will be further explored in the next section. These particles may have different toxicological implications since the Aitken mode may be more biologically persistent [25,26] and therefore it is of interest to explore how each mode contributes to overall particle numbers and peak concentrations. The study of Brauner [26] determined that following the exposure of volunteers to ambient particulate concentrations over 6 to 24 hour periods, greater DNA strand breakage was observed after exposure to Uf particles, with the soot mode (57nm) causing the most damage. This soot mode falls within the Aitken mode discussed here.

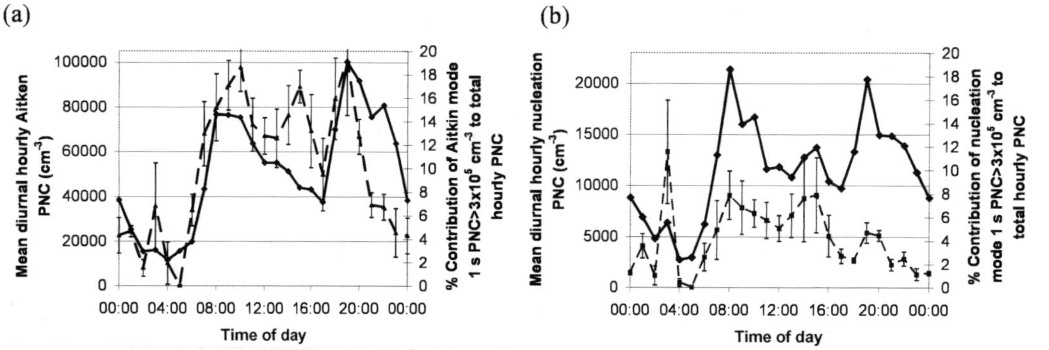

Figure 4. Mean diurnal hourly PNC (solid lines) and % contribution of 1 s exceedences > $3x10^5$ cm^{-3} to hourly PNC (dashed line) of (a) Aitken mode and (b) nucleation mode particles.

The mean modal particle counts for the Aitken and nucleation modes and their % contribution to peak excursions over 3 x 10^5 cm^{-3} are shown in figure 4. Peak events (1 s PNC>3 x 10^5 cm^{-3} compared to diurnal means of ~ 1 x 10^5 cm^{-3}) in the Aitken mode dominate transient PNC increases for most of the day, contributing to a fifth of hourly total PNC during the morning and evening rush hours and during the middle of the day. In contrast nucleation mode events contribute from 7 to 12% of hourly PNC during the early morning and mid afternoon hours. Higher contributions are seen at night where ambient temperatures and pre-existing Aitken mode particle concentrations are lower, potentially favouring nucleation of semi-volatile gases over condensation onto existing particles [17,18].

The domination of the Aitken mode within the transient peak events reflects the transient nature of exhaust emissions. Commonly, it is assumed that the larger, internally mixed Aitken mode particles would be predominantly sourced from diesel vehicles and are soot based in nature. In order to explore this point, particle concentration data and video footage of the concurrent traffic activity passing the sampling site were combined in order to provide a source apportionment of the peaks in particle concentration. Figure 5 shows representative particle size distributions for a series of peak particle concentration events obtained from night-time data 4th March 2005. On the whole, the events associated with the passage of HGVs and LGVs show size distributions typical of the Aitken mode, with count mean diameters (CMDs) in the region of 28-42 nm, although there is also a single large nucleation mode event with number concentrations far in excess of the Aitken mode events. This may relate to rapid cooling and dilution of a diesel exhaust resulting in the saturation ratio of semi-volatile compounds becoming great enough for nuclei formation. Casati et al., [25] showed a strong dependence of nucleation mode concentrations with dilution temperature and vehicle speeds for a EURO-III diesel passenger vehicle. For passenger cars there are a number of nucleation mode events as well as low numbers of Aitken mode type distributions. This is in broad agreement with previous suggestions that it is diesel vehicles that dominate Aitken mode emissions and that SI engines can produce comparable numbers of particles compared to diesels but with smaller peak diameters [27].

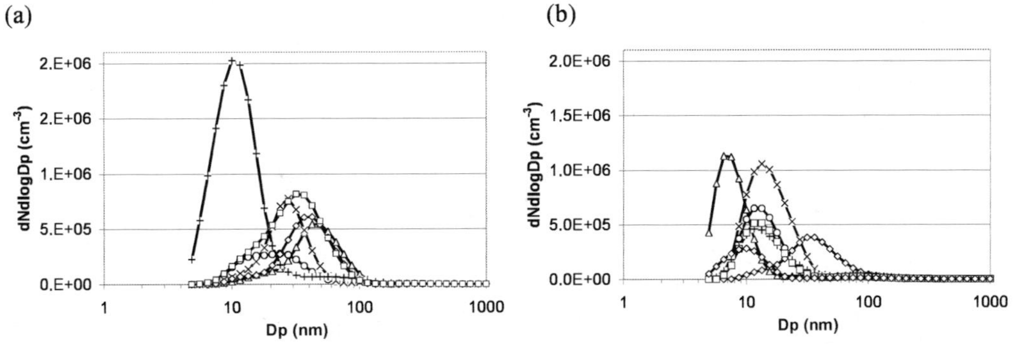

Figure 5. Example particle size distributions at peak of events associated with (a) LGV's, vans, and HGV's and (b) passenger cars. 1 s averages.

The proportion of nucleation mode to Aitken mode particles is highly variable over short timescales and therefore it is worth asking whether a number based standard would adequately reflect the concentrations of the potentially more toxic Aitken mode particles, particularly when a decrease in Aitken mode particles can be accompanied by an increase in nucleation mode numbers. Total PNC measurements do not represent these subtleties. Representative number concentration measurements are difficult to perform and depend strongly on many factors including instrument scan time, transient emissions and dispersion processes, atmospheric temperature and dilution conditions. Total PNC measurements with no size distribution information would not represent the diverse nature of particle sources and their atmospheric processing. Furthermore, whilst the commonly used SMPS gives size distributions, its long scanning time appears to lead to an underestimation of total PNC. In addition, whilst total PNC represents both nucleation and Aitken mode particles, perhaps these two types of particle mixing state are not equally relevant from the toxicological point of view. Nucleation mode particles could be excluded from PNC measurements through absorption or volatilisation techniques but this would have to be achieved consistently and achievable in the roadside environment.

In order to explore the issue of relevant particle measurement metrics, the relationship between nucleation and Aitken mode number concentrations and mass based measurements at the site was explored. Previous roadside studies e.g. [28] have found a poor correlation between $PM_{2.5}$ and Uf number concentrations. However, figure 6a shows that the Aitken mode diurnal hourly average concentrations correlated well with $PM_{2.5}$ at the Narborough Road site, although the nucleation mode did not. This may explain why $PM_{2.5}$ has often been found to be a poor indicator of Uf particle counts, since there is potentially so much variability in nucleation mode concentrations depending on traffic and ambient conditions. However, if $PM_{2.5}$ is adopted as a future standard, the correlation between $PM_{2.5}$ and Aitken mode concentrations may be more important, since these particles are potentially more toxicologically relevant. There have also been suggestions by the scientific community in both Europe and the USA that a PM_1 standard would be a better mass standard than $PM_{2.5}$, although a study in the Los Angeles Basin found that complex seasonally dependant relationships existed between PM_1 and $PM_{2.5}$ [29]. The potential usefulness of using PM_1 as a standard was investigated for the Narborough Road measurements. The mean hourly particle volume between 5 and 1000nm was calculated from size distribution measurements and is used to represent PM_1, based on assumptions of sphericity and constant density. The normalised mean hourly Aitken mode PNC plotted against the normalised estimated total volume is shown in figure 6b. Strong correlations ($R^2 = 0.92$) are seen and are stronger than the correlation between $PM_{2.5}$ and normalised Aitken particle count. Given the assumptions made in calculating the particle volume, this is only indicative of the possible relationship that might exist, but shows the potential usefulness of a PM_1 standard in representing number concentrations of the potentially more toxic Aitken mode particles. In contrast the R^2 value for the equivalent relationship for the nucleation mode is 0.79. The infrastructure change to adopt a PM_1

standard would not be too significant, since the measurement would be TEOM based, although the problems associated with the loss of semivolatile material from the TEOM may well be more significant for the lower cut-off size of 1 μm.

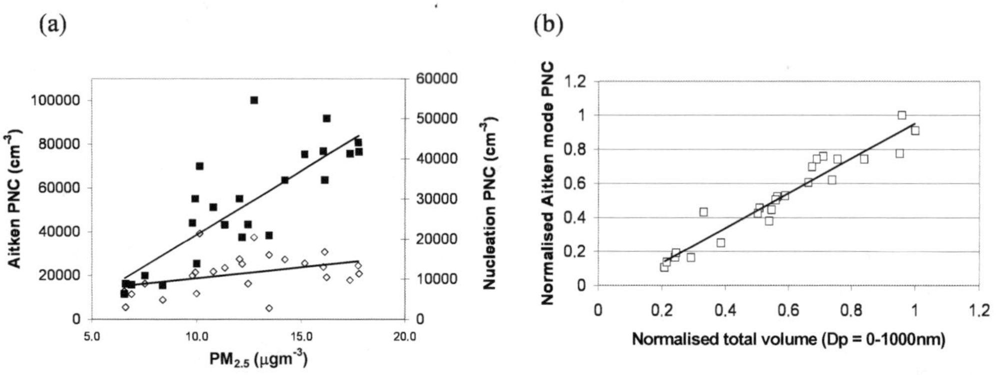

Figure 6. (a) Weekday relationship between hourly diurnal average Aitken and nucleation mode PNC and $PM_{2.5}$ on Narborough Road. ■ – Aitken mode (R^2 = 0.65), ◊ – nucleation mode (R^2 = 0.16). (b) Relationship between normalised diurnally averaged hourly Aitken mode PNC and estimated normalised volume on Narborough Road (R^2 = 0.92).

3.3. Further investigation of the nature of nucleation and Aitken modes using volatility studies.
Possible differences in particle mixing state between the different size fractions of the whole distribution were suggested in the previous section. This was explored further in a separate study at the Kirkstall Road roadside location in Leeds. Figure 7 represent contour plots of number concentrations across the size distributions for diurnally averaged data across the sampling period for 3 different TD temperatures. Here the concentrations have been adjusted based on simultaneous measurements of total PNC from a CPC in order to account for week to week variability in concentrations. For ambient temperatures it is clear that there are high concentrations across the size distribution from below 10 nm up to 90 nm for much of the day, with particular peaks in the morning rush hour period. The size distributions appear to be continuous with no distinct modes obvious. However, through fitting sums of log normal distributions using the programme R-Mix [30], individual modes can be identified with the hourly diurnal average count mean diameters (CMD) shown on the plots. CMDs of around 10 nm (nucleation mode), 30nm (small Aitken mode) and 90nm (larger Aitken mode) are present within the distribution but there is clearly overlap of these modes.

Particular peaks in nucleation mode concentrations are seen during the early morning rush hour and also in the early afternoon (possibly with a secondary aerosol contribution). A distinct peak in the Aitken mode is seen in the morning rush hour. When the sample is heated to 50°C using the TD it is noticeable that the afternoon peak in nucleation mode numbers almost disappears, whilst there is less impact on the morning nucleation mode peak. This suggests that the material contributing to nucleation mode particles in the afternoon is more volatile than that contributing to nucleation mode particles in the morning which might be expected to have primary emission sources from peak traffic numbers. There is a clear drop in nucleation mode numbers in the afternoon following heating to 50°C, indicating that these particles are droplets in nature, possibly comprising semi-volatile alcohols and carbonyl compounds [31]. There are no apparent differences in Aitken mode particle numbers on heating to 50°C, indicating that these are composed of lower volatility components in a possibly internally mixed structure. This is also confirmed by figure 8 where daily average size distributions are presented for the different TD temperatures. The figure includes either normalisation or adjustment of total number concentrations with respect to total PNC from the CPC measurements in order to account for week to week variability. At 50°C there is a distinct drop in the nucleation mode compared to ambient temperature but no obvious difference for the larger particles.

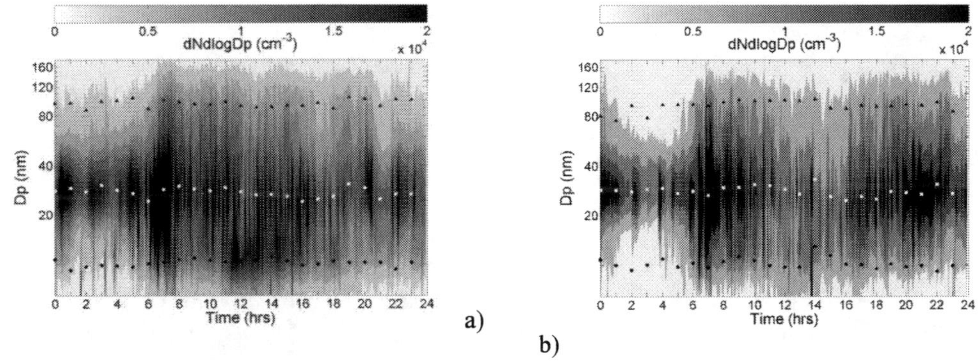

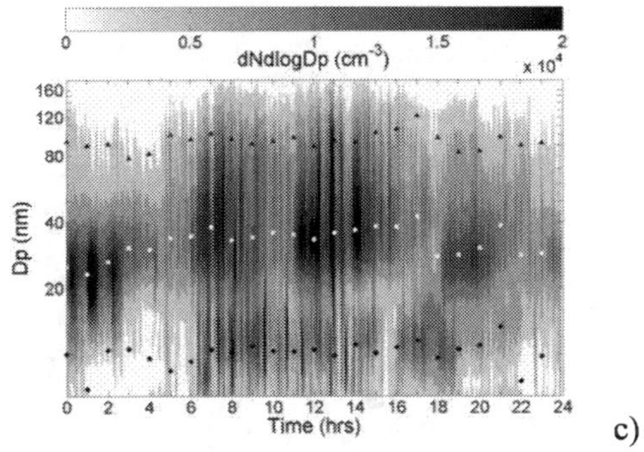

Figure 7. Contour plots of the diurnal variation in number concentration size distributions for diurnally averaged data across the sampling period for (a) ambient temperature (average 14 °C), (b) TD = 50°C, (c) TD = 100°C. Also shown is the diurnal variation in hourly average count mean diameter for ● Mode I, □ Mode II, and ▲ Mode III.

Further heating to 100°C leads to a significant change in the diurnal size distribution pattern. Now there is a distinct drop in Aitken mode particle numbers (figure 8) whilst no further reduction in the nucleation mode is observed. This is perhaps surprising since one might expect that further heating would remove further nucleation mode particles of slightly lower volatility than those removed at TD = 50°C. The contour plots of size distribution reveal further interesting features which may help to explain this result. At TD = 100°C (figures 7c,8) distinct modes appear, with peak concentrations more obviously coinciding with the mean CMDs. A possible explanation is that a proportion of the Aitken mode is made up of either agglomerated low volatility particles also incorporating a semi-volatile component, or agglomerated semi-volatiles. On heating, the semi-volatiles are driven off leaving lower volatility particles in the lower size fractions. If these are solid particles, there are potential implications for inhaled particle toxicity, since larger agglomerated particles in the ambient atmosphere could decoagulate, leaving smaller particles that would be classified within the nucleation mode but are not composed of highly volatile droplets. This explanation implies that there is significant coagulation of primary particles in the near field of the exhaust, which is possible given the high concentrations found there during slow moving rush hour periods where initial dispersion due to vehicle movement would be limited. A second explanation is that a proportion of the Aitken mode is comprised of background aerosol which has been processed within the atmosphere and is therefore likely to contain agglomerates. However, in periods of high traffic volume and low traffic speeds the roadside enhancement of PNC is likely to dominate over background sources which would become more significant during night-time periods.

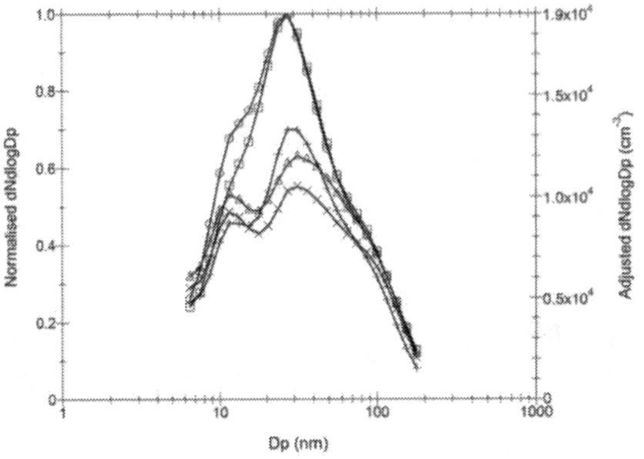

Figure 8. Normalised (left) and CPC adjusted (right) average particle size distributions for TD temperatures: ○ $T_{ambient}$, □ 50°C, Δ 75°C, + 100°C and × 125°C.

4. Conclusions

High time resolution, size segregated studies of Uf particle number concentrations were performed at two UK roadside locations using a DMS500 particle spectrometer with and without a thermodenuder. Intercomparisons between the DMS500, an SMPS and a CPC, indicate that the SMPS can underestimate total number concentrations compared to the DMS500 and CPC, due to the transient nature of excursions in concentrations and the longer scan time of the instrument. Short time scale periods of high number concentrations were found to make considerable contributions (up to 25% using a threshold of 3×10^5 cm^{-3}) to hourly average total PNC. The duration of these events was often of the order of a few seconds. The transient nature of changes in PNC may have implications for the type of exposure experienced in the roadside environment, as well as for indoor outdoor exchange. It also potentially affects instrument selection for size distributed number concentration measurements.

The transients were dominated by Aitken mode particles during high traffic periods, with the Aitken mode shown to be internally mixed particles with a low volatility core (thought to be primarily soot) and a surface of condensed and absorbed material, some of which was lost on heating with the thermodenuder. Transient periods of high number concentrations of nucleation mode particles were also found, and contributed up to 12% of hourly total number concentrations. They were relatively more important during night-time and afternoon periods, potentially due to secondary formation processes and the lower concentration of pre-existing particles favouring nucleation over condensation or absorption onto existing particles. The TD studies indicated the presence of highly volatile liquid droplets in the nucleation mode which were lost on heating. Such complexities in size distributions and particle mixing state within the Uf fraction, and the discrepancies between instruments, make the potential adoption of a total PNC based standard more complicated than first appears. Numbers of nucleation mode particles are highly dependant on atmospheric conditions such as dilution and temperature as well as pre-existing particle concentrations. They may also be toxicologically less significant if they are predominantly liquid droplets. Although techniques could be developed to remove the nucleation mode from total PNC measurements, their practicality in a roadside environment requires assessment. Aitken mode particles have been shown in previous studies to have relatively high toxicological significance within the Uf fraction [26], and in this study they were seen to dominate the size distribution following heating to >50°C. Their hourly average PNC correlatedwell with $PM_{2.5}$ and estimated PM_1, with lower significance of the equivalent correlation of nucleation mode particles. Mass based measurements may easier to undertake on a UK wide basis within existing infrastructure and may well give a reasonable representation of numbers of Aitken mode particles. Issues related to spatial and temporal variability in particle volatility would also affect mass measurements however, suggesting further studies of this kind at different types of locations would be useful.

References

[1] Kittelson D B 1998 *J. Aerosol Sci.* **29** 575 – 588
[2] Lingard J J N, Agus E L, Young D T, Andrews G E and Tomlin A S 2006 *J. Environ. Monitor.* **8** 1203–1218
[3] Ketzel M, Wåhlin P, Berkowicz R, Palmgreen F 2003 *Atmos. Environ.* **37** 2735 – 2749
[4] Oberdörster G, Ferin J, Lehnert B E 1994 *Environ. Health Persp. Suppl.* **102** 173 – 179
[5] Anderson HR, Bremner SA, Atkinson RW, et al. 2001, *Occ. Environ. Med.* **58** 504-510
[6] de Hartog JJ, Hoek G, Peters A, et al., 2003 *Am. J. Epid.* **157** 613-623
[7] Kettunen J, Lanki T, Tiittanen P, et al., 2007, *Stoke* **38** 918-922
[8] von Klot S, Wolke G, Tuch T, Heinrich J, Dockery DW, Schwartz J, Kreyling WG, Wichmann HE, Peters 2002, *Eur. Resp J.* **20** 691-702
[9] Schwartz J, Neas LM, 2000 *Epidemiology* **11** 6-10
[10] Osunsanya T, Prescott G, Seaton A, 2001 *Occ. Environ. Med.* **58** 154-159
[11] Allen G, Sioutas C, Koutrakis P, Reiss R, Lurmann FW, Roberts PT 1997 *J. Air Waste Man.* **47** 682-689
[12] Charron A, Harrison RM, Moorcroft S, et al. 2004 *Atmos. Environ.* **38** 415-423
[13] Ayers GP, Keywood MD, Gras JL 1999 *Atmos. Environ.* **33** 3717-3721
[14] Sakurai H, Tobias H J, Park K, Zarling D, Docherty K S, Kittelson D B, McMurry P H and Ziemann P J 2003 *Atmos. Environ.* **37** 1199-1210
[15] Kittelson D B, Watts W F and Johnson J P 2006 *J. Aerosol Sci.* **37** 913-930
[16] Shi J P and Harrison R 1999 *Environ. Sci. Technol.* **33** 3730-3736
[17] Agus E L, Young D T, Lingard J J N, Smalley R J, Tate J E, Goodman P S and Tomlin A S 2007 *Sci. Tot. Environ.* **386** 65-82
[18] Charron A and Harrison R 2003 *Atmos. Environ.* **37** 4109-4119
[19] Gerde P, Muggenburg B A, Lundborg M and Dahl A R 2001 *Carcinogenesis* **22** 741 – 749.
[20] Harris S, Hodges N, Jenkins H, Ellison J, Bell M C, Goodman P, Namdeo A and Häggkvist K 2003 Project Deliverable 8.11. EU 5th Framework HEAVEN project. Project Ref: IST-1999-11244.
[21] Cambustion 2005 DMS500 Fast Particulate Spectrometer, User Manual. Cambustion, Cambridge.
[22] Agus E L, Lingard J J N, Andrews G E, Tate J E, Ropkins C and Tomlin A S 2008 Using volatility analysis to investigate aerosol processing in a roadside environment, in prep.
[23] Oberdörster G, Oberdörster E and Oberdörster J 2005 *Environ. Health Persp.* **113** 823-839
[24] Michaels R A and Kleinman M T 2000 *Aerosol Sci. and Technol.* **32** 93-105
[25] Casati R, Scheer V, Vogt R and Benter T 2007 *Atmos. Environ.* **41** 2125-2135
[26] Bräuner E V, Forchhammer L, Møller P, Simonsen J, Glasius M, Wåhlin P, Raaschou-Nielsen O and Loft S 2007 *Environ. Health Persp.* **115** 1177-1182
[27] Kittelson D B, Watts W F, Johnson J P, Schauer J J and Lawson D R 2006 *J. Aerosol Sci.* **37** 931-949
[28] Molnar P, Janhill S and Hallquist M 2002 *Atmos. Environ.* **36** 4115-4123
[29] Geller M D, Fine P M and Sioutas C 2004 *J. Air & Waste Man. Assoc.* **54** 1029-1039
[30] Leys J, McTanish G, Koen T, Mooney B and Strong C 2005 *Earth Surface Processes and Landforms* **30** 579-590
[31] Ning Z, Geller M D, Moore K F, Sheesley R, Schauer J J, Sioutas C, 2007. *Environ. Sci. Technol.* **41** 6000 – 6006

Home outdoor models for traffic-related air pollutants do not represent personal exposure measurements in Southern California

R Ducret-Stich[1], R J Delfino[2], T Tjoa[2], A Gemperli[1], A Ineichen[1], J Wu[2], H C Phuleria[1], L-J S Liu1[3]

[1] Institute of Social and Preventive Medicine, University of Basel, Switzerland

[2] Department of Epidemiology, School of Medicine, University of California, Irvine, CA, USA

[3] Department of Environmental and Occupational Health Sciences, University of Washington, Seattle, WA, USA

E-mail: regina.ducret@unibas.ch

Abstract. Recent studies have used measurements or estimates of traffic-related air pollutants at home or school locations to link associations between exposure and health. However, little is known about the validity of these outdoor concentrations as an estimate for personal exposure to traffic. This paper compares modelled outdoor concentrations at home with personal exposure to traffic air pollution of 63 children in two areas in Los Angeles in 2003/2004. Exposure monitoring consisted of sixteen 10-day monitoring runs, with each run monitoring 4 subjects concurrently with the active personal DataRAM for particulate matter <2.5 μm ($PM_{2.5}$), elemental carbon (EC) and organic carbon (OC). One child per run had concurrent indoor/outdoor home monitoring. Measurements at central sites (24-hr $PM_{2.5}$, EC, OC) were taken daily and concentrations of PM_{25}, EC, and OC from traffic sources were calculated using the CALINE4 model for individual residences. We modelled outdoor concentrations of $PM_{2.5}$, EC and OC with multilinear regression including GIS and meteorological parameters and adjusted for auto-correlation between repeated measurements. The model fit (R^2) for home outdoor estimates was 0.94, 0.74 and 0.80 for $PM_{2.5}$, EC and OC, respectively. Comparisons between these outdoor estimates and the personal measurements showed a good agreement for $PM_{2.5}$ (R^2=0.65-0.70) with a mean bias of –0.7±11.8μg for the smog receptor area, and 18.9±16.2μg for the traffic impacted area. However the outdoor estimates were not related to personal exposure for EC (R^2=0.01-0.29) and OC (R^2=0.03- 0.14). **Conclusions:** Predictions of outdoor concentrations can be used as approximations of personal exposure to $PM_{2.5}$. However, they are not appropriate for estimating personal exposure to traffic-related air pollutants including EC and OC in studies of acute exposure-response relationships.

1. Introduction

Many children living in the traffic dominated Los Angeles area have an increased exposure to traffic air pollution due to their homes and schools being located close to major roads or highways [1]. A birth cohort study in this area found an increased risk of low birth weight and preterm birth in association with residential proximity to heavy-traffic roadways [2]. Positive associations between exposure to traffic-related air pollution and asthma [3] and negative effects on lung growth independent of background air pollution levels [4, 5] were found in the Southern California Children's Health Study.

Assessing effects or exposure of traffic exhaust is more complicated than assessing ambient air pollution in general due to the significant intra-urban spatial variation of traffic air pollution [6-8]. Therefore individual traffic index or dispersion models were used in many air pollution epidemiologic studies to link traffic-related air pollution to health effects [2-5, 9-11]. Only few studies used personal measurements to link acute respiratory effects to personal measurements of $PM_{2.5}$ [12-16]. Defino et al. [17, 18] presented one of the few studies associating personal exposure to traffic related pollutants to respiratory effects. He found associations between acute asthma outcomes and personal exposure to traffic pollutants but not with ambient $PM_{2.5}$ or OC. There may be other more toxic pollutants from traffic-related sources or other PM constituents, which are not sufficiently represented by home outdoor measurements that might be attributed to these acute asthma outcomes.

The current state-of-the-art exposure assessment using measured or modeled concentrations of NO_2, black smoke, and/or $PM_{2.5}$ outside homes or schools remains as the most affordable exposure assessment option for studies linking traffic pollution to health effects [4, 9, 11, 19-23]. Although these outdoor concentration estimates are widely used, only one study had examined the validity of using home or school outdoor concentrations to represent personal exposure to traffic pollution [24-26]. Van Roosbroeck and coworkers reported that the indices "living on a busy road" and "traffic intensity" at the street of residence predicted reasonably well long-term personal exposure to soot, but not exposure to $PM_{2.5}$ or nitrogen oxides. It remains unknown whether personal exposure to traffic exhaust can be represented by the outdoor estimates from the land-use regression or dispersion models.

This study evaluates the validity of using home outdoor measurements and estimates from land-use regression models for personal short-term exposure to $PM_{2.5}$, EC, and OC. Exposure models for $PM_{2.5}$, EC, and OC outside homes were constructed using geographic parameters and outdoor measurements. Predictions from these outdoor models were then compared to personal exposure measurements from children with asthma in two communities with different ambient pollution source profiles.

2. Methods

2.1. Study Design

This work was part of a panel study evaluating acute health effects of 63 children with asthma living in two areas of the Los Angeles air basin [17, 18]. Between August and December 2003 31 subjects were followed in Riverside, a smog receptor site, and in Whittier, a site immediately downwind of vehicular emission sources, 32 subjects were monitored from July to November 2004. Each child completed a time-place-activity diary every two waking hours during the 56 days of follow-up. Eight 10-day exposure monitoring runs were conducted in each area. Each run consisted of 4 subjects with concurrent personal monitoring of $PM_{2.5}$, EC and OC. Concurrent $PM_{2.5}$, EC and OC measurements were also taken at one child's residence (indoor and outdoor) as well as at the central site. The CALINE4 dispersion model [27] was used to estimate traffic-source related $PM_{2.5}$, EC and OC at residences.

2.2. Exposure Measurements

The personal DataRAM (MIE pDR-1200; Thermo Electron Corp., Franklin, MA), an integrated nephelometer, was carried in a backpack during waking hours to measure one-minute average $PM_{2.5}$. 24-hr integrated $PM_{2.5}$ was collected on a 37 mm quartz filter (Whatman Inc, Florham Park, NJ) using a 2.5-μm sharp-cut cyclone (BGI model GK 2.05, KTL cyclone, GI Inc., Waltham MA) operated at 4 L/min upstream to the pDR. Relative humidity (RH), and temperature loggers were used at 1-min intervals (Onset Computer Corp., Pocasset, MA). Data were downloaded and checked for quality and compliance daily. All pDR data were adjusted for RH effect [28,29]. Detailed description and validation of the pDR can be found elsewhere [28].

One subjects' home per run had simultaneous indoor and outdoor monitoring concurrent with the personal measurements. Harvard Impactors (Air Diagnostics and Engineering, Inc., Naples, ME) [30] were used to collect 24-hour PM_{10} and $PM_{2.5}$ on Teflon and quartz filters at home and central sites. The central site in Riverside was the South Coast Air Quality Management District (SCAQMD) monitoring site, while the one in Whittier was located at one subject's residence. All quartz filters were analyzed for EC and OC using the thermal manganese dioxide oxidation technique [31].

Residences were geocoded using the TeleAtlas Eagle Geocoding service (TeleAtlas, Redwood City, CA). Distance to different types of roadways (freeway, arterial, and collector roads) were calculated in ArcGIS 8.3 (ESRI, Redlands, CA) based on TeleAtlas MultiNet™ USA roadway network. Annual average daily traffic (AADT) count data in 2000 were obtained from the California Department of Transportation (Caltrans), assigned to TeleAtlas roadway links, and adjusted to represent the years of 2003 and 2004 based on a statewide vehicle-miles-traveled growth (2.4% per year from 2000 to 2004) [32].

Traffic-source related $PM_{2.5}$, EC and OC at residences were estimated using CALINE4, a Gaussian dispersion model which predicts traffic-specific pollutant concentrations for receptors near roadways [27, 33]. The original CALINE4 model was modified to include contributions from road segments within 5 km to a receptor.

2.3. Analysis

For personal, home indoor, home outdoor and central site traffic air pollution concentrations summary characteristics and the Pearson correlation coefficients were calculated by area. Separate models were constructed for estimating 24-h averages of home outdoor $PM_{2.5}$, EC and OC using pooled data from both areas and stepwise linear regression with a 0.1 significance level for entering and retaining predictors. A random effect was added to the final models to account for data clustering within homes. In order to minimize the temporal variation we also used the same modeling approach to predict the ratios of home outdoor to central site measurements.

All models were examined for reliability using the leave-one-out cross validation. Each observation was removed from the dataset and evaluated against its model prediction, which was calculated using the remaining observations. To compare predicted home outdoor concentrations with measured personal exposure, R^2 from the linear regression and the bias (the differences between the measured and modeled values) were reported. All statistical analyses were computed with SAS 9.1 (SAS Institute Inc., Cary, NC).

3. Results

3.1. Quality control

The data collection rate for personal measurements of $PM_{2.5}$, EC and OC ranged between 76% to 89% in Riverside and 94% to 95% in Whittier. Data from 4 of the 31 subjects in Riverside were excluded from analysis, including one with home monitoring, due to inadequate geocoding data. From the 16 runs, outdoor measurements from 7 Riverside homes (excluding one above) were pooled with those from 8 Whittier homes. A total of 131 $PM_{2.5}$ and 129 EC/OC measurements at the home outdoor sites with matched central site measurements were valid for modeling.

3.2. Exposure data

In Table 1 daily averages of $PM_{2.5}$, EC and OC concentrations are summarized. Central-site and home outdoor $PM_{2.5}$, EC and OC concentrations were higher in Riverside than in Whittier, whereas average personal $PM_{2.5}$ EC and OC were higher in Whittier than in Riverside. Indoor $PM_{2.5}$ concentrations at

Table 1. Daily averages of air pollution measurements and GIS parameters at subjects' homes by sample type and location.

	Riverside Panel (N=27)				Whittier Panel (N=32)			
	N (missing[a])	Mean (Std Dev)	Median	Min / Max	N (missing[a])	Mean (Std Dev)	Median	Min / Max
Daily air measurements (24-hr)								
Central Site								
PM_{25} ($\mu g/m^3$)[b]	76 (3)	28.5 (20.1)	22.1	6.1 / 87.2	74 (5)	16.7 (10.2)	14.9	2.8 / 66.8
PM_{25} EC ($\mu g/m^3$)	70 (9)	1.59 (0.82)	1.35	0.36 / 5.04	76 (3)	0.69 (0.44)	0.59	0.14 / 2.95
PM_{25} OC ($\mu g/m^3$)	70 (9)	6.09 (2.74)	5.89	1.55 / 19.73	76 (3)	3.89 (1.50)	3.73	1.64 / 8.82
Personal Exposure								
PM_{25} ($\mu g/m^3$)[c]	196 (70)	26.8 (19.0)	21.1	5.8 / 98.4	299 (17)	36.4 (26.8)	29.1	7.6 / 220.0
PM_{25} EC ($\mu g/m^3$)	234 (32)	0.38 (0.98)	0.24	0.00 / 14.64	295 (21)	0.76 (1.32)	0.47	0.00 / 17.20
PM_{25} OC ($\mu g/m^3$)	237 (29)	4.81 (2.82)	3.86	0.98 / 17.32	301 (15)	6.83 (3.41)	6.43	2.18 / 31.50
Outdoor home								
PM_{25} ($\mu g/m^3$)[b]	65 (4)	23.2 (16.4)	18.0	3.8 / 71.8	70 (9)	17.5 (10.4)	15.8	3.2 / 64.6
PM_{25} EC ($\mu g/m^3$)	63 (6)	1.01 (0.44)	0.97	0.25 / 2.23	78 (1)	0.84 (0.51)	0.70	0.21 / 3.10
PM_{25} OC ($\mu g/m^3$)	63 (6)	5.14 (1.82)	5.44	1.19 / 9.19	78 (1)	4.50 (1.87)	4.22	2.05 / 10.28
Indoor home								
PM_{25} ($\mu g/m^3$)[b]	66 (3)	12.6 (7.6)	10.4	2.9 / 33.8	69 (10)	16.2 (7.3)	16.3	3.6 / 46.9
PM_{25} EC ($\mu g/m^3$)	61 (8)	0.66 (0.32)	0.67	0.07 / 1.55	78 (1)	0.79 (0.92)	0.62	0.14 / 7.75
PM_{25} OC ($\mu g/m^3$)	61 (8)	5.35 (2.11)	5.02	1.42 / 11.57	78 (1)	5.96 (2.39)	5.38	2.64 / 13.53
CALINE4 estimates (24-hr)								
Home outdoor								
PM_{25} ($\mu g/m^3$)	266 (0)	4.4 (3.1)	3.8	0.4 / 17.0	316 (0)	5.1 (2.6)	4.4	1.8 / 17.2
PM_{25} EC ($\mu g/m^3$)	266 (0)	0.46 (0.32)	0.38	0.04 / 1.69	316 (0)	0.51 (0.28)	0.43	0.18 / 1.79
PM_{25} OC ($\mu g/m^3$)	266 (0)	2.28 (1.62)	2.06	0.22 / 9.68	316 (0)	2.63 (1.19)	2.37	0.93 / 7.67
GIS variables								
Population density (1/km^2)	27 (0)	1701 (1072)	1696	141 / 3782	32 (0)	3577 (1551)	2952	978 / 6575
Minimal distance to highways (m)	27 (0)	2684 (1993)	2175	460 / 6938	32 (0)	1372 (937)	1084	121 / 3339
Minimal distance to arterial roads (m)	27 (0)	487 (543)	310	6 / 2593	32 (0)	335 (320)	248	20 / 1130
Minimal distance to collector roads (m)	27 (0)	496 (347)	401	26 / 1275	32 (0)	291 (328)	144	4 / 1238
Minimal distance to local roads (m)	27 (0)	14 (11)	12	2 / 67	32 (0)	19 (19)	12	10 / 97

[a] missing and excluded together.
[b] 24-hr PM2.5 collected with Harvard Impactors.
[c] 24-hr averages of 1-min pDR readings.

Riverside homes were lower than outdoor and central-site concentrations, while home outdoor and indoor concentrations were comparable to the central site levels in Whittier. The CALINE4 model estimates for $PM_{2.5}$, EC and OC from local mobile sources were expectedly lower than measurements from all sources. Based on the CALINE4 estimates, about 30% of outdoor $PM_{2.5}$ and 60% of EC/OC came from local traffic in Whittier, while in Riverside the local traffic contribution was only 20% and

45% for $PM_{2.5}$ and EC/OC, respectively. This is in accordance with the GIS data showing that subjects in Whittier lived closer to major roads than subjects in Riverside (Table 1).

3.3. Correlation between personal, central site and home measurements

Home outdoor and personal concentrations of $PM_{2.5}$ showed good correlations to those at the central, home outdoor and home indoor sites (Table 2). Weaker but still strong correlations were found for OC between home outdoor and central site measurements. Correlations were smaller between indoor and outdoor OC measurements and between personal and central site OC measurements. The smallest correlation coefficients were found for EC, between the home outdoor and central sites and between home indoor and outdoor sites. Personal EC measurements in Riverside were significantly correlated with only home outdoor measurements, whereas in Whittier a strong correlation was observed with measurements at both the home indoor and the central sites.

Table 2. Pearson correlations between the concentrations at different locations by pollutant.

		Riverside			Whittier			Pooled		
		Central Site	Home outdoor	Home indoor	Central Site	Home outdoor	Home indoor	Central Site	Home outdoor	Home indoor
$PM_{2.5}$	Home outdoor	0.95**		0.80**	0.97**		0.58**	0.95**		0.64**
	Personal	0.78**	0.81**	0.84**	0.83**	0.77**	0.74**	0.60**	0.65**	0.79**
OC	Home outdoor	0.83**		0.48**	0.79**		0.65**	0.79**		0.55**
	Personal	0.16*	0.48**	0.25*	0.22**	0.53**	0.54**	0	0.26**	0.39**
EC	Home outdoor	0.60**		0.52**	0.32**		0.38**	0.26**		0.39**
	Personal	0.05	0.33**	0.13	0.29**	0.18*	0.89**	-0.01	0.17**	0.80**

* = p-value <0.05.
** = p-value <0.001.

3.4. Models for home outdoor pollution

Table 3 shows the final models for home outdoor measurements of $PM_{2.5}$, OC, and EC for both areas. All 3 models were dominated by the central site measurements and included minimal distance to highways. The $PM_{2.5}$ model (R^2=0.94) was dominated by central site $PM_{2.5}$. The OC model (R^2=0.80) was also influenced by the population density and wind direction. The EC model, with the smallest R^2 value among all 3 models (R^2=0.74) was also influenced by the population density, minimal distance to highways, and wind direction. The CALINE4 estimates did not enter the OC or the EC model.

Models for the ratios of home outdoor to central site measurements had smaller R^2 values than the concentration models because of the adjustment for the seasonal variation. The major predictors in the model for the $PM_{2.5}$ ratio included minimal distance to highways, wind direction, and temperature (R^2=0.28). The major predictors in the OC ratio model were distances to highways and collector roads, wind direction, and temperature (R^2=0.62). For the EC ratio model minimal distance to highways to the EC model was replaced by the CALINE4 estimate, which was the major predictor, followed by wind direction and RH (R^2=0.38).

Predictions from the concentration and ratio models showed comparable performance for predicting home outdoor concentrations, after the ratio estimates were multiplied by the central-site measurements. Subsequently we used the concentration models for the following evaluation of their representativeness to personal exposure measurements.

Table 3. Results of multilinear regression modeling for home outdoor air pollution measurements.

Variable	Estimate		SE	% of model R^2 explained	Adj. Model R^2
$PM_{2.5}$ Model (N=131)					0.94
Central site $PM_{2.5}$ measurement (μg/m^3)	0.92	**	0.01	98	
Minimal distance to highway (m)	-3.4E-04		2.3E-04	<1	
Average temperature at central site (°F)	-0.04		0.03	<1	
Weekly time term (sin)	0.51	*	0.24	<1	
$PM_{2.5}$ CALINE4 estimate at home (μg/m^3)	0.29		0.15	<1	
OC Model (N=129)					0.80
Central site OC measurement (μg/m^3)	0.83	**	0.04	77	
Population density (per km^2)	3.3E-04	**	5.1E-05	16	
Average wind direction from W to N	0.27		0.21	4	
Minimal distance to highway (m)	-1.6E-04	**	5.4E-05	<1	
Average wind direction from E to S	0.02		0.25	<1	
Average wind speed (miles/hour)	-0.20	*	0.10	<1	
Frequency of wind direction from E to S	-0.08	**	0.02	<1	
Relative humidity at central site (%)	0.01		0.01	<1	
Daily time term (sin)	-0.19		0.11	<1	
Frequency of wind direction from S to W	-0.03	*	0.02	<1	
EC Model (N=129)					0.74
Central site EC measurement (μg/m^3)	0.66	**	0.04	64	
Population density (per km^2)	8.7E-05	**	1.5E-05	23	
Minimal distance to highway (m)	-5.0E-05	**	1.7E-05	7	
Frequency of wind direction from E to S	-0.03	**	6.8E-03	4	
Relative humidity at central site (%)	5.8E-03	**	1.7E-03	<1	
Frequency of wind direction from S to W	-0.01		4.5E-03	<1	
Average wind direction from N to E	-0.20	*	0.09	<1	

* $p<0.05$.
** $p<0.01$.
SE: Standard error

3.5. Comparisons between measured personal exposure and outdoor model predictions

65% of the variation in personal $PM_{2.5}$ measurements in Riverside and 70% in Whittier were explained by the estimates of home outdoor $PM_{2.5}$ (Figure 1A). The prediction bias (measured minus predicted concentration) was larger in Whittier than in Riverside. Although the models were constructed using outdoor measurements from only 25% of the subjects' homes, the model predicted equally well for all subjects regardless of the availability of home monitoring data (Figure 1B).

The predictions of the home outdoor OC model explained very little of the variation in the personal exposure to OC in both Riverside (R^2=0.03) and Whittier (R^2=0.14, two outliers excluded). The prediction bias was again larger in Whittier (2.2 μg/m^3) than in Riverside (-1.1 μg/m^3). Outdoor OC estimates at monitored homes seemed to explain a slightly higher percentage of the variation in personal OC (R^2=0.06 vs. 0.003, respectively) (results not shown).

Similarly, home outdoor EC estimates did not relate to personal exposure to EC (Figure 2). Analysis without four outliers increased the model fit to 29% (from 7%) in Whittier. The prediction bias was -0.8 μg/m^3 in Riverside and -0.2 μg/m^3 in Whittier. As with the OC model, the EC model predicted better for the subjects with home monitoring (Figure 2B).

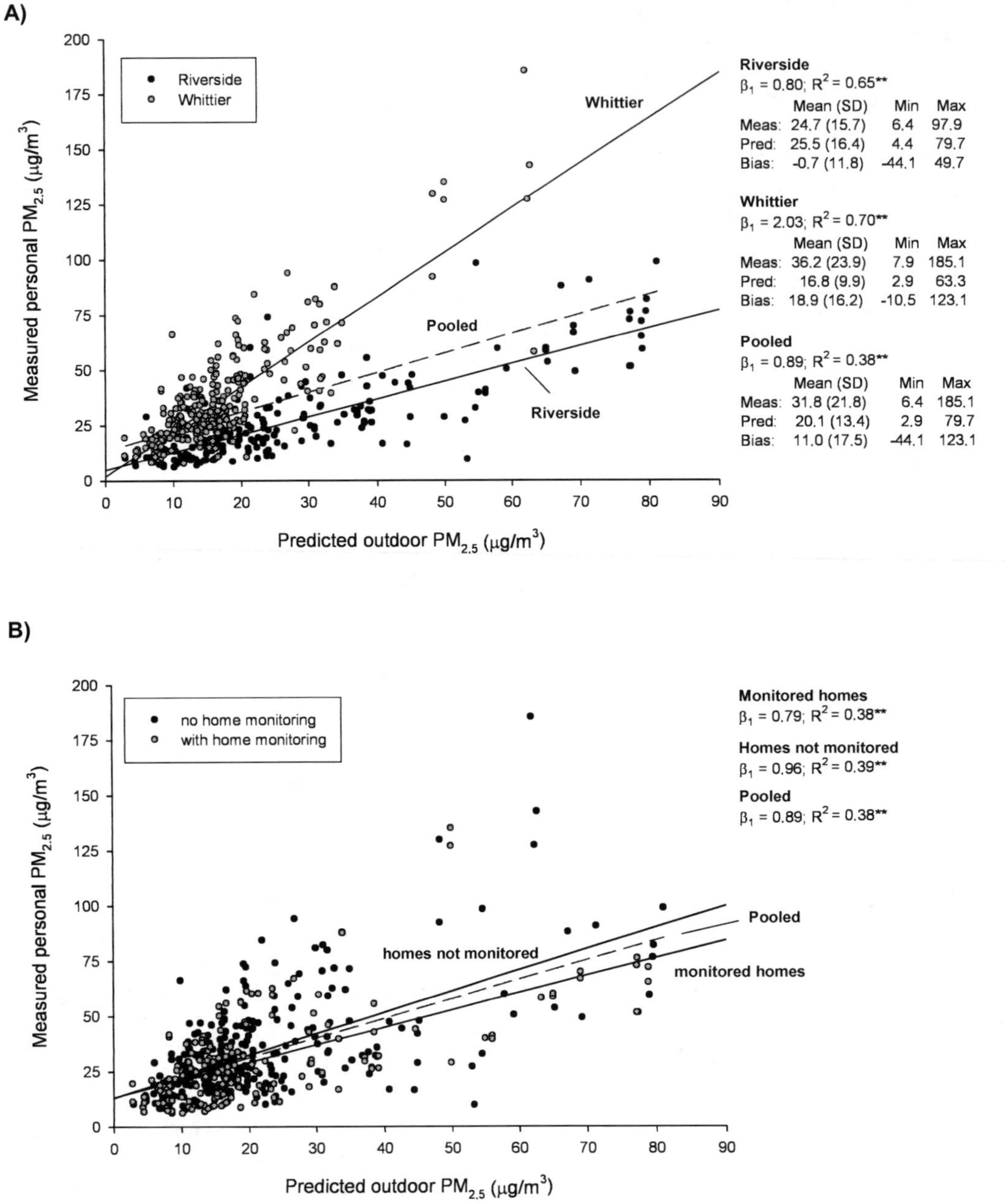

Figure 1. Measured personal $PM_{2.5}$ vs. predicted home outdoor $PM_{2.5}$ (concentration model with random effect):(A) by city, (B) by homes monitored and not monitored (β_1 = estimate of regression slope; **p<0.01). Statistics for measured (Meas), predicted (Pred) and the difference (Bias = Meas-Pred) $PM_{2.5}$ are also shown next to the Figure.

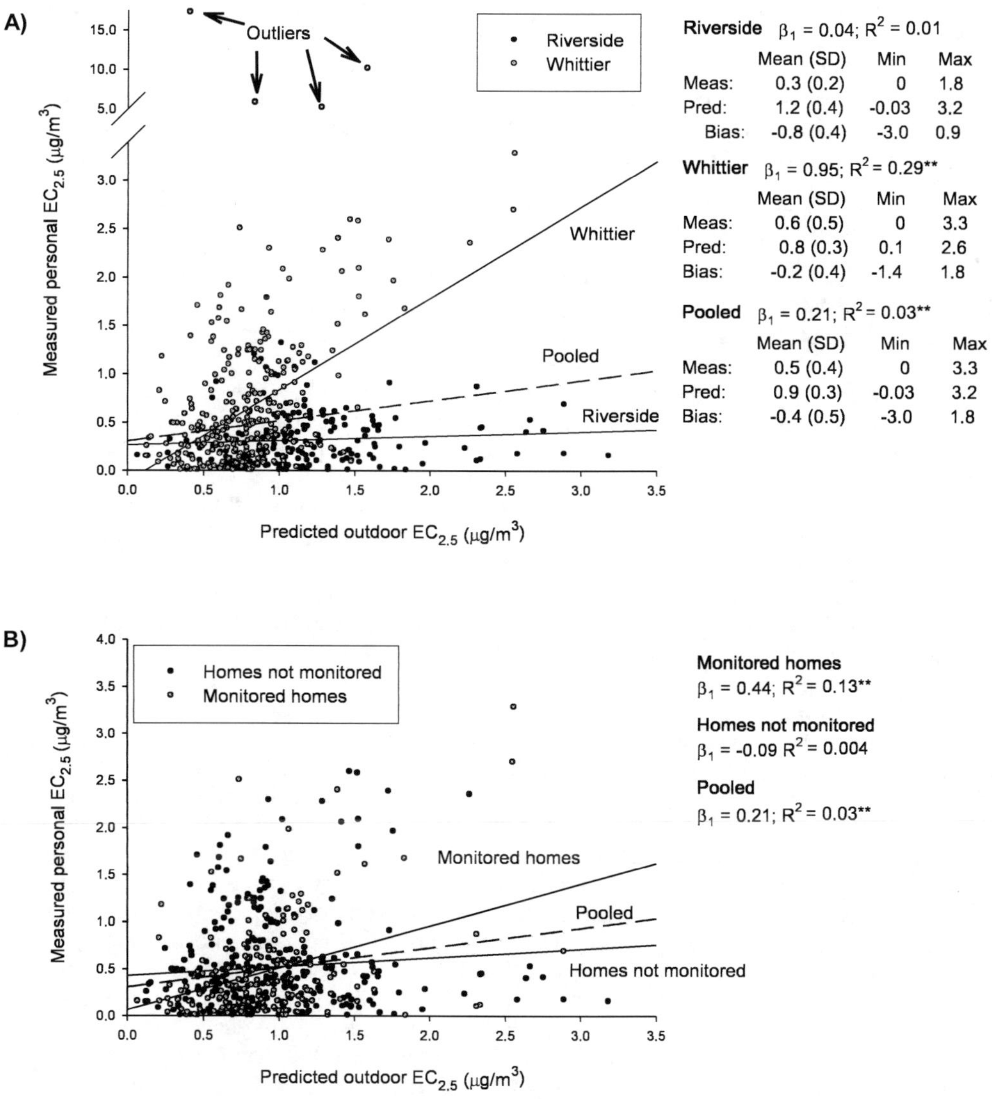

Figure 2. Measured personal EC vs. predicted home outdoor EC (concentration model with random effect):(A) by city, (B) by homes monitored and not monitored, without 4 outliers (β_1 = estimate of regression slope; $^{**}p<0.01$). Statistics (without outliers) for measured (Meas), predicted (Pred) and the difference (Bias = Meas-Pred) $PM_{2.5}$ are also shown next to the Figure.

3.6. Sensitivity analysis

To test for an improvement of the prediction by smoothing out the temporal variation, we averaged the personal measurements and modeled home outdoor values over the 10 monitoring days for each subject/home and re-examined the relationships. The R^2 for $PM_{2.5}$ increased in Riverside from 0.65 (daily averages) to 0.77 (10-day averages). However, some large prediction biases in Whittier had a larger influence on the 10-day mean and reduced the R^2 value from 0.70 to 0.48. Using the 10-day averaged EC data, there was a modest increase in the association from 1% to 24% in Riverside and from 7% to 17% in Whittier. Similar results were observed for OC with an increase from 3% to 21% in Riverside and from 9% to 13% in Whittier.

Analysis stratified by "Near source" days, meaning days with at least one 15-min entry of near smoking or cooking, revealed only a significant difference for $PM_{2.5}$ in Whittier. The R^2 for modeled home outdoor $PM_{2.5}$ and personal $PM_{2.5}$ exposure was 0.74 for the days without PM sources and 0.55 for days with any source.

4. Discussion

One finding of this study is the importance of area-specific modeling as indicated by the reduced predictive power of the home outdoor models when the two areas were pooled. Overall results for the predictions of air pollution outside homes indicate that $PM_{2.5}$ is spatially homogeneous within the two areas, and while this is the case for OC to some extend, it is not the case for EC.

The contribution from traffic to the home outdoor concentrations was higher in Whittier due to the subjects living closer to highways and arterials, although the central site and home outdoor concentrations were higher in Riverside, the smog receptor site. There PM mass concentrations were influenced by sources other than local traffic exhaust. This is in agreement with results from Kim et al. identifying Riverside as a receptor site, affected by secondary air pollution and local traffic emissions while Downey, located near central Los Angeles (10 km west of Whittier) was identified as a source site primarily affected by local traffic emissions [34].

Personal exposure to EC was correlated only slightly with home outdoor measurements and model predictions in Riverside and strongly with home indoor measurements in Whittier. However, the outdoor EC model could not predict well the exposure of subjects whose homes were not monitored. In Riverside, this might have been partially due to measuring errors. In Whittier, this points to personal exposure to EC being sensitive to local sources which may not be captured by the home outdoor measurements or estimates alone. Correlations between personal $PM_{2.5}$ and indoor concentrations were stronger than between personal $PM_{2.5}$ and home outdoor or central site concentrations due to the relatively higher fraction of time spent at home indoors (>70%). This is in accordance with results in studies in southern California [16] and other cities [30, 35-38]. However, the performance of the $PM_{2.5}$ home outdoor model for personal exposure was good, indicating that most personal exposure to $PM_{2.5}$ came from outdoor sources. The correlations between personal OC exposure and home outdoor OC were significant. However, the associations with home indoor and central site measurements differed by region. The stronger personal-home outdoor correlations likely indicated a significant influence of local outdoor sources of OC on personal exposure in both regions.

Central-site measurement was an important predictor for all our prediction models for home outdoor $PM_{2.5}$, EC and OC concentrations, although the degree of importance varied. Predictors indicating local sources and meteorological influences overwhelmed the CALINE4 estimates for the EC and OC models. CALINE4 estimates were not a significant predictor in the EC model, which is surprising as CALINE4 estimates and EC were considered specific for local traffic sources. In CALINE4 seasonal averages of meteorological parameters were used as the inputs, which might not reflect the variations in our daily measurements. In addition, heavy-duty truck count data were not available for all potentially important roadways in the study areas, which likely influenced the accuracy of the predictions. In contrast to the EC concentration model, CALINE4 became the main predictor in the EC ratio model. This indicates that the CALINE4 model is adequate in estimating spatial variations of traffic pollution in some degree.

Although many epidemiologic studies of traffic exhaust made the assumption of concentrations outside homes or schools being good indicators of personal exposure to traffic, we could show that this applies only to $PM_{2.5}$, reflecting the high correlations between personal and outdoor $PM_{2.5}$ from regional sources, and not to EC or OC, reflecting local sources. This may be partly why $PM_{2.5}$ associations with airway inflammation [17] and lung function [18] in children with asthma were seen with personal $PM_{2.5}$ measurements but not when ambient $PM_{2.5}$ measurements were used. Although, independent effects from $PM_{2.5}$ and gaseous copollutants or from specific $PM_{2.5}$ constituents on respiratory health can not be ruled out either. Therefore these models might not be adequate for specific sources. It should also be noted that measurement errors might have been involved, especially in Riverside.

We expected an improvement of the model fit with longer averaging durations and/or simultaneous sampling across sites, as reported in longer-term exposure and chronic health effects studies [24, 26]. However, home outdoor EC and OC models averaged over 10 days still could not predict more than 25% of the variation in the personal measurements. Thus, outdoor models for EC or OC could not fully capture the temporal and spatial variation in short-term personal exposure to EC or OC

As subjects spend most of the time indoors, we expected a negative influence of indoor sources on the performance of the home outdoor models for personal exposure. This was true for $PM_{2.5}$ in Whittier but it did not apply to $PM_{2.5}$ in Riverside or EC or OC in either region. This could be due to the limited number of records for indoor source reporting, reporting errors, and most likely, the relatively short occurrence of these indoor sources over the 24-hr sampling period. Similar results were found by [24] with no effect of indoor sources on personal soot.

5. Conclusions

Modeling of traffic-related air pollution at different outdoor locations is regularly used in epidemiological studies to represent personal exposures. In our study daily personal exposure of $PM_{2.5}$ was in good agreement with predicted $PM_{2.5}$ concentrations at home, validating the use of these outdoor estimates to assess the acute health impacts of short-term exposures to $PM_{2.5}$ for children. However, personal short-term exposure to OC or EC can not be approximated by the home outdoor estimates, mainly due to the contributions from sources other than the traffic exhaust outside the home. Future work should focus on building personal exposure models that incorporate additional information on personal activities and locations.

References

[1] Künzli N, McConnell R, Bates D, Bastain T, Hricko A, Lurmann F, Avol E, Gilliland F and Peters J 2003 Breathless in los angeles: The exhausting search for clean air *Am J Public Health* **93** 1494-9

[2] Wilhelm M and Ritz B 2003 Residential proximity to traffic and adverse birth outcomes in los angeles county, california, 1994-1996 *Environ Health Perspect* **111** 207-16

[3] McConnell R, et al. 2006 Traffic, susceptibility, and childhood asthma *Environ Health Perspect* **114** 766-72

[4] Gauderman W J, Avol E, Lurmann F, Kuenzli N, Gilliland F, Peters J and McConnell R 2005 Childhood asthma and exposure to traffic and nitrogen dioxide *Epidemiology* **16** 737-43

[5] Gauderman W J, et al. 2007 Effect of exposure to traffic on lung development from 10 to 18 years of age: A cohort study *Lancet* **369** 571-7

[6] Goswami E, Larson T, Lumley T and Liu L J 2002 Spatial characteristics of fine particulate matter: Identifying representative monitoring locations in seattle, washington *J Air Waste Manag Assoc* **52** 324-33

[7] Liu L J, Curjuric I, Keidel D, Heldstab J, Kunzli N, Bayer-Oglesby L, Ackermann-Liebrich U and Schindler C 2007 Characterization of source-specific air pollution exposure for a large population-based swiss cohort (sapaldia) *Environ Health Perspect* **115** 1638-45

[8] Zhu Y, Hinds W C, Kim S, Shen S and Sioutas C 2002 Study of ultrafine particles near a major highway with heavy-duty diesel traffic *Atmospheric Environment* **36** 4323-35

[9] Oftedal B, Brunekreef B, Nystad W, Madsen C, Walker S E and Nafstad P 2008 Residential outdoor air pollution and lung function in schoolchildren *Epidemiology* **19** 129-37

[10] Zmirou D, et al. 2004 Traffic related air pollution and incidence of childhood asthma: Results of the vesta case-control study *J Epidemiol Community Health* **58** 18-23

[11] Nordling E, Berglind N, Melen E, Emenius G, Hallberg J, Nyberg F, Pershagen G, Svartengren M, Wickman M and Bellander T 2008 Traffic-related air pollution and childhood respiratory symptoms, function and allergies *Epidemiology* **19** 401-8

[12] Ebelt S T, Wilson W E and Brauer M 2005 Exposure to ambient and nonambient components of particulate matter - a comparison of health effects *Epidemiology* **16** 396-405

[13] Trenga C A, Sullivan J H, Schildcrout J S, Shepherd K P, Shapiro G G, Liu L J S, Kaufman J D and Koenig J Q 2006 Effect of particulate air pollution on lung function in adult and pediatric subjects in a seattle panel study *Chest* **129** 1614-22

[14] Koenig J Q, Mar T F, Allen R W, Jansen K, Lumley T, Sullivan J H, Trenga C A, Larson T and Liu L J 2005 Pulmonary effects of indoor- and outdoor-generated particles in children with asthma *Environ Health Perspect* **113** 499-503

[15] Strand M, Vedal S, Rodes C, Dutton S J, Gelfand E W and Rabinovitch N 2006 Estimating effects of ambient pm(2.5) exposure on health using pm(2.5) component measurements and regression calibration *J Expo Sci Environ Epidemiol* **16** 30-8

[16] Delfino R J, Quintana P J, Floro J, Gastanaga V M, Samimi B S, Kleinman M T, Liu L J, Bufalino C, Wu C F and McLaren C E 2004 Association of fev1 in asthmatic children with personal and microenvironmental exposure to airborne particulate matter *Environ Health Perspect* **112** 932-41

[17] Delfino R J, Staimer N, Gillen D, Tjoa T, Sioutas C, Fung K, George S C and Kleinman M T 2006 Personal and ambient air pollution is associated with increased exhaled nitric oxide in children with asthma *Environ Health Perspect* **114** 1736-43

[18] Delfino R J, Staimer N, Tjoa T, Gillen D, Kleinman M T, Sioutas C and Cooper D 2008 Personal and ambient air pollution exposures and lung function decrements in children with asthma *Environmental Health Perspectives* **116** 550-8

[19] Kim J J, Smorodinsky S, Lipsett M, Singer B C, Hodgson A T and Ostro B 2004 Traffic-related air pollution near busy roads: The east bay children's respiratory health study *Am. J. Respir. Crit. Care Med.* **170** 520-6

[20] Brunekreef B, Janssen N A, de Hartog J, Harssema H, Knape M and van Vliet P 1997 Air pollution from truck traffic and lung function in children living near motorways *Epidemiology* **8** 298-303

[21] Janssen N A, Brunekreef B, van Vliet P, Aarts F, Meliefste K, Harssema H and Fischer P 2003 The relationship between air pollution from heavy traffic and allergic sensitization, bronchial hyperresponsiveness, and respiratory symptoms in dutch schoolchildren *Environ Health Perspect* **111** 1512-8

[22] Brauer M, Hoek G, Smit H A, de Jongste J C, Gerritsen J, Postma D S, Kerkhof M and Brunekreef B 2007 Air pollution and development of asthma, allergy and infections in a birth cohort *Eur Respir J* **29** 879-88

[23] Morgenstern V, Zutavern A, Cyrys J, Brockow I, Gehring U, Koletzko S, Bauer C P, Reinhardt D, Wichmann H E and Heinrich J 2007 Respiratory health and individual estimated exposure to traffic-related air pollutants in a cohort of young children *Occup Environ Med* **64** 8-16

[24] Van Roosbroeck S, Hoek G, Meliefste K, Janssen N A H and Brunekreef B 2008 Validity of residential traffic intensity as an estimate of long-term personal exposure to traffic-related air pollution among adults *Environ. Sci. Technol.* **42** 1337-44

[25] Van Roosbroeck S, Jacobs J, Janssen N A H, Oldenwening M, Hoek G and Brunekreef B 2007 Long-term personal exposure to pm2.5, soot and nox in children attending schools located near busy roads, a validation study *Atmospheric Environment* **41** 3381-94

[26] Van Roosbroeck S, Wichmann J, Janssen N A, Hoek G, van Wijnen J H, Lebret E and Brunekreef B 2006 Long-term personal exposure to traffic-related air pollution among school children, a validation study *Sci Total Environ* **368** 565-73

[27] Benson P E 1992 A review of the development and application of the caline3 and 4 models *Atmospheric Environment. Part B. Urban Atmosphere* **26** 379-90

[28] Chakrabarti B, Fine P M, Delfino R and Sioutas C 2004 Performance evaluation of the active-flow personal dataram pm2.5 mass monitor (thermo anderson pdr-1200) designed for continuous personal exposure measurements *Atmospheric Environment* **38** 3329-40

[29] Wu C. F., Delfino R J, Floro J N, Samimi B S, Quintana P J, Kleinman M T and Liu L J 2005 Evaluation and quality control of personal nephelometers in indoor, outdoor and personal environments *J Expo Anal Environ Epidemiol* **15** 99-110

[30] Liu L J, Box M, Kalman D, Kaufman J, Koenig J, Larson T, Lumley T, Sheppard L and Wallace L 2003 Exposure assessment of particulate matter for susceptible populations in seattle *Environ Health Perspect* **111** 909-18

[31] Fung K, Chow J C and Watson J G 2002 Evaluation of oc/ec speciation by thermal manganese dioxide oxidation and the improve method *J Air Waste Manag Assoc* **52** 1333-41

[32] Wu J., Funk T H, Lurmann F W and Winer A M 2005 Improving spatial accuracy of roadway networks and geocoded addresses *Transactions in GIS* **9** 585-601

[33] Wu J., Lurmann F, Winer A, Lu R, Turco R and Funk T 2005 Development of an individual exposure model for application to the southern california children's health study *Atmospheric Environment* **39** 259-73

[34] Kim S, Shen S, Sioutas C, Zhu Y F and Hinds W C 2002 Size distribution and diurnal and seasonal trends of ultrafine particles in source and receptor sites of the los angeles basin *Journal of the Air & Waste Management Association* **52** 297-307

[35] Turpin B, Weisel C, Morandi M, Colome S, Stock T, Eisenreich S and Buckley B 2007 Relationships of indoor, outdoor, and personal air (riopa): Part ii. Analyses of concentrations of particulate matter species *Res Rep Health Eff Inst* 1-77; discussion 9-92.

[36] Crist K C, Liu B, Kim M, Deshpande S R and John K 2008 Characterization of fine particulate matter in ohio: Indoor, outdoor, and personal exposures *Environmental Research* **106** 62-71

[37] Meng Q Y, et al. 2005 Influence of ambient (outdoor) sources on residential indoor and personal pm2.5 concentrations: Analyses of riopa data *J Expo Anal Environ Epidemiol* **15** 17-28

[38] Janssen N A, Lanki T, Hoek G, Vallius M, de Hartog J J, Van Grieken R, Pekkanen J and Brunekreef B 2005 Associations between ambient, personal, and indoor exposure to fine particulate matter constituents in dutch and finnish panels of cardiovascular patients *Occup Environ Med* **62** 868-77

Acknowledgments

The project described was supported by grant number ES11615 from the National Institute of Environmental Health Sciences (NIEHS), U.S. National Institutes of Health (NIH), and its contents are solely the responsibility of the authors and do not necessarily represent the official views of the NIEHS, NIH. The authors declare they have no competing financial interests.

Impacts of Air Pollution Exposure on the Allergenic Properties of Arizona Cypress Pollens

Y Shahali[1, 2], Z Pourpak [1], M Moin [1], A Zare [1] and A Majd [2]

[1]Immunology, Asthma and Allergy Research Institute, Medical Sciences/ University of Tehran, I. R. Iran

[2]Department of Biology, Faculty of Sciences, Islamic Azad University, North Tehran Branch, I. R. Iran

E-mail: youcef.shahali@espci.fr

Abstract: Epidemiological studies have demonstrated that urbanization and high levels of vehicle emissions correlated with the increasing trend of pollen-induced respiratory allergies. Numerous works have investigated the role of pollutants in the pathogenesis of respiratory diseases but impacts of anthropogenic pollution on pollen allergenic properties are still poorly understood. The objective of this survey was to evaluate impacts of the traffic-related pollution on the structure and allergenic protein content of Arizona cypress (*Cupressus arizonica,* CA) pollens, recognized as a rising cause of seasonal allergy in various regions worldwide. According to our results, traffic-related air pollution by its direct effects on the elemental composition of pollens considerably increased the fragility of the pollen exine, causing numerous cracks in its surface and facilitating pollen content liberation. Pollen grains were also covered by numerous submicronic orbicules which may act as effective vectors for pollen-released components into the lower regions of respiratory organs. On the other hand, this study provides us reliable explications about the low efficiency of standard commercial allergens in the diagnosis of the Arizona cypress pollen allergy in Tehran. Although traffic related pollution affects the allergenic components of CA pollens, the repercussions on the respiratory health of urban populations have yet to be clarified and need further investigations.

1. Introduction

Pollen grains are the male gametophyte generation of seed plants. Anemophilous pollens (wind pollinated) dominate the pollen rain of any area, representing major seasonal carriers of allergens. Of all the sources of allergy, pollen is undeniably one of the most pervasive; carried on the wind, it can travel over very long distances and interact with numerous biological and chemical agents present in the atmosphere.

The importance to study the interaction between pollens and environmental pollution has been put forward two decades ago by Ishizaki *et al.* [1], who demonstrated the relationship between pollutants and a higher prevalence of pollinosis in urban areas. Since then, effects of pollution on the molecular and developmental biology of pollen have been the subject of interesting studies revealing the accumulation of numerous inorganic elements such as sulphur, cadmium, and lead in pollen grains [2-5] and the acidification of pollen surfaces by the absorption of acid gases such as nitric acid in polluted areas [6]. Several studies also proved that pollutants by attaching to the surface of pollen grains and sub-pollens particles can modify the morphology of these antigen-carrying agents and alter their allergenic potential [4-10].

The city of Tehran with more than 2 millions cars emitting about 1.5 million tons of pollutants annually is a good model to study interactions between pollens and traffic-related pollution. The geographic situation of this town, hemmed in by the Alborz Mountains to the north, favors the absence of wind, causing the increasing volume of pollutants to become trapped.

In winter, the inversion phenomenon frequently happens and the amount of gaseous and particulate air pollutants attains critical levels. These peaks of pollution correlate with the early pollination period of Arizona cypress trees (mid-winter time), abundantly planted for reforestation, wind barrier and landscaping programs in numerous parts of the city [11]. These facts have prompted us to investigate the impacts of the traffic-related pollution on the structure and allergenic protein content of Arizona cypress (*Cupressus arizonica,* CA*)* pollen, currently recognized as a rising cause of winter allergies in various regions worldwide.

2. Materials and Methods

2.1. Sampling

Polluted areas of the city were selected according to data provided by Air Quality Control Company related to the National Department of Environment (table 1).

In the first sampling, pollens were collected just after microsporangia (male cones, figure 2) bursting, and the second sampling was performed two weeks later, in order to permit a natural pollen exposure to the urban air pollution.

Pollens were sifted by passage through mesh and the purity checked by light and scanning electron microscopy. The mode of sampling and storage was exactly the same for both samples.

2.2. Scanning electron microscopy (SEM)

Pollen grains were dried prior to gold coating. Samples were coated by physical vapor deposition (Sputter coater SCDOOS, BAL-TEC. Co, Switzerland) and then, were observed by using a scanning electron microscope (Philips XL30).

Elemental composition of exposed (EP) and non-exposed (NEP) pollens to air pollution were compared by using Energy Dispersive X-ray microanalysis (EDX). The percentage of different elements was determined by ZAF quantification method.

Figure 1. Arrows indicate the ranges of cypress trees planted alongside one of the most important highways of the city (Chamran highway). *Cupressus arizonica* and *Cupressus sempervirens* represent the most common species used in Tehran green spaces belonging to the Cupressaceae family.

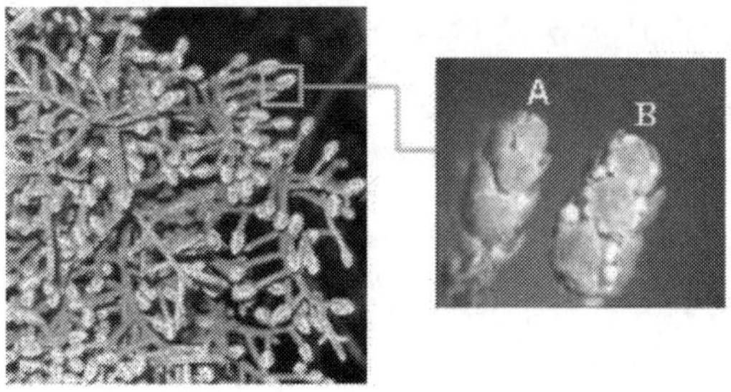

Figure 2. Mature male cones of two main varieties of CA planted in Tehran: *var. arizonica* (A) and *var. glabra* (B).

Table 1. Average amounts of air pollutants in the highest polluted areas and a non-polluted area situated outside the city. Data have been obtained from 3 air pollution measurement stations situated in the polluted areas of the city and a control station localized outside the city. Standard amounts of air pollutants have been adopted from Nation Air Quality Standards, US.

Pollutants	Carbon monoxide CO (ppm)	Sulfur dioxide SO_2 (ppb)	Nitrogen dioxide NO_2 (ppb)	Hydrocarbons (ppm)	Respirable particles (μg/m^3)
Polluted areas	8.5	52	63	4.8	124
Non-polluted area	1.1	7.6	16	2.2	50
Standard amounts	9.5	35	54	-	51

2.3. Protein studies

For the preparation of pollen extracts, an equal amount of each samples, 50 mg per ml, were extracted in 0.01 M PBS, pH 7.4 for 14 h at 4 °C under stirring. The suspensions were centrifuged at 14,000 g for 1 h at 4 °C and the supernatant were dialyzed against distilled water.

Total protein contents of *C. arizonica var. arizonica* (var. a) and *C. arizonica var. glabra* (var. g) pollens were determined and compared before and after air pollution exposure according to Bradford protein assay [12] by using bovine serum albumin (BSA) as standard. Extracted proteins were separated by 10% SDS-polyacrylamide gel electrophoresis (90 μl of each pollen extract per well) and detected by Coomassie blue staining. The molecular weight was calculated by comparison with known markers. Protein bands of each extract were quantitatively compared by densitometric analysis.

2.4. Immunoblotting for the detection of IgE reactivity

After electrophoresis, proteins were transferred from the gel onto membrane polyvinylene difluoride [13] and membranes were blocked for 1 h at room temperature with 0.05% Tween20 and 3% Bovine Serum Albumin (BSA) in PBS. The membrane was washed three times for 10 min in PBS-Tween20 and incubated overnight with 1:15 diluted serum of allergic subject. After incubation, the membrane was washed and incubated for 1 hour with 1:20000 dilution of anti-human IgE tagged with horse radish peroxidase in PBS. The blots were immersed in an enhanced chemiluminescence's solution (Amersham Kit) and the specific IgE-binding bands were revealed after 1-min exposure to a X-OMAT Kodak film.

3. Results and Discussion

3.1. Pollen

Pollen grains are sealed in a double-layered wall. Its outer layer, the exine, is composed out of sporopollenin, a lipophilic polymer of carotenoid, which is extremely resistant to enzymatic and chemical degradations. Our SEM analysis illustrated that exposure to the ambient air pollution considerably increased the fragility of the pollen exine, causing numerous cracks and collapses in its surface (figure 4) and facilitating pollen content liberation (Figure 5).

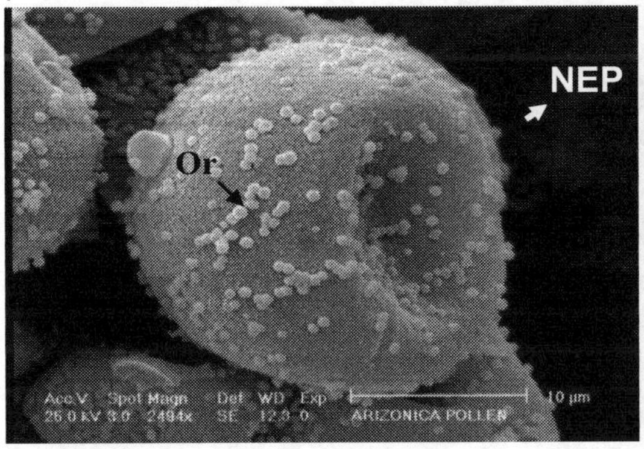

Figure 3. Scanning micrograph of a non-exposed pollen to pollution (NEP). Pollen is spherical with approximately 24µm in diameter. Numerous orbicules (Or), sat over the pollen surface.

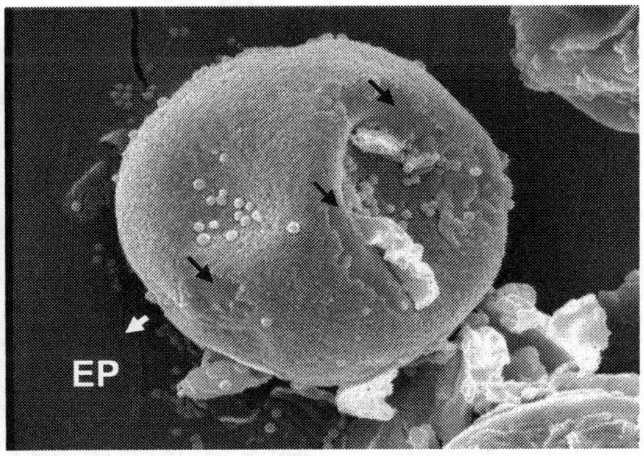

Figure 4. Scanning micrograph of a CA pollen exposed to air pollution (EP). Exine degradations are shown by black arrows.

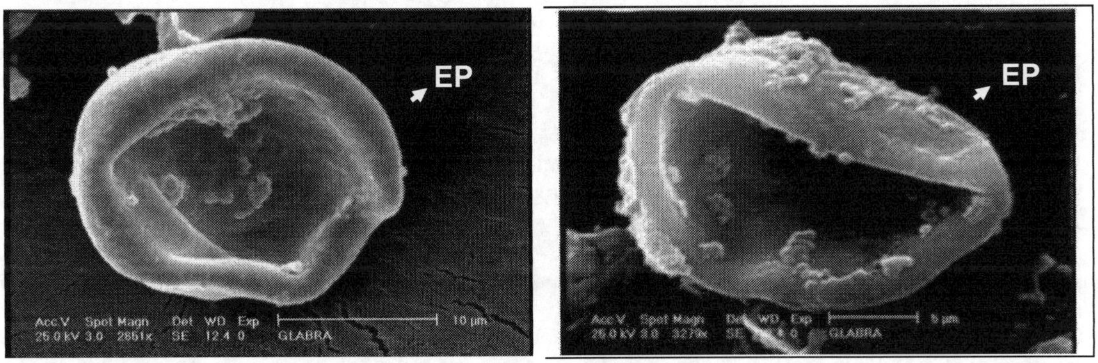

Figure 5. CA pollen exposed to pollution (EP): before their dispersion from male cones, many of polluted pollens were empty of their content.

Due to their sizes (> 5 µm), intact cypress pollens can be easily removed by the protective mechanisms of nasal and upper tracheobronchial passages and they are unable to reach the bronchial tree. That is probably the reason why cypress pollinosis tends to particularly affect conjunctivae and upper respiratory tracts. In polluted areas, pollutants by affecting the exine structure and permeability of cypress pollens may modify two important physical factors influencing their clinical significance:

the particle sizes, by increasing the premature release and dispersion of pollen derived particles smaller than pollen in the atmosphere, and the rapidity with which antigen molecules diffuse or leach out the pollen. Rapidly released allergens may induce immediate allergic reactions and may be adsorbed through the respiratory mucosa before the pollen grains are removed or swallowed. One of the major mechanisms of allergen release in CA pollens is described in the figure 6.

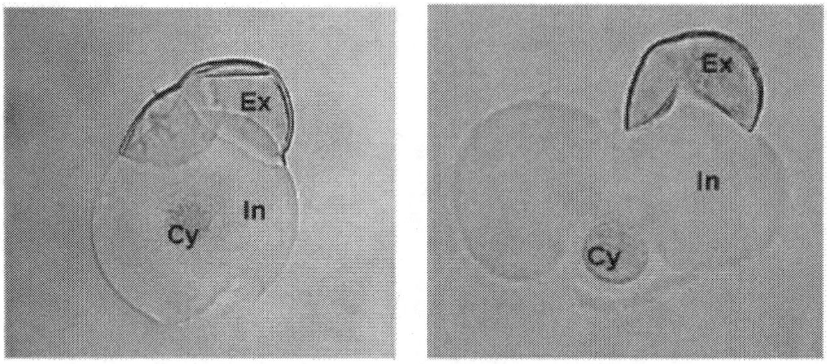

Figure 6. In contact with any source of humidity, which can be in the atmosphere, on a female flower, or on the mucosal surface of the upper respiratory tract, pollen metabolic activity is reengaged. Pollen hydration rapidly results to an important dilatation of the pollen intine (In), the layer situated beneath the exine (Ex) and essentially made of hydrophilic components such as pectin. Following pollen hydration, the exine is progressively pushed out (A). The dilatation of the intine may continue till its rupture (B) causing the liberation of the cytoplasm (Cy).

3.2. Orbicules

Cypress pollen grains are covered by numerous submicronic orbicules (350-600 nm) which may act as effective vectors for pollen-released components into the lower regions of respiratory organs (figure 7). Aerobiological studies demonstrated that small pollen-derived particles (< 2.5µm) can bind to fine aerosols in polluted air, such as DEP (diesel exhaust particles), forming an association of inhalable particles and facilitating allergen transmission [14-15].

The interactions between these cypress nanoparticles and pollutants may explain the increasing frequency of asthma reported in sensitized patients to cypress pollens, reaching 18% in Italy [16] and 20% in Izmir, Turkey [17].

3.3. EDX analysis

EDX analysis showed that Si, S, F, Co, Ni, Cu, Zn, Br, Ag , Mg, Al, Ca, K, Cl and Pb represent 15 essential elements composing the pollen tectum. The comparison of the elemental composition of pollens collected before and after exposure to air pollution showed that most important augmentations were in the ratios of nickel, sulfur, iron and copper respectively. In polluted pollens, sulfur (49%) and copper (11%) were the dominant elements and amounts of nickel and iron were respectively fourteen-fold and fourfold higher than control pollen (table 2).

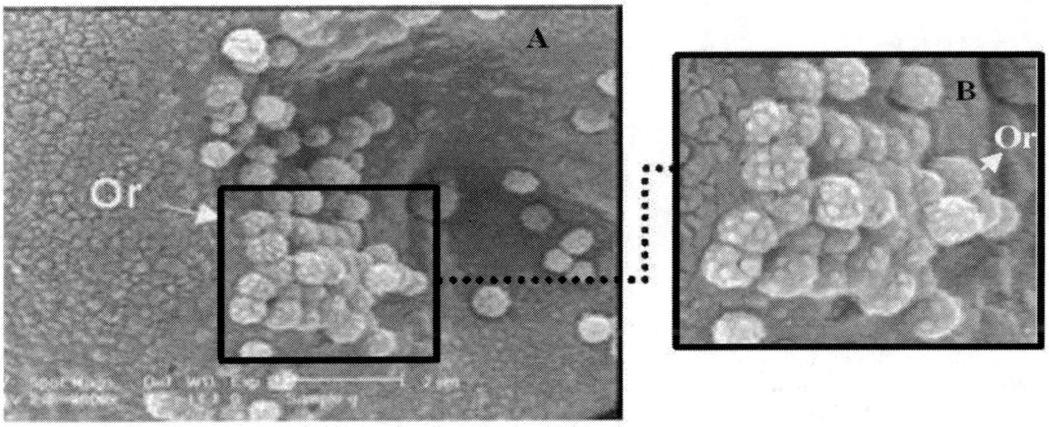

Figure 7. Scanning micrograph of orbicules (Or), A: magnified 8000x. The surface of these spherical nanoparticles of sporopollenin is irregular and gemmate, similar to the pollen exine B: Orbicules magnified 15000x.

Table 2. Energy dispersive X-ray microanalysis: elemental composition of exposed (EP) and non-exposed pollens (NEP) to the urban pollution. Levels of silica (Si), sulphur (S), cobalt (Co), nickel (Ni), zinc (Zn), bromide (Br), argent (Ag), copper (Cu), iron (Fe), magnesium (Mg), aluminium (Al), calcium (Ca), potassium (K), chloride (Cl) and lead (Pb) were compared.

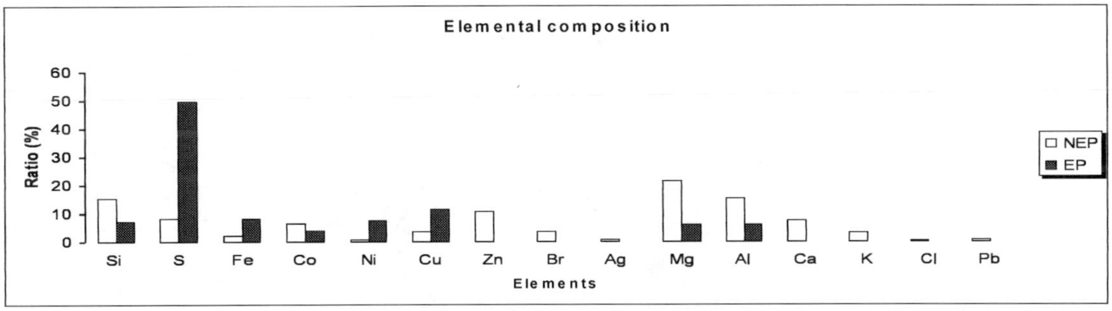

Elements	Si	S	Fe	Co	Ni	Cu	Zn	Br	Ag	Mg	Al	Ca	K	Cl	Pb
NEP (%)	15.18	8.04	2.08	6.56	0.58	3.60	10.72	3.65	0.85	21.5	15.32	7.64	3.34	0.37	0.57
EP (%)	7.30	49.48	8.35	3.88	7.57	11.39	-	-	-	6.05	5.5	-	-	-	-

3.4. Protein Studies

Bradford protein assay showed an important decrease of the total protein content in polluted pollens, from 220 μg/ ml to 180 μg/ ml for var. a and from 250 μg/ ml to 150 μg/ ml in var. g.

The comparison of the protein components of CA mature pollens collected just after microsporangia bursting with those collected 2 weeks later (exposed to the urban environment) revealed that in both var. a and var. g pollens, the amount of Cup a 1 (a pectate lyase protein with an approximate molecular weight of 45 kDa which is currently recognized as the major allergen of Arizona cypress pollen extracts), considerably decreased to only 14.6% and 14.9% of all protein content, respectively, and a new protein of 35 kDa represented more than 50% of the total protein content (figure 8).

3.5. Immunoblottings

Western blotting revealed that serum specific IgE of our CA allergic subject showed a high reactivity to the new 35 kDa major allergen in all tested extracts. On the other hand, no IgE-binding to Cup a 1 (45 kDa) has been detected in the lanes 3 and 4, representing extracts of pollens exposed to the traffic-related pollution of Tehran (figure 9).

The lack of this new 35kDa allergen in the established commercial extracts intended to in vivo and in vitro clinical tests considerably brought into question their performance in the diagnosis of the CA pollen allergy in Tehran.

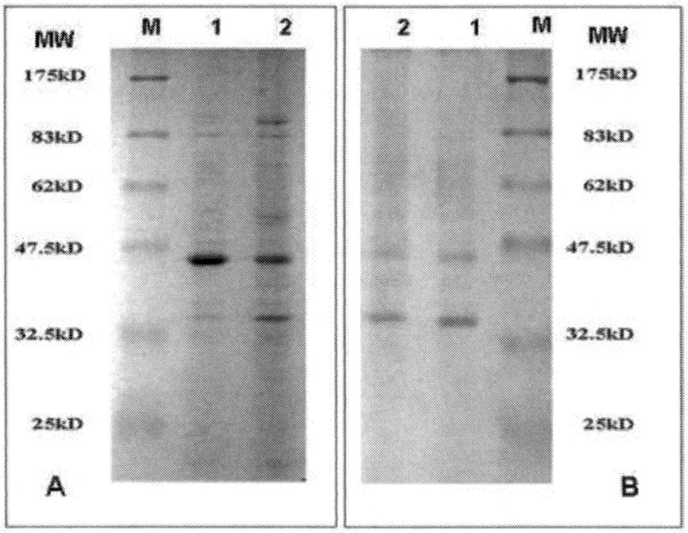

Figure 8. SDS-PAGE profile of CA pollen extracts obtained just after microsporangia bursting (A) and 2 weeks later (B). MW indicates molecular weights, M, marker, lanes 1 and 2, extracts of respectively var. a and var. g pollen collected from trees planted in Tehran.

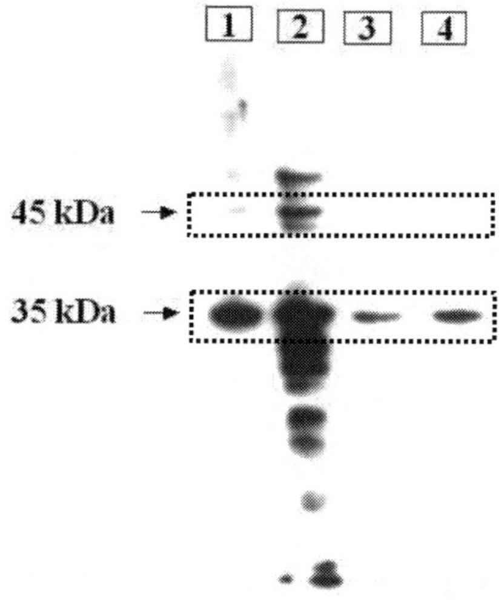

Figure 9. Immunoblotting: specific IgE-binding protein bands to respectively var. a and var. g pollen extracts obtained just after microsporangia bursting (lane 1 and 2) and after 2 weeks exposure to the urban pollution (lane 3 and 4). Immunoblots were probed with the serum of a CA allergic patient.

4. Conclusion

According to our results, traffic-related air pollution by its direct effects on the elemental composition of pollens considerably increases the fragility of the pollen exine, causing numerous cracks and collapses in its surface and facilitating pollen content liberation. Cypress pollen grains are also covered by numerous submicronic orbicules which may act as effective vectors for pollen-released components into the lower regions of respiratory organs.

The present study also demonstrates that in polluted areas, Arizona cypress pollens are not only important carriers of allergens but also of pollutant elements, such as sulphur and copper, able to increase mucosal hyper reactivity in susceptible subjects.

Although the influence of the traffic related pollution on the allergenic components of CA pollens is evident, the repercussions on the respiratory health of urban populations have yet to be clarified and need further investigations.

Acknowledgments

This study was supported in part by research grants from Tehran University of Medical Sciences

References

[1] Ishizaki T, Koizumi K, Ikemori R, Ishiyama Y and Kushibiki E 1987 Studies of prevalence of Japanese cedar pollinosis among residents in a densely cultivated area *Ann. Allergy* **58** 265

[2] Cerceau-Larrival M, Nilsson S, Cauneau-Pigot A, Berggren B, Derouet L, Verhille A Mand Carbonnier-Jarreau M C 1991 The influence of environmental (natural and experiment) on the mineral composition of the exine of allergenic pollen with deposit of mineral polluting particles *Grana* **30** 532

[3] Sawidis T 1997 Accumulation and effects of heavy metals in Lilium pollen *Acta. Horticult.* **437** 153

[4] Behrendt H, Becker W M, Friedrichs K H, Darson V and Tomingas R 1992 Interaction between aeroallergens and airborne particulate matter *Int. Arch. Allergy Immunol.* **99** 425

[5] Behrendt H, Becker W M, Friedrichs K H and Ring J 1997 Pollution and allergy *Int. Arch. Allergy Immunol.* **113** 69

[6] Okuyama Y, Kiyoshi M, Okochi H and Igawa M 2007 Adsorption of air pollutants on the grain surface of Japanese cedar pollen *Atmospheric Environment* **41** 253

[7] Knox B and Suphioglu C 1996 Environmental and molecular biology of pollen allergens *Trends in plant science (Elsevier Science)* **5** 156

[8] D'Amato G 2000 Urban air pollution and plant-derived respiratory allergy *Clin. Exp. Allergy* **30** 628

[9] D'Amato G, Liccardi G, D'Amato M and Cazzola M 2002 Outdoor air pollution, climatic changes and allergic bronchial asthma *Eur. Respir. J.* **20** 763

[10] Chehregani A, Majd A, Moin M, Gholami M, Shariatzadeh MA and Nassiri H 2003 Increasing allergy potency of Zinnia pollen grains in polluted areas *Ecotoxicol. Environ. Saf. (Elsevier)* **58** 267

[11] Shahali Y, Majd A, Pourpak Z, Tajadod G, Haftlang M and Moin M 2007 *Iran J. Allergy Asthma Immunol.* **6(3)** 123

[12] Bradford M 1976 A rapid and sensitive method for the quantfication of microgram quantities of protein utilizing the principle of protein-day binding *Anal. Biochem.* **72** 248

[13] Towbin H, Staehelin I, Gordon J 1979 Electrophoretic transfer of proteins from polyacrylamidegels to nitrocellulose sheets: Procedure handsome applications *Proc. Natl. Acad. Sci. USA* **76** 4350

[14] Knox R B, Suphioglu C, Taylor P, Desai R, Watson H C, Peng J L and Bursill L A 1997 Major grass pollen allergen Lol p 1 binds to diesel exhaust particles: Implications for asthma and air pollution *Clin. Exp. Allergy* **27** 246

[15] Namork E, Johansen B V and Lovik M 2006 Detection of allergens adsorbed to ambient air particles collected in four European cities *Toxicology Letters* **165** 71

[16] Italian Association of Aerobiology 2002 An epidemiological study of Cupressaceae pollinosis in Italy *J. Investig. Allergol. Clin. Immunol.* **12** 287

[17] Sin A Z, Ersoy R, Gulbahar O, Ardeniz O, Gokmen N M, Kokuludag A 2008 Prevalence of Cypress Pollen Sensitization and Its Clinical Importance in Izmir, Turkey, With Cypress Allergy Assessed by Nasal Provocation *J. Investig. Allergol. Clin. Immunol.* **18(1)** 46

Development of a bio-mathematical model in rats to describe clearance, retention and translocation of inhaled nano particles throughout the body.

L MacCalman[1], C L, Tran[1] and E Kuempel[2].
[1]Institute of Occupational Medicine, Edinburgh

[2]National Institute for Occupational Safety and Health, Cincinnati, Ohio, USA

laura.maccalman@iom-world.org

Abstract. Studies of the translocation of inhaled nanoparticles from rodent lungs to different target organs provide data to model the inhalation and translocation of nanoparticles. A compartmental model was developed based on biological information about the major organs and how they interrelate. This model, which is quantified by a set of differential equations 'describing' the passage of nanoparticles through the body, gives estimates of the particle mass present in each organ. Optimal parameter estimates were found by minimising the model mean squared error. Data from two different studies in rats (one endotracheal instillation and one inhalation exposure) were used to calibrate the model. Most of the nanoparticle mass remained in the lungs. The overall fit of the model to the total measured particle mass in the body was very good for both studies (R^2=0.98-0.99), although different parameter estimates were sometimes required to fit the study-specific data. The model fit to the measured particle mass by organ was very good for the lungs, brain, and spleen (R^2=0.81-0.98) in both studies, but not as good for the liver and kidney (R^2=0.32-0.53). While this model describes the retention and translocation of nanoparticles from the lungs reasonably well, further model evaluation and validation is needed using additional data.

1. Introduction

There is a need to quantify the distribution of internalised particles in the body for hazard and risk assessment. Traditionally, the quantification process is done by mathematical modelling. For micron size particles, the route of exposure is usually via inhalation and the target organ is the lungs. Models have been developed to describe the exposure-dose-response relationship for a range of particles, for example the paper by Tran, Kuempel and Castranova [1]. Nanoparticle (NP), exposure can be via many routes (inhalation, ingestion and dermal) and the target organs can be beyond the portal of entry. We introduce a new mathematical model describing the exposure-dose-response relationship for internalised NP, with the exposure dose part of the model being described here and the extension to include dose-response presented in a later paper.

2. Model Development

The model to be used is an extension to an earlier model describing the retention and clearance of particles in the lungs [1] and [2]. The original model consists of compartments representing the mass of particles (i) on the alveolar surface; (ii) inside macrophages; (iii) in the interstitial space; (iv) in the lymph nodes; (v) in the olfactory; (vi) in the upper airways region.

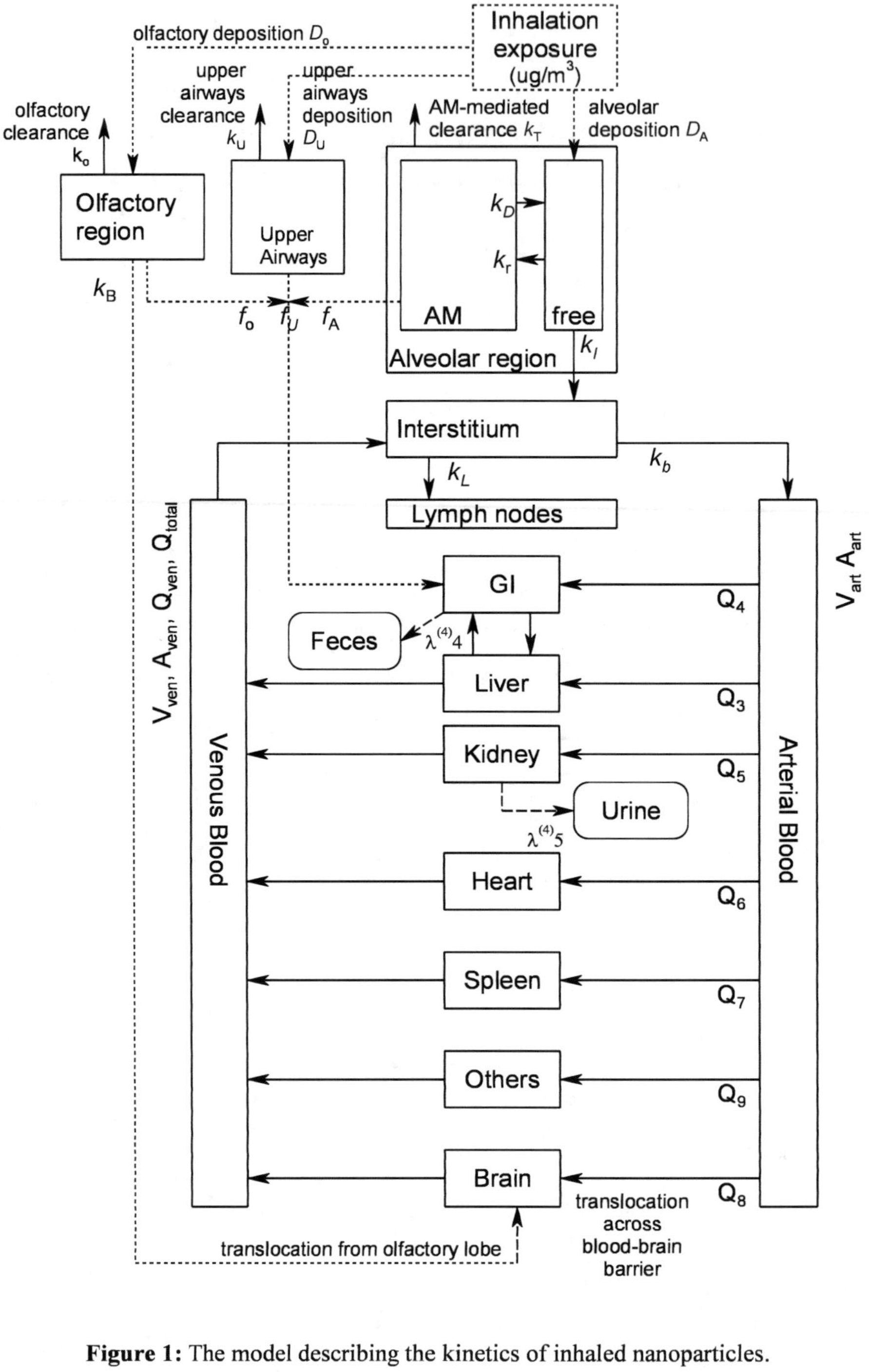

Figure 1: The model describing the kinetics of inhaled nanoparticles.

It is assumed that NP can move into the extra-pulmonary compartments because of their size. The model is therefore extended by adding the translocation route from the interstitium to the venous blood and by adding a model of the circulation of blood around the major organs in the body (Figure 1). The major organs considered here are brain, heart, liver, kidney, spleen, gastrointestinal tract and 'other' which represents other tissue sites not specifically named in the model.

First-order kinetic processes were assumed to dictate the retention and clearance of NP from each of these organ tissues. Each organ was represented by a compartment which contains sub-compartments for the tissue, capillary and sequestration as outlined in Figure 2.

Using a series of differential equations to represent the movement of NP throughout the body the mass of NP in each organ was estimated, given information about the organ characteristics and experimental conditions. The details of the differential equations and the estimates of the parameters obtained through the model calibration will be reported in a later paper.

The model requires information regarding the volume of the organ tissue, the volume of the organ capillary and the blood flow rate to each organ. The required information was collected from a number of papers; [3], [4], [5], [6], [7].

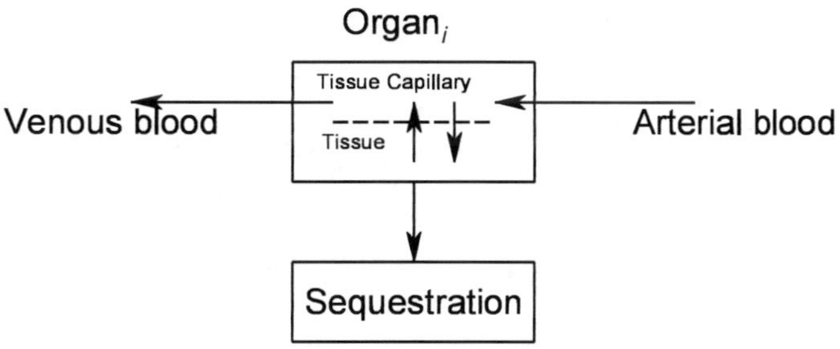

Figure 2: Extra pulmonary organ retention and excretion

3. Methods for Model calibration

The first step was to obtain data with which to compare the results of our model and therefore enable model calibration. The data was extracted by examining plots from papers published on translocation of ultrafine particles and confirmed through communication with the authors of the papers themselves. The model calibration was based on data from two different experiments.

3.1. Data

The first paper from which data was obtained was Semmler *et al.* [8]. This paper was chosen for two reasons: (1) There was data available for the translocation of iridium NPs to a number of organs; lung, brain, spleen, liver and kidney, and (2) The translocation was measured up until 6 months after the initial exposure.

The data were initially obtained from each paper but the actual data were later provided by the authors. The data obtained from Semmler *et al.* [8] were transformed from the fraction of the initial dose given in the paper to the estimated mass of NPs. The initial lung burden was given in the paper as 11.21μg, but this was later modified to 3.5μg [10]. The authors confirmed that the new lung burden was the more accurate one.

As this study used an endotracheal route of exposure there was no olfactory deposition and therefore a number of the parameters in our model did not need to be estimated (i.e. f_o, k_o and k_B) as there were no particles moving to the other organs from the olfactory region.

The second set of data was obtained from a study by Takenaka *et al.* [9]. This study differed from the first in particle type (silver), route of exposure (whole body inhalation) and length of follow-up (Table 1). We used the same model as for the first paper with changes to some of the parameters. On examining the data it appeared that the inhaled silver NPs were deposited in the nasal and upper airway regions of the respiratory tract, but not in the alveolar region of the lung; for this reason the parameters for the alveolar region were set to be zero in this model.

A program, Mutiple-Path Particle Dosimetry (MPPD) [11], was used to determine the deposition fractions in the olfactory, upper airways and alveolar regions. This program is designed to calculate deposition and clearance of aerosols in rats and humans given information about the particle characteristics, exposure conditions, and breathing patterns. The program is free and can be downloaded from www.thehamner.org/technology-and-development/technology-transfer/index.html

Table 1: The experimental conditions of both of the studies used for model calibration.

	Semmler	Takenaka
Particle Type	Iridium	Silver
Aerosol Concentration	200 $\mu g / m^3$	133 $\mu g / m^3$
Particle Diameter	15 to 20 nm	17.1 nm (MMD), 1.38 (GSD)
Duration of exposure	1.5 hours	6 hours
Breathing Frequency	45 min^{-1}	Not Given
Inspiratory Fraction	75 – 80 %	Not Given
Breathing Scenario	Endotracheal	Inhalation
Length of Follow-up	6 months	7 days

3.2. Obtaining optimal estimates of the parameters

Using sensible initial parameter estimates a constrained minimisation of the sum of squares was carried out to obtain the best parameter estimates possible, using MATLAB vs 7.0.

A vector of the list of all of the actual data obtained from the papers is denoted as y. The mass of particles estimated by the model to be in each organ, at each time point for which there are actual data is then denoted by x. The parameters to be estimated can be denoted by the vector θ, where $\theta = [\theta_1,\ldots., \theta_n]$. We then carried out a search over many possible estimates of the parameters and choose the vector θ which minimises the mean square error (MSE). Where the MSE is a measure of the amount by which the estimated, y_i, mass of NPs differs from the true mass of NPs, x_i, and is calculated by:

$$\sum_i (x_i - y_i)^2 .$$

The R^2 value was calculated as a measure of how good a fit to the model gave to the actual data. This was calculated by:

$$R^2 = \frac{\left[\sum_i (x_i y_i) - \left(\sum_i x_i \sum_i y_i\right)/n\right]^2}{\left[((\sum_i x^2 - (\sum_i x)^2/n)(\sum_i y^2 - (\sum_i y)^2/n))^{1/2}\right]^2}$$

The mean square error (MSE) is a measure of the amount by which the estimated, y_i, mass of NPs differs from the true mass of NPs, x_i, and is calculated by:

$$\frac{1}{n}\sum_i (x_i - y_i)^2$$

4. Results

4.1. Model for Semmler *et al.*
After calibration the mass of NP estimated by the model was very close to the actual mass but the estimates for the other organs were not. This is because the majority of the deposited particles were in the lung and the minimisation appears to 'concentrate' on minimising this difference.

Fixing the lung parameters and re-running the model gave a better fit for all of the organs, but with a slightly poorer fit for the lung than in the initial run. Adjusting the model, by changing the translocation rate to the lymph nodes slightly, gave a compromise between the two. Transforming the estimated mass to fractions of the initial lung burden enabled the standard errors of the original data to be shown. Figures 3 - 5 show the estimated fraction of NP in each organ from the resulting model and the actual fractions given in the paper, along with their standard errors.

Table 2 shows the R^2 values for each organ. The overall R^2 of 0.99 suggests that the data obtained from the final model is a very good fit to the data. Considering the organs separately however, there is a suggestion that the model does not fit the data very well for both the kidney and the liver.

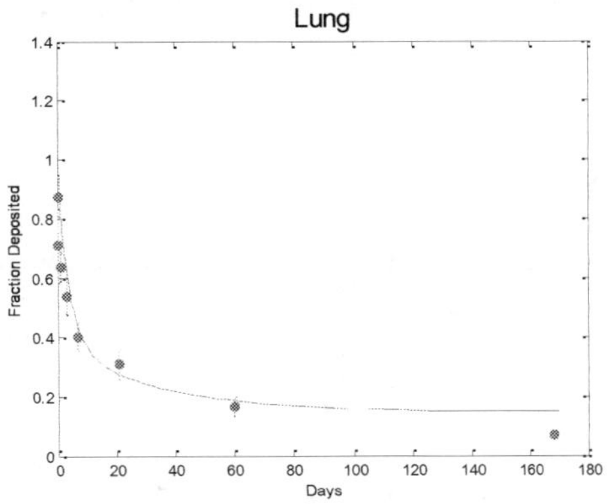

Figure 3: The fraction deposited in the lung estimated by the final model as well as the actual data points and their error bars.

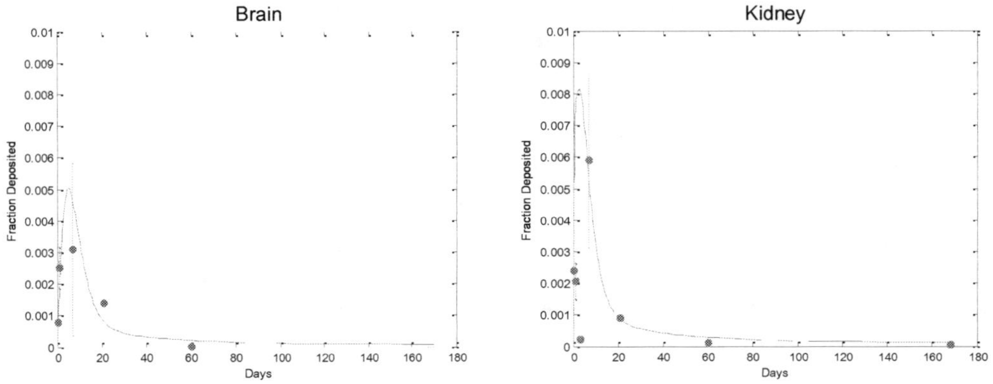

Figure 4: The fraction deposited in the brain, left, and kidney, right, estimated by the final model as well as the actual data points and their error bars.

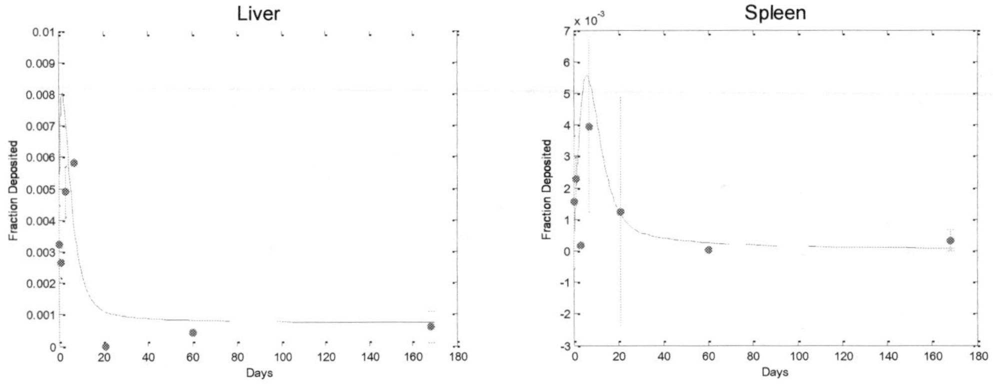

Figure 5: The fraction deposited in the liver, left, and spleen, right, estimated by the final model as well as the actual data points and their error bars.

Table 2: The R^2 and mean square error (MSE) values obtained when comparing the fitted data to the actual data for each organ separately and for the body as a whole.

	R^2	MSE
Brain	0.81	0.0000055
Kidney	0.43	0.000031
Liver	0.32	0.000038
Spleen	0.85	0.0000075
Lung	0.98	0.34
Overall	0.99	0.067

4.2. Model for Takenaka e*t al.*

There were a few differences in how we classified the compartments to match the data supplied by Takenaka *et al.* [9]. The lymph nodes were no longer included in the total lung as data were available to enable a direct comparison to these compartments separately. The brain compartment was adjusted so that the nasal (or olfactory) compartment was no longer included in the brain as data were also available for these separate compartments. Figures 6 to 9 illustrate the mass of nanoparticles estimated to be in each organ, by our model, and the actual mass measured in the study.

Table 3 illustrates the MSE values for the overall model and for each organ individually. For a number of the organs the R^2 value could not be calculated as there were only two data points. The overall model was again a good fit to the data with an overall, $R^2 = 0.98$. Of the 2 organs for which there were more than two data points one was a good fit to the data, lung with $R^2 = 0.98$, and one did not fit as well, liver with $R^2 = 0.53$.

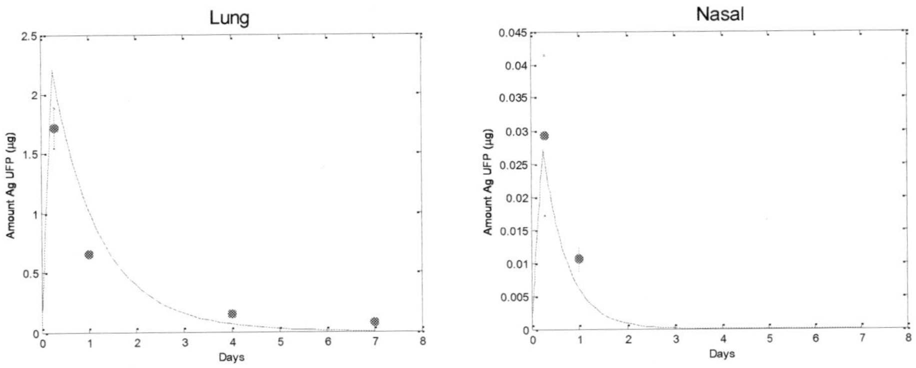

Figure 6: The mass deposited obtained using the parameter estimates for the lung, left, and nasal, right. The points are the actual data obtained from Takenaka *et al.*

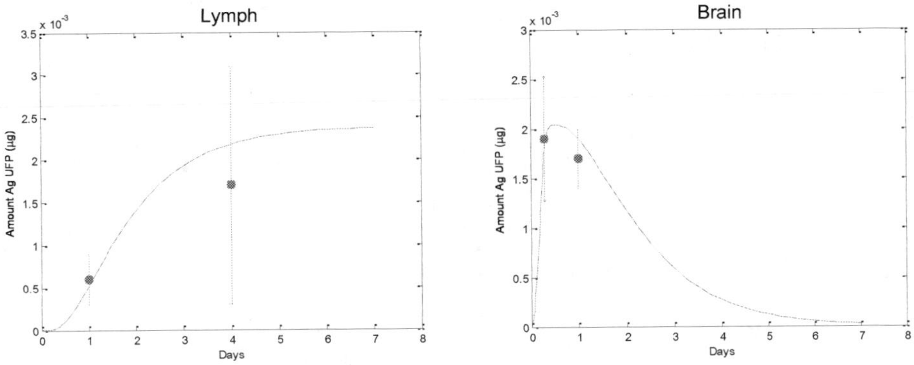

Figure 7: The mass deposited obtained using the parameter estimates for the lymph nodes, left, and brain, right. The points are the actual data obtained from Takenaka *et al.*

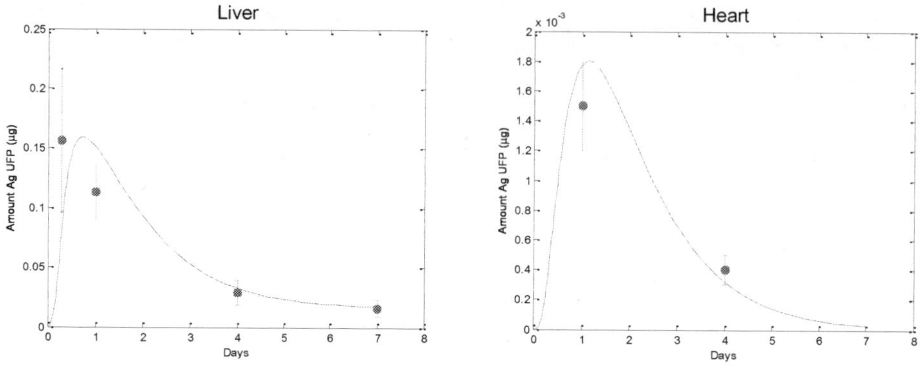

Figure 8: The mass deposited obtained using the parameter estimates for the liver, left, and heart, right. The points are the actual data obtained from Takenaka *et al.*

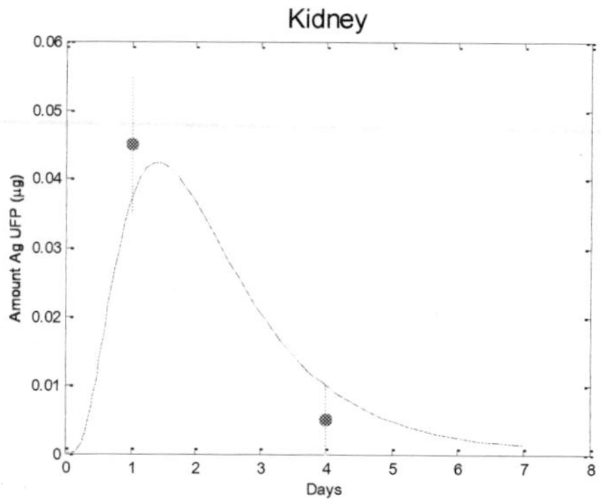

Figure 9: The mass deposited obtained using the parameter estimates in for the kidney. The points are the actual data obtained from Takenaka *et al.*

Table 3: The R^2 and mean square error (MSE) values obtained when comparing the fitted data to the actual data for each organ separately and for the body as a whole.

	R^2	MSE
Brain		0.000020
Nasal Olfactory		0.000040
Lymph		0.00000012
Kidney		0.000042
Liver	0.53	0.0017
Heart		0.000000042
Lung	0.98	0.073
Overall	0.98	0.017

5. Discussion

In this paper, we have described a biologically-based model of nanoparticle clearance from the lungs and translocation to the major organs of the body over time. Model calibration showed that the optimal parameter estimates may be dependent on the route of exposure and the physical-chemical characteristics of the particles.

The parameters associated with the compartments through which the NPs were introduced to the body (olfactory, alveolar and upper airways) are dependent on the method of exposure; olfactory deposition was eliminated when endotracheal instillation was used, while exposure through inhalation eliminated any alveolar deposition. The initial differences in the exposure methods resulted in quite different optimal parameter estimates, despite the fact that the measured mass of NPs in certain organs was quite similar.

It was hoped that the data from the second experiment could have been used to validate the model built using the first data set. However, due to the differences in the experimental conditions, particularly the exposure methods, this was not possible. The next step would ideally be to validate the model using data from further experiments. The main limitation for doing this is that there are rarely two experiments in the published literature with the same conditions. Some attempt is being made to validate the model, however, and will be reported in a later paper.

As noted earlier the mathematical detail of the model and the resulting parameter estimates will be reported at a later date along with an extension of the model to include dose-response estimation.

Acknowledgements

The authors would like to acknowledge that the work carried out in this paper was supported by a grant from the National Institute for Occupational Safety and Health (NIOSH).

References

[1] Tran CL, Kuempel ED, Castranova V (2002). A rat lung model of exposure, dose, and response to inhaled silica. Ann Occup Hyg. 46(suppl. 1):14-17.

[2] Tran CL, Graham M, Buchanan D (2001). A biomathematical model for rodent and human lung describing, exposure, dose, and response to inhaled silica. Institute of Occupational Medicine Technical Memorandum TM/01/01.

[3] Brown, R.P., Delp, M.D., Lindstedt, S.L., Rhomberg, L.R. & Beliles, R.P. (1997). Physiological parameter values for physiologically based pharmacokinetic models. Toxicol and Ind Health. 13(4):407-484.

[4] Luu, H.M.D., Hutter, J.C. & Bushar, H.F. (1997). Physiologically based pharmacokinetic model for 2, 4-toluenediamine leached from polyurethane foam-covered breast implants. Environmental Health Perspectives. 106:393-400.

[5] Kohn, M.C. & Melnick, R.L. (2000). The privileged access model of 1,3-Butadiene disposition. Environmental Health Perspectives, 108(Suppl 5):911-917.

[6] Kreyling, W.G., Semmler-Behnke, M., Erbe, F., Mayer, P., Takenaka, S. Schulz, H., Oberdörster, G. & Ziesenis, A. (2002). Translocation of ultrafine insoluble iridium particles from lung epithelium to extrapulmonary organs is size dependent but very low. J of Toxicol and Environ Health, Part A. 65:1513-1530.

[7] Bernareggi, A. & Rowland, M. (1991). Physiologic modelling of cyclosporine kinetics in rat and man. J Pharmaco Biopharm. 19:21-50.

[8] Semmler, M., Seitz, J., Erbe, F., Mayer, P., Heyder, J., Oberdörster, G. & Kreyling, W.G. (2004). Long-term clearance kinetics of inhaled ultrafine insoluble iridium particles from the rat lung, incluing transient translocation into secondary organs. Inhal Toxicol. 16:453-459.

[9] Takenaka, S., Karg, E., Roth, C., Schultz, H., Ziesenis, A., Heinzmann, U., Schramel, P. & Heyder, J. (2001). Pulmonary and systemic distribution of inhaled ultrafine silver particles in rats. Environ Health Perspect. 109(Suppl 4):547-551.

[10] Semmler-Behnke, M., Takenaka, S., Fertsch, S., Wenk, A., Seitz, J., Mayer, P., Oberdörster, G. & Kreyling, W.G. (2007). Efficient elimination of inhaled nanoparticles from the alveolar region: evidence for interstitial uptake and subsequent reentrainment onto airways epithelium. Environmental Health Perspectives. 115:728-733.

[11] CIIT and RIVM. (2006). Multiple-path particle dosimetry model (MPPD V 2.0, 2002-2006). Research Triangle Park, NC: Centres for Health Research (CIIT) and The Netherlands: National Institute for Public Health and Environment (RIVM).

Changes to the structure of blood clots formed in the presence of fine particulate matter

Sofian Metassan[1,2], Robert A S Ariens[2], D Julian Scott[2] and Michael N Routledge[1]

[1]Molecular Epidemiology Unit and

[2]Cardiovascular and Diabetes Research Division, Leeds Institute for Health, Genetics and Therapeutics, The LIGHT Laboratories, University of Leeds, Leeds LS2 9JT

Email: umphsp@leeds.ac.uk

Abstract. Both long-term and short-term exposure (one to two hours) to particulate matter are associated with morbidity and mortality caused by cardiovascular diseases. The underlying mechanisms leading to cardiovascular events are unclear, however, changes to blood coagulability upon exposure to ultrafine particulate matter (UFPM, the smallest of which can enter the circulation) is a plausible mechanism. Objectives: This study aims to investigate the direct effects of particulate matter on fibrin polymerization, lateral aggregation and the formation of fibrin network structure. Methods: Standard Urban Particulate Matter (PM) was suspended in Tris buffer centrifuged and filtered with <200nm filter to obtain ultrafine PM or their water-soluble components. Purified normal fibrinogen was made to clot by adding thrombin and calcium chloride in the presence of varying concentrations of PM. Permeation properties (Darcy constant [Ks]) and turbidity of clots were measured to investigate the effects on flow-rate, pore size, and fibrin polymerization. In addition, confocal microscopy was performed to study detailed clot structure. Results: Total PM increased the Ks of clots in a dose dependant manner (Ks = 4.4, 6.9 and 13.2 x 10-9 cm2 for 0, 50 and 100 μg/ml total PM concentrations, respectively). Filtered PM also produced a significant increase in Ks at PM concentration of 17 μg/ml. Final turbidity measurements at 20min were obtained for varying concentrations of PM. Maximum optical density (OD) for 1 mg/ml fibrinogen at 0, 50, 100 and 200 μg/ml total PM concentrations were 0.39, 0.42, 0.45 and 0.46, respectively. The maximum OD for 0, 17, 34 and 68 μg/ml filtered PM concentrations were 0.39, 0.42 0.47 and 0.51, respectively, suggesting an increase in fibre diameter with increasing particulate concentration. The lag phase was significantly shorter and the rate of polymerisation was significantly faster in the presence of 68 μg/ml filtered PM. Confocal microscopy results showed decrease in fibre density without a significant increase in fibre diameter in the presence of total PM and filtered PM. Conclusion. The results indicate that total PM and filtered PM are capable of causing alterations to the fibrin polymerization and network structure as shown by the changes in permeation properties, the turbidity experiments as well as by confocal microscopy.

1. Background and objectives

Epidemiological studies have shown that exposure to air pollution is associated with increased hospitalisation and mortality due to cardiovascular disease [1-3]. Peters *et al.* showed that recent exposure to air pollution was associated with myocardial infarction with an odds ratio of 2.92 (95% Confidence Interval (95% CI), 2.22 to 3.83, $p<0.001$) [4]. Air pollution constitutes gaseous pollutants and particulate matter (PM) of varying sizes. There is increasing evidence that suggests that exposure to varying concentrations of PM is associated with myocardial infarction (MI), coronary heart disease (CHD), arrhythmia and cerebrovascular disease [5, 6]. The main source of traffic related PM is from diesel and petrol combustion engines, although road surface, tyre and brake wear all contribute to PM. PM has been classified according to size. PM_{10}, $PM_{2.5}$ and $PM_{0.1}$ are particles less than 10 μm, 2.5 μm and 0.1 μm in aerodynamic diameter, respectively. The particles consist mainly of elemental carbon (carbon black or soot), with small amounts of trace metals (such as lead, zinc, iron and aluminium), sulphate salts and nitrate salts. There may also be traces of varying compositions of polycyclic aromatic hydrocarbons. The size of the particles determines their ability to penetrate deep into the lungs, finer particles reaching the bronchioles and the alveoli. In order to enter the blood circulation, particles have to pass through the air-blood barrier, which is only 0.5 μm thick, and is where gaseous exchange takes place. Ultrafine particles (UFP; <0.2 μm) have been found capable of crossing the cellular membrane of the alveoli without the need for endocytosis or any form of receptor-mediated, actin-based process. In the lungs, the particles have been found in the tissue compartments, cells and within capillaries [7]. The presence of UFP in blood has been shown in vivo [8, 9].

One of the most important plasma proteins during the coagulation process is fibrinogen. The final step in the formation of clot is the conversion of fibrinogen to fibrin which polymerizes into a three-dimensional network. The polymerization of fibrinogen is initiated by the action of thrombin to cause the conversion of fibrinogen to fibrin. The fibrin clot is broken down by the process of fibrinolysis through the action of plasmin. During clot formation, plasminogen is bound to fibrin. Tissue plasminogen activator (t-PA) converts plasminogen to plasmin, initiating fibrinolysis. The rate of fibrinolysis may depend on fibrin fibre diameter and fibrin clot structure. Fibrin clot with a dense structure made of thin fibres dissolved more slowly than coarse fibrin clot structure made of thicker fibres. The formation of abnormal structure of fibrin clot has been linked to the thrombosis[10-12]. People with myocardial infarction are more likely to have tighter and more rigid fibrin clot network with lower permeability compared to normal subjects [13]. According to confocal microscopy data, plasma clots from coronary artery disease (CAD) patients were made of fibrin fibres that were shorter, thinner and more numerous than clots from normal control subjects. Furthermore, the fibres from the CAD patients were more resistant to lysis with a slower rate of lysis than that of control subjects [14]. In this study we have studied the characteristics of blood clots formed in vitro in the presence of PM and UFP.

2. Study description

2.1. Particulate matter

The PM used was a Standard Reference Material (SRM) 1648 purchased from National Institute of Standards and Technology (NIST). The SRM was collected from St Louis, MO, United States, accumulated over a period of over 12 months. The SRM mainly consists of carbon particles with various elements such as aluminium, iron, potassium, lead, sodium and zinc. Two portions of the SRM were suspended in Tris buffer (0.05M Tris-HCl, 0.1M NaCl, pH7.5) at concentration of 5mg/ml. One portion was kept aside as total particulate matter (TPM). Another portion was centrifuged at 13,400 rpm for 30 minutes and the supernatant filtered with a serological 0.22 μm filter with a hydrophilic polyethersulfone membrane (Millex GP, Millipore). This portion was called filtered particulate matter (FPM).

2.2. Fibrin permeation

Purified fibrinogen (100 μl) was made to clot by adding 10 μl of activation mixture (thrombin and calcium chloride). The final concentrations were 1 mg/ml, 1 U/ml and 10 mM for fibrinogen, thrombin and calcium chloride, respectively. Immediately after mixing, 100 µl of the clotting mixture was transferred to a plastic pipette tip that had been prepared by carefully cutting 4.5cm off the tip and making the interior surface rough. The clotting mixture was kept in a moist chamber for 2 hours at room temperature. The plastic tip was then connected through a tube to a syringe containing Tris buffer. The clot was washed by allowing 6 drops of the Tris buffer to pass through. The time taken for 6 more drops were recorded and the weight of the drops measured. The Darcy's constant (Ks), which is a measure of the pore size of fibrin network through which liquid may flow [11] was calculated.

2.3. Turbidity measurements

The stages of fibrin polymerisation were followed by turbidity measurements. The Lag phase measures the protofibril formation. Rate of polymerisation (Vmax) measures the lateral aggregation and fibre formation. The maximum optical density (OD) gives the information on fibre diameter.

2.4. Laser scanning confocal microscopy

Confocal microscopy was performed on clots formed in a confocal microscope slide chamber. A Leica TCS-SP2 confocal system was used with an inverted microscope (Leica DMIRE2), equipped with a 63X water immersion objective lens using the reflection mode. The scan format was set at 512x512 pixels. The pinhole set to one Airy unit to obtain maximum resolution in the z-plane. The fibre diameters were measured by using the software (Quantification Tools - Profile). The scans were taken using 8x zoom-in mode (30 x 30 µm). A line was drawn across the fibre avoiding any junction and the length of the line was recorded. Five fibre diameters were measured for each scan. The fibre density for each scan was calculated as follows: the scans were taken at 4x zoom-in mode (60 x 60 µm). A line was drawn across the scanned image, from right to left and the number of fibres crossing the line were counted.

2.5. Statistical methods

Statistical significance was determined by ANOVA followed by Bonferroni's test. The level of significance was taken as $p<0.05$. Results were express as mean ± 95% confidence interval (95% CI). Some results were also expressed as mean ± standard error mean (SEM).

3. Results

3.1. Fibrin permeation

As it is thought that UFP can enter the blood, the TPM was centrifuged and filtered to remove particles larger than 220 nm (FPM). Both FPM and TPM were lyophilised and the weight measured. Distilled water was used as a blank control. The percentage of the FPM was calculated to be 34% of the TPM. This value was used to calculate the actual concentrations of filtered PM. Ks of TPM at concentration of 100 µg/ml was significantly different than control ($p<0.05$). In a similar experiment as for TPM, FPM caused a significant effect on the fibrin clot permeability. As FPM was increased from 0 µg/ml to 17 µg/ml, there was a significant increase in the mean Ks (3.74 and 8.29 x 10-9 cm2 for 0 and 17 µg/ml FPM, respectively, $p<0.05$) (Figure 1). The graph was plotted as Darcy constant (Ks) against total PM concentration. Every point represents the mean and 95% confidence interval. (A) 0, 50 and 100 µg/ml total PM. (B) 0, 17 and 34 µg/ml filtered PM.

3.2. Turbidity measurements

When fibrinogen was incubated with increasing concentrations of FPM, there was also a corresponding increase in the maximum OD (mean= 0.39, 0.42, 0.47 and 0.51 for control, 17, 34 and 68 µg/ml of FPM, respectively; Fig 2). The OD at concentrations of 34 and 68 µg/ml were significantly different from control ($p<0.05$). Similar results were seen for TPM. The lag phase, the rate of protofibril formation, was measured and found to be significantly reduced in the presence of 68 mg/ml FPM (39.4 vs 12 sec for control vs filtered PM). Vmax, the rate of polymerisation, was also measured and shown to be increased in the presence of 68 mg/ml FPM (1.4 vs 2.3 x 10^{-3} sec^{-1}) for control vs FPM) (Table 1). Therefore, the presence of FPM caused faster formation of half-staggered double-stranded protofiblil as well as the formation of lateral aggregation of the protofibrils leading to the formation of fibrin fibres.

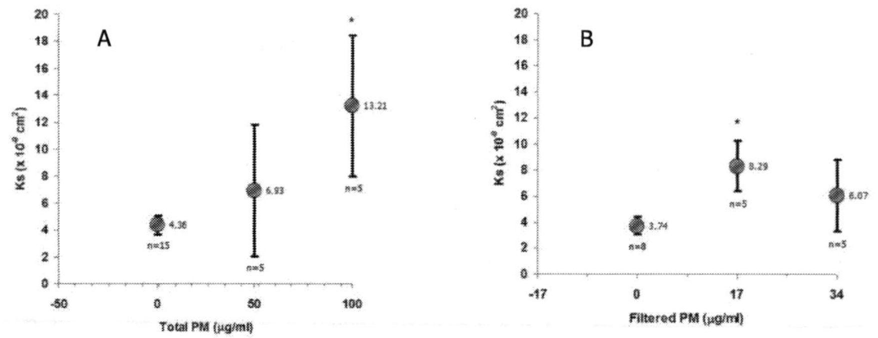

Figure 1 Permeation of 1 mg/ml fibrinogen clot in the presence of PM.

Table 1 Lag phase and rate of polymerisation of fibrin clot in the presence of total PM and filtered PM

	Lag phase (s)	Vmax (10^{-3} s^{-1})
Control	39.4 ± 6.3 (n=4)	1.4 ± 0.1 (n=10)
200 µg/ml TPM	21.0 ± 9.0 (n=4)	1.7 ± 0.2 (n=5)
68 µg/ml FPM	12.0 ± 0.0 (n=4)	2.3 ± 0.3 (n=5)

The values were mean ± SEM. The Lag phase and Vmax of 68 µg/ml Filtered PM are statistically significant from control ($p<0.05$).

The changes to the fibrin network structure were investigated in detail by digital measurements of fibre density and fibre diameter using LSCM performed at increasing concentrations of TPM and FPM (fig 3). The mean of fibre density was calculated from the LSCM scans and plotted against total PM concentration. Total PM caused decrease in the fibre density (mean = 9.1, 8.2, 8.3, 7.2 fibres/60µm for control, 50, 100 and 200 µg/ml of TPM). Fibre density at 200 µg/ml TPM concentration was significantly decreased from control ($p< 0.05$). FPM also caused a significant decrease in fibre density.

Fibre densities for 17, 34 and 340 µg/ml of FPM were statistically significant ($p<0.05$). In contrast to fibre density, there was no significant change in the fibre diameter with increasing concentration of either TM or FM. Also, the presence of high concentration of FPM (340 µg/ml) did not cause any significant change to the fibre diameter.

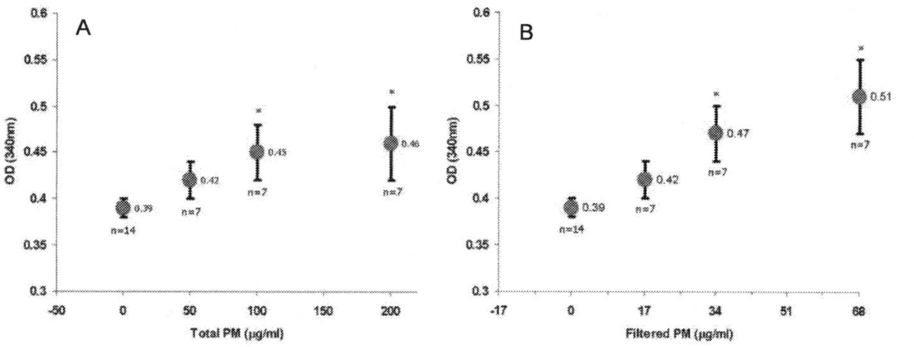

Figure 2 Effects of filtered PM fibrin polymerization, measured by turbidity. Clots were formed with 1 mg/ml fibrinogen, 0.25 U/ml thrombin and 10 mM calcium chloride. Each point represents mean and 95% CI. (A) 0, 50, 100 and 200 µg/ml total PM. (B) 0, 17, 34 and 68 µg/ml filtered PM. The OD for total PM concentrations of 100 and 200 µg/ml as well as filtered PM concentrations of 34 and 68 µg/ml were statistically significant from control ($p < 0.05$).

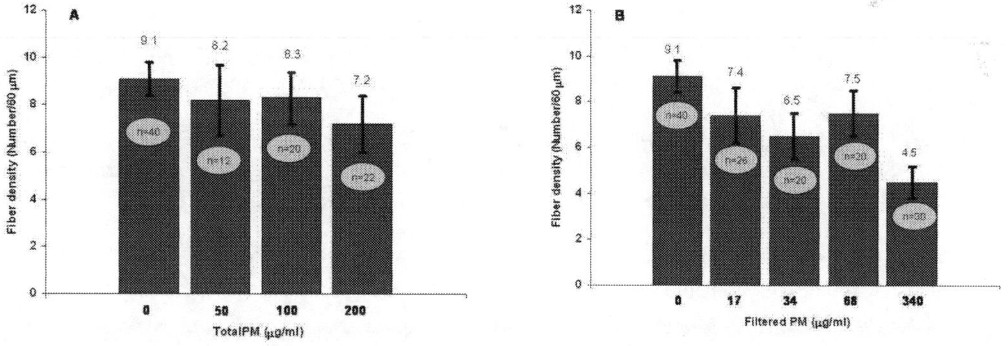

Figure 3. Fibre density in the presence of total PM and filtered PM, measured from confocal microscopy scans.Clots were formed with 1.0 mg/ml fibrinogen, 1 U/ml thrombin and 10 mM calcium chloride. (A) Presence of total PM. (B) Presence of filtered PM. Each point was plotted as mean and 95% CI.

4. Conclusions

Fibrinogen is central to the formation of thrombus. Alteration to the fibrinogen structure and function, changes to fibrin polymerisation, formation of heterogeneous clot structure due to the formation of clustered fibrin fibres might all add up to the formation of an unstable thrombus. Our findings indicate that filtered PM which contained soluble components (such as metal ions) as well as UFP particles of less than 0.22 μm in diameter were capable of causing alterations to fibrin network properties, with changes in fibre density, network structure and clot permeability . However, independent effects of UFP have not been studied. Some of these changes may contribute to the observed health effects of airborne PM on cardiovascular function.

Acknowledgements
Sofian Metassan is funded by a grant from the Sultanate of Brunei.

References

[1] Peters A, Dockery DW, Muller JE & Mittleman MA (2001) Increased particulate air pollution and the triggering of myocardial infarction. *Circulation, 103*, 2810-2815.

[2] Pope CA III, Burnett RT, Thurston GD, Thun MJ, Calle EE, Krewski D & Godleski JJ (2004) Cardiovascular mortality and long-term exposure to particulate air pollution: epidemiological evidence of general pathophysiological pathways of disease. *Circulation, 109*, 71-77.

[3] Pope CA III, Muhlestein JB, May HT, Renlund DG, Anderson JL & Horne BD (2006) Ischemic heart disease events triggered by short-term exposure to fine particulate air pollution. *Circulation, 114*, 2443-2448.

[4] Peters A, von Klot S, Heier M, Trentinaglia I, Hormann A, Wichmann HE & Lowell H (2004) Exposure to traffic and the onset of myocardial infarction. *N Engl J Med, 351*, 1721-1730.

[5] Miller KA, Siscovick DS, Sheppard L, Shepherd K, Sullivan JH & Anderson GL (2007) Long term exposure to air pollution and incidence of cardiovascular events in women. *N Engl J Med, 356,* 458.

[6] Toren K, Bergdahl IA, Nilsson TK & Jarvholm B (2007) Occupational exposure to particulate air pollution and mortality due to ischemic heart disease and cerebrovascular disease. *Occup Environ Med, 64,* 515-519.

[7] Geiser M, Rothen-Rutishauser B, Kapp N, Schurch S, Kreyling W, Schulz H , Semmler M, Hof VI, Heyder J & Gehr P (2005) Ultrafine particles cross cellular membranes by nonphagocytic mechanisms in lungs and cultured cells. *Environ Health Perspect, 113*, 1555-1560.

[8] Nemmar A, Hoylaerts MF, Hoet PHM, Dinsdale D, Smith T, Xu, HY, Vermylen J & Nemery B (2002) Ultrafine particles affect experimental thrombosis in an in vivo hamster model. *Am J Respir Crit Care Med, 166*, 998-1004.

[9] Nemmar A, Vanbilloen H, Hoylaerts MF, Hoet PH, Verbruggen A & Nemery B (2001) Passage of intratracheally instilled ultrafine particles from the lung into the systemic circulation in hamster. *Am J Respir Crit Care Med, 164,* 1665-1668.

[10] Ajjan R & Grant PJ (2006) Coagulation and atherothrombotic disease. *Atherosclerosis, 186*, 240-259.

[11] Lim BCB, Ariens RAS, Carter AM, Weisel JW & Grant PJ (2003) Genetic regulation of fibrin structure and function: complex gene-environment interactions may modulate vascular risk. *Lancet, 361,* 1424-1431.

[12] Standeven KF, Ariens RA & Grant PJ (2005) The molecular physiology and pathology of fibrin structure/function. *Blood Rev, 19,* 275-288.

[13] Fatah K, Silveira A, Tornvall P, Karpe F, Blombäck M & Hamsten A (1996) Proneness to

formation of tight and rigid fibrin gel structures in men with myocardial infarction at a young age. *Thromb Haemost, 76,* 535-540.

[14] Collet JP, Allali Y, Lesty C, Tanguy ML, Silvain J, Ankri A, Blanchett B, Dumaine R, Gianetti J, Payot, L, Weisel JW & Montalescot G (2006) Altered fibrin architecture is associated with hypofibrinolysis and premature coronary artherothrombosis. *Arterioscler Thromb Vasc Biol, 26,* 2567-2573.

Journal of Physics: Conference Series **151** (2009) 012030 doi:10.1088/1742-6596/151/1/012030

Imaging carbon nanoparticles and related cytotoxicity

C Cheng[1], A E Porter[1], K Muller[2], K Koziol[3], J N Skepper[2], P Midgley[3] and M Welland[1]

[1]The Nanoscience Centre, University of Cambridge, 11 JJ Thompson Ave, Cambridge CB3 OFF, United Kingdom, [2] Multi-imaging Centre, Dept of Physiology, Development and Neuroscience, Anatomy Building, University of Cambridge, Downing St, Cambridge, CB2 3DY, United Kingdom, [3]Dept of Materials Science and Metallurgy, University of Cambridge, Pembroke St, Cambridge, CB2 3QZ, UK

Email: mew10@cam.ac.uk

Abstract. Carbon-based nanoparticles have attracted significant attention due to their unique physical, chemical, and electrical properties. Numerous studies have been published on carbon nanoparticle toxicity; however, the results remain contradictory. An ideal approach is to combine a cell viability assay with nanometer scale imaging to elucidate the detailed physiological and structural effects of cellular exposure to nanoparticles. We have developed and applied a combination of advanced microscopy techniques to image carbon nanoparticles within cells. Specifically, we have used EFTEM, HAADF-STEM, and tomography and confocal microscopy to generate 3-D images enabling determination of nanoparticle spatial distribution in a cell. With these techniques, we can differentiate between the carbon nanoparticles and the cell in both stained and unstained sections. We found carbon nanoparticles (C_{60}, single-walled carbon nanotubes (SWNT), and multi-walled carbon nanotubes (MWNT)) within the cytoplasm, lysosomes, and nucleus of human monocyte-derived macrophage cells (HMM). C_{60} aggregated along the plasma and nuclear membrane while MWNTs and SWNTs were seen penetrating the plasma and nuclear membranes. Both the Neutral Red (NR) assay and ultra-stuctural analysis showed an increase in cell death after exposure to MWNTs and SWNTs. SWNTs were more toxic than MWNTs. For both MWNTs and SWNTs, we correlated uptake of the nanoparticles with a significant increase in necrosis. In conclusion, high resolution imaging studies provide us with significant insight into the localised interactions between carbon nanoparticles and cells. Viability assays alone only provide a broad toxicological picture of nanoparticle effects on cells whereas the high resolution images associate the spatial distributions of the nanoparticles within the cell with increased incidence of necrosis. This combined approach will enable us to probe the mechanisms of particle uptake and subsequent chemical changes within the cell, essential for identifying the toxicological profiles of carbon nanoparticles.

1. Introduction

Carbon-based nanoparticles (CNP) have attracted significant attention due to their unique physical, chemical, and electrical properties. Of specific interest are C60 fullerenes, single-walled (SWNT), and multi-walled carbon nanotubes (MWNT). Potential applications are extending into everyday settings: C60 as structural reinforcers [1], building blocks in photovoltaics and optoelectronics [2], chemotherapeutic agents [3], antioxidants, enzyme inhibitors, and imaging contrast agents [4]; SWNTs as field-emission transistors [5], imaging probes [6], gene and drug delivery vehicles [7][8], and anti-tumour agents[9]; and MWNTs as fillers in composite materials to improve structural and electrical properties [10], sensors [11], microcatheters [12], tissues engineers [13], and drug carriers [14]. Consequently, attention has focused on their potential toxicity to human health.

Numerous studies have been published on CNP toxicity; however, the results remain contradictory. Table 1 highlights some of the recent toxicological findings for C60s, SWNTs, and MWNTs. Generally, C60s and SWNTs have been found to exhibit both cytotoxic and non-cytotoxic responses. Some studies have suggested unpurified SWNTs are more toxic than MWNTs and C60 [15] while others have found no inflammation [16] and fast systemic excretion [17]. MWNTs seem fairly cytotoxic, generating free radicals [18], inflammation [19], and DNA damage [20]. Behaving similarly to asbestos, longer nanotubes can cause frustrated phagocytosis, where digestive enzymes can leak from the phagosomes of incompletely phagocytosed nanoparticles [21], while shorter ones can yield little to no toxicity [22]. No clear uptake mechanism has been identified; however, some have suggested a mechanical nanoneedle piercing of the cell membrane [7].

Clearly, cell viability assays are not sufficient tools alone for assessing CNP toxicity as they are subject to variability and only reveal broad toxicological effects. Similarly, imaging alone provides a visual depiction that can be subjectively interpreted positively or negatively. An ideal approach is to combine the cell viability assays with nanometer scale imaging to elucidate the detailed physiological and structural effects of cellular exposure to nanoparticles. Imaging, however, remains a challenge due to the difficulty of attaining high spatial resolution with sufficient contrast to distinguish between nanoparticles and various cell organelles.

Transmission electron microscopy (TEM) is invaluable for imaging such small particles; however, the traditional techniques developed for inorganic imaging limit what can be seen when applied to organic systems. Standard bright-field imaging yields little contrast amongst cell structures; consequently, heavy stains are used. Yet potential problems arise when differentiating between the stains and nanoparticles. Additionally, conventional sample thicknesses of 70 nm or greater can cause misinformation as 3D data is projected into a 2D image [23].

We have developed and applied a combination of advanced microscopy techniques that address these issues and allow us to image CNPs within cells. One technique is energy filtered TEM (EFTEM), where images are created by filtering out inelastically-scattered electrons. Cell structures and CNPs have different energy loss spectrums, and as stains may alter these spectrums, EFTEM is ideal for mapping nanoparticles in cells unstained. Furthermore, electron energy loss spectroscopy (EELS) can be used to identify CNPs [24].

Another technique is high angle annular dark field scanning TEM (HAADF-STEM), where electrons scattered to high angles by differing carbon atom densities provide image contrast. As the intensity of scattering is proportional to the mass-density, HAADF-STEM is sensitive to the atomic number and is ideal for imaging SWNTs and MWNTs because of the stronger scattering from the iron catalysts. HAADF-STEM can be further combined with tomography to generate 3-D images enabling determination of nanoparticle spatial distribution in a cell without the effect of chromatic aberration as described in Porter et al [24].

As TEM is a non-viable technique, confocal microscopy offers the advantage of generating 3-D images from living cells. Cells are stained to visualise the plasma membrane and nucleus, while the CNPs are seen via reflectance from either the metal catalysts or post-filled heavy metals [25].

In this paper, we present data of C60, SWNT, and MWNT effects on human macrophage cells from cell viability assays and various imaging techniques in order to understand the potential toxic effects of CNPs on cells. We will also highlight the limitations of each tool when used alone and the benefits when combined.

Table 1. Summary of recent toxicity studies for C60, SWNTs, and MWNTs.

	C60	TOX[a]	SIZE[b]	MODEL	CITATION[c]
Toxicity	Inhibited cell proliferation and differentiation	+	Not specified	Mice midbrain micromass	Tsuchiya T 1996 *FEBS Lett.* **393** 139-45
	Water derived nano-C_{60} caused cell membrane damage and generated O^{2-}	+	60nm	Human dermal fibroblast & liver carcinoma	Sayes CM 2004 *Nano. Lett.* **4** 1881-7
	Caused no lethal, acute, or subacute toxicity; produced hypertrophy & hyperplasia but no fibrosis	-	200-1000 nm	Male Wistar rats	Gharbi N 2005 *Nano. Lett.* **5** 2578-85
	Pristine C_{60}: rapid ROS-dependent, caspase-independent necrotic cell death; $C_{60}(OH)_n$: delayed, ROS-independent, caspase-mediated apoptosis	+	96.3nm	Mouse fibrosarcoma, rat & human glioma	Isakovic A 2006 *Toxicol. Sci.* **91** 173-83
	Water derived nano-C_{60} & $C_{60}(OH)_{24}$ yielded little to no toxic difference compared to control	-	160±50 nm	Rats: intratracheal instillation	Sayes CM 2007 *Nano. Lett.* **8** 2399-406
	Minimal toxicity: no gross or microscopic lesions; C_{60} nanoparticle lung burden greater than microparticle; lung half-lives = 26-29 days	-	C_{60} nano & micro particles	Male rats: aerosol inhalation	Baker GL 2008 *Toxicol. Sci.* **101** 122-31
Uptake	Caused lipid peroxidation; damaged fish brain cell membranes; increased detoxification enzymes (ie. GSH, P450)	+	30-100nm	Juvenile Largemouth Bass	Oberdorster E 2004 *Environ. Health Perspect.* **112** 1058-62
	Stongly bound to single-strand DNA and minor grooves of double-stranded DNA		~1nm	Simulation	Zhao X 2005 *Biophys. J.* **89** 3856-62
	Found C60 along plasma and nuclear membranes and within cytoplasm, lysosomes, and nucleus		900±100 nm	Human macrophages	Porter AE 2006 *Acta Biomater.* **2** 409-19
	Pristine C_{60} entered lipid bilayer easily and remained inside creating micropores; functionalised C_{60} ($C_{60}(OH_{20})$) attached to bilayer surface only		1.2 nm	Simulation	Qiao R 2007 *Nano. Lett.* **7** 614-9

	SWNT	TOX[a]	LENGTH[d]	MODEL	CITATION[c]
Toxicity	Altered nucleus, mitochondria, tonofilaments, organelles, F-actin	+	Not specified	Human keratinocyte	Shvedova AA 2003 *Free Rad. Oxid. Stress.* 91-103
	Inhibit cell proliferation, increased apoptosis genes, and decreased cell adhesion time and dose dependently; caused chromatin condensation, nucleosomal DNA fragmentation	+	Not specified	Human embryo kidney	Cui D 2005 *Toxicol. Lett.* **155** 73-85
	More toxic than MWNTs and C_{60}	+	~1μm	Rat alveolar macrophages	Jia G 2005 *Environ. Sci. Technol.* **39** 1378-83
Inflammation	Thickened alveolar walls, formed granulomas, increased inflammation	+	Not specified	Mouse: pharyngeal aspiration	Shvedova AA 2005 *Am. J. Physiol. Lung Cell Mol. Physiol.* **289** L698-708
	No O^{2-} or NO^-, ingestion, nor apoptosis	-		Mouse macrophages	
	Caused greatest platelet aggregation (70%) vs MWNTs and C_{60}	+	Not specified	Human platelets	Radomski A 2005 *Brit. J. Pharm.* **146** 882-93
Iron	Both unpurified and purified did not generate O^{2-} or NO^-	-	Not specified	Rat macrophages	Kagan VE 2006 *Toxicol. Lett.* **165** 88-100
	Induced ROS at higher extent than asbestos	+	0.5-1μm	Human mesothelioma	Wick P 2006 *Phys. Stat. Sol.* **243** 3556-60
	Dose dependent ROS generation corresponds to Fe content	+	Not specified	Mouse keratinocyte	Grabinski C 2007 *Carbon* **45** 2828-35

Length	Estimated threshold length of uptake at 189±17 nm		189±17 nm	Human lung fibroblasts	Becker ML 2007 *Adv. Mater.* **19** 939-45
Functionalise	Less toxic than unfunctionalised; distributed through body quickly, blood half-life = 3.5 hours	-	Not specified	Male KM mouse	Wang HF 2004 *J. Nanosci. Nanotechnol.* **4** 1019–24
Uptake	Entered via active-uptake (phagocytosis)		~1μm	Mouse peritoneal macrophage	Cherukuri P 2004 *J. Am. Chem. Soc.* **126** 15638-9
	Functionalised not enter via endocytosis		300-1000 nm	Human & mouse fibroblast	Pantarotto D 2004 *Chem. Commun.* **1** 16-17
	Found in cytoplasm, lysosomes, and nucleus with dose dependent toxicity		Not specified	Human macrophages	Porter AE 2007 *Nature Nanotech.* **2** 713-7

	MWNT	**TOX**[a]	**LENGTH**[d]	**MODEL**	**CITATION**[c]
Toxcity	Oxidized more toxic than pristine; time and dose dependent	+	1-5μm	Human T lymphocyte	Bottini M 2006 *Toxicol. Lett.* **160** 121-6
	Agglomerates more toxic than well-dispersed	+	Not specified	Human mesothelioma	Wick P 2007 *Toxicol. Lett.* **168** 121-31
Inflammation	Increased IL-8, macrophages, lymphocytes, neutrophils; multifocal granuloma around NTs; mild to no peribronchial fibrosis	+	Not specified	Guinea pigs: intratracheal instillation	Grubek-Jaworska H 2006 Carbon **44** 1057-63
	DNA damage by unpurified and purified; dose dependent	+	Not specified	Mouse stem embryonic	Zhu L 2007 *Nano. Lett.* **7** 3592-7
Length	Unground (5.9μm) harder to clear than ground (0.7μm); both increased inflammation	+	0.7um 5.9um	Sprague-Dawley rats	Muller J 2005 *Tox. Appl. Pharmacol.* **207** 221-31
	Ground (not unground) caused dose-dependent toxicity and increased TNF-α	+		Rat peritoneal macrophage	
	220nm caused less inflammation than 825nm; 220nm inflammation disappeared after 4 wks while 825nm remained same throughout	- +	220nm 825nm	Human acute monocytic leukemia	Sato Y 2005 *Mol. BioSyst.* **1** 176-82
Functionalise	Glucosamine functionalisation: No renal or severe acute toxicity; blood half-life longer than SWNTs	-	Tons of μm	Kunming mice	Guo J 2007 *Nucl. Med. Biol.* **34** 579-83
	Functionalising with carbonyl (C=O), carboxyl (COOH), & hydroxyl (OH) increases toxicity	+	160-180nm	Lung cancer H596	Magrez A 2006 *Nano. Lett.* **6** 1121-5
Iron	Dose dependent ROS generation corresponds to Fe content	+	Not specified	Mouse keratinocyte	Grabinski C 2007 *Carbon* **45** 2828-35
Uptake	Penetrated cell by 2 hour, internalized by 6 and 24 hours		70μm	Mouse microglyia	Kateb B 2007 *NeuroImage* **37** S9-S17
	Found in vacuoles, cytoplasm, close to nucleus, pierce nuclear membrane		< 3.6μm	Human epidermal keratinocytes	Monteiro-Riviere NA 2005 *Tox. Lett.* **155** 377-84
	Observed frustrated phagocytosis with dose dependent TNF-α and O^{2-} production	+	> 50μm	Human peripheral mononuclear	Brown DM 2007 *Carbon* **45** 1743-56
	Nanoneedle hypothesis, not via energy-dependent process; crossed membrane with no cell death	-	20nm 200nm	HeLa	Pantarotto D 2004 *Angew. Chem. Int. Ed.* **43** 5242-6

[a] Quick assessment of toxic potential: '+' exhibits toxicity, '-' exhibits little to no toxicity
[b] Average particle size in diameter
[c] Only the first author is listed
[d] Average length

2. Materials and Methods

2.1. CNPs

C60s were purchased from Aldrich, Dorset, UK, and were characterised via TEM. SWNTs produced via the HiPco process were purchased from Carbon Nanotechnologies, Inc., and were characterised by high-resolution TEM and EELS. MWNTs were synthesised via chemical vapour deposition (CVD) from a solution of 5.6 wt% ferrocene in toluene injected into a furnace at 760 °C under an argon atmosphere; they were characterised by scanning electron microscopy (SEM), TEM, energy dispersive x-ray microanalysis (EDX), and inductively coupled plasma atomic emission spectroscopy (ICP-AES).

2.2. In vitro study

2.2.1. *Cell culture.* Mature human monocyte-derived macrophage cells (HMM) were isolated from human buffy coat residues (National Blood Transfusion Service, Brentwood, UK). Buffy coat residues were washed once with phosphate-buffered saline (PBS), and the resulting cell sediment was mixed with an equal volume of fresh PBS. Thirty milliliters of diluted buffy coat residue were layered onto 15 ml of LymphoPrep (Axis-Shields, Oslo, Norway), and, after centrifugation at 20 °C for 30 min at 600g, the opaque interphase of mononuclear cells was removed and washed three times with PBS containing 4 mg/ml bovine serum albumin (BSA) to remove platelets. Monocytes were then enriched by an additional centrifugation step in a Percoll gradient (24). Mononuclear cells were re-suspended in 4 ml of PBS and mixed with 8 ml of Percoll: Hanks' balanced salt solution (10× concentrate) (6:1, at pH 7.0). After centrifugation at 20 °C for 30 min at 450g, the monocytes were collected from the top of the gradient, washed in PBS/BSA, and seeded in 24 and 48-well tissue culture plates at 1–2×10^6 cells/well and $0.5–1 \times 10^6$ cells/well respectively using macrophage serum-free medium (Mϕ-SFM, Invitrogen) unless otherwise stated. After incubation for 1 h at 37 °C, any remaining non-adherent cells were removed by washing twice with PBS. Adherent monocytes were cultured at 37 °C in humidified air/5% CO_2 using Mϕ-SFM for at least 6–7 days prior to experiments unless stated otherwise, renewing the culture medium twice a week.

2.2.2. *CNP treatment.* C60 was dispersed in 10ml of freshly distilled, filtered, and de-gassed tetrahydrofuran (THF) and left in a vial with a screw cap to stir for 24 hours at ambient temperature; when added to cells, the final THF concentration was < 0.1% THF in Mϕ-SFM. SWNTs were dispersed in 0.1% of THF and Mϕ-SFM and water bath sonicated for 30 min. MWNTs were first sterilized in ethanol then dispersed in a final volume concentration of 250 μg/ml of BSA in Mϕ-SFM; the solution was sonicated in a water bath for ~1 hour. Concentration for imaging was 5 μg/ml for all CNPs unless otherwise stated.

2.2.3. *Cell assays.* Cell viability was assessed via Neutral Red (NR), which measured lysosomal accumulation of NR dye in viable cells. For each treatment, three experiments each in triplicate were performed and statistically analysed using one-way ANOVA and least significance difference (LSD) with significance set at $p < 0.05$.

2.3. Imaging

2.3.1. *Electron microscopy.* For electron microscopy studies, samples were fixed in 4% gluteraldehyde in piperazine-1,4-bis(2-ethanesulphonic acid) (PIPES) buffer, dehydrated, embedded, and sectioned with an ultramicrotome at 40 nm for EELS and high-resolution STEM (HR-STEM), 70 nm for EFTEM studies, and 300 nm for 3D electron tomography. TEM sections were deposited on lacy carbon film grids and bare 600-mesh copper TEM grids. Selected sections were bulk stained with osmium tetroxide and post-stained with uranyl acetate and lead citrate for 5 min in each to enhance contrast from cell. For freeze-drying, HMMs seeded and treated on formvar coated gold grids were cryo-fixed by rapid immersion into liquid propane cooled by liquid nitrogen. Samples were freeze-dried by raising the temperature to room temperature over 24 hours at a pressure of $5x10^{-6}$ bar [26].

EFTEM studies were performed on a Philips CM300 operated at 300 kV with a Gatan imaging filter (GIF) model 2002 using 10μm objective aperture. To optimize contrast, zero-loss images of unstained sections were taken using a 3eV slit while stained sections were taken using a 20eV slit. Low-loss energy filtered series were recorded from 0 to 30 eV using a 2 eV slit and 2 eV step size. EELS was extracted from the EFTEM series using IDL image processing software. HAADF-STEM of stained sections was performed on an FEI Tecnai F20 operated at 200 kV using a 30 mm condenser aperture and a camera length of 150-200 mm. EFTEM tomography was performed on the stained and unstained 150 nm thick sections and HAADF-STEM tomography on unstained freeze-dried whole cells. An image series was acquired using the FEI Tecnai F20 operated at 200kV. Datasets were acquired over a tilt range of –68° to +56° (EFTEM) and –70° to +70° (STEM) using a step size of 2°. 3D reconstruction was carried out using weighted back-projection [27] and a simultaneous iterative reconstruction technique (SIRT) [28].

Cell death was also quantitatively differentiated between necrosis and apoptosis via morphology [29]. Necrotic cells were characterised as electron-translucent with cytoplasmic and nuclear contents appearing to be leached out while apoptotic cells appeared shrunken with heavily capped chromatin. 100 cells were indexed for each group.

2.3.2. *Confocal microscopy.* HMMs were seeded and treated on coverslips and imaged on a Leica TCS SP2 confocal microscope. Hoerchst and FM-464 stained the nuclei and plasma membrane, respectively. CNPs were visualised via reflectance from the iron catalyst in the MWNTs and the AgI filled in the SWNTs. Z-plane stacks were taken to create a 3-D image.

3. Results

3.1. Characterisation

Average diameter of C60 clusters was 900±100 nm. HiPco-produced SWNTs have diameters from 0.5 to 3 nm with ~10 wt% ash content [30]. MWNTs had an average diameter of 68 nm and length of 26 μm with ~5 wt% Fe_2O_3.

3.2. Cell viability

No alteration in cell viability was found in the THF vehicle control; none was performed for BSA due to its non-toxic nature. At 2 days, the CNPs exhibited little or inconsistent toxicity: C60 was not toxic, SWNT was significantly toxic at 0.31, 1.25, 2.5, and 5 μg/ml, and MWNT at 1.25 and 2.5 μg/ml. At 4 days, the SWNTs were significantly toxic at 0.31, 0.625, and concentrations greater than 2.5 μg/ml, while MWNTs at 10 μg/ml (Figure 1).

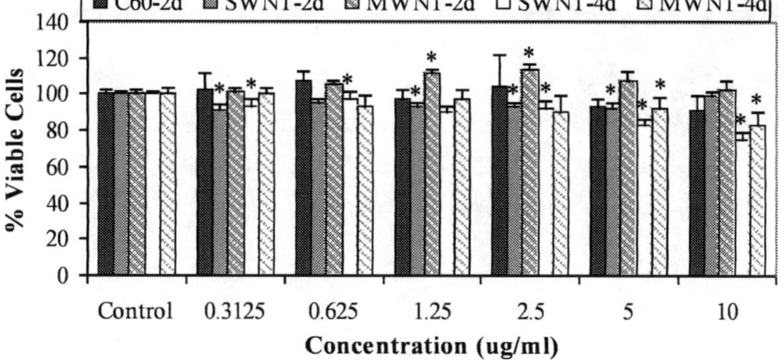

Figure 1. Toxicity of CNPs after 2 and 4 day treatment as measured by NR assay. Values represent mean ± SE of three experiments each performed in triplicates (ANOVA with LSD, $p < 0.05$). Significance within each sample labelled with asterisk.

Journal of Physics: Conference Series **151** (2009) 012030 doi:10.1088/1742-6596/151/1/012030

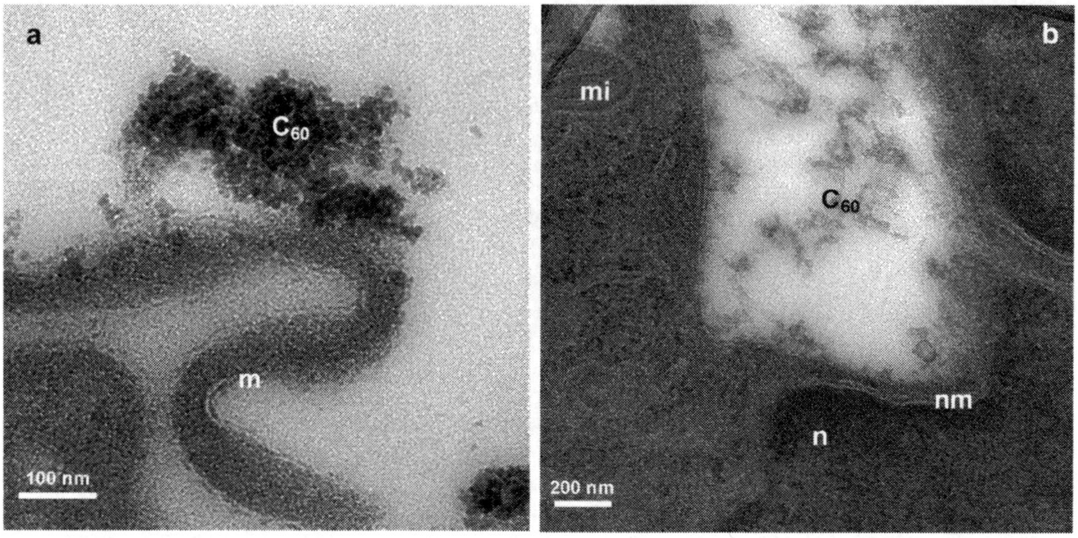

Figure 2. Zero-loss EFTEM of C60 in HMMs: (a) C60 clusters along plasma membrane, (b) C60s inside a lysosome. (m = membrane, mi = mitochondria, n = nucleus, nm = nuclear membrane)

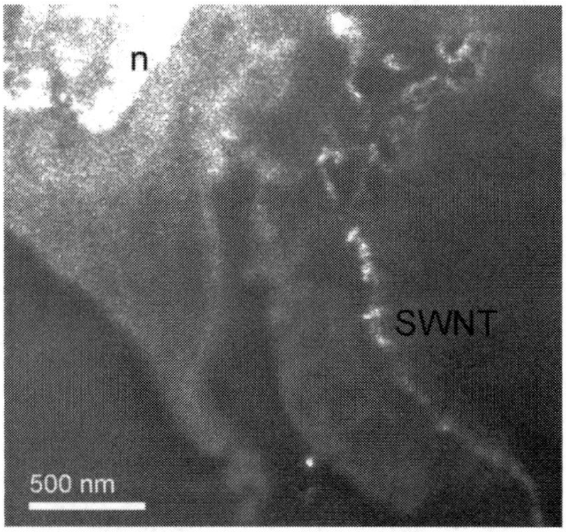

Figure 3. Zero-loss dark-field EFTEM of SWNT aligned along plasma membrane in the cytoplasm. (n = nucleus)

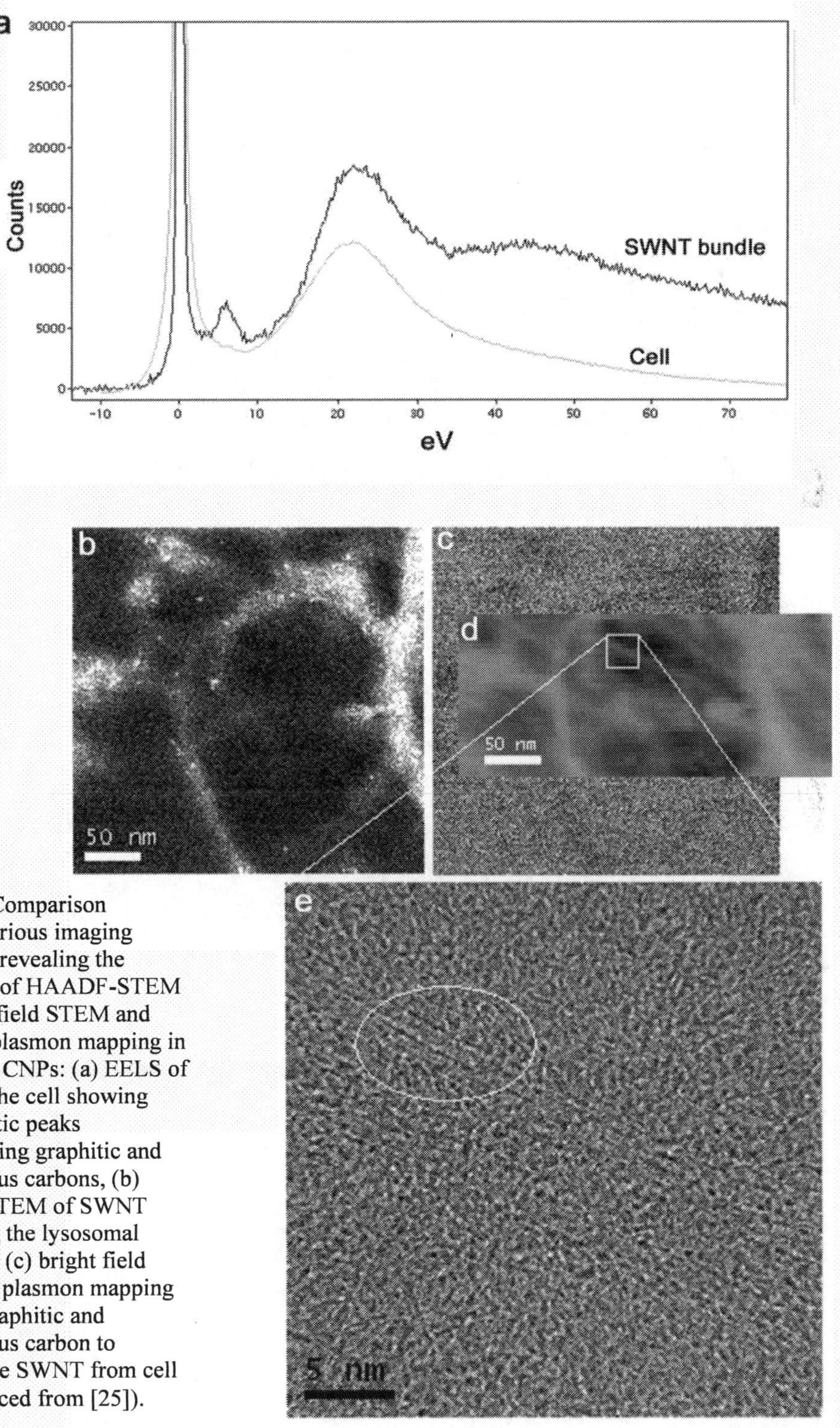

Figure 4. Comparison between various imaging techniques revealing the limitations of HAADF-STEM and bright field STEM and benefit of plasmon mapping in identifying CNPs: (a) EELS of SWNT in the cell showing characteristic peaks differentiating graphitic and amorphorous carbons, (b) HAADF-STEM of SWNT penetrating the lysosomal membrane, (c) bright field STEM, (d) plasmon mapping between graphitic and amorphorous carbon to differentiate SWNT from cell (e, reproduced from [25]).

3.2.1.

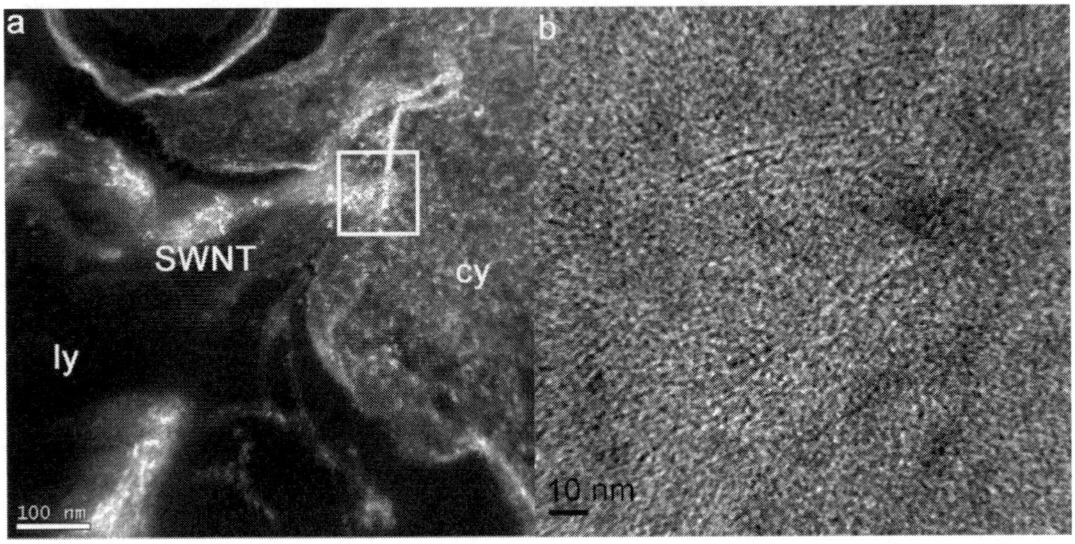

Figure 5. HAADF-STEM of SWNT: (a) in the lysosome and pushing up against lysosomal membrane, (b) bright field STEM of boxed region from (a) showing SWNTs. (ly = lysosome, cy = cytoplasm)

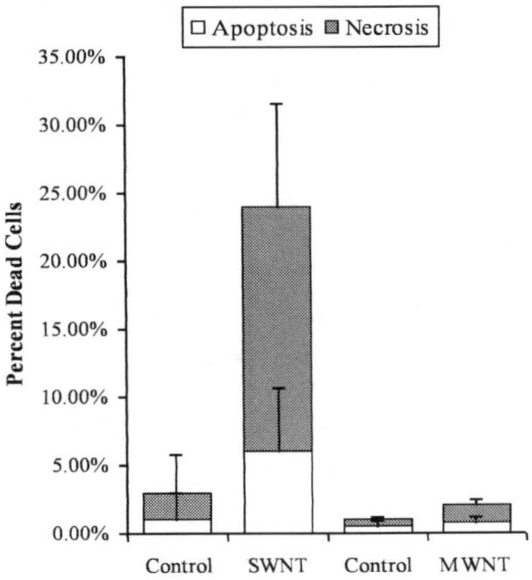

Figure 6. TEM analysis of cells treated with SWNTs and MWNTs at 5 μg/ml for 4 days showing proportion of necrotic to apoptotic cells. Values represent mean ± SE (ANOVA with LSD, $p < 0.05$).

3.3. Imaging

3.3.1. *2D.* With bright-field TEM, cells were stained with lead citrate and uranyl acetate stains in order obtain the contrast needed to visualise mitochondria, ribsomes, and nuclei. CNPs were found within cells; however, it was difficult to determine whether the CNPs were inside or on the surface. With zero-loss EFTEM, after 2 days, C60 clusters were found aligned along the plasma membrane Figure 2a) and inside lysosomes and nuclei (Figure 2b). SWNTs were found in lysosomes after 2 days, while after 4 days they were found with the long axes aligned parallel to the plasma membrane translocating the lysosomal membrane (Figure 3). MWNTs were found translocating the plasma membrane into the cytoplasm and, at times, nuclei after 4 days. Low-loss EELS confirmed identification of the CNPs (Figure 4). HAADF-STEM images of stained cells were taken for SWNTs and MWNTs to further compare CNP internalisation and cell morphology alterations. SWNTs were found in lysosomes and nuclei (Figure 5). Cells remained healthy after 2 days; however, increased cell mortality was correlated with higher SWNT and MWNT concentration. Furthermore, cell death via necrosis or apoptosis were differentiated and found that both SWNTs and MWNTs were correlated with greater necrotic death (Figure 6).

3.3.2. *3D.* Zero-loss EFTEM tomography of C60s confirmed alignment along the plasma and nuclear membrane as well as localisation in secondary lysosomes and the nucleus; HAADF-STEM tomography of a freeze-dried whole cell revealed C60 alignment along the nuclear membrane but no nuclear internalisation (Figure 7). EFTEM tomography of SWNT-treated cells confirmed alignment along and penetration of the nuclear membrane (Figure 8). HAADF-STEM of a whole cell revealed MWNT internalisation (Figure 9); confocal microscopy further confirmed membrane translocation into the cytoplasm and nucleus with live cells (Figure 10).

4. Discussion

After 2 days incubation of HMMs with C60s, SWNTs, and MWNTs, the NR assay revealed inconsistent toxicity patterns which may result from assay variability. Increased viability values may be due to increased dye uptake from lysosomal distension by particle accumulation or adsorption of the dye to the CNPs causing false increased viability [31]. At 4 days, results were more consistent with SWNT toxicity increasing significantly at concentrations greater than 2.5 μg/ml. The NR assay is widely used for its fast, inexpensive, and simple procedure for screening viability in large number of samples; however, the assay is subject to variability and does not discriminate between the routes of cell death. As a result, we wanted to use imaging to further understand the relationship between CNP localisation and cell mortality.

In traditional TEM, heavy metal stains are required to generate the contrast to differentiate various cell components; however, these metal salts can obscure CNP identification. As a result, we pursued EFTEM using the low loss region of the energy-loss spectrum because: (i) acquisition times and beam damage could be minimized from the considerably lower energy-scattering magnitudes for core-loss, and (ii) the ($\pi + \sigma$) volume plasmon excitation which dominates the low-loss spectrum is sensitive to any slight differences in the electronic structure of the carbon species. Graphite-like carbonaceous material such as C60 and SWNTs exhibit a bulk plasmon energy at ~26 eV while amorphorous carbonaceous material such as those in a cell are ~23 eV. Furthermore, graphitic carbon has an interband $\pi\rightarrow\pi^*$ transition at 6 eV which is weak to non-existent in cell carbon material. Using an energy filter slit of 2 eV, images surrounding 20 eV and 26 eV could be recorded and an image ratio of 26 / 20 eV used to generate images that differentiated CNPs from cell material [32]. As graphitic carbon material has the extra 6 eV $\pi\rightarrow\pi^*$ transition, EELS showing 2 peaks—6 and 26 eV—could confirm CNP identification.

Two-dimensional zero-loss EFTEM can be extended to tomography for 3D images; similar ratio images taken at every tilt can then be reconstructed. EFTEM tomography can be further combined with EELS to characterise CNPs; however, the volume information is limited by a specimen thickness of less than 100 nm. HAADF-STEM tomography allows imaging of thicker specimens or even whole cells up to 300 nm thick due to no post-specimen imaging lens, but CNPs cannot be characterised. Confocal microscopy allows imaging of live whole cells with 3D image reconstruction; however, resolution limits visualisation of smaller CNPs and CNPs cannot be characterized. With such differing benefits, we used all three in parallel to determine and confirm CNP localisation in cells.

We were able to confirm the presence of CNPs within the cell cytoplasm, lysosomes, and nuclei. C60s were found clustered along the plasma and nuclear membranes, and SWNTs fused along the plasma membrane; this supports the implications that C60s and SWNTs can cause lipid peroxidation [33][34][35]. Furthermore, peroxidation of phagosomal and lysosomal membranes could release digestive enzymes into the cell itself and result in cell mortality.

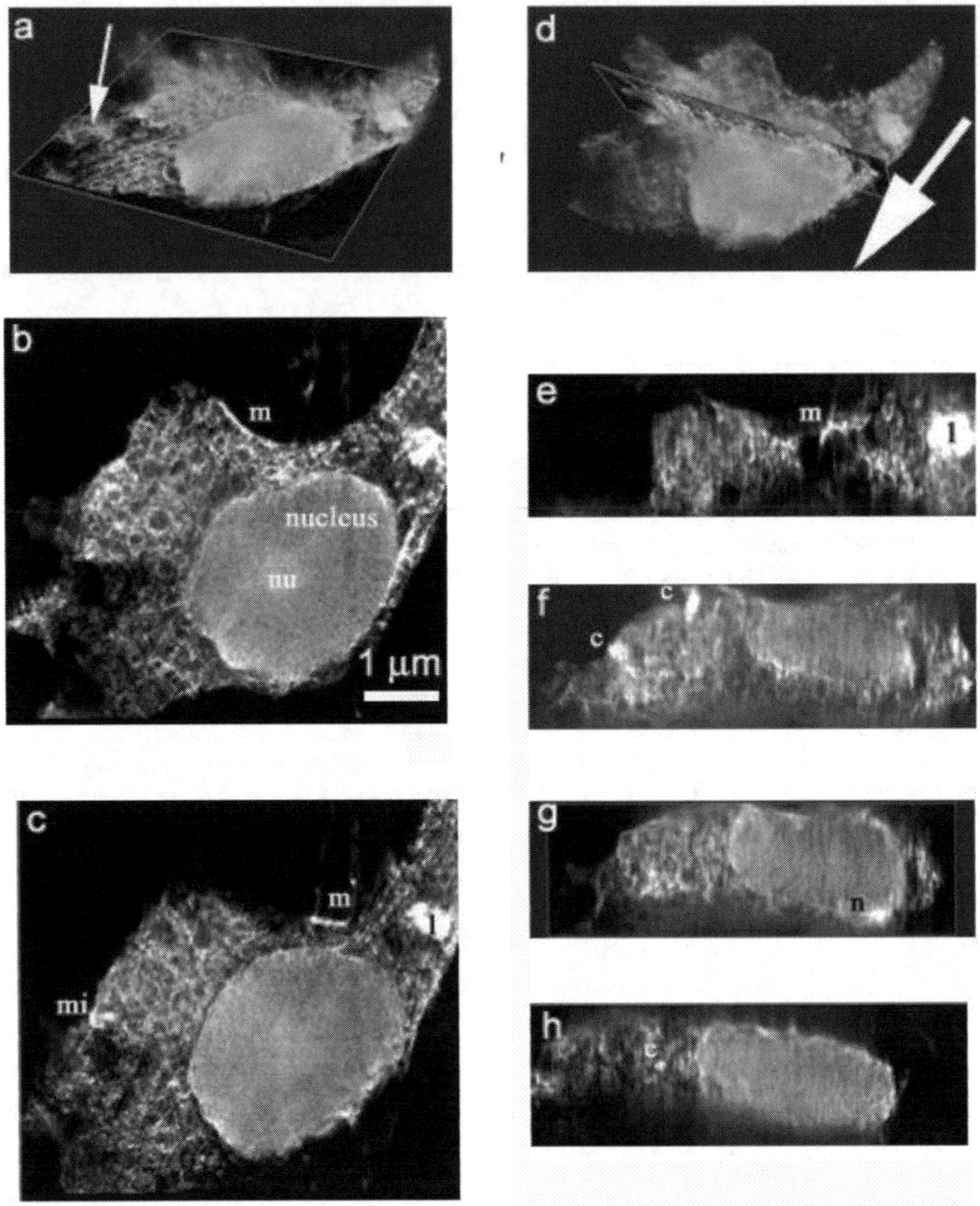

Figure 7. HAADF-STEM tomography of a freeze-dried cell treated with C60 (bright spots in b-h): (a, d) voltex reconstructions with a horizontal (a) and vertical slice (d) bisecting the reconstruction (arrow indicates direction perpendicular to orthoslice), (b-c) horizontal slices at different heights through the reconstruction – (e-h) vertical slices through the reconstruction (m = membrane, nu = nucleus, l = lysosome, c = cytoplasm, mi = mitochondria); reprinted with permission from [24]

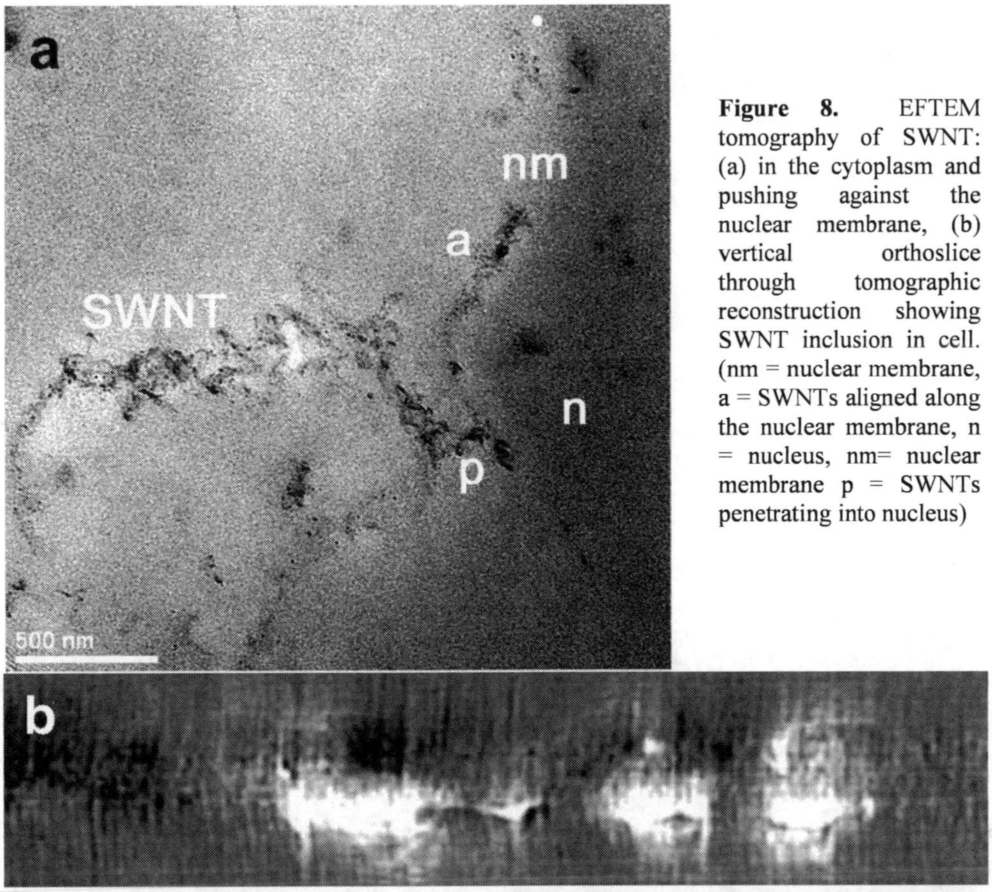

Figure 8. EFTEM tomography of SWNT: (a) in the cytoplasm and pushing against the nuclear membrane, (b) vertical orthoslice through tomographic reconstruction showing SWNT inclusion in cell. (nm = nuclear membrane, a = SWNTs aligned along the nuclear membrane, n = nucleus, nm= nuclear membrane p = SWNTs penetrating into nucleus)

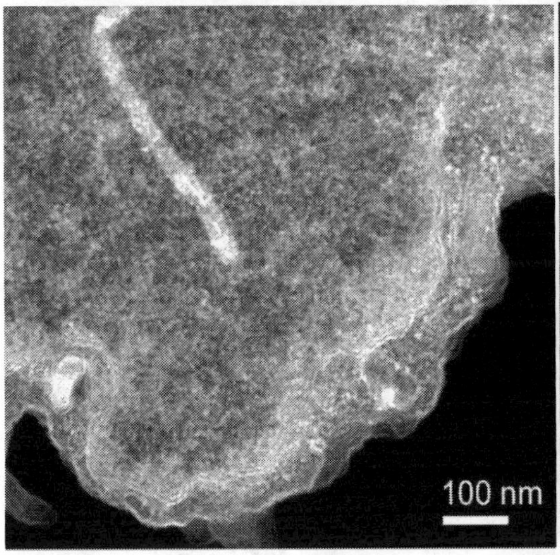

Figure 9. HAADF-STEM tomography projection of a MWNT in the nucleus.

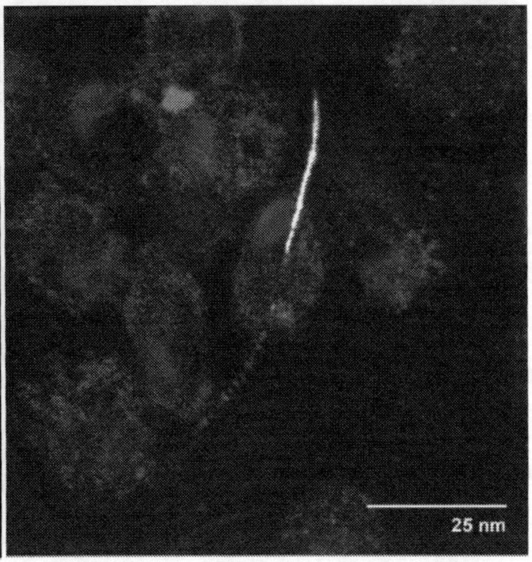

Figure 10. Confocal microscopy 3D reconstruction of a MWNT penetrating the plasma membrane of a living cell.

Some SWNTs and MWNTs were found translocating the plasma, lysosomal, and nuclear membranes supporting the nanoneedle hypothesis as well as suggesting potential cytotoxicity from damaged membrane integrity or undesired fluid and ion exchanges [7]. The similarity between MWNTs and asbestos also suggested the possibility of frustrated phagocytosis [21]. Additionally, CNP presence within the nucleus could potentially cause DNA damage [33][35][36].

We further elucidated CNP effects on cells by directly combining cell viability with imaging. The high resolution of TEM imaging allows visual differentiation between apoptotic and necrotic death. For both SWNTs and MWNTs, we found increased cell death with longer exposure time; additionally, there was a proportionally greater increase in necrotic than apoptotic cells, suggesting a preference for necrotic death from CNP exposure.

5. Conclusion

Clearly, further research is required; however, this study demonstrates that only a combination of cell viability assays with high resolution imaging will help to fully understand the uptake mechanism, internal trafficking, and toxicological profile of CNPs.

References

[1] Zheng S S, Zeng L, Che S L, Deng G Z, Zha C G and Zhang L 2006 Experimental study on high strength and high performance concrete applied to steel reinforced concrete structures *Advances in Structural Engineering: Theory and Applications (Fuzhou and Xiamen, China, 18-21 August 2006)* 1794-1800

[2] Mateo-Alonso A, Guldi D M, Paolucci F and Prato M 2007 Fullerenes: Multitask components in molecular machinery *Angew. Chem. Int. Ed.* **46** 8120-8126

[3] Yamakoshi Y, Umezawa N, Ryu A, Arakane K, Miyata N, Goda Y, Masumizu T and Nagano T 2003 Active Oxygen Species Generated from Photoexcited Fullerene (C60) as Potential Medicines: O2- versus 1O2 **125** 12803-12809

[4] Bosi S, Da Ros T, Spalluto G and Prato M 2003 Fullerene derivatives: an attractive tool for biological applications *Eur. J. Med. Chem.* **38** 913-923

[5] Popov V N 2004 Carbon nanotubes: properties and application *Mat Sci Eng R.* **43** 61-102

[6] Hafner J H, Cheung C L, Woolley A T and Lieber C M 2001 Structural and functional imaging with carbon nanotube AFM probes *Prog. Biophys. Mol. Biol.* **77** 73–110

[7] Pantarotto D, Briand J P, Prato M and Bianco A 2004 Translocation of bioactive peptides across cell membranes by carbon nanotubes *Chem Commun.* 16-17

[8] Li X, Peng Y and Qu X 2006 Carbon nanotubes selective destabilization of duplex and triplex DNA and inducing B–A transition in solution *Nucl. Acids Res.* **34** 3670-3676

[9] Kam N W S, O'Connell M, Wisdom J A and Dai H 2005 Carbon nanotubes as multifunctional biological transporters and near-infrared agents for selective cancer cell destruction *PNAS* **102** 11600-11605

[10] Miaudet P, Badaire S, Maugey M, Derré A, Pichot V, Launois P, Poulin P and Zakri C 2005 Hot-drawing of single and multiwall carbon nanotubes fibers for high toughness and alignment *Nano. Lett.* **5** 2212-2251

[11] Shekar C, Giri R, Malik S K and Srivastava O N 2007 Improved critical current density of MGB2-carbon nanotubes composite *J. Nanosci. Nanotechnol.* **7** 1804-1809

[12] Barone P W, Baik S, Heller D A and Strano M S 2005 Near-infrared optical sensors based on single-walled carbon nanotubes *Nat. Mater.* **4** 86-92

[13] Koyama S, Haniu H, Osaka K, Koyama H, Kuroiwa N, Endo M, Kim Y A and Hayashi T 2006 Medical application of carbon-nanotube-filled nanocomposites: the microcatheter *Small* **2** 1406-1411

[14] Harrison B S and Atala A 2007 Carbon nanotube applications for tissue engineering *Biomat.* **28** 344-353

[15] Jia G, Wang H, Yan L, Wang X, Pei R, Yan T, Zhao Y and Guo X 2005 Cytotoxicity of carbon nanomaterials: single-wall nanotube, multi-wall nanotube, and fullerene *Environ. Sci. Technol.* **39** 1378-1383

[16] Shvedova A A et al 2005 Unusual inflammatory and fibrogenic pulmonary responses to single-walled carbon nanotubes in mice. *Am. J. Physiol. Lung Cell Mol. Physiol.* **289** L698-708

[17] Lacerda L, Bianco A, Prato M and Kostarelos K 2006 Carbon nanotubes as nanomedicines: from toxicology to pharmacology *Adv. Drug. Deliv. Rev.* **58** 1460-1470

[18] Grabinski C, Hussain S, Lafdi K, Braydich-Stolle L and Schlager J 2007 Effect of particle dimension on biocompatibility of carbon nanomaterials Carbon 45 2828-2835

[19] Muller J, Huaux F, Moreau N, Misson P, Heilier J F, Delos M, Arras M, Fonseca A, Nagy J B and Lison D 2005 Respiratory toxicity of multi-wall carbon nanotubes. *Toxicol. Appl. Pharmacol.* **207** 221-231

[20] Zhu L, Chang D W, Dai L and Hong Y 2007 DNA damage induced by multiwalled carbon nanotubes in mouse embryonic stem cells *Nano. Lett.* **7** 3592-3597

[21] Brown D M, Kinloch I A, Bangert U, Windle A H, Walter D M, Walker G S, Scotchford C A, Donaldson K and Stone V 2007 An in vitro study of the potential of carbon nanotubes and nanofibres to induce inflammatory mediators and frustrated phagocytosis *Carbon* **45** 1743-1756

[22] Sato K et al 2005 Influence of length on cytotoxicity of multi-walled carbon nanotubes against human acute monocytic leukemia cell line THP-1 in vitro and subcutaneous tissue of rats in vivo *Mol. Biosyst.* **1** 176-182

[23] Porter A E, Gass M, Muller K, Skepper J, Midgley P and Welland M 2007 Visualizing the uptake of C60 to the cytoplasm and nucleus of human monocyte-derived macrophage cells using energy-filtered transmission electron microscopy and electron tomography *Environ. Sci. Technol.* **41** 3012-3017

[24] Porter A E, Muller K, Skepper J, Midgley P and Welland M 2006 Uptake of C60 by human monocyte macrophages, its localization and implications for toxicity: studied by high resolution electron microscopy and electron tomography *Acta Biomater.* **2** 409-419

[25] Porter A E, Gass M, Muller K, Skepper J, Midgley P and Welland M 2007 Direct imaging of single-walled carbon nanotubes in cells *Nature Nanotech.* **2** 713-717

[26] Warley A and Skepper J N 2000 Long freeze-drying times are not necessary during the preparation of thin sections for X-ray microanalysis *J. Microsc.* **198** 116-123

[27] Reimer L 1994 Energy-filtering transmission electron microscopy. *Springer Ser Opt Sci* **71** 373–375

[28] Kak A C and Slaney M 1988 *Principles of computed tomography imaging* (New York, NY: IEEE)

[29] Hardwick S J, Hegyi L, Clare K, Law N S, Carpenter K H, Mitchinson M J and Skepper J N 1996 Apoptosis in human monocyte-macrophages exposed to oxidized low density lipoprotein. *J Pathol.* **179** 294-302

[30] Hafner J H, Bronikowski M J, Azamian B R, Nikolaev P, Rinzler A G, Colbert D T, Smith K A and Smalley R E 1998 Catalytic growth of single-wall carbon nanotubes from metal particles *Chem. Phys. Lett.* **296** 195–202

[31] Casey A, Herzog E, Davoren M, Lyng F M, Byrne H J, and Chambers G 2007 Spectroscopic analysis confirms the interactions between single walled carbon nanotubes and various dyes commonly used to assess cytotoxicity *Carbon* **45** 1425-1432

[32] Gass M H, Koziol K K, Windle A and Midgley P 2006 Four-dimensional spectral tomography of carbonaceous nanocomposites *Nano Lett.* **6** 376-379

[33] Cui D, Tian F, Ozkan C S, Wang J and Gao H 2005 Effect of single wall carbon nanotubes on human HEK293 cells *Toxicol. Lett.* **155** 73-85

[34] Oberdörster E 2004 Manufactured nanomaterials (Fullerenes, C_{60}) induce oxidative stress in brain of juvenile largemouth bass *Environ. Health Perspect.* **112** 1058-1062

[35] Shvedova A A, Murray A R, Kisin E R, Kagan V E, Gandelsman V Z and Castranova V. Exposure of human bronchial epithelial cells to carbon nanotubes caused oxidative stress and

cytotoxicity *Free Radicals and Oxidative Stress: Chemistry, Biochemistry and Pathophysiological Implication (Ioannina, Greece, 26-29 June 2003)* 91-103

[36] Zhao X, Striolo A and Cummings P T 2005 C60 binds to and deforms nucleotides *Biophys. J.* **89** 3856-3862

In vivo Study with Quartz-Containing Ceramic Dusts: Inflammatory Effects of Two Factory Samples in Lungs after Intratracheal Instillation in a 28-Day Study with Rats

O Creutzenberg[1], C Ziemann[1], T Hansen[1], H Ernst[1], P Jackson[2], D Cartlidge[2], R Brown[3]

1 Fraunhofer Institute of Toxicology and Experimental Medicine, Hannover, Germany

2 CERAM Research Ltd., Stoke-on-Trent, UK

3 TOXSERVICES, Stretton, UK

E-mail: otto.creutzenberg@item.fraunhofer.de

Abstract. As various quartz polymorphs react differently in lungs, a differentiation of effects is needed while setting occupational exposure levels. The objective of this European Collective Research Project SILICERAM was to characterize differences in biological activity of four quartz species, i.) 2 quartz-containing materials collected at typical ceramic manufacturing sites (Tableware granulate, TG and Tableware cast, TC) versus ii.) a designed ceramic dust sample (Contrived Sample, CS) and iii.) ground quartz DQ12 (well-characterised standard quartz (Positive Control, PC) and TiO_2 (negative control). TG and TC had been selected as the most promising two candidates based on a preceding *in vitro* screening of 5 factory samples. Total doses of 5 mg per rat of the TG and TC, 1.1 mg of the CS and 0.33 mg of the PC corresponding to 0.29, 0.16, 0.29 and 0.29 mg quartz per rat, respectively, were administered to rats by intratracheal instillation. After 3 days, bronchoalveolar lavagate (BAL) analysis resulted in polymorphonuclear neutrophil (PMN) levels of 15%, 25%, 0.6% and 25% in the TG, TC, CS and PC groups, respectively. At 28 days, the values were 29%, 20%, 7% and 45%. Histopathologically, the TG and TC groups showed very slight to slight effects, the PC group, however, stronger effects after the same period. In conclusion, the following ranking was found: PC > TG > TC > CS > TiO_2 > Vehicle Control. Thus, a clear differentiation of effects for TG and TC, CS and PC was found. From a regulatory point of view, the substance-specific toxic potentials of TG and TC may need to be considered when devising occupational exposure limits.

1. Introduction

The traditional ceramics industry is divided into several subsectors: tiles, tableware, sanitary ware, refractory, bricks and roofing tiles. In order to manufacture these products powdered crystalline silica is used. The inhalation of respirable crystalline silica (RCS) can lead to silicosis, which ultimately results in to ill-health through breathing difficulties and even death. A new input to the debate has given the International Agency on Research in Cancer (IARC) in 1997 when classifying inhaled quartz or cristobalite from occupational sources as a category 1 carcinogen (human).

At present different EU states have different limits for RCS. Within Europe the Scientific Committee on Occupational Exposure Limits (SCOEL) is working towards identifying a common EU position on mg/m^3 levels. One of the difficulties associated with setting common exposure limits for workers is the fact that different forms of crystalline silica appear to pose different threats. A Social Dialogue between trade-union and employers agreed on that i.) quartz concentrations at various workplace sectors should be held as low as technically possible and that ii.) toxicological experiments should continue to generate data allowing a differentiation of industrial quartz varieties. Setting a single low limit to encourage continual improvement, but allowing concessions based on proven reduced risks associated with certain RCS forms is seen as a pragmatic way forward.

Against the above background, the overall aim of this project was to provide legislators with useful data for defining RCS in air limits.The project focused on different industries of the ceramic sector: tiles, tableware, sanitary ware, refractory, bricks and roofing tiles.

Objectives: This non-clinical *in vivo* rat study was designed to expand the database obtained in preceding *in vitro* screening assays using sensitive cytotoxic and genotoxic endpoints (see accompanying extended abstract by Ziemann et al.):

- To validate the endpoints established in an *in vitro* screening in a 28-day *in vivo* study.
- To characterize dust-dependent lung effects by using bronchoalveolar lavage (BAL) on day 3 and day 28 and histopathological parameters on day 28 of the post-instillation period.
- To determine if there were differences in the toxicological profiles of the TG and TC sample, the CS sample and the well-characterized reference quartz DQ12 and to rank the observed effects.

2. Materials and Methods

2.1. Particle samples

2.1.1. *Test materials* Two high-volume sampling apparatus were developed for RTDs' use and dust amounts of approximately 1000 mg were collected at various industrial sectors (Table 1). Two test items were investigated in this study: TG and TC have been collected at various workplaces of the tableware sector under the responsibility of the Instituto de Tecnología Cerámica (ITC), Spain and the Société Francaise de Céramique (SFC), France, respectively.

2.1.2. *Reference materials* The reference materials are shown in Table 1. Three reference items were investigated in this study, TiO2, the contrived sample (CS) and the positive control quartz DQ12 (PC). CS was prepared in a wet process mixing DQ12, feldspar and clay and lyophilising the mixture subsequently (tables 1 and 2).

2.2. Test system

Male Wistar WU rats [strain Crl:(WI)WU] were purchased from Charles River Deutschland, Sulzfeld, Germany. The age of the animals at the start of exposure was 8 weeks and the weight approx. 300 grams. The rats were exposed to the test/reference items by intratracheal instillation. Males are preferred to females because of literature data that indicate a higher sensitivity with regard to silicotic response for males [2].

2.3. Dosing scheme and investigations

TG and TC served as the basis for dosing with 5 mg per rat. These fractions consist of 5.8% and 3.1% quartz, respectively. To reach a similar quartz dose in the quartz DQ12 and in the CS groups as compared to the TG and TC samples, the contained quartz percentage was adapted. This allows a direct comparison of quartz polymorphs at similar quartz doses (see Table 2).

Each total dose was administered by intratracheal instillation of 4 equal aliquots on 4 consecutive days. Animals were sacrificed 3 days (--> BAL) and 28 days (--> BAL, histopathology) post-treatment for inflammatory endpoints. The experimental set-up is shown in Table 2.

Table 1 Particle Size Distribution of Quartz-Containing Factory and Reference Samples

Name/code	Intended use / Collected by	Geometric Mean (µm) (geom. STD) weighting by mass	Geometric Mean (µm) (geom. STD) weighting by number	Quartz Content (weight-% - XRD)
TiO_2 Bayertitan T	Negative Control	1.80 (1.90)	0.44 (1.58)	---
Tableware (granulate) TG	ITC Spain	5.97 (1.83)	0.52 (1.95)	5.8
Tableware (cast) TC	SFC France	4.36 (1.76)	0.53 (2.15)	3.1
Contrived sample [a] CS	Surrogate for a Ceramic Dust	3.77 (1.86)	0.87(1.66)	26.1
Quartz, type DQ 12 PC mid-size Dörentrup [1]	Positive control	3.01 (1.53) 87% α quartz, 13% amorphous	1.72 (1.56) 87% α quartz, 13% amorphous	87

[a] 30% DQ12, 20% feldspar, 50% china clay

Table 2 Dosing Scheme

Treatment (17 animals per group)	Cumulative Particle Dose (mg per rat)	Quartz Dose (mg per rat)
Saline (vehicle control)	- (0.3 ml saline)	-
TiO2 - Bayertitan T (negative control)	5	-
Tableware (granulate) TG	5	0.29
Tableware (cast) TC	5	0.16
Contrived sample [a] CS	1.1	0.29
Quartz DQ12 mid-size PC	0.33	0.29

[a] 30% DQ12, 20% feldspar, 50% china clay

Note: Dosage of Quartz DQ12 Mid-Size has been specified on the basis of known quartz contents of the two ceramic samples ---> the quartz dose of the CS and DQ12 group was the same to allow a direct comparison.

Journal of Physics: Conference Series **151** (2009) 012031 doi:10.1088/1742-6596/151/1/012031

3. Results

TG and TC BAL data, 3 days after the end of treatment, showed a moderate inflammatory response. The intensity of the initial response did either not change or decreased after 28 days. In contrast, the quartz DQ12 group showed a strong response which continued to intensify during the subsequent 25 days. At 3 days, bronchoalveolar BAL analysis resulted in polymorphonuclear neutrophil levels (PMNs) of 15%, 25%, 0.6% and 25% in the TG, TC, CS and PC groups, respectively. At 28 days, the values were 29%, 20%, 7% and 45%. Histopathologically, the TG and TC groups showed very slight to slight effects, the PC group, however, stronger effects after the same period. Results of endpoint analyses are given as normalized data in Table 3.

4. Discussion

The scientific approach in this study was to characterize differences in the biological activity of various quartz species showing a similar particle size. It is well-known that such differences in toxicological potency exist depending on age, treatment, composition, surface contamination and other factors representing the multifactorial toxicity aspects of quartz and quartz-containing particles [3], e.g. for crushed quartz and aged quartz [4] [5].

DQ12 mid-size used as the positive control in this study is a crushed quartz that has been widely accepted as a biologically active standard in similar toxicological studies [1]. The particularity of quartz DQ12 samples is their lack in any aging effect. Most of the DQ12 fractions nowadays used have been produced in the eighties of the last century and there exists a more than 20-year experience in various labs regarding the persistence of the intrinsic toxicity of these samples.

CS was prepared in a wet process mixing DQ12, feldspar and china clay and mimics the typical composition of a ceramic dust with high quartz content (here DQ12). The toxicological results reveal that feldspar and china clay have a strong masking effect on the toxic properties of DQ12 (Table 3).

5. Conclusions

From the results of the present study the following conclusions can be derived:

- The vehicle control did not induce lung inflammation.
- The positive control quartz DQ12 produced significant and progressive lung inflammation.
- The Tableware (granulate) and the Tableware (cast) samples induced a weaker inflammatory response than the positive control (statistically significant) and this response, albeit statistically significant compared to the vehicle control, was not progressive.
- The Tableware (cast) sample showed a (partially statistically significantly) weaker response than the Tableware (granulate) sample.

Overall, the above-mentioned biological effects of the test items can be ranked in the following order combining the endpoints BAL (LDH, β-glucuronidase, total protein, PMN) and histopathology findings (interstitial mononuclear cell infiltration, alveolar inflammatory cell infiltration, granulomatous alveolitis, perivascular granulocytic infiltration): quartz DQ12 mid-size > Tableware (granulate) > Tableware (cast) ≥ contrived sample > TiO_2 (Bayertitan T) > vehicle control. Thus, a differentiation of effects for TG and TC, CS and PC was found. TG, CS and PC showed differences in their toxic potential albeit dosed with the same quartz amount. A substantial decrease was observed for CS as compared to PC. Irrespective of the contained DQ12 quartz, the wet mixing process with feldspar and china clay had drastically reduced the toxicity of the final CS. From a regulatory point of view, the substance-specific toxic potentials of TG and TC may need to be considered when devising occupational exposure limits.

Table 3 Summary of main effects after 3 and 28 days (normalized data as compared to controls)

Test Materials - not significant (+),+,++,+++ effect grading	**TiO_2**		**Tableware granulate**		**Tableware cast**		**Contrived sample**		**Quartz DQ12**	
BAL	Day 3	28	3	28	3	28	3	28	3	28
Lactate Dehydrogenase (LDH)	-	-	3.3 ### ++	3.5 ###	2.1 # +§§	2.1 ##+++ §§§	1.2	1.3	2.0 #	3.3 ###
β-Glucuronidase	-	-	2.6 ###	2.5 ###	1.1 §§§	1.6 +++§	0.6	1.0	1.6	3.2 ###
Total Protein	-	-	2.8 ###	1.9	1.6 §§§	1.4 ###+++ §§§	1.1	1.1	1.8 #	1.8 ###
PMN[a] (% in differential cell count)	1.8	4.9	14.5 ###+	28.6 ###+++	2.3 +++§§	20.4 ###+++	0.6	6.7	25.0 ###	44.8 ###
Lung Weights Relative	-	-	###++	##	###	-	-	-	-	#
Histopathology	Day 28		28		28		28		28	
Inflammation	not significant		very slight to slight		very slight to slight		very slight		slight to moderate	
Interstitial Fibrosis	not significant		very slight		not significant		not significant		very slight	

Statistically significantly different as compared to vehicle controls (#, ##, ###), DQ12 (+, ++, +++) and TG (§, §§, §§§) (Tukey's test)

[a] Total PMN numbers resulted in relative values of 0.01 (VC), 0.06 (TiO_2), 0.52 (TG), 0.43 (TC), 0.08 (CS) and 1.00 (PC), thus, the outcome was similar as compared to the percentage values. TC shows the same number as TG, i.e. non-quartz is contributing to the effect observed.

References

[1] Robock K 1973 Ann. Occup. Hyg. 16, 63
[2] Honnons and Porcher JM 2000 J. Environm. Path. Tox. Onc. 19, 391
[3] Fubini B 1998 Health effects of silica The Surface Properties of Silicas ed AP Legrand (John Wiley & Sons Ltd) chapter 5 pp 415-464
[4] Castranova V, Vallyathan V, Ramsey DM, McLaurin JL, Pack D, Leonhard S,. Barger MW, Ma JYC, Dalal NS and Teass A 1997 Environm Health Persp 105 (Suppl 5), 1319
[5] Vallyathan V, Castranova V, Pack D, Leonhard S, Shumaker J, Hubbs AF, Shoemaker DA, Ramsey DM, Pretty JR and McLaurin JL Am. J. Respir. Crit. Care Med. 152, 1003

Acknowledgement

The work was supported by the EC through the COLL-CT-2003-500 896 project SILICERAM.

Inhaled Particles X, (23–25 September 2008, Manchester) IOP Publishing
Journal of Physics: Conference Series **151** (2009) 012032 doi:10.1088/1742-6596/151/1/012032

Biological effects of nano-nickel in rat lungs after administration by inhalation and by intratracheal instillation.

A Ogami[1] Y Morimoto[1] M Murakami[1] T Myojo[1] T Oyabu[2] and I Tanaka[2]

1 Department of Occupational Pneumology,

2 Department of Environmental Health Engineering,

Institute of Industrial Ecological Sciences, University of Occupational and Environmental Health, Japan, 1-1 Iseigaoka Yahatanishiku Kitakyushu, Japan 807-8555

E-mail: gamisan@med.uoeh-u.ac.jp

Abstract. We examined the biological effects of the nickel oxide (NiO) nanoparticles by inhalation and instillation study. Wistar male rats were exposed to NiO nanoparticles (nNiOs) for 4 weeks (6 hrs/day). The nNiOs was black-colored NiO (99.8%) and average particle size (APS) was 20 nm. The geometric mean diameter of the particles in the chamber and the daily average exposure concentration were 139 ± 12 nm and $1.0 \pm 0.5 \times 105$ particles/cc, respectively. The deposited amount of nNiOs in the rat lungs at 4 days after the inhalation exposure ended was 29 ± 4 µg. Although nNiOs exposure group showed temporal significant increase in the number of total cells in brochoalveolar lavage fluid (BALF) at 4 days after the exposure end, the difference were not seen at one month after an exposure end compared with control group. The histopathological change was not severe just after the inhalation nor throughout the observation time. Elemental mappings of nickel showed that nickel particles were located in agglomeration at the pulmonary macrophages.

1. Introduction

Nanoparticles have specific optical and electric properties. The production of highly functional materials by utilizing such properties has already begun; however, the effects of nanoparticles on humans have not been elucidated. Concerns about the biological effects of these nanoparticles being produced for use in industrial products have arisen since epidemiologic data have shown a correlation between airborne nanoparticles, as typified by $PM_{2.5}$ derived from the combustion of fossil fuel, and cardiovascular diseases[1]. There is a need for inhalation exposure tests that simulate human exposure to nanoparticles to observe their health effects experimentally. In general, nanoparticles have a tendency to agglomerate; therefore, it is necessary to prevent the agglomeration of the nanoparticles while determining the particle size during the inhalation exposure. Nickel compounds are classified as carcinogenic substances (Group 1) in the International Agency for Research on Cancer (IARC) categories. Nickel oxide is a nickel compound with extremely poor solubility that is considered to have high biopersistence in the lung[2]. With its physiological and chemical characteristic usefulness, NiO is preferable referential particle for evaluation of the pulmonary effect of other nanoparticles. In this study, we evaluated histopathological changes in the lungs of rats following inhalation exposure to nanoparticles. The inhalation experiment was conducted after determining that the generated nanoparticles were dispersed stably in the chamber.

2. Material and Methods

2.1. Nickel oxide particles

The NiO nanoparticles (nNiOs) were black-colored NiO (99.8%) and average particle size (APS) was 20 nm (Nanostructured & Amorphous Materials Inc.).

2.2. Animals

Wistar rats (males, 9 weeks old, 30 animals in each group) had been fed standard diet, and were maintained in an inhalation chamber. All of the experimental procedures were approved by the Experimentation Committee of the University of Occupational and Environmental Health, Japan.

2.3. Inhalation procedure

The generation and inhalation system of nNiOs is described elsewhere[3]. Briefly, the generation system was composed of an ultrasonic nebulizer and diffusion dryers. The particles were generated in the gas phase by spraying suspension. Thirty rats were exposed by inhalation to nNiOs for 4 weeks (6 hrs/day). Control rats were exposed to fresh air only. Ten rats from each group were randomly sacrificed at 4 days, 1 month and 3 months after the 4 weeks exposure. The geometric mean diameter of the particles in the chamber and the daily average exposure concentration were 139 ± 12 nm and 1.0 ± 0.5×10^5 particles/cc, respectively (Fig. 1). The deposited amount of nNiOs in the rat lungs at 4 days after the inhalation was 29 ± 4 μg.

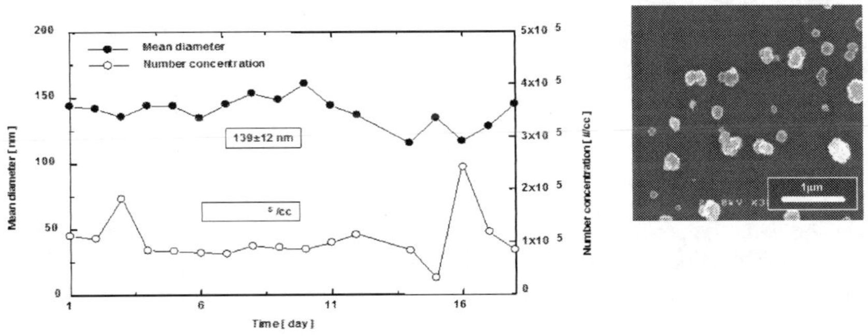

Figure 1. nNiOs particle concentration in chamber after nebulization.

2.4. Bronchioalveolar lavage fluid (BALF) and Histopathological examination

The left lung was clamped and 50 ml of bronchoalveolar lavage fluid (BALF) was collected from the right lung. The number of inflammatory cells in the collected BALF was determined by an automatic cell counter (Celltac MEK 5204 Nihon Koden, Tokyo, JAPAN) and the neutrophil fraction was counted from the cytospin sample stained by Diff-Quik kit (International Reagent Co., Kobe, Japan). After extracting the BALF, the left lung was inflated and fixed by intratracheal infusion with 4 % paraformaldehyde at 25 cm H_2O pressure for one night and then embedded in paraffin. The paraffin sections of 3μm thickness were stained with Hematoxylin and Eosin. After each specimen was HE-stained, 6 random digital images, focusing mainly on the alveoli and excluding large vessels and trachea, were taken per lung section with a digital camera (DS-5M, Nikon Instech Co. Ltd., Kanagawa, Japan) under light microscopy. A 300-point grid was placed over each image on the computer screen and we examined pulmonary inflammation in each using the point counting method (PCM) [4].

2.5. Ni mapping of lung tissue by intratracheal instillation

One mg of nNiOs was suspended in each 0.4 ml of physiological saline aliquots, sonicated thoroughly, and intratracheally instilled into the animals (Wistar rats, males, 10 weeks old). At 3 days after instillation, the left lung was inflated and fixed as same procedures as the inhalation study. The paraffin sections of 3μm thickness were stained with HE. The elemental mappings of Ni in the HE-stained lung tissue or pulmonary macrophages on the cytospin samples were analyzed directly by scanning electron microscopy and Energy Dispersive X-Ray Micro Analyzer (HORIBA).

3. Results

3.1. BALF findings and Histopathological findings

Figure 2 shows the representative image of rat lung sections with HE stain at 4 days after nNiOs inhalation. The histopathological change was not severe throughout the observation time.

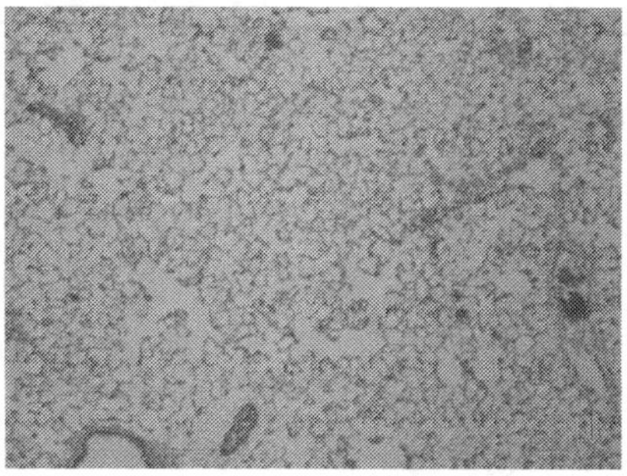

Figure 2. Representative image of rat lung sections with HE stain at 4 days after nNiOs inhalation. (x100)

Although nNiOs showed temporal significant increase in the number of total cells in brochoalveolar lavage fluid (BALF) at 4 days after the exposure end ($p< 0.01$), the difference were not seen at one month after an exposure end compared with control group (Fig.3). The inflammation areas by PCM are also shown in Figure 3. In nNiOs, inflammation areas significantly increased at 4 days after inhalation period ended. No significant differences in inflammation areas were found at 1month and 3 months.

3.2. Nickel mapping on lung tissue by intratracheal instillation

Elemental mappings of nickel showed that nickel particles were located in agglomeration at the mild inflammatory lesion or pulmonary macrophages.

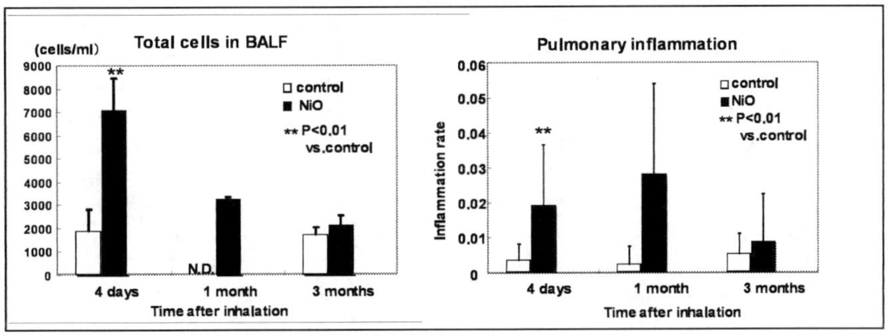

Figure 3. Total cell count in BALF (left) and pulmonary inflammation (right) after inhalation of nNiOs.

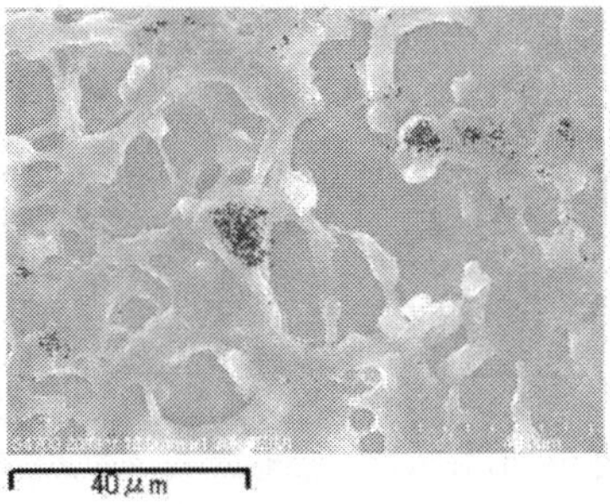

Figure 4 Nickel mapping image on the lung tissue after intratracheal instillation of nNiOs. Black dots are nickel-positive sites by scanning electron microscopy and Energy Dispersive X-Ray Micro Analyzer.

4. Discussion

This study was designed to evaluate the biological effects of the nickel oxide (NiO) nanoparticles in an inhalation study.

Using the exposure system in this study, we confirmed stable exposure of NiO and the deposited amount of NiO nanoparticles in the rat lungs at 4 days after the inhalation was 29 ± 4 μg. We also calculated that the biological half-time of NiO nanoparticle from the lungs was 62 days ($r^2 = 0.90$) [3].

The distribution and location of inhaled nanoparticles are important for evaluation of clearance. We tried to visualize the location of particles in the alveoli after entering the airway. Using the scanning electron microscopy and Energy Dispersive X-Ray Micro Analyzer, we are able to map the nickel elements directly on the lung tissue. Using standard formalin embedded section, this technique is relatively convenient without glutaraldehide fixation. Although single nanoparticle was unable to detect for its sensitivity, the mapping of Ni element on the lung tissue after intratracheal instillation of nNiOs showed the most of the particles were located in the alveolar macrophage as an aggregated form.

Histopathological findings in the nNiOs group and evaluation by point counting method (PCM) showed slight temporal inflammatory features and the temporal increase pattern of inflammation rate in the lung at 4 days. This temporal inflammatory trend was also observed in the increase of total cell numbers in bronchoalveolar lavage fluid (BALF) after inhalation. These inflammatory responses were very low compared with the other inhalation studies on particles with micrometer size. This minimum or mild histopathologic change observed in the lungs was consistent with the findings that initial deposited amounts of nanoparticles in the lungs were small and the clearance from the lungs was not impaired. The slight inflammation seen in this study in spite of the small deposition and normal clearance may be due to the toxicity of NiO.

In conclusion, we exposed rats to NiO nanoparticles by inhalation at a stable particle size and concentration by using an ultrasonic nebulizer for 4 wk. Although NiO exposure group showed temporal significant increase in the number of total cells in brochoalveolar lavage fluid (BALF) at 4 days after the exposure end, the difference were not seen at one month after an exposure end compared with control group. The histopathological change was not severe just after the inhalation nor throughout the observation time. Elemental mappings of nickel following the intratracheal instillation showed that nickel particles were located in agglomeration at the pulmonary macrophages.

Acknowledgments
This study was supported by a Grant of the New Energy and Industrial Technology Development Organization (NEDO) of Japan.

References

[1] Brunekreef B and Forsberg B 2005 Epidemiological evidence of effects of coarse airborne particles on health *Eur Respir J* **26** 309-318
[2] NTP Toxicology and Carcinogenesis Studies of Nickel Oxide (CAS No. 1313-99-1) in F344 Rats and B6C3F1 Mice (Inhalation Studies), 1996.
[3] Oyabu T, Yamato H, Ogami A, Morimoto Y, Akiyama I, Ishimatsu S, Hori H, and Tanaka I 2006 The effect of lung burden on biopersistence and pulmonary effects in rats exposed to potassium octatitanate whiskers by intratracheal instillation *J Occup Health* **48** 44-48
[4] Ogami A, Morimoto Y, Yamato H, Oyabu T, Kajiwara T, and Tanaka I 2004 Patterns of histopathological change determined by the point counting method and its application for the hazard assessment of respirable dust *Inhal Toxicol* **16** 793-800

Cell shape imaging analysis: a fast and reliable technique for the investigation of internalised carbon nanotubes in flat macrophages

F Tian[1], A Prina-Mello[2], G G Estrada,[3]A Beyerle,[1] W Kreyling[1] and T Stoeger[1]

1 Institute for Inhalations Biology, German Research Centre for Environmental Health, Helmholtz Zentrum München (GmbH), Munich, Germany.

2 Centre for Research on Adaptive Nanostructures and Nanosciences, Trinity College, Dublin, Ireland.

3 Institute for Bioinformatics, German Research Centre for Environmental Health, Ingolstaedter Landstr. 1, D-85764 Neuherberg, Munich, Germany.

email: tobias.stoeger@helmholtz-muechen.de

Abstract: The aim of this work is to elucidate the mechanisms involved in the morphological adaptation and regulation of macrophages in the presence of internalised materials. This development will accelerate the toxicology assessment of novel nanomaterials and subsequently reduce their environmental and health exposure. For this purpose, we adapted our established in vitro culture system to investigate and measure cell shape changes with and without functionalized carbon nanotubes (CNTs). Two nanomaterials, such as fluorescent polystyrene (PS) beads and functionalized CNTs, were employed to track the material location under confocal microscopy, light microscopy and Transmission Electron Microscopy (TEM). It was found that particles equally spread throughout the entire cytoplasm in spherical macrophage; whereas when macrophages where forced to adhere to the substrate, via fibronectin coating, the accumulation of particles and tubes was limited to the vicinity of the nucleus due to the modified cellular micro architecture. TEM analysis also confirmed these findings and demonstrated that CNTs of about 5 μm laid at the bottom of adherent cells. Therefore, this cell shape analysis and manipulation may result very important for the quantification of internalised novel materials with high aspect ratio like nanotubes, nanorods and nanowires.

1. Introduction

Carbon nanotubes (CNTs) are one of the most well-studied nanomaterials in the literature since they were first reported back in 1991 [1]. The development of carbon nanotubes (CNTs) as delivery systems for nucleic acids, proteins, and drug molecules into mammalian cells and living tissues has attracted much attention [2]. Surface functionalization of carbon nanotubes is an effective way of improving the solubility and dispersion of the nanotube. Functionalization of carbon nanotubes has been proposed in various biological applications, such as DNA and protein biosensors, biocatalysts, and bioseparators Nanotubes also find potential applications in the fields of cancer diagnosis and therapy. Pantarotto and coworkers reported on the internalisation of fluorescently labelled CNTs into human T cells [3]. Recently, Kam et al. reported single-wall carbon nanotube (SWCNT)-streptavidin conjugates into human promyelocytic leukaemia (Hela or HL60), Chinese hamster ovary (CHO), and MC3T3 osteoblast cell line [4]. Getting the number of fluorescent SWCNT has been described in lung epithelium cell in the work of Lacerda et al [5]. However, it is necessary to calibrate a number of factors to get the right number of internalised nanomaterials.

It is well known that phagocytes, and in particular macrophages, play a key role in the cellular uptake and response to foreign bodies, such as particles, fibres or tubes. Alveolar macrophages in lungs represent the first defence against various hazardous environmental exposures such are particles and fibres when inhaled. To better describe in vivo environmental interactions and in vitro models, we should go beyond simple membrane filopodia to full actin-filaments cross-link which provides sufficient substrate for macrophage spreading.

Over the past decade, several methods have described how to enhance the macrophage adhesion by modifying the substrate surface-chemistry with immobilized ligands or proteins [6,7]. Current in vitro models use the interactions of macrophages with extra-cellular matrix such as fibronectin (Fn). This interaction is known to occur mainly via the amino acid sequence RGD [6]. Fn has been largely used to enhance actin polymerization rapidly in neutrophils, monocytes and macrophages [8]. Therefore an accurate concentration of Fn is essential for the formation of receptor clusters and enhancement of cell adhesion on substrates. In this paper, different concentrations of Fn were used to coat glass substrates to maximize alveolar macrophage adhesion and subsequently control cell, shape such as cell height and width.

To validate this shape control, well-defined commercially available 1 μm polystyrene particles were used as reference material. Polystyrene (PS) beads are regarded as reference material when it comes to their geometrical regularity, shape and low toxicity. It is known that due to their tube-shape geometry, in the nanoscale domain, nanotubes are not easy to focus and be detected under fluorescent microscopy. In this work, fluorescently conjugated labels onto CNTs were used for their visualization by confocal microscopy of the CNT post-internalisation and cell interaction. We speculate on the mechanisms that regulate the interaction between the macrophage cytoskeleton and the internalised CNTs. This may lead to develop a quantitative assay to measure the internalised mass of the nano-structured materials into macrophages.

Our research hypothesis is that cell shape can orientate CNTs when the cell height is less than the CNT length. This paper will show how the cytoskeletal reorganisation of geometrically-tubular internalised CNTs can be seen by light microscopy, confocal microscopy and TEM. Here, the macrophage cell shape is used as a tool to orientate the carbon nanotube and make them easier to detect under any of the microscopy techniques employed. From the results presented in this work we can confirm that this technique can be very useful for the toxicology assessment of tube shaped nano materials.

2. Materials and Methods

2.1. Particles and Nanotubes

Fluosphere particles of diameter 1.0 μm (Molecular probes F-8823) were purchased from Invitrogen (Invitrogen Ltd, UK). Particles were supplied at a concentration of 2.0 $\%_w$ suspended in 50 mM Na_3PO_4, 50 mM NaCl, pH 7.5. The number for particles, N, per ml of suspension can be determined from the following equation: $N = C\ (6 \times 10^{12}) / (\rho\pi\phi^3)$, where C is the concentration of suspended beads in g/ml (0.02 g/ml for 2 % suspension); diameter ϕ of particles in micrometres (1 μm), and density ρ of polystyrene in g/mL (1.05 g/cm^3).

Multiwalled carbon nanotubes (MWCNTs) of 50 nm wide and 5 μm long were purchased from IIJIN (Diamond Co. Ltd, Korea). CNTs were functionalized with amino acids, according to the work of Pantarotto and co-workers [3].

2.2. Substrate preparations with Fn

Fn coating of substrates for cell counting were achieved on purposely prepared glass cover slips (ϕ = 10 mm, MatTek Corporation, MA., USA) by using 20 μl coating of 2.5×10^{-3} mg/ml of fibronectin III (Sigma Chemical Co. St. Louis, MO.).

2.3. Culture

A murine macrophage cell line RAW264.7 was grown in Dulbecco's minimal Eagle's medium (DMEM) containing 10% foetal bovine serum (FBS), 1% L-glutamine and 1% penicillin–streptomycin at 37°C, 5% CO_2 and 95% relative humidity (Incubator model: MCO-20AIC, SANYO, Japan). Macrophages and PS beads/CNTs are incubated in the plastic tubes. The tubes were then rotated at 38°C (rotator model ST, New Brunswick Scientific Company, New Brunswick, New Jersey) for 2 hours. Under these conditions more than 90 per cent of nanoparticles were ingested after 30 minutes of rotation.

2.4. Cell staining and confocal laser microscopy

First, cells were seeded onto coverslips and incubated for 2 hours in complete DMEM then were fixed with 4% paraformaldehyde in PBS at room temperature for 20 minutes. Cells were then washed again with PBS and permeated with 0.2% Triton X-100 for 5 minutes at room temperature. The samples were subsequently stained with mouse anti-actin (A4700, Sigma, Taufkirchen, Germany) and anti-α-tubulin (T4026, Sigma) antibodies by a FITC conjugated secondary anti-mouse antibody (F0257, Sigma) and Chromeo™ 642 Goat anti-Mouse IgG (ACTIVE MOTIF, Germany). Then the cell nucleus was stained by 10 mM Hoechst for 1 minute. Finally the stained coverslips were mounted for fluorescence microscopic examination. Data were obtained using a Zeiss Laser Scanning Microscope (LSM 510 META, Zeiss, Germany). The LSM was based on the Axiovert 100M inverted research microscope equipped with 4 laser sources (Argon, Helium-neon 1, Helium-neon 2 and Titanium : Sapphire).

2.5. Transmission Electron Microscopy (TEM) sample preparation

For TEM analysis the cell-CNT samples were prepared following the same protocol used for the fluorescent investigation. Macrophages with CNT at 0.05 μM concentration were incubated for 2 hours. After the incubation, cells were seeded onto fibronectin coated (2.5 $\times 10^{-3}$ mg/ml) glass slides for 2 hours. Afterwards, the cells were washed with 0.01M phosphate buffer (PBS) and fixed for 2h by 2.5% glutaraldehyde, which was previously dissolved in PBS (pH 7.4). The cells were embedded into 0.1% agar at 37 °C for 5min. This agar was fixed by 2.5% glutaraldehyde in PBS at 4 °C, for at least 2h. The samples were washed with PBS, and then fixed by 1% osmium tetroxide at 4 °C for 2h. Cells were dehydrated in graded series of ethanol and later embedded in epoxy resin, as described elsewhere [9]. The cells were cut by vertical and horizontal section. Ultra-thin cross sections (~50nm) of cells were observed under a Transmission Electron Microscope, Philips CM10.

3. Results

3.1. Measurements of cell geometry and particle location

After 2h exposure to polystyrene particles, RAW264.7 macrophages were successfully plated for 2 hours on fibronectin-coated glasses and untreated borosilicate glasses. This resulted in intrinsically induced morphological cell-shape changes of the plated macrophages, both at the cellular and subcellular level. In particular, the subcellular particle distributions varied a lot within the two different cell shapes, here call "spherical" and "flat", as shown from the confocal fluorescence image (Figure 1). It can be seen that for so-call spherical cells, plated on untreated substrates, the particles are equally distributed through the cytoplasm (Figure 1A). Where as for the fibronectin-coated substrate the flat cells (cell height = 1.5 ±0.2 μm) showed a preferential accumulation of the particles near the nucleus (Figure 1B). This is attributed to the modified cytoskeletal micro architecture network induced by the fibronectin coating. Thus, the extracellular matrix coating has a direct impact on the cell focal adhesion and cell geometrical aspect ratio (diameter/height = 20). This high aspect ratio implies that spherical macrophages can be rearranged in their cytoskeletal scaffold into a nearly flat, 2-dimensional (2D) shape (Figure 1B). We thus call this cell shape 'flat'. Enhanced imaging resolution can be immediately gained from the comparison of the two confocal images (Figure 1A and Figure 1B). Figure 1C illustrates parameters, such as height and diameter of each nucleus, heights and diameter of all the accumulated particles. All the geometrical information, quantification and localisation of accumulated particles are necessary for the 3-dimensional rendering in confocal microscopy.

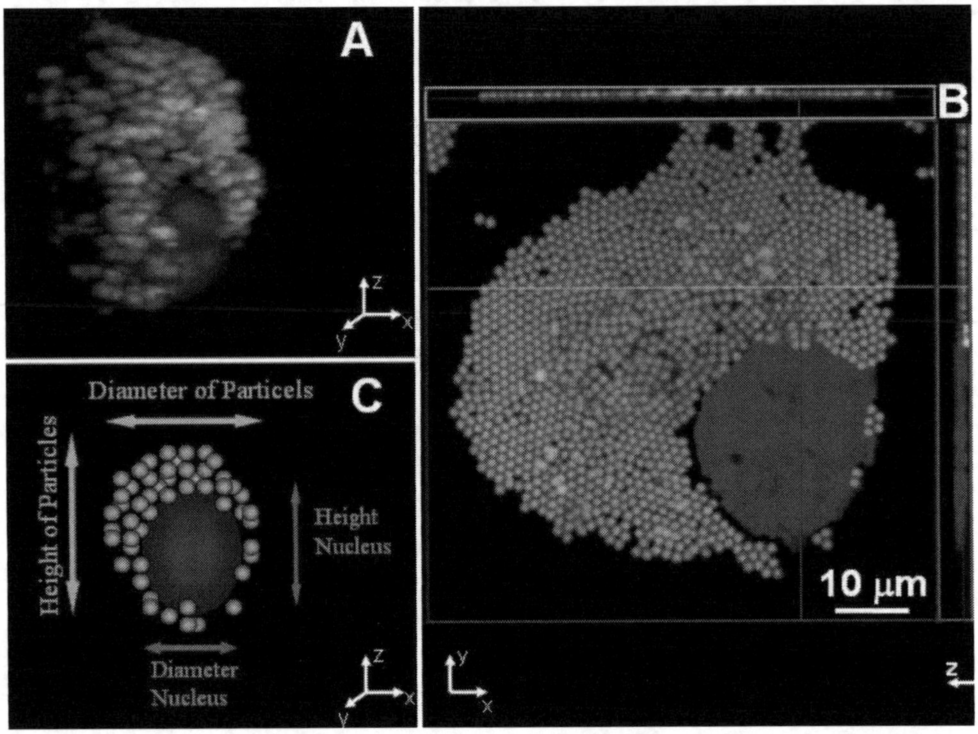

Figure 1. Morphology of a 'spherical' and 'flat' macrophages with internalised 1 μm fluospheres. (A) Particle distribution in a spherical macrophage. Uniform distribution throughout the cytoplasm of a spherical cell; 2D image. (B) X-Y and Z-plane section of a macrophage adherent to fibronectin-coated substrate. The Z-plane section (right side of B image) showed a flat cell with an height of 1.5 μm. (C) 3D schematic image illustrating a macrophage with internalised 1 μm fluorescent beads (in green) and nucleus (in blue).

3.2. Cell adaptation to particles: cytoskeletal reorganisation

From our observations it was found that spherical macrophages do not show any streched cytoskeletal structural organisation. Instead, a rather diffuse organisation of the microtubule- and actin filaments was observed, as shown by Figure 2A. This is highlighted by the recurrent colour merging between the green fluorescent polystyrene beads and the yellow fluorescent stained microtubules, as it is shown in Figure 2A.

On the other hand, when the flattened macrophages were analysed and imaged, it was clearly distinguishable that microtubules and actin filaments were spreading throughout the cytoplasm. This cytoskeletal organisation identifies a filamentous intracellular network. Actin-bundle structures appeared in straight filaments; whereas microtubules occurred as subtle structural networks, radiating throughout the cytoplasm. In this study the simplified structural scaffold is only limited to actin filaments and microtubules since they are the main active filaments involved during cell active cytoskeletal adaption to the ingested particles. They filamentous scaffold results so organised that the green fluorescence of the polystyrene particles can be seen entangled with the yellow microtubules (Figure 2B). A schematic representation of what we believe is intervening during this structural cell-scaffolding rearrangement on flattened macrophages is shown in Figure 2C. Our hypothesis is that the cell membrane (in violet) and the actin filaments (in red) are constricting the cytoplasm structure and moved the rearrangement of the cytoskeletal components (i.e., actin and microtubules) which subsequently force the particles to redistribute in organised layers underneath the filamentous structures. This constrains an effective accumulation of particles to the immediate vicinity of the nucleus. By our microscopic analysis we also observed that when high concentration of particles were forced towards the nucleus, they rearranged into a dense 2-dimensional packed clusters, as shown in Figure 1B. These combined evidences lead us to assume that the relatively "bulky" 1 μm polystyrene particles were simply accumulated around the perinuclear region of the cytoplasm due to the cell lack of space. Speculations on a tighter packing order can be drawn by the presence of cytoskeletal forces rearranging the cell shape.

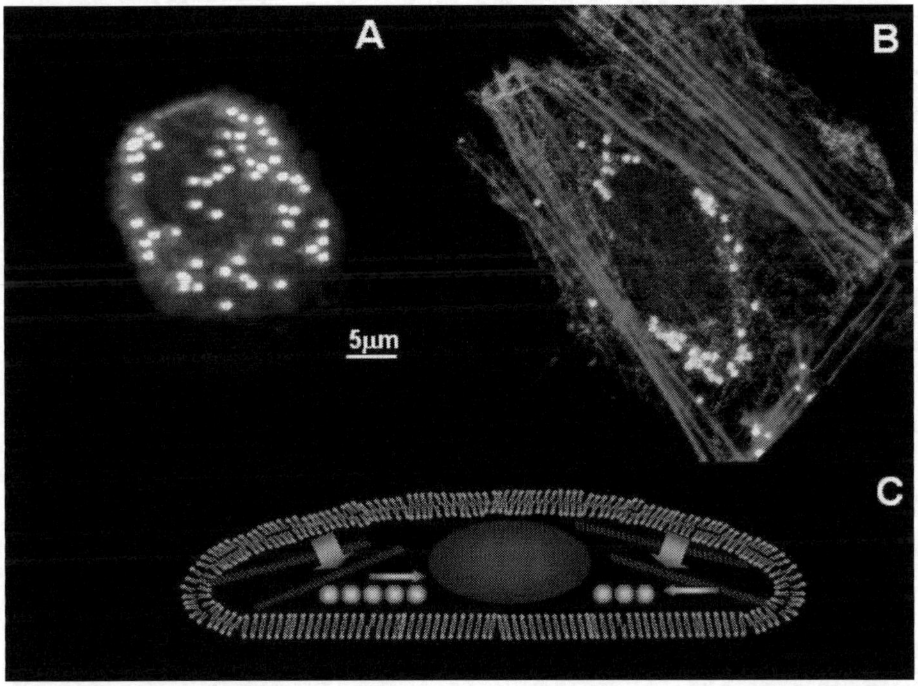

Figure 2. Particle distribution and macrophage morphological geometry: images and schematic representation; (A) Top view of spherically rounded macrophage (adhesion on untreated glass substrate); (B) Top view of flattened macrophage (adhesion on fibronectin coated glass substrate); (C) Schematic illustration of X-Z plane section of flattened macrophage (from Figure B); The nucleus is stained blue, microtubules are stained yellow, and actin is stained red. Green fluorescent particles are moved by the microtubule and actin cytoskeletal reorganisation.

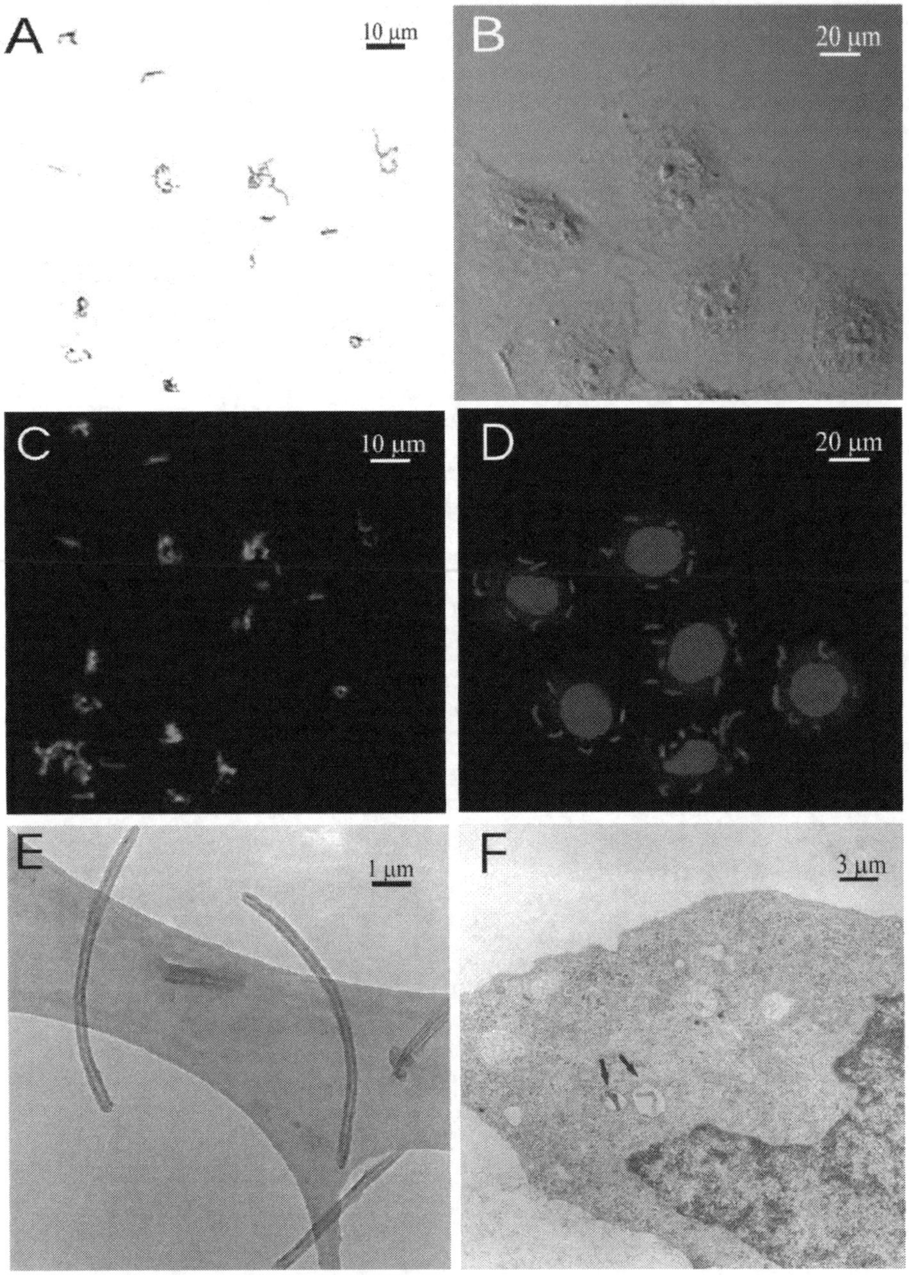

Figure 3. Micrographs of carbon nanotubes (CNTs) and cells with internalised CNTs; (A) and (C) highlight a heterogeneous distribution of CNTs onto a glass substrate under phase contrast and dark-field fluorescent microscopy, respectively; (B) and (D) show cells with internalised CNTs under light (phase contrast) and fluorescent (dark-field) microscopy (CNTs: green, nucleus: blue); (E) shows a HRTEM image of entangled MWCNTs with a membrane; (F) shows TEM contrast details of the internalised CNTs inside a macrophage, their localization and storage inside vesicles.

3.3. Determination of ingested CNTs in flat macrophages

A direct consequence of flattening macrophages is the possibility to image individual CNTs with a standard fluorescent microscope. After ingesting functionalized carbon nanotubes (CNT), macrophages were plated on fibronectin-coated substrates and the internalised CNTs were observed using three different microscopy techniques: phase contrast (Figure 3A and B), fluorescence (Figure 3C and D) and high resolution transmission electron microscopy (HRTEM) (Figure 3E and F). The left column of Figure 3 (Figure 3A, 3C, and 3E) shows images of separated CNTs; whereas the right column of Figure 3 shows cultured macrophages with internalised CNTs (Figure 3B, 3D, and 3F). The phase contrast micrographs show CNTs separated and spread on the glass slides (Figure 3A). From the matching fluorescent microscopy image (Figure 3C) the green fluorescent tube-shape structures exactly matched the CNTs imaged by phase contrast technique. Therefore, we achieved individual carbon nanotube imaging by fluorescence microscopy.

The TEM image in Figure 3E shows a 10^4-fold magnification on three individual CNTs. There, the CNTs appear well separated, as single structure on the carbon film substrate. The right column of the images shows cells grown on fibronectin-treated glass slides. According to the phase contrast image (Figure 3B), the cells appear as flat-shaped and well spread. The smooth surface of the cell indicates that there are not any CNTs on the outer side of the cell membrane. By using the fluorescence microscopy, the fluorescent CNTs are shown as green tubes arranged around the nucleus (Figure 3D). There, the arrangement of the CNTs appears in a 2-dimensional orientation. This allow for the geometrical sizing of the internalised CNT lengths at around 5 μm each. The TEM image in Figure 3F shows two internalised CNTs, engulfed and stored inside individual vesicles. Furthermore, the high staining contrast between cell membrane and cytoplasm made it also possible to visually quantify the number of CNTs internalised in each cell.

4. Discussion

The total account of "potentially hazardous" materials such as nanoparticles or CNTs requires the establishment of their potential bio-effect, uptake and subcelullar localisation. Many studies have reported on the cell uptake of carbon nanotube by confocal microscopy, fluorescent microscopy and TEM [2, 3]. In those works, bulky CNTs resulted very difficult to observe because nanotubes were entangled into the cell-cytoplasm and overlapped with all the others internalised tubes. In our work, it was possible to described how the cytoskeleton would oriented the carbon nanotube inside the cytoplasm by using FN coating to flatten the cells.

From our results, we also found that in well-adherent flat-shaped macrophages, fluorescent particles were systematically accumulated around the perinuclear region in a geometrically regular packing order. Hence, we suggested that the cell membrane bending forces, exerted by the recruited actin filaments traction, may confine the large particles to the region where a maximal void space is available. A similar explanation has also been suggested in the study of Kodali and co-workers [10]. To further support of our findings, it has been suggested that, after their cellular uptake, these particles follow the dynamic instability of the microtubules and are directed towards the nucleus [11,12]. In our experimental settings, the preferential accumulation of particles, closer to the nucleus, was observed only when the cells reached their maximum flattened shape, which represents an almost 2-dimensional condition. There, we highlighted that the geometry of the cytoskeleton micro architecture limits the accumulation of particles near the nucleus. This cell response is most likely caused by the constricted cell height which directs particles toward an ordered arrangement. On the basis of this, the intracellular particle accumulation model could also be applied to explain the agglomeration of carbon particles and paclitaxel etanidazole nanoparticles closer to the cytoplasmic nucleus [13,14].

From the phase contrast images we also showed the flat shape of macrophages grown on fibronectin treated glass slides. There, the arrangement of the CNTs appears in a 2-dimensional orientation. The lack of cell membrane coarse rugosity indicates that the CNTs did not perforate the cell membrane; this was also shown by fluorescence microscopy (Figure 3D). Where from the TEM image in Figure 3F it was reported that each CNT were neutralised and stored within the cell cytoplasm [3]. Furthermore, the high contrast made it possible to quantify and localize low density number of internalised CNTs per cell. Orientation of the nanotubes is possible in flat-shaped cells. However, it has been shown that there is a limit for localisation and orientation of CNTs. This has been reported in Lacerda's work, at a concentration of 500 μg/ml [5]. On the other hand, another work reported the imaging of individual SWCNTs, inside macrophage, by using EFTEM (Energy Filtered Transmission Electron Microscopy. In their work, Porter and co-workers reported a of 5 μg/ml which is far lower than the one described of Lacerda et al. [5, 14]

In our work, CNTs appeared well separated, as single tubes. This has been shown by the TEM image and suggested a high geometrical aspect ratio. However, the main limitation of TEM imaging technique is the applicability to very low density of CNTs. Therefore, the imaging technique here proposed has an advantage over the other techniques, by mapping the particles into a 2D plane. This also facilitates the use of light transmission microscopy (or similar techniques) for the quantification and localization of the internalised nanoparticles.

This implies that the detection and quantitative measurement of CNTs can be carried out from a small amount of cells (< 10,000 cells). This is commonly occurring in clinical medicine and occupational health when the cell samples are obtained from biopsies or bronchoalveolar lavages (BAL). Therefore, if the quantitative technique is not sensitive enough the toxicological analysis of the alveolar macrophages will be negatively biased. The technique here presented would facilitate and accelerate the experimental preparation for BAL investigations, and increase the laboratory resource capacities by simply reducing the cost per experiment due to the low sample volume employed. Finally, we can expect that this technique can be very useful for the toxicology assessment of any tube-like and spherically shaped nanomaterials.

Acknowledgements

We thank Dr. Shinji Takenaka and Dr. Peter Hutzler for excellent technical advice in the confocal microscopy. This work was supported by Helmholtz-DAAD (Germany), National Council of Science and Technology (CONACyT, Mexico) and CRANN-Science Foundation Ireland (SFI).

References

[1] Cui D, Tian F, Ozkan C S, Wang M and Gao H 2005 *Toxicol. Lett.* **155** 73
[2] Monthioux M and Kuznetsov V L 2006 Carbon **44** 1621
[3] Pantarotto D, Briand J P, Prato M, and Bianco A 2004 *Chem. Comm. (Camb).* **7** 16
[4] Kam N W S, Jessop T C, Wender P A and Dai H 2004 *J. Am. Chem. Soc.* **126** 6850
[5] Lacerda L, Pastorin G, Gathercole D, Buddle J, Prato M, Bianco A and Kostarelos K. 2007 *Adv. Mater.* **19** 1480
[6] Reaven E P and Axline S G. 1973 *J. Cell Biol.* **59** 12
[7] Hill H S, Augustine N H, Williams P A, Brown E J and Bohnsack J F. 1993 *Infect Immun* **61** 2334
[8] Kao WJ, Lee D, Schense J C and Hubbell J A. 2001. *J. Biomed. Mater. Res.* **55** 79
[9] Tian F, Cui D, Schwarz H, Estrada G G and Kobayashi H 2006 *Toxicol. In Vitro* **20** 1202
[10] Kodali V K, Roos W, Spatz J P, Curtis J E. 2007 *Soft Matter* **3** 251
[11] Brodbeck W G, Patel J, Voskerician G, Christenson E, Shive M S, Nakayama Y, Matsuda T, Ziats N P and Anderson J M 2002 *Proc. Natl. Acad. Sci. USA* **99** 10287
[12] Jung T, Koneberg R, Hungerer K D and Kissel T. 2002 *Int. J. Pharm.* **234** 75
[13] Jin C, Bai L, Wu H, Tian F and Guo G. 2007 *Biomaterials* **28** 3724
[14] Porter A, Gass M, Muller K, Skepper J N, Midgley PA and Welland M. 2007 Nature nanotech 2 713

Screening strategy to avoid toxicological hazards of inhaled nanoparticles for drug delivery: the use of α-quartz and nano zinc oxide particles as benchmark

Andrea Beyerle[1,2], Holger Schulz[1], Thomas Kissel[2], Tobias Stoeger[1]

[1] Institute of Inhalation Biology, Helmholtz Zentrum München, German Research Center for Environmental Health (GmbH), Ingolstädter Landstraße 1, D - 85764 Neuherberg, Germany

[2] Department of Pharmaceutical Technology and Biopharmacy, Philipps- University Marburg, Germany

e-mail: tobias.stoeger@helmholtz-muenchen.de

Abstract: Nanotechnology is a broad, revolutionary field with promising advantages for new medicine. In this context the rapid development and improvement of so called nanocarriers is of high pharmaceutical interest and some devices are already on the market. In our project we aim to develop well characterized nanoscaled drug delivery systems for an inhalative application. To this end, we focus on the most adverse side-effects within the lung, the cytotoxic and the proinflammatory responses to these nanoparticles (NPs). Before performing any animal experiments, we start with an in vitro screening for analyzing the cytotoxic and proinflammatory effects of the investigated particles on two murine lung target cell lines, the alveolar epithelial like typ II cell line (LA4) and the alveolar macrophage cell line (MH-S). Three different endpoints were estimated, (i) cellular metabolic activity, determined by the WST-1 assay, (ii) membrane integrity, by detection of LDH release and hemolytic activity, and (iii) secretion of inflammatory mediators. To analyze the relative particle toxicity we choose two reference particles as benchmarks, (i) fine α-quartz, and (ii) ultrafine ZnO particles. The investigation of dose-response and kinetics of proinflammatory and toxic effects caused to the named cell lines provide an insight to a close evaluation of our cell based screening strategy. α-quartz is well known for its inflammatory and toxic potential caused by inhalation, and nanosized ZnO particles - used in a broad field of nanotechnology like electronics, but also cosmetics and pharmaceuticals - is to a high degree cytotoxic and proinflammatory in vitro. Preliminary experiments indicated not only particle and cell specific inflammatory responses, but also different susceptibilities of the cell types being exposed to our benchmark particles regarding their size and surface activities. Exposure to the μm-sized α-quartz particles affected the viability of epithelia cells less than that of macrophages, pointing to the impact of particle uptake by phagocytosis. In contrast, the nanosized ZnO particles caused much stronger decrease in cell viability and higher levels of LDH in the macrophage cell line compared to epithelial cells, even though the hemolytic activity was much higher for the α-quartz particles than for the nanosized ZnO. For the proinflammatory effects, we observed a clear dose-dependent release of acute phase cytokines (TNF-α, IL-6, G-CSF> CXCL10>CCL2) for both alveolar cell lines after Min-U-Sil exposure. After ZnO treatment the cytokine responses were negligible compare to control cells. In conclusion, our data attach value to the use of different cell types to detect different pathways of toxicity generated by different particle properties.

Journal of Physics: Conference Series **151** (2009) 012034 doi:10.1088/1742-6596/151/1/012034

Therefore, we will establish both lung target cell lines for an in vitro screening to analyze proinflammatory and cytotoxicity effects of nanocarriers. The implementation of the two reference particles facilitate the validated classification of the cytotoxic responses caused by the NPs investigated.

1. Introduction:

In the field of nanotechnology the emerging developments make demands more and more on considerations of 'nano-hazards' from nano-devices, -application or combustion-derived nanoparticles (NPs). To ensure the safety of such a material or device and to categorize nanomaterial regarding its toxic potential, many researchers focus now on the most crucial issues, namely cytotoxicity and proinflammatory effects, especially for the lung exposure [1]. These issues are of particular interest since instilled non-biodegradable nanospheres have been shown to induce pulmonary inflammation [2]. In addition, upon pulmonary delivery, for similar particles even systemic effects like enhanced thrombosis have been observed [3]. The commonly accepted model suggests, that the toxicity of particulate matter (PM) is related to their ability to generate free radicals and induce oxidative stress and inflammation. Inflammation sounds crucial for PM-exposure related diseases like exacerbations of airways disease and cardiovascular disease [4]. Since sub-micron sized particles could inherently be more toxic due to their increased surface area, this drawback has to be kept in mind [5].

In our project we aim to develop well characterized nanoscaled drug delivery systems for an pulmonary application. A lot of studies dealt with the physicochemical characterization and application behavior of these material [6], but only a few studies focus on the risk assessment [7].

In the literature it has been well described that particle size and surface area are the most important and toxicity driving parameters, and it is thus generally accepted, that at a given mass, smaller sized particles are more toxic than larger sized particles.

In this study, we choose two particles as benchmarks, (i) fine α-quartz, and (ii) ultrafine ZnO particles to estimate the related particle toxicity. Min-U-Sil-5 possess well characteristized physico-chemical properties, is of high purity (98.3% silicium dioxide), and has over many years been extensively studied for its acute inflammatory and cytotoxic lung response [8-11]. Since this fine sized quartz might not reflect the toxic aspects inherent to nanomaterials we additionally included 70nm ZnO particles, a well known lung-toxic nanoparticle from occupational medicine [12-14], as reference. Both particles are form high purity and well characterized by the manufacturers for their physico-chemical properties, all these attributes mandatory for any 'reference material' being applied as positive control. To minimize animal experiments we start with an *in vitro* screening of the two different reference particles described above for better understanding of the involved toxicity driven pathways.

2. Materials and Methods:

2.1. Cell culture

Cell culture experiments were carried out using the murine alveolar epithelial – like type II cells (LA4; ATCC No. CCL-196™) and the murine alveolar macrophages (MH-S; ATCC No. CRL-2019). LA4 cells were grown in HAM's F12 medium with stable Glutamax containing 15% fetal bovine serum (FBS, Gibco, Germany) and 1% non essential amino acids and 100U/ml penicillin and 100mg/ml streptomycin, MH-S cells were cultured in Dulbecco's modified Eagle's medium with stable Glutamax supplemented with 10% fetal bovine serum (FBS, Gibco, Germany) and 100U/ml penicillin and 100 mg/ml streptomycin at 37°C and 5% CO_2. All cells were passaged every 2-3 days. All reagents were obtained from Biochrom AG, Seromed, Germany or otherwise signed.

2.2. Particle preparation
Min-U-Sil 5 (α-quartz) was obtained from U.S. Silica Company, Berkerly Springs, WV, USA, with a median diameter of 1.7μm declared on the datasheet from the manufacturer. Zinc oxide (CAS-No: 1314-13-2) was obtained from Alfa Aesar (A Johnson Matthey Company, Karlsruhe, Germany) and the manufacturer declares a avarage diameter of 70nm. Stock suspensions (10mg/ml) of each particle were prepared in sterile, double-distilled water. Each suspension was sonicated for 15min. prior to exposure and subsequently diluted in sterile, double-distilled water to reach a range of particle concentration from 20-1000μg/ml.

2.3. Cell viability
Cell viability was determined using the Cell Proliferation Reagent WST-1 (Roche Diagnostics, Germany) according to the method of Mosmann [15]. Briefly, LA4, MH-S cells were seeded at a density of 0.25*106 cells/well/2cm² in 24-well-plate in cell culture medium containing FBS and grown overnight in an incubator at 37°C and 5%CO2. For the treatment, the cell culture medium was replaced by freshly, pre-warmed, serum-reduced (2% FBS) cell culture medium without antibiotics. Particles were exposed to the cells for 2h, 6h and 24h and the relative viability [%] related to control samples (untreated cells) was calculated by following equation: Cell viability = (ODsample/ ODcontrol)*100. All data represent at least three independent experiments.

2.4. Cytotoxicity
For detection of the cytosolic enzyme lactate dehydrogenase (LDH) characteristically for membrane damage effects we used the Cytotoxicity Detection Kit (Roche Diagnostics, Germany) according to the manufacturer protocol. The experiments were carried out according to the conditions at the WST-1 assay. After three different time points (2h, 6h, and 24h, the LDH concentration in the cell culture supernatant was spectrophotometrically determined in an ELISA reader (Labsystems iEMS Reader MF) at a wavelength of 492nm. As control served cells treated with 2% (w/v) Triton X-100 according to the manufacturer protocol and set as maximum of LDH release (100%). The relative LDH release is defined by the ratio of LDH released over total LDH in the intact cells (high control). Less than 10% LDH release were regarded as non-toxic effect level in our experiments [16]. All data represent at least three independent experiments.

2.5. Hemolysis test
The hemolytic activity was determined according to [17].Fresh blood were taken from human volunteers and collected in heparinized tubes. Blood was centrifuged at 700g at 4°C for 10min. and washed several times with PBS until the supernatant was colorless. The supernatant was removed and a 2.5% (v/v) suspension of erythrocytes was prepared. 500μl of the 2.5%(v/v) suspension of erythrocytes was mixed with 500μl of the appropriate particle dilution in eppendorf cups. After 60min. incubation time in a shaking water bath, the red blood cells were removed by centrifugation, and the supernatants were investigated spectroscopically at a wavelength of λ=540nm for detection of the release of hemoglobin. As references served PBS (negative) and 0.2% Triton X-100 solution (positive). All experiments were performed in triplicates.

2.6. Enzyme-linked immunosorbent assay
22 cytokines/chemokines were detected simultaneously in the cell culture supernatant by using Luminex technology (Linco Research, St. Charles, MO). In this study, the secretion of following cytokines/chemokines was investigated: IL-1α, IL-1ß, IL-4, IL-5, IL-6, IL-7, IL-9, IL-10, IL-12p70, IL-13, IL-15, IL-17, TNF-alpha, INF-γ, G-CSF, GM-CSF, CXCL1 CXCL10, CCL2, CCL3, and CCL5. The assay was performed as described previously [18]. The mean fluorescence intensity (MFI) was detected by the Multiplex plate reader (Luminex System, Bio-Rad Laboratories, Germany) for each sample (50μl) with a minimum of 50 beads per region being analyzed. The raw data (MFI) were

Journal of Physics: Conference Series **151** (2009) 012034 doi:10.1088/1742-6596/151/1/012034

captured using the Multiplex plate reader software (Bioplex Manager, Version 2.0). For data analysis, a 4-parameter logistic curve fit was applied to each standard curve and sample.

2.7. Statistics

All values are presented as means±standard error (SEM) of at least three independent experiments. Significant differences between two groups were evaluated by Student's t-test or between more than two groups by one-way ANOVA followed by Tukey's multiple comparison test. Statistical analysis was performed using the program STATGRAPHICS PLUS Version 5.0 (Statpoint, Inc., Virginia, US).

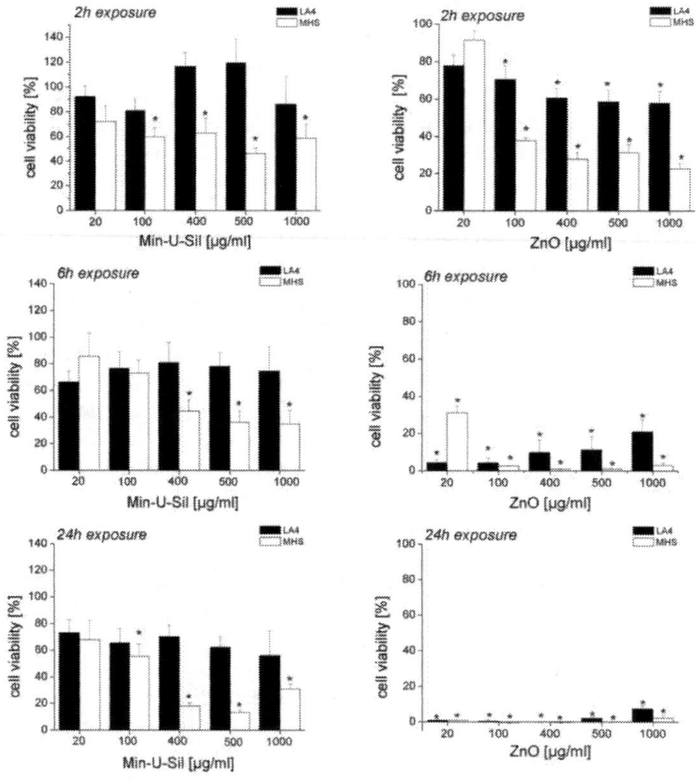

Figure 1: Cell viability as determined by WST-1 Cell Proliferation Assay. Values are expressed as means±SEM (n=4-5). Five different concentrations (20, 100, 400, 500, and 1000µg/ml) of Min-U-Sil 5 (left figures), and ultrafine ZnO (right figures) were exposed to alveolar epithelial cells (LA4, black bars) and alveolar macrophages (MH-S, open bars). Differences to control (untreated cells) with (p)<0.05 were considered statistically significant and marked with an asterisk.

3. Results and Discussion:

3.1. Cell viability

In both alveolar cell lines the cell viability decreased in a time- and dose-dependent manner, which was more prominent in the alveolar macrophages (MH-S). Regarding the two different particle exposures, namely the α-quartz particle with a mean diameter of 1.2μm and the ultrafine ZnO particles with an average diameter of 70nm, we observed a statistically significant higher decrease in metabolic activity after ZnO treatment (Figure 1).

Already after 2h particle exposure the cell viability decreased statistically significant in both cell lines in a dose-dependent manner. After α-quartz treatment the cell viability in the alveolar epithelial cells remained constantly around 80%, even after 24h the cell viability was around 60% with the highest quartz dose. Nevertheless, in the alveolar macrophages we detected a statistically significant decrease in cell viability with a dose more than 100μg/ml Min-U-Sil. Remarkably, after 6h and 24h exposure of the two particles to the alveolar cell lines we could distinguished between a dramatically loss of metabolic activity after ultrafine ZnO treatment and a sustained decrease in cell viability after α-quartz exposure, especially for the alveolar epithelial cells. In all cases the alveolar macrophages (MH-S) were more susceptible than the alveolar epithelial cells. The different particle effects could be explain in general because of the different sizes and surface areas of these paticles.

3.2. Cytotoxicity (LDH release)

In agreement to the WST-1 data, the α-quartz exposure caused much less membrane damage detected by LDH release in the supernatants than the ultrafine ZnO exposure and again, the alveolar macrophages were more sensitive to these particles (Figure 2).

After both particle treatments LDH release was negligible after 2h, less than 10%, which is considered as a threshold for membrane toxic behavior [16]. Upon α-quartz exposure the LDH release remained under 10%, but yielded some statistically significance compare to control (untreated cells). In contrast, after the ultrafine ZnO treatment high levels of LDH were observed in the alveolar macrophages after 6h and 24h, and the maximum was reached after 6h, but not in a dose-dependent manner. In the alveolar epithelial cells the LDH release was under the 10% threshold for membrane toxicity and were therefore negligible similar to the α-quartz exposure. Again, we could show that the ultrafine ZnO particles caused a much higher toxicity, which we correlate to the higher surface area of these particles in comparison to the α-quartz particles according to [19].

Journal of Physics: Conference Series **151** (2009) 012034 doi:10.1088/1742-6596/151/1/012034

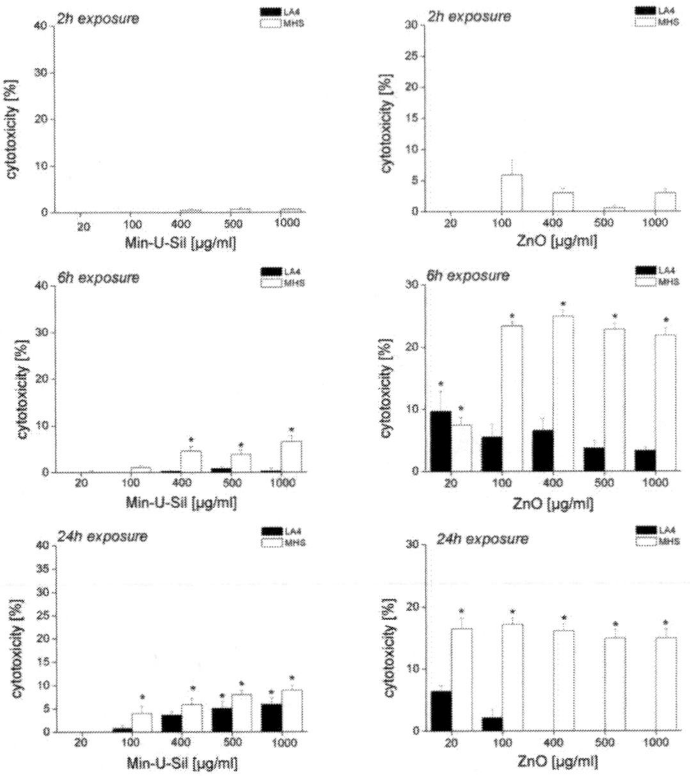

Figure 2: Influence of cell membrane integrity was observed by measuring LDH in the cell culture supernatant after three different time points (2h, 6h, and 24h) in the alveolar epithelial cell line (LA4, black bars) and the alveolar macrophages (MH-S, open bars). Cells were exposured to Min-U-Sil 5 (left figures) and ultrafine ZnO (right figures) with five different concentrations (20, 100, 400, 500, and 1000µg/ml). Values are expressed as means±SEM (n=4-5). Propability (p)<0.05 was considered as significant compared to control (untreated cells) and indicated with an asterisk.

3.3. Hemolytic activity

In addition, we investigated the hemolytic activity of both particles. After 60min. treatment of human red blood cells we observed a dose-dependent elevated release of hemoglobin upon α-quartz treatment, whereas the ultrafine ZnO particles indicated a sustained release of hemoglobin independent from the dose. Both particles caused significantly high hemolytic activity, but in a different manner with regard to the dose.

Remarkably, the hemolytic activity of Min-U-Sil particles was much higher and dose-dependent, whereas Min-U-Sil particles released less LDH, which indicated only few membrane damage effect. In contrast, the ultrafine ZnO particles caused very high sustained LDH release already after 6h as well as a very high sustained release of hemoglobin, but much lower compare to the hemolytic activity of the α-quartz particles. The different underlying mechanism regard to cell death need to be observed in further experiments.

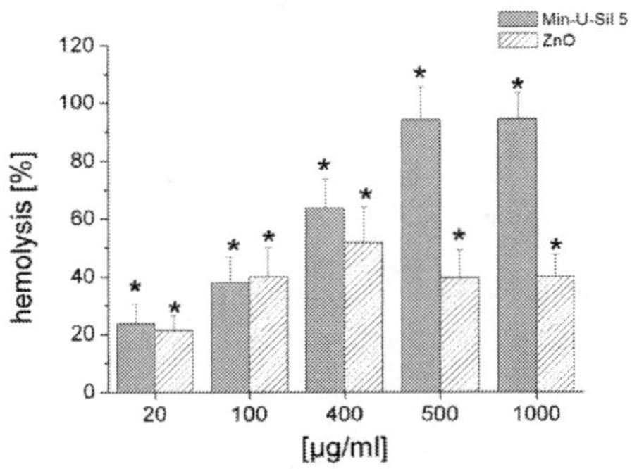

Figure 3: Measurement of the hemolytic activity after 60min. treatment of human red blood cells (2.5% (v/v)) with five different concentration of Min-U-Sil particles (grey bars) and ultrafine ZnO particles (hatched bars). Values are represented the mean±SEM of three independent experiments. Propability (p)<0.05 was considered as significant compared to control (untreated cells) and indicated with an asterisk.

3.4. Cytokine response

The cytokine secretion of 22 proinflammatory cytokines were detected by multiplex technique. From this cytokine profiling five cytokines, mainly acute phase cytokines (TNF-α, IL-6, G-CSF, CXCL10, and CCL2) were induced more than 2-fold compare to controls (untreated cells), see table 1. For Min-U-Sil we observed the most significant changes in a dose-dependent manner in both cell lines. In contrast, ZnO exposure caused only slightly elevated cytokine levels, but without any statistically significance. Due to very high cytotoxic effects we observed the proinflammatory effects of ZnO particles only for two non toxic concentrations (1.25µg/ml and 2.5µg/ml), where the cell viability is more than 80% and the LDH release is negligible after 24h treatment (data not shown).

Table 1: Relative cytokine secretion in LA4 and MH-S cells after 24h treatment of Min-U-Sil (20µg/ml,100µg/ml, and 400µg/ml) and ultrafine ZnO (1.25µg/ml, and 2.5µg/ml) determined by multiple analyte detection immunoassay. Means were normalized to control cells (untreated cells). Values represent mean ± SEM, n=2-3; b.d.l. = below detection limit; statistically significant (p<0.05) changes compared to control levels (untreated cells) are marked with an asterisk.

	TNF-a		**IL-6**		**G-CSF**		**CXCL10**		**CCL2**	
	LA4	MH-S	LA4	MH-S	LA4	MH-S	LA4	MH-S	LA4	MH-S
Min-U-Sil										
20µg/ml	1.2±0.9	1.5±0.2	1.2±0.8	1.3±0.5	1.3±0.4	3.2±1.3	2.6±0.6	1.1±0.3	1.3±0.7	1.6±0.3
100µg/ml	3.4±2.5	3.5±1.3	4.6±2.3	2.2±0.4	2.0±0.6	5.9±1.7	3.6±0.2	1.5±0.5	1.6±0.8	1.5±0.4
400µg/ml	5.6±2.9	*****15.4±5.4**	7.2±3.0	2.2±0.3	3.1±0.9	4.1±1.9	1.9±1.3	2.0±1.0	1.0±0.4	0.7±0.2
ZnO										
1.25µg/ml	b.d.l.	0.9±0.2	0.4	0.8±0.0	b.d.l.	2.1±0.8	3.9	0.4±0.1	6.3	0.4±0.0
2.5µg/ml	b.d.l.	1.2±0.3	0.5±0.4	1.2±0.3	b.d.l.	2.1±0.9	1.6±1.2	1.1±0.2	4.9±3.5	1.2±0.8

4. Conclusion:

In this study, we detected different time- and dose-dependent cytotoxic and proinflammatory responses of the two different particles. The ultrafine ZnO particles caused high, but sustained membrane damage effects confirming to a high decrease in metabolic activity, whereas the α-quartz particles revealed much less decrease in cell viability and no significant LDH release, but higher hemolytic activity in a dose dependent manner. Regarding the two different cell lines cytotoxic effects were more prominent in the alveolar macrophage cell line (MH-S). For the proinflammatory effects, we observed a clear dose-dependent release of acute phase cytokines (TNF-α, IL-6, G-CSF, CXCL10, CCL2), and to a higher extent in the macrophages, and only after Min-U-Sil exposure. After ZnO treatment the cytokine responses were negligible compare to control cells, but this might be because of the very low doses (1.25μg/ml and 2.5μg/ml).

In summary, we aimed to etablished two positive reference particles for our *in vitro* assessment of nanoscaled pulmonary drug delivery and discovered high cytotoxic effects in a dose- and time-dependent manner for both particles, most prominent in macrophages and for the ultrafine ZnO particles. Interestingly, the proinflammatory effects seem to be more pronounced after Min-U-Sil treatment. Thus, we concluded that for further *in vitro* assessment of nanocarriers for pulmonray applications it is necessary to estimated additionally the effects of two positive reference particles such as fine Min-U-Sil and ultrafine ZnO.

References

[1] [Donaldson, K. and V. Stone, *Current hypotheses on the mechanisms of toxicity of ultrafine particles.* Ann Ist Super Sanita, 2003. **39**(3): p. 405-10.

[2] Brown, D.M., Wilson M.R:, MacNee W., Stone V., Donaldson K., *Size-dependent proinflammatory effects of ultrafine polystyrene particles: a role for surface area and oxidative stress in the enhanced activity of ultrafines.* Toxicol Appl Pharmacol, 2001. **175**: p. 191-199.

[3] Nemmar, A.H.M.F., Hoet Peter H.M., Vermylen Josef, Nemery Benoit, *Size effect of intratracheally instilled particles on pulmonary inflammation and vascular thrombosis.* Toxicol Appl Pharmacol, 2003. **186**: p. 38-45.

[4] Duffin, R., N.L. Mills, and K. Donaldson, *Nanoparticles-a thoracic toxicology perspective.* Yonsei Med J, 2007. **48**(4): p. 561-72.

[5] Oberdorster, G., et al., *Principles for characterizing the potential human health effects from exposure to nanomaterials: elements of a screening strategy.* Part Fibre Toxicol, 2005. **2**: p. 8.

[6] Mao, S., et al., *Influence of polyethylene glycol chain length on the physicochemical and biological properties of poly(ethylene imine)-graft-poly(ethylene glycol) block copolymer/SiRNA polyplexes.* Bioconjug Chem, 2006. **17**(5): p. 1209-18.

[7] Fischer, D., Li Y. , Ahlemeyer B., Krieglstein J., Kissel T., *In vitro cytotoxicity testing of polycations: influence of polymer struture on cell viability and hemolysis.* Biomaterials, 2003. **24**: p. 1121-1131.

[8] Albrecht, C., et al., *Inflammatory time course after quartz instillation: role of tumor necrosis factor-alpha and particle surface.* Am J Respir Cell Mol Biol, 2004. **31**(3): p. 292-301.

[9] IARC, *Silica.* IARC Monogr Eval Carcinog Risks Hum, 1997. **68**: p. 41-242.

[10] Sayes, C.M., K.L. Reed, and D.B. Warheit, *Assessing toxicity of fine and nanoparticles: comparing in vitro measurements to in vivo pulmonary toxicity profiles.* Toxicol Sci, 2007. **97**(1): p. 163-80.

[11] Warheit, D.B., et al., *Pulmonary bioassay studies with nanoscale and fine-quartz particles in rats: toxicity is not dependent upon particle size but on surface characteristics.* Toxicol Sci,

2007. **95**(1): p. 270-80.

[12] Fine, J.M., et al., *Metal fume fever: characterization of clinical and plasma IL-6 responses in controlled human exposures to zinc oxide fume at and below the threshold limit value.* J Occup Environ Med, 1997. **39**(8): p. 722-6.

[13] Brunner, T.J., et al., *In vitro cytotoxicity of oxide nanoparticles: comparison to asbestos, silica, and the effect of particle solubility.* Environ Sci Technol, 2006. **40**(14): p. 4374-81.

[14] Karlsson, H.L., et al., *Copper Oxide Nanoparticles Are Highly Toxic: A Comparison between Metal Oxide Nanoparticles and Carbon Nanotubes.* Chem Res Toxicol, 2008.

[15] Mosmann, T., *Rapid colorimetric assay for cellular growth and survival: application to proliferation and cytotoxicity assays.* J Immunol Methods, 1983. **65**(1-2): p. 55-63.

[16] Choksakulnimitr Suthummar, M.S., Tokuda Hideaki , Takakura Yoshinobu, Hashida Mitsuru, *In vitro cytotoxicity of macromolecules in different cell culture systems.* J Control Release, 1995. **34**: p. 233-214.

[17] Parnham, M.J. and H. Wetzig, *Toxicity screening of liposomes.* Chem Phys Lipids, 1993. **64**(1-3): p. 263-74.

[18] Prabhakar, U., et al., *Validation and comparative analysis of a multiplexed assay for the simultaneous quantitative measurement of Thl/Th2 cytokines in human serum and human peripheral blood mononuclear cell culture supernatants.* J Immunol Methods, 2004. **291**(1-2): p. 27-38.

[19] Stoeger, T., et al., *Instillation of six different ultrafine carbon particles indicates a surface area threshold dose for acute lung inflammation in mice.* Environ Health Perspect, 2006. **114**(3): p. 328-33.

Development of a Bolus Injection System for regional deposition studies of nanoparticles in the human respiratory system

V Koujalagi[1], S L Ramesh[2], G P P Gunarathne[2], S Semple[1] and J G Ayres[1]

1 Dept. of Environmental and Occupational Medicine, Liberty Safe Work Research Centre, University of Aberdeen, Foresterhill Road, Aberdeen AB25 2ZP, UK

2 School of Engineering, The Robert Gordon University, Schoolhill, Aberdeen AB10 1FR, UK

Email: g.gunarathne@rgu.ac.uk

Abstract. This study presents the work carried out in developing a precision bolus injection system in order to understand the regional deposition of nanoparticles (NP) in human lung. A real-time control system has been developed that is capable of storing graphite NP, assessing human breathing pattern and delivering a bolus of the stored NP at a pre-determined instance of the inhalation phase of breathing. This will form the basis for further development of a system to deliver radioactive nanoparticles to enable 3-dimensional lung imaging using techniques such as positron emission tomography (PET). The system may then be used to better understand the actual regional deposition in human lung, which could validate or challenge the current computational lung models such as that published by the International Commission for Radiation Protection (ICRP-1994). A dose related response to inhaled PM can possibly be shown, which can be used to review the current workplace exposure limits (WELs).

1. Introduction

Particulate matter (PM) deposited in human lung are known to cause adverse health effects both in normal subjects and in patients with chronic obstructive pulmonary disease (COPD), asthma and cardiovascular diseases. PM is composed of both fine (micrometer) and ultrafine (or nano-particles, NP) particles and it is hypothesised that the NP may cause greater health hazards than their larger counterparts [1].

The advent of nano-technology means that engineered nano-particles are now being produced in large amounts which might represent a risk to the workforces involved. At present there is no clear idea of how these might have an impact on the health and therefore there are no current Workplace Exposure Limits (WEL) for nano-materials. Also, the amount of exposure in an individual is also associated with rate of respiration i.e., the physical effort, and hence the dose. Hence, the system measuring the impact should take the inhalation patterns and lung volume into consideration.

In this study a method to achieve a controlled delivery of bolus of graphite NP into the lungs, at different instances in the inhalation cycle was developed. The prototype bolus injection system comprising of an electro-mechanical system and a real-time software control system was used to control the entire operation of bolus delivery.

2. Background

Some of the previous studies are based on the hypothesis that different size fractions of the PM, preferentially deposit in different regions of lung [2]. There is some data available on the mechanism of deposition; however, they mostly deal either with lung models [3], computer simulations [4] or mathematical calculations [2, 5]. The (1994) ICRP model of lung deposition [6], shown below in figure 1, is primarily based on the principles of computational fluid dynamics. The accuracy of such a model has been questioned due to inaccurate assessment of Brownian motions with respect to particle deposition [7, page 26].

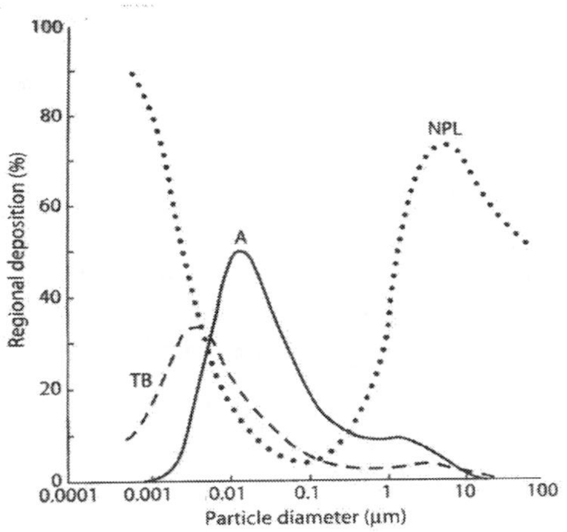

A - Alveolar deposition;
TB - Tracheo-bronchial deposition;
NPL- deposition in nose pharynx and larynx

Source: Occupational Hygiene [6]

Figure 1. ICRP model for lung deposition of particles (1994)

Hence, to understand the exact method of NP deposition and their distribution within the lung it is necessary to carry out the study *in vivo*. This involves:

- A mechanism for storage and controlled delivery of recently-generated NP to lung.
- Ability to select different PM size distributions to study preferential deposition.
- Identifying the areas where the NP actually deposited by means of imaging such as Positron Emission Tomography (PET).

This study partially fulfils the 1st and to a lesser extent 2nd requirement and will form the basis for development of 3rd stage which will involve the design of a clinical system for controlled delivery of radioactive labelled NP (such as Technegas) required for PET lung imaging.

3. System Design and Implementation

The first prototype system has been developed with the general assumption that the particles inhaled at the beginning of inhalation has a greater tendency to deposit in distal lung (e.g. alveolar region) and the ones inhaled towards the end of inhalation deposit more proximally (e.g. tracheo-bronchial region).

So, if short pulses of bolus were to be injected at a given point in the inhalation cycles as in figure 2, then cumulative accumulation of particles in a specific region could be achieved.

As shown in the ICRP model (figure 1), there is also the tendency for size dependant regional deposition. This can now be observed by using different size fractions, introduced as short pulses to compare the size dependant deposition at a specified target site. In this way, the need for continuous inhalation of large doses to study regional deposition may be avoided.

3.1. System Requirements

To achieve the above objectives the first prototype system was designed to comply with the following main end-user requirements [8].

1. Breath activation – the system should be able to detect the start of breathing cycle and activate the data acquisition and the control system.
2. Breath Pattern detection – Examine breath pattern histograms and identify the maximum (Vmax) and minimum (Vmin) lung volumes on the basis of a prescribed quantitative method, such as moving averages.
3. Programmable bolus size – Once programmed, the system should determine the parameters necessary to deliver the bolus as in equation (1) below.
4. Deliver a bolus of NP aerosol at a selected point of the inhalation cycle.
5. Programmable site of deposition in the lung as determined by the time of release.

The desired deposited dose *W (mg)* of a bolus of PM may be written as:

$$W(deposited)mg = \Delta w_{(T^0,t,p)} \times D_f \times \Delta t \times n \quad (1)$$

Where, $\Delta w_{(T^0,t,p)}$ is the weight of the PM of a chosen size distribution, released during a time window of Δt; D_f is the deposition factor based on inhalation and exhalation concentration, and n is the chosen number of breath cycles over which the total weight of the bolus is dispensed [9]. The subscripts (T^0, t, p) denote that Δw is dependent on the temperature, elapsed time of storage of the aerosol and the pressure (p) at which the aerosol was stored. The whole operation can be represented as shown in figure 2 below.

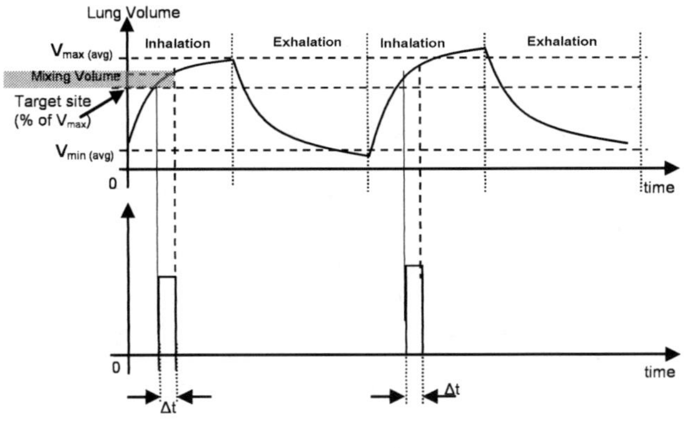

Figure 2. Specified lung target (% of Vmax(avg) – Vmin (avg)) and bolus delivery period (Δt) [7]

3.2. System Description

The system block diagram is shown in Figure 3 below. An external particle generator (GFG1000, PALAS - not shown) generates the graphite NP and is let into the system via the pump, which fills the chambers C_2 and C_3. The system is ready to deliver the NP once the chambers reach specified pressures. This pressure is maintained at the same level throughout the operation. Storing the PM under pressure would enable the release of a higher concentration of aerosol within a short time window (Δt), for a given bolus size. This improves target resolution as it allows for a finer control of the site of delivery in the lungs.

Inhaled Particles X, (23–25 September 2008, Manchester) IOP Publishing
Journal of Physics: Conference Series **151** (2009) 012035 doi:10.1088/1742-6596/151/1/012035

The data acquisition and control system measures the breathing flow rate using the mass flow meters (MFM) and also calculates the lung volume and average tidal volume. A bolus of aerosol is injected during inhalation at the required lung volume by opening valve S_3 for a fixed amount of time specified by the user.

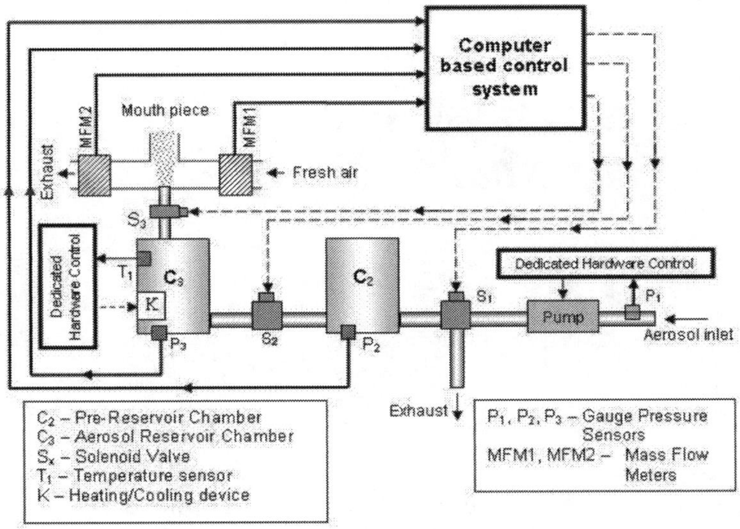

Figure 3. System block diagram

3.3. The software

The control system was initially implemented using National Instruments LabVIEW software (ver.8.0). However, significant jitter in the release valve timing was found due to non-deterministic execution delay, affecting the trigger point at which the bolus was released. Hence the control system was redesigned employing a Real-time Compact Reconfigurable I/O embedded controller (cRIO-9004, National Instruments) that runs the LabVIEW code in real time. The graphical user interface is shown in the figure 4. As shown, the controls on the front panel are used for controlling different operations and are self explanatory.

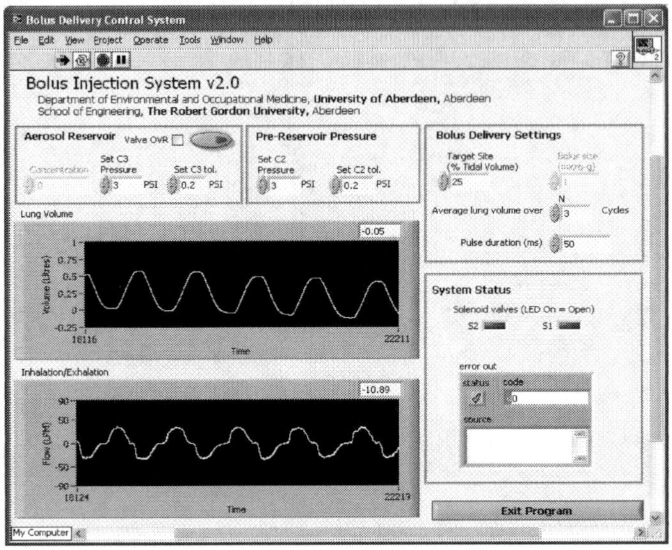

Figure 4. User Interface for control system

The block level architectural diagram and the code for the main program are shown in figure 5 and 6 respectively. The cRIO runs the actual control software in real-time and the PC resident software simply acts as a host which provides graphical user interface and display. With this approach, the problem of uncertainty of time (jitter) was resolved.

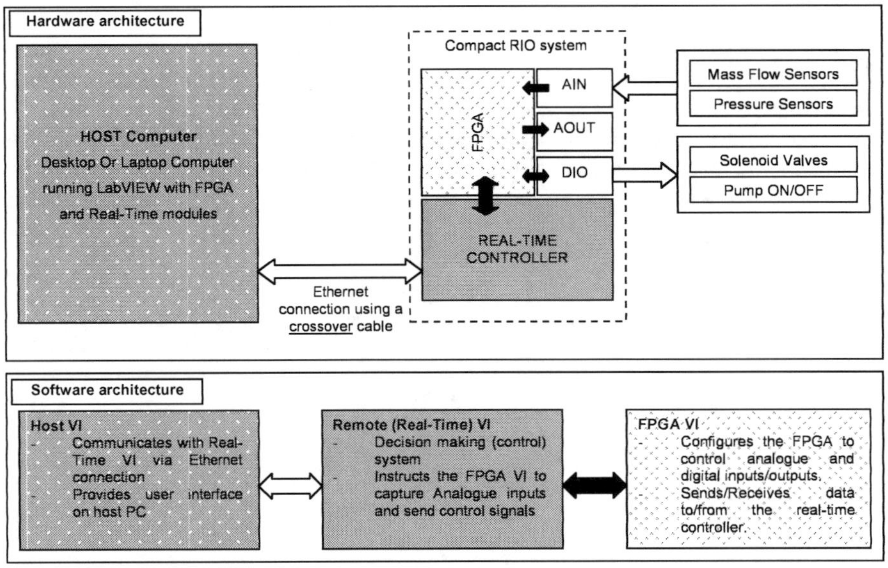

Figure 5. Block level architectural diagram

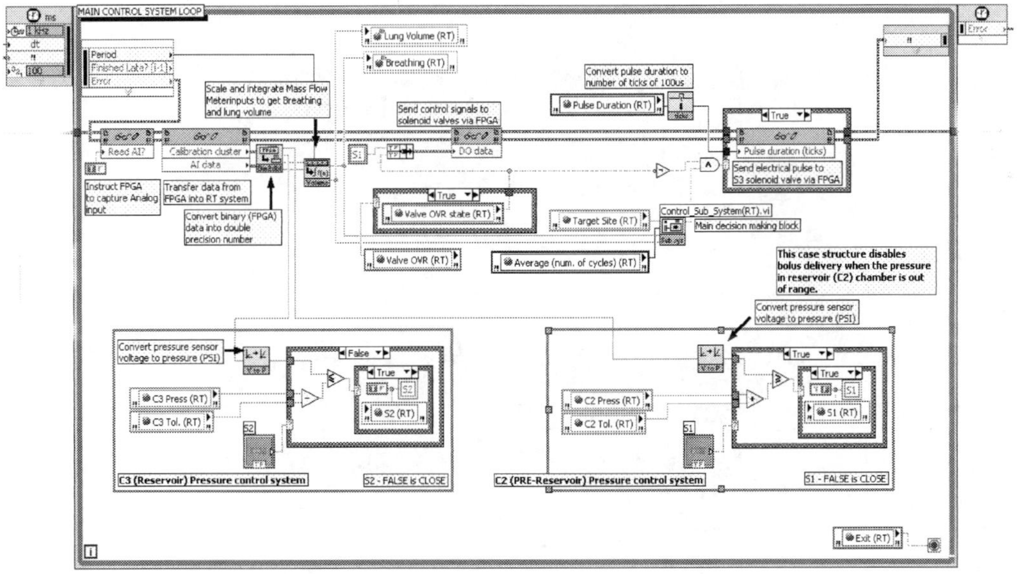

Figure 6. LabVIEW code for main program

3.4. Hardware Modifications
It was found that the results using the original mouthpiece led to some errors in the lung volume calculations by the software (constant drifting). This can be seen from figure 8(a) in which the experiment was carried out after attaching the mouthpiece to a sinusoidal air pump. The problem of drifting was found to be due to flow disturbances introduced during the exhalation phase. It was corrected by introducing laminar flow using an additional smoothing filter on the mouthpiece as shown in figure 7 below.

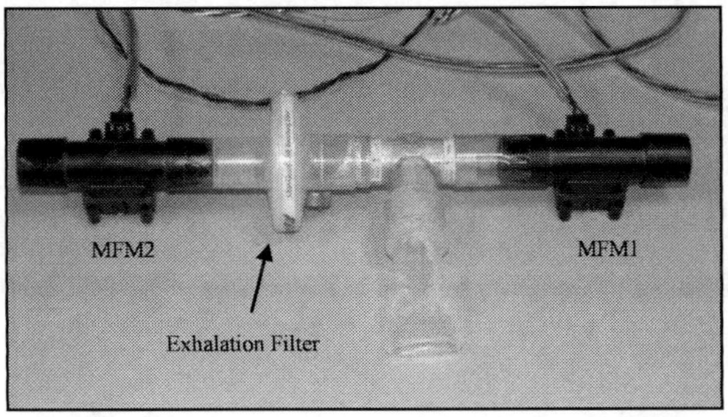

Figure 7. Modified mouthpiece assembly with Filter

Figure 8(b) shows the current response of the flow system after the above modification.

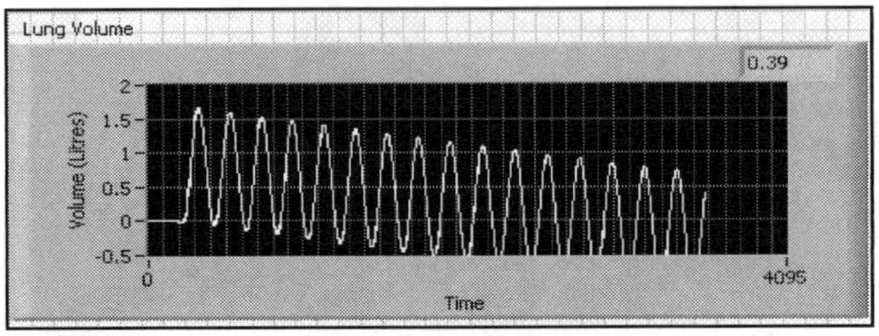

(a)

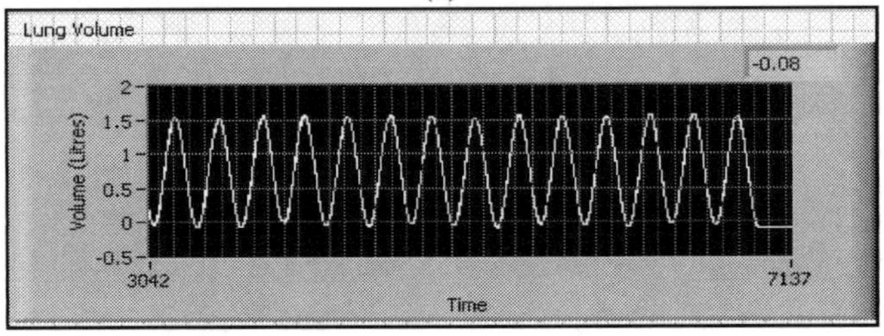

(b)

Figure 8. Flow sensor response (Volume) graphs (a) before and (b) after installation of smoothing filter

In addition to the above modifications, it was found out that if the system were to be used without any imposed breath pattern, then the changes in each breath cycle makes the volume integrator base line to continue shifting upwards or downwards depending on the lung volume calculations. For this reason it was decided to add a facility in the software that automatically resets the integrator for each cycle to enable constant base-line to be maintained as shown in figure 9 below.

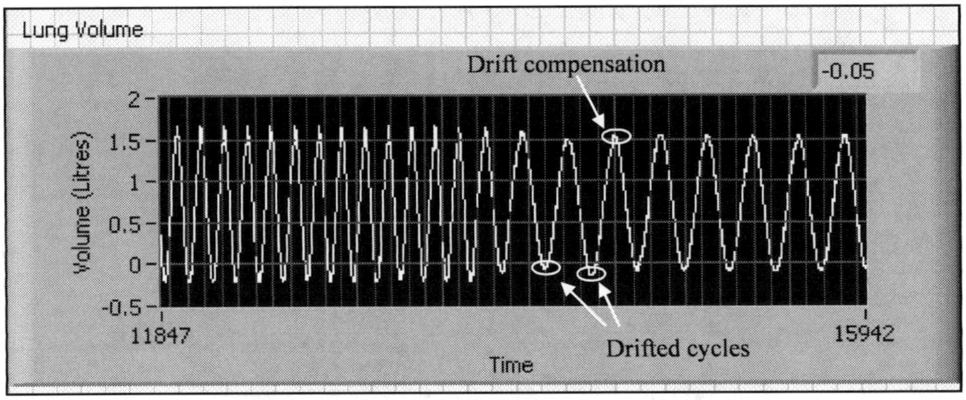

Figure 9. Ilustration of integrator reset function

4. Discussion of results

The modified system with real time hardware was successfully demonstrated as seen from figure 4, 8(b) and 9. With regard to the properties of PM passing through the system it was observed that the PM size distribution increased with storage time, due to likely causes such as agglomeration and aggregation. This can be seen by comparing figure 10(a) and (b) below. Figure 10(a) shows the size distribution when the generated aerosol is free flowing through the system without being stored. In contrast, figure 10(b) represents typical distribution graph when the aerosol is stored at 1.3PSI for 1min 30sec. It is evident from the graphs that there were:

- Increase in the mean particle size of PM during storage.
- Reduction in peak values of particle numbers and mass. (This may be attributed to deposition or settling within the storage chambers.)

The Mass Summation curve can be used to see whether a particular mass distribution remains within a required limit of the maximum size distribution. For example in figure 10 (a), over 98% of particles are still within a maximum of 100nm range, whereas in figure 10(b) this has been reduced to about 92%.

Further experimentation revealed that in this particular arrangement, introducing agitation within the storage chamber and the shearing forces at the valve openings, there is a possibility to influence the PM size distribution. The effect of agitation was observed by installing a miniature fan inside chamber C_3. The effect of switching the fan ON, on the particle distribution can be observed by comparing the figure 10(b) and (c). There is an observable reduction in peak particle size distribution and particle mass distribution. It should be mentioned that only one type of PM (graphite) was used, and further experimentation with other type of agitations is required to validate this result, as the fan may itself introduce complications.

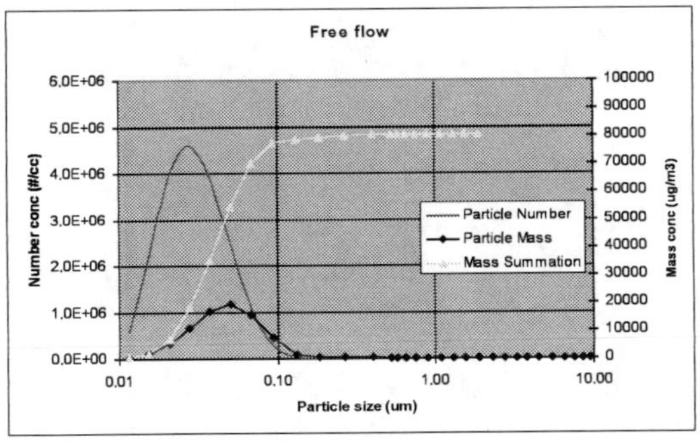

(a)

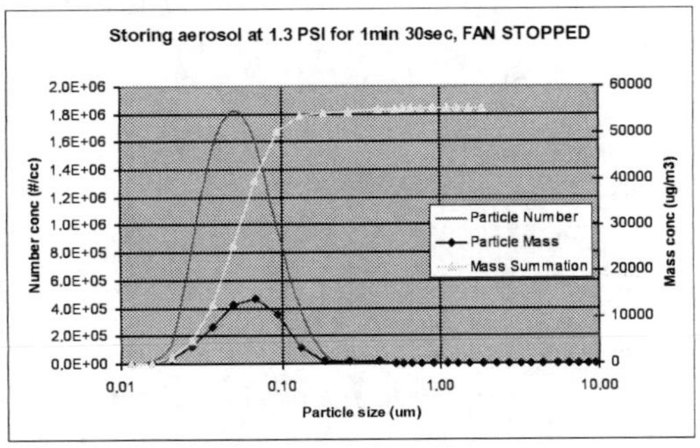

(b)

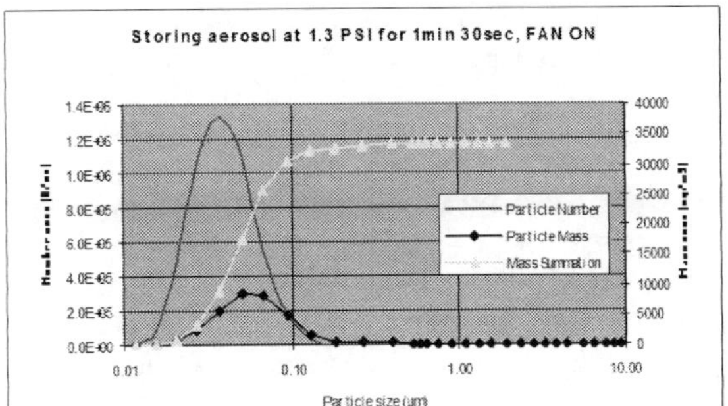

(c)

Figure 10. Variation of particle size distribution with pressure and elapsed time

5. Conclusion

A prototype system for delivery of controlled bolus of PM has been designed. The main control functionalities according to original design specifications have been demonstrated.

Some of the factors affecting the changes in PM size distribution have been observed and the possibility of maintaining the size distribution by impeding the process of particle growth was also observed for one type of NP (graphite), under specified conditions.

From this first experimental model, a number of parameters for re-designing the system towards a clinical prototype have been established.

6. Further Work

Further experimentation is needed to quantify the effect of introducing a disturbance into the storage chamber and increasing shear forces at the point of discharge. This has to be repeated for different aerosols in order to justify that it is possible to break at least in part, the aggregated/agglomerated NP to smaller sizes in order to maintain a required size distribution.

The system would then have to be tested with radioactive labelled NP in a precise manner in safe conditions before assessing the regional deposition in human lung. Techniques for selecting a particular particle size range may also be incorporated. The new design is likely to consist of a much smaller and easily interchangeable discharge cell to enable the use of different aerosols. This work is in progress.

Acknowledgement

This project is the result of a collaborative venture between the School of Engineering of the Robert Gordon University, UK and the Department of Environmental and Occupational Medicine of the University of Aberdeen, UK. Authors wish to express their gratitude to all those who helped facilitate this work.

References

[1] Donaldson K, Tran L, Jimenez LA, Duffin R, Newby DE, Mills N, et al. 2005 Combustion-derived nanoparticles: A review of their toxicology following inhalation exposure *Particle and Fibre Toxicology* **2005(2)** 10

[2] Kim CS, Hu SC, DeWitt P and Gerrity TR 1996 Assessment of regional deposition of inhaled particles in human lungs by serial bolus delivery method *J. Appl. Physiol*. **81(5)** 2203

[3] Chan TL and Lippmann M 1980 Experimental measurements and empirical modelling of the regional deposition of inhaled particles in humans *Am. Ind. Hyg. Assoc. J.* **41(6)** 399

[4] Zhang Z and Kleinstreuer C 2002 Cyclic micron-size particle inhalation and deposition in a triple bifurcation lung airway model *J. Aerosol Sci.* **33(2)** 257.

[5] Martonen TB 1992 Deposition patterns of cigarette smoke in human airways *Am. Ind. Hyg. Assoc. J.* **53(1)** 6

[6] Harrington JM and Gardiner K (editors) 1995 *Occupational Hygiene* 3rd ed. (Cambridge, MA: Blackwell Science)

[7] Koujalagi V 2006 Development of a New Method for the Study of Exposure to Nanoparticles within the Respiratory System in order to Better Define and Understand Health Risks *MSc Dissertation* University of Aberdeen, Aberdeen, UK.

[8] Koujalagi V, Gunarathne GPP, Semple S and Ayres JG 2008 Development of a Bolus Injection System to achieve storage and controlled delivery of nanoparticles to human lung *Preprint IAEng Journal special issue ICEEE_57*

[9] Gunarathne GPP, Koujalagi V, Semple S and Ayres JG 2008 A Real-time Instrumentation System for the Study of Regional Deposition of Airborne Particles and Drug Delivery in the Human Respiratory System *IEEE Int. Instr. Meas. Tech. Conf. Proc.* 2015

Generation of Nanoparticle Agglomerates and their Dispersion in Lung Serum Simulant or Water

B A Wong[1], D G Nash[2], O R Moss[1]

[1]The Hamner Institutes for Health Sciences, PO Box 12137, Research Triangle Park, NC 27709

[2]Biostatistics, University of North Carolina, Chapel Hill, NC 27599

bwong@thehamner.org

Abstract. Nanoparticles released into the atmosphere, due to their high diffusivity, will likely begin to agglomerate. The state of agglomeration upon inhalation and the potential to disperse back into nanoparticles may affect the toxicity of the inhaled material. In order to investigate particle dispersion, a system was set up to generate aggregates from agglomerates. Primary particles, composed of zinc, were generated using zinc rods in a spark generator (Palas GFG-1000, Karlsrhue, Germany). These particles formed agglomerates which were passed through a room temperature aging chamber or through a tube furnace (Carbolite HST, Derbyshire, UK). Agglomerate size was measured with a scanning mobility particle sizer (SMPS model 3936, TSI Inc., Shoreview, MN). When furnace temperature was set near the zinc coalescence temperature, instead of decreasing in size, agglomerate size increased up to 30%; a percentage increase duplicated with the room temperature aging chamber. Starting with an aerosol of primary zinc particles, equal concentrations of agglomerate and aggregrate aerosol were produced. The extent of breakup and dispersion of agglomerates and aggregates to individual nanoparticles in lung serum simulant will be assessed using transmission electron microscopy.

1. Introduction

With the rapid emergence of nanotechnology and increased opportunity for exposure to nanomaterials, questions regarding the toxicity of nanoparticles and agglomerates continue to emerge as well. Initial studies investigating the toxicity of nanoparticles suggest that these particles possess enhanced toxicity beyond that of larger particles composed of the same species [1,2,3].

However, the question of nanoparticle toxicity is not as simple as examining individual particles only. Due to their high diffusivity, nanoparticles should exist as individual particles for only a short time and will agglomerate rapidly. Therefore, it is essential to understand the fate of these agglomerates upon inhalation to ascertain whether subsequent toxicological effects are attributable to the nanoparticle physical properties or is a function of their chemical composition. To date, studies on the degree of nanoparticle agglomeration have involved examining nanoparticle powders in solution [4] but not in the aerosol state.

In this study, we used zinc to generate agglomerates of constituent nanoparticles. Zinc is a relevant species to study as humans are exposed to it in different forms and from varying sources. For one, zinc is found bound to ligands in red meat and animal protein that facilitate its absorption [5]. Zinc has also been identified as a component of industrial emissions [6]. It is also detected in ambient air samples at higher levels than many other metals [7,8]. Exposure to zinc fumes in the workplace have shown a tendency to cause pulmonary inflammation [9, 6], thus making it an appropriate species to use for these initial agglomerate studies.

2. Experimental Procedure

Primary zinc nanoparticles were generated using zinc rods in a spark generator (GFG-1000, Karlsrhue, Germany). Before exiting the spark generator, these primary particles form agglomerates due to their high diffusivity. The degree of this agglomeration was controlled by adjusting the flow of dilution air both at the point of nanoparticle generation, as well as immediately after exiting the generator (Figure 1). Upon exiting the spark generator, agglomerates were sent into a glass tube for aging and dilution. Nanoparticles and agglomerates could then be sent through an aging volume for further agglomeration. Next, agglomerates were sent through a furnace (Carbolite HST, Derbyshire, UK) to be sintered to varying degrees by adjusting the furnace temperature, or sent through a bypass line for unsintered agglomerates. At this point, the agglomerates were characterized with a scanning mobility particle sizer (SMPS, model 3936, TSI Inc., Shoreview, MN) and collected on copper TEM grids using a nanometer aerosol sampler (NAS, model 3089, TSI Inc., Shoreview, MN) for TEM imaging.

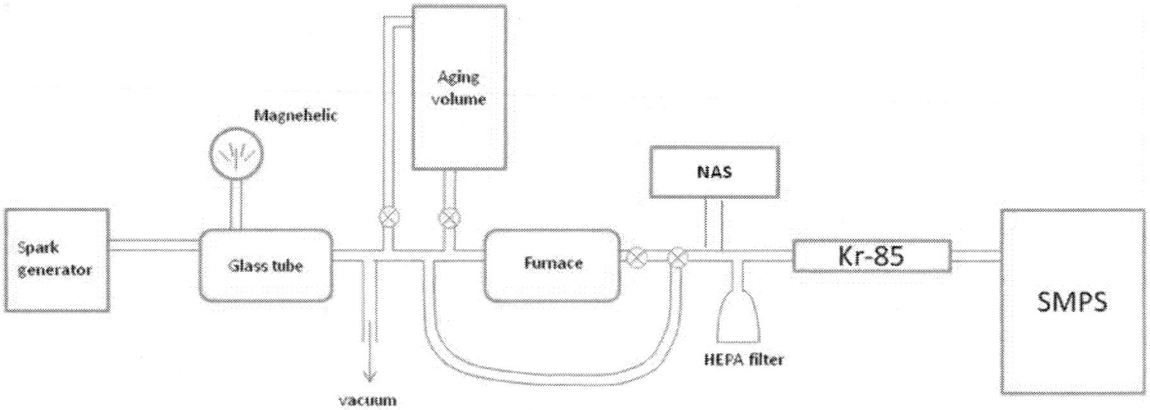

Figure 1: Experimental set up to produce agglomerated zinc nanoparticles. In this figure, NAS = nanometer aerosol sampler, and SMPS = scanning mobility particle sizer

3. Imaging Agglomerate dispersion and translocation

In order to view agglomerates and the effect of immersion in a liquid, a drop of a biologically relevant fluid such as deionized (DI) water or lung serum simulant (LSS) [10] was then placed on the TEM grid. After allowing the liquid to evaporate, TEM images of the grids were again taken to examine the degree to which agglomerates that had been sintered to varying degrees had dispersed in the liquid.

Another method was also employed to exposed agglomerates to pure water. Zn agglomerates that had been collected on TEM grids and shadow coated were then exposed to water. Humidified air passed over the grids placed on a cooled surface (Figure 2). Moisture condensed onto the grid, wetting the surface without significant movement of liquid.

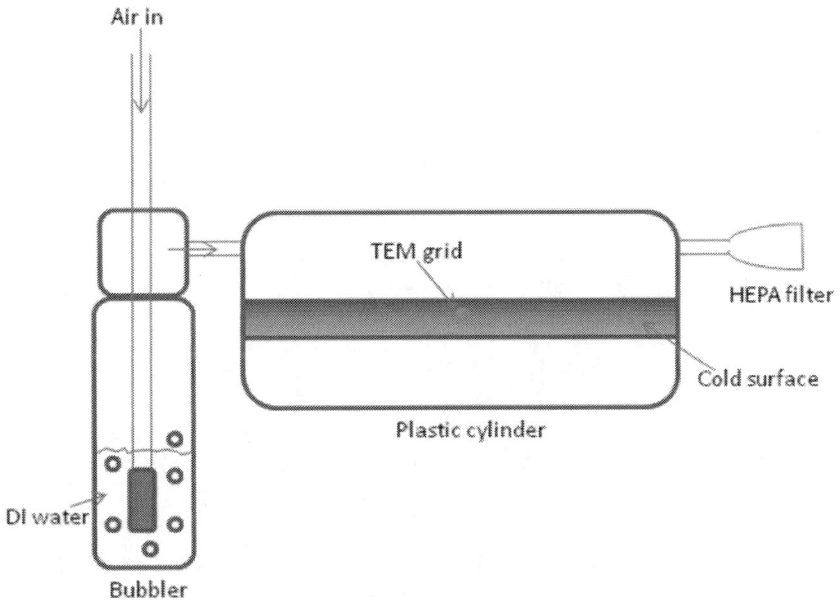

Figure 2: Schematic of DI water application method involving humidified air

4. Results and Discussion

Initially, TEM grids, on which agglomerates that had been heated in a furnace to varying degrees had been deposited, were exposed to lung serum simulant (LSS) to observe its effect on the extent of agglomeration. An example of TEM images taken before and after exposure to LSS are shown in Figure 3. Comparing the images in Figure 3, it appears that the Zn agglomerates indeed dispersed following the application of LSS as evidenced by the lack of chain-like agglomerates in Figure 3b.

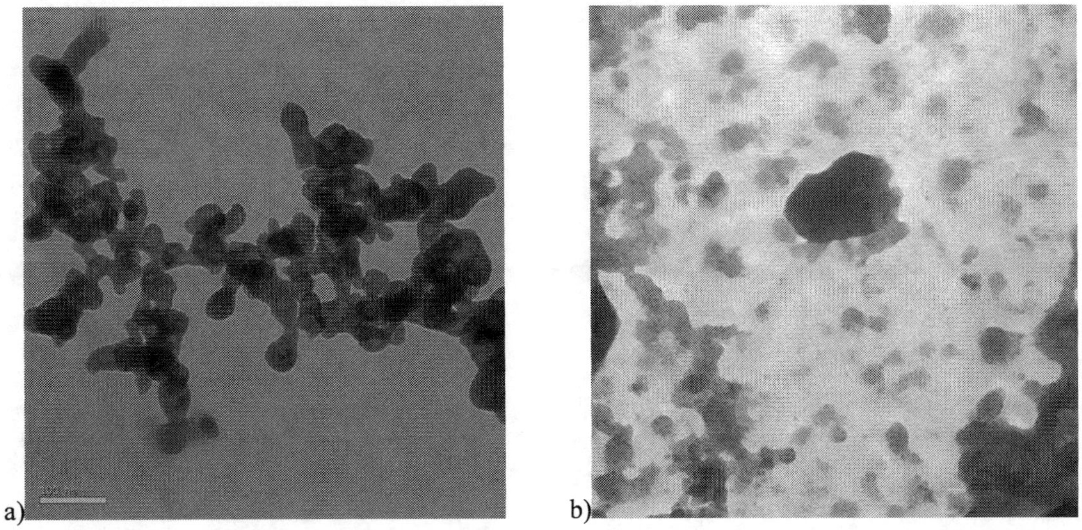

Figure 3: TEM images of unsintered Zn agglomerates a) before and b) after exposure to LSS

Journal of Physics: Conference Series **151** (2009) 012036 doi:10.1088/1742-6596/151/1/012036

One difficulty with this technique, however, was that upon evaporation, compounds dissolved in the LSS precipitated out, making it difficult to see the Zn nanoparticles or distinguish them from precipitated LSS components, as shown in Figure 4. Given this particular difficulty, application of DI water only to grids coated with Zn agglomerates was examined and is shown in Figure 5. The images in Figure 5 suggest that not only dispersion, but perhaps also dissolution of the agglomerates is occurring. In figure 5a, agglomerates are uniformly present across the grid, while in figure 5b, after application of DI water they are non-existent with what appears to be diffuse particle remnants from dissolution of the original agglomerates. While zinc is insoluble in water, it is slightly soluble in slightly acidic conditions. Since DI water is indeed slightly acidic (pH ~ 6.7), it is possible that these nanoparticles do dissolve to some extent.

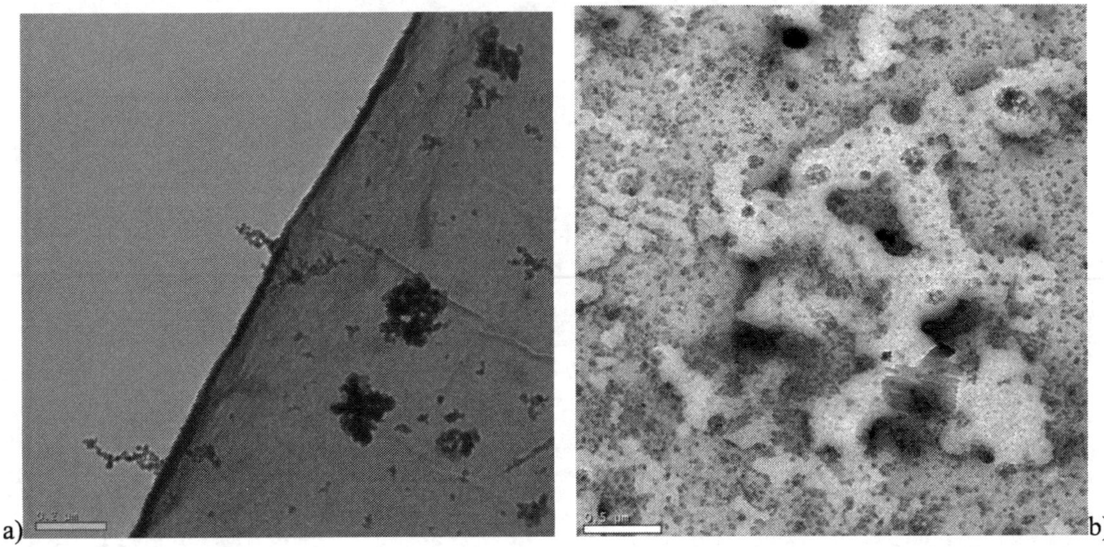

Figure 4: TEM images a) before and b) after application of LSS. Zn particles are difficult to distinguish from precipitated LSS components.

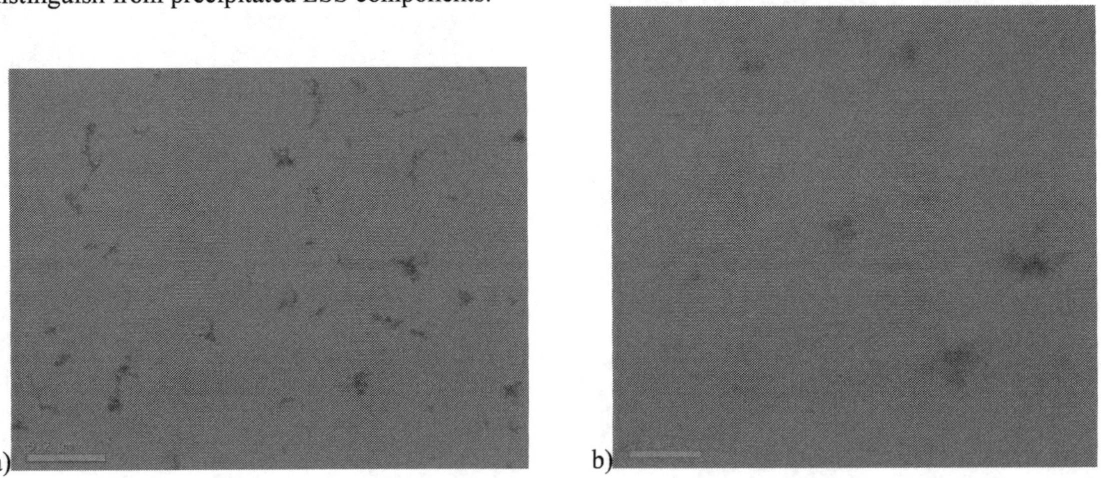

Figure 5: TEM images of room temperature agglomerates a) before and b) after application of DI water.

Another possible explanation for what appeared to be dispersion and possible dissolution is the manner in which water was applied to the grids. Initially, this was done by placing a drop on each grid using a pipette. One thought was that perhaps the force of the water droplet from the pipette was forcing agglomerate dispersion to occur. An alternative method was develop in which water was condensed onto the grid. In order to detect the movement of deposited particles, the grids were first shadow coated (Figure 6), showing the shadows (in white) cast by the zinc agglomerates.

a) b)

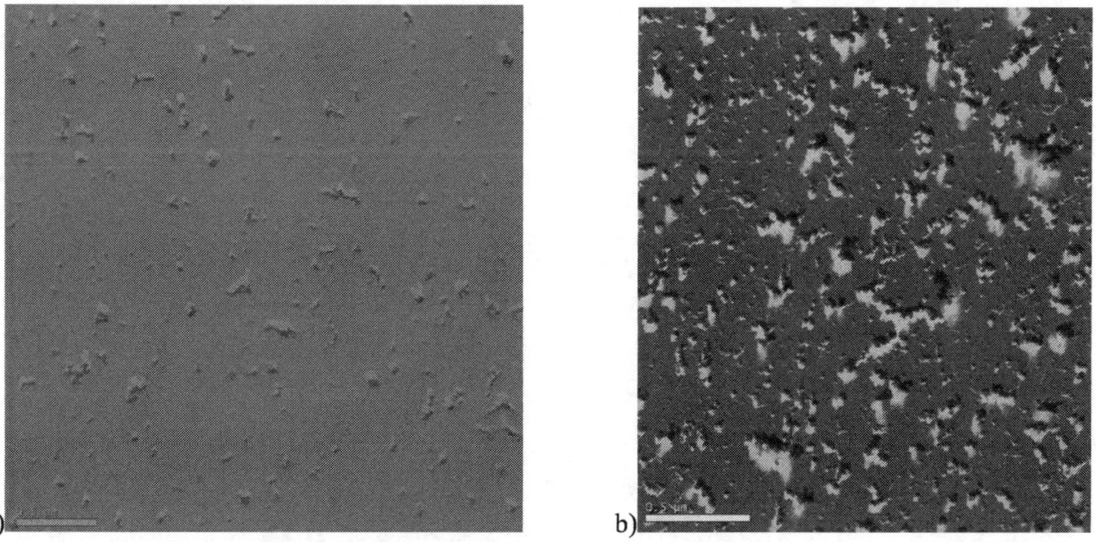

Figure 6: TEM images of Zn agglomerates that have been shadow coated

a)

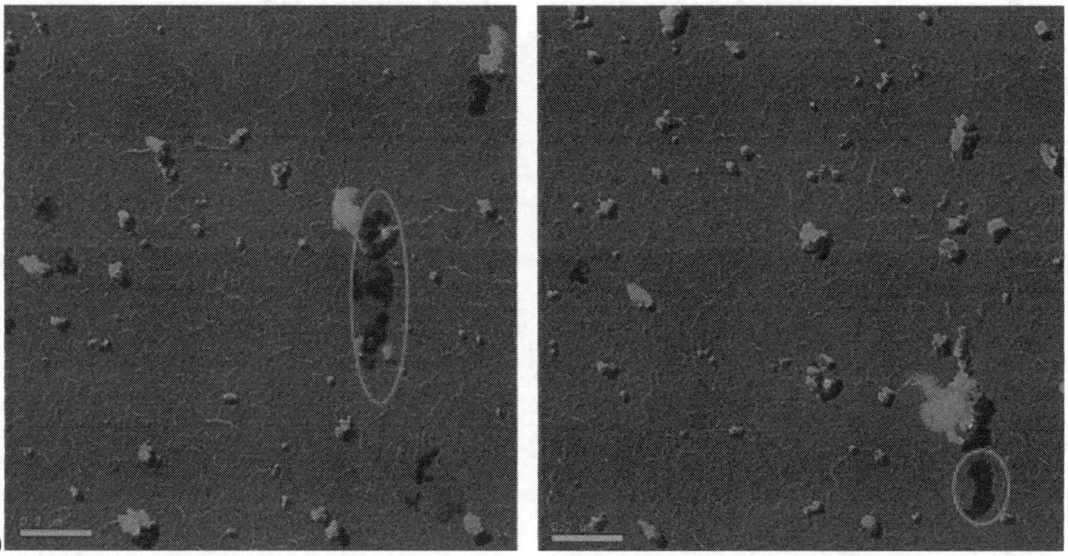

Figure 7: a) TEM images of shadow coated Zn agglomerates exposed to DI water for 1hr. Inside the circled area of each image is a large Zn agglom ssociated" with any shadow.

Figure 7 shows an example of the results obtained when unsintered Zn agglomerates were shadow coated and exposed to DI water using the humidified air method. There are a few very important observations to be made regarding Figure 7. First, as shown in the circles drawn in Figure 7a, there are several instances of larger agglomerates that appear unassociated with any shadow, suggesting that they have moved, not necessarily dispersed, and certainly not dissolved, with the application of DI water. This can be contrasted with the images in Figure 5a, taken of agglomerates that hadn't yet been exposed to DI water with each remaining with an associated shadow. It appears that mostly the larger agglomerates that moved. Also in Figures 7a and b as compared to Figure 5, is that smaller, more compact agglomerates appear not to move, instead remaining adjacent to their respectively cast shadow even after the application of DI water via the humidified air method. This is an interesting result that suggests a possible factor in the degree to which agglomerates are transported is both how they land and their shape factor, or how spherical versus spread out they are. For instance, it would be expected that long, branched agglomerates with high dynamic shape factors, elevated above the surface upon which they are deposited, experience a greater amount of torque for example from condensation of DI water onto the grid surface, making them more likely than smaller, more compact agglomerates to be moved by such a process.

b)

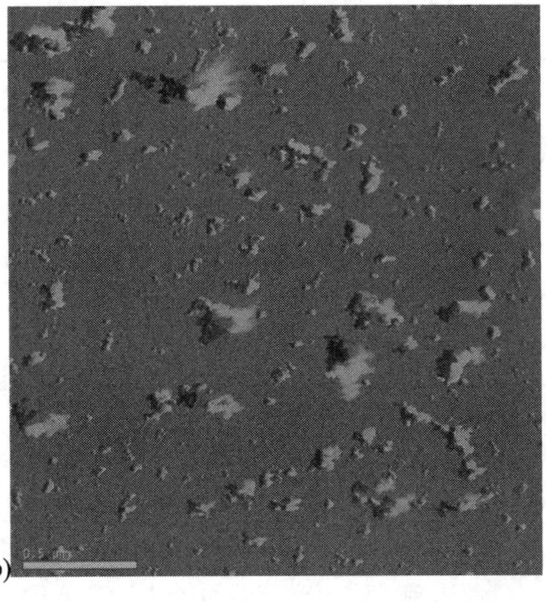

Figure 7b) TEM image of shadow coated Zn agglomerates exposed to DI water for 3hrs.

These techniques of agglomerate generation and imaging require further refinement in order to be able to quantitate the size and number of the agglomerates before and after application of liquid. Agglomerates of nanoparticles that break up into individual nanoparticles may expose more respiratory tissues to nanoparticles, while larger agglomerates may be cleared via mucociliary action. On the other hand, there may be an agglomerate size or shape that is less likely to be broken up into its nanoparticle constituent parts. The transformation of agglomerates of nanoparticles in biological fluids requires further study to be able to assess the toxicological impact of nanoparticles.

References

[1] Donaldson K, Stone V, MacNee W. 1999 The toxicology of ultrafine particles. In: Particulate Matter: Properties and Effects Upon Health (Maynard RL, Howards CV, eds). Oxford: BIOS Scientific Publishers; 115-129.

[2] Ferin J, Oberdorster G, Penney D P. 1992 Pulmonary retention of ultrafine and fine particles in rats. American Journal of Respiratory Cell and Molecular Biology 6(5): 535-42.

[3] Li X Y, Gilmour P S, Donaldson K, MacNee W. 1996 Free radical activity and pro-inflammatory effects of particulate air pollution (PM10) in vivo and in vitro. Thorax 51:1216-1222.

[4] Grassian V H, O'Shaughnessy P T, Adamcakova-Dodd A, Pettibone J M, Thorne P S. 2007 Inhalation Exposure Study of Titanium Dioxide Nanoparticles with a Primary Particle Size of 2 to 5 nm. Environ Health Perspect 115(3):397-402.

[5] Vasto S, Mocchegiani E, Candore G, Listi F, Colonna-Romano G, Lio D, Malavolta M, Giacconi R, Cipriano C, Caruso C. 2006 Inflammation, genes and zinc in ageing and age-related diseases. Biogerentology 7:315-27.

[6] Kodavanti U P, Schladweiler M CJ, Ledbetter A D, Hauser R, Christiani D C, Samet J M, McGee J, Richards J H, Costa D L. 2002 Pulmonary and Systemic Effects of Zinc-Containing Emission Particles in Three Rat Strains: Multiple Exposure Scenarios. Toxicol Sci 70:73-85.

[7] Adamson I YR, Prieditis H, Hedgecock C, Vincent R. 2000 Zinc Is the Toxic Factor in the Lung Response to an Atmospheric Particulate Sample. Toxicol Appl Pharmacol 166:111-119.

[8] Balachandran S, Meena B R, Khillare P S. 2000 Particle size distribution and its elemental composition in the ambient air of Delhi. Environ Int 26(1-2):49-54.

[9] Fine J M, Gordon T, Chen L C, Kinney P, Falcone G, Sparer J, Becket W S. 2000 Characterization of clinical tolerance to inhaled zinc oxide in naïve subjects and sheet metal workers. J Occup Environ Med 42:1085-1091.

[10] Moss, O R 1979 Simulants of lung interstitial fluid. Health Physics, 36(3), 447-8.

Journal of Physics: Conference Series **151** (2009) 012037 doi:10.1088/1742-6596/151/1/012037

A good practice guide for safe work with nanoparticles: the Quebec Approach

C Ostiguy[1], B Roberge[1], L Ménard[2] and C A Endo[3]

[1]: IRSST, 505 De Maisonneuve Blvd. West, Montréal, Québec, Canada, H3A 3C2

[2]: CSST, De Bleury, Montréal, Québec, Canada

[3]: NanoQuébec, Montréal, Québec, Canada

E-mail: claude.ostiguy@irsst.qc.ca

Abstract: new industrial revolution has already begun around nanotechnologies, letting us anticipate major scientific breakthroughs that will affect each economic activity sector and whose expected global economic impacts will exceed $1000 billion annually by 2012. Simultaneously, many studies reveal that nanoparticles represent different occupational health and safety (OHS) risks unique to them and that often differ from the risks related to the same chemical substances with larger dimensions. As the number of potentially exposed workers increases and much uncertainty persists about OHS risks, this extended abstract proposes a framework for occupational risk management with the objective of controlling exposure to NPs in a context of a major lack of specific data related to the hazards of these substances and to the level of occupational exposure. The framework takes into consideration the equal representation of both the employers and workers in the Québec legislation and accounts the potential routes of exposure and focuses on a structured approach dealing with hazard identification, exposure characterization, risk assessment and risk management through different control methodologies. These are included in a prevention program that must be followed up, once it has been implemented, and refined through an iterative approach as new data become available.

1. Introduction

The interest of nanotechnology (NT) is based on the fact that many specific properties of nanoparticles (NPs) exist only at their small size (<100 nm). On the one hand, the incorporation of these NPs into NT-based products could substantially improve their performances, promote economic development, improve the quality of life, and contribute to the protection and rehabilitation of the environment [1]. On the other hand, the development of new products through research, their industrial synthesis and the incorporation of these new particles into various products, processes, and workplaces raise important health and safety concerns since some workers could be exposed to these NPs through inhalation, skin absorption or ingestion, during synthesis, packing, storage, shipping, transportation, spills, etc [1-3]. This is of particular interest since many expert groups or research organizations have already completed reviews of the scientific literature and identified various potential risks related to engineered NPs [1-10]. Specific data on hazards are still extremely limited. There are many uncertainties as to what extent the unique properties of engineered nanomaterials, which underpin their commercial potential, pose specific occupational health risks. Among these uncertainties, the actual difficulty of determining which metric of exposure (mass, surface area, number of particles…) is best related to toxicity and the complexity of evaluating the occupational exposure are also contributing to the challenge of establishing quantitative risk assessment. Another concern with NPs relates to their very high specific surface area per unit of mass, with some highly reactive or easily oxidizible NPs possibly posing a threat of uncontrolled reactions, explosions and fires.

With such uncertainties and the increasing number of workers potentially exposed to NPs, it becomes essential that safe work practices be developed, implemented and validated in research laboratories and industrial plants. This extended abstract proposes an approach to hazard identification; to exposure, toxicological and risk assessment; and to risk management in the specific context of limited information. These data will constitute the basis for the implementation of a prevention program in research laboratories and in industrial workplaces.

2. Risk Assessment

Risk assessment is based on the hazard (toxicity or safety) that can cause harm and the probability that it happens (level of exposure). The risk associated to the toxicity is associated with the inherent toxicity of the material, the magnitude and duration of the exposure, the persistence of the material in the body and, for a specific worker, that person's susceptibility or health status. The toxicity is usually well correlated to the absorbed mass of the substance. Studies with NPs clearly show that the measured biological effects are not well correlated to the mass of the NPs absorbed, but instead to other parameters such as the specific surface area, the number of particles, the size, the granulometric size distribution, the surface reactivity or the chemical composition, and the solubility of the NPs [3-5, 9-12]. There is actually no international consensus on the metric or metrics that best correlate effects and doses. In addition, most studies conclude that, at equal mass, NPs are more toxic than larger particles of the same chemical composition. Important uncertainties arise because of gaps in knowledge about the factors that are essential for predicting health risks. These factors include routes of exposure, translocation of materials once they enter the body, and interaction of the materials with the body's biological systems [3-7, 9-14].

With a very high specific surface area per unit of mass and under specific conditions, some chemically-reactive or combustible NPs could have a potential for catalytic reactions, fires or explosions. Although insufficient information exists to predict these physical risks associated with NPs, nanoscale combustible materials might present a higher risk than coarser materials of similar composition and quantity. Decreasing the particle size of combustible materials could reduce the minimum ignition energy and increase the combustion rate [1, 13, 15, 16]. The dustiness of the NPs, dust explosiveness, and the minimum ignition energy and ignition temperature are essential parameters for the characterization of the safety-relevant aspects of dusts. Some industrial processes and manipulations could create conditions that could promote explosion or fire: confinement, concentration of combustible NPs in the air within the limits of explosivity, sufficient concentration of oxygen, and an ignition source. Inadequate ventilation, spills, leaks, or poor housekeeping using inappropriate methods are examples of situations that could promote the formation of NP clouds, and increase the probability of explosions and exposure of workers, mainly through inhalation [1, 14-16].

It is apparent from the above discussion that parameters such as the specific surface area, number of particles, size distribution and composition are essential for measuring exposures to nanoaerosols [1, 13, 17, 18]. Many instruments can be used to document exposure: mass concentration (cascade impactors, piezoelectric microbalances, tapered element oscillating microbalance, electrical low pressure impactor (ELPI), scanning mobility particle sizer (SMPS)); surface area (diffusion charger, direct-reading instruments, SMPS, transmission electron microscopy (TEM)); number concentrations (condensation particle counter, electrometers, SMPS, ELPI, scanning electron microscopy (SEM), TEM); granulometric distribution (SMPS, differential electrical mobility sizer, cascade impactors, ELPI, SEM, TEM), chemical composition (laboratory techniques, TEM, SEM). One important factor is that the evaluation of all these parameters in the worker's breathing zone is actually impossible. In fact, specific detection of NPs is actually a challenge because natural and anthropomorphic nano-sized particles are already present in the ambient air. The sensitivity and specificity of existing instruments could also be insufficient since low concentrations are expected in many environments, and most existing instruments are cumbersome, nonspecific, expensive and difficult to use in workplaces. Despite the actual limitations of the available instrumentation, it seems essential to document the workers' exposure by measuring a maximum of parameters that could be correlated to adverse health effects or safety risks. The mass concentration should also be evaluated as it is the only parameter that can often be used to compare results to previously available data [1, 13, 17, 19].

Quantitative risk assessment normally relies on the availability of exposure limits based on dose-response relationships and the quantitative evaluation of the exposure. The risk assessment then triggers the risk management measures to be implemented [20, 21]. In the case of NPs, neither the toxic effects (or the safety hazards) nor the exposure level can be quantified, thus leading to major uncertainties in risk assessment (Figure 1). In the absence of data for utilizing traditional quantitative risk assessment methods to quantify the risk, expert judgments can be used.

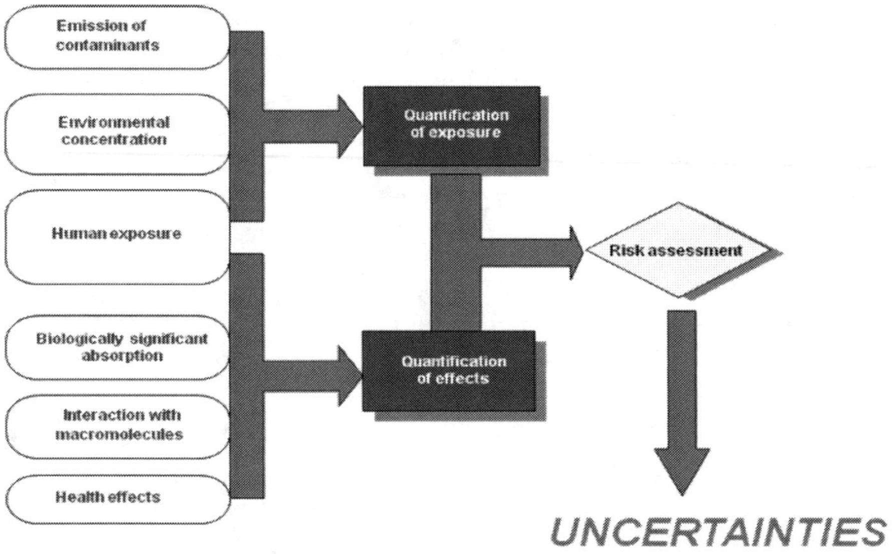

Figure 1: Risk assessment of nanoparticles results in uncertainties

3. Risk management and prevention program

In such a situation of multiple uncertainties, the authors recommend that a risk management program be implemented through an iterative prevention program (Figure 2) that includes a regular re-assessment of available hazard and exposure information [22]. A precautionary approach should be adopted until the risk can be quantified and managed. Control banding could be extremely useful as it helps to identify the level of prevention to implement [22]. The right two columns of that Figure 2 detail different items to take into account during the implementation step.

The Québec legislation is based on the assumption that an essential condition for the successful implementation of a prevention program is the employer's strong leadership and commitment, and total employee involvement [1, 22]. When these essential conditions are in place, the next step is to interpret all the existing information (nature of NPs; hazards; phase (solid, suspension…); processes, equipment and volumes involved; flow of material; sources of potential exposure; worker's position relative to airstreams and emission source; movement of workers (tasks, activities); work methods; sites of potential leaks and emission; handling; transportation; storage; housekeeping; ventilation;

equipment maintenance/decontamination; prevention measures; emergency procedures; exposure data, etc. Based on this information, decisions are made on how to control these risks at an acceptable level, and action is taken.

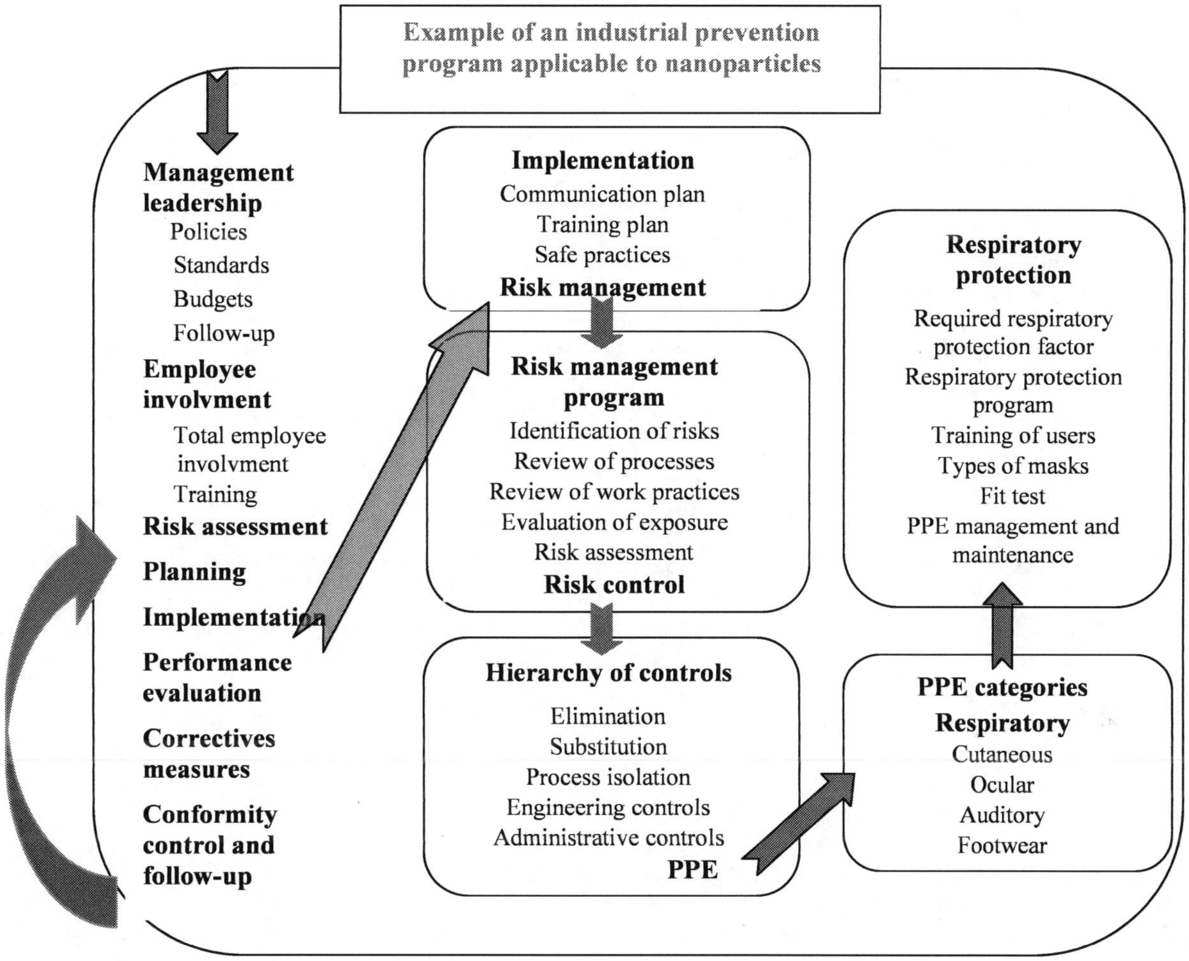

Figure 2. Proposed prevention program applicable to nanoparticles detailing major elements of the implementation step

Due to the many uncertainties in the case of NPs, the expert's judgment becomes a critical element in risk management. The OHS prevention program should be an integral part of the risk management program and it should also be integrated into the overall management of the organization [1, 13, 16, 21, 22]. Each step of the program is essential. Nevertheless, with the uncertainties related to NPs, all the elements of this program should be continuously reviewed in an iterative process as new data become available. The classical hierarchy of controls should be efficient against NPs. Our experience in workplace interventions suggests that one of the key elements in optimizing the impact of the prevention program is related to the validation of the efficiency of the prevention measures and to the maintenance of the prevention equipment. The use of respirators should be limited to cases where all the other prevention measures do not succeed in controlling the concentration of NPs in the air at the desired level, or during specific high-risk exposure operations (equipment maintenance, emergency following spillages and accidental releases). Because the health impacts of worker exposure to NPs are not known, it could be useful to implement biological monitoring and medical surveillance, whenever possible, but this approach is extremely limited as little information is available at this time to define the appropriate parameters of biological monitoring. Medical screening and surveillance could be useful for identifying health problems that might be linked to NP absorption.

4. Conclusion

Nanotechnology is a rapidly developing but the different hazards associated with NPs are only partially known and the level of air contamination is difficult to relate to a significant metric. In most situations, the risks cannot be quantified. Nevertheless, the development and implementation of an efficient prevention program is possible despite the many uncertainties related to NPs.

Bibliography

[1] Ostiguy C, Lapointe G, Ménard L, Cloutier Y, Trottier M, Boutin M, Antoun M, Normand C, 2006. *IRSST Report* # R-470, Montréal, 86 p
[2] Ostiguy C, Soucy B, Lapointe G, Woods C, Ménard L 2008. Second Edition, *IRSST Report* # R-558, 112 p
[3] Ostiguy C, Lapointe G, Trottier M, Ménard L, Cloutier Y, Boutin M, Antoun M, Normand C, 2006. *IRSST Report* #R-469
[4] Borm PJA, Robbins D, Haubold S, Kuhlbusch T, Fissan H, Donaldson K, Schins R, Stone V, Kreyling W, Lademann, J, Krutmann J, Warheit D, Oberdörster E, 2006. *Particle and Fibre Toxicology* **3** 11
[5] Donaldson K, Aitken R, Tran L, Stone V, Duffin R, Forrest G, Alexander A, 2006. *Toxicological Sciences* **92** (1) 5-22
[6] Fiorito S, 2007. *In* Nanotoxicology, interactions of nanomaterials with biological systems, Edited by Zhao Y, Nalwa HS, *American Scientific Publishers*, 167-180
[7] Hardman R, 2006. *Environmental Health Perspectives* **114** (2) 165-172
[8] Lam CW, James JT, McCluskey R, Arepalli S, Hunter RL, 2006. *Critical Reviews in Toxicology* **36** (3) 189-217
[9] Monteiller C, Tran L, MacNee W, Faux S, Jones A, Miller B, Donaldson K, 2007. *Occupational and Environmental Medicine*, **64** (9) 609-615
[10] Oberdorster G, Stone V, Donaldson K, 2007. *Nanotoxicology* **1** (1) 2-25
[11] Wittmaack K, 2007. *Environ Health Perspect* **115** (2) 187-194
[12] Moss OR, 2006. *CITT Activities* **26** (2) 1-7
[13] National Institute for Occupational Safety and Health – [Online] Available at http://www.cdc.gov/niosh/topics/nanotech/safenano/pdfs/approaches_to_safe_nanotechnology_28november2006_updated.pdf (Accessed April 20, 2008)
[14] Wittmaack K, 2007. *Environ Health Perspect* **115** (2) 187-194
[15] Biswas P, Wu CY, 2005. *Journal of the Air & Waste Management Association* **55** 708-746
[16] Pritchard DK, 2004. *Health and Safety Laboratory*, HSL/2004/12
[17] International Standards Organisation (ISO), 2007. *ISO/TR* 27628
[18] Chow JC, Watson JG, 2007. Aerosol and Air Quality Research 7 (2) 121-173
[19] International Standards Organisation (ISO), 2008. Technical Committee TC 229 - WG3, Work in progress.
[20] Kuempel ED, 2006. Ist International Conference on Nanotoxicology: Biomedical Aspects, February 1, Miami, Florida
[21] Schulte P, Geraci C, Zumwalde R, Hoover M, Kuempel E, 2008. J. Occup Env Hyg 5 239-249
[22] Ostiguy C, Roberge B, Ménard L, Endo CA, 2008. A guide to safe work with nanoparticles in occupational settings. IRSST Report, In press
[23] Paik SY, Zalk DM, Swuste P, 2008. Application of a pilot control banding tool for risk assessment and control of nanoparticle exposures. Ann Occup Hyg 52 (6) : 419-428.

The UCB Particle Monitor: A tool for logging frequency of smoking and the intensity of second-hand smoke concentrations in the home

Sean Semple, Andrew Apsley, Gill Moir, George Henderson, Jon Ayres

Department of Environmental & Occupational Medicine, Liberty Safe Work Research Centre, University of Aberdeen, Foresterhill Road, Aberdeen AB25 2ZP, UK

E mail sean.semple@abdn.ac.uk

Abstract. Second-hand tobacco smoke (SHS) exposure generates a large public health burden. Recent legislation has moved to prohibit smoking in public places and there are concerns that this may lead to an increase in exposures in private homes. Measurement of SHS aerosol has tended to use active pumped samples or longer-term diffusive badges. Pumped methods are noisy and poorly tolerated in home settings while diffusive badges do not provide real-time data. The UCB particle monitor (UCB-PM) is a modified smoke-alarm device capable of logging changes in airborne particulate matter over extended periods and has been used successfully to measure biomass fuel smoke concentrations in developing world settings This study has examined the use of the UCB-PM to measure SHS aerosol in both controlled laboratory conditions and a pilot field trial over a 7 day period in a smoker's home. Comparisons with a pumped sampler (TSI Sidepak Personal Aerosol Monitor) indicate good agreement over a range of exposure concentrations but there is evidence of a threshold effect at approximately 0.5 mg/m^3 of fine particulate measured as $PM_{2.5}$. While this threshold effect undermines the ability of the device to provide useful data on the time-weighted average SHS concentration, the field trial indicates that that the UCB-PM has a sensitivity of about 71% and a specificity of 98%. The device has many advantages including zero noise operation, low cost and long battery life and may be a useful tool in quitting and smoke-free home intervention studies.

1. Introduction:

Indoor air pollution from second-hand smoke (SHS) has been linked to a range of health effects including lung cancer [1], exacerbations of asthma [2] and cardiovascular disease including acute myocardial infarction [3]. Recent smoke-free legislation in many countries has moved to prohibit smoking in enclosed public spaces. Legislation in the UK in 2006 and 2007 has banned smoking in most workplaces including pubs and restaurants and restricted smoking to outdoors and in private spaces. There are concerns that the introduction of these restrictions would lead to a displacement of smoking activity from bars to private homes and a consequent increase in exposure to SHS among children and others living in homes where smoking is permitted. Initial data examining the levels of salivary cotinine (a metabolite of nicotine) in schoolchildren in Scotland suggests that displacement of smoking to the home has not occurred [4] but there have been few direct studies of SHS levels in home environments.

One of the major problems in collecting data on SHS levels in private domestic environments is that active sampling devices tend to produce noise that can be irritating to study participants and are generally only tolerated for short periods. Diffusive methods, particularly for nicotine or vapour–phase components, are available but these then suffer from the disadvantage of only providing average concentrations over the whole sampling period. There is a need for a diffusive-based system that can provide real-time data on SHS concentrations within a home.

The UCB Particle Monitor (UCB-PM) has been developed by scientists at the University of California, Berkeley [5] for use in studies looking at exposure to smoke generated from the use of biomass fuels in developing world environments. The device is based on a modified smoke-alarm and uses the voltage changes generated by particles passing across a photometric chamber to express airborne concentrations of fine particulate matter [6]. The device has been used successfully to measure particulate in homes burning wood, crop residues, dried cow-dung and charcoal in a number of countries [7, 8].

This pilot study aimed to determine if the UCB-PM could be used as a device to provide information on SHS concentrations and the frequency of smoking activity in domestic settings where tobacco smoking was the primary source of fine particulate matter concentrations.

2. Methods

2.1. Chamber experiments

The first series of tests compared the response from two UCB-PM devices and a co-located TSI AM510 Sidepak Personal Aerosol Monitor (TSI Inc., Shoreview, MN, USA) fitted with a $PM_{2.5}$ impactor, within a 9 m^3 exposure chamber. The Sidepak used a 0.295 calibration factor for SHS aerosol as described in previous studies [9, 10].

The UCB-PM devices require a zero-ing period prior to use and so were set to log and then sealed inside a ziplock bag for 30 minutes prior to the start of the chamber experiment. This complete, the UCB-PM devices were removed from their bags and placed alongside the other instruments within the chamber. The UCB-PM and Sidepak instruments were set to log data every 1 minute. A cigarette was lit within the chamber and allowed to smoulder for a short period (between 1 and 5 minutes burn time) until the real-time display on the Sidepak device indicated airborne concentrations of $PM_{2.5}$ between 0.5 and 5.0 mg/m^3. Airborne particle concentrations then decayed over time and at various timepoints the extract ventilation was switched on to remove the SHS and bring $PM_{2.5}$ levels back to the baseline for the room. The experiment was repeated between 2-4 times each day for a total of 11 runs. The Sidepak device was zero calibrated each day during the study and the flow rate set to 1.7 l/ min.

On completion of the experiment, data from the Sidepak was downloaded using TSI Software (Trackpro V3.6.2), UCB data was downloaded using the UCB-PM 2.2 software using the monitor manager facility.

A graph of the output from a typical experimental run is illustrated in figure 1.

2.2. Use in the home

The field-testing of the devices took place in a volunteer's house over a period of 7 days. Two UCB-PM devices were co-located with a Sidpeak monitor in the living-room area of the house. This room was approximately 20 m^3 in volume and was the location where most smoking activity took place within the home. The devices were placed on a side-table at a height of about 1 metre above the floor. The instruments were zeroed, set to log and downloaded using a similar protocol to that described for the chamber experiments. The participant was asked to record the number and approximate timing of smoking activity during the whole 7 day period.

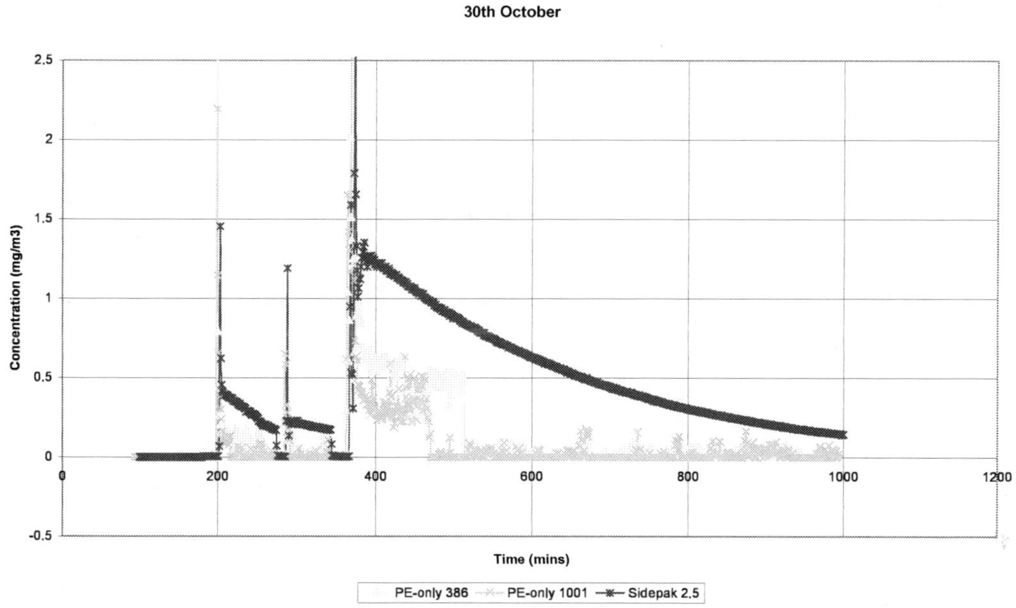

Figure 1. Particulate matter levels measured during one chamber experiment. The graph shows the concentration in mg/m^3 measured by two UCB-PMs (PE-only 386 and PE-only 1001) and from a Sidepak measuring $PM_{2.5}$. The experiment lasted over 15 hours and had three 'smoking' events during this time.

2.3. Statistical analysis

SPSS v 15 and Microsoft excel software packages were used to analyse the data. Time weighted average concentrations were compared using Pearson correlation coefficients to compare UCB data with Sidepak data and UCB devices with each other.

In addition we also analysed the length of time that each instrument indicated a concentration above two threshold values. These values (65 and 250 μg/m3) are used by the US Environmental Protection Agency in their Air Quality Index for PM2.5 levels in outdoor air to represent an 'unhealthy' and 'hazardous' level respectively [11].

3. Results

3.1. Chamber experiments

Across these 11 chamber experiments the correlation between the TWA levels measured by SP and UCB-PM devices was high with a Pearson correlation co-efficient of 0.95 ($p<0.001$) for the UCB1001 and 0.92 ($p<0.001$) for the UCB386. Correlation between the two UCB-PM devices is also highly significant at 0.91 ($p<0.001$). Using the duration above 250 mg/m3 as the exposure metric the relationship is less strong. The Pearson correlation coefficient for the UCB386 and Sidepak is 0.69 ($p=0.020$), between the UCB1001 and Sidepak is 0.68 ($p=0.021$) and between both UCB-PM devices is 0.57 ($p=0.067$).

Figure 2 shows the plot of TWA concentrations measured by each device for all 11 experiments.

A

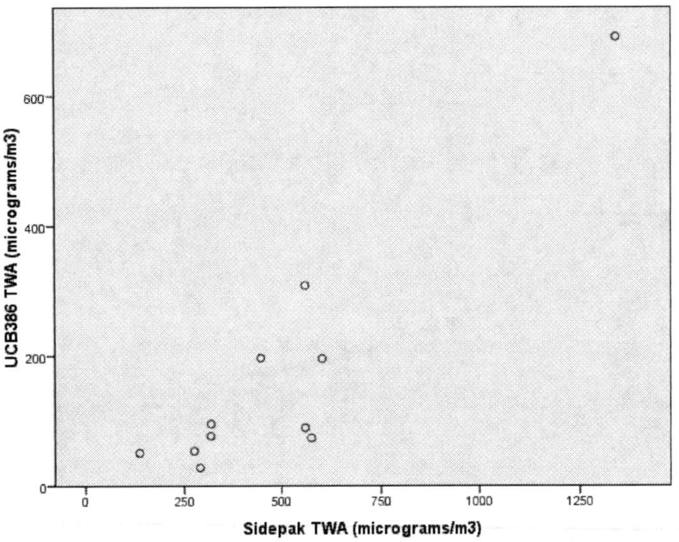

B

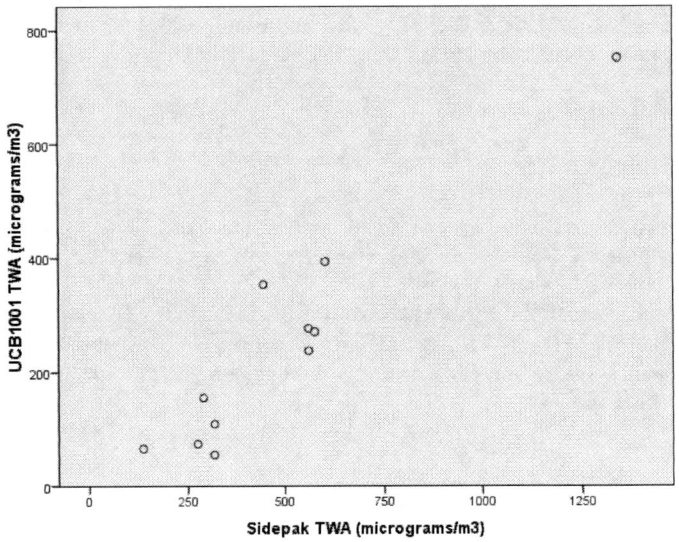

Figure 2. TWA particulate matter levels measured during eleven chamber experiment. (A) presents the data from UCB386 and Sidepak devices. (B) presents the data from UCB1001 and Sidepak devices.

3.2. Use in the home

The TWA exposure concentrations and duration of exposure above the two defined threshold levels (65 and 250 μg/m^3) as measured by the three devices is provided in table 1.

Table 1: Results from 3 devices used to measure SHS in a smoker's home over 7 day period.

Sampler	*TWA concentration (μg/m^3)*	*Duration (mins) above 65 μg/m^3*	**Duration (mins) above 250 μg/m^3**
Sidepak ($PM_{2.5}$)	42	2298	61
UCB386	28[a]	288	3
UCB1001	57[a]	1568	32

[a]The output for UCB386 had a minimum value of 25 μg/m^3 and UCB1001 had a minimum level of 49 μg/m^3 throughout the sampling period and when this is removed the TWAs are reduced to 3 and 8 μg/m^3 respectively. Figure 2 and 3 below corrects this systematic zero-shift in values by subtracting 25 μg/m^3 (UCB386) or 49 μg/m^3 (UCB1001) from each data-point.

Figures 3 and 4 illustrate the particulate matter concentrations registered by the Sidepak and the two UCB-PM (386 and 1001) devices over the 7-day sampling period.

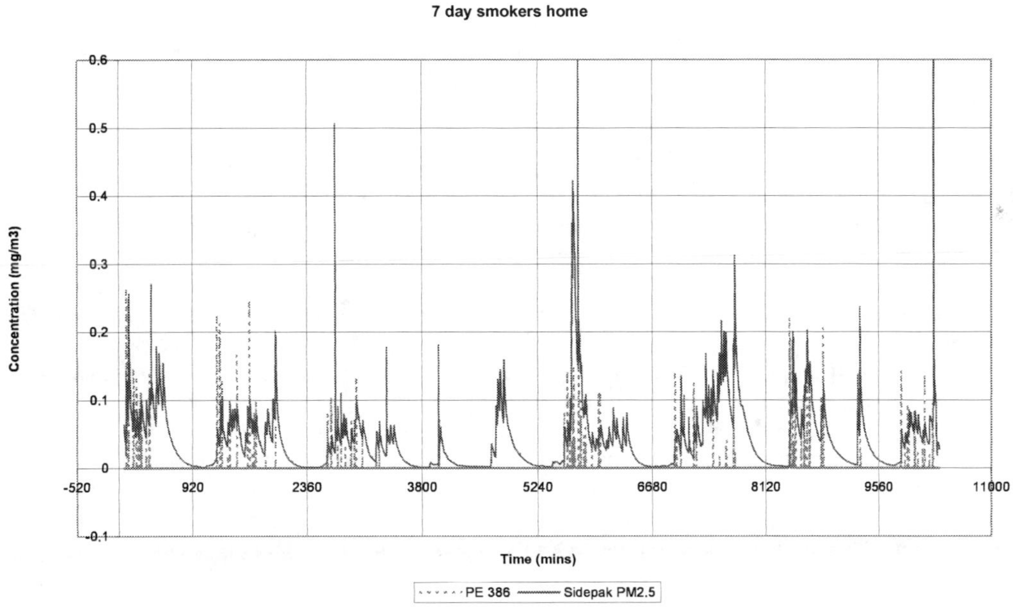

Figure 3. Particulate matter levels measured during a seven day period in the home of a smoker. The graph shows the concentration in mg/m^3 measured by a UCB-PM (PE386) and from a Sidepak measuring $PM_{2.5}$.

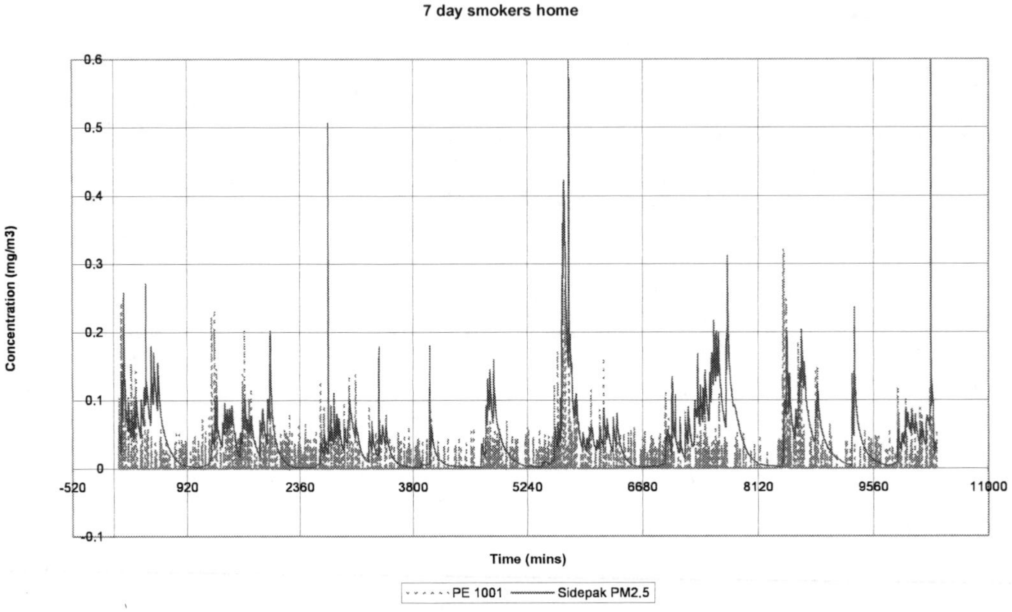

Figure 4. Particulate matter levels measured during a seven day period in the home of a smoker. The graph shows the concentration in mg/m^3 measured by a UCB-PM (PE1001) and from a Sidepak measuring $PM_{2.5}$.

The output from UCB386 appears to closely match that generated by the Sidepak device however that generated by UCB1001 demonstrates much poorer agreement. There appear to be many small peaks of approximately 50 μg/m^3 in intensity even at night-time periods when smoking does not take place. This may be due to instrument noise or some processing artefact.

According to the self-completed record form the number of cigarettes smoked in the house over the 7 day period totalled 106. From this data the number of cigarettes smoked each day was compared with identifiable peaks as indicated by the Sidepak (n=112) and the UCB386 device (table 2). The sensitivity and specificity of the UCB386 were calculated by comparing the number of concurrent Sidepak and UCB-PM peaks and assuming that the Sidepak peaks are representative of true smoking events (i.e. the participant failed to identify 6 occasions when smoking took place). This process indicated that the UCB386 device has a sensitivity of 71% and a specificity of 98% to detect smoking within this environment. That is to say that the UCB-PM device records an identifiable peak, indicative of smoking, for nearly 3 out of every 4 cigarettes smoked while very rarely (less than 1 in 50 smoke-free periods) registering a false smoking event when smoking does not appear to be occurring.

Table 2: Self-reported cigarette smoking and device peaks during field-trial.

Day	*Self-reported # cigarettes*	*Sidepak peaks*	*Concurrent UCB386 peaks*	**Non-concurrent UCB386 peaks**
1	11	13	8	0
2	17	18	13	0
3	12	15	11	0
4	9	6	1	0
5	19	17	12	1
6	14	20	13	0
7	15	10	10	0
8	9	13	12	1
Total	**106**	**112**	**80**	**2**

4. Discussion:

To the best of our knowledge this is the first assessment of the use of the UCB-PM to provide information on SHS levels. In chamber experiments the device performed well but did demonstrate a threshold effect at various levels between 0.5 and 1.0 mg/m^3 of $PM_{2.5}$ aerosol- when levels fell below this threshold, particularly in the decay of aged SHS aerosol, the UCB-PM devices responded by returning to baseline zero. Clearly this has important implications for the calculation of time-weighted average exposures using these devices.

From the field-trial there were similar issues with the UCB-PM instruments and in addition to a zero-shift problem the TWA concentrations from both devices were much lower than those recorded by the Sidepak. One device, the UCB-PM (1001), also appeared to suffer from some 'noise' that caused voltage changes unrelated to smoking activity throughout the 7-day period. The other UCB-PM (386) produced data that was much more closely correlated to that generated by the Sidepak. This device demonstrated a moderate degree of sensitivity (71%) and a high specificity (98%) to detecting smoking events. In a house where twenty cigarettes are smoked each day the device will be able to detect approximately fourteen of these smoking events and can determine with 98% accuracy when smoking does not occur.

One drawback of the device at present is the variation between two UC PM devices when monitoring SHS in the same environment. The devices variation can range from a factor of six to unity in high concentration environments. In situations where the concentration of SHS was low this device variation ranged from 1 to 1.4. Why this variation between the two devices tested in our experiments occurs is unknown at present

We acknowledge that these UCB devices have previously been used to monitor aerosols generated by burning biomass fuel and the concentrations in these studies are generally very much higher than those found during cigarette smoking activities in homes [7-8]. It is possible that modifications to these devices may allow them to measure aerosols at the lower concentrations found in homes where smoking takes place. There are however several areas where the device, as it stands, may be adequate to measure SHS concentrations. Our next trials will look at the performance of the UCB-PM device in a car. SHS concentrations in cars where smoking takes place may be sufficiently high to allow the instrument to perform at a higher sensitivity.

Despite the limitations of threshold and moderate sensitivity identified in this study, the UCB-PM device has many advantages over currently available methods for measuring SHS in domestic environments. The relative low cost of the device (350 euros), noise-free operation and long battery-life together with an ability to provide a real-time log of changing SHS levels are some of the major benefits of the device over pumped sampling or diffusive nicotine badge methods.

In its present form the device could be used to provide feedback to smokers about SHS levels and the frequency of smoking within their home and may be a useful tool to support quitting or smoke-free home intervention programmes. The UCB-PM devices may also be useful in quantifying SHS levels in cars and other vehicles.

Acknowledgements

We wish to acknowledge the help and assistance we received from Professor Kirk Smith at the University of California, Berkley and from Charles Litton from the National Institute of Occupational Safety and Health in Pittsburgh.

Reference list

[1] Hacksaw AK, Law MR, Wald NJ. The accumulated evidence on lung cancer and environmental smoke. BMJ 1997;315:980-88.

[2] Eisner M. Environmental tobacco smoke and adult asthma. Exp Lung Res 2005; 31 Suppl 1:8-14.

[3] Pell JP, Haw S, Cobbe S, et al. Smoke-free legislation and hospitalizations for acute coronary syndrome. N Engl J Med. 2008 ;359:482-91.

[4] Akhtar PC, Currie DB, Currie CE, et al. Changes in child exposure to environmental tobacco smoke (CHETS) study after implementation of smoke-free legislation in Scotland: national cross sectional survey. BMJ. 2007 ;335:545.

[5] Chowdhury Z, Edwards RD, Johnson M, et al. An inexpensive light-scattering particle monitor: field validation. J Environ Monit. 2007;9:1099-106.

[6] Litton CD, Smith KR, Edwards R, et al. Combined optical and ionization measurement techniques for inexpensive characterization of micrometer and submicrometer aerosols, Aerosol Science and Technology. 2004. 38:1054–1062.

[7] Smith KR, Dutta K, Chengappa C, et al. Monitoring and evaluation of improved biomass cookstove programs for indoor air quality and stove performance: Conclusions from the Household Energy and Health Project. Energy for Sustainable Development. 2007; 15: 5-18.

[8] Chengappa C, Edwards R, Bajpai R, et al. Impact of improved cookstoves on indoor air quality in the Bundelkhand Region in India. Energy for Sustainable Development. 2007; 15: 33-44.

[9] Repace J. Air Pollution in Virginia's hospitality industry. Final report 2006. http://www.repace.com/pdf/VAAQSurvey.pdf [Accessed 15th July 2008]

[10] Semple S, Maccalman L, Naji AA, et al. Bar workers' exposure to second-hand smoke: the effect of Scottish smoke-free legislation on occupational exposure. Ann Occup Hyg. 2007;51:571-80.

[11] US Environmental Protection Agency air quality guidelines. http://airnow.gov/index.cfm?action=aqibroch.index [Accessed 24th July 2008]

Personal exposure to airborne ultrafine particles in the urban area of Milan

A Cattaneo[1], G Garramone[2], M Taronna[2], C Peruzzo[2,3] and D M Cavallo[2]

1 Department of Occupational Health, University of Milan, via S.Barnaba 8, 20122 Milan, Italy

2 Department of Chemical and Environmental Sciences, University of Insubria, c/o Clinica del Lavoro "L.Devoto", via S.Barnaba 8, 20122 Milan, Italy

3 Occupational and Preventive Health Division, Macchi Foundation Hospital, Viale Borri 57, 21100 Varese, Italy

E-mail: domenico.cavallo@uninsubria.it

Abstract. The relevance of health effects related to ultrafine particles (UFPs; aerodynamic diameter < 100 nm) can be better evaluated using high-resolution strategies for measuring particle number concentrations. In this study, two different portable Condensation Particle Counters (CPCs) were used to measure personal exposure to UFPs in the central area of Milan for one week period during spring, with three sampling sessions per day. Experimental data were continuously collected along an established urban pathway, moving afoot or by different private and public means of transport. Correlation analysis between data measured by two CPCs was performed and general results showed a good agreement, especially at concentrations lower than 2×10^5 particles /cm^3. UFPs measures were divided on the basis of crossed environments or micro-environments, days of the week and day time (hours). The highest measured mean concentrations and data variability were observed during walking time and moving on motorized vehicles (bus and car), indicating that the highest exposure to UFPs can be reached near motorized traffic. The lowest exposures were observed in green areas and in office microenvironments. An appreciable difference between working and non-working days was observed. Concentration patterns and variation by days of the week and time periods appears related to time trends in traffic intensity.

1. Introduction

Ultrafine particles (UFPs; aerodynamic diameter < 100 nm) are generated by homogeneous nucleation of condensable gases, including photochemical production from precursors and emissions of condensable gases from high temperature combustion, or combustion processes with direct emission of soot particles. Thus, the UFPs production is often associated with high temperature phenomena occurring in natural processes such as volcanic activities or wild fires and in anthropogenic activities as industrial combustion processes, domestic activities (smoking, burning candles, cooking) and automotive mobility. As a matter of fact, combustion is the main anthropogenic source of fine particles, with the vast majority of them included in the sub-micrometer range. These particles contain a host of organic and inorganic materials [1]. Automotive traffic and in particular diesel engines are considered the major sources of ultra-fine particles in urban environments [2]. Diesel exhausts are mainly constituted by particles in the 20-130 nm dimensional range [3], while spark ignition vehicles emit particles belonging to the 20-60 nm range [4].

Human exposure in private and public means of transport is of particular interest because many citizens and commuters spend a substantial component of their outdoor time in these microenvironments. In addition, exposures to urban atmospheric pollutants in several means of transport are often highly elevated in comparison to elsewhere so that individuals may gain a significant contribution to their daily total exposure [5].

After release into the atmosphere, UFPs are subjected to complex transformation processes, as dilution with background air, coagulation, deposition and condensation [6]. Some field data taken near freeways have shown that the size distribution of emitted particles change substantially within a few hundred meters of the emission source. Both atmospheric dispersion and coagulation appears to contribute to the rapid decrease in particle number concentration and change in particle size distribution with increasing distance from the traffic source [7]. Typical particle growth rates were determined in the range 1–20 nm per hour in mid-latitudes, depending on the temperature and the availability of condensable vapours [8].

In addition, UFPs can be also generated indoors by other combustion processes such as gas cooking [9], burning of wood and candles, tobacco smoke or gas-to-particle conversion [10].

Indoor particle concentrations can greatly differ from outdoors, even in the absence of indoor sources. Therefore, it can be important to understand the relationships between indoor and outdoor concentrations in order to assess the source and transport pathways of UFPs indoors. Some recent studies have examined this issue [11-13]. These studies show that, in absence of strong indoor sources, particles in indoor air are mainly of outdoor origin but, in the presence of significant indoor sources, the indoor/outdoor ratio can exceed unity [14].

The scientific and public interest in assessing the exposure and the risk due to UFP inhalation is growing in time, from the moment in which epidemiological studies have suggested that exposure to ultra-fine particles may cause pulmonary diseases, cardiovascular health effects and impairment of the immune system [15,16]. As a matter of facts, UFPs differ from larger particles with respect to deposition and alveolar clearance [17] and particles in the range 10-100 nm have a high probability of deposition into the deeper parts of the respiratory tract [18]. It was also found that UFPs contain higher levels of trace elements and toxins such as polycyclic aromatic hydrocarbons and some mutagenic substances [19,20]. Health effects rising from exposure to UFPs were first recognized through pulmonary toxicological studies, which have reported that UFPs can have a greater potency to cause adverse health effects than larger particles [21-25]. There are indications that particle number or surface area rather than mass may be more important factors when health implications are considered [26,27]. Currently, however, only the mass concentrations of fractions including particles with larger aerodynamic diameter (PM_{10} and $PM_{2.5}$) are regulated in Europe.

In general the study of human exposure during daily activities in different microenvironments should be conducted using high-resolution strategies, taking into account at the same time the duration, location and magnitude of exposure in order to accurately assess environmental health risks and so correctly manage and reduce the risk [28]. In the case of UFPs, the high spatial variability and the sudden variations in exposure even more suggest the use of continuous monitoring instruments.

The aim of this paper is to evaluate the personal exposure to UFPs as a function of space, investigating some microenvironments in the city Milan, and time, considering the different days of the week and periods of the day. The obtained results can be useful for future risk assessment of populations living or working in urban environments, as accurate (personal) human exposure assessment is one of the crucial steps in the risk assessment process. Until now no studies on personal exposure to UFPs in the metropolitan area of Milan (almost 3 millions inhabitants) have been reported.

2. Materials and Methods

Two hand-held Condensation Particle Counters were used to measure personal outdoor and indoor exposure to UFPs: a TSI Condensation Particle Counter model 3007 and a TSI P-Trak Ultrafine Particle Counter model 8525.

The CPC model 3007 enables number concentration measurements of particles >0.01 μm and is equipped with a continuous inlet flow rate, controlled by a pressure drop sensor which allows to minimize errors caused by flow rate variability. The P-Trak is capable of measuring particle ranging from 0.02 to > 1 μm in size. This instrument is not equipped with the flow control system present in the model CPC 3007.

It is important to underline that there is not yet a standard method or instrument for the measurement of true UFPs air concentrations in relation to health effects, particularly in the case of personal exposure, which requires the use of portable devices with reduced dimensions and less reliable performance. Among two adopted instruments, however, CPC 3007 will be considered to be the reference, owing to its wider operating particle size range and its more reliable technical features. Number concentration values reported in the following sections, then, if not differently specified, have to be considered as measured by CPC 3007.

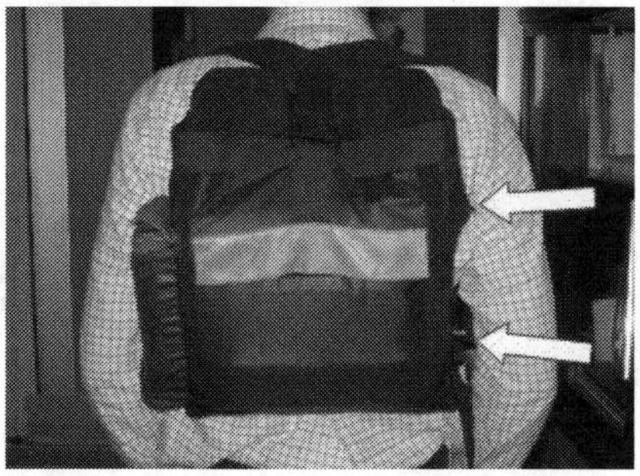

Figure 1. The personal exposure pack used for personal monitoring.

Both CPCs operated with six AA batteries, at an inlet flow rate of 700 cm^3 min^{-1}. 100 cm^3 min^{-1} of the total sampled volume were automatically analyzed. Measured concentration values were recorded by internal data loggers with an acquisition time of 1 minute. Data were downloaded using the provided software. The working alcohol used for both instrument was isopropyl alcohol, with a purity grade of at least 99.5%.

The instruments were placed into a small backpack with sampling inlets on the same side (figure 1). This personal monitoring pack was carried by trained team members.

Experimental data were collected within the central area of Milan during a week in April 2005.

The city of Milan is divided into four areas of different size by two-lane, two-way concentric streets. The inner ring was chosen as sampling area for the study. The sampling strategy involved data collection whilst walking in busy and minor streets, travelling by bus (number 94, diesel), car (petrol with three-way catalyst) and metro (MM1). UFP concentrations inside car were collected while driving along a very busy street, with closed windows, ventilation system turned on and air recycling deactivated. Moreover, the car was equipped with an anti-particulate filter. Measurements were made along the route represented in figure 2, according to a sequential protocol that started and finished at the Department of Occupational Health, also used as office sampling area. The protocol comprised also the investigation of a green area and a great multi-road intersection (Cadorna square).

The sampling protocol was established in order to collect data for at least 15 minutes in each studied environment or micro-environment. Data were collected at three timings: morning (7.00 - 11.00 am), lunchtime (11 am - 3 pm) and afternoon (4 - 8 pm). A time-activity diary was completed in order to accurately separate the exposure data as a function of the different monitored environments.

The statistical analysis of collected data was performed by means of Microsoft Excel 2003 and SPSS 13.0.

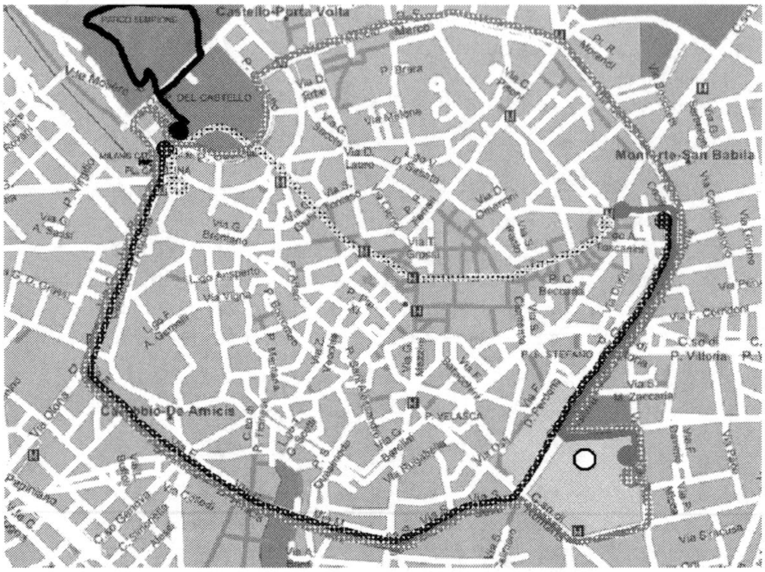

Figure 2. Sampling route map (Grey = Walking, Dotted white = Metro, Black = Park, Dotted black = Bus, Dotted grey = Car, White circle = Office).

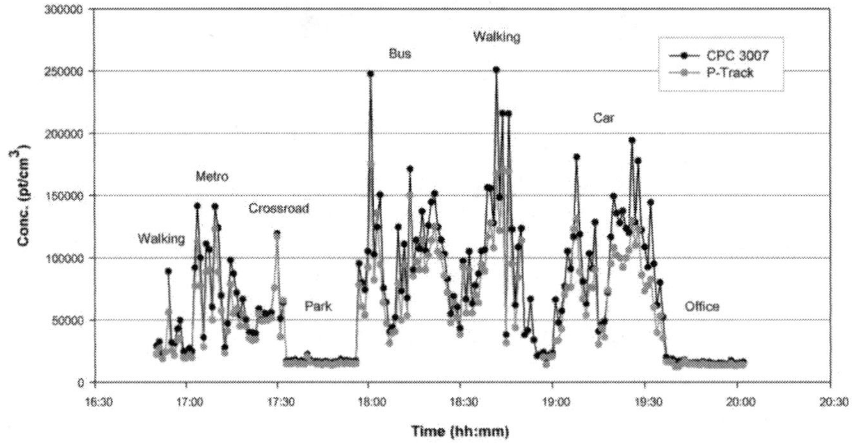

Figure 3. An example of UFP concentration trend measured by CPC 3007 and P-Trak during one sampling session.

3. Results

In total, N=4005 1-min data were collected using the two instruments of which respectively N=3728 were acceptable for CPC 3007 and N=3884 for P-Trak, since some samples were lost due to tilting errors or alcohol wick depletion. Figure 3 displays an example of particle number concentration trend detected by CPC 3007 and P-Trak during one session of measurements. Peaks corresponding to each sampled environment or micro-environment are labeled. Observed trends indicated that both CPCs are able to detect short-time UFP variations, even if P-Trak showed a lack in detecting high concentration peaks. The best agreement between the two instruments was found for concentrations lower than 200000 particles (pt)/cm^3.

3.1. Correlation analysis between data collected by CPC 3007 and P-Trak

A linear correlation analysis was performed between time-paired concentrations measured by CPC 3007 and P-Trak (figure 4). This comparison yielded a Pearson correlation coefficient (r=0.965) close to unity (P<0.001) and the slope of the regression line was m=0.717. The relative accuracy improved (m>0.8) when considering low concentration data (< 50000 pt /cm^3).

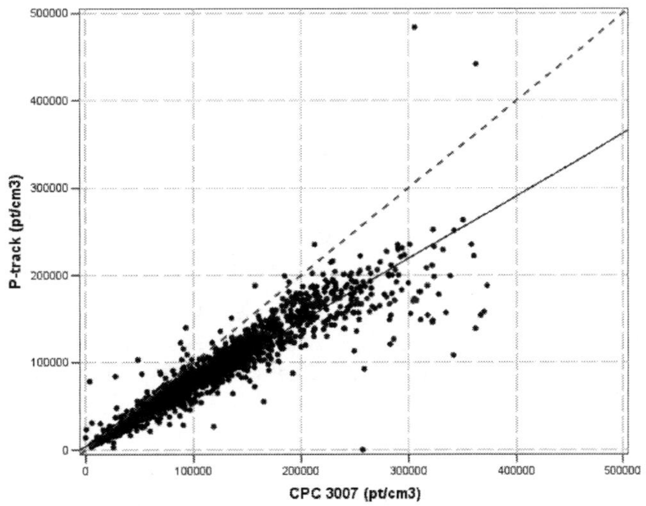

Figure 4. Linear regression plot between time-paired concentrations measured by P- Track and CPC 3007 (dotted line: y=x).

3.2. Exposure analysis

UFP personal exposure measurements were analyzed on the basis of the considered environments or micro-environments, days of the week and time (hours), and statistically significant differences were always found ($P_{ANOVA} < 0.001$). Only the statistical outputs resulted from data collected by CPC 3007 were used to quantify UFP exposure, as a consequence of the higher efficiency showed by this device in detecting the highest peak excursions (> 200000 pt/cm^3). Particle counts measured by P-Trak, however, are plotted in all graphical representations for comparison purposes.

The analysis of data showed that personal exposure magnitude and concentration variability were specific for each sampled environment or micro-environment. The highest measured mean concentrations and data variability [mean; standard deviation] were observed during walking time [100200 pt/cm^3; 80900 pt/cm^3] and moving by means of motorized vehicles like bus [117600 pt/cm^3; 59800 pt/cm^3] and car [107000 pt/cm^3; 64200 pt/cm^3]. On the contrary, the lowest exposures were observed in green areas [23400 pt/cm^3; 10800 pt/cm^3] and office [13500 pt/cm^3; 4700 pt/cm^3].

Graphical representations of data distributions according to the different considered environments and microenvironments are displayed in figure 5.

An appreciable difference between working and non-working days was observed. Concentrations measured during working days showed higher median and mean values, as well as wider data distributions, in comparison with those collected in non-working days. The highest concentrations were observed during the monitoring session of Wednesday [86000 pt/cm^3; 73200 pt/cm^3] and Thursday [87300 pt/cm^3; 73600 pt/cm^3], while lower values were measured on Saturday [40800 pt/cm^3; 37200 pt/cm^3] and Sunday [39200 pt/cm^3; 40000 pt/cm^3]. The data distribution as a function of days of the week is represented in figure 6.

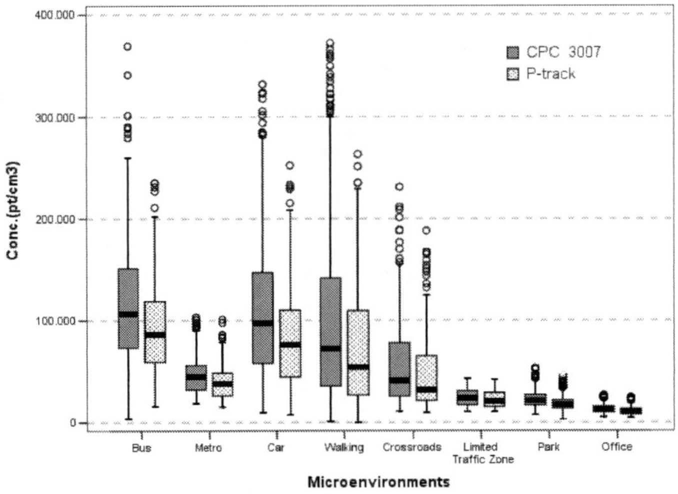

Figure 5. Box plots of data collected by CPC 3007 and P-Trak in sampled environments and micro-environments. (The central box comprises values between the 25th and 75th percentiles, the whiskers show the range of values that falls within 1.5 times the interquartile range beyond the box).

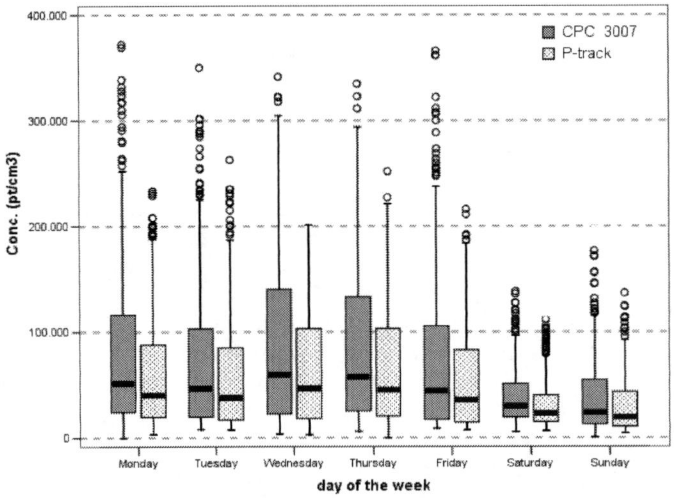

Figure 6. Box plots of data collected by CPC 3007 and P-Trak in different days of the week.

The analysis of concentrations collected in the three daily sampling sessions (7.00 - 11.00 a.m.; 11.00 a.m. - 3.00 p.m.; 4.00 - 8.00 p.m.) revealed some differences between the early morning session and the rest of the day. In fact, the most wide-spread distribution of measures and the highest mean values were obtained from 7.00 to 11.00 a.m. [82200 pt/cm^3; 77400 pt/cm^3]. The variability and mean concentrations obtained during lunchtime and afternoon were generally lower and comparable. These results are graphically displayed in figure 7.

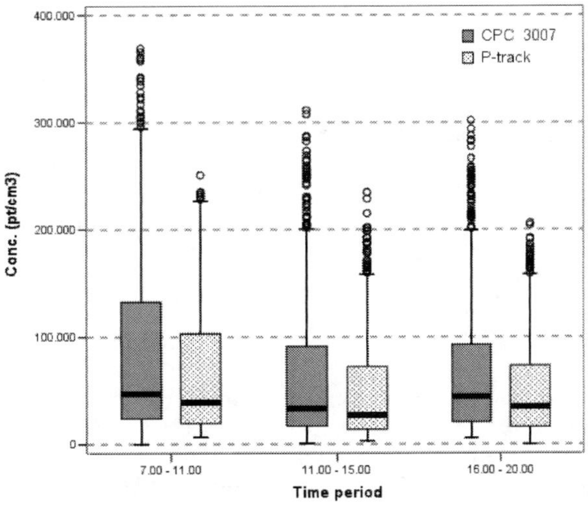

Figure 7. Box plots of particle number concentrations measured by CPC 3007 and P-Trak in three different periods of the day.

4. Discussion

The two portable instruments used in this study showed a good capacity of explaining the short time variability of UFP concentrations (R^2=0.93), even if P-Trak exhibits a lack in the detection of concentrations above 200000 pt/cm^3. Of course, a difference in readings between devices was expected owing to the lower threshold detection of CPC 3007 (> 0.01 μm) compared to P-Trak (> 0.02 μm). Nevertheless, figure 4 clearly shows a lacking of sensitivity that characterizes P-Trak respect to CPC 3007, evidenced by the displacement from the regression line when considering particle concentrations above the indicated limit. Moreover, it should be specified that, for concentration values higher than 200000 pt/cm^3, CPC 3007 also exhibits a weak loss in counting efficiency when compared with a more reliable method (electrometer TSI 3068) [29]. Therefore, P-Trak can be properly used in studies on the detection of UFP variations and in environments with low expected concentrations, whereas CPC 3007 is highly preferable in applications that require a better accuracy.

In general the highest UFP background levels were observed in southern European urban areas [30], with average values in Rome and Barcelona four times higher than those collected in Augsburg, Helsinki and Stockholm. This behaviour is confirmed in Milan, showing mean background levels (green areas) more than double than those found in northern European cities.

The personal concentrations measured in vehicles with instruments and in conditions similar to those adopted in this study (driving along a very busy street with closed windows) are comparable to the ones reported in similar international researches. A recent study in Athens [31] reported a mean value of 94000 pt/cm^3 compared with 107000 pt/cm^3 measured in Milan. These high values are related both to the meteo-climatic conditions of the considered city and to the traffic intensity along the examined streets [32]. Another study demonstrated that the highest UFP exposures occurred while driving in traffic congestions or when behind a heavy diesel-driven vehicle [33]. It is likely that higher concentrations than those observed would be expected if tunnel microenvironments were considered: here UFP concentrations higher than 10^6 pt/cm^3 have been measured [33], due to the absence of atmospheric dispersion.

Analysis of temporal variability of collected measures within the week reveals relevant differences between weekdays and the rest of the week. Sunday is characterized by lower concentration values, while UFPs levels measured on Saturday lie in an intermediate position between Sunday and the working days. A similar behaviour have been observed for data variability: the most wide-spread distributions have been found during weekdays. The concentrations observed in different days of the week confirms a relationship with traffic trends, similarly to what described in other international research studies [30, 34]. The impact of traffic intensity on exposure is also highlighted in figure 7, where the distributions of UFP concentrations according to three different diurnal time periods are displayed. Indeed, more wide-spread distributions and higher particle number concentrations were measured during rush hours, when an increase of UFP emissions is caused by the elevated number of circulating vehicles. Similarly to other European cities [30, 31, 35], peaks of concentrations were found in the morning, between 7.00 and 11.00 a.m., while in the evening measured UFP levels are slightly higher than during the central hours of daytime.

5. Conclusions

Our observations indicate that the highest exposure to UFPs occurs while moving along busy streets or in their immediate proximities, either afoot or by motorized vehicles. Here exposure to UFPs is higher than elsewhere (7.4 to 8.8 times higher than in office micro-environments) so that individuals may gain a significant contribution to their daily total exposure, especially who spend a lot of time in these high-polluted environments. This is the case of traffic wardens, bus and taxi drivers, or those moving during rush hours (commuters), when the increase of UFP concentration is related to the elevated number of circulating vehicles.

References

[1] Morawska L, He C, Hitchins J, Gilbert D and Parappukkaran S 2001 The relationship between indoor and outdoor airborne particles in the residential environment *Atm. Environ.* **35** 3463–73

[2] Kittelson D B 1998 Engines and nanoparticles: a review *J. Aerosol Sci.* **29** 575–88

[3] Morawska L, Bofinger N D, Kocis L and Nwankwoala A 1998 Submicrometer and supermicrometer particles from diesel vehicle emissions *Environ. Sci. Technol.* **32** 2033-42

[4] Ristowski Z D, Morawska L, Bofinger N D and Hitchins J 1998 Submicrometer and supermicrometer particles from spark ignition vehicles *Environ. Sci. Technol.* **32** 3845-52

[5] Kaur S, Clark R D R, Walsh P T, Arnold S J, Colvile R N and Nieuwenhuijsen M J 2006 Exposure visualisation of ultrafine particle counts in a transport microenvironment *Atm. Environ.* **40** 386-98

[6] Ketzel M and Berkowicz R 2004 Modelling the fate of ultrafine particles from exhaust pipe to rural background: an analysis of time scales for dilution, coagulation and deposition. *Atmos. Environ.* **38(17)** 2639-52

[7] Zhu Y, Hinds W C, Kim S, Shen S and Sioutas C 2002 Study of ultrafine particles near a major highway with heavy-duty diesel traffic. Atmos. Environ. 36 (27), 4323–4335.

[8] Kulmala M, Vehkamäki H, Petäjä T, Dal Maso M, Lauri A, Kerminen V M, Birmili W, McMurry P H 2004 Formation and growth rates of ultrafine athmospheric particles: a review of observations. *J. Aerosol. Sci.* **35** 143-76

[9] See S W and Balasubramanian R 2006 Risk assessment of exposure to indoor aerosols associated with Chinese cooking **102(2)** 197-204

[10] Rohr A C, Weschler C J, Koutrakis P and Spengler J D 2003 Generation and quantification of ultra-fine particles through terpene/ozone reaction in a chamber setting *Aerosol Sci. Technol.* **37(1)** 65–78

[11] Franck U, Herbarth O, Wehner B, Wiedensohler A and Manjarrez M 2003 How do the indoor size distributions of airborne submicron and ultra-fine particles in the absence of significant indoor sources depend on outdoor distributions? *Indoor Air* **13** 174–81

[12] Koponen I K, Asmi A, Keronen P, Puhto K and Kulmala M 2001 Indoor air measurement

campaign in Helsinki, Finland 1999—the effect of outdoor air pollution on indoor air. *Atmos. Environ.* **35** 1465– 77

[13] Wallace L and Howard-Reed C 2002 Continuous monitoring of ultra-fine, fine, and coarse particles in a residence for 18 months in 1999–2000 *J. Air Waste Manage Assoc.* **52** 828–44

[14] Matson U 2005 Indoor and outdoor concentrations of ultrafine particles in some Scandinavian rural and urban areas *Sci. Total Environ.* **343** 169–76

[15] Dockery D W and Pope A C 1994 Acute respiratory effects of particulate air pollution *Ann. Rev. Public Health* **15** 107–32

[16] Hagdnagy W, Stiller-Winkler R, Kainka R, Ranft U and Idel H 1998 Influence of urban air pollution on the immune system of children *J. Aerosol Sci.* **2** 997–8.

[17] Jaques P A and Kim C S 2000 Measurement of total lung deposition of inhaled ultra-fine particles in healthy men and women *Inhal. Toxicol.* **12** 715–31

[18] Oberdörster G, Oberdörster E and Oberdörster J 2005 Nanotoxicology: An Emerging Discipline Evolving from Studies of Ultrafine Particles *Environ. Health Persp.* **113(7)** 823-39

[19] Kiss G, Vargapuchony Z, Rohrbacher G, Hlavay J 1998. Distribution of polycyclic aromatic hydrocarbons on atmospheric aerosol particles of different sizes *Atmos. Res.* **46(3–4)** 253–61

[20] Xia T, Korge P, Weiss J N, Li N, Venkatesen M I, Sioutas C and Nel A 2004 Quinones and Aromatic Chemical Compounds in Particulate Matter Induce Mitochondrial Dysfunction: Implications for Ultrafine Particle Toxicity *Environ. Health Persp.* **112(14)** 1347-58

[21] Ferin J, Oberdörster G and Penney D P 1992 Pulmonary retention of fine and ultra-fine particles in rats *Am. J. Resp. Cell. Mol. Biol.* **6** 535-42

[22] Donaldson K, Li X Y and MacNee W 1998 Ultra-fine (nanometre) particle mediated lung injury *J. Aerosol Sci.* **29** 553–60

[23] Johnston C J, Finkelstein J N, Mercer P, Corson N, Gelein R, Oberdörster G 2000 Pulmonary effects induced by ultra-fine PTFE particles *Toxicol. Appl. Pharmacol.* **168** 208–15

[24] Oberdörster G 2001 Pulmonary effects of inhaled ultra-fine particles *In.t Arch. Occup. Environ. Health* **74** 1–8

[25] Stone V, Shaw J, Brown D M, MacNee W, Faux S P and Donaldson K 1998 The role of oxidative stress in the prolonged inhibitory effect of ultra-fine carbon black on epithelial cell functions *Toxicol. InVitro* **12** 649-59

[26] Peters A, Wichmann H, Tuch T, Heinrich J and Heyder J 1997 Respiratory effects are associated with the number of ultrafine particles *Am. J. Respir. Crit. Care Med.* **155** 1376-83

[27] Beck-Speier I, Dayal N, Karg E, Maier K L, Roth C, Ziesenis A, Heyder J 2001 Agglomerates of ultrafine particles of elemental carbon and TiO2 induce generation of lipid mediators in alveolar macrophages *Environ Health Perspec* **109** 613-8

[28] Wagener D K, Selevan S G and Sexton K 1995 The importance of human exposure information: a need for exposure related data bases to protect and promote public health *Annu. Rev. Pub. Health* **16**, 105–21

[29] Hämeri K, Koponen I K, Aalto P P and Kulmala M 2002 The particle detection efficiency of the TSI-3007 condensation particle counter *Aerosol Sci.* **33** 1469-69

[30] Aalto P, Hämeri K, Paatero P, Kulmala M, Bellander T, Berglind N, Bouso L, Castaño-Vinyals G, Sunyer J, Cattani G, Marconi A, Cyrys J, von Klot S, Peters A, Zetzsche K, Lanki T, Pekkanen J, Nyberg F, Sjövall B and Forastiere F 2005 Aerosol particle number concentration measurements in five european cities using TSI-3022 condensation particle counter over a three-year period during health effects of air pollution on susceptible subpopulations *J. Air Waste Manage. Assoc.* **55** 1064–76

[31] Diapouli E, Chaloulakou A and Spyrellis N 2007 Levels of ultrafine particles in different microenvironments - Implications to children exposure *Sci. Total Environ.* **388** 128-36

[32] Westerdahl D, Fruin S, Sax T, Fine P M and Sioutas C 2005 Mobile platform measurements of ultrafine particles and associated pollutant concentrations on freeways and residential streets in Los Angeles *Atmos. Environ.* **39** 3597-610

[33] Weijers E P, Khlystov A Y, Kos G P A and Erisman J W 2004 Variability of particulate matter concentrations along roads and motorways determined by a moving measurement unit *Atmos. Environ.* **38** 2993-3002

[34] Wåhlin P, Palmgren F and Van Dingenen R 2001 Experimental studies of ultrafine particles in streets and the relationship to traffic *Atmos. Environ.* **35** (Suppl. 1) S63-S69

[35] Marconi A, Cattani G, Cusano M, Ferdinandi M, Inglessis M, Viviano G, Settimo G and Forastiere F 2007 Two-years of fine and ultrafine particles measurements in Rome, Italy *J. Toxicol. Env. Heal. A* **70** 213-21

Journal of Physics: Conference Series **151** (2009) 012040 doi:10.1088/1742-6596/151/1/012040

Occupational Exposure to Mineral Oil Metalworking Fluid (MWFs) Mist: Development of New Methodologies for Mist Sampling and Analysis. Results from an Inter-laboratory Comparison

C Khanh Huynh[1], H Herrera[1], J Parrat[2], R Wolf[3] and V Perret[4]

1. Institut Universitaire Romand de Santé au Travail (IST), CH-1005 Lausanne, Switzerland

2. Service des Arts et Métiers et du Travail (LIST), CH-2800 Delémont and Laboratoire intercantonal de santé au travail (LIST), CH-1034 Peseux, Switzerland

3. SUVA, CH-6002 Luzern, Switzerland

4. Service cantonal de toxicologie industrielle et de protection contre les pollutions intérieures (STIPI), CH-1211 Geneve, Switzerland

chhuynh@hospvd.ch

Abstract. Metalworking Fluids (MWFs) are largely used in the sector of undercutting, a large professional activity in Switzerland, in particular in the fine mechanic and watch making industry. France proposes a Permissible Exposure Limit (PEL) of 1 mg.m^{-3} of aerosol. The American Conference of Governmental Industrial Hygienists (ACGIH) sets its value at 5 mg.m^{-3} but a proposal to lower the standard ("intended changes") to 0.2 mg.m^{-3} of aerosol is pending since 2001. However, it has not become a recognized threshold limit value for exposure. Since 2003, the new Swiss PEL (MAK) recommendations would be 0.2 mg.m^{-3} of aerosol (oil with boiling point > 350°C without additives) and/or 20 mg.m^{-3} of oil aerosol + vapour for medium or light oil. To evaluate evaporative losses of sampled oil, the German "Berufsgenossenschaftliches Institut für Arbeitssicherheit" (BGIA) recommends the use of a XAD-2 cartridge behind the filter. The method seems to work perfectly for MWFs in a clean occupational atmosphere free from interference of light vapour cleaning solvent such as White Spirit. But, in real situation, machine shop atmosphere contaminated with traces of White Spirit, the BGIA method failed to estimate the MWFs levels (over-estimation). In this paper, we propose a new approach meant to measure both oil vapours and aerosols. Five inter-laboratory comparisons are discussed, based on the production of oil mist in an experimental chamber under controlled conditions.

1. Introduction

Metalworking Fluids (MWFs) are largely used in the metallurgical industry, especially in undercutting, a widely exercised professional activity in the Swiss watch making industry. Various respiratory pathologies are caused by the production and dispersion of oil mists in the workplace atmosphere. Despite the use of hooding systems around the machines, providing ventilation and filtration in order to reduce oil mists exposure, there are among exposed workers increasing cases of chronic cough, irritation of the airways, bronchitis and asthma. Several epidemiological studies, undertaken since the eighties, showed a higher frequency of respiratory symptoms in subjects exposed to the MWF. In 1995, Ameille et al. [1] studied chronic respiratory symptoms among 308 workers in a French automobile manufacturing plant. Those exposed to straight oil MWFs had a significant higher prevalence of chronic cough, chronic phlegm and dyspnea. These observations were made for exposure levels of 2.6 $mg.m^{-3}$, half the recommended values of 5 $mg.m^{-3}$.

There is no current statistic on the tonnage used, the companies concerned and the number of workers exposed in Switzerland. Nevertheless, adapting a French study [2] carried out in 2001 to the Swiss situation by keeping a proportion of 1:10 between Switzerland and France, we can estimate that approximately 20'000 Swiss workers, out of the 120'000 registered workers, are exposed to oil mists, to a quantity of approximately 0.36 tons of oil per year and per worker. Pulmonary function tests used in occupational medicine are sensitive enough to study the dose-effect relationship. Indeed, Eisen et al. [3], in 2001, showed that, in a group of 1’811 employees of an automobile factory of Michigan, the additional exposure to mineral oil mists at the thoracic level of 1 mg/m3 during one year causes as much effect on the Forced Ventilation Capacity (FVC) as smoking one pack of cigarettes per day during more than one year.

Presently, the national occupational exposure limits concerning oil mists are not homogeneous and standardization at the international level would be welcome. The German standard (BGIA) [5] is 10 $mg.m^{-3}$ of oil aerosol + vapour. France (INRS) [6] proposes a recommended value of 1 $mg.m^{-3}$ of aerosol only, while the USA (ACGIH) keeps the value at 5 $mg.m^{-3}$. ACGIH have also proposed a new Threshold Limit Values (TLV®) for mineral oil of 0.2 $mg.m^{-3}$, with the notation A2 (suspected human carcinogen) for poorly refined mineral oils and A4 (not classified as a human carcinogen) for highly refined mineral oils. However, a proposal to lower the ACGIH standard (“intended changes”) to 0.2 $mg.m^{-3}$ of aerosol is pending since 2001, without becoming a recognized threshold limit for exposure because of lack of more thorough knowledge. Since 2003, the new Swiss recommendations for PEL (MAK)(4) would be 0.2 $mg.m^{-3}$ for heavy oil with Boiling Point (BP) >350°C of aerosol and/or 20 $mg.m^{-3}$ of oil aerosol + vapour for medium or light oil. The National Institute for Occupational Safety and Health (NIOSH), as well as the Occupational Safety and Health Administration (OSHA), recommend an 8 hours Time Weighed Average (TWA) exposure limit of 0.5 $mg.m^{-3}$ for oil mist [7].

Various techniques are used to determine the weight of oil mists collected on the filters: gravimetry [6], spectrophotometry using ultraviolet wavelength [8], or infrared spectrophotometry [5]. However, with all these methods, evaporation can occur from oil mist droplets collected on filter membranes, because the droplets remain in contact with flowing air during sampling [9-12].

Conventional methods recommend the use of a filter to quantify oil mists but this is only reliable for common use MWFs that form stable airborne mists. Oil with viscosities inferior to 18 centi-Stokes (cSt) at 40°C, which is commonly used in Switzerland, looses more than 70% of its weight on the filter for periods up to 6 h at 2 $l.min^{-1}$ of standard sampling conditions [11]. The losses from the MWFs were mostly aliphatic hydrocarbons (C_{12}-C_{24}), but additives such as alkyl benzenes, esters, phenols and terpenes were also lost. The problem of exposure underestimation and inappropriate exposure sampling could be widespread.

To prevent evaporative loss, the BGIA standard [5] involves the placing of a XAD-2 absorbent cartridge behind the filter. The method seems perfect for MWFs in an occupational interference free from light vapour solvent such as White Spirit (C_5-C_{11}), which has a high Swiss PEL (525 $mg.m^{-3}$). In fact, in machine shops atmosphere usually contamined with traces of White Spirit, the BGIA method over-estimated the MWFs levels. Simpson [13] in another recent study proposed a similar method using both filter and cartridges of XAD-2 to measure the total airborne oil (i.e. mist and vapour). Any mist sample that is vaporized should be trapped as vapour by the absorber charcoal tube and XAD-2 cartridge. A gas chromatographic with flame ionization detector (GC-FID) technique was used to determine the vapour fraction.

In this paper, we suggest a new approach designed to overcome the drawbacks of both techniques.

2. Materials and Methods

2.1. Oil mist generation

Oil mist was generated with light, medium and heavy mineral oils based MWF (Somentor 29, Blaser VP1006 or Blasomil 22 and Blaser 220 respectively) in a home-made glass nebulizer [17] using compressed-air inside first, a small cylindrical experimental chamber (130 cm x 50 cm ∅id) during a previous isocyanate study [16, 17] and second, in a experimental chamber of 10 m^3. The compressed air flow rate was automatically controlled between 0 and 10 $l.min^{-1}$ by a mass flow control valve (Hi-tech MFC model F200). The size characteristics of the aerosol were determined using optical method and gravimetric method: an optical particle analyser (Climet Model 208A, USA) equipped with its dilution system (Climet CI-294-1, USA) which counted the particles in the diameter range of <0.3 to 10 μm and classified them into seven categories; an impactor (Ambient Particle Sizing Sampler, model 2000, Andersen Inc., USA) equipped with 8 glass fibre filter stages and a backup filter (Glass fibre filter 934AH, ∅ 81 mm, Whatman) which collected aerosols in the aero-dynamic diameter range of <0.4 to >11 μm. The granulometry distribution of aerosol was computerized using "in house" software, based on Andersen individual calibrated penetration curve of each stage.

To test the performance of oil mist methods, interference was introduced in the test chamber during aerosol generation: Industrial solvent vapour of White spirit (Indurei) at 60-100 $mg.m^{-3}$, 1/5 of PEL (Swiss MAK: 100 ppm or 525 $mg.m^{-3}$). This interference induced no effect on standard aerosol-on-filter methods, but gives an over-estimation by the BGIA-like methods, using absorbers cartridges and IR method. To check the real target value, one round of pure oil mist was generated and during a second round, with identical level of oil mist, White Spirit interference was introduced. In round 5, we also introduced in the experimental chamber other interference, inert dust (glass polydispersed spherical particles, Spheriglass, 0-15 μm, 5 $mg.m^{-3}$ level).

2.2. Proposal methods

The principal aim of this work was to investigate sampling and analyze methods for measuring both mist and vapour, to comply with the new Swiss regulation (2003) and to demonstrate the robustness of the proposals during a mini inter-laboratory comparison. Our laboratory (IST) developed two new methods and proposed to other participants for adaptation during an "inter-laboratory comparison":

2.2.1. *Gravimetric determination method.* Based on the French INRS method [6] for particular fraction and BGIA method [5] for vapour fraction, the new proposal method tries to determine the oil vapour fraction with controlled evaporation under nitrogen and correct the interference from vapour solvent. As to the analytical method, we developed a new technique because the Infrared (IR) method used in Germany, BGIA method [5] does not distinguish the vapours from White Spirit or from the others cleaning solvents. This fact results in an overestimation of the true exposure level to the oil vapours.

The new method consists in initially extracting the filter and XAD-2 cartridge with CH_2Cl_2, then determining the residual fraction by gravimetric measurement after evaporation of the extracting solvent according to French method from INRS [6]. A detailed attention is devoted to the level of evaporation to make it possible the elimination of CH_2CL_2 and other solvents, the cleaning solvents or collected White Spirit on XAD2 cartridges, without loss of oil vapour. After extraction of XAD-2 cartridge by organic solvents, CH_2Cl_2 or tetrachlorethylene (PER), a GC-FID or gravimetric method of evaporated residues were used to determine the vapour oil fraction. The figures 1 to 3 show the determination of CH_2Cl_2, White Spirit and vapour oil fraction by gravimetric with gentle evaporation of extract solution, repetitive weigh of residue on vial and calculate the theoretical amount of volatile fraction. The real amount of volatile fraction was calculated using regression method and intercept at time t=0 (start time, 0 min). This approach was better than conventional constant weighing technique, avoiding loss of oil evaporation. By this technique, we can confirm the possibility to analyse the vapour oil fraction without interference, even at very high level. This technique was proposed for validation in an "inter-laboratory comparison" as shown in this study, and can be applied by any laboratories without special instruments or costly apparatus.

2.2.2. GC-FID or GC-MS method. Based on ISO16703 [18] for the determination of hydrocarbon content in soil, the total signal of FID detector in the retention range of White Spirit against oil volatile fraction or the Single Ion Monitoring (SIM) signal of ion m/z 57 of respective fractions, representing the hydrocarbon molecules were integrated and computed. Quantification was done by comparing the total signal of sample to the standard, in this case, a solution of oil diluted in CH_2Cl_2 or PER. The figure 4 shows the determination of vapour oil fraction by GC-MS method with quantification by summing aliphatic chromatographic area peaks corresponding to oil vapour fraction (C_{12} to C_{24} in this case). The interference compound, the White Spirit fraction of C_5 to C_{11} is separated and discarded. This technique was also validated in an inter-laboratory comparison as shown in this study, but it required more costly instruments like GC-MS or GC-FID.

2.3. Participants

The principal aim of this work was to prove the inefficiency of actual official methods to measure the oil mist and to suggest Swiss official laboratories for adaptation of new proposals. There is not a real inter-laboratory comparison with certified methods but a tentative work-team to check the feasibility of new proposals. To achieve the objectives, five inter-laboratory comparisons were organised based on generation of oil mist in an experimental chamber, under controlled conditions.

In the beginning, each participant used its own conventional and official method, without modifications. By the introduction of interference during the oil mist generation, we can point out the lacks of official methods against interferences and the loss-by-evaporation during samplings. After a few round, each participant will be asked for modification if necessary to overcome the loss-by-evaporation during sampling or White Spirit interference problem. IST laboratory proposed two methods: gravimetry with controlled evaporation rate (method C) and GC-MS (method D).

The adopted sampling techniques for aerosol and oil vapour were: glass fibre filters + XAD-2 adsorbent cartridges [7]. Gravimetric, Fourier Transform Infrared Spectroscopic (FTIR), gas chromatography with flame ionisation detection (GC-FID) and gas chromatography with mass spectrometric detection (GC-MS) analytical methods were considered for quantitative determination of collected oil mist and vapour. The sampling time (2 to 8 hours) was determined in function of the concentration level of oil mist, the capacity load of aerosol samplers and the Limit of Quantification (LOQ) of the methods (at least 0.2 mg) on the filters for accuracy gravimetric determination. Briefly, the different adopted methods by participants were described herewith, and summarized in table 1:

2.3.1. Method BGIA. Standard BGIA method [5] with GSP samplers were used with minor modifications for lab 5: 1 l/min sampling flow rate instead of 3.5 l/min with a modified entry orifice of ∅ 2 mm. This modification tries to minimize the evaporation loss of droplet oil collected on the filters, but no effects was observed and finally, lab 5 reused the original BGIA GSP sampler. After sampling, the filter and the cartridge of absorber (3g of XAD-2) were eluted with 10 ml of PER and the total oil mist weight was done by measuring IR absorbance at 3000-2800 cm^{-1} of desorbed solution with 10 mm path length. Quartz cell comparing to standard solution of oil in PER freshly prepared. Blank correction was done with extracted blank glass fibre filter solution of PER.

2.3.2. *Method BGIA mod.* The determination of volatile oil fraction can also be done by gentle evaporation of extract PER solution, elimination of interfering solvents and determination of the amount of volatile oil fraction by FTIR. Before analyzing the samples, the evaporation method is optimized with known mixtures of the used MWF and the interfering solvents in PER. With this adjusted method all extracts of filters and adsorbents and all calibration solutions are treated in the same way, advantageous simultaneously by a multi-rotavapor device. The quantitative determination is done by FTIR.

2.3.3. Method B. Standard NIOSH method 5026 [20] was applied, using PVC filters with modification for volatile fraction by adding sorbent cartridge (1g of XAD-2) as backup vapour sampler. IOM heads were also tested at 2 l/min during few rounds. Gravimetric determinations of oil mist on the filters were done after conditioning during 24 hours in controlled humidity box (52 % RH ± 5 %) before weighing by microbalance (Model M5P, Sartorius). The determination of volatile fraction on XAD-2 cartridges was done by similar procedure described in method C.

2.3.4. Method C. Based on French INRS method [6] for particular fraction and BGIA method [5] for vapor fraction, the new proposal method tries to correct the interference from vapour solvent. The sampling train was ∅ 37 mm closed-face polystyrene cassette containing glass fibre filters, (GF/B, Whatman), coupled with a Solid Phase Extraction (SPE) cartridge (6 ml Polypropylene tube and frits, ref 57242 and 57181, Supelco) containing 1 g of XAD-2 (ref 1-0357, Supelco). The flow rates were 2 l/min for all round tests. For particular fraction, gravimetric methods were done with filters conditioned before and after sampling in the same glow box as method B. Extraction of oil mist on the filter with 20 ml CH_2Cl_2 followed by evaporation under gentle nitrogen stream (200 ml/min of N_2 at 40°C), and gravimetric determination of extracted fraction was also applied to avoid non-oil mist dust in the real situation. Volatile fraction on XAD-2 cartridges was extracted by elution of 20 ml of distilled CH_2Cl_2. After evaporation to small volume (5 ml), under gentle nitrogen stream, the final solution was transferred into a small vial and gravimetric determination of volatile oil fraction can also be done by gentle evaporation of extract solution, repetitive weigh of residue on vial and calculate the theoretical amount of volatile oil fraction (see result section).

2.3.5. *Method D.* Volatile fraction on XAD-2 cartridges was extracted by elution of 20 ml of distilled CH_2Cl_2. After evaporation to small volume (5 ml), under gentle nitrogen stream, a amount of extract was injected into a GC-MS for determination of interference (White Spirit, C_5-C_{11}) and volatile oil fraction, C_{12}-C_{24}). The method was based on ISO 16703 [18] for the determination of hydrocarbon content in soil. Briefly, 1-2 µl of extracted solution were injected in a GC-FID or GC-MS instrument and the oil volatile fraction (C_{12}-C_{24}) fraction was quantified by using the total signal of FID detector or the Single Ion Monitoring (SIM) signal of ion m/z 57 of GC-MS detector in their retention range were integrated and computed. Fraction representing White Spirit (C_5-C_{11}) was ignored. Calibration curve was computed with standard solution of oil at different concentrations.

It seem that there are a plenty of methods applied by participants, but in fact, the general approach was based on both IST proposals: gravimetric with controlled evaporation under nitrogen and by GC technique.

2.4. Quality control

The comparison procedure was based on the International Organisation for Standardization [14] (ISO 5725), to provide information about the repeatability (within-laboratory variation) and the reproducibility of measurements (within- and between-laboratory variation). To gauge the acceptability of the method performances, the results were compared with the requirements of the European Standard EN 482 Workplace atmospheres-General requirements for the performance of procedures for the measurement of chemical agents [15]. It requires the Relative Overall Uncertainty (ROU) or bias plus twice the standard deviation, for comparison with limit values to be less than 30 percent, when used in range 0.5 to 2 times a limit value, including sampling and analytical errors. The reference level was computerized with all individual data of each participating lab.

3. Results and discussions

3.1. Generated WMFs oil mist performances

Oil mist was generated with light, medium and heavy mineral oils based MWF (Somentor 29, Blaser VP1006 or Blasomil 22 and Blaser 220 respectively) in a home-made glass nebulizer [17] using compressed-air in a experimental chamber of 10 m^3. The generated range of oil mist can be regulated in the range of 0.1 to > 20 mg.m^{-3}. After a 15-min stabilization time, the aerosol concentration of commercial machine cutting oil (Blaser VP1006) can be held constant for more than 6 hr within a deviation standard of < 5 %. The spatial uniformity of the aerosol inside the test chamber was in the range of 2-3 % reported by a 5 points test. This clearly illustrates the good stability and spatial uniformity of the aerosol concentration, ready for an inter-laboratory comparison. The granulometry distribution of aerosol was computerized and show an aerodynamic diameter means of 2.74 μm ± 1.8 μm. This size distribution of oil mist corresponds to the observed aerosol distribution in machining shop atmosphere.

3.2. Loss of oil mist during sampling

Table II shown the loss of oil vapour fraction spiked on filter in function of the Boiling Point (BP) or the Flash Point of respectively light, medium and heavy oil during the sampling, using a flow rate of 2 l/min and a total air volume of 480 l through the filters. The amount of evaporated vapour oil was collected on XAD2 cartridges mounted in serial after the filter. With light oil (flash point: 74°C), all the aerosol fraction was evaporated and moved to XAD2 cartridge, with medium oil, the vapour-liquid repartition is 50% and with the heavy oil (flash point: > 250°C), only 25% vapour fraction was evaporated during the sampling. Depending the nature of oil, the problem of exposure under-estimation and inappropriate exposure sampling could be widespread.

To prevent evaporative losses of sampled oil, the BGIA standard [5] proposed a XAD-2 absorbent cartridge behind the filter. The method seem perfect for MWFs in an occupational interference free of light vapour solvent such as White Spirit (C_5-C_{11}) with a very high Swiss PEL (525 mg.m^{-3}). In reality, in machine shops atmosphere, we uncounted usually high level of White Spirit or others cleaning solvents, and the BGIA method overestimates the MWFs levels.

3.3. Gravimetric determination of oil vapour

As to the analytical method, we developed a new technique because the IR method used in Germany, BGIA method [5] does not distinguish the vapours from White Spirit or from the others cleaning solvents. This fact results in an overestimation of the true exposure level to the oil vapours.

The new gravimetric method consists in initially extracting the filter and XAD-2 cartridge with CH_2Cl_2, then determining the residual fraction by gravimetric measurement after evaporation of the extracting solvent according to French method from INRS [6]. A detailed attention is devoted to the level of evaporation to make it possible the elimination of CH_2CL_2 and other solvents, the cleaning solvents or collected White Spirit on XAD2 cartridges, without loss of oil vapour. Contrarily to the French method, we are not going to evaporate until dryness to constant weight, but we try to monitor the evaporation rate of the oil fraction to avoid losses. The determination of vapour oil fraction by gravimetric with gentle evaporation of extract solution was done by repetitive weigh of residue on vial and calculation of the theoretical amount of volatile oil fraction. The real amount of vapour oil was calculated using regression method and intercept at time t=0. After extraction of XAD-2 cartridge by organic solvents (CH_2Cl_2 or PER), a GC-FID or gravimetric method of evaporated residues were used to determine the vapour oil fraction. The figures 1 to 3 show the determination of CH_2Cl_2 (theoretical added value 3.98 g, founded: 3.656 g), White Spirit (theoretical added value: 100 mg, founded: 105.2 mg) and vapour oil fraction (theoretical added value: 10 mg, founded: 9.58 mg) by gravimetric with gentle evaporation of extract solution, repetitive weigh of residue on vial and calculate the theoretical amount of volatile fraction. The precision of the method was better than ± 20 μg by weighing for oil determination. The errors for interference White Spirit and solvent CH_2Cl_2 were much more higher but this is not important for MWFs determinations. Other conventional sampling methods (Coconut Charcoal tubes, GC method) can be used simultaneously to perform White Spirit level in air. This approach was better than conventional constant weighing technique, avoiding loss of oil evaporation. By this technique, we can confirm the possibility to analyse the vapour oil fraction without interference, even at very high level. This technique was validated in an inter-laboratory comparison as shown in this study, and can be applied by any laboratories without special instruments or costly apparatus.

3.4. GC-FID or GC-MS determination of oil vapour

Based on ISO16703 [18] for the determination of hydrocarbon content in soil, the total signal of FID detector in the retention range of White Spirit against oil volatile fraction or the Single Ion Monitoring (SIM) signal of ion m/z 57 of respective fractions, representing the hydrocarbon molecules were integrated and computed. With the high resolution of the GC capillary columns, the separation of dissolvent and oil vapour fraction was made without difficulties. Quantification was done by comparing the total signal of sample to the standard, in this case, a solution of oil diluted in CH_2Cl_2 or PER. The figure 4 shows the determination of vapour oil fraction by GC-MS method with quantification by summing aliphatic chromatographic area peaks corresponding to oil vapour fraction (C_{12} to C_{24} in this case). The interference compound, the White Spirit fraction of C_5 to C_{11} is separated and discarded. This technique was also validated in an inter-laboratory comparison as shown in this study, but it required more costly instruments like GC-MS or GC-FID.

3.5. Round-Robin test results

The figure 5 shows the experimental chamber of 10 m^3 during Round Robin inter-laboratory sampling exercise. Each participant was asked to bring their own sampling devices, pumps, sampling heads, cartridges, etc. All participants applied the combined sampling devices (filters+cartridges XAD2) to eliminate the loss of oil vapour. Table III shown the all laboratory results in term of sum of particulates and vapour of oil mist generated over 5 rounds. First of all, as show the fig 6, we observed that at 24.18 mg/m^3 level of pure oil mist (Switzerland 2003 PEL = 20 $mg.m^{-3}$), all laboratories complied with the ROU<30%, no matter the sampling heads used (Ø47 mm BGIA sampling head, Ø25 mm IOM sampling head, and Ø37 mm closed-face polystyrene cassette), nor the used methods (IR, gravimetry or GC-MS). The overall performance of all laboratories remained very good for lower levels: 14.75, 3.32 mg/m^3, excepted lab 5, with a ROU slightly higher than 30. The potential explanations for slightly high ROU (56.3) of lab 5 in this round were high blank level, contamination and eventually technical problems. At 3.32 $mg.m^{-3}$ of pure oil mist level of round 4, we introduced interference, White Spirit at 1/5 of PEL level (100 $mg.m^{-3}$), We observed, as show the fig 7, a bias on results of lab 4 and 5 using standard BGIA method and IR technique to measure the total oil mist. The interference, a family of aliphatic hydrocarbons, absorbs in a same IR range than oil mist and the results were overestimated. Other labs (1 to 3) using gravimetric with controlled evaporation under nitrogen or GC-MS technique, found the correct result. 4. At 1.64 $mg.m^{-3}$ level of pure oil mist + interferences (White Spirit and Spheriglass) of round 5, lab 3 lost vapour oil during evaporation and weighing. Lab 4 used BIA mod method and found good results. For the determination of volatile fraction of oil, GC-FID, GC-MS or gravimetric methods worked well, except that some special oils containing polar or thermo sensible "non-chromatographed" compounds by GC. In this case, only gravimetric method was applied.

4. Conclusions

In this study, we propose two new and validated methods of measuring mixed mist and vapour oil and to overcome the difficulties of loss of vapour oil during sampling and interference problems: by determining separately the aerosol fraction on filter and by adding a absorber cartridge (XAD2) for trapping vapour fraction of oils, completed by a selective determination of vapour oil fraction (C_{12}-C_{24}) against interferences such as cleaning solvent fraction White Spirit like (C_5-C_{11}).

Two techniques to quantify the oil vapour fraction can be used: by gravimetric with controlled evaporation under nitrogen, by FTIR, by GC-FID and by GC-MS:

1. The gravimetric method, using controlled evaporation under nitrogen technique was proposed as standard method. This method is easy to apply successfully by any laboratories without the needs of special instruments or costly apparatus.
2. Alternative techniques, using GC-FID or GC-MS methods were also validated in this inter-laboratory comparison.

For low viscosity oils, it may be appropriate to measure both the mist and vapour, as request the new Swiss regulation. The inter-laboratory circuit to validate sampling and analytical approaches was helpful and the accuracy results of developed methods give more reproducible results suitable for assessment or survey of real oil mist exposure for futures epidemiological studies.

Acknowledgments
The authors would like to thank Philippe Boiteux, Pierre Alain Porchet, Christine Nicole, Corine Burla, Olivier Schenk, Konrad Schelbert and Maria Barbey for their greatly appreciated technical assistance.

References

[1] Ameille J, Wild P, Choudat D, Ohl G, Vaucouleur JF, Chanut JC and Brochard P **1995** *American J. of Ind. Med.* **27**: 247-256
[2] Lafontaine M **2001** *Travail et Sécurité* **No. 607**, 36-40
[3] Eisen EA, Smith TJ, Kriebel D, Woskie SR, Myers DJ, Kennedy SM, Shalat S and Monson RR **2001** *American J. of Ind. Med.* **39**: 443-453
[4] SUVA **2007** : *Valeur limites d'exposition aux postes de travail (Permissible Exposure Limit at working places)*, Switzerland
[5] BIA **1997** *Kühlschmierstoffe 7750*, BIA-Arbeitsmappe 19. Lfg XI/1997, BG-Code 07432
[6] INRS **2003** *Brouillards d'huile entière, détermination gravimétrique (Mineral oil mist, gravimetric determination)* Fiche **006**, INRS, France
[7] Harper M 2002 *American Int Hyg Assoc J* **63** 488-492
[8] Turner D, Baxter HG, Davies GM, Eyres AR, Lees GL, Mitchell J, Sanderson JT, Steel J and Moore JP **1975** *Ann. Occup. Hyg.* **18** 293-297
[9] Park D, Kim SB and Yoon CS **2003** *American Int Hyg Assoc J* **64** 837-841
[10] Raynor PC, Volckens J and Leith D **2000** *Appl. Occup. Hyg.* **15** 90-96
[11] Simpson AT, Groves JA, Unwin J and Piney M **2000** *Ann. Occup. Hyg.* **44** 165-172
[12] McAneny JJ, Leith D and Boundy MG **1995** *Appl. Occup. Environ. Hyg.* **10** 783-787
[13] Simpson AT **2003** *Appl. Occup. Environ. Hyg.* **18** 865-876
[14] International Organisation for Standardization **1994** *Accuracy (trueness and precision) of measurement methods and results* **Part 2** Basic method for the determination of repeatability and reproducibility, ISO 5725-2, Geneva
[15] Comité Européen de Normalisation **1994** *Workplace atmospheres-General requirements for the performance of procedure for the measurement of chemical agents* **EN482** CEN, Brussels
[16] Huynh CK, Vu Duc T and Savolainen H **1992** *American Int Hyg Assoc J* **53** 157-162
[17] Huynh CK, Vu Duc T and Savolainen H **1989** *Ann. Occup. Hyg.* **33** 573-581
[18] International Organisation for Standardisation **2004** *Soil quality – Determination of content of hydrocarbon in the range C10 to C40 by gas chromatography* **ISO 16703**, Geneva
[19] Lorberau C and Glaser R **1996** *National Institute for Occupational Safety and Health (NIOSH)* Oil mist, mineral Method **5026** Issue 2. Manual of Analytical Methods (NMAM), 4th edition. P.C. Schlecht and P.F. O'Connor (eds.). DHHS (NIOSH) Pub no. 94-113. Cincinnati, Ohio: Department of Health and Human Services, Public Health Service, Centers for Disease Control, NIOSH.

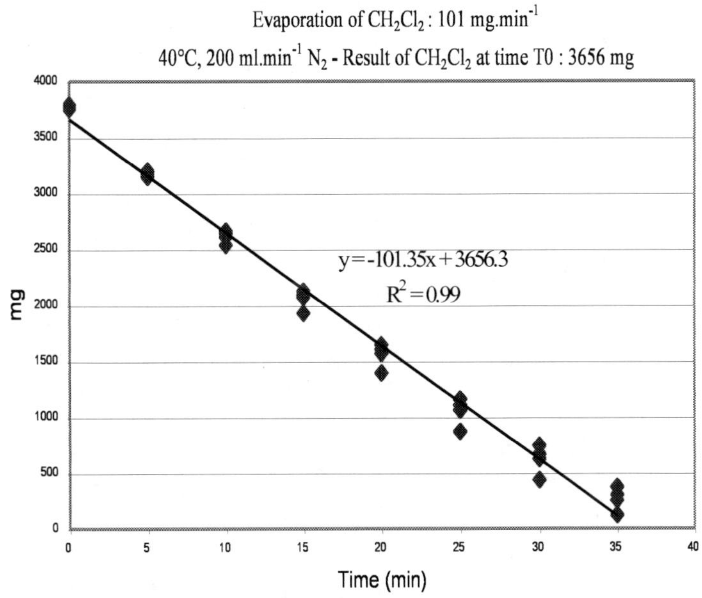

Figure 1. Gravimetric determination of CH_2Cl_2 extracting solvent fraction was presented by gentle evaporation of 3ml of CH_2Cl_2 solution and repetitive weighing of residue on vial (N=4). The theoretical value of CH_2Cl_2 was 3.98 g and the experimental value was 3.656 g. This experiment suggests gentle evaporation of CH_2Cl_2 extracting solvent to avoid losses, without exact determination of solvent amount. Under 200 ml.min^{-1} flow rate of N_2 at 40°C, the evaporation rate of CH_2Cl_2 was 101 mg.min^{-1}.

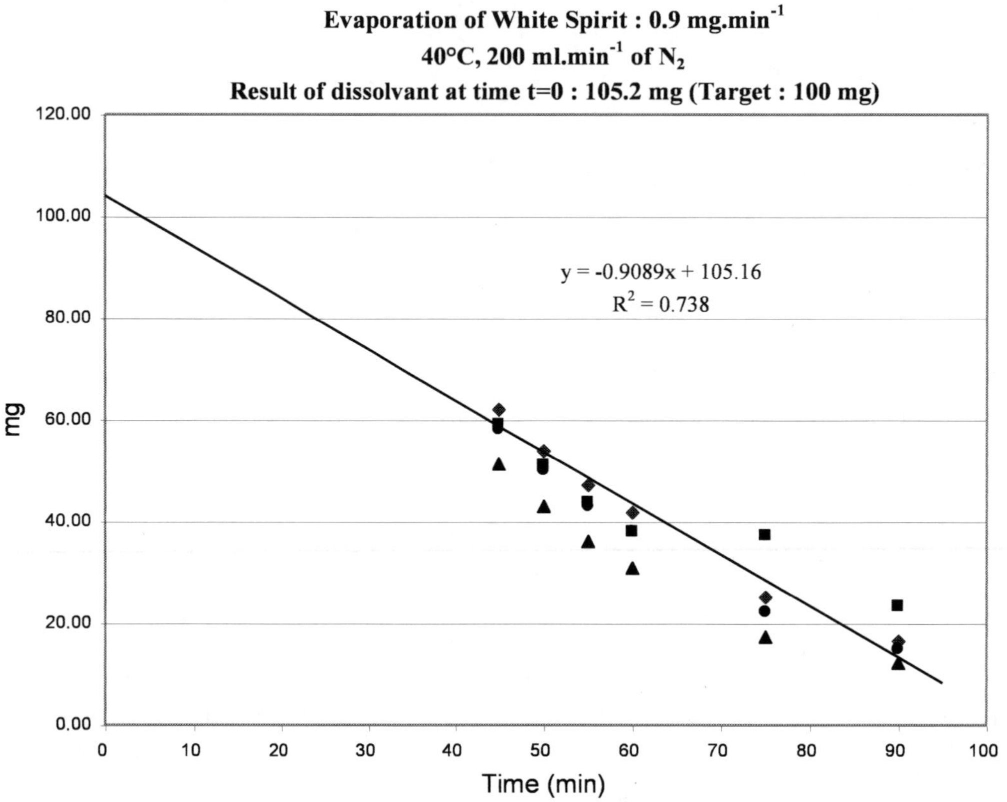

Figure 2. Determination of interference White Spirit fraction by gravimetric with gentle evaporation of extract solution, repetitive weigh of residue on vial and calculate the theoretical amount of White Spirit fraction. The real amount of White Spirit (100 mg) was calculated using regression method and intercept at time T=0 (N=4) and the experimental value was 105.2 mg. The evaporation rate of White Spirit was 0.9 mg.min^{-1}. Again, this experiment tends to suggest gentle evaporation of White Spirit interference to avoid losses, without exact determination of White Spirit amount

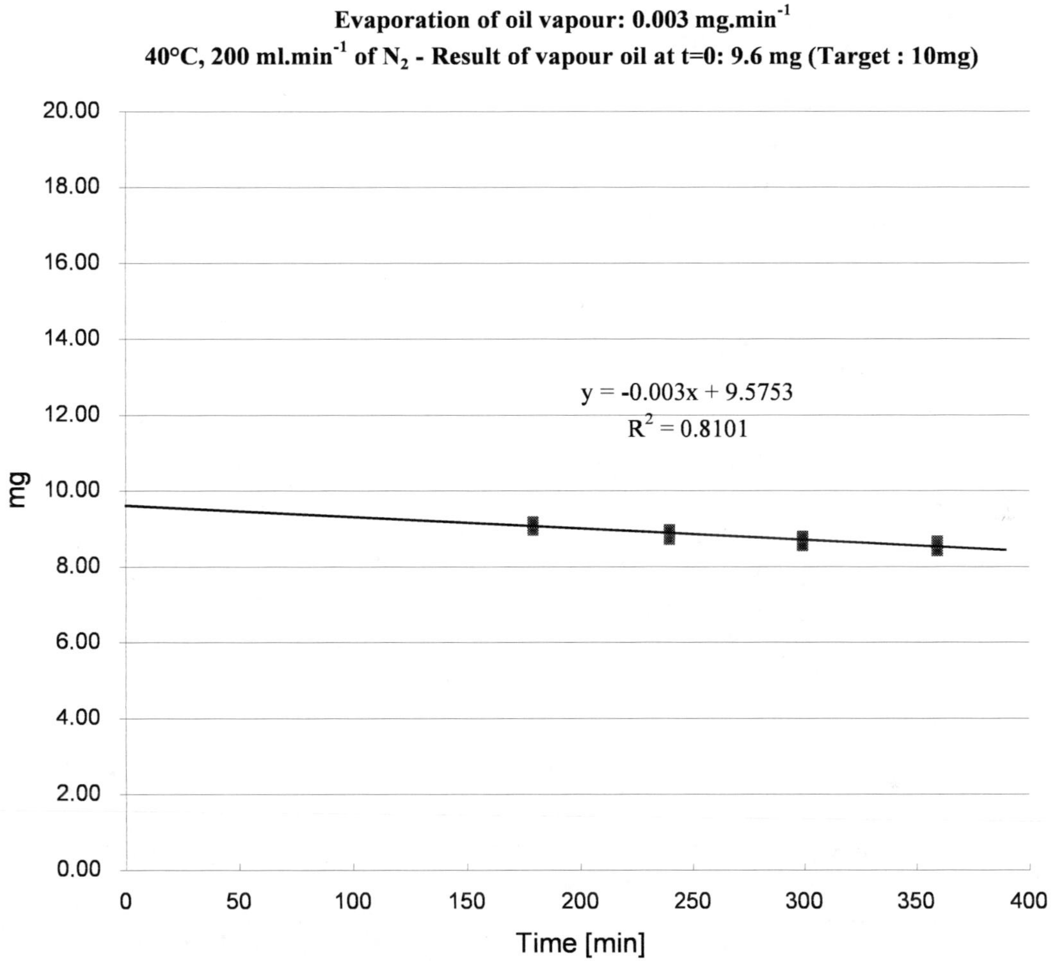

Figure 3. Determination of vapour oil fraction by gravimetric with gentle evaporation of extract solution, repetitive weighing of vial residue and calculating the theoretical amount of volatile oil fraction. The real amount of vapour oil (10 mg) was calculated using a regression method and intercept at time t=0 (N=4) and the experimental value was 9.58 mg. The evaporation rate of oil was very low, 0.003 $mg.min^{-1}$, but not negligible. Contrary to the INRS method assuming constant residual weight over time, the determination of the oil fraction was made using the slope of evaporation.

Figure 4. Determination of vapour oil fraction by GC-MS method with quantification by summing aliphatic chromatographic area peaks corresponding to oil vapour fraction (C_{12} to C_{24} in this case). The interference compound, the White Spirit fraction of C_5 to C_{11} is separated and discarded.

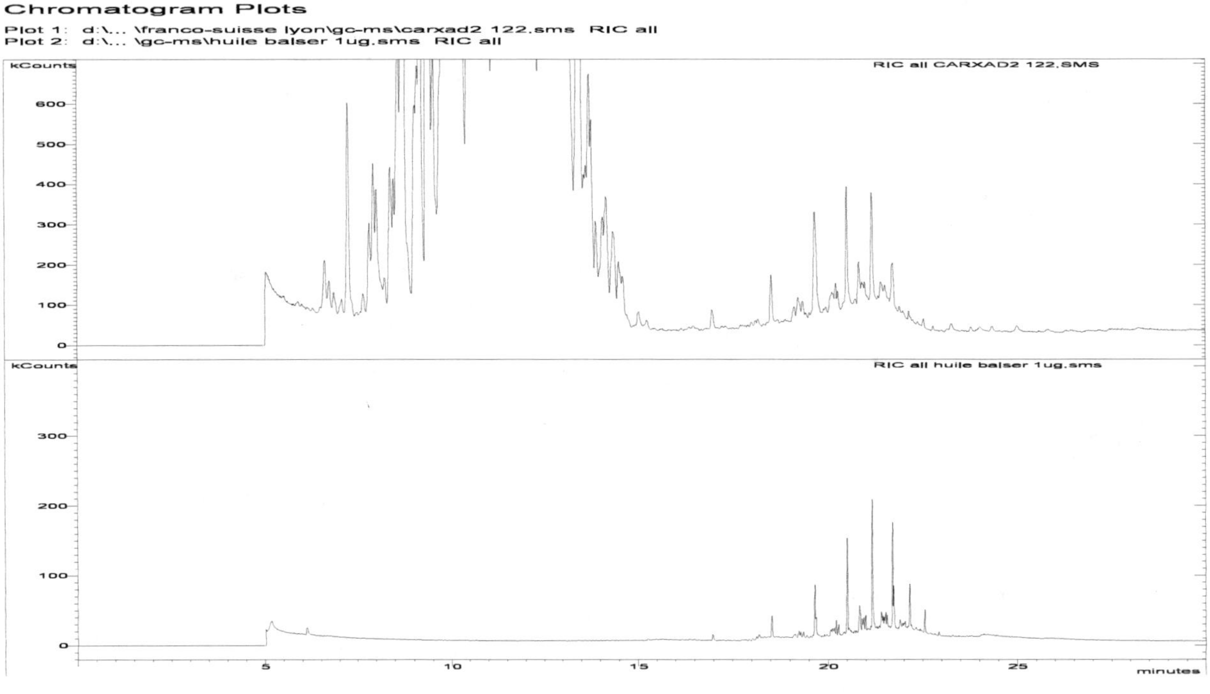

Figure 5. The picture shows the various sampling devices inside the experimental chamber of 10 m^3, during inter-laboratory sampling exercise. Oil mist from light, medium and heavy mineral oils based MWF (Somentor 29, Blaser VP1006 or Blasomil 22 and Blaser 220 respectively) was generated in a homemade glass nebulizer. The generated range of oil mist can be regulated in the range of 0.1 to > 20 $mg.m^{-3}$ and can be held constant at lest for 6 hr within an deviation standard of < 5 %. The spatial uniformity of the aerosol inside the test chamber was in the range of 2-3 % reported by 5 test points. Each participant used five to six sampling heads to perform reproducible results.

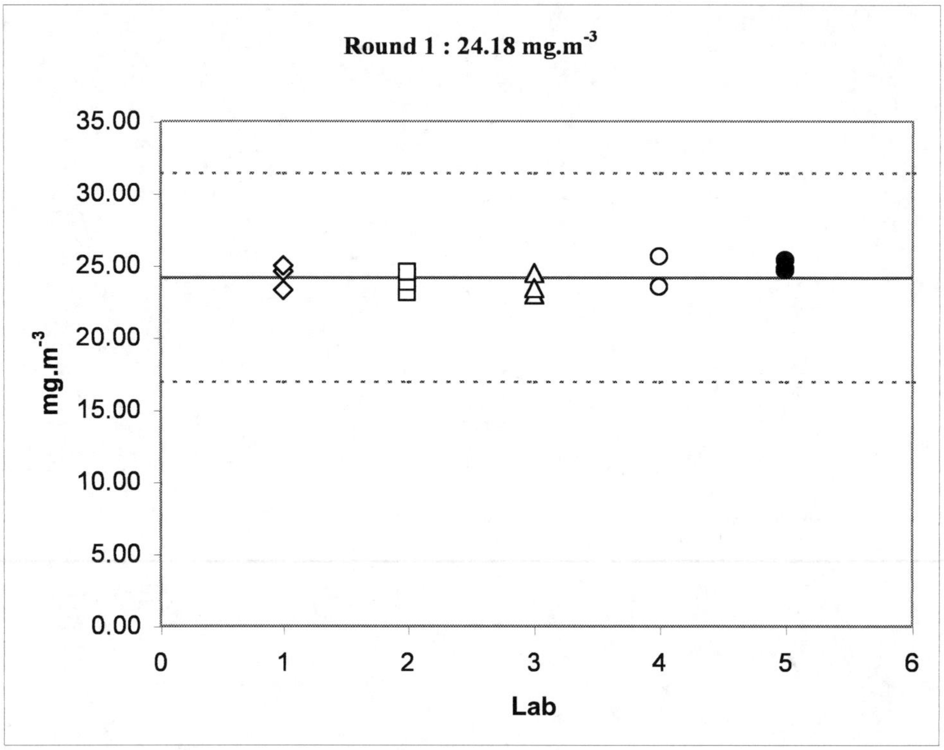

Figure 6. All laboratories round 1 results of medium oil mist generated at 24.18 $mg.m^{-3}$. We observed an excellent correlation of results independently of laboratory method and sampling technique, dotted lines mean ±30% of the Relative Overall Uncertainty (ROU) or bias plus twice the standard deviation, for comparison with limit values to be less than 30 percent, when used in range 0.5 to 2 times a limit value, including sampling and analytical errors.

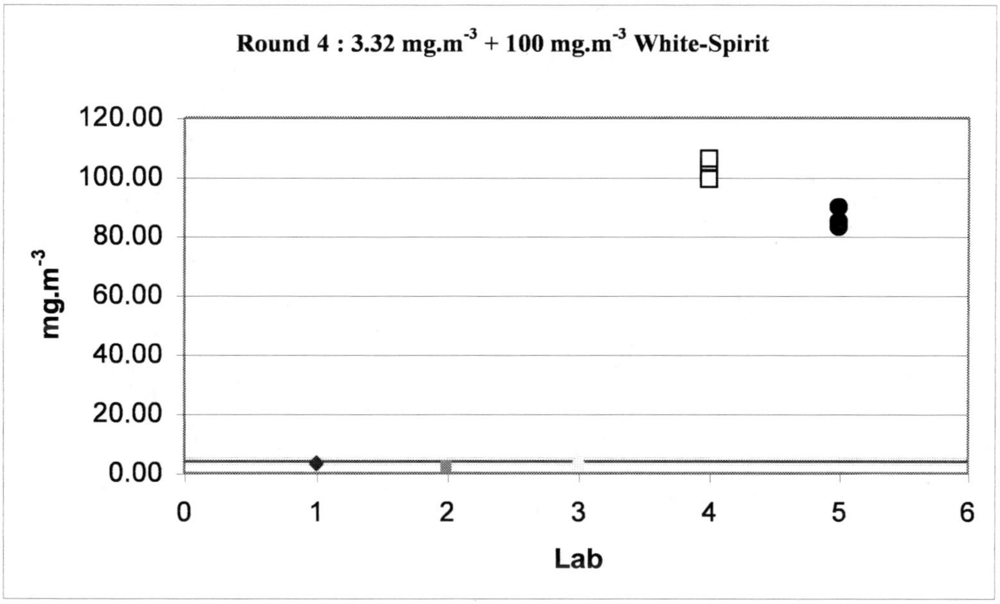

Figure 7. All laboratories round 4 results of medium oil mist generated at 3.32 mg/m^3 in presence of interference, White Spirit at 1/5 of PEL level (100 mg.m^{-3}). We observed a bias on results of lab 4 and 5 using BGIA method and IR technique to measure the total oil mist. The interference, a family of aliphatic hydrocarbons, absorbs in a same IR range than oil mist and the results were over estimated. Other labs (1 to 3) using gravimetric with controlled evaporation under nitrogen or GC-MS technique, found the correct results.

Table 1. Summary of methods used by participating laboratories (N=5).

Method	Lab	Sampling devices	Flow rate l/min	Filter nature Ø mm	XAD-2 g	Extraction solvent	Measurement
BGIA	4, 5	GSP sampler	3.5	Glass fibre filter Ø 47	3	PER	FTIR
BGIA mod	5	GSP sampler	3.5	Glass fibre filter Ø 47	3	PER	FTIR + multi-rotavapor
B	2	IOM	2	PVC Ø 37	1	CH_2Cl_2	Gravimetry
C	1, 3	Closed-face polystyrene cassette	2	Glass fibre filter Ø 37	1	CH_2Cl_2	Gravimetry
D	1	Closed-face polystyrene cassette	2	Glass fibre filter Ø 37	1	CH_2Cl_2	Gravimetry + GC-MS

Table 2. Loss of oil vapour fraction collected on filter in function of the Boiling Point (BP) or the Flash Point of respectively light, medium and heavy oil during the sampling. With light oil, the entire aerosol fraction is evaporated and moved to XAD2 cartridge, with medium oil, the repartition is 50% and with the heavy oil, only small vapour fraction is evaporated. Since 2003, the new Swiss recommendations for PEL (MAK) [4] would be 0.2 mg.m^{-3} for heavy oil with Boiling Point (BP) >350°C of aerosol and/or 20 mg.m^{-3} of oil aerosol + vapour for medium or light oil. In some cases, we have not founded the BP, because the thermo degradation of heavy oils and Flash Point unit was more usually reported in oil industries.

Oil	Flash Point	Oil amount (mg) spiked on filter (N=3)	480 l of air at 2 l/min	
			Filter (mg)	XAD2 (mg)
Light oil (Somentor 29)	74°C	4.91 ± 0.05	-	5.04 ± 0.17
Medium oil (Balser VP1006 or Blasomil 22)	180°C	2.05 ± 0.05	0.96 ± 0.08	0.82 ± 0.03
Heavy oil (Blaser 220, BP>350°C)	>250°C	0.22 ± 0.05	0.16 ± 0.002	0.08 ± 0.006

Table 3. All laboratory Round-Robin results of generated oil mist.

Laboratory	Method	N	Mean	STD	ROU
Round 1:	**Medium oil, reference value: 24.18 ± 0.69 mg.m^{-3}**				
1	C	5	24.72	0.86	9.3
2	B	4	23.43	0.91	10.6
3	C	3	23.63	0.76	8.5
4	BGIA	3	24.15	1.21	10.1
5	BGIA	3	24.87	0.38	6.0
Round 2:	**Medium oil, reference value: 14.75 ± 0.44 mg.m^{-3}**				
1	C	5	14.58	0.80	12.1
2	B	4	14.38	1.83	27.3
3	C	2	11.58	0.48	27.9
4	BGIA	3	15.96	0.18	10.7
5	BGIA	3	16.43	1.82	**36.0**
Round 3:	**Medium oil, reference value: 3.32 ± 0.10 mg.m^{-3}**				
1	C	4	3.18	0.10	10.5
2	B	4	3.17	0.31	23.2
3	C	3	2.84	0.23	28.6
4	BGIA	3	3.35	0.30	18.6
5	BGIA	3	4.17	0.51	**56.3**
Round 4:	**Medium oil, reference value: 3.32 ± 0.10 mg.m^{-3} + interference 100 mg.m^{-3} of White Spirit (1/5 of Swiss PEL)**				
1	C	5	3.09	0.28	24.2
2	B	2	1.90	0.02	**44.2**
3	C	3	4.08	0.58	**58.0**
4	BGIA	3	103.12	3.65	**2838.0**
5	BGIA	3	85.57	3.52	**2364.5**
Round 5:	**Medium oil, reference value: 1.64 ± 0.33 mg.m^{-3} + interference 100 mg.m^{-3} of White Spirit (1/5 of Swiss PEL) + 5 mg.m^{-3} inert dust (Spheriglass)**				
1	C	5	1.74	0.17	26.6
1	D	5	1.66	0.05	7.3
2	B	5	1.69	0.20	**38.5**
3	C	5	0.95	0.04	**46.8**
4	BGIA mod	5	1.72	0.03	9.3

Remarks:

1. Round 1, at 24.18 mg.m^{-3} level of pure oil mist (PEL = 20 mg.m^{-3}, Switzerland 2003), all laboratories complied with the ROU<30%, no matter the sampling heads used (∅47 mm BGIA head, ∅25 mm IOM head, and ∅37 mm closed-face polystyrene cassette), nor the used methods (IR, gravimetry or GC-MS).
2. The overall performance of all laboratories remained very good for lower levels, rounds 2 and 3: 14.75, 3.32 mg.m^{-3}, excepted lab 5, with a ROU slightly higher than 30.
3. At 3.32 mg.m^{-3} of pure oil mist level of round 4, we introduced interference, White Spirit at 1/5 of PEL level (100 mg.m^{-3}). We observed a bias on results of lab 4 and 5 using standard BGIA method and IR technique to measure the total oil mist.
4. At 1.64 mg.m^{-3} level of pure oil mist + interferences (White Spirit and Spheriglass) of round 5, lab 3 lost vapour oil during evaporation and weighing. Lab 4 used BGIA mod method and found good results.

A mini-sampler for welding aerosol mounted in close vicinity of the mouth/nose

Göran Lidén[1] and Jouni Surakka[2]

[1] Department of Environmental Science, Stockholm University, Stockholm, Sweden
[2] Swedish Work Environment Authority, Solna, Sweden

eMail: Goran.Liden@itm.su.se

Abstract. A small personal aerosol mini-sampler to be used inside modern welding visors has been developed. The main object of the mini-sampler has been to sample manganese. The sampler is based on commercially available 13 mm filter holders but modified to incorporate an inlet nozzle made of aluminium. The nominal flow rate of the mini-sampler is 0,75 l/min. The sampler is to be worn mounted on a headset, modified from professional microphone headsets. The headset mounting arrangement was accepted by the welders. The sampling bias of the mini sampler versus the IOM sampler depends on the coarseness of the sampled aerosol. At the lowest concentration ratio of the open-face 25 mm filter holder to the IOM sampler equal to 0,65, the bias of the mini sampler is approximately -26% versus the IOM. The RMS sampling bias of the mini sampler versus the IOM sampler for manganese is -4,6%. The inhalable fraction of welding aerosol mass consists only of 25-55% of welding fume. The rest of the mass is made up of spatter particles and grinding particles. For manganese generally more than 65% is found in the fume.

1. Introduction

Personal aerosol sampling at workplaces is performed with the samplers mounted in what is termed the breathing zone, generally defined as a hemisphere with its centre between the nose and mouth and with a 25 or 30 cm radius. This definition accepts samplers mounted both at the lapel/collar and at the top of the head. Many scientific reports have demonstrated that the concentration of a workplace air pollutant in general is not constant within this zone [1-8]. Several suggestions have been presented over the years to sample aerosols closer to the mouth/nose [9-14]. Of these only HSL's measurement procedure for resin acids in solder fume seems to have gained wider application [14].

In welding sampling has for a long time preferably been carried out with samplers mounted at the lapel/collar bone. Since the early seventies only a few investigations have been published on the possible concentration bias between samplers mounted behind the welders visor and at the lapel/collar [15, 16]. With the adoption of the international standard on the sampling of welding fumes [17] several countries have begun changing the mounting of samplers. Chung et al. [18] investigated the relationship between the concentration sampled by several different samplers mounted behind a welders visor and the concentration inhaled by a breathing (inhalation only) mannequin during welding.

In this paper the design and validation of a headset-mounted mini-sampler for manganese in welding aerosols is described.

2. Background

The Swedish Work Environment Authority (SWEA) decided to reduce the occupational exposure limit (OEL) for manganese sampled as total dust from 400 µg/m^3 to 200 µg/m^3. This change became effective as of 1 January 2007. Welders were considered as practically the only occupation for which the new OEL might require measures to reduce the exposure. Due to the low value of the new OEL it was considered necessary to investigate whether the sampling position could be changed from close to the lapel/collar bone to inside the welder's visor. This would make it more reasonable to assume that the sampler actually sampled from the air actually inhaled by the welder. This would also comply with the ISO standard for welding fume, ISO 10882-1. [17]

3. Methods and materials

The collection efficiency of the mini sampler was determined by static parallel sampling with the IOM sampler and the open-face 25 mm filter holder at five welding plants. The practical usability of the headset-mounted mini sampler was evaluated by personal sampling at three welding plants.

3.1. The mini-sampler

The mini sampler is based on the 13 mm Swinnex filter holder. The filter holder has been modified from being a closed-face filter holder to being a open-face filter holder by incorporating an entry nozzle in aluminium with an inner diameter of 10 mm and a length of 9 mm. In the work reported here Millipore 13 mm mixed cellulose ester (MCE) filters with a nominal pore size of 8 µm (SCWP) have been used. In this work the nominal flow rate has been 0,75 l/min, generally as low as the pumps used were able to operate for 6-8 h.

3.2. Headset mounting systems

Four different commercial headsets with professional microphones intended for musicians, TV commentators etc. were bought; IsoMax HH (Countryman), WH20QTR (Shure), HT2 (AUDIX) and C420^{III}L (AKG). These models were all hung over the ears with a mounting around the back of the head, but not over the top of the head, and with a beam holding the microphone close to the mouth. The mounting around the back of the head is needed to increase the vertical and lateral stability of the beam. The headsets were stripped of the microphone and the electrical cables. A PVC tubing was fastened with shrinking tubing onto the beam of each headset. The headset mounting system allowed the welder to work with or without the visor, with the visor in the up or down position, and to wear ear protectors (either inside the ear canal or as an ear-muff), and still the sampler could be sturdy positioned close to the mouth/nose.

3.3. Visited plants

The mini sampler has been evaluated at five manufacturing companies (A – E) based on welding. Static rig sampling was performed at all five plants. Personal sampling was performed on 5-10 of the welders at plants C – E. MIG welding was the main type of welding performed at the plants.

3.4. Aerosol sampling

Static rig parallel sampling was performed with the mini-sampler, the IOM sampler, the Millipore open-face 25 mm filter holder (FH25OF) and SIMPEDS cyclones operated at the flow rate 5 l/min in order to obtain information on the fraction of particles smaller than 2 µm. The air flow was supplied by personal sampling pumps of various makes, for the mini-sampler the AirChek2000 (SKC) was used.

Personal sampling was carried out only with headset-mounted mini-samplers behind the welder's visor. Apex pumps (Casella) were employed for the personal sampling with the mini-sampler. In total, 43 personal samples were taken on 23 welders over 34 work shifts.

Journal of Physics: Conference Series **151** (2009) 012041 doi:10.1088/1742-6596/151/1/012041

3.5. Static rig sampling

The static rig sampling was performed with horizontal rods mounted on a vertical rod. The horizontal rods were mounted approximately in the height of the breathing zone of a standing man. On each rod were two samplers mounted on each side of the vertical rod. The rig was not rotated but at some workplaces was turned at regular intervals. The total number of samplers of the different types mounted in the rig varied between plants. In each test 5-12 mini-samplers were compared to five samplers each of FH25OF and IOM samplers. All filter samples were weighed and selected samples additionally analysed by inductively coupled plasma/mass spectrometry (ICP/MS) and/or by portable X-ray fluorescence spectrometers (XRFS).

The concentration ratio FH25OF/IOM is used as an indicator of the coarseness of the sampled aerosol, and be termed the "aerosol coarseness ratio".

3.6. Laboratory analyses

3.6.1. *Filter weighing.* The collection substrates were weighed before and after sampling in an air-conditioned room using a MT5 balance. The temperature is controlled to 21±0,5°C and the relative humidity to 50±2%. Tests were performed to determine the weighing uncertainty of 13 mm filters having been mounted in the mini-sampler for 2-3, 10 and 15 weeks.

3.6.2. *Analyses for manganese by ICP/MS and XRFS.* For each run with the rig one sample per cyclone flow rate and two filter samples each for the IOM sampler and the FH25OF were selected for analysis with ICP/MS. Some static samples and most of the personal samples were analysed by XFRS, Niton XLt-700.

3.6.3. *Transport test of loaded 13 mm filters mounted in mini-samplers.* Loss of the deposit from filters mounted in mini-samplers by ordinary mail has been determined. Twenty samples each from plants C – E, were collected by static sampling with five different nominal flow rates. After sampling the mini-samplers were transported to the weighing lab, weighed and mounted anew in the original mini-samplers. Then the samplers were packed in a padded envelope and sent by ordinary mail a distance of approximately 500 km and back again. Upon return the filters were weighed anew. The transport loss was determined as the change in sample weight between the second and the first weighing.

4. Results

4.1. Static sampling

At the first run at plant B the concentration was approximately 20 mg/m3, and almost all pumps had stopped within one hour. With the exception of run B#1, the measured inhalable welding aerosol concentrations are in the range 0,7-3,3 mg/m3. The standard deviations for the IOM sampler, the FH25OF and the mini sampler are generally in the range 2-10%, apart from the first run at plant A.

The biases between the mini-sampler, the IOM sampler and the FH25OF are generally twice as large for welding aerosol mass as for manganese.

Proper welding fume (submicron particles or at least particles smaller than 2 µm) at these workplaces did not constitute more than 25-55% of the inhalable fraction, the rest is particles generated by spattering, grinding or slagging. Proper manganese fume for all but one run at the workplaces at plants C – E constituted more than 88% of the inhalable manganese.

4.2. Laboratory evaluations of the mini-sampler

4.2.1. *Precision of weighing for 13 mm MCE filters with nominal pore size 8 µm mounted in mini-samplers for a period of two to fifteen weeks.* The pooled standard deviation for weighing 13 mm filters mounted in mini-samplers was 2,2 μg. The standard deviation did not increase with longer shelf life between weighing. This corresponds to a limit of quantification of 22 μg.

4.2.2. *Transport loss.* The maximum filter loadings obtained for this experiment were 0,9, 1,5 and 0,8 mg, for plants C – E respectively, and the lowest loading was 10 μg (plant E). The minimum and maximum values of the transport loss are -5 and +19 μg, respectively. The average transport loss is 3,7 μg with a standard error of 0,6 μg. The RMS transport loss is 6,0 μg.

4.3. Bias of the mini-sampler versus the IOM sampler

Figure 1 shows the dependence of the concentration ratio of the mini sampler to the IOM sampler for welding aerosol mass and manganese on the aerosol coarseness ratio for each substance. The mini sampler undersamples relative to the IOM sampler, significantly for aerosol mass, but much less for manganese as the coarseness ratio for manganese is closer to unity. The regression line plotted is forced through $(x; y) = (1;1)$ and has been determined by the model

$$\left(1-\frac{C_{mS}}{C_{IOM}}\right)=k\left(1-\frac{C_{FH25OF}}{C_{IOM}}\right)$$

where C_{mS}, C_{IOM} and C_{FH25OF} are the concentration (welding aerosol mass and manganese respectively) for the mini-sampler, the IOM sampler and the open-face 25 mm filter holder, respectively. The obtained value for the regression coefficient (k) is 0,750, and for the residual standard deviation 0,051, respectively.

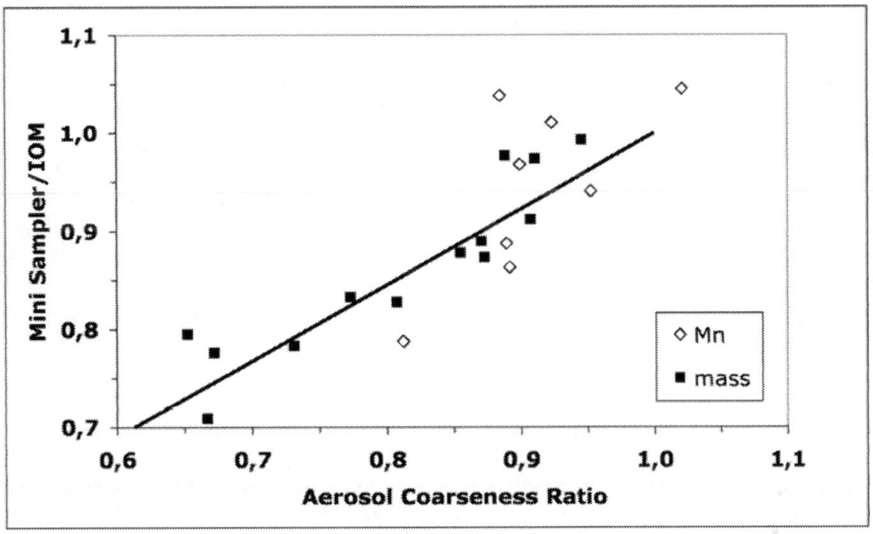

Figure 1. Concentration ratio mini sampler/IOM sampler for welding aerosol mass and manganese versus the aerosol coarseness ratio for each substance. NB! Only the filter of the IOM sampling cassette has been analysed for manganese.

4.4. Personal sampling

Samplers mounted on a modified headset were accepted by the welders. No big problems were encountered by the welders, even though they at rare occasions had to take them off or on by themselves, for example sometimes when going to the lavatory or at a coffee break. None of the headsets broke during this project. Distributions of personal samples of welding aerosol mass and manganese are shown in figure 2. The limits of quantification for mass (gravimetric) and manganese (XRFS) were

0,022 and 0,004 mg, respectively. Only the lowest manganese sample was below the corresponding limit of quantification. For five of the welders did the exposure on eight of nine sampled shifts exceed the current Swedish OEL for manganese, 0,2 mg/m^3.

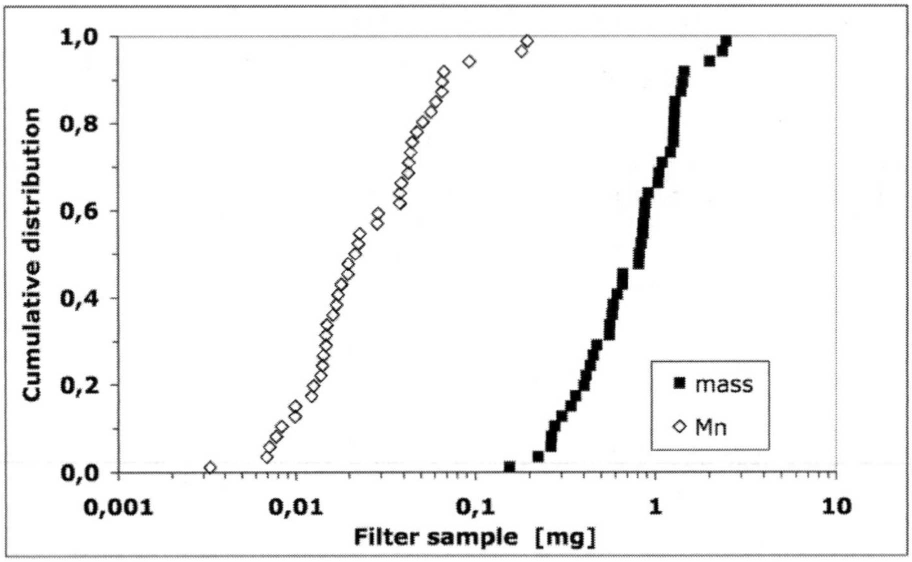

Figure 2. Distribution of personal 43 samples for welding aerosol mass and manganese

5. Discussion

5.1. The headset
It has been shown that aerosol sampling by a mini sampler mounted on a headset behind a welders visor is possible and does not unduly disturb the welders. Of the four tested versions, three were possible to mount in such a way that they were not felt uncomfortable by the welders. Headset-mounted samplers is preferable to the mounting arrangements shown in the year 2000 edition of the EN ISO 10882-1 standard on sampling airborne particles during welding[17] as the sampler is positioned close to the mouth/nose whether the visor or other face shield is worn or not, and whether it is in the up or down position.

5.2. Uncertainties of sample analyses versus sample transport loss
The weighing uncertainty for 13 mm filters was found to be 2,2 µg, considerably less than the uncertainty due to transport by ordinary mail, 6,0 µg. Both of these uncertainties are small compared to the amount of welding aerosol mass obtained by personal sampling in this project, which are at least 25 times larger than the transport uncertainty. For the personal samples, manganese constituted on average 3,5% of the welding aerosol mass collected, and therefore the transport loss is negligible also for manganese.

5.3. Sampling bias of the mini sampler versus the IOM sampler

5.3.1. Welding aerosol mass. At personal sampling for welding aerosol larger particles can be expected than encountered at static rig sampling. Within a reasonable coarseness ratio range of [0,50;1,00] for the welding aerosol mass, i.e. implying that welding aerosol is not coarser than for example flour dust, the average bias of the mini-sampler, based on the regression model above, would be -0,193. This bias is so large that in order to use the mini-sampler for sampling inhalable dust, it would first need to somehow be optimized to that sampling convention.

5.3.2. *Manganese.* Manganese mainly exists as fume and to a minor fraction as spatter particles, but constitutes a negligible fraction of grinding dust. Therefore the range of coarseness ratios to be expected at personal sampling of manganese will be similar to that was encountered in our static sampling, {0,88;1,02}. Within a coarseness ratio range of {0,85;1,00} for manganese, the average bias of the mini sampler, based on the regression model above, would be -0,056. This bias less than 10% need not be further reduced by optimization of the sampler. This bias is of the same order as Chung et al. [18] found among samplers mounted behind a welder's visor.

6. Conclusions

- Headsets for mounting a mini sampler behind a welder's visor have been manufactured from commercial headsets and were found to be acceptable by the welders.
- At the visited plants the inhalable fraction of welding aerosol mass only consisted of 25-55% of welding fume. The remainder was welding spatter, grinding dust and slagging dust. However, more than 65% of the manganese is found in the fume proper.
- The mini sampler exhibited for welding aerosol mass a negative bias versus the IOM sampler. The value of the sampling bias depended on the coarseness of the welding aerosol sampled. For a coarser welding aerosol the absolute value of the bias increased. For a welding aerosol with an aerosol coarseness ratio of 0,65, the bias of the mini sampler versus the IOM sampler was approximately -0,26.
- The weighing precision for 13 mm MCE filters was found to be 2,2 µg and the RMS transport losses of 13 mm filters mounted in mini samplers transported by mail 500 km there and back was 6,0 µg. Both of these values are small compared to the range of personal samples collected.
- The sampling bias of the mini sampler versus the IOM sampler for manganese was for all eight sampled welding aerosols numerically less than -0,14, and does not depend on the aerosol coarseness ratio of the sampled manganese aerosol. The average sampling bias of the mini sampler versus the IOM sampler for manganese, (with a negative RMS value of -0,046) is not statistically significant.

Acknowledgements
We would here like to thank everybody who has helped us in this project.

Leif Bäcklin has manufactured the mini sampler in several versions and rebuilt the commercial headsets for our purposes. Gunnel Sundström has weighed all filters. Jüri Waher has performed several of the laboratory tests of the mini sampler. Karin Holm has performed all analyses with ICP/MS. Villy Glas, Nils Kulle and Bengt Christensson have participated in the workplace sampling campaigns.

We would also like to thank those who have lend us their sampling pumps; IVL Swedish Environmental Research Institute, (the now shut down) Swedish National Institute for Working Life and the Occupational and Environmental Health, Stockholm Centre for Public Health, Stockholm County Council.

Last, but not least, we would like to thank management and the workers at the plants we have visited and which have helped us with our sampling.

References

[1] Martinelli C A, Harley N H, Lippmann M and Cohen B S 1983 *Am. Ind. Hyg. Assoc. J.* Monitoring real-time aerosol distribution in the breathing zone **44** 280-285
[2] Bull R K, Stevens D C and Marshall M 1987 *J. Aerosol Sci.* Studies of aerosol distributions in a small laboratory and around a humanoid phantom **18** 321-335

[3] Parker R C, Bull R K, Stevens D C and Marshall M 1990 *Ann. occup. Hyg.* Studies of aerosol distributions in a small laboratory containing a heated phantom **34** 35-44

[4] Steel, J and Maican A 1994 The measurement of exposure *Second International Occupational Hygiene Association Conference* (Hong Kong: HKIOEH Co. Ltd)

[5] Rosén G, Andersson I-M, Niemelä R, and Welling I 1997 Air transport phenomena in the breathing zone and its implication on exposure assessment *Ventilation '97* (Ottawa: The Canadian Environment Industry Association)

[6] Brohus, H 1997 Measurement of personal exposure using a breathing thermal manikin *Ventilation '97* (Ottawa: The Canadian Environment Industry Association)

[7] Malek R F, Daisy J M and Cohen B S 1999 *Appl. Occup. Environ. Hyg.* Breathing zone concentration variations in the reinforced plastic industry – field measurements in a boat manufacturing plant **14** 777-784

[8] Welling I, Andersson I-M, Rosén G, Räisänen G J, Mielo T, Marttinen K and Niemelä R 2000 *Ann. occup. Hyg.* Contaminant dispersion in the vicinity of a worker in a uniform velocity field **44** 219-225

[9] Bloor W A, Eardley R E and Dinsdale A 1968 *Ann. occup. Hyg.* A gravimetric personal dust-sampler **11** 81-86

[10] ter Kuile W M 1982 *Ontwikkeling van de totaalstof helm (in Dutch - Development of the Total Dust Helmet)* (Delft: IMG-TNO)

[11] Chien P T 1992 *Am. Ind. Hyg. Assoc. J.* A proposed sampling device for personal exposures **53** 527

[12] Allen J, Bellinger E G and Higgins R I 1981 *Proc. Instn. Mech. Engrs.* A full shift true breathing zone air sampler and its application to lead workers **195** 325-328

[13] Pengelly M I, Groves J A, Foster R D, Ellwood P A and Wagg R M 1994 *Ann. occup. Hyg.* Development of a method for measuring exposure to resin acids in solder fume **38** 765-776

[14] Health and Safety Executive 1997 *MDHS83 Resin Acids in Rosin (Colophony) Solder Flux Fume* (London:HSE)

[15] Goller J W and Paik N W 1985 *Am. Ind. Hyg. Assoc. J.* A comparison of iron oxide fume inside and outside of welding helmets **46** 89-93

[16] Liu D, Wong H, Quilan P and Blanc P D 1995 *Am. Ind. Hyg. Assoc. J.* Welding helmet airborne fume concentrations compared to personal breathing zone sampling **56** 280-283

[17] International Organization for Standardization 2000 *EN ISO10882-1 Health and Safety in Welding and Allied Processes – Sampling of Airborne Particles and Gases in the Operator's Breathing Zone -- Part 1: Sampling of Airborne Particles* (Geneva:ISO)

[18] Chung K Y K, Carter G J and Stancliffe J D 1999 *Appl. Occup. Environ. Hyg.* Laboratory evaluation of a protocol for personal sampling of airborne particles in welding and allied processes **14** 107-118

Characterisation of fume from hyperbaric welding operations

John A S Ross[1,5], Sean Semple[1], Rodger Duffin[2], Frank Kelly[3], Joerg Feldmann[4] and Andrea Raab[4].

[1] Environmental and Occupational Medicine, University of Aberdeen.

[2] ELEGI Colt Laboratory, University of Edinburgh.

[3] Lung Biology Group, Kings College, University of London.

[4] Trace Element Speciation Laboratory, University of Aberdeen.

[5] Environmental and Occupational Medicine, University of Aberdeen, Aberdeen AB25 2ZP,

email j.a.ross@abdn.ac.uk, 01224 558197

Abstract. We report preliminary work characterising dust from hyperbaric welding trials carried out at increased pressure in a helium and oxygen atmosphere. Particle size and concentration were measured during welding. Samples for quartz and metal analysis and toxicity assessment were taken from a filter in the local fume extraction system. The residue of dust after metal extraction by nitric acid in hydrogen peroxide predominantly a non-metallic white powder assumed to be dust from welding rod coatings and thermal insulation material. Metallic analysis showed predominantly calcium, from the welding rod coating, and period 4 transition metals such as iron, manganese, magnesium and titanium (inductively coupled mass spectrometry, Agilent 7500c). The presence of zirconium indicated a contribution from grinding. The fume was nanoparticulate in nature with a mean particle diameter of 20-30 nm (MSI Inc WPS 1000XP). It showed an intermediate level of oxidative potential regarding the low-molecular weight respiratory tract lining fluid antioxidants ascorbate and glutathione and caused release of the inflammatory marker IL-8 in a human lung A 549 epithelial cell culture with no indication of cytotoxicity. The study findings have strong implications for the measurement techniques needed to assess fume exposure in hyperbaric welding and the provision of respiratory protection.

1. Introduction

In the offshore oil and gas industry there is a requirement to carry out welding on underwater pipelines. While in-water welding is possible it does not produce an adequate quality of weld and pipeline welding operations are carried out in a hyperbaric welding habitat. The worksite is enclosed in a metal container which is filled with a breathable gas mixture of helium and oxygen. The oxygen level in the habitat is held at between 30 and 40 kPa and so although oxygen partial pressure is higher than in air the concentration is very much lower at 2.7-3.6% at an illustrative depth of 100 metres of sea water. The divers who conduct the weld live on board a diving support vessel in chambers pressurised with helium and oxygen to the same pressure as the underwater workplace. They usually work an eight hour shift and are transported from the vessel to the welding habitat in a diving bell. Occupational exposures are very difficult to assess under such conditions due to the remoteness of the work site. Each pipeline weld, however, is preceded by a trial weld conducted in a hyperbaric simulator onshore in order to demonstrate the competence of the welding team for insurance purposes. Such trial welds offer a practical opportunity to study hyperbaric workplace exposures.

Fume generation is an inevitable consequence of welding operations and such fume was considered a "nuisance" dust. More recently, however, it has been understood that welding fume does have a significant level of toxicity and UK legislation requires that its individual constituents are considered when calculating tolerable workplace exposures. High manganese exposure can lead to a Parkinsonoid condition of manganism [1] and welders are at risk from pneumonia [2] and ischaemic heart disease [3]. Divers who weld are also more likely to express cognitive symptoms [4]. Accordingly, we have carried out some work to characterise hyperbaric welding fume and report on the metallic content and cytotoxicity of dust from hyperbaric welding operations and its oxidant potential with preliminary work on the assessment of particle size.

2. Methods

The methods were applied during several different welding trials (Table 1).

Table 1. Analysis methods used for each welding trial

Sample	WF001	WF002	WF003	WF006	n/a
Welding trial	1	2	3	4	5
Silica	Yes	Yes			
Metal	Yes	Yes	Yes	Yes	
IL-8	Yes	Yes	Yes	Yes	
Cytotoxicity	Yes	Yes	Yes	Yes	
Oxidant potential	Yes	Yes	Yes	Yes	
Particle size					Yes

2.1. Welding

While a limited amount of tungsten inert gas welding was carried out the greater part of the welding was conducted by manual metal arc welding (MMAW). Grinding was carried out using zirconium containing abrasive wheels. The weld piece was heated and lagged externally when heating pads were in use. Local fume extraction was used. The extraction system recirculated chamber atmosphere through a series of filters and chemical absorbents to remove the particulate and gaseous components of the welding fume. In addition to the local extraction system, welders wore respiratory protection equipment which was either a positive pressure respirator (Aga) or a particulate respirator (3M) and reflected the appreciation of risk of two different diving companies. The welding habitat was a cylindrical chamber with domed ends having nominal measurements of 3 m diameter, 8 m length and 63 m^{-3} empty volume.

2.2. Sampling

Samples for metal analysis or toxicity testing were taken from the dust collected on a filter in the local extraction system which had a nominal pore size of 30 μm. The sample collected, therefore, represented all the dust present in the chamber atmosphere and not just welding fume. To provide a sample for online monitoring of atmospheric dust particle size, a penetration in the chamber hull was plumbed with a quarter turn valve on the outside of the chamber which was adjusted to give a free flow at atmospheric pressure of 27 litre minute-1 into an open ended 2m length of 19 mm diameter tungum pipe. Inside the chamber, the gas inlet for the sample line was at waist level and sampled the general chamber atmosphere. The length of pipe descended on the outside of the chamber and allowed gas sampling from it into the analysers. The analyser sample tubes were inserted into the open end of the tungum pipe. Sampling was therefore performed at one atmosphere pressure.

2.3. Metal analysis and toxicity tests

Metal analysis was carried on dust samples from three welding trials. The dust was digested on a hotplate in a one to five mixture of nitric acid in hydrogen peroxide. Analysis in duplicate was by inductively coupled mass spectrometry (Agilent 7500c) using an internal standard of 10 ppb rhodium. A full set of calibration and quantification of most elements was carried out with additional use of quality control certified reference material.

Silica analysis was carried out on the dust from two welding trials at the commercial laboratories in the Institute of Occupational Medicine, Edinburgh. Portions of the samples were finely ground to create particles of uniform size. These were scanned qualitatively using routine inductively coupled mass spectrometry. The resultant diffraction patterns were then compared with standard reference materials and search match indices.

The oxidative potential of dust samples from four welding trials was assessed by their capacity to deplete (oxidatively consume) the low-molecular weight respiratory tract lining fluid antioxidants ascorbate, glutathione and urate from a synthetic respiratory tract lining fluid at a range of particle concentrations. Samples were suspended in ultra pure (18Ω) Chelex-resin treated water (pH adjusted to 7.0 with Chelex treated 1M HCl) at concentrations of 27.78, 55.56 and 111.11μg/ml. Samples were then vortexed for 10 minutes followed by 30 seconds of sonication on ice at an amplitude of 15 microns to produce an homogenous suspension. Control residual oil fly ash (ROFA – oxidatively active control) and carbon black (M120 – oxidatively inactive control) particulate samples were prepared in an identical fashion at a concentration of 55.56μg/ml. The oxidative activity of these suspensions was assessed using a synthetic respiratory tract lining fluid (RTLF) containing the antioxidants ascorbate (AA), urate (UA) and glutathione (GSH), each at 200μM. PM suspensions were added to achieve a final particle concentration of 25, 50 and 100μg/ml. Samples were then incubated for 4 hours at 37°C at which time the samples were acidified to stop oxidation reactions. Particles were then removed by centrifugation and the resultant supernatants assessed for remaining concentrations of ascorbate, urate by HPLC [4]. Glutathione was detected using a spectrophotometric method [5], which allows the simultaneous detection of oxidised (GSSG), total (GSx) and reduced (GSH) glutathione. For comparison, residual oil fly ash (ROFA) and inert carbon black (M120) were used as positive and negative controls respectively. Parallel control incubations were also carried out in the absence of dust with samples taken at 0 and at 4 hours to determine background rates of auto-oxidation.

Cytotoxity testing and assessment of inflammatory potential were carried out on dust samples from four separate welding trials. Nimrod c276 welding fume and DQ12 quartz c were also included in all experiments as control particles: At 24hrs post-exposure, cytotoxicity was measured by lactate dehydrogenase (LDH) release and inflammatory potential by the cells' ability to produce the pro-inflammatory cytokine interleukin 8 (IL-8). A549 Type II lung epithelial cells, were seeded into 24-well plates at a concentration of 71,250 cells/well in cell culture media containing 10% foetal calf serum (FCS) and cultured over-night. The following day, the cell culture media was removed, cells washed with phosphate buffered saline (PBS) and cultured for a further 24hrs in cell culture media containing 2% FCS. Cells were then treated for a further 24 hrs with the various particle types, at the various treatment doses, suspended in fresh 25 FCS cell culture media. Following the different cell treatments, the tissue culture plate were centrifuged at 250xg, for 10 minutes and the supernatants removed for LDH and IL-8 quantification. LDH was measured using a standard Roche LDH Cytotoxicity Assay. IL-8 was measured using a standard R&D IL-8 ELISA kit.

2.4. Welding fume particle size assessment

Particle size measurement was carried out during a single welding shift when the local extraction ventilation was described as ineffective by the welding team who were performing manual arc welding. The welders were wearing positive pressure respirators. A MSP Corporation Wide Spectrum Particle Spectrometer 1000XP was used to quantify particles of 10 to 200 nm in diameter.

The measurement was expressed as total particles ml^{-1} and also in 47 grades of diameter. Particle concentration was also measured using a TSI Inc. P-Trak Model 8525 Ultrafine Particle Counter. Values were corrected for the effects of decompression by multiplication by chamber pressure in atmospheres absolute. Concurrent events in the chamber were taken from the welding log compiled by the welding supervisor who worked in close communication with the welders who were also monitored on closed circuit television.

3. Results

3.1. Metal and silica analyses

The metal analysis is summarised in table 2. Although calcium is the predominant element this measurement would have been subject to interference from the presence of titanium in the sample and is an overestimation. Nevertheless, calcium is a major constituent of welding rod coatings and high levels were expected. Period four transition metals were next best represented. Seven transition elements out of the ten in this period are in the table. The remaining three, scandium, vanadium and cobalt, however, were also detected although at levels marginally below the table cut point. Zirconium was detected and the source of this was almost certainly the grinding wheels used to shape the weld. Quartz and cristabolite levels were below the method's detection limits of 0.3% in both samples tested.

Table 2. Summary of Metal Analyses.
Units are $\mu g\ g^{-1}$ of sample and are expressed as the mean of the measurements made. Only elements present at >1$\mu g\ g^{-1}$ in all three samples are tabulated. Elements are in periodic table categories and transition metals are ordered by period.

Element category	Element	WF002	WF003	WF006
Alkali metals	Lithium	50	72	660
Alkali earth metals	Magnesium	348	362	329
	Calcium[a]	8,558	26,913	21,058
	Barium	2.0	16	10
Period 4 transition metals	Titanium	100	330	290
	Chromium	2.2	4.9	4.2
	Manganese	460	746	663
	Iron	1,656	6,331	4,581
	Nickel	10	19	27
	Copper	1.4	4.6	3.6
	Zinc	13	22	11
Period 5 transition metal	Zirconium	54	205	151
	Molybdenum	1.4	4.2	2.7
Other metals	Aluminium	70	154	107
	Lead	1.4	1.5	1.5
	Bismuth	2.0	1.8	2.
Metalloids	Boron	8.0	17	10
	Silicon	63	98	110

[a] The values for calcium are not corrected for titanium interference in the method of analysis used.

3.2. Oxidant potential assessment

The dust elicited significant and dose dependent losses of ascorbate ($P<0.05$) compared with background oxidation associated with the particle free controls after the 4hr exposure (figure 1). At the lowest dose of 25μg/ml the addition of all 4 particles led to a loss of ascorbate from the system. At the 50μg/ml particle concentration all 4 WPM samples were less reactive than ROFA, whilst significantly more reactive than M120. Dose responses were seen with four of the particle samples with the exception of WPM003.

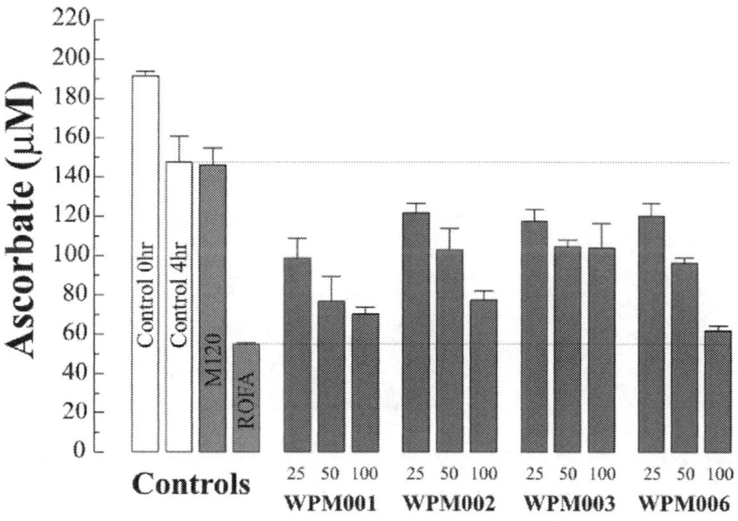

Figure 1. Ascorbate remaining in the synthetic RTLF after a 4-hour incubation with 25, 50 and 100μg/ml of WPM (blue bars). Control values are also displayed; particle-free controls at 0 and 4 hrs (white bars) and M120 and ROFA (grey bars) at 50μg/ml. Data are illustrated as mean (SD) of 3 separate experiments. WPM001-006 are the same samples as WF001-006.

All dust samples were less reactive towards glutathione with only sample 6 at the highest dose of 100μg/ml depleting significantly more GSH (-37.31μM) after the 4hr exposure than the particle free control. The loss in GSH observed with sample 6 was due to its oxidation to GSSG and did not reflect adsorption to the particle surface as no significant change was seen with total GSx concentrations. No depletion of urate was noted under any incubation condition.

3.3. Cytotoxicity and inflammatory potential assessment

After 24 hours exposure of A549 cells to welding fume at doses up to 500 μg ml^{-1} there was little indication of cytotoxicity. In contrast, cytotoxicity rose rapidly with dose for Nimrod welding fume and DQ12 and, at 250 μg ml^{-1} averaged 15% for each of the two agents.

IL-8 release in the untreated preparation was 370 pg ml^{-1}. At lower doses of fume, hyperbaric welding fume had significant effects (figure 2). The four hyperbaric welding fume samples induced a higher IL-8 response than the control particles but this result then reversed as the dose increased with the toxic control particles (DQ12 and Nimrod) becoming more active than the welding fumes. Increased activity was sustained up the full dose of fume for DQ12 but, at doses of more than 150 μg ml^{-1} Nimrod fume was inhibitory. At a dose of 250 μg ml^{-1} IL-8 release was significantly suppressed by Nimrod welding fume at a mean level of 187 pg ml^{-1} and stimulated by DQ at a mean level of 723 pg ml^{-1} while it was essentially no different to the untreated control for the hyperbaric welding fume samples.

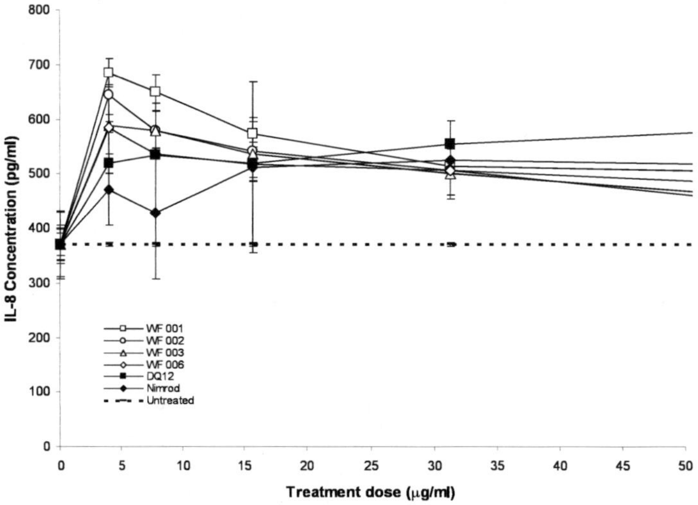

Figure 2. IL-8 Release 24 hours after exposure with treatment dose 0-50 μg ml^{-1}

3.4. Welding fume particle size assessment

Data gathered by the WPS during a welding session are shown in figure 3. It is clear that the fume is largely nanoparticulate in nature. Total values of particles from the same welding session are shown in figure 4 as measured by both the WPS and P-Trak. The P-Trak substantially underestimated particle concentration while welding was being performed. The overall distribution of particle size during the welding session is shown in figure 5. Over the period monitored, mean particle diameter was 29 nm (sd 17 nm) with a distribution mode of 22 nm. From the distribution plot it seems probable that there is a substantial amount of fume of a diameter less than 10 nm.

4. Discussion

Hyperbaric welding fume, while not being demonstrably cytotoxic in this study does have the potential to cause oxidative stress and induce inflammation at relatively low levels. This may be attributable to the representation of transition metals in the fume. Such metals are oxidatively active and have been associated with the inflammatory effect of other types of welding fume [6]. Under the conditions encountered by hyperbaric welders this gives some cause for concern as they breath a raised partial pressure of oxygen of 30-40 kPa for the duration of their work at pressure. An active hyperbaric welder may spend 90-150 days at pressure in any one year. Under such conditions any oxidative stress may be amplified and this justifies further consideration and investigation.

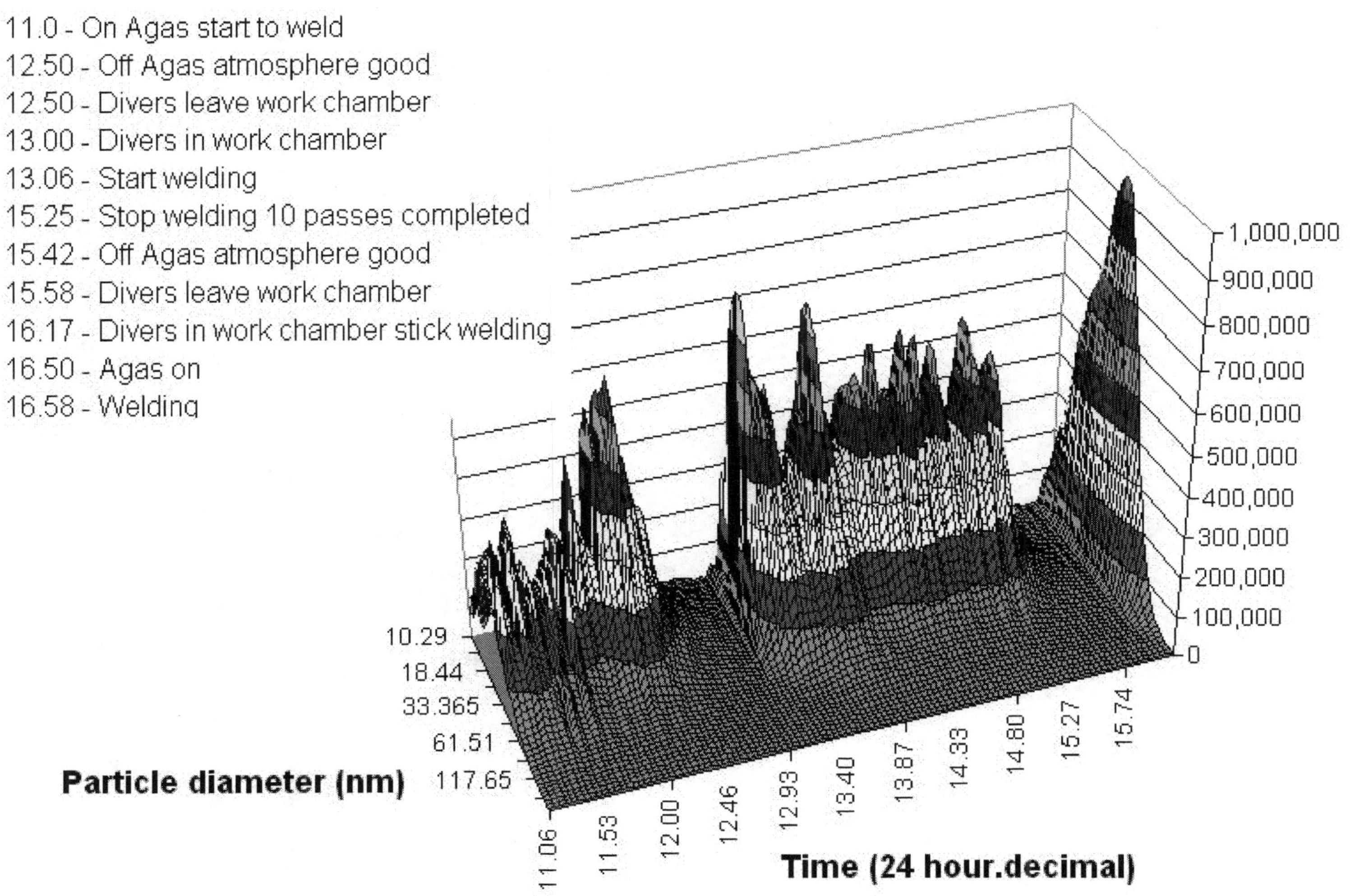

Figure 3. Three dimensional plot of WPS data during manual arc welding with indication of background events

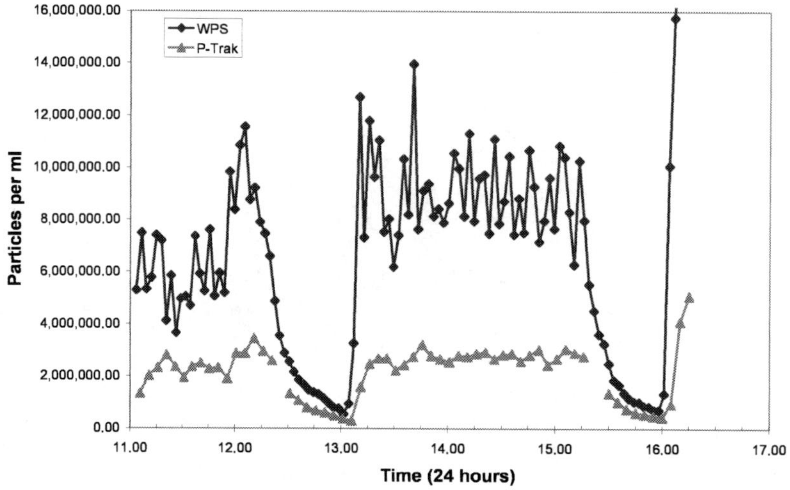

Figure 4. Total particle count over the same period as in figure 2 measured by both the WPS and P-Trak.

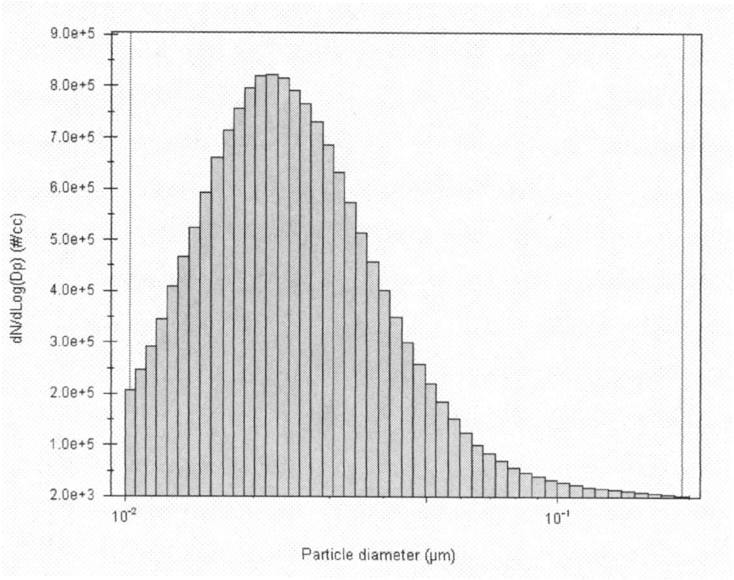

Figure 5. The average distribution of 110 samples analysed during the welding session from figure 2.

The particle diameter identified for hyperbaric welding fume was very low in this study. The fume generated by welding which employs flux contain a mixture of chainlike and spherical structures which suggest that condensation of vapour into primary particles and this is followed by a complex and dynamic process which includes coagulation, condensation and oxidation [7]. Agglomeration of particles occurs and it is assumed that this occurs before the fume reaches the welder's breathing zone and there is work to support this for CO_2 arc welding [8]. This assumption regarding agglomeration is carried into the specification for particle respirators for use during welding which work to a particle size of 300 nm and above. At first sight this would seem to be an unjustified assumption for hyperbaric welding. There are factors that may influence the fume generated in hyperbaric welding. The atmosphere is largely composed of helium and contains a low concentration of oxygen. Vapour condensation may be affected as the thermal properties of helium produce higher cooling rates and there may be less oxidation of the fume constituents. The observation of a high concentration of nanoparticles in this study does not preclude the presence of larger agglomerated particles since these were not studied. The presence of such small particles, however, does give rise to some cause for concern. The filter respirators used by welders are not tested against such dust and as particle diameter falls there is a greater chance of inward leakage round the edge of any face mask so reducing the protection offered.

The mean particle diameter identified was 29 nm. At this diameter, approximately 80% of fume is retained in the lung with 40% being deposited in the alveolar region. Unagglomerated ultrafine particles are not phagocytosed efficiently by macrophages and this can lead to a major accumulation of particles if exposure is continued and ultrafine particles can pass through the epithelium to the interstitial tissues. At lower particle size increasing fractions of particles are deposited in the extrathoracic and tracheobronchial airways. The effects of such ultrafine dusts have recently been comprehensively reviewed [9]. Such low particle sizes are difficult to assess by simple online analysis and this is demonstrated here by the performance of the P-Trak analyser which measures form 20 nm in contrast to the WPS unit which measures from 10 nm. A small difference in capability lead to a very large discrepancy in measurement. With adequate characterisation of the fume of a particular operation, however, it may be possible to introduce correcting factors so that any risk is not overlooked. Online analysis to assess risk with such small particles seems appropriate since at quite high concentrations there may be no perception of a hazard in the immediate environment and welders in this study felt able to remove respiratory protection at particle levels of 1,000,000-2,000,000 particles ml^{-1} (figures 3 and 4)

This study has weakness and strengths. Particle sizes greater than 200 nm were not studied in detail and so the presence of larger agglomerated particles was not identified. Also such particles may be held together by weak attracting forces which might have been sheered apart in the orifice of the quarter turn valve controlling sample flow from the chamber. This effect would lead to a lowering of observed particle size. This point remains to be addressed by the use of a laminar flow sampling system. It is tempting to assume that the fume observed during welding operations is entirely fume coming from the welding process itself. Grinding, however, is an important process in the type of weld monitored in this project and the welders here were used high speed grinders as much as welding torches. This is reflected by the finding of zirconium in the metal analysis. The study examined, therefore, fume and dust from combined welding and grinding operations and not just welding alone. Strengths of the study lie in the ability to observe welding operations at increased ambient pressure. For analysis, the sample was decompressed and, consequently diluted due to expansion during decompression. This gives the ability to observe higher concentration of particles by on-line analysis than is possible at atmospheric pressure. For example the P-Trak has a measurement range of 0-500,000 particles ml^{-1} and in this study was used to identify particle concentrations of up to 5,000,000 particles ml^{-1}.

In conclusion, this study has identified the predominance of nanoparticulate as a result of hyperbaric welding operations. The fume has the potential at relatively low concentrations to cause oxidant stress and inflammation and these properties have the potential to interact with other factors in the hyperbaric diving environment.

References

[1] Racette BA, McGee-Minnich L, Moerlein SM, Mink JW, Videen TO and PerlmutterJS 2001 Welding-related parkinsonism: clinical features, treatment, and pathophysiology Neurology **56** 8-13.

[2] Palmer KT, Poole J, Ayres JG, Mann J, Burge PS and Coggon D *Exposure to metal fume and infectious pneumonia* Am. J. Epidemiology. **157** 227-33

[3] Sjogren B, Gyntelberg F and Hilt B. *Ischemic heart disease and welding in Scandinavian studies* Scand. J. Work, Environ. Health **Suppl 2** 50-3

[4] Macdiarmid JI, Ross JAS, Taylor CL, Watt SJ, Adie W, Osman LM, Adie W, Godden D, Murray AD, Crawford JR and Lawson A 2004 *Co-ordinated investigation into the possible long term health effects of diving at work. Examination of the long term health impact of diving*. Research Report 230. (Sudbury: HSE Books)

[5] Baker MA, Cerniglia GJ, and Zaman A 1990.*Microtiter plate assay for the measurement of glutathione and glutathione disulfide in large numbers of biological samples* Anal Biochem **190** 360-5

[6] Mudway IS, Stenfors N, Duggan ST, Roxborough H, Zielinski H, Marklund SL, Blomberg A, Frew AJ, Sandstrom T, Kelly F 2004 A*n in vitro and in vivo investigation of the effects of diesel exhaust on human airway lining fluid antioxidants* Arch Biochem Biophys **423** 200-12

[7] McNeilly JD, Heal MR, Beverland IJ, Howe A, Gibson M.D, Hibbs LR, MacNee W, Donaldson K 2004. *Soluble transition metals cause the pro-inflammatory effects of welding fumesin vitro* Toxicol. Appl. Pharmacol. **196** 95–107

[8] Zimmer AT 2002 *The influence of metallurgy on the formation of welding aerosols* J. Environ. Monit. **4** 628-632

[9] Saito H, Ojima J, Takaya M, Iwasaki T, Hisanaga N, Tanaka S and Arito H 2000 *Laboratory measurement of hazardous fumes and gases at a point corresponding to breathing zone of welder during* CO_2 *arc welding* Ind. Health **38** 69-78

[10] Ostiguy C, Lapointe G, Trottier M, Menard L, Cloutier Y, Boutin M, Antoun M and Normand C 2006 *Health effects of nanoparticles* (Montreal: IRSST – Communications Division)

Journal of Physics: Conference Series **151** (2009) 012043 doi:10.1088/1742-6596/151/1/012043

Real-time measurement of dust in the workplace using Video Exposure Monitoring: farming to pharmaceuticals

P T Walsh, A R Forth, R D R Clark, K P Dowker and A Thorpe

Health & Safety Laboratory, Harpur Hill, Buxton, Derbyshire, SK17 9JN, UK.

peter.walsh@hsl.gov.uk

Abstract. Real-time, photometric, portable dust monitors have been employed for video exposure monitoring (VEM) to measure and highlight dust levels generated by work activities, illustrate dust control techniques, and demonstrate good practice. Two workplaces, presenting different challenges for measurement, were used to illustrate the capabilities of VEM: (a) poultry farming activities and (b) powder transfer operations in a pharmaceutical company. For the poultry farm work, the real-time monitors were calibrated with respect to the respirable and inhalable dust concentrations using cyclone and IOM reference samplers respectively. Different rankings of exposure for typical activities were found on the small farm studied here compared to previous exposure measurements at larger poultry farms: these were mainly attributed to the different scales of operation. Large variations in the ratios of respirable, inhalable and real-time monitor TWA concentrations of poultry farm dust for various activities were found. This has implications for the calibration of light-scattering dust monitors with respect to inhalable dust concentration. In the pharmaceutical application, the effectiveness of a curtain barrier for dust control when dispensing powder in a downflow booth was rapidly demonstrated.

1. Introduction

Real-time dust monitors are employed by occupational hygienists for tasks such as walk-through surveys, background sampling, site dust measurements, assessment of the effectiveness of dust control systems and measurement of indoor air quality [1]. They also form part of Video Exposure Monitoring (VEM), the combination of video and synchronized real-time exposure data [2], to illustrate dust control techniques and highlight dust levels generated by work activities. VEM can also play an important role in exposure risk communication by demonstrating good practice.

In VEM, dust monitors are used qualitatively and semi-quantitatively. For qualitative measurements, the relative effect of changing controls on exposure is measured. For quantitative measurements, the real-time monitors must be calibrated with respect to the respirable and, if required, inhalable dust concentrations [3,4,5]. Quantitative data can place the exposure in context by reference to control and exposure limits, although the exposure data will only be semi-quantitative because of uncertainties due to sampling and, particularly for aerosols, calibration associated mainly with the real-time instrument response which is a function of particle characteristics, e.g. size distribution and other scattering properties [6].

Two workplace examples, at opposite ends of the spectrum in relation to the working environment, are described here to illustrate the role of VEM with real-time dust monitors for the measurement of respirable dust concentration and, by inference, inhalable dust concentration, arising from (a) poultry farming activities and (b) powder transfer operations in a pharmaceutical company. They present different challenges for measurement, and particularly for VEM: low light, dusty farming conditions with a highly mobile workforce contrasting with ultra-clean pharmaceutical operations where dust levels are typically very low.

2. Poultry farming

2.1. Background

Recent research into the incidence of ill health in agriculture in Great Britain [7] concluded that there is very little current information available on the incidence or prevalence of occupational ill health in the agricultural industry. However the data indicated that respiratory disease generally and upper respiratory tract infection symptoms in particular were high, and reported by just under 40% of farm workers exposed to organic dusts. The prevalence of chronic bronchitis amongst agricultural workers is reported to be 6.5%. The agricultural workforce totals about 400,000 with the majority exposed to organic dust during the course of their work. The figures would therefore suggest that up to approximately 162,000 workers could be suffering from respiratory symptoms at any one time and 27,000 suffering from bronchitis.

A specific group of farm workers, those in poultry farming, can be exposed to significant amounts of dust produced from poultry dander, feathers and dry wastes during their work activities. While other studies [8] have focused on identifying the scale of the risk through more comprehensive exposure measurements, in this work VEM was used to provide information on the hazard and its control which could form part of a package of electronic training material. Such risk management guidance for managers and safety representatives of large poultry farms and trade associations can then be disseminated in a suitable form to employees and contract workers.

2.2. Poultry farm activities

Various tasks in a commercial poultry farm, identified from previous work [8], were targeted for action as part of the investigation. The cycle of poultry rearing can be simplified to:

- Laying new litter;
- Repopulation (introducing new chicks);
- Growing the birds to adult size;
- Depopulating (removing the birds);
- Litter removal; and
- Cleaning and disinfecting.

All the above activities, apart from the growing phase, were identified as producing significant dust levels and suitable for VEM. At the farm, these activities can be described as follows:

- Laying new litter: The central part of the Rearing Shed is fenced off and bags of white wood shavings are opened and spread over the concrete floor using forks.
- Repopulation: Recently hatched chicks are introduced into the Rearing Shed under subdued lighting (the chicks are attracted to and congregate under hanging red lights, hanging gas burners heat the same areas).
- Depopulation: Removal of all of the young adult birds. Batches of a few hundred birds are herded to one end of the Rearing Shed, corralled and collected by poultry catchers, inoculated, then removed to cages on a lorry to be transported away.
- Litter Removal: Once the shed is emptied of birds, the used litter is removed. A 'Bobcat' type vehicle (a small, quick, manoeuvrable loader) is used to shovel up the litter into a pile at one end of the shed, then load it into a tractor/trailer. Shovelling and brushing by hand is needed

around the edges of the shed and the supporting pillars. A mechanical rotating brush attached to the 'Bobcat' shovel enables the floor to be swept clean.

- Cleaning and disinfecting: A power-washer with hand 'gun' is used to wash everything from the top down (roof, roof mounted ventilation, walls, bird feeder containers and floor). Following washing, a disinfectant is mixed with the water and the same equipment (with a long lance replacing the hand 'gun') and technique is used to disinfect all of the surfaces.

2.3. Materials and methods

The VEM technique (Exposure Level Visualization - 'ELVis') involves real-time, personal monitoring while simultaneously videoing the worker and combining the exposure profile with the video image on a computer; it has been described previously [2]. Radio telemetry was not used in this case for transmission of the real-time monitor data in order to simplify the kit and minimize the weight carried by the subject. The data was simply stored on the real-time monitor's datalogger. The real-time monitor (1 s logging period) and the camera clocks were synchronized with the PC clock and a digital radio-controlled clock respectively. Then the real-time monitor's clock was filmed alongside the radio-controlled clock to synchronize them to 'real' time. The video and data files were joined to display on one screen as illustrated in figure 1. Here, one subject was monitored and carried the monitor in a harness with gravimetric samplers located immediately adjacent to the monitor.

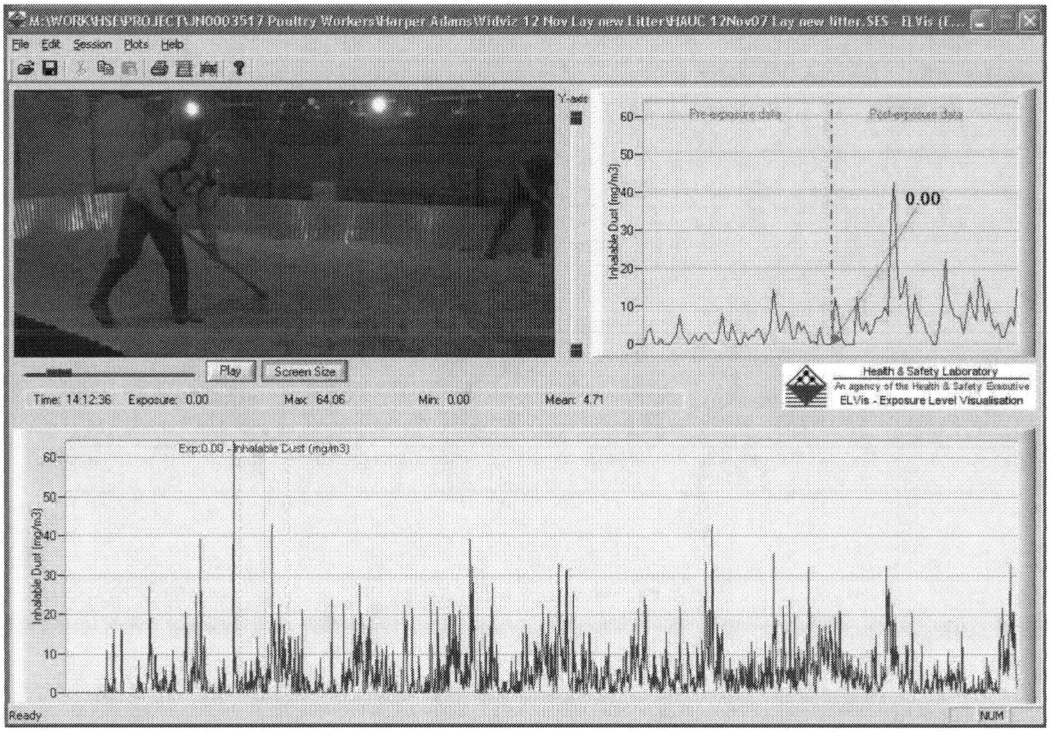

Figure 1. Example of the VEM screen display showing spreading of new litter over the concrete floor of the rearing shed. The complete exposure profile is shown at the bottom and a detailed window, either side of the moveable cursor, is displayed at the upper right of the screen. The real-time monitor is calibrated with respect to inhalable dust.

A camcorder (Sony Handycam) was used to video activities. Although it was able to work in low light, depopulation and repopulation occurred in very low light levels and it was not possible to use natural light. Satisfactory footage was obtained by illuminating the subjects with infra-red from portable infra-red lights (Sony Battery IR Light) situated next to each camera, and by setting the cameras to 'night-shot'. Reflective tape was affixed to the real-time monitor and the harness to further improve identification of the subject and instrument being filmed.

The real-time monitors (Thermo Personal DataRAMs - PDR) were located in a harness attachment sampling in the breathing zone towards the top of the chest (the poultry catchers bent forward a lot during their activities). Up to three real-time monitors were used during a visit: two subjects were monitored, each with a real-time monitor and gravimetric samplers; and a further real-time monitor with adjacent gravimetric samplers recorded the background dust level.

Personal gravimetric samplers (IOM head and cyclone samplers for inhalable and respirable fractions respectively) were also employed to obtain exposure data and retrospectively calibrate the real-time monitors. The samplers were operated according to the HSE reference method MDHS14/3 [9]. Each time-series of real-time monitor data was compared with its corresponding gravimetric samplers and the relevant calibration factors applied.

2.4. Results and Discussion

2.4.1. Inhalable dust concentrations

The exposure levels to inhalable dust determined from gravimetric measurements using the IOM head for three activities are shown in table 1. The workers were monitored for periods between 0.5 – 2 hr to determine exposure concentrations for the activities.

Table 1. Exposure to inhalable dust (IOM head) for three poultry farm activities.

Activity	n	Maximum ($mg\ m^{-3}$)	Median ($mg\ m^{-3}$)	Minimum ($mg\ m^{-3}$)
Depopulation	3	39.4	23.4	22.8
Repopulation	2	4.4	4.4	4.3
Litter/manure removal	3	7.9	3.0	0.9

Three other activities (routine cleaning, laying new litter and cleaning/disinfecting) were also measured but only one measurement was taken for each (this data was used in section 2.4.2).

The ranking of the activities in table 1 (with median values of exposure in $mg\ m^{-3}$ in parentheses), albeit based on limited data, in terms of the inhalable dust exposure is:

depopulation (23.4) >> repopulation (4.4) ~ litter/manure removal (3.0).

Previous work [8] at another poultry farm reported different exposure rankings of activities (median values in $mg\ m^{-3}$):

litter/manure removal (35.4) > repopulation (5.5) ~ depopulation (4.6).

There are various differences between the farm studied here and those used for the earlier work. In this work there were fewer data collected (11 samples compared to 60), the farm was a much smaller enterprise, processes were less mechanised and there were possibly less time constraints.

Larger farms require more materials and there is more product and waste to be removed. During litter removal at larger farms there may be two or more vehicles collecting manure compared with the one vehicle used in this study which would confer some control over where the dust is generated. Also, collecting damp with dry manure can reduce dust, as was performed at the farm studied here.

Petrol-engined blowers used at larger farms to clear remaining dust around stanchions can raise clouds of dust. At this farm, new litter was manually broken away from an opened bag, then distributed using a pitchfork. Laying new litter at larger farms is, however, largely mechanised and can include using a 'hay-turner' attachment to spread the litter.

Depopulation resulted in high inhalable dust concentrations in this study but fairly low concentrations at the other larger farms. Different types of bird feathers (eg from chickens, turkeys and ducks) may result in differing dust levels, as well as differences between the bird catchers' technique (e.g. farm employees or contracted staff may operate differently). The use of powered and natural ventilation at larger farms tends to prevent accumulation of dust inside the shed. In other respects both depopulation and repopulation were carried out similarly at this and larger farms.

2.4.2. Comparison of monitors

The use of inhalable and respirable samplers and the PDR at the farm allowed comparisons to be made between the measurements; the results are shown in table 2.

Table 2. Mean inhalable, respirable and PDR concentration ratios (standard deviations in parentheses)

Activity	n	Gravimetric inhalable/ Gravimetric respirable	Gravimetric inhalable/ PDR	Gravimetric respirable/ PDR
Depopulation	3	7.6 (1.1)	6.6 (1.5)	0.9 (0.2)
Routine cleaning	1	4.8	4.6	1.0
Repopulation	2	21.8 (0.4)	18.4 (5.1)	0.9 (0.2)
Laying new litter	1	12.1	26.1	2.2
Litter removal	3	5.2 (2.1)	9.7 (0.9)	2.1 (0.8)
Clean/Disinfect	1	2.9	14.4	5.1

There is a wide variation in gravimetric inhalable/gravimetric respirable ratio for the activities - a factor of approximately 8 between the maximum and minimum. The dust particle size profiles are therefore different for the activities measured, which necessitates the application of separate calibration (response) factors for each activity to each PDR. Repopulation and laying litter are the two activities which generate the greatest proportion of inhalable dust. The gravimetric inhalable/PDR ratios do not follow the same ranking as the gravimetric inhalable/gravimetric respirable ratios, although the statistics do not allow detailed analysis. The range of gravimetric respirable/PDR ratios is not as great as that for gravimetric inhalable/PDR (especially if the cleaning/disinfecting result is considered anomalous, ie the low gravimetric inhalable/gravimetric respirable ratio and the high gravimetric respirable/PDR ratio). Nevertheless, the respirable dust from the different activities still requires different PDR calibration factors, as is found for most dusts other than 'standard' silica (eg Arizona road dust) used to factory calibrate the instruments [4,5].

The accuracy of real-time dust monitors for workplace (and environmental) monitoring relies on the validity of the calibration, which is typically based on gravimetric samplers. Photometric (light scattering) dust monitors, which are commonly used for workplace air monitoring, including the PDR used in this study, have a response as a function of particle size quite similar to the respirable sampling convention and therefore to the standard cyclone sampler [6]. The real-time monitor can, however, only have an average response factor applied over the duration of the activity. Consequently, if there are any variations in the dust particle characteristics (size distribution, refractive index etc) over this period, then at any given instant there may be significant deviation from the average value. This situation could be exacerbated when the photometric ('quasi-respirable') real-time monitor is calibrated with respect to an inhalable reference derived from a gravimetric sampler. Here the respective particle size profiles of the monitors are very different which increases the chance of wider deviations of the real-time monitor values from the 'true' values. For example, generated particles that are greater than 10 μm will be captured by the reference sampler, but not measured by the real-time monitor. Therefore, as the particle size increases above 10 μm the real-time monitor will increasingly underestimate the inhalable concentration.

The use of a portable TEOM [10] should help in deriving more accurate response factors for a real-time monitor. This is because its response is not subject to uncertainties in measurement caused by changes in particle size, colour, or shape and responds purely as a function of mass of dust sampled. The TEOM's response time can be quite long for VEM depending on the concentration of aerosol being measured, typically tens of seconds compared to seconds for photometers including the PDR. It is, however, much shorter than the gravimetric sampling period (of the order of an hour). The use of the TEOM should thus be able to confirm that the response factor of the photometer monitor remains essentially constant over the measurement period or, if not, provide more accurate values for the periods where the response factor is significantly different. Further work is necessary to validate this approach for various types of dust and workplace activity.

2.4.3. VEM output

The results from the VEM were processed (using Adobe Flash media tools) for the occupational hygienists as shown in figure 2. The activities were highlighted on the chart and the corresponding exposure and video are viewed by moving the cursor (highlighted in figure 2 by arrow) over the chart and exposure profile. This information is suitable for adapting and incorporating into guidance such as Toolbox talks [11] which are short talks focused around specific health and safety issues and allow workers, safety professionals and managers to explore risks and develop strategies for dealing with them. It is the intention that Toolbox talks can help to demystify health and safety and to show the relevance of specific topics to particular jobs.

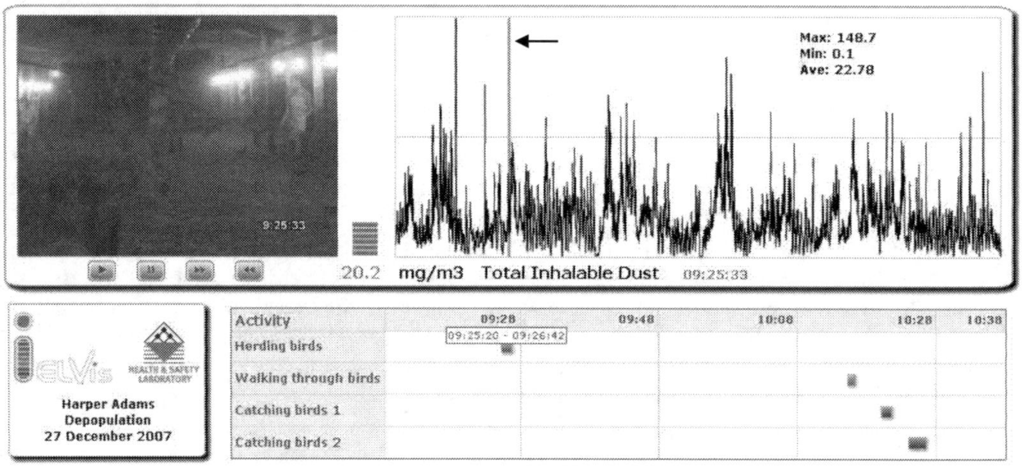

Figure 2. VEM depicting poultry farming activities (moveable cursor highlighted by arrow)

3. Pharmaceutical powder transfer operations

3.1. Background

In the pharmaceutical industry, there is potential for exposure to active agents during powder transfer operations. Various control measures can be adopted to minimise exposure risk, in particular, the use of a curtain barrier when dispensing in a downflow booth. VEM, using a non-pharmaceutically active powder (xanthan gum), was undertaken to show how improved control further reduced personal exposure by monitoring with and without a curtain barrier. Moreover, this offered the opportunity for disseminating training information using the 'ELVis' software.

3.2. Materials and methods

The ultraclean environment required that all equipment be checked and wiped, and operators wore coveralls. Also, note that for clean applications such as this, separate instruments are used from those for 'dirty' applications as described in section 2. Both wireless instruments (radio telemetry - Satel Radiomodems) and wireless video (transmitter/receiver – Astrotel Communication Corp.) were employed. A dust lamp (HSE Tyndall beam dust lamp) was used to enhance the video footage by showing the suspended dust particles. Two cameras were employed for wide angle and close-up views. The real-time monitor (Casella Microdust) was chosen to allow periodic checks of the span as well as zero during the measurement period. A harness was adapted to fit the sampling probe (25 cm long) in the breathing zone of the worker. The monitor (based on the photometric light scattering principle) was factory calibrated with respect to total suspended particulate (TSP), which approximates to the inhalable fraction [4].

The activity to be monitored was simple and used as a test case for the technique. Powder was transferred in a downdraft booth from one receptacle to an adjacent one using a scoop (figure 3), when full the contents were bagged and transferred back and the process repeated. Monitoring was performed before and after installation of a curtain barrier - a clear plastic sheet fixed between opposite walls of the booth with two holes cut centrally for insertion of the operator's arms. The operator carried out the same activities but behind the barrier.

Figure 3. Powder transfer operations using a scoop

3.3. Results and Discussion

Task analysis was undertaken, identifying three activities: scooping, emptying the bag and bulk transfer of the contents, and the results were analysed and displayed as in figure 4.

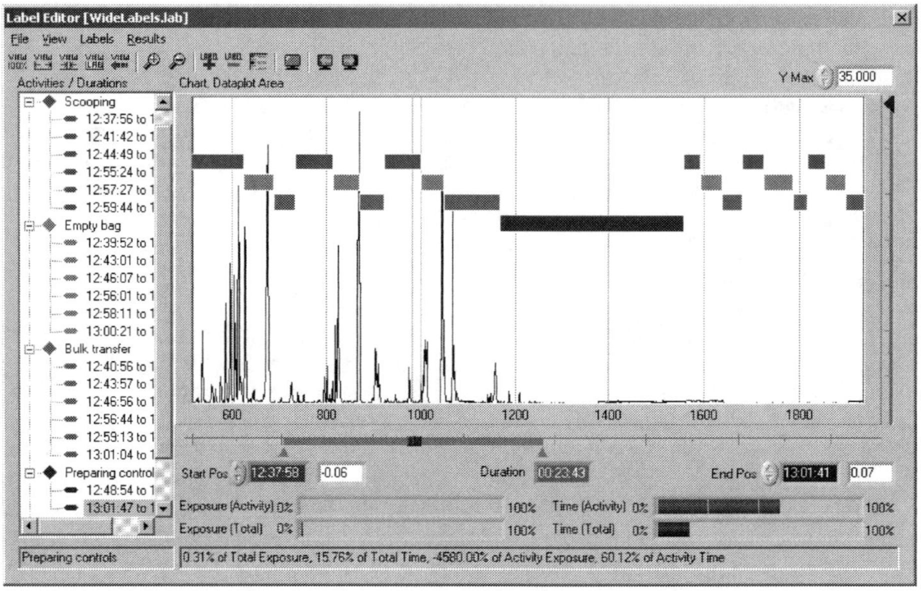

Figure 4. Task analysis for powder transfer operations derived from VEM using a Microdust dust monitor.

These data were then summarized and displayed as bar graphs as shown in figure 5. Emptying the bag resulted in the highest exposure while scooping and bulk transfer exposures were roughly similar.

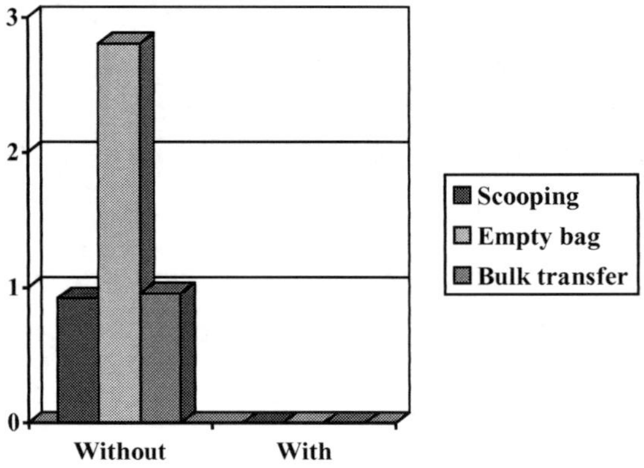

Figure 5. Comparison of exposure (mg m^{-3} TSP) with and without the curtain barrier for the three activities.

The beneficial effect of the simple curtain barrier can be clearly seen and was demonstrated in a very short time – the exercise took a half-day with the actual measurement taking less than one hour.

4. Conclusions

Two contrasting examples of workplace environment have been used to illustrate how VEM with real-time dust monitors, calibrated with reference samplers, can transform task analysis data into useful information on the hazards of dust (poultry waste and pharmaceutical powder) and its control. The different environments placed different demands on the VEM measurement equipment, eg low light levels and very mobile workers at the poultry farm and the ultra-clean, sterile environment at the pharmaceutical production facility.

The calibration of the real-time monitors, which are essentially responsive to the respirable fraction, such that they read in inhalable dust concentration has been highlighted as an area where further work may be needed to (a) test the validity of such calibration over the measurement period for various dusts and activities and (b) develop real-time, portable inhalable monitors, possibly based on the TEOM.

References

[1] Maynard A D and Jensen P A 2001 Aerosol measurement in the workplace *Aerosol Measurement*, ed P A Baron and K Willeke (New York: Wiley-Interscience) ch 25 pp 792-3.

[2] Rosén G, Andersson I-M, Walsh P T, Clark R D R, Säämänen A, Heinonen K, Riipinen H and Pääkkönen R 2005 A review of video exposure monitoring as an occupational hygiene tool. *Ann. Occup. Hyg* 49 201-17.

[3] Thorpe A 2007 Assessment of personal direct-reading dust monitors for the measurement of airborne inhalable dust. *Ann. Occup Hyg.*, **51**, 97 – 112.

[4] Thorpe A and Walsh P T 2002 Performance testing of three portable, direct-reading dust monitors. *Ann. Occup. Hyg.* **46** 197-207.

[5] Thorpe A and Walsh P T 2007 Comparison of portable, real-time dust monitors sampling actively, with size-selective adaptors, and passively. *Ann. Occup. Hyg.* **51** 679-691.

[6] Gebhart J 2001 Optical direct-reading techniques: light intensity systems *Aerosol Measurement*, ed P A Baron and K Willeke (New York: Wiley-Interscience) ch 15 pp 430-1.

[7] Cowie H A, Graveling R A G, Cherrie J W, Soutar C A, Cattermole T J, Graham M K, Mulholland R E 2005 Baseline incidence of ill health in agriculture in Great Britain. *HSE Research Report RR370/2005*. (Sudbury: HSE Books).

[8] Crook B, Easterbrook A and Stagg S 2008 Exposure to dust and bioaerosols in poultry farming; summary of observations and data. *HSL Report OH2008/12*. HSL.

[9] HSE 2000 General methods for sampling and gravimetric analysis of respirable and inhalable dust MDHS 14/3. www.hse.gov.uk/pubns/mdhs/pdfs/mdhs14-3.pdf

[10] Patashnick H, Meyer M and Rogers B 2002 Tapered element oscillating microbalance technology. Proc of N American/Ninth U.S. Mine Vent Symp. June 2002. 625-631.

[11] HSE Toolbox Talks (accessed June 2008) www.hse.gov.uk/involvement/planning.htm

Acknowledgements

We are grateful for the assistance of Harper Adams University College, especially David Chadwick, with the poultry farming work; and GSK, especially Russell Collins, Debra Stewart and Erica Joseph, with the pharmaceutical work. We also thank our colleagues in the Health and Safety Executive for their help and support: Neil Davey, Len Morris, Nigel Black and Robert Hadway.

Alveolar macrophage accumulation rates, for 28 nm and 250 nm PSL, are mediated by separate mechanisms.

O R Moss and V A Wong

The Hamner Institutes for Health Sciences, Research Triangle Park, NC 27509-2137, USA;

moss@thehamner.org

Abstract. When macrophages accumulate 28 nm and 250 nm diameter polystyrene latex (PSL) beads, the accumulation rates should reflect differences in molecular and cellular function. We used a confocal microscope to measure the accumulation rates of nanoparticles by F344-rat-alveolar macrophages (~25,000 cells adhered to a 0.7 cm^2 surface). Over the cells were layered 0.1 ml of media, and 0.1 ml of media-with-beads. Fresh cells were introduced for each exposure scenario. The maximum possible individual macrophage exposures were as follows: $8x10^6$, $8x10^5$, and $8x10^4$ 28 nm beads per macrophage; and $8x10^4$ and $1.12x10^4$ 250 nm beads per macrophage. Accumulation rates were estimated over 23 minutes. The increase in bead accumulation-rate matched changes in bead-availability: 7x increase for 250 nm beads; 100x increase for 28 nm beads; and 700x increase for all bead availabilities. The maximum sustained 28 nm bead accumulation rate was > 30,000 /min (for 5 min). Increases in bead accumulation could be explained by two mechanisms: bead-diffusion; and, for the macrophage, macropinocytosis. Also for the highest concentrations of 28 nm beads, we saw a colligative threshold -- possibly due to beads masking the cell surface or obstructing cellular mechanisms.

1. Introduction:

Human exposures to nanoparticles occur through injection, ingestion, skin contact, or inhalation: The initial interaction is between nanoparticles and cells. In particular, during inhalation, an airway-system cell may come into contact with nanoparticles of different composition. Each nanoparticle may impact different mechanistic pathways; each may have a different toxicity. Yet these differences may be overshadowed by the cell's mechanistic and toxic responses to the colligative properties of nanoparticles.

Colligative properties are properties that depend on number density. For example, inside a sealed container, the pressure on the walls is due to the number of gas molecules (actually the rate that the gas molecules bounce off of the container walls). Likewise, although the term "pressure" may not be adequate, for nanopartaicles, the number of nanoparticles may impact cells in three ways: the number nanoparticles inside a cell may exert a pressure on (or impede) molecular machinery; the total cross-sectional area of nanoparticles per cellular surface area my impede basic functions of the cell surface; and the total volume of nanoparticles per cell volume may impede basic cell mobility and response. In each case, the number of nanoparticles, a colligative property, impacts efforts to compare and rank the cellular-toxicity of different types of nanoparticles.

The inverse should also be true: When colligative effects are kept negligable, comparisons can be made of the mechanism of action and toxicity of different nanoparticles.

Journal of Physics: Conference Series **151** (2009) 012044 doi:10.1088/1742-6596/151/1/012044

We use a laser scanning confocal microscope to measure the rates, that 26 nm, 28 nm, and 250 nm diameter polystyrene latex (PSL) beads are accumulated by macrophages. In our previous work (Moss, 2008 [(1)]) we showed that the confocal laser scanning microscope could be used to measure, in alveolar macrophages, the accumulation-rate of 26 nm polystyrene latex beads.

In this current work, our objectives are threefold: first, to duplicate our previous work (this time with 28 nm and 250 nm diameter beads); second, to use the rate that nano-beads are accumulated by macrophages as an early indicator of the colligative effects of nanoparticles; and third, to associate the accumulation rate with changes in the molecular processes involved in endocytotic and cellular function. The nano-bead accumulation rates are expected to be impacted by three separate mechanisms: nanoparticle diffusion, macropinocytotic vesicle transport, and cell surface modification.

What we found was that, for the F344 rat alveolar macrophage exposed to PSL nano-beads, a 700x increase in bead availability was matched by the same increase in bead accumulation-rate. The maximum sustained 28 nm bead accumulation rate was > 30,000 /min (for 5 min). The 700x increase in bead accumulation could be explained by two mechanisms: bead-diffusion; and, for the macrophage, macropinocytosis.

2. Materials and Methods

Our approach was to measure the rate that freshly harvested (not activated; not synchronized) rat alveolar macrophages accumulate fluorescent Polystyrene Latex (PSL) beads (28 nm and 250 nm in diameter). Bead concentrations were selected to avoid, within the macrophage, possible colligative effects due to number of beads accumulated per cell, total bead cross sectional surface per cell surface, and total bead volume per cell volume.

The three concentrations of 28 nm PSL beads [High (H), Medium (M); and Low (L)] increased by factors of 10. The two concentrations of 250 nm PSL beads [Low (L); and Very low (V)] increased by a factor of 7. For both sizes of beads, the Low concentration was the same. The concentration range between the V and H concentrations was a factor of 700x. Bead-accumulation by the macrophage was measured for up to 30 minutes or until the fluorescence signal saturated or exceeded the maximum fluorescence of the calibration curves. From the slope of the accumulation curve, we calculated the accumulation rate.

2.1. Materials:

2.1.1. *PSL beads:* The nano-particles used in this study were 28 nm and 250 nm diameter, fluorescent Polystyrene Latex (PSL) beads (Duke Scientific Corp., Palo Alto, CA). In the fluid covering the macrophages, concentrations of the 28 nm beads were 10^{12}, 10^{11}, 10^{10} beads/ml -- corresponding to the high (H), medium (M), and low (L) concentrations. For the 250 nm beads, concentrations were 10^{10} and $1.4x10^9$ beads/ml -- corresponding to the low (L) and very low (V) concentrations (For only the L concentration of 250 nm beads, non-fluorescent beads of the same size were used to dilute the fluorescence; for this dilution we used a ratio of 1 fluorescent bead to 9 non-fluorescent beads.).

2.1.2. *Maximum PSL beads available per macrophage:* The concentration of PSL beads (beads/ml) does not provide sufficient information about individual macrophage exposure. Such exposures are better expressed as the maximum number of PSL beads available per macrophage. This alternative expression is calculated from the number of cells (~25,000), the volume of liquid layered over the cells (0.2 ml), and the concentration (beads/ml). From this perspective the maximum possible individual macrophage exposures were $H=8x10^6$, $M=8x10^5$, $L=8x10^4$, and $V=1.12x10^4$ PSL beads/macrophage.

This estimation is still not ideal. The maximum number of beads available per cell does not indicate the rate at which beads come into contact with the cell membrane. In this study we do not know the rate of contact (the tap-tap rate in contacts/second). However, if we assume that, for every

Journal of Physics: Conference Series **151** (2009) 012044 doi:10.1088/1742-6596/151/1/012044

change in bead concentration there is a proportional change in the tap-tap rate, then we can test whether the magnitude of this change is duplicated in the magnitude of the change in the rate that macrophages accumulate beads.

2.1.3 PSL bead exposure limits: The four exposure levels (H, M, L, and V) were selected to minimize the possibility of PSL beads covering the surface of the macrophage: That is, for the maximum number of PSL beads available per cell, we minimized the possibility of the total cross-sectional area of nanoparticles being greater than the surface area of the cell (For a 12 micron diameter "idealized-spherical" macrophage, and for 28 nm diameter beads, the number is approximately 10^6 beads per macrophage; for 250 nm diameter beads, the number is ~ 10^4 beads per macrophage.). The four exposure levels were also selected to assure that the total volume of nanoparticles per macrophage would be less than 6% of the volume of the macrophage: For 28 nm beads this number is ~ $5x10^6$ beads per macrophage; for 250 nm beads the number is ~ $7x10^3$ beads per macrophage (Moss, 2008 [1]).

2.2. Methods:

2.2.1. *Animals*. This study was conducted under federal guidelines for the care and use of laboratory animals (National Research Council, 1996) and was approved by the Hamner Institutes' Institutional Animal Care and Use Committee. For this study, eight to ten week old F344 female rats were purchased from Charles River Laboratories (Raleigh, NC); the rats were housed in standard cage racks in an AAALAC-accredited facility. Animals were given NIH-07 cereal-based diet (Zeigler Brothers, Gardner, PA) and water ad libitum. Room temperature was maintained at 18 to 26°C; relative humidity was maintained at 30 to 70%.

2.2.2. *Collection of Macrophages from BAL.* Bronchioalveolar lavage (BAL) fluid was collected from approximately 12-week-old F-344 female rats (weight ~230 g). Rat lungs were lavaged with ice cold phosphate-buffered saline (PBS; Invitrogen, Grand Island, NY). Total cells from the BAL were resuspended in medium (Dulbecco's modified Eagles medium, DMEM-F12K without phenol red; Invitrogen) with 1% fetal bovine serum (FBS; Invitrogen) and 10mM HEPES (N-2-hydroxyethylpiperazine-N'-2-ethanosulphonic acid; Sigma, St. Louis, MO). Lavage fluid was counted (97-99% macrophages; Bermudez et al. 2003) and then diluted to concentrations of $6.6x10^5$ macrophages/ml.

2.2.3. *Imaging of particles in cells.* Uptake of 28 nm green fluorescent polystyrene latex (PSL) beads was measured using a confocal laser scanning microscope (LSM 510 META confocal microscope; Zeiss, Thornwood, NY). Visualization of particle uptake by macrophages was accomplished by adjusting the focal plane to step, 800 nm thick optical-slice by optical-slice, through the entire volume of a macrophage (Figure 1).

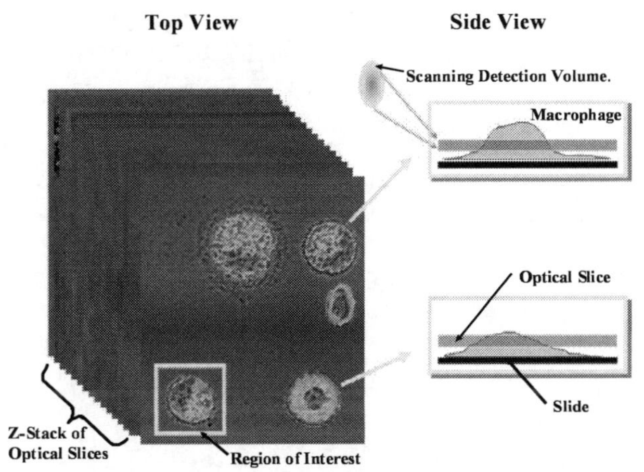

Figure 1. Confocal lasar Scanning Microscopy of bead-accumulation in alveolar macrophages.

For measuring the accumulation of PSL particles in a macrophage, macrophages from the BAL were diluted to $5x10^5$ macrophages/ml. Fifty thousand macrophages (100 µl) were placed in each chamber of an 8-chambered cover slip (0.7 cm^2/chamber; Lab-Tek II) and allowed to adhere at 37°C. Non-adherent cells (~50%) were removed and the chambers were washed 2 times each with medium. Segmented nuclei were not observed among the remaining adherent cells. Adherent macrophages were layered with 100 µl of medium and placed on the heated stage of an LSM 510 META microscope (Zeiss, Thornwood, NY). In each chamber, one cell was examined using a Plan Achromatic 100x objective (NA= 1.4). A Z-stack of the cell was imaged with a zoom setting of 1, an area of 729 μm^2 and an optical slice height of 0.8 µm (volume of 583 μm^3) and at intervals of 800 nm. To the chamber were added 100 µl of a continuously mixed fluorescent bead suspension. Z-stack images were acquired as rapidly as possible for up to 30 minutes. Calibration consisted of matching relative flourescence units to bead-concentration; the calibration was carried out several times during each day.

2.2.4. Calculation of beads accumulated per macrophage: From an optical slice captured by the confocal microscope at least six (6) macrophages were selected. Selection was based on whether the macrophage was fully contained within the Z-stack of optical slices. Around each macrophage, a box was drawn in order to define a region of interest. This region of interest was extended upward through each optical slice of the Z-stack. From these images we extracted four types of relative fluorescence (RF): the RF of the medium before bead addition; the RF of the medium after bead addition; the RF of the medium and macrophage before bead addition; and the RF of the medium and macrophage after bead addition. These values allowed an estimation to be made of the number of beads accumulated within a macrophage. By repeating this process for the selected macrophages, we calculated at each time point the average number of beads per macrophage.

2.2.5. *Calculation of bead accumulation rates.* From the time-series of the average number of beads per macrophage, we calculated, at each time point, the slope of the curve (beads accumulated per minute per macrophage).

3. Results

By using the confocal laser scanning microscope, we measured the rate that freshly harvested (not activated; not synchronized) rat alveolar macrophages accumulated fluorescent Polystyrene Latex (PSL) beads (28 nm and 250 nm in diameter). Bead exposures were selected to avoid possible colligative effects due to number of beads per cell, total bead cross sectional surface per cell surface, and total bead volume per cell volume. Even with these restrictions, the concentration of beads -- and thus the availability of beads to each macrophage -- varied by a factor of 700x. Our results were as follows:

3.1. Calculation of beads per macrophage:

We found that, in spite of the macrophages not being activated or synchronized, at each PSL bead exposure level and at each time-point, the average number of beads accumulated was reproducible. By 13 minutes, the macrophages, exposed to the high (H) exposure level of 28 nm diameter PSL beads, had accumulated an average of 380,000 beads; while, during the same time period, the macrophages exposed to the very low (V) exposure level of the larger, 250 nm diameter beads, had accumulated only 99 beads (Figures 2 and 3).

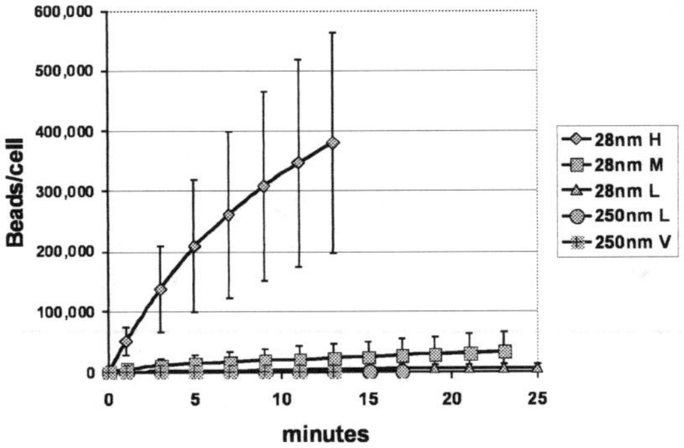

Figure 2 Alveolar macrophage accumulation of 28 nm and 250 nm diameter PSL beads. Exposure levels, as maximum number of beads available per macrophage, were H($8x10^6$), M($8x10^5$), L($8x10^4$), and V($1.12x10^4$). Shown are means plus standard deviations for n equal to or greater than 6.

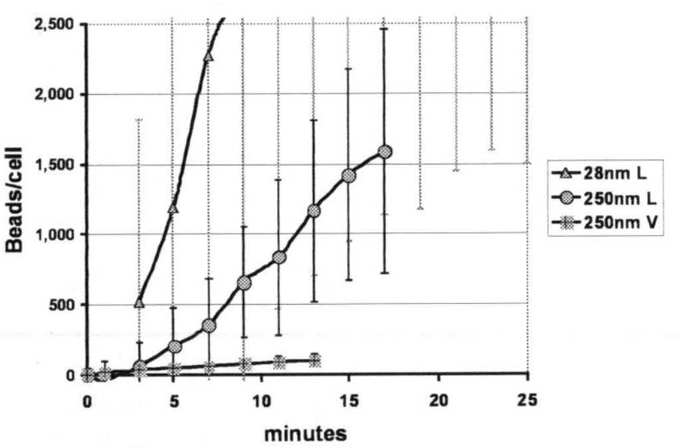

Figure 3 Expansion of Figure 2 -- Alveolar macrophage accumulations of 28 nm and 250 nm diameter PSL beads. Exposure levels, as maximum number of beads available per macrophage, were L($8x10^4$) and V($1.12x10^4$). Shown are means plus standard deviations for n equal to or greater than 6.

3.2. Calculation of bead accumulation rates.

In every case, during the first five minutes, when the exposures of PSL beads changed by a factor of 7 or 10, the average bead-accumulation rate by the macrophages changed by the same factor (Figure 4). However during this same time period, for the low (L) exposures of 28 nm or 250 nm beads, the average rate that macrophages accumulated the 250 nm beads was six (6) times slower than the accumulation rate for the 28 nm beads (73 vs. 438 beads per minute per macrophage). After five minutes, for the high (H) exposures of 28 nm beads, the average bead-accumulation rate began to drop from a sustained rate of 31,000 beads per minute per macrophage, reaching approximately half this value by the eleventh minute. Moreover, for the low (L) exposures of 250 nm beads, after five minutes, the bead accumulation rate began to increase from 73 beads per minute per macrophage, doubling the rate by the thirteenth minute.

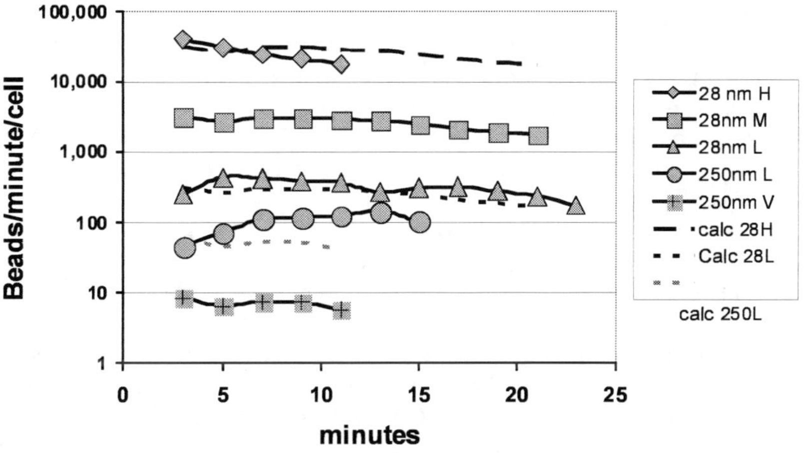

Figure 4: Average rates that alveolar macrophages accumulate 28 nm and 250 nm diameter PSL beads. Exposure levels, as maximum number of beads available per macrophage, were H($8x10^6$), M($8x10^5$), L($8x10^4$), and V($1.12x10^4$). The dashed line "cal 28H" is calculated by multiplying the accumulation rate for "28nm M" by a factor of 10; "cal 28L" is calculated by dividing the accumulation rate for "28nm M" by a factor of 10; and "cal 250L" is calculated by multiplying the accumulation rate for "250nm V" by a factor of 7.

4. Discussion:

In these experiments we showed that, for 28 nm and 250 nm diameter Polystyrene Latex (PSL) beads, as the maximum possible number of beads available per macrophage increased, the parallel -- and of equal magnitude -- change in the rate that macrophages accumulated beads could be explained by the difference in the rate at which beads were presented to the surface of the cells. For the 28 nm diameter beads, as the bead exposure changed by a factor of 10, the rate that macrophages accumulated the beads also increased by a factor of 10. Likewise, for the 250 nm diameter beads, as the exposure changed by a factor of 7, the accumulation rate also increased by a factor of 7. In these experiments, the macrophages had sufficient capacity to handle an overall increase of 700x in the rate that beads were accumulated.

The low (L) exposures (a possible maximum of $8x10^4$ beads per macrophage) provide data that confirms predictions by Teeguarden [2]. The 28 nm diameter PSL beads were predicted to move by diffusion at a rate 9x faster than that of the 250 nm diameter PSL beads. However, the 250 nm beads move by sedimation at approximately the same rate as they move by diffusion. The net effect is a prediction that, compared to the 250 nm beads, the rate that the 28 nm diameter beads move towards a macrophage would be 4x to 5x faster: In our measurements (Figure 4), when the starting concentration of beads was the same, the difference in the rate that macrophages accumulated the two sized beads was between 2x and 6x. For the low (L) exposures of 28 nm and 250 nm beads, the 2-to-6x difference in the rate that macrophages accumulated the beads can be explained by the respective differences in the rate of bead- diffusion and sedimentation.

We saw a 700x increase in the rate that macrophages could accumulate PSL beads. This capacity could be explained by a specific nano-particle transport mechanism -- macropinocytosis (Radcliffe [3]). A macropinocytotic vesicle capable of carrying one 250 nm diameter PSL bead would also be capable of carrying approximately 700 of the 28 nm diameter PSL beads. The macropinocytotic vesicle number and transport rate would not have to increase in order to produce the observed bead accumulation rates -- including the highest rates (30,000 to 40,000 beads/minute/macrophage) seen for the 28 nm diameter beads. The capacity was sufficient to handle the increase in the rate that the beads interacted with the macrophage surface membrane. However, after 5 minutes at an accumulation rate of 30,000 to 40,000 beads per minute per macrophage, the rate at which the macrophage accumulated the 28 nm beads began to drop -- as predicted due to colligative effects of the total cross sectional area of the beads approaching the area of the surface membrane of the macrophage. If, for the 28 nm PSL beads, the macrophage was able to sustain the observed accumulation rate of 30,000 beads per minute, then, in 165 minutes the beads would occupy 6% of the macrophage volume: Moreover, at this bead-accumulation rate, in 32 minutes there would be enough beads to cover the surface membrane of an idealized spherical macrophage.

There was one unexpected result for the low exposures of 250 nm diameter PSL beads (a possible maximum of $8x10^4$ beads per macrophage). From minute five to minute thirteen, the average rate that the macrophages accumulated beads doubled. Either the capacity doubled -- more macropinocytotic vesicles; or there was a doubling of the rate that the 250 nm diameter PSL interacted with the macrophage surface membrane; or additional endocytotic processes were initiated. However, as discussed above, we expected the 28 nm beads to approach the macrophage surface membrane at a rate that was 4-to-5x faster than that for the 250 nm beads. For the 250 nm beads, the doubling of the accumulation rate most likely represented a statistical variation in diffusion and sedimentation.

Because the freshly collected rat-alveolar macrophages were not synchronized and not specifically activated, our observations primarily provide an indication of the interaction between nanoparticles and macrophages -- a worst case demonstration that reflects the status of macrophages in the body. If the macrophages were activated and synchronized, the standard deviations of the bead accumulations (Figures 2 and 3) would be significantly smaller. The measurement of bead accumulation rates provides a means for future investigation of the initiation, limits, and obstruction of endocytotic processes; it also provides a means for developing and evaluating high throughput, *in vitro* tests for ranking the toxicity of nanomaterials.

5. Summation.

In this work we measured, in non synchronized, non activated alveolar macrophages, the average rates that macrophages accumulate 28 nm and 250 nm PSL beads. We draw three conclusions:

First, The volume of macropinocytotic vesicles capable of transporting 250 nm beads is sufficient to explain the broad capacity of the rate that the macrophage accumulated 28 nm diameter beads. As the exposure (the maximum possible number of beads available per macrophage) increased, the rate of accumulation also increased: We saw a 7x increase for 250 nm beads; a 100x increase for 28 nm beads; and a 700x increase for all bead availabilities.

Second, a colligative threshold -- possibly due to beads masking the cell surface or obstructing cellular mechanisms -- could explain the observed decrease in the maximum rate that macrophages accumulated 28 nm diameter beads. The maximum sustained 28 nm bead accumulation rate was > 30,000 /min/macrophage (for 5 min). However, after 5 minutes, this accumulation rate began to decrease, as expected for accumulations of greater than 150,000 28 nm diameter beads per macrophage.

Third, the dose-rate (or rate of accessibility) of the beads at the macrophage surface dominated the observed rate of accumulation. For equal exposures of 28 nm and 250 nm PSL beads (a possible maximum of $8x10^4$ beads per macrophage), during the first 5 minutes, the difference in the macrophage rate of accumulation of 250 nm and 28 nm diameter beads (a factor of 2x to 6x) is of the same order of magnitude as the combined difference in bead-diffusion and -sedimentation (a factor of 4x to 5x).

Acknowledgments:
This study was supported as part of a grant from the Long Range Research Initiative of the American Chemistry Council.

References:

[1] Moss, O. R. (2008) Insights into the health effects of nanoparticles: Why numbers matter. International Journal of Nanotechnology 5(1):3-14

[2] Teeguarden, J. G., Hinderliter, P. M., Orr, G., Thrall, B. D., Ponds, J. G. (2007) Particokinetics *In Vitro*: Dosimetry considerations for *In Vitro* Nanoparticle Toxicity Assessments, Toxicological Sciences 95(2): 300-312.

[3] Radcliffe, P. M., Wong, V. A., Moss, O. R., and Dorman, D. C. (2008), Uptake of Fine and Ultra-fine Polystyrene Latex beads in A549 Cells, The Toxicologist 92(S-1): 1511.

Dissolution and reactive oxygen species generation of inhaled cemented tungsten carbide particles in artificial human lung fluids

A B Stefaniak, S S Leonard, M D Hoover, M A Virji and G A Day

National Institute for Occupational Safety and Health, Morgantown, WV, USA, 26505

E-mail: AStefaniak@cdc.gov

Abstract. Inhalation of both cobalt (Co) and tungsten carbide (WC) particles is associated with development of hard metal lung disease (HMD) via generation of reactive oxygen species (ROS), whereas Co alone is sufficient to cause asthma via solubilization and hapten formation. We characterized bulk and aerodynamically size-separated W, WC, Co, spray dryer (pre-sintered), and chamfer grinder (post-sintered) powders. ROS generation was measured in the murine RAW 264.7 cell line using electron spin resonance. When dose was normalized to surface area, hydroxyl radical generation was independent of particle size, which suggests that particle surface chemistry may be an important exposure factor. Chamfer grinder particles generated the highest levels of ROS, consistent with the hypothesis that intimate contact of metals is important for ROS generation. In artificial extracellular lung fluid, alkylbenzyldimethylammonium chloride (ABDC), added to prevent mold growth during experiments, did not influence dissolution of Co (44.0±5.2 vs. 48.3±6.4%); however, dissolution was higher ($p<0.05$) in the absence of phosphate (62.0±5.4 vs. 48.3±6.4%). In artificial macrophage phagolysosomal fluid, dissolution of Co (36.2±10.4%) does not appear to be influenced ($p=0.30$) by the absence of glycine (29.8±2.1%), phosphate (39.6±8.6%), or ABDC (44.0±10.5%). These results aid in assessing and understanding Co and W inhalation dosimetry.

1. Introduction

Cemented tungsten carbide (CTC) refers to a class of alloys containing 80 to 95% tungsten carbide (WC) bound in a matrix of 5 to 20% cobalt (Co), with additives such as nickel, chromium, vanadium, titanium carbide, and tantalum carbide to impart performance-specific properties. Cemented tungsten carbide tools exhibit extreme hardness and wear resistance at the tool and work piece interface during cutting, which is retained at elevated temperatures. As such, applications of CTC include cutting tools for metals; and masonry, concrete, coal, and rock drills and cutters.[1,2]

Inhalation of dusts generated during production of CTC is associated with development of hard metal lung disease (HMD) and occupational asthma. Cases of HMD have been reported among workers exposed to pre-sintered material [3] and among workers exposed to post-sintered CTC material.[4] Using pre-sintered powders, investigators have demonstrated greater generation of reactive oxygen species (ROS) for Co in the presence of WC relative to ROS generation among the individual constituents.[5] Thus, available data suggest that WC particles in association with Co particles, rather than WC or Co particles alone, should be considered a specific toxic entity in

Journal of Physics: Conference Series **151** (2009) 012045 doi:10.1088/1742-6596/151/1/012045

development of HMD. Additionally, workplace studies have identified cases of occupational asthma among employees who handle pre-sintered material (e.g., powder mixers, pressers, and shapers) and among those who grind and manipulate sintered CTC parts.[3-6] To date, a clear toxicological mechanism for cobalt-induced asthma is lacking.[7] Evidence indicates, however, that Co is soluble in artificial lung fluids [8-9] and that ionic cobalt may conjugate with proteins to form a hapten which may elicit an immune response.[6,11]

Using well-characterized particles from representative CTC manufacturing processes, [12] we evaluated the hypothesis that aerosol physicochemical properties are important exposure factors for HMD and asthma. We measured particle solubility in variations of artificial extracellular lung fluid and lung alveolar macrophage phagolysosomal fluid; evaluated solvent effects on observed dissolution rates; and determined particle ROS generation using an in vitro cellular model.

2. Materials and Methods

2.1. Study powders

Five representative bulk powders that span the spectrum of powders used in the manufacture of CTC were chosen for study: feed-stock W, WC, and Co powders, pre-sintered powder from a spray dryer, and dust from a chamfer grinder used for final machining of post-sintered CTC product. The morphology and chemistry of these bulk powders are similar to that observed for airborne particles in the work areas where the powders are handled. Thus, these powders are thought to be representative of aerosols to which workers are potentially exposed during CTC production.[12] A sixth material was prepared as an admixture by mechanically blending aerodynamically size-separated (see section 2.2) WC and Co powders at a ratio of 94:6 by mass. The admixture is not an actual industrial powder, but a mechanical mixture of feedstock powders prepared for the purpose of comparing free radical generation of loosely-associated cobalt and tungsten carbide particles to the individual feedstock powders.

2.2. Aerodynamic size separation of study powders

Each bulk powder was aerodynamically size-separated in the laboratory using a dry powder disperser (Model 175, DeVilbiss, Somerset, PA) and five-stage aerosol cyclone operated at 10 L min^{-1}.[13,14] The 50% aerodynamic cutoff diameters (D_{50}) of the aerosol cyclone were >10, 4.3, 2.9, 2.0, and 1.2 μm for stages 1 to 5, respectively. For the purposes of this study, we evaluated ROS generation and solubility in artificial lung fluids for the materials collected in stages 1, 2, and 3 of the aerosol cyclone. (There were insufficient masses of materials collected in stages 4 and 5 of the aerosol cyclone to perform the material characterization assays and ROS and solubility studies for all study powders.) Material collected in stage 1 ($D_{50} > 10$ μm) have particle sizes capable of penetrating into the conducting airways (i.e., thoracic-sized particles). Materials collected in stage 2 ($D_{50} = 4.3$ μm) and stage 3 ($D_{50} = 2.9$ μm) have particle sizes capable of penetrating into the alveolar region of the lung (i.e., respirable-sized particles).[15]

2.3. Characterization of study powders

Physicochemical properties of the aerodynamically-size separated powders were characterized using scanning electron microscopy to evaluate the morphology and size of powders [12]; Auger spectroscopy to determine surface elemental composition (down to 100Å thickness); x-ray diffraction to identify crystalline constituents; helium pycnometry (Multipycnometer, Quantachrome Instruments, Boynton Beach, FL) to determine material density; and nitrogen gas adsorption (NOVA 2000e, Quantachrome) by the Brunauer, Emmett, and Teller (BET) method to determine specific surface area.

2.4. Electron spin resonance (ESR) studies

ESR spin trapping was used to detect both carbon radicals and short-lived free radical intermediates. Hydroxyl radicals were measured using the addition-type reaction of a short-lived radical with a paramagnetic compound (spin trap) to form a relatively long-lived free radical product (spin adduct), which can then be studied using conventional ESR. RAW 264.7 mouse peritoneal monocytes (ATTC, Rockville, MD, USA) were used because they respond to particle exposure in a manner similar to primary alveolar macrophages.[16,17] RAW 264.7 cells were cultured in Dulbecco's modified eagles medium (DMEM) with 10% fetal bovine serum, 2 mM L-glutamine, 100 units mL^{-1} penicillin, and 0.1 mg mL^{-1} streptomycin at 37 °C in a 5% CO_2 incubator. For each study powder, a sample of material collected in stage 1, 2, or 3 of the aerosol cyclone was weighed, mixed with artificial extracellular lung fluid, [18] vortexed for 10 sec, and then immediately pipetted into the reaction system of RAW 264.7 cells (1 x10^6) and DMPO spin trap (200 mM). The final powder concentration in each reaction system was 5 mg mL^{-1}. The cells were incubated for 3 min at 37 °C and then transferred to a flat cell for ESR measurement using a Bruker EMX spectrometer (Bruker Instruments Inc. Billerica, MA). Hyperfine couplings were measured (to 0.1 G) directly from magnetic field separation using potassium tetraperoxochromate (K_3CrO_8) and 1,1-diphenyl-2-picrylhydrazyl (DPPH) as reference standards.[19] The relative radical concentration was estimated by multiplying half the peak height by $(\Delta Hpp)^2$, where ΔHpp represents peak-to peak width. The Acquisit program was used for data acquisitions and analyses (Bruker Instruments Inc).

2.5. Artificial lung fluid solubility studies

A known mass of each powder was weighed on a 0.025-µm pore size 47-mm diameter mixed cellulose ester (MCE) filter (Millipore, Bedford, MA) and then covered with a second filter. The filter "sandwich" was secured tightly in a static dissolution chamber (InTox Products, Moriarity, NM) using four nylon screws on the chamber periphery, providing a particle-tight seal around the edge of the sandwich. Each chamber was placed in a polypropylene plastic beaker and immersed in 80 mL of artificial extracellular lung fluid (see section 2.5.1) or phagolysosomal simulant fluid (see section 2.5.2) and maintained at 37 °C for the duration of a static dissolution study. For each dissolution experiment, triplicate samples were prepared for each size fraction of each study powder. At each time point, the chamber was removed from its beaker, the fluid transferred to a borosilicate glass jar, and fresh equilibrated fluid was added to the sample beaker. All samples were stored frozen at -5 °C and then thawed prior to quantification of dissolved Co and W in accordance with U.S. OSHA Method 213 (inductively coupled plasma-atomic emission spectroscopy).

2.5.1. *Artificial extracellular lung fluid.* Artificial extracellular lung fluid termed serum ultrafiltrate (SUF) was prepared using the recipe described by Finch et al.[18] The pH of SUF was maintained at 7.3±0.1 for the duration of the dissolution experiments by blanketing the headspace in each sample beaker with CO_2 (5%) and air (95%). SUF contains 5 mmol L^{-1} of glycine, an important constituent in the dissolution of Co [20] and 1.2 mmol L^{-1} phosphate, which may react with dissolved cobalt to form insoluble cobalt phosphate.[21] Although endogenous to extracellular lung fluid, 50 ppm of ABDC was added to the SUF to prevent mold growth [16] during this short-term experiment and to evaluate whether this constituent could be included in long-term (i.e., 30 day) studies. To evaluate whether these fluid constituents influenced dissolution, experiments were performed using only bulk feedstock powder to evaluate the dissolution of Co in SUF (with and without glycine or phosphate) and to evaluate the dissolution of W, WC, and Co in SUF (with and without ABDC). Samples were collected at 1, 4, 8, 12, 18, 24, 36, 48, and 72 hr to measure the mass of dissolved Co or W.

2.5.2. *Artificial macrophage phagolysosomal fluid.* Artificial lung alveolar macrophage phagolysosomal fluid (PSF) was prepared using the recipe described by Stefaniak et al.[22] The pH of the PSF solution was maintained at 4.55±0.10 for the duration of the dissolution experiments. PSF contains 6 mmol L^{-1} of glycine, which is intended to represent the organic acid constituents present in phagolysosomal fluid, 1 mmol L^{-1} phosphate, and 50 ppm ABDC. As with SUF, experiments were performed using bulk feedstock powders to evaluate whether these fluid constituents influenced dissolution. Samples were collected at 1, 3, 6, 12, and 24 hr to measure the mass of dissolved Co or W.

3. Results and Discussion

As summarized in Table 1, feedstock powders were homogeneous, whereas the pre-sintered powder and the post-sintered CTC dust were heterogeneous mixtures of WC/Co powders. Cobalt was not identified on the surfaces of pre-sintered spray dryer powder, whereas W, C, and Co were each present on the surface of post-sintered chamfer grinder powder.

Table 1. Physicochemical properties of aerodynamically size-separated feedstock powders, process intermediary powders, and post-sintered cemented tungsten carbide powders.

Powder	Aerodynamic Size (μm)	Surface Constituents[a]	Crystalline Constituents[a]	Density (g/cm^3)	Specific Surface area (m^2/g)
Tungsten	>10	O, W, C	W	18.9±0.2	0.61±0.04
	4.3	O, W, C	W	19.8±0.3	0.87±0.05
	2.9	O, W, C	W	18.9±0.2	1.04±0.11
Tungsten carbide	>10	O, W, C	WC, W_2C	15.1±0.1	0.55±0.02
	4.3	W, O, C	WC, W_2C	15.6±0.1	0.65±0.03
	2.9	O, W, C	WC, W_2C	14.8±0.1	0.82±0.06
Cobalt	>10	O, Co	Co	7.7±0.0	6.05±0.87
	4.3	O, Co	Co	7.6±0.2	6.19±0.61
	2.9	O, Co	Co	7.7±0.1	6.02±0.35
Admixture	>10	O, W, C, Co	WC, W_2C, Co	13.8±0.1	0.74±0.08
	4.3	O, W, C, Co	WC, W_2C, Co	14.4±0.1	0.84±0.10
	2.9	O, W, C, Co	WC, W_2C, Co	13.9±0.1	1.14±0.05
Spray dryer	>10	C	WC, W, W_2C, Co	10.4±0.1	0.79±0.09
	4.3	C	--[b]	--	0.79±0.11
	2.9	C, O	--	--	1.43±0.08
Chamfer grinder	>10	C, O, W, Co	WC	12.9±0.1	0.49±0.07
	4.3	O, C, W, Co	--	--	0.61±0.09
	2.9	C, O, W, Co	--	--	1.11±0.05

[a] The order of constituents is from most to least abundant; O = oxygen, W = tungsten, C = carbon, WC = tungsten carbide, W_2C = tungsten carbide
[b] Insufficient mass of aerodynamically size-separated material available for this analysis

3.1. Results of electron spin resonance studies

On an equal powder mass basis, ROS generation in the RAW 264.7 cell line increased as particle size decreased for all powders except Co powder, which did not generate radicals (see figure 1). When dose was normalized to surface area, ROS generation was independent of particle size, which suggests that surface chemistry may be important for radical generation (see figure 2). On both a mass and surface area basis, chamfer grinder particles generated the highest levels of radicals. This observation may be explained as follows: during sintering, the Co binder phase becomes liquid and, upon cooling, the WC grains become embedded in the binder to form CTC. Thus, the more intimate chemical bonds among surface metals in heterogeneous post-sintered alloy material relative to the weaker and less extensive physical bonds encountered among particles in pre-sintered material appears important for ROS generation.

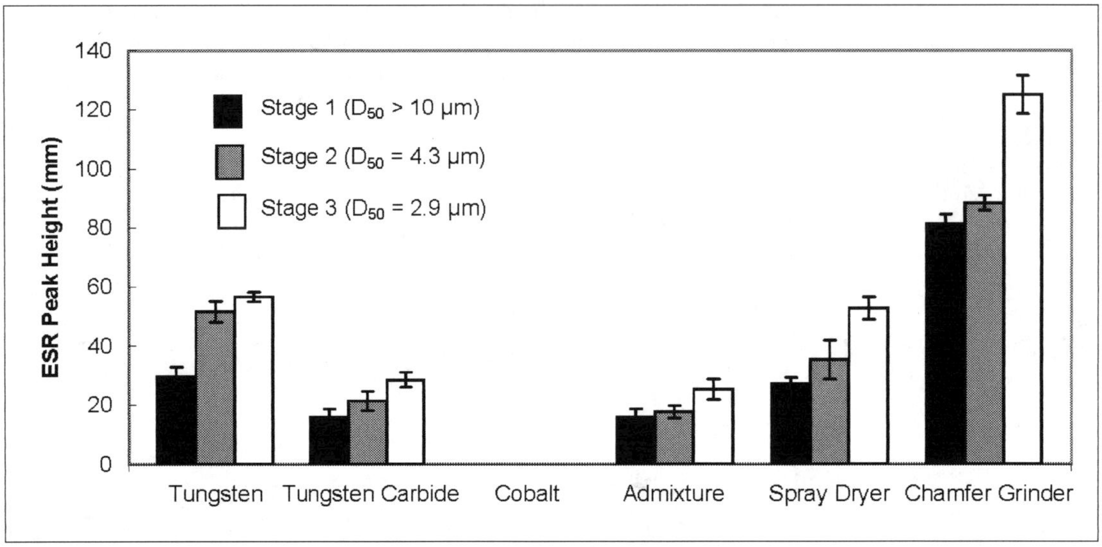

Figure 1. Plot of the influence of powder type on electron spin resonance (ESR) peak height in artificial extracellular lung fluid+RAW 264.7 cell line system for feedstock and process sampled powders. On a mass basis, all powders except cobalt metal exhibited a trend of increased radical production with decreased particle size. Cobalt metal powder alone did not generate radicals.

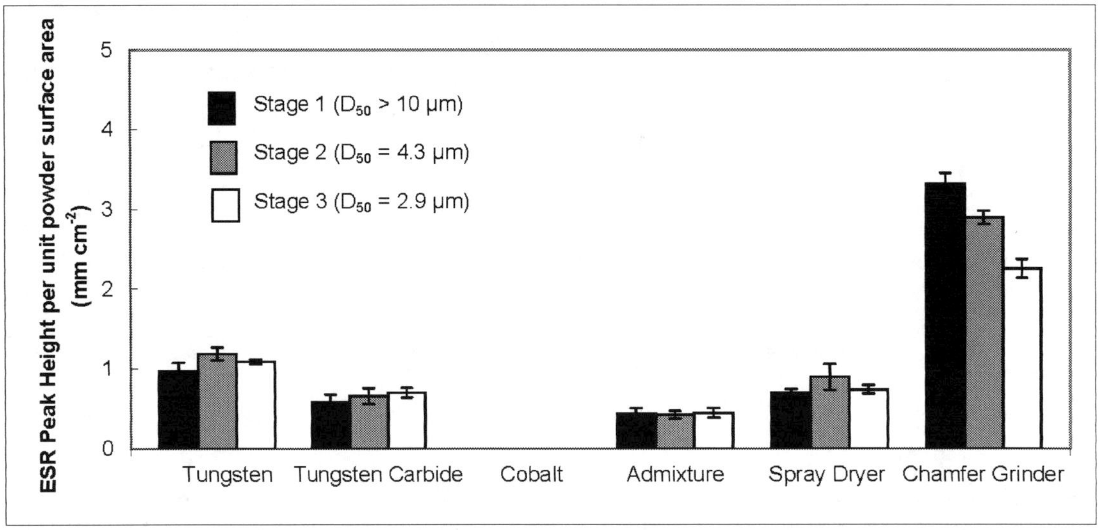

Figure 2. Plot of the influence of particle type on electron spin resonance (ESR) peak height in artificial extracellular lung fluid+RAW 264.7 cell line system for feedstock and process sampled powders. On a surface area basis, radical production was independent of particle size. Cobalt metal powder alone did not generate radicals.

3.2. Results of artificial lung fluid solubility studies

At the time of this writing, quantification of dissolved Co and W in samples of size-fractionated powders was in progress; therefore, only results from studies of bulk powders are reported herein. Figure 3 summarizes the total masses of metals dissolved in SUF without ABDC, glycine, or phosphate relative to SUF during a 72 hr exposure period. Dissolution of cobalt was higher ($p<0.05$)

in the absence of phosphate suggesting that cobalt ions may have reacted with phosphate to form cobalt phosphate $Co_3(PO_4)_2$, a very insoluble salt.[21]

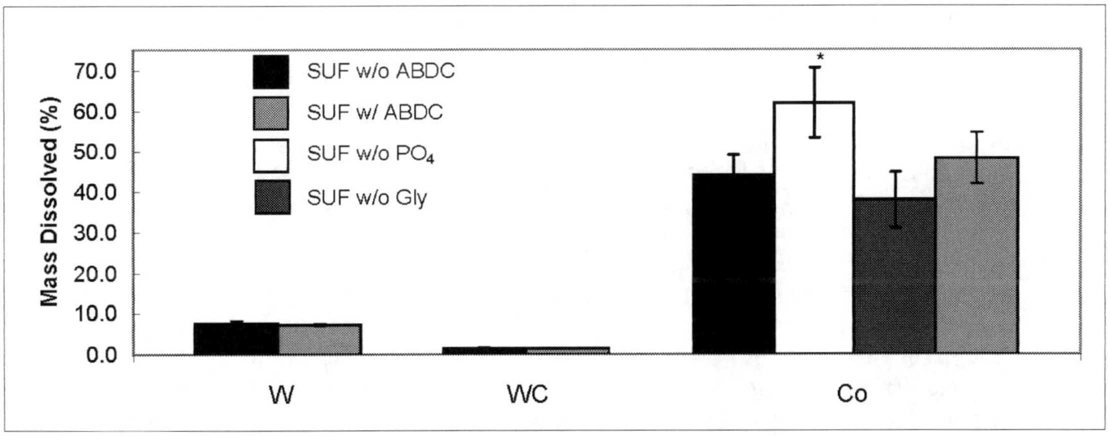

Figure 3. Total mass fraction of tungsten, tungsten carbide, and cobalt dissolved from artificial extracellular lung fluid (SUF) during a 72 hr exposure to solvent. * = mass fraction dissolved was significantly different ($p < 0.05$) from SUF.

Figure 4 summarizes the total masses of metals dissolved in PSF during a 24 hr exposure period. Dissolution of bulk W and WC powders were higher in the absence of ABDC. The phosphate concentration of PSF is similar to that of SUF; however, $Co_3(PO_4)_2$ did not form in PSF ($p=0.30$) which may be due its equilibrium solubility in acid.[23] The mass fraction of tungsten dissolved from W and WC powders was lower in PSF (24 hr exposure) relative to SUF (72 hr exposure). Some glass fibers are also known to be less soluble in macrophages relative to extracellular lung fluid.[24] These dissolution data for W and WC are consistent with results from microanalyses of lung tissue samples collected from CTC workers up to four years after cessation of exposure which demonstrated the presence of W but not Co in lung tissues.[25-27] Thus, W and WC particles that are deposited in lung and engulfed by macrophages may exhibit slower dissolution than if exposed to extracellular lung fluid alone for the same period.

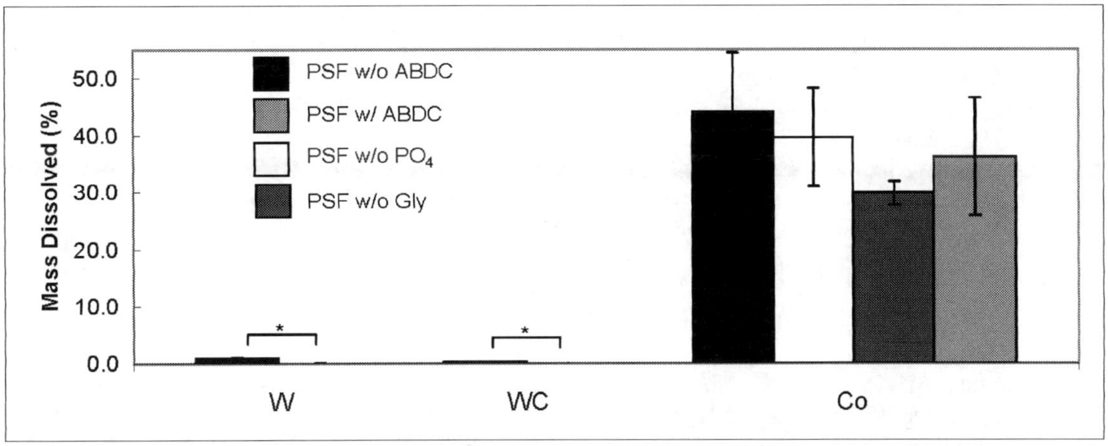

Figure 4. Total mass fraction of tungsten, tungsten carbide, and cobalt dissolved from artificial lung alveolar macrophage phagolysosomal fluid (PSF) during a 24 hr exposure to solvent. * = mass fraction dissolved was significantly different ($p < 0.05$) from PSF.

Collectively, the implication of these data for HMD is that the total exposure of the lung to ROS may be enhanced by retention of particles in macrophages. The extent of this effect will depend on the rate of physical clearance of engulfed particles from the lung by transport of macrophages up the muco-ciliary escalator, as compared to chemical dissolution and clearance of the particle constituents from the extracellular lung environment.

4. Summary

Size-dependent differences in ROS generation were observed for all study powders, except Co, which did not generate radicals in the cellular model. Chamfer grinder particles generated the highest levels of ROS in the RAW 264.7 cell line. Collectively, these results suggest an important role for particle physicochemical properties such as surface chemistry and bonding contact of Co and WC (i.e., pre- or post-sintered material) in generation of ROS. In artificial extracellular lung fluid and lung alveolar macrophage phagolysosome models, the rank order of dissolution from bulk feedstock powders was: Co> W> WC. The mass fractions dissolved were slower in PSF compared to SUF, suggesting that particles may persist longer in the lung if phagocytized and undergo dissolution than if cleared by chemical dissolution in the extracellular airway lining fluid. Results from studies of artificial lung fluid composition on observed dissolution rates indicate that careful attention to endogenous lung fluid constituents (e.g., phosphate) and exogenous model constituents (e.g., ABDC) is necessary when performing dissolution studies of W, WC, and Co powders.

References

[1] Schneider G 1989 *Principles of Tungsten Carbide Engineering* 2nd edition (Materials Park, OH: ASM International) pp 1-1 to 1-15
[2] Santhanam AT 1995 *Kirk-Othmer Encyclopedia of Chemical Technology* 4th edition vol 4 ed M Howe-Grant (New York: John Wiley & Sons, Inc.) pp 848-860
[3] Meyer-Bisch C, Pham QT, Mur J-M, Massin N, Moulin J-J, Teculescu D, Carton B, Pierre F, and Baruthio F 1989 *Br. J. Ind. Med.* **46** 302-09
[4] Sprince NL, Oliver LC, Eisen EA, Greene RE, and Chamberlin RI 1988 *Am. Rev. Respir. Dis.* **138** 1220-26
[5] Lison D, Carbonnelle P, Mollo L, Lauwerys R, and Fubini B 1995 *Chem. Res. Toxicol.* **8** 600-06
[6] Shirakawa T, Kusaka Y, Fujimura N, Goto S, Kato M, Heki S, and Morimoto M 1989 *Chest* **95** 29-37
[7] Kusaka Y, Iki M, Kumagai S, and Goto S 1996 *Occup. Environ. Med.* **53** 188-93
[8] Cirla AM 1994 *Sci. Total Environ.* **150** 85-94
[9] Johansson A, Lundborg M, Hellstrom P-A, Camner P, Keyser TR, Kirton SH, and Natusch DFS 1980 *Environ. Res.* **21** 165-76
[10] Stopford W, Turner J, Cappellini D, and Brock T *J. Environ. Monit.* **5** 675-80
[11] Shirakawa T, Kusaka Y, Fujimura N, Goto S, and Morimoto M 1988 *Clin. Allergy.* **18** 451-60
[12] Stefaniak AB, Day GA, Harvey CJ, Leonard SS, Schwegler-Berry DE, Chipera SJ, Sahakian NM, and Chisholm WP 2007 *Ind. Health.* **45** 793-803
[13] Smith WB, Wilson RR, and Harris DB 1979 *Environ. Sci. Technol.* **13** 1387-92
[14] Hoover MD, Castorina BT, Finch GL, and Rothenberg SJ 1989 *Am. Ind. Hyg. Assoc. J.* **50** 550-53
[15] Lippman M 1999 Rationale for particle size-selective aerosol sampling *Particle Size-Selective Sampling for Particulate Air Contaminants* ed Vincent JH (Cincinnati, OH: American Conference of Governmental Industrial Hygienists) p 18
[16] Leonard SS, Wang S, Shi X, Jordan BS, Castranova V, Dubick MA 2000 *Toxicology.* **150** 147-57
[17] Leonard SS, Roberts JR, Antonini JM, Castranova V, Shi X 2004 *Mol. Cell. Biochem.* **255** 171-

79
[18] Finch GL, Mewhinney JA, Eidson AF, Hoover MD, and Rothenberg SJ 1988 *J. Aerosol Sci.* **19** 333-42
[19] Janzen EG, Blackburn BJ 1968 *J. Am. Chem. Soc.* **90** 5909-5910
[20] Takahashi H and Koshi K 1981 *Ind. Health.* **19** 47-59
[21] Lison D and Lauwerys R 1992 *Toxicol. Letters.* **60** 203-10
[22] Stefaniak AB, Guilmette RA, Day GA, Hoover MD, Breysse PN, Scripsick RC 2005 *Toxicol. In Vitro.* **19** 123-34
[23] 2000 *CRC Handbook of Chemistry and Physics* 81st edition (Properties of the Elements and Inorganic Compounds) ed DR Lide (Boca Raton, FL: CRC Press) p 4-56
[24] Davis JMG 1994 *Environ. Health Perspect.* **102(Suppl 5)** 113-17
[25] Rüttner JR, Spycher MA, and Stolkin I 1987 *Br. J. Indust. Med.* **44** 657-60
[26] Abraham JL, Burnett BR, and Hunt A 1991 *Scanning Microsc.* **5** 95-108
[27] Auchincloss JH, Abraham JL, Gilbert R, Lax M, Henneberger PK, Hietzman ER, and Peppi DJ 1992 *Br. J. Ind. Med.* **49** 832-36

The role of macrophage mediators in respirable quartz-elicited inflammation

D van Berlo[1], C Albrecht[1], AM Knaapen[2]*, FJ van Schooten[2], and RPF Schins[1]

[1]Institut für Umweltmedizinische Forschung (IUF) at the Heinrich Heine University Duesseldorf, Germany.

[2]Nutrition and Toxicology Research Institute Maastricht (NUTRIM), Department of Health Risk Analysis and Toxicology, University of Maastricht, The Netherlands. * current address: Department of Toxicology and Drug Disposition, Schering-Plough, 5340 BH Oss, The Netherlands

roel.schins@uni-duesseldorf.de

Abstract. The instigation and persistence of an inflammatory response is widely considered to be critically important in quartz-induced lung cancer and fibrosis. Macrophages have been long recognised as a crucial player in pulmonary inflammation, but evidence for the role of type II epithelial cells is accumulating. Investigations were performed in the rat lung type II cell line RLE and the rat alveolar macrophage cell line NR8383 using Western blotting, NF-κB immunohistochemistry and qRT-PCR of the pro-inflammatory genes iNOS and COX-2, as well as the cellular stress gene HO-1. The direct effect of quartz on pro-inflammatory signalling cascades and gene expression in RLE cells was compared to the effect of conditioned media derived from quartz-treated NR8383 cells. Conditioned media activated the NF-κB signalling pathway and induced a far stronger upregulation of iNOS mRNA than quartz itself. Quartz elicited a stronger, progressive induction of COX-2 and HO-1 mRNA. Our results suggest a differentially mediated inflammatory response, in which reactive particles themselves induce oxidative stress and activation of COX-2, while mediators released from particle-activated macrophages trigger NF-κB activation and iNOS expression in type II cells.

1. Introduction

Exposure to high concentrations of respirable crystalline silica is well known to cause progressive fibrosis and lung cancer in rats. In humans, respirable quartz induces silicosis, a disease that is associated with an increased risk of lung cancer [1]. Based on chronic inhalation studies in the rat and epidemiological data, the ability of quartz to elicit a marked and persistent inflammation is believed to be a crucial factor for the development of these severe pathogenic conditions [2, 3].The cell types orchestrating the initial inflammatory response after particle exposure are thought to be the alveolar macrophage and epithelial cells, two main airway cell types likely to interact with inhaled particles; however, the contribution of each is not fully known and the mechanisms involved in their activation are at present not clearly understood. On a cellular level, the central regulator activating cells in response to quartz is considered to be transcription factor nuclear factor κB (NF-κB) [4]. In its dormant state, NF-κB, typically present as a dimer of the RelA (p65) and p50 subunits, is bound in the cytosol to its inhibitor protein, which is usually IκBα. In the classical NF-κB activation pathway, an

NF-κB activating stimulus reaches the cell, upon which the inhibitor protein IκBα is phosphorylated at serines 32 and 36, ubiquitinated at lysines 21 and 22, and subsequently degraded by the 26S proteasome. This liberates the NF-κB unit from the nuclear export signal of IκBα, and allows for its migration into the nucleus where it activates the transcription of many pro-inflammatory genes [4, 5]. Previously, we and others have shown that crystalline silica can activate the NF-κB pathway in vivo [6, 7, 8]. When considered together, these sudies are indicative of NF-κB activation within alveolar macrophages, but also in bronchiolar as well as alveolar epithelial cells. Macrophages have been long recognised as a crucial player in pulmonary inflammation, but evidence for the role of epithelial cells is accumulating [9]. Pro-inflammatory activation of lung epithelial cells in quartz-exposed lungs principally may involve two principal pathways, i.e. (1) direct activation of the epithelium by quartz particles or (2) activation of the epithelium by mediators released from quartz-interacting macrophages. Recently we provided further support for such an indirect mechanism by showing that bronchoalveolar lavage fluids from quartz-exposed rats can trigger NF-κB activation in lung epithelial cells in vitro [8]. The present work was undertaken to further investigate the role of macrophages in quartz-induced inflammation, focusing on quartz vs. macrophage mediated activation of the NF-κB pathway in lung epithelial cells. Moreover, we evaluated the mRNA expression of inducible Nitric Oxide Synthase (iNOS) and cyclooxygenase-2 (COX-2) because of their discussed involvement in (quartz-induced) lung disease [6, 10, 11, 12, 13], as well as of heme oxygenase-1 (HO-1) which is considered a senstive marker of particle-induced oxidative stress [13, 14].

2. Methods

For the present study, two cell types were used, i.e. the rat alvoeolar macrophage cell line NR8383 [15] and the rat lung epithelial cell line RLE-6TN (RLE) [16]. Dörentruper quartz (DQ12, batch 6 IUF, mean diameter of 0.96μm) was used as a standard respirable quartz sample. In order to inactivate any endotoxins that might be present at the particle surface, DQ12 was baked at 220°C for 24h. For cell treatments, DQ12 was suspended in reduced RLE medium (Ham's F12 containing 0.1% FCS, 1% γ-glutamine, 1% penicillin and 1% streptomycin) or normal NR8383 medium (F12-K with 15%FCS, 1% γ-glutamine, 1% penicillin and 1% streptomycin), respectively. Particle suspensions were prepared immediately prior to treatment and added to the cells after sonicated for 5 minutes in a water-bath sonicator (60Watt, 35Hz, Sonorex TK 52, Schaltech, Mörfelden-Walldorf, Germany). The experimental approach of the study is shown in Figure 1.

2.1. Culture and treatment of NR8383 cells and RLE cells

NR8383 macrophages were seeded in 60-mm culture dishes at a concentration of 0.5×10^6/ml and then cultivated for 3 days at 37°C and 5% CO until they reached 60% confluence. Subsequently, cells were treated with DQ12 at concentrations and time intervals as indicated in the results section. Control cells received fresh NR8383 medium. After incubation, supernatants were harvested for treatment of RLE cells. Obtained supernatants destined for RLE treatment were spun at low speed (200g for 5 min) to remove non-adherent cells, and spun once more at 18000g for 5 min to remove cell debris and particles. RLE cells were seeded in 60mm or 100mm culture dishes at a concentration of 3×10^4 cells/cm^2 and cultivated in respectively 2 or 6ml complete Ham's F12 medium. At 90% confluence, medium was replaced by reduced medium containing 0.1% FCS for 24h for cell synchronization. Cells were incubated for various time intervals with DQ12 particles at a concentration of 40μg/cm^2 (direct treatment), a concentration selected on the basis of prior concentration range finding experiments. To compare this direct treatment with the indirect, macrophage-mediated induction, cells were treated with 1:6 diluted supernatants of untreated or DQ12-treated NR838 cells. Recombinant rat Tumor Necrosis Factor-α (TNF-α) and Interleukin-1β (IL-1β), as wel as TNF-α and IL-1β neutralizing antibodies (R&D Systems, Germany) were used to determine the role of the inflammatory cytokines Tumor Necrosis Factor-α (TNF-α) and Interleukin-1β (IL-1β) on NF-κB pathway activation by DQ12 conditioned macrophage media.

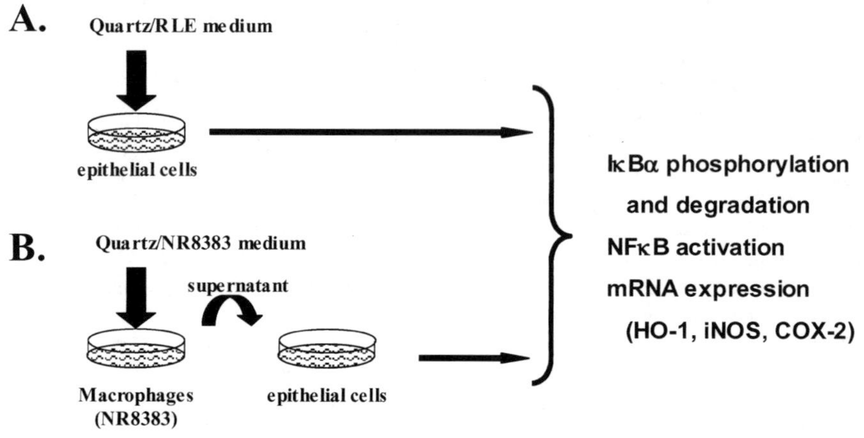

Figure 1. Experimental approach: RLE cells were exposed either to DQ12 quartz (A) or to the conditioned medium obtained from untreated or quartz treated NR8383 cells (B). For both models, the RLE cells were then evaluated for NF-κB activation by evaluation of phosphorylation and degradation of its cytosolic inhibitor IκBα and translocation of its RelA (p65) subunit, as well as for mRNA expression of iNOS, COX-2 and HO-1.

2.2. Immunocytochemistry

RLE cells were cultured to near confluence on 4-chamber culture slides (BD Falcon, Heidelberg Germany). After treatment, cells were rinsed twice with PBS and fixed in 4% paraformaldehyde/PBS (pH 7.4) followed by endogenous peroxidase inactivation with 0.3% H_2O_2 for 30 min. After permeabilisation in citrate buffer (pH 6.0) by microwave treatment, slides were blocked with normal goat serum (1:65) for 1 h to prevent nonspecific binding. Incubation with a rabbit polyclonal antibody against NF-κB RelA (Santa Cruz Biotechnology, USA) was followed by a secondary goat anti-rabbit fluorescent antibody (MFP555, 1:200, MoBiTec, Göttingen, Germany) and nuclear counterstaining with Hoechst 33342 (1μg/ml Sigma, Deisenhofen, Germany). After dehydration slides were covered in DePex (Serva, Heidelberg, Germany). As a negative control, slides of each treatment were incubated with rabbit IgG instead of the primary antibody at the same IgG concentration. Slides were analyzed using a fluorescence microscope (Olympus BX60) at an original magnification of ×1000.

2.3. Western blotting

RLE cells were harvested by gentle scraping and incubated for 30 min in 200μl lysis buffer (1% NP-40, 0.5% sodium deoxycholate, 0.1% sodium dodecyl sulfate in PBS) at 4°C containing freshly added protease inhibitors (Complete cocktail; Roche, Penzberg, Germany). Subsequently, lysates were centrifuged for 20min at 15000g and 4□ C. Protein concentrations were determined using the bicinchoninic acid (BCA) assay and Equal amounts (20μg) were electrophoresed in 10% sodium dodecyl sulfate (SDS)-polyacrylamide gels and transferred onto nitrocellulose membranes. Non-specific protein binding was blocked with 5% dried milk powder and 0.1% Tween-20 in PBS. IκBα protein was detected using a rabbit polyclonal antibody (1:1000, C-21, Santa Cruz Biotechnology) and an anti-rabbit-IgG whole protein HRP-conjugated secondary antibody (1:3000, Sigma). To detect IκBα phosphorylated at serine 32 and/or 36, a monoclonal antibody was used (1:2000) (ser32/36, clone 16A6, Cell Signaling Technologies, Danvers USA) in combination with an anti-mouse-IgG whole protein HRP-conjugated secondary antibody (1:2000, Sigma). Band formation was visualized using an ECL Plus reagent/detection system (Amersham Bioscience Europe, Freiberg, Germany).

2.4. Quantitative rtPCR analysis of gene expression in RLE cells

RLE cells were transferred into Trizol (Invitrogen, Karlsruhe Germany). Subsequently RNA was extracted according to the manufacturer's instructions. The RNeasy® mini kit coupled to DNAse treatment (RNAse-free DNAse set, Qiagen) was used to purify total RNA from salts and residual DNA. Quantity and purity of RNA were evaluated using spectrophotometry. cDNA was synthesized using the iScript cDNA Synthesis kit (BioRad, CA, USA), starting from 0.5 µg of RNA. cDNA was diluted 15x in water before use. rtPCR primers for rat HO-1, iNOS, COX-2 and the housekeeping gene HPRT were designed using Primer express software (Applied Biosystems) and obtained from Operon. Primer sequences for iNOS were 5'-AGG AGA GAG ATC CGG TTC ACA GT-3' (forward) and 5'-ACC TTC CGC ATT AGC ACA GAA-3' (reverse), for COX-2 5′-GCA CAA ATATGA TGT TCG CAT TCT-3′ (forward) and 5′-GAA CCC AGG TCC TCG CTT CT-3′ (reverse), for HO-1 5'-GGG AAG GCC TGG CTT TTTT -3' (forward) and 5'-CAC GAT AGA GCT GTT TGA ACT TGGT -3' (reverse), and for HPRT 5'-GCC CTT GAC TAT AAT GAG CAC TTC A-3' (forward) and 5'-TCT TTT AGG CTT TGT ACT TGG CTT TT-3' (reverse). Real time PCR was performed with a MyiQ Single Colour real time PCR detection system (BioRad) using SYBR© Green Supermix (Biorad, CA, USA.), 5 µl diluted cDNA, and 0.3 µM primers in a total volume of 25 µl. PCR was conducted as follows: a denaturation step at 95°C for 3 min was followed by 40 cycles at 95°C (15 seconds) and 60°C (45 seconds). Melt curves (60-95°C) were produced for product identification and purity. PCR efficiency of all three primer sets, as assessed by the use of cDNA dilution curves, was 90-100%. Data were analyzed using the MyiQ Software system (BioRad) and were expressed as relative gene expression (fold increase) using the 2-ΔΔCt method [17].

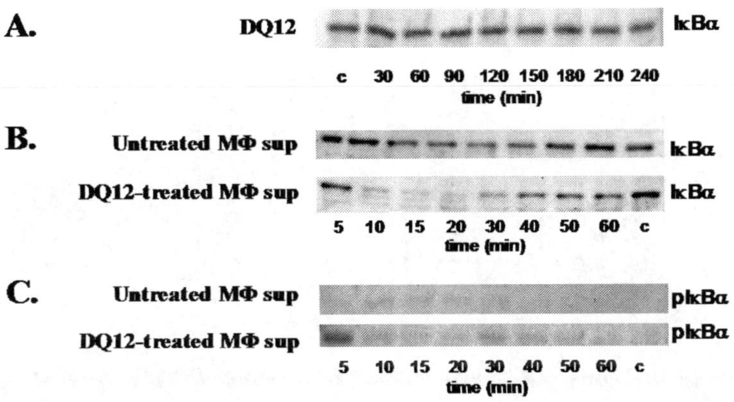

Figure 2. Western blots of IκBα degradation and phosphorylation in RLE cells. A: Effect of DQ12 on IκBα degradation in RLE cells. Cells were treated with 40µg/cm2 DQ12 for the indicated time intervals. B: Effects of conditioned medium from untreated vs. DQ12-treated NR8383 macrophages on IκBα degradation (B) and IκBα Ser-32/36-phosphorylation. RLE cells were treated for the indicated time intervals with a 1:6 dilution of supernatants from untreated NR8383 cells or NR8383 cells that were treated for 24h with 40 µg/cm^2 DQ12.

3. Results

To compare NF-κB activation in RLE cells induced by DQ12 particles (direct effect) with that as induced by the supernatants of quartz treated macrophages (indirect, macrophage-mediated effect), we first investigated IκBα phosphorylation and degradation by Western blotting analysis (Figure 2). Direct treatment of RLE cells with quartz, did not result in a significant degradation of IκBα degradation (Fig. 2A). In concordance with these findings no phosphorylation of IκBα could be detected (not shown). In contrast, conditioned media obtained from quartz treated NR8383 cells were found to cause a fast and strong degradation of IκBα (Fig. 2B). The degradation of this NF-κB inhibitor protein was found to be preceded by its phosphorylation (Fig. 2C). The fast phosphorylation and degradation of IκBα was followed by a rapid restoration (30 to 60 min), indicative of its de novo synthesis. In RLE cells that were treated with supernatants from untreated NR8383 cells also a mild IκBα reduction was observed, but in the absence of any detectable phosphorylation (Figs 2B and C).

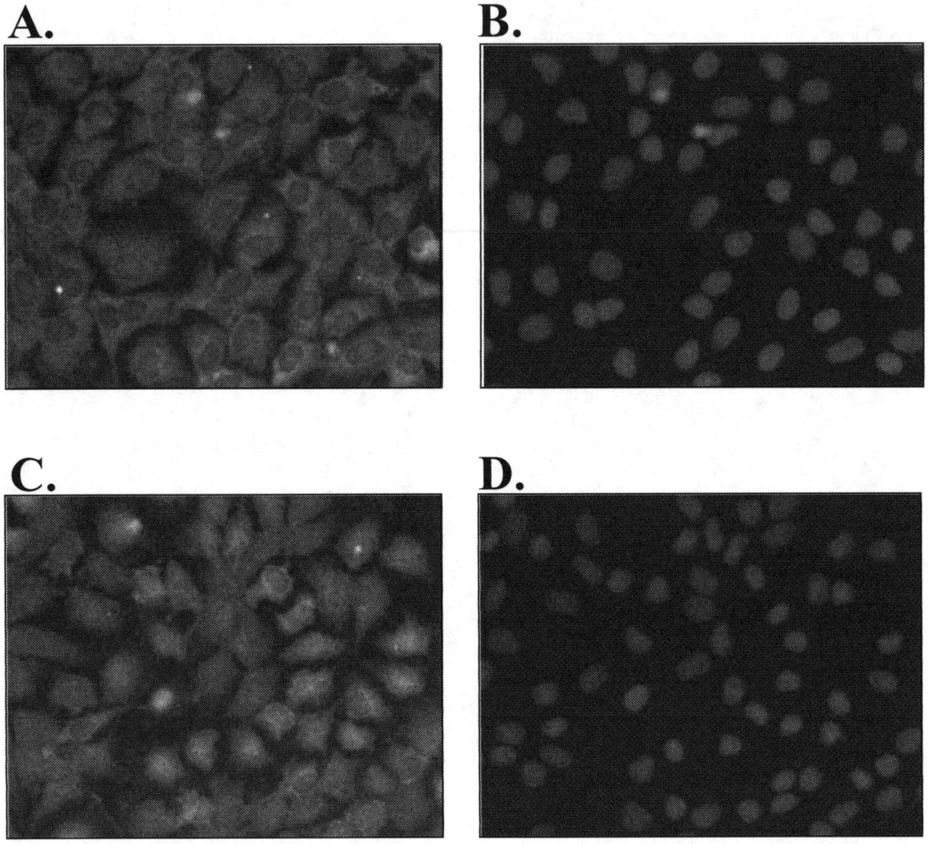

Figure 3. Immunocytochemical analysis of the nuclear translocation of the NF-κB subunit RelA in RLE lung epithelial cells after treatment with conditioned medium from untreated NR8383 cells (A, B) or conditioned medium from DQ12-treated NR8383 cells (C, D). Cells were treated for 10 minutes with a 1:6 dilution of supernatants obtained from untreated NR8383 cells or from NR8383 cells that were treated for 6 hours with $40 \mu g/cm^2$ DQ12. Staining for RelA is shown in panels A and C. For the verification of its nuclear translocation, counterstaining with Hoechst 33342 was used, shown in panels B and D for the respective treatments.

Supernatants from DQ12 treated macrophages were also found to elict a marked nuclear translocation of of RelA in RLE cells, while superantants from untreated NR8383 cells failed to cause such clear nuclear translocation (Figure 3). These fidings were in concordance with their respective effects on phosphorylation and degradatation of IκBα as observed obtained in the Western blot analysis (Figure 2). In RLE cells that were treated directly with DQ12 quartz, RelA nuclear translocation was also absent (not shown). When taken together, the Western blot and immunocytochemical analysis suggest that NF-κB activation induced in RLE cells by quartz, occurs via the indirect mechanism, mediated by macrophage products, rather than directly through the quartz particles themselves. The data also indicate that this NF-κB activation occurs through the classical, IκB-degradation-mediated pathway, involving phosphorylation and degradation of the most common IκB molecule, IκBα.

The potential role of TNFα and/or IL-1β in the indirect, macrophage mediated NF-κB activation we investigated the effects of these cytokines on RLE cells combined with analysis of effects of their inhibitors. As shown in Fig. 4, TNFα was induced a clear IκBα degradation followed by a rapid restoration. However, anti-TNFα neutralizing antibodies failed to block activation of the classical NF-κB pathway in RLE cells as induced by supernatants from DQ12-treated macrophages, while completely blocking the effect of TNFα (Figure 4). Such findings were also obtained for IκBα ser32/36 phosphorylation and RelA nuclear translocation, as well as in further blocking experiments applying IL-1β neutralizing antibody or even a combination of both TNFα and IL-1β inhibitors (not shown). Taken together, these results indicated that the acute inflammatory cytokines TNFα and IL-1β do not play a crucial role in the indirect macrophage-mediated NF-κB activation in RLE cells.

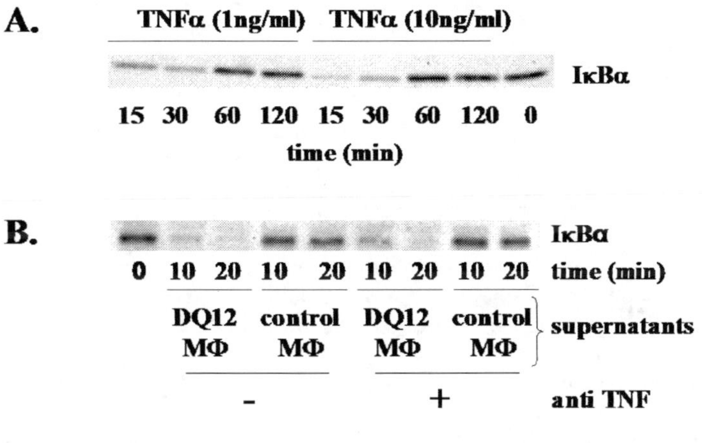

Figure 4. Western blot analyses on the investigation of the involvement of TNFα in indirect, macrophage mediated IκBα degradation in RLE cells. A. Degradation of IκBα degradation in RLE cells upon treatment with TNFα at indicated concentration and time points. B. Expression of IκBα protein in RLE cells after treatment with NR8383 supernatants that were pretreated with a TNFα neutralizing antibody. For these experiments, supernatants from untreated as well as DQ12-treated NR8383 macrophages were incubated for 1h with anti-TNFα at a concentration of 5μg/ml and then added to RLE cells in a 1:6 dilution..

To further investigate and compare the direct versus indirect pro-inflammatory effects of DQ12 quartz on lung epithelial cells, RLE cells were also evaluated on the mRNA expression levels for of iNOS, COX-2 and HO-1 by quantitative rtPCR. The findings of these experiments are summarised in table 1. The supernatants from the DQ12 treated NR8383 macrophages were found to be capable of inducing iNOS almost 2000-fold in the RLE cells, whereas direct DQ12 treatment elicited an approximate 50-fold increase of the same transcript. In contrast, COX-2 mRNA expression was found to be induced more strongly in the RLE cells upon direct treated with DQ12 than upon treatment with the macrophage supernatants. In line with COX-2, also HO-1 mRNA expression was found to be relatively stronger increased in the RLE cells upon direct quartz treatment (see Table 1).

Table 1. Induction of HO-1, iNOS and COX-2 mRNA expressions, in RLE lung epithelial cells after treatment with treatment with DQ12 quartz versus DQ12-treated NR8383 macrophages.

	iNOS (fold induction)	**COX-2 (fold induction)**	**HO-1 (fold induction)**
Treatment with conditioned media from DQ12-treated MΦ	1845.8 ± 1038.9 (2h)	12.9 ± 8.1 (2h)	2.7 ± 1.7 (2h)
Treatment with DQ12	55.0 ± 12.4 (8h)	164.9 ± 10.5 (4h)	15.7 ± 4.5 (8h)
ratio of fold induction (MΦ supernatants vs. DQ12)	*33*	*0.08*	*0.17*

The fold induction compared to untreated cells is shown upon treatment of RLE cells with DQ12 for 4 or 8 hours with supernatants from DQ12-treated NR8383 cells for 2 hours. The selected time points were those showing the highest induction (indicated between brackets) of the respective genes as determined from the kinetic investigations (2 - 8 hours).

4. Discussion and conclusion

In this study we found that conditioned media from quartz-treated macrophages activated the NF-κB signalling pathway and induced a far stronger upregulation of iNOS mRNA than quartz itself in epihelial cells. In contrast, quartz particles elicited a stronger, progressive induction of COX-2 and HO-1 mRNA, in the absence of any notable NF-κB activation. The gene encoding iNOS is known to be NF-κB regulated [18], and has been implicated in silica induced inflammation and fibrosis in the basis of studies with iNOS knockout mice [10, 11]. Our current in vitro findings show that the effect of macrophage products secreted in response to quartz on NF-κB activation in alveolar epithelial cells is greater than the effect of the quartz particles directly, providing further support for the macrophage as crucial mediator of (quartz-induced) inflammation [9, 19, 20]. Our findings also showed that TNFα and IL-1β do not play a crucial role in quartz-induced NF-κB activation in RLE cells, but that other factors are at least involved or possibly more important.

The observed induction of COX-2 mRNA expression by DQ12, however, also suggests a role for direct particle-epthelial interactions in triggering pro-inflammatory effects. COX-2 is involved in the pathogenesis of various inflammatory lung diseases [21]. Although COX-2 is known to be regulated by NF-κB [22], in the present study DQ12 treatment did not lead to any notable induction of this pathway in the RLE cells. Since this treatment also induced the mRNA expression of the oxidative stress marker HO-1, it appears that quartz particles induce COX-2 expression via the activation of NF-κB independent, yet oxidative stress-dependent, cell signalling. Finally, the observed profiles of HO-1 mRNA expression in our study is also of interest for two further reasons. On the one hand, the failure of macrophage mediators to strongly activate this gene indicates that the indirect (macrophage mediated) epithelial NF-κB activation occurs rather in an oxidative stress-independent manner. On the other hand, the ability of quartz to directly activate HO-1 in epithelial cells provides further support for direct quartz-driven pro-inflammatory effects, as silica-induced inflammation in mice was recently shown to depend on HO-1 expression in the lung [23]. In conclusion, our results suggest a differentially mediated inflammatory response, in which reactive particles themselves induce oxidative stress and activation of COX-2, while mediators released from particle-activated macrophages trigger NF-κB activation and iNOS expression in type II cells.

Acknowledgements.
This study was supported by a grant from the German Federal Ministry of Environment (BMU)

References

[1] Steenland K, 't Mannetje A, Boffetta P, Stayner L, Attfield M, Chen J, Dosemeci M, DeKlerk N, Hnizdo E, Koskela R and Checkoway H 2001 Cancer Causes Control **12** 773
[2] Kuempel ED, Attfield MD, Vallyathan V, Lapp NL, Hale JM, Smith RJ and Castranova V 2003 J. Biosci **28** 61
[3] Driscoll KE 1996 Inhal. Toxicol. **8** 139
[4] Chen F and Shi X 2002 **234-235**(1-2) 169
[5] Schins RPF and Donaldson K 2000 Inhal. Toxicol. **12** 317
[6] Porter DW, Ye J, Ma J, Barger M, Robinson VA, Ramsey D, McLaurin J, Khan A, Landsittel D, Teass A and Castranova V 2002 Inhal Toxicol **14** 349
[7] Hubbard AK, Timblin CR, Shukla A, Rincón M and Mossman BT 2002 Am. J. Physiol. Lung. Cell. Mol. Physiol. **282** 968
[8] Albrecht C, Schins RPF, Höhr D, Becker A, Shi T, Knaapen AM and Borm PJA 2004 Am. J. Respir. Cell. Mol. Biol. **31** 292
[9] Cheng DS, Han W, Chen SM, Sherrill TP, Chont M, Park GY, Sheller JR, Polosukhin VV, Christman JW, Yull FE and Blackwell TS 2007 J. Immunol. **178** 6504
[10] Srivastava KD, Rom WN, Jagirdar J, Yie TA, Gordon T and Tchou-Wong KM 2002 Am. J. Respir. Crit. Care. Med **165** 527
[11] Zeidler P, Hubbs A, Battelli L and Castranova 2004 J. Toxicol. Environ. Health. A. **67** 1001
[12] Choi JK, Lee SG, Lee YJ, Nam HY, Lee WK, Lee KH, Kim HJ and Lim Y 2005 J. Environ. Pathol. Toxicol. Oncol. **24** 163
[13] Li H, van Berlo D, Shi T, Speit G, Knaapen AM, Borm PJ, Albrecht C and Schins RPF 2008 Toxicol. Appl. Pharmacol. **227** 115
[14] Suzuki K and Hei TK 1996 Carcinogenesis **17** 661
[15] Helmke RJ, Boyd RL, German VF and Mangos JA 1987 In Vitro Cell. Dev. Biol. **23** 567
[16] Driscoll KE, Carter JM, Iype PT, Kumari HL, Crosby LL, Aardema MJ, Isfort RJ, Cody D, Chestnut MH, Burns JL and LeBoeuf RA 1995 In Vitro Cell. Dev. Biol. Anim. **31** 516
[17] Livak KJ, Schmittgen TD 2001 Method. Methods **25** 402
[18] Taylor BS, de Vera ME, Ganster RW, Wang Q, Shapiro RA, Morris SM Jr, Billiar TR and Geller DA 1988 J. Biol. Chem. **273** 15148
[19] Lentsch AB, Czermak BJ, Bless NM, Van Rooijen N and Ward PA 1999 Am. J. Respir. Cell Mol. Biol. **20** 692
[20] Elder AC, Gelein R, Oberdörster G, Finkelstein J, Notter R, and Wang Z 2004 Exp. Lung Res. **30** 105
[21] Park GY and Christman JW 2006 Am. J. Physiol. Lung. Cell. Mol. Physiol. **290** 797
[22] Jobin C, Morteau O, Han DS, and Balfour Sartor R 1998 Immunology **95** 537
[23] Sato T, Takeno M, Honma K, Yamauchi H, Saito Y, Sasaki T, Morikubo H, Nagashima Y, Takagi S, Yamanaka K, Kaneko T and Ishigatsubo Y 2006 Am. J. Respir. Crit. Care. Med. **174** 906

Journal of Physics: Conference Series **151** (2009) 012047 doi:10.1088/1742-6596/151/1/012047

Building an industry-wide occupational exposure database for respirable mineral dust – experiences from the IMA Dust Monitoring Programme

Remko Houba[1], Jelle Vlaanderen[2], Richard Jongen[1] and Hans Kromhout[2]

[1] Arbo Unie, Expert Centre for Chemical Risk Management, Nijmegen, The Netherlands

[2] Institute for Risk Assessment Sciences, Utrecht University, The Netherlands

Corresponding author: h.kromhout@uu.nl

Abstract: Building an industry-wide database with exposure measurements of respirable mineral dust is a challenging operation. The Industrial Minerals Association (IMA-Europe) took the initiative to create an exposure database filled with data from a prospective and ongoing dust monitoring programme that was launched in 2000. More than 20 industrial mineral companies have been collecting exposure data following a common protocol since then. Recently in 2007 ArboUnie and IRAS evaluated the quality of the collected exposure data for data collected up to winter 2005/2006. The data evaluated was collected in 11 sampling campaigns by 24 companies at 84 different worksites and considered about 8,500 respirable dust measurements and 7,500 respirable crystalline silica. In the quality assurance exercise four criteria were used to evaluate the existing measurement data: personal exposure measurements, unique worker identity, sampling duration not longer than one shift and availability of a limit of detection. Review of existing exposure data in the IMA dust monitoring programme database showed that 58% of collected respirable dust measurements and 62% of collected respirable quartz could be regarded as 'good quality data' meeting the four criteria mentioned above. Only one third of the measurement data included repeated measurements (within a sampling campaign) that would allow advanced statistical analysis incorporating estimates of within- and between-worker variability in exposure to respirable mineral dust. This data came from 7 companies comprising measurements from 23 sites. Problematic data was collected in some specific countries and to a large extent this was due to local practices and legislation (e.g. allowing 40-h time weighted averages). It was concluded that the potential of this unique industry-wide exposure database is very high, but that considerable improvements can be made. At the end of 2006 relatively small but essential changes were made in the dust monitoring protocol and the data collection sheet. In addition a system of quality control was set up and each new set of data is thoroughly investigated before inclusion into the database. Recently, it became apparent that more than 80% of the measurement data collected since winter 2005/2006 is of high quality. The IMA Dust Monitoring Programme Database contains personal measurements of more than 2,000 monitored workers who are representative of in total 5,000 workers in the industrial minerals production. This unique prospective exposure database will prove to be very valuable when health effects due to exposure to respirable mineral dust among these workers will be evaluated in the future.

1. Introduction

Building an industry-wide database with occupational exposure measurements of silica dust is a challenging operation. The Industrial Minerals Association (IMA-Europe) took the initiative to create an exposure database filled with data from a prospective and ongoing dust monitoring programme that was launched in 2000.

The IMA hygiene or dust monitoring programme aims at collecting representative and comparable dust exposure data at the level of the industrial minerals sector across Europe to be able to discuss on both industry level and on company level on the compliance with current occupational exposure limits, to develop prevention strategies to reduce exposure, to check the efficiency of implemented control measures (e.g. by studying time trends in silica dust exposure), and to be able to discuss with authorities on setting future occupational exposure limits. In addition, the exposure database could be used in the future to elaborate a job-exposure matrix for a potential epidemiological study of health effects of respirable crystalline silica.

This extended abstract describes the requirements, experience, and first results of the IMA dust monitoring database.

2. Methods

2.1. Dust monitoring protocol minimal requirements

The occupational exposure database should only contain comparable and high quality data. Therefore, all data should comply with minimal requirements to be included in the database:

- Collect 6 dust samples per job function per site per campaign (the number of six samples was chosen as a compromise between statistical needs and practical feasibility for participating companies)
- The protocol includes a list of thirteen common job functions in the industrial minerals sector (see table 1)
- Recommendation to start with two campaigns per year (Summer and Winter);
- All data should be based on personal monitoring only;
- Respirable dust fraction;
- Sampling duration should correspond to a full shift (6-8 hours)
- Allocate a unique code to each individual worker, to allow appropriate statistical analysis of time trends and cumulative exposure following repeated measurements;
- Sampling equipment must be in conformity with the European standard EN 481. Several respirable dust samplers are allowed by the protocol (e.g. CIP-10, Respicon particle sampler, Dorr-Oliver 10 mm nylon cyclone, Higgins-Dewell cyclone). Based on an international comparison study for respirable dust samplers, the allowance of different sampling equipment will introduce only small differences in absolute measured values [1]. In addition, corrections are possible in the statistical analyses by introducing sampling equipment as fixed effect in the linear mixed models;
- Provide field blanks for each sampling campaign (for quality control and for calculation of limits of detection);
- Analytical technique: Either X-ray diffraction or Fourier Transform Infrared Spectroscopy. Information on the limit of detection of the analytical techniques should be available and the laboratories involved should join an inter-laboratory round exercise;
- Record observations of work activities and factors affecting exposure during each sampling period;
- Transmit the data through a standardised MS Excel® collection sheet with standardized variable codes

A detailed protocol on all these items is available for participating companies and periodically training sessions have been and are organized for participants to learn and discuss the dust monitoring protocol.

Table 1. List of job functions within the protocol

	Job function
1	Quarry operator
2	Crusher operator
3	Wet process operator
4	Dry process operator
5	Miller operator
6	Bagging operator
7	Transport / bulk loading
8	Foreman / plant management staff
9	Maintenance
10	Multi-skilled operator
11	Laboratory workers
12	Research & Development
13	Plastification

2.2. Quality control procedure

Before inclusion in the final database, all data is extensively checked to verify if data meet all requirements and if all information is complete and internally consistent. If relevant, this step is followed by communication with company representatives and changes to the data will be made, if needed. The quality control procedure was intensified during the 11th campaign, when Arbo Unie and IRAS took over the management of the Dust Monitoring Programme.

2.3. Participation

A total of 85 different worksites of 24 industrial mineral companies from 13 countries (Belgium, Denmark, Finland, France, Germany, Greece, Italy, the Netherlands, Norway, Portugal, Spain, Sweden and the United Kingdom) have submitted exposure monitoring data since the start of the project. This represents a population of approximately 5,000 workers, whose job functions are covered by the protocol. The companies produce bentonite, clays, feldspar, kaolin, silica, talc, and other industrial minerals. Several participants were small and medium sized enterprises (SME's). Both large and small companies very well managed to run the sampling campaigns within the programme.

2.4. Feed back of sampling results to companies and industry

All companies are receiving site-specific reports per campaign containing the statistical evaluation of the measurements to check with national exposure limits, to identify problem areas and job titles with high exposure levels within their company in order to develop prevention strategies to lower exposure levels. This feed-back on company level is one of the benefits stimulating companies to participate. Once every year an industry-wide report is being produced for IMA-Europe (and the participating companies) containing more sophisticated analyses on all available data. Importantly, however, all company data are treated confidentially and individual companies cannot be identified in this report.

3. Results

3.1. Available data and data quality

Recently in 2008 Arbo Unie and IRAS evaluated the quality of collected exposure data up to summer 2007 (14 sampling campaigns). A substantial body of data has been collected since the first campaign in 2002: a total of 10,207 respirable dust samples, with 8,533 of these being analysed for crystalline silica are present in the IMA-DMP database. Approximately 1,000 – 1,500 new personal exposure measurements are currently being collected annually. Table 2 present an overview of the number of samples per country currently available.

Review of existing exposure data in the IMA dust monitoring programme database showed that 83% of collected respirable dust measurements and 80% of collected respirable quartz could be regarded as 'good quality data' meeting at least three crucial quality criteria (personal exposure measurement, sampling duration in accordance with protocol requirements and availability of information on limit of detection). The percentage of 'good quality data' considerably improved over time after the introduction of the strict quality control procedures. Problematic data was collected in some specific countries and to a large extent this was due to local practices and legislation (e.g. allowing 40-h time weighted averages).

Only 34% of the measurement data included repeated measurements of identifiable workers (within a sampling campaign) that would allow advanced statistical analysis incorporating estimates of within- and between-worker variability in exposure to respirable (crystalline silica) dust. This data came from 18 companies comprising measurements from 10 countries. Due to the improved quality control procedures, exposure data suitable for advanced statistical analysis has increased sharply to 85% in the last two campaigns.

Table 2. Number of data in the IMA DMP database per country

	Respirable dust	Respirable quartz
Belgium	841	623
Denmark	26	26
Finland	209	115
France	2,335	1,949
Germany	1,287	1,287
Greece	87	55
Italy	440	327
The Netherlands	834	395
Norway	110	103
Portugal	99	99
Spain	829	550
Sweden	108	108
United Kingdom	3,002	2,896
TOTAL	**10,207**	**8,533**

3.2. Trends in exposure concentrations over time

Time trends were studied on a subset of the total database (about one-third of the data), using data that met all the necassary requirements (personal sampling, sampling duration, unique worker codes available, ≥ 5 samples per cell, ≥ 2 workers per cell, and number of observations divided by number of workers ≥ 1.25). Time trends were studied using linear mixed effects models. For several minerals (e.g. silica, clay, mixed minerals), considerable and statistically significant downward trends (up to

14.5% decrease in exposure per campaign) in both respirable dust and respirable quartz concentrations could be found, As an example, the trend in respirable dust concentrations is showed in figure 1, for sites working with the mineral silica. The estimated time trend per campaign was -13.0 % (95%-confidence interval: 11.6% - 14.5%).

The number of groups of workers (defined by function and site) with overexposure were a factor of three higher for respirable quartz than for respirable dust, indicating that achieving compliance with existing OEL's for respirable quartz is more difficult than achieving compliance with OEL's for respirable dust. Interestingly, this number has declined rapidly in the first part of the programme and leveled off in the more recent campaigns.

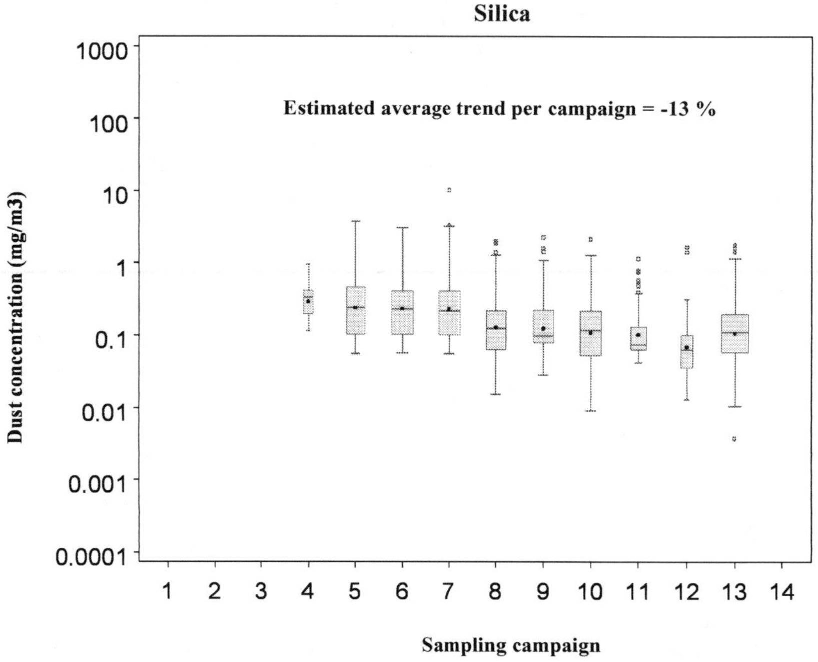

Box plots represent distribution, median and mean of observed dust concentrations. The estimated time trend in dust concentrations is based on the observed dust concentrations and corrected for influences of site and job title.

Figure 1. Trend in respirable dust concentrations for mineral silica

4. Conclusions

It was concluded that the potential of this unique industry-wide exposure database is very high. Both large companies and SME's are able to make contributions to the exposure database. Importantly, however, a stringent quality control of incoming data is crucial. At the end of 2006 relatively small but essential changes were made in the dust monitoring protocol and the data collection sheet. In addition, a system of quality control was set up and each new set of data is thoroughly investigated before inclusion into the database. Recently, it became apparent that more than 80% of the measurement data collected since winter 2005/2006 is of high quality.

The IMA Dust Monitoring Programme Database contains personal measurements of more than 2,000 monitored workers who are representative for a total of 5,000 workers in the industrial minerals production in Europe. This unique prospective exposure database will prove to be very valuable when health effects due to exposure to respirable crystalline silica among these workers will be evaluated in the future. This study shows that it is possible to build a multi-national industry-wide exposure database in which data from SMEs is included which may serve as an example for other industries.

References

[1] Gömer P, Wrobel R, Micka V, Skoda V, Denis J and Fabriès 2001 Study of fifteen respirable aerosol samplers used in occupational hygiene *Ann Occup Hyg* **45** 43-45

Sanding dust from nanoparticle-containing paints: physical characterisation.

I K Koponen, K A Jensen, T Schneider

National Reseach Centre for Working Environment (NRCWE), Lersø Park Alle 105, DK-2100 Copenhagen, Denmark

E-mail : ikk@nrcwe.dk

Abstract. Increasing use of nanoparticles in different industrial applications has raised a new potential health risk to the workers as well as to the consumers. This study investigates the particle size distributions of sanding dust released from paints produced with and without engineered nanoparticles. Dust emissions from sanding painted plates were found to consist of five size modes; three modes under 1 μm and two modes around 1 and 2 μm. We observed that the sander was the only source of particles smaller than 50 nm and they dominated the number concentration spectra. Mass and surface area spectra were dominated by the 1 and 2 μm modes. Addition of nanoparticles caused only minor changes in the geometric mean diameters of the particle modes generated during sanding of two paints doped with 17 nm TiO2 and 95 nm Carbon Black nanoparticles as compared to the size modes generated during sanding a conventional reference paint. However, the number concentrations in the different size modes varied considerably in between the two NP-doped paints and the reference paint. Therefore, from a physical point of view, there may be a difference in the exposure risk during sanding surfaces covered with nanoparticle-based paints as compared to sanding conventional paints.

1. Introduction

Nanotechnology enables the design of particle-based chemicals and materials at the molecular and atomic scale. These new nanomaterial-based products have tremendous potential benefits for the society. However, nanomaterials are also associated with significant uncertainties regarding possible hazards to human health and the environment. There is evident risk of exposure to persons who produce or handle nanoparticles (NPs) in their production. The paint and lacquer industry is a potential high tonnage user of NPs. In paint and lacquers vast amounts of conventional particulate, pigments, fillers and binders are used. The new use of NPs has been aiming at developing special paint properties, e.g. water repellence, scratch resistance and improved durability. Direct or indirect nanoparticle exposure during production and application of this product group as well as during subsequent finishing and renovation may reach a large number of persons in the future [1].

Sanding painted surfaces is often one of the major activities when renovating buildings. Sanding produces a large amount of dust, and could result in an extensive exposure to NP containing dust particles or even individual nanoparticles released from the paint and lacquer matrix during sanding. Therefore, more detailed information on the particle sizes of sanding dust and whether NPs are changing the physico-chemical properties of sanding dust is needed in order to assess the potential health risks from exposure by inhalation.

In the past few years, numerous studies have investigated and discussed the potential health risks of nanomaterials [e.g. 2, 3]. Fewer studies have studied the exposure risks during handling nanoparticles [4,5,6,7,8]. There seems to be no studies available characterizing the potential exposure to nanoparticles from sanding nanoparticle-doped materials. Nevertheless, a few scientific studies have characterised the exposure to dust from sanding conventional materials in the field as well as in the laboratory. Scholz et al. [9] studied 11 job sites in order to asses the exposure to dust and lead from sanding lead-bearing paint. Total dust was sampled at 2 lpm using closed face filter cassettes. Thirty-minute total dust exposures were 1.6 (SD=2) , 6.7 (SD=3.6), and 14 (SD=13) mg/m^3 for HEPA exhausted power sanding, manual dry sanding and uncontrolled power sanding, respectively. The lead-content in the airborne dust was generally lower than in the paint itself. For comparison, Choe et al. [10] studied the dust generation during scraping and dry machine sanding of wood, covered with leaded paint at laboratory conditions. The tests were performed in a 24.3 m^3 test room representing the size of a typical residential room. They measured particle concentrations with a Grimm model 1.108 optical particle counter, using data in the size range 0.4-20 µm only. Immediately after sanding there was one size mode around 3 µm and one at or below the lower size limit, 400 nm. Emission of nanosized particles from the sanding machine was not addressed. However, electrically driven sanding machines can be a prominent source for ultrafine particles in the sanding operation. Szymaczak et al. [11] reported electrical motors to be a major source of particles; especially for particles below 100 nm size. In Szymaczak's study a professional vacuum cleaner engine was used as the source of particles. The total number concentration produced by the motor operated at full power in the test room (4.2 m^3 with air exchange rate 0.021 m^{-1}) saturated rapidly at 3.0×10^{11} particles m^{-3} within 15 min of operation.

Other relevant studies include Teitsworth and Sheehan [12] who measured the effectivity of a specific shrouded tool. Personal samples of total and respirable dust were taken during grinding, sanding, trimming, and removing surplus resin. When using un-shrouded tools, the personal exposure to total or respirable dust ranged from values below the limit of detection (1.6-1.7 mg m^{-3}) to 50.8 mg m^{-3}. Carlton et al.,[13] tested a pneumatic, random orbital type sanders in a glove box during sanding of aluminium panels coated with aircraft epoxy primer and polyurethane paint, using grit size 180 and 240. No difference was found in inhalable dust generation between grit size 180 and 240.

Noteworthy, none of the previous studies have focused on a detailed characterisation of the airborne particle size distribution of dust generated by sanding painted surfaces. Complete knowledge of the particle size-distribution enables a better assessment of exposure risk and effectiveness of personal protective equipment. In this paper we characterise the airborne particle size distribution spectra generated during sanding of painted MDF plates. The sanding particles were generated in a set-up designed for simultaneous electrostatic collection of the sanding particles for subsequent physicochemical and toxicological analysis. We characterise the contribution from both the sander and paint and discuss the results in terms of number size distribution as well as surface area and mass size distribution. The presented work is generated as part of a larger project effort with the overall aim to complete a detailed physico-chemical and toxicological characterisation of NPs, which currently are in the promising candidates for use in commercial paints and lacquers, and of the dust particles generated during sanding paints produced with and without NPs.

2. Methods

2.1. Experimental set up and Instrumentation

The experiment was conducted in a 20.6 m^3 stainless steel human exposure chamber, which has a nominal air exchange rate 9.2 ± 0.8 h^{-1} [14]. HEPA-filtering of the laboratory air assured low total background particle concentrations (<100 cm^{-3}). Figure 1 shows the experimental set-up, which consists of a sander unit, a sampling tube, a 0.03 m^3 dust reservoir made of plastic, and a commercial electrostatic precipitator (ESP), previously characterized by Sharma et al., [15], for particle sampling.

The dust particles were sampled for completion of subsequent psysico-chemical and toxicological analysis.

The particle size distribution was measured from the chamber using an APS Model 3321 (Aerosol Particle Sizer, TSI Inc.), which measures the size distributions of particles from 0.542 to 19.81 μm (aerodynamic diameter). Due to the high concentrations we used a TS1 diluter model 3302A with dilution ratio of 100 to 1. An FMPS Model 3091 (Fast scanning Mobility Sizer, TSI Inc.), which measures the particle size distribution from 5.6 nm to 542 nm in 32 channels (mobility diameter), was used for sizing smaller particles. The APS and FMPS data were exported at a 10 seconds time resolution, which was sufficient to observe the relatively rapid changes in the aerosol spectra.

Sanding was performed using a commercial hand-held orbital sander (Metabo Model FSR 200 Intec) with an internal fan for dust removal. Grit size 240 sanding paper was used as recommended by the paint and lacquer manufactures. This grit size is usually applied for finishing. By using a fine grit size, we also expected an enhanced production of small particles. However, the actual enhancement is uncertain [13, 16]. Charlton et al., [13] did not observe differences in the particle size distribution using grit size 180 and 240, whereas Thorpe and Brown [16] observed an inconsistent effect of grit size on the particle size in the dust. The Metabo sander was delivered with a filter bag attached to filter exhaust air. For our purpose the filter bag was removed and the outlet modified to connect a 60 cm long and 32 mm ID flexible plastic tube after the sander to lead the dust to the dust reservoir chamber from which sampling by the APS and FMPS was performed through 10 mm, TSI Conductive Silicone Tubing. The ESP was attached at the side of the chamber sampling air through a 21 cm deep, 37 cm wide and 15 cm high tunnel made in aluminum (see Fig.1).

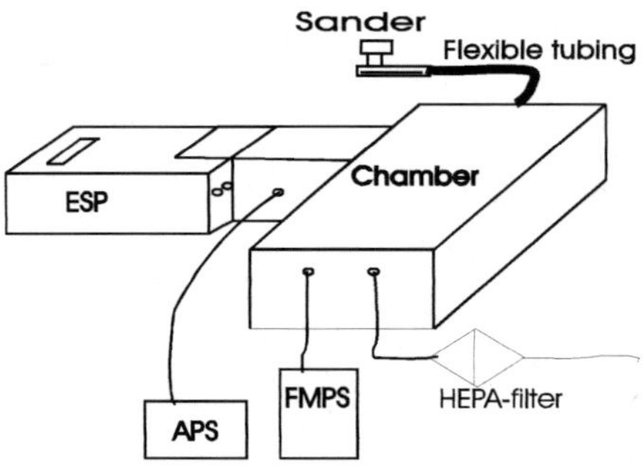

Figure1. Schematic illustration of the measurement set up.

The air-flow from the sander to the chamber and the flow from the chamber to ESP were measured with a pitot tube and a micro manometer (TT 570SV, DP Measurement UK). The volume flow to the ESP influences the cut size and collection efficiency of the ESP sampler and will be described elsewhere. The volume flow from the sander was 100 m3/hour measured at the end of the tube leading dust to the sampling chamber. The volume flow of the HEPA-filtered make-up air was 20 l/min, resulting in a 20 to 834.4 l/min ratio between the sample flow and the dilution air, which is less than 3 % of the total sampled air.

2.2. Description of the paints

For use in the large project, the Danish paint-and lacquer industry provided 13 plates, three painted with reference paints i.e. not containing NP, six painted with NPs, one lacquer with and one without NP. All the plates had a few millimeter thick layer of paint. In addition, we received two plates covered with two different fillers. In this paper we will focus on two paints containing NPs and their reference paint, all from the same manufacturer and intended for indoor use. We denoted the NP containing paints as Paint 1 and Paint 2, respectively. According to the supplier Paint 1 contained 17 nm TiO_2 NP and Paint 2 contained 95 nm Carbon Black NP.

2.3. Measurement procedure

Each experiment was conducted following the same procedure. First the ventilation system in the human exposure chamber was turned on to clean the background air while measuring the particle concentrations in the chamber by the APS and FMPS. This lasted until the background concentration was less than 300 particles cm^{-3}. After checking the background number concentrations the instruments and sander were connected to the small sampling chamber and the size distribution from the sampling chamber was measured. The total particle concentration was always below 10 cm^{-3} inside the chamber before starting the sanding experiments. Sander emissions were measured for a 2 to 3 minute period before each experiment (see spectra Fig. 2). After measuring the sander emissions some time was allowed for the total particle number concentration in the chamber dust reservoir to be reduced to background concentrations of a few hundred particles cm^{-3}. Then the ESP was turned on and sanding of the painted MDF plates started. Sanding was carried out for one or two minute periods until sufficient amounts of material was collected on the ESP collection plates, as assessed visually. This took 15 to 30 minutes depending on the type of paint. After sanding, all the parts from the measurement set-up were dismantled and cleaned with a vacuum cleaner.

2.4. Data-analysis

In order to evaluate the characteristics of the measured aerosol number size distributions, we fitted a multi modal log-normal distribution to the measured spectra. Fitting gave three parameters; the geometric mean diameter (GMD), the number concentration (N) and the geometric standard deviation (GSD) of each mode.

Several studies have discussed that particle parameters such as size, number concentrations and available surface area to a large extent may determine the toxicity of low-solubility particles [e.g. 3, 17]. However, the mass of the airborne particles may still be an important descriptor and is still used as a toxicological metric in characterising dose. Consequently, both the mass and surface area weighted distributions were calculated from the number size distribution assuming the particles are spherical with a density of 1 kg/m3.

3. Results

3.1. Sanding machine emissions

The sanding machine emissions were checked before each sanding experiment. The continued check of particle emissions from the sander was important, as the sander was new at the start of the experiments and we wanted to control whether the sander emissions changed over time. Even-though the sander was cleaned after every experiment, some dust apparently remained from the previous experiments. Therefore, prior to each sanding experiment, the sander was allowed to run until the total number concentration and aerosol spectra were constant. Figure 2a presents the average number concentration spectrum calculated from all sander background measurements. The spectrum shows a multimodal size distribution, which will be analyzed in detail below.

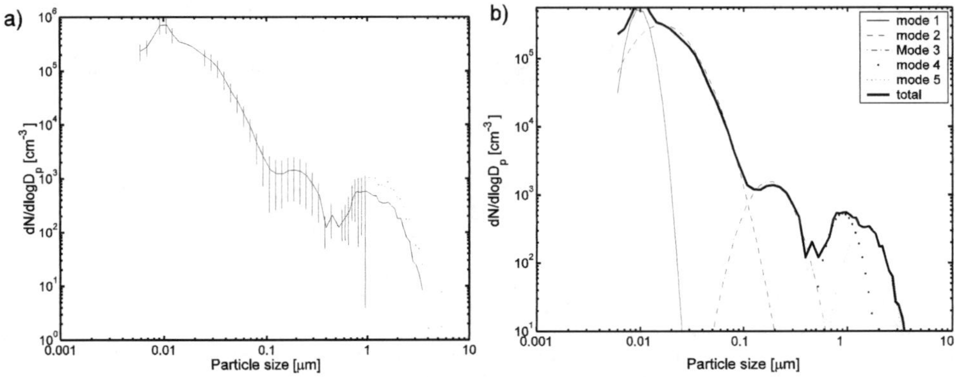

Figure 2: a) The average number size distribution spectrum measured from the grinding machine and standard deviations of each point. b) The average number size distribution spectrum measured from the grinding machine and five-mode fitted log-normal size distribution modes.

Electrical motors produce both mechanical wear and spark particles. Szymaczak et al. [11] reported in they study, that spark generated particles dominated the emissions of ultrafine particles (UFP). They also conducted PIXE (proton induced X-ray analysis) analysis, which showed that Cu dominated the mass concentrations in the UFP size range. The number size distributions of the sander emissions were similar those found by Szymaczak et al. [11]. To enable a more in-depth understanding of the background emissions from the sander, the average number size distribution spectra was fitted, assuming five log-normal size distribution modes (Figure 2b). The 1st mode at 9.8 nm had 1.18e5 (6.06e4) cm^{-3}, the 2nd mode at 16 nm had 1.85e5 (1.27e5) cm^{-3}, the 3rd mode at 179 nm had 664.5(506) cm^{-3}, the 4th mode at 904 nm had 141 (125) and the 5th mode at 1.32 (μm) had 89 (114) cm^{-3}, where the numbers in parenthesis are standard deviations of the fitted data. Our measurements indicated that the particle size distribution in the sander emissions was very reproducible for particles smaller than 100 nm. In mode 3 and mode 4, the variations in particle concentrations were notably higher in between the measurements than the particle concentration variations in between mode 1 and mode 2. That is probably due to paint dust residuals inside the sander, chamber and tubes. It must be concluded that the system could not be cleaned completely in between the experiments.

3.2. Paint dust emissions

3.2.1. Number size distribution modes. Several separate size modes were observed in the particle size distribution measurements. When including sander emissions, we observed five size modes and when subtracting the sander emissions we observed three modes (Figure 3). It is clear that there is no significant particle contribution from the sanded paint under 50 nm, however both mode 3 (the smallest mode when sander emissions are subtracted) 4 and 5 also contain sander emissions.

Table 1 shows the modal concentrations and geometric mean diameters of the observed modes corrected for sander emissions. Comparison of particle emissions from the sander and from the paint dust shows that mode 3 in the paint dust has slightly smaller GMD than mode 3 in sander emissions. There is a large difference in GMD's in modes 4 and 5. In the sander emission, mode 4 and 5 occurred at 0.90 and 1.32 μm, respectively. Sanding dust mode 4 and 5 were found at 1.02 and 2.02 μm, respectively. GMDs of the emissions from the sander and from sanding painted plates are similar. This indicates that mode 3, 4 and 5 are mixtures of sander emissions and paint dust particles. Sanding plates painted with different paints did not change sanding emission GMDs. Therefore, the size of the observed number concentration modes most likely depends in other factors such as the sander paper grit size and the sander engine rotation speed. The modal number concentrations in the three paints varied (see Table 1). By comparing these variations with the standard deviations calculated from the sander emissions it is evident that the emissions varied with the type of paint.

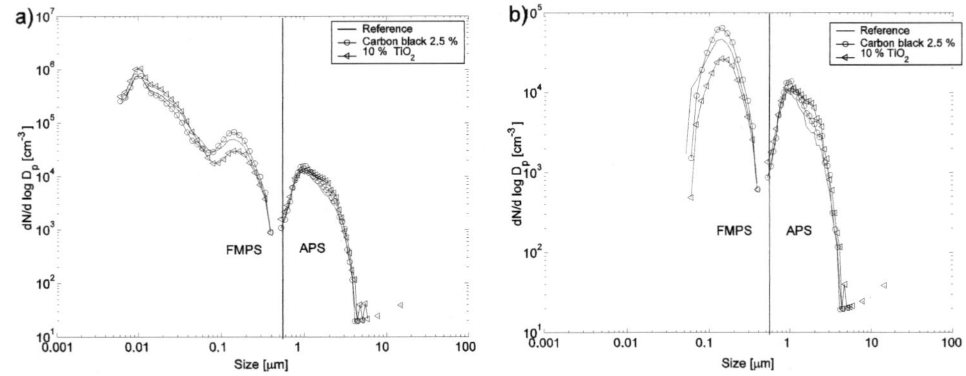

Figure 3. Particle number concentration spectra measured during sanding: (a) with sanding machine emissions and (b) corrected for sander emissions.

Table 1: Averages of the six fitting parameters obtained from the fitting procedure (geometric standard deviation is not presented): geometric mean diameter GMD_p (μm), number concentration in mode N_i (cm^{-3}). The standard deviation of each quantity is given in parentheses. Paint 1 contains TiO_2 and Paint 2 contains carbon black nanoparticles.

	GMDp (μm)	N1 (cm-3)	GMDp (μm)	N2 (cm-3)	GMDp (μm)	N3 (cm-3)
Ref.	0.140(0.05)	1.81*1e4(4248)	1.02(0.01)	2257(461)	2.02(0.06)	482(114)
Paint 1	0.152(0.03)	1.03*1e4(740)	1.02(0.02)	3005(629)	1.84(0.05)	1182(249)
Paint 2	0.144(0.03)	2.38*1e4(4000)	1.02(0.03)	3194(1119)	2.16(0.01)	599 (240)

3.2.2. *Surface- and mass distributions.* In Figure 4, the mass and surface area weighted distributions calculated from number size distributions, are presented. Figures 4a,b present distributions corrected for the sander emissions and 4c,d sander emissions are included. Figures 4a,b show that most of the mass and surface are is above 1 μm and Figures 4c,d illustrate that despite high influence in the number size distribution spectra, the sander emissions only play a minor role on mass and surface area distribution spectra.

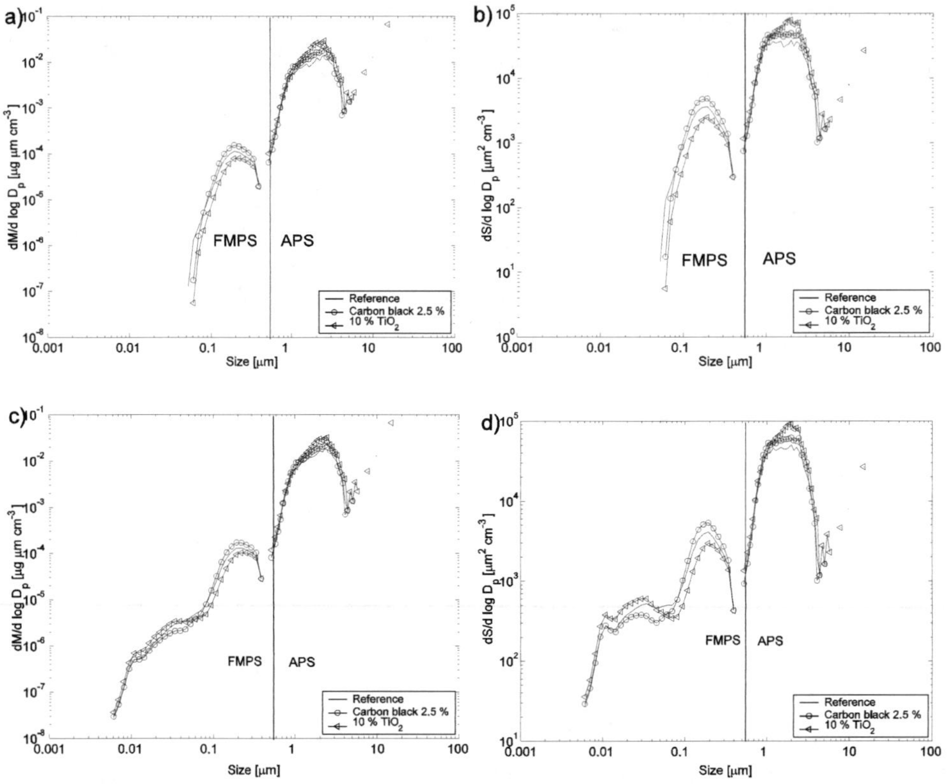

Figure 4. Mass, and surface size distribution spectra calculated from the sanding emission number size distribution measurements. a) and b) are presented when corrected for sander emissions and c) and d) with sander emissions. For mass calculation density =1 kg/m^3 was used.

4. Conclusions

The objective was to investigate if adding NPs to paints would result in emissions of free NP and/or change the particle size distribution spectra of dust produced by sanding. Dust was produced by sanding MDF plates covered with several mm of paint. Two plates were painted with NP-based paint namely with TiO_2 and carbon black and one plate was painted with conventional paint, which was also our reference paint. In order to evaluate the properties of the released dust, we measured the aerosol size distribution from 5.6 nm to 19.8 μm using a FMPS and an APS with 10 second time resolution. Three to five modal log-normal distributions were fitted to measured data.

The sander was a main source of particles smaller than 50 nm. Total emission number size distributions were clearly five modal. The First two modes had peaks below 20 nm and these were identified as sander emissions. Mode 3 was found at about 200 nm, which contained particles from the sander and paint dust emissions. Mode 4 and 5 were at 1 and 2 μm and they contained mostly dust released from the paints. While the number size distribution spectra were dominated by particles below 50 nm, mass and surface area distributions showed that particles above 1 μm might have an important rule on these measures. Consequently, these characteristics may be important when investigating toxicity of the sanded paint dust. However, when focusing on the possible health risks caused by particles generated from sanding paint, one also has to take emissions from the sander into account when assessing the real exposure situation.

Parameters obtained from the log-normal fitting showed that the GMDs of the modes did not depend on whether NPs were present in the paints or not. The GMDs were probably more controlled by other factors of the sanding process. However, even-though the positions of the modes were the same in all spectra, the modal number concentrations varied in between all three paints. Therefore NPs may have an impact to properties of the paints. It is evident that NPs released during the sanding do not occur in individual particles. Therefore accurate physical and chemical characterization is needed for obtaining more information on particle characteristics and the potential toxicity of the dust.

References

[1] Balbus M., Florini K., Denison R. A. and Walsh S.A. (2006), Protecting workers and environment: An environmental NGO's perspective on nanotechnology. Journal of Nanoparticle Research, 9,11-22

[2] Oberdorster G., Oberdorster E., Oberdorster J., (1995). Nanotoxilogy: An Emerging Discipline Evolving Studies of Ultrafine Particles Environ. Health Perspectives, 113:823-839.

[3] Wallace W.E., Keane M.J., Murray D.K., Chisholm W.P., Maynard A.D. and Ong T-M. (2006), Phospholipid lung surfactant and nanoparticle surface toxicity: Lessons from diesel soost silicate dusts. Journal of nanoparticle research, 9:23-38.

[4] Schneider T. and Jensen K.A. (2007) Combined Single-Drop and Rotating Drum Dustiness Test of Fine to Nanosize Powders Using a Small Drum. Ann.Occup.Hyg.2008;52(1):23-34

[5] Bohgard M, Anderberg MT, Eklund A (1994) Source characteristics of emitted aerosols in the size range 0.02 to 20 μm at pigment handling in dye manufacturing. J Aerosol Sci 25(Suppl 1):S461–S462

[6] Kuhlbusch TAJ, Neumann S, Fissan H (2004) Number size distribution, mass concentration, and particle composition of PM1, PM2.5, and PM10 in bag filling areas of carbon black production. J Occup Environ Hyg 1:660–671

[7] Maynard A.D., Baron PA, Foley M et al (2004) Exposure to carbon nanotube material: aerosol release during the handling of unrefined single-walled carbon nanotube material. J Toxicol Environ Health A 67:87–107

[8] Jensen K.A., Koponen I.K., Clausen P.A. and Schneider T., (2008), Dustiness behaviour of loose and compacted Bentonite and organoclay powders: What is the difference in exposure risk? Journal of Nanoparticle Research, doi : 10.1007/s11051-008-9420-1.

[9] Scholz P.F., Materna B.L., Harrington D, Uratsu C. (2002), Residential and commercial painters' exposure to lead during surface preparation. Am Ind Hyg Assoc J; 63: 22-28

[10] Choe KT, Trunov M, Grinshpun SA, Willeke K, Harney J, Trakumas S, Mainelis G, Bornschein R, Clark S, Friedman W. (2000), Particle settling after lead-based paint abatement work and clearance waiting period. Am Ind Hyg Assoc J; 61: 798-807.

[11] Szymczak W., Menzel N., Keck L.(2007), Emission of ultrafine copper particles by universal motors controlled by phase angle modulation. Journal of Aerosol. Sci., 38, 520-531.

[12] Teitsworth JE, Sheehan MJ. (1998), The effectiveness of local exhaust-ventilated (shrouded) hand power tools used for grinding/sanding composite materials. Am Ind Hyg Assoc J; 59: 689-693.

[13] Carlton GN, Patel KB, Johnson DL, Hall TA. (2003), The effectiveness of handheld ventilated sanders in reducing inhalable dust concentrations. Appl Occup Environ Hyg; 18: 51-56.

[14] Lennert A., Nielsen F, Breum N.O.,(1997) Evaluation of evaporation and concentration distribuyton models and a test chamber study. Am. Occup. Hyg. Vol 41, No. 6, 625-641.

[15] Sharma A. K., Wallin H., Jensen K. A. (2007) High volume electrostatic field-sampler for collection of fine particle bulk samples. Atm. Env., 41, 369-381.

[16] Thorpe A, Brown RC. Measurements of the effectiveness of dust extraction systems of hand sanders used on wood. Ann Occup Hyg 1994; 38: 279-302.

[17] Wittmaack Klaus,(2006), In Search of the Most Relevant Parameter for Quantifying Lung Inflammatory Response to Nanoparticle Exposure: Particle Number, Surface Area or What?Env. Health Perspectives, 115, 2.

[18] Donaldson, K., Stone, V., Gilmour, P. S., Brown, D. M., & MacNee, W. (2000). Ultrafine particles: Mechanisms of lung injury. Philosophical Transactions of the Royal Society of London B, 358, 2741–2749.

Journal of Physics: Conference Series **151** (2009) 012049 doi:10.1088/1742-6596/151/1/012049

Toxicity of Inhaled Traffic Related Particulate Matter

Miriam E Gerlofs-Nijland[1]; Arezoo Campbell[2]; Mark R Miller[3]; David E Newby[3]; Flemming R Cassee[1]

1 National Institute for Public Health and the Environment, Bilthoven, Netherlands.

2 Western University of Health Sciences, Pomona, CA, USA.University of Edinburgh,

3 Edinburgh, United Kingdom.

Email: miriam.gerlofs@rivm.nl

Abstract. Traffic generated ultrafine particulates may play a major role in the development of adverse health effects. However, little is known about harmful effects caused by recurring exposure. We hypothesized that repeated exposure to particulate matter results in adverse pulmonary and systemic toxic effects. Exposure to diesel engine exhaust resulted in signs of oxidative stress in the lung, impaired coagulation, and changes in the immune system. Pro-inflammatory cytokine levels were decreased in some regions of the brain but increased in the striatum implying that exposure to diesel engine exhaust may selectively aggravate neurological impairment. Data from these three studies suggest that exposure to traffic related PM can mediate changes in the vasculature and brain of healthy rats. To what extent these changes may contribute to chronic neurodegenerative or vascular diseases is at present unclear.

1. Introduction

Epidemiological studies have shown associations between both short-term and chronic exposure to particulate matter (PM) and shortening of life expectancy [1, 2]. In addition, several studies revealed that emissions from highways result in inflammation in the lung and cardiovascular changes [3, 4]. Ultrafine particulates (UF) which are part of diesel soot may enter the systemic circulation and might trigger systemic effects [5]. However, little is known about adverse effects from prolonged exposure. Therefore, we hypothesized that repeated exposure to PM results in exacerbation of adverse pulmonary effects which will lead to systemic toxic effects.

To test this hypothesis PM effects were studied in the presence of a mild inflammation status, induced by ozone. The assumption is that PM by itself is not causing (significant) health effects, but can induce or worsen them when some inflammation is present. Exposure to particulates could then result in oxidative stress which can not be handled adequately. As a result various processes may be initiated which eventually lead to negative consequences on overall health status.

2. Material and methods

Male Fischer F344/CrlBR rats of 15-16 weeks old were randomly allocated to either control or exposure groups (n=15/group). Three independent experiments were performed in which animals were exposed to different PM exposures in addition to HEPA-filtered air (clean air).

2.1. Exposure

All animals were exposed to 0.4 ppm ozone for 12 hours prior to (concentrated) PM exposure to induce a mild pulmonary inflammation at the onset of exposure. Thereafter, animals were exposed for 4 weeks, 5 days/week for 6 hours a day to clean air or the following various PM exposures:

A) concentrated ambient particles (0.15-2.5 µm) at an urban background (CAPs background) using an Ambient Fine Particle Concentrator (AFPC);
B) diluted diesel engine exhaust (DEE) obtained by mixing a small flow of freshly generated diesel engine exhaust taken from a stationary (1500 rpm) diesel engine (35 KVA Genset) into conditioned and purified air or;
C) concentrated PM (<2.5 µm) near a freeway (CAPs freeway) generated by a Versatile Ambient Concentration Enrichment System (VACES).

2.2. Effect parameters

Different parameters were measured to assess the effect of PM exposure on inflammation, cell damage and oxidative stress. In addition, pathological and haematological changes were examined. Vascular function was assessed in arteries using the myograph system. Different brain sections were taken to investigate the inflammatory response in different brain regions. All effects were investigated 24 hours after termination of the exposure.

3. Results

3.1. Exposure

Particle mass levels ranged between 150 and 650 µg/m3 with the highest concentrations (and variation) observed for the freeway study. Particle number concentrations were relatively low for the urban background location (26,000 cm^{-3}) and substantially higher for the DEE (430,000 cm^{-3}) and the freeway (260,000 cm^{-3}) exposures due to a larger contribution of freshly emitted particles.

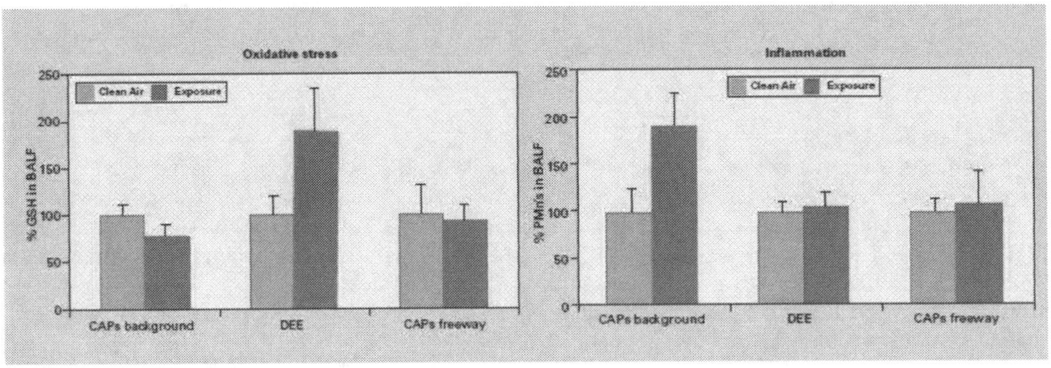

Figure 1. Oxidative stress and inflammation measured in the bronchioalveolar lavage fluid (BALF) of rats exposed to different PM exposures or clean air. Data are expressed as percentages relative to clean air exposed animals (mean ± SEM), $p \geq 0.05$.

3.2. Effects

There were indications of mild oxidative stress determined by a non-significant increase in the levels of glutathione (GSH) in bronchioalveolar lavage fluid (BALF) of rats exposed to DEE only (figure 1). Although a small non-significant increase in polymorphonuclear cells (PMNs) was detected in BALF after exposure to urban background PM (figure 1), levels of tumor necrosis factor-alpha (TNF-α) were significantly reduced (clean air 8.22 ± 1.26 versus CAPs background 6.06 ± 0.99; $p < 0.001$).

Exposure to DEE and freeway PM did not induce inflammation (figure 1). In addition, no significant changes were seen in markers related to cytotoxicity. Examination of the lung did not reveal any pathological effect after 4-week exposures.

Exposure to PM from an urban background and DEE resulted in mild increase in von Willebrand factor (vWF) after 1-week exposure and vWF was slightly reduced after 4-week exposure to DEE (figure 2). However, besides a reduction in vWF levels no major effect was detected on the coagulation cascade (fibrinogen, figure 2). Interestingly, white blood cell (WBC) numbers were decreased in animals exposed to DEE (figure 2).

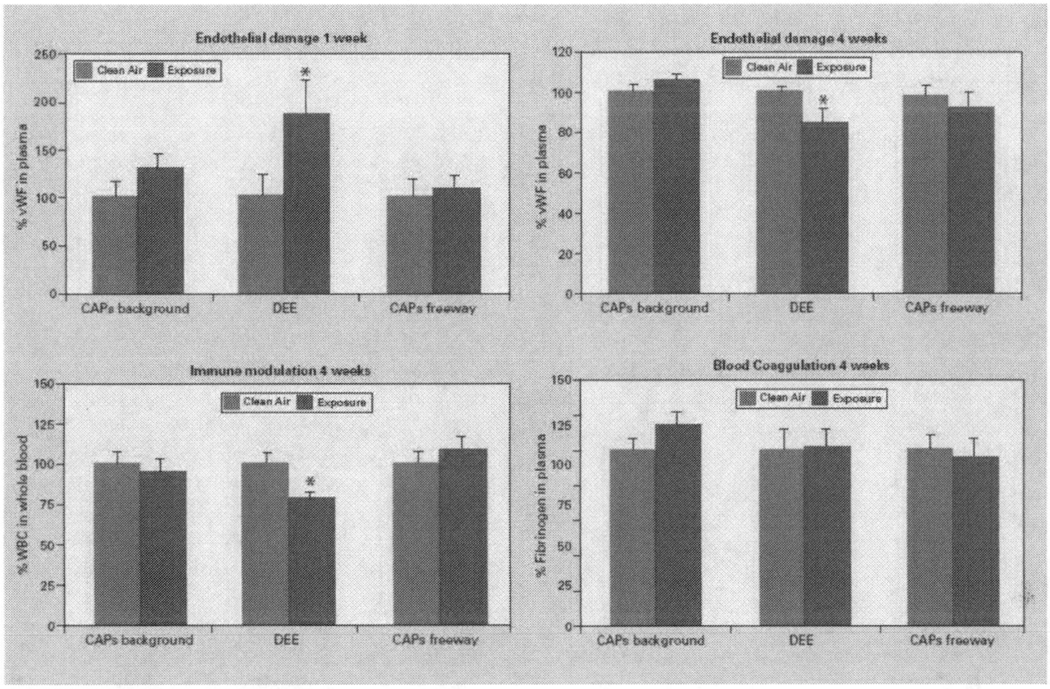

Figure 2. Endothelial damage, immune modulation and blood coagulation measured in blood of rats exposed to different PM exposures or clean air. Data are expressed as percentages relative to clean air exposed animals (mean ± SEM), $^*p \leq 0.05$.

No changes were observed in vascular function in arteries as measured by *ex vivo* responses to vasodilators in rat aortic rings (figure 3). No significant effect on responses to the endothelium-dependent vasodilator, acetylcholine (figure 3), the endothelium-independent vasodilator, sodium nitroprusside, or the NO-independent vasodilator, isoprenaline, was observed after exposure to different traffic related PM exposures.

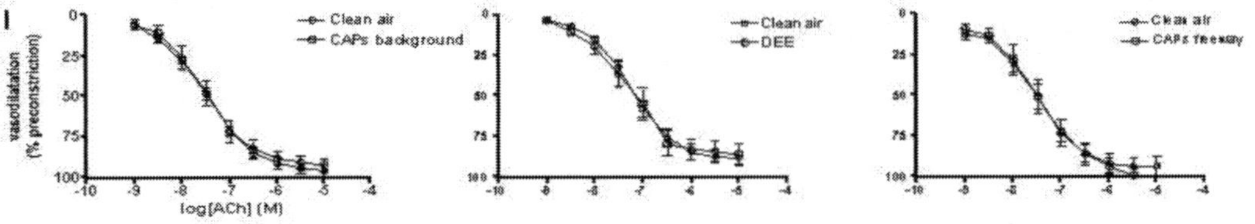

Figure 3. *Ex vivo* response to the endothelium-dependent vasodilator acetylcholine in rat aortic rings after exposure to different PM exposures or clean air. Vasodilator response is expressed as percentage of the maximal contraction to phenylephrine, $p \geq 0.05$.

A decrease in pro-inflammatory cytokine levels was noticed in some parts of the brain of animals exposed to urban background PM and DEE while these cytokines were enhanced in the striatum after DEE exposure (figure 4).

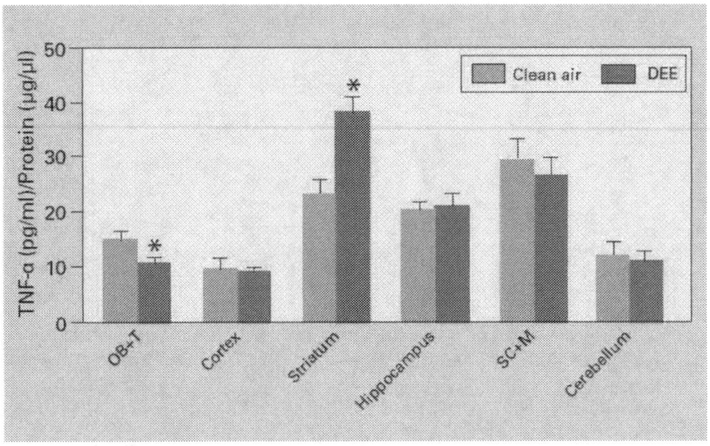

Figure 4. Pro-inflammatory marker TNF-α in different brain regions of rats exposed to different PM exposures or clean air. Abbreviations: OB+T - olfactory bulb and tubercles; SC+M – spinal cord and medulla. Data are expressed as mean ± SEM, $*p \leq 0.05$.

4. Conclusions

The presented studies suggest that exposure to traffic related PM can mediate changes in the vasculature and brain of healthy rats. The fact that only very mild effects were detected may be a consequence of the relatively healthy animal model that has been used in these studies, despite the fact that pulmonary inflammation was induced at the onset of the PM exposures.

Prolonged exposure to PM (including gaseous co pollutants) mainly from traffic at levels roughly 10 times higher than ambient did not result in significant toxic effects in rats suggesting that a threshold for PM induced effects may exist. In general, no clear signs for a more harmful effect of DEE compared to ambient PM was noted, irrespective the contribution of any gaseous component. DEE seems to be a more powerful oxidant to which the defence system has been adapted.

Reduced numbers of white blood cells were observed after exposure to diesel engine exhaust suggesting that the immune system is affected and can not react adequately or these cells might adhere to the vessel wall. The increase in pro-inflammatory markers in the striatum implies that exposure to diesel engine exhaust may selectively aggravate neurological impairment. Data on CAPs freeway are not available yet. To what extent the changes in the vasculature and brain may contribute to chronic neurodegenerative or vascular diseases is at present unknown.

References

[1] Boldo E, Medina S, LeTetre A, Hurley F, Mücke H G, Ballester F, Aguilera I, Eilstein D and Apheis Group 2006 *Eur. J. Epidemiol.* **21** 449-58

[2] Rabl A. 1: J Air Waste Manag Assoc. 2003 Jan;53(1):41-50

[3] Kleinman M T, Hamade A, Meacher D, Oldham M, Sioutas C, Chakrabarti B, Strom D, Froines J R and Cho A K 2005 *J. Air E Waste Manag.Assoc.* **55** 1277-88

[4] Elder A, Couderc J P, Gelein R, Eberly S, Cox C, Xia X, Zareba W, Hopke P, Watts W, Kittelson D, Frampton M, Utell M and Oberdörster G 2007 *Inhal. Toxicol.* **19** 1-12

[5] Frampton M W, Utell M J, Zareba W, Oberdörster G, Cox C, Huang L S, Morrow P E, Lee F, Chalupa D, Framer L M, Speers D M and Stewart J 2004 *Res. Rep. Health Effect Inst.* **126** 1-47; discussion 49-63

Mortality in an extended follow-up of British coal workers.

MacCalman, L and Miller, B G

Institute of Occupational Medicine, Edinburgh.

laura.maccalman@iom-world.org

Abstract. The Pneumoconiosis Field Research (PFR) programme was established in the 1950s, to evaluate effects of coal mining exposures on the health and mortality of British coal workers. Surveys of working miners were carried out at 5-yearly intervals, initially in 24 collieries but later concentrating on 10, collecting detailed work histories and health information for each recruit. Here we report on cause-specific mortality in a cohort of almost 18,000 men from 10 British collieries, followed up for periods up to 47 years, yielding over 516,000 life-years of follow-up. External analyses compared cause-specific death rates in the cohort to those of the population of the regions in which the collieries were situated, using Standardised Mortality Ratios (SMRs). The causes investigated included lung cancer, stomach cancer, non-malignant respiratory disorders and cardiovascular disorders. SMRs showed evidence of an initial healthy worker effect diminishing over time. Several causes, including non-malignant respiratory disease and lung cancer, showed a significant deficit of mortality at the start of the study period with an excess in the latter part of the follow-up period. In these results, effects of working conditions are likely to be confounded with smoking habits. Overall, we believe our results may be generalised to the British coal industry since nationalisation.

1. Background

The Pneumoconiosis Field Research (PFR) programme of the then National Coal Board began in the early 1950s. The aims were "to determine how much and what kinds of dust cause pneumoconiosis, and to establish what environmental conditions should be maintained if mineworkers are not to be disabled by the dust they breathe during the course of their work". The research began in the 1950s in 24 representative British collieries, later reduced to 10. All industrial-grade workers at these collieries were invited to regular health surveys (around every five years) including chest radiographs, lung function tests and questionnaires on respiratory symptoms and smoking habits. Response rates were high, generally well above 90%. A separate programme of measurement quantified typical concentrations of respirable dust and quartz at the locations of a wide variety of colliery occupations. These were linked to data from the colliery payroll systems, on the times worked by each worker in the same occupations, to produce individual and period-specific estimates of exposure to respirable dust and quartz.

Study of coalminers' mortality within the PFR began in 1970. Several phases of tracing led to the inclusion in a mortality database of all workers surveyed during the PFR. Several reports ([1],

[2], [3], [4], [5]) have compared mortality experienced in the cohort with that expected on the basis of local reference rates, and have shown increased mortality risks in those with severe pneumoconiosis at the start of follow-up, and a general increase in risk from stomach cancer. Analyses (based on subsets of the cohort with appropriate data) have also shown exposure-response relationships for risks of various respiratory causes of death with increasing exposure to dust, but little evidence of increased cancer risks from dust or quartz exposures.

The present study is based on an extended period of follow-up in a previously analysed cohort of almost 18,000 workers from the 10 continuing PFR collieries. Exposure estimates are the same as before, since the collieries were all closed by the early 1980s.

2. Methods

The entire PFR cohort had been flagged for death notification in the National Health Service Central Registers (NHSCR) administered by the Office of National Statistics (ONS) and the General Register Office (Scotland) (GRO(S)), but death notification had been suspended some years previously. Notification was reactivated, and the database used in the most recent PFR study report [3] was updated with the new death details. These included the underlying cause of death coded according to the International Classification of Diseases and Causes of Death (ICD) published by the World Health Organisation (WHO), in the contemporary revision ([6], [7], [8], [9]).

To create regional reference rates for selected causes of death, data were obtained from the national authorities on numbers of deaths from those causes and on the mid-year male populations, by age and calendar year, for each of the relevant administrative regions of Britain. In a few cases where death data for specific cause groups were not available for a small number of early years, we extrapolated or interpolated rates from years for which data existed, using regression models.

Comparisons with regional reference data used indirect standardisation to calculate expected deaths from selected cause groups, and to compare observed numbers with these, stratified by age group and calendar year within region, calculating standardised mortality ratios (SMRs) with 95% confidence intervals assuming Poisson variation. All the calculations were carried out using the table manipulation facilities of the statistical package Genstat [10].

3. Results

Miller *et al.* (1997)[3] analysed mortality in 18,166 men from 10 collieries within the PFR cohort. On reactivation of the cohort tracing, vital status could not be ascertained for 346 of these men, and the final analysis cohort consisted of 17,820 men.

SMR calculations were based on follow-up to the end of 2005, yielding 516,431 person years and 10,698 deaths from all causes. The resulting SMRs and 95% confidence intervals are displayed in Table 1. (Table 1 includes the numbers of deaths described as from pneumoconiosis and silicosis, but no SMRs are calculated for these causes. Since these are diseases associated entirely with occupational exposure, there are no meaningful background rates with which to compare the observed incidences.)

Table 1 Summary results of comparisons of mortality in cohort with external reference rates. The table shows, for chosen cause groups, numbers of deaths, age- year- and region-standardised mortality ratios (SMR) and 95% confidence interval.

Cause of death	Observed deaths	SMR %	Confidence bounds Lower	Confidence bounds Upper
All causes	10,698	100.9	99	102.9
All external causes	278	87.5	77.8	98.4
All internal causes	10,421	103.7	101.7	105.7
Tuberculosis	16	77.8	47.6	126.9
All cancer	2,732	98.0	94.4	101.8
Stomach Cancer	318	129.0	115.6	144.0
Lung Cancer	958	98.7	92.6	105.1
Cardiovascular Disease:	4,890	97.8	95.1	100.6
Ischaemic Heart Disease	3,298	100.2	96.8	103.7
Acute Pulmonary Heart Disease	28	71.1	49.1	102.9
Non-Malignant Respiratory Disease	1,966	138.2	132.3	144.5
COPD	849	115.5	108.0	123.6
Chronic Bronchitis	500	138.9	127.3	151.7
Emphysema	70	164.4	130.1	207.8
Pneumoconiosis	288	NA	NA	NA
CWP	222	NA	NA	NA
Silicosis	10	NA	NA	NA

The all-cause SMR was 100.9% (95% confidence interval – CI – 99.0, 102.9), while that for all internal causes was 103.7% (101.7, 105.7). SMRs for tuberculosis, lung cancer, external causes and cardiovascular disease showed no overall excess, while those for stomach cancer and non-malignant respiratory diseases showed significant excesses.

The SMRs for chronic obstructive pulmonary disease (COPD) overall, 116%, and subcategories chronic bronchitis, 139%, and emphysema, 164% were all significantly different from 100%. Miller *et al.* (1997)[3] also found a significant excess of mortality from chronic bronchitis, but did not calculate SMRs for any other NMRDs. As the fashions of cause attribution change over time it is more sensible to consider the broad category of NMRD when considering changes in the SMRs over time.

The above statements on uncertainty in the overall SMRs should be interpreted with caution, since in this case average results conceal important temporal patterns. For example, the current lung cancer SMR of 99% is considerably higher than the previously reported 87% [4] and 86% [3] and a similar rise is apparent in our all-cause SMR of 101%.

All the previous reports on this cohort identified the low all-cause mortality as indicating a 'healthy worker' effect. The cohort was defined initially by employment, so a healthy worker effect was to be expected. However, we should also expect that as the cohort aged (and the coalmining industry disappeared) this effect would diminish and perhaps disappear. Figure 1 shows the time course of all-cause SMRs calculated for successive 5-year periods. It is clear that the initial deficit had been diminishing steadily throughout the follow-up, and that the SMR had in fact been above 100% since the mid-1970s. The size of the apparent dip in the most recent period could be consistent with normal fluctuations, but the evidence seems clear that not only has the healthy worker effect long dissipated, the recent experience of this group is of higher all-cause mortality than in the general populations in their regions.

Figure 2 shows the time-course of lung cancer SMRs calculated for successive 5-year periods, and again shows a clear rising trend over the follow-up, more than doubling from around 60% at the beginning to around 130% recently. This pattern is seen again, although with an earlier rise, for NMRD (Figure 3), where the SMR levels out at around 140%. This was also repeated for the majority of the other causes of death considered here (Miller et al, 2007 [11]).

In contrast, the excess of stomach cancer seen here is consistent with that found by Miller *et al.* (1997)[3] who calculated an SMR of 124% (95% CI 110-141), suggesting that the healthy worker effect was not relevant to these stomach cancers. Figure 4 shows no convincing evidence of a healthy worker effect during the follow-up period.

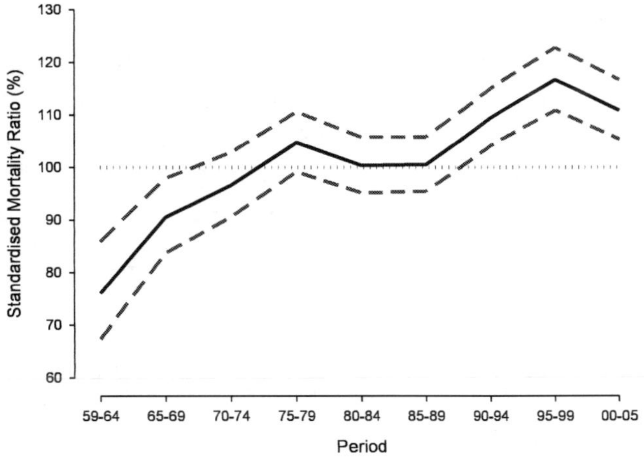

Figure 1 Standardised Mortality Ratio (SMR) for all internal causes over the length of the follow-up period where year is grouped. The solid line is the SMR while the dashed lines represent the 95% confidence interval. The dotted line shows the SMR equal to 100%.

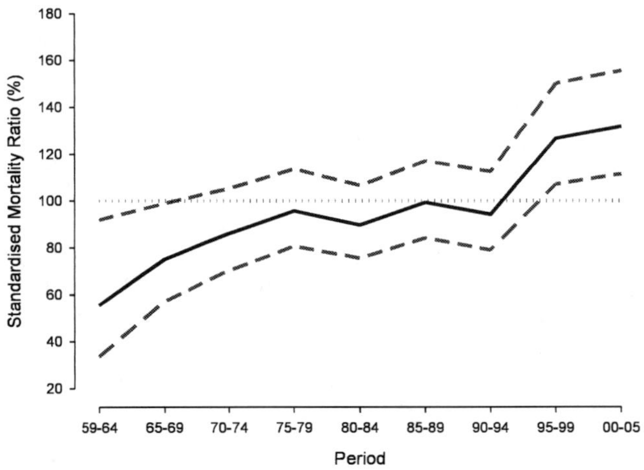

Figure 2 Standardised Mortality Ratio (SMR) for lung cancer over the length of the follow-up period where year is grouped. The solid line is the SMR while the dashed lines represent the 95% confidence interval. The dotted line shows the SMR equal to 100%.

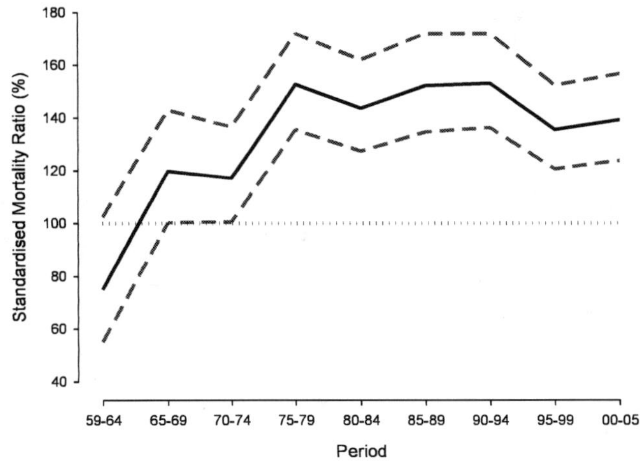

Figure 3 Standardised Mortality Ratio (SMR) for non-malignant respiratory disorders (NMRD) over the length of the follow-up period where year is grouped. The solid line is the SMR while the dashed lines represent the 95% confidence interval. The dotted line shows the SMR equal to 100%.

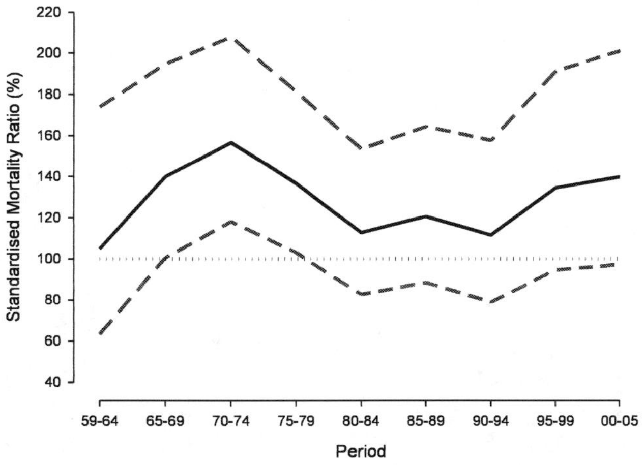

Figure 4 Standardised Mortality Ratio (SMR) for stomach cancer over the length of the follow-up period where year is grouped. The solid line is the SMR while the dashed lines represent the 95% confidence interval. The dotted line shows the SMR equal to 100%.

4. Discussion

Our data on coal workers' mortality are unique in their combination of response rates at the studied collieries, detailed exposure characterisation and length of follow-up. The original collieries were chosen to represent the range of coal types and regions in the British coalfields, and the continuing 10 collieries retained this coverage. The response rates for the health surveys were high because screening for pneumoconiosis was accepted generally by workers and their representatives. Non-ascertainment of vital status was low, as is typical of British registration systems. Overall, we believe our results to be representative of the British coal industry since nationalisation.

Our extended follow-up has permitted examination of the time-trends in the comparison of mortality rates with regional reference rates. We interpret many of the early deficits as evidence of 'healthy worker' selection effects, because all cohort members had to be healthy enough to be employed in coalmining at study entry. However, the five-year breakdown of SMRs shows clearly that this effect was temporary, and the later SMRs for NMRD, lung cancer and all-cause mortality were in clear excess. Given such trends, interpretation of summary SMRs for the whole follow-up period is problematic and potentially misleading.

It is not surprising that there is a significant excess of mortality from NMRD in the cohort as coal mine dust has long been associated with respiratory disorders, specifically pneumoconiosis and COPD: and respiratory symptoms might well have been taken into account in the recruitment process, so a healthy-worker effect for this cause is highly plausible. However, not all of the trends found can be due to the selection of healthy men to work in coal mines. Lung cancer, with its poor cure rate and short expectation of survival from diagnosis, would not be expected to show strong healthy-worker selection effects, but the early deficits are evident in these data. These deficits in lung cancer mortality were not explainable by the prevalence of smoking as reported at the health surveys, which in this cohort was higher than typical of that era. A more likely explanation is in the strict prohibition on smoking within coalmines, limiting total cumulative doses in those coalworkers who were regular smokers. That prohibition disappeared with the dissolution of British coalmining industry around 1980, and presumably the workers made redundant then had on average greater freedom to indulge their tobacco habit. This could partly explain the steep rise in NMRD deaths from the late 1970s onwards, and latency for lung cancer would be consistent with the later rise in those deaths. However, we have no data on actual cigarette consumption after the collieries closed.

Our results for stomach cancer add to a body of data that has demonstrated consistently elevated risks in coal workers. Other studies have shown similarly elevated risks in non-miners and in miners' wives in coalmining communities. Possible causes have been suggested in diet, other lifestyle factors, coal-based local air pollution and infectious agents such as *helicobacter pylori*. Whatever the cause of the excess, the quality of our exposure data, in analyses of the deaths contribution to the earlier excesses, has given us confidence that the cause is not occupational exposure to respirable dust or quartz ([3], [11]).

Now that death notification has been reactivated, death events will continue to be notified until the cohort is extinct, so there are opportunities for future analyses over even longer follow-up periods. Analyses relating cause-specific mortality to individual exposures to respirable dust and quartz will be reported separately.

Acknowledgements
The authors gratefully acknowledge financial support from Bundesanstalt für Arbeitsschutz und Arbeitsmedizin (Forschungsprojekt F2213).

References

[1] Jacobsen M. (1976). Dust exposure, lung diseases, and coalminers' mortality. PhD Thesis. Edinburgh: University of Edinburgh.
[2] Miller BG, Jacobsen M, Steele RC. (1981). Coalminers' mortality in relation to radiological category, lung function and exposure to airborne dust. Edinburgh: Institute of Occupational Medicine. (IOM Report TM/81/10).
[3] Miller BG, Jacobsen M. (1985). Dust exposure, pneumoconiosis and coalminers' mortality. British Journal of Industrial Medicine; 42: 723-733.
[4] Maclaren WM. (1992). Coalminer's mortality in relation to low level exposures to radon and thoron daughters. Edinburgh: Institute of Occupational Medicine. (IOM Report TM/92/06).
[5] Miller BG and Buchanan D. (1997) The effects of exposure to diesel fumes, low-level radiation, and respirable dust and quartz, on cancer mortality in coalminers. (IOM Report TM/97/04).
[6] World Health Organisation. (1957). International classification of diseases. Manual of the international statistical classification of diseases, injuries and causes of death. Based on the recommendations of the Seventh Revision Conference 1955, and adopted by the Ninth World Health Assembly under the WHO Nomenclature Regulations. Geneva: WHO.
[7] World Health Organisation. (1967). International classification of diseases. Manual of the international statistical classification of diseases, injuries and causes of death. Based on the recommendations of the Eighth Revision Conference 1965, and adopted by the nineteenth World Health Assembly. Geneva: WHO.
[8] World Health Organisation. (1977). International classification of diseases. Manual of the

international statistical classification of diseases, injuries and causes of death. Based on the recommendations of the Ninth Revision Conference 1975, and adopted by the twenty-ninth World Health Assembly. Geneva: WHO.

[9] World Health Organisation. (1990). International statistical classification of diseases and related health problems. Tenth revision. Based on the recommendations of the Tenth Revision Conference 1989, and adopted in the UK by 2001. World Health Organisation, Geneva.

[10] Payne RW. (2004). The Guide to GenStat Release 8. Rothamsted: Lawes Agricultural Trust.

[11] Miller BG, MacCalman L and Hutchison PA. (2007). Mortality over an extended follow-up period in coal workers exposed to respirable dust and quartz. (IOM Report TM/07/06).

Trends in pneumoconiosis mortality and morbidity for the United States, 1968–2005, and relationship with indicators of extent of exposure

M D Attfield, K M Bang, E L Petsonk, P L Schleiff, J M Mazurek
Division of Respiratory Disease Studies, National Institute for Occupational Safety and Health, Centers for Disease Control, 1095 Willowdale Road, Morgantown, WV, 26505, USA

E-mail: mda1@cdc.gov

Abstract. This surveillance report examines trends in selected pneumoconioses in the U.S. for 1968–2005 and their relationship with past indicators of extent of exposure. Numbers of deaths with asbestosis, silicosis, and coal workers' pneumoconiosis (CWP) were tabulated by time and age at death. Worker monitoring CWP prevalence data were tabulated by tenure group. Information on indicators of extent and intensity of exposure were obtained from various sources. Asbestosis deaths from 1968–2005 closely followed the historical trend in asbestos consumption, and appear to be declining in most age groups. Given appropriate exposure control, asbestosis could be eliminated by 2050. Silicosis deaths decreased substantially from 1968–2005, but levelled off after 1998 in all age groups, indicating a continuing occupational risk. In the anthracite coal region, CWP mortality has been declining rapidly. If there is no resurgence in the industry, CWP could disappear in that region by 2030. In the much larger bituminous region, deaths have declined over time but may be increasing among younger individuals. In addition, although CWP prevalence in working coal miners declined substantially from 1970 to 1994, it increased from 1995 to 2006. This indicates the need for increased vigilance in dust control in underground coal mining.

1. Introduction

Although silicosis and coal workers' pneumoconiosis (CWP) have been recognized occupational diseases for many centuries, the industrialization led to a substantial rise in the prevalence of these diseases and associated mortality. Similarly, the increasing use of asbestos starting in 1900 led to the epidemic of asbestosis and other asbestos-related diseases. From 1968–2005 alone, over 100,000 individuals died in the U.S. with mention of at least one of the three conditions on the death certificate [1]. The belated recognition of the burden and costs of these diseases on individuals and the state motivated the introduction of workplace control measures and airborne dust standards in the 1970s in the U.S., thus reducing exposure levels. Now that almost 40 years have elapsed from the introduction of the first major U.S. federal exposure limit (for coal dust in 1969), it is time to assess what progress has been made in eliminating pneumoconiosis and what the future might hold. Accordingly, this paper presents information on the temporal patterns of mortality in the U.S. for the three principal pneumoconioses: asbestosis, silicosis, and CWP, and examines their trends with indicators of extent exposure as well as discussing the likely impact of the compliance dust standards introduced around the 1970s. Mortality was selected as the principal outcome variable because there are no satisfactory national morbidity data on asbestosis and silicosis. National morbidity data on CWP do exist and are included in this analysis. The findings presented here should be considered preliminary.

2. Methods

2.1. Mortality data
U.S. national data from death certificates for 1968–2005 were employed. These data were supplied from by the National Center for Health Statistics, with the cause of death coded using the International Classification of Diseases (ICD) system in force at time of death. The following ICD codes were used for each cause: asbestosis – 515.2 (ICD-8), 501 (ICD-9), and J61 (ICD-10); silicosis – 515.0 and 010 (ICD-8), 502 (ICD-9), and J62 (ICD-10); and CWP – 515.1 (ICD-8), 500 (ICD-9), and J60 (ICD-10). All mentions on the entity axis of the death certificate, i.e., both underlying and contributing causes of death, were included. Industry information, extracted from the death certificates and available on a subset of deaths from 1985–1999 for certain states was employed to aid in interpretation of the data on silicosis. All of the mortality information was derived using the National Institute for Occupational Safety and Health (NIOSH) web-based interactive mortality query system [1].

2.2. Morbidity data
NIOSH has operated a federally-mandated worker monitoring program for underground coal miners since 1970 [2]. Every working underground coal miner is eligible for free periodic chest radiographs. These are processed by NIOSH, and read by at least 2 NIOSH readers using the ILO system for classifying the pneumoconioses. Data are presented here for 1970 – 2006 by tenure group for prevalence of category 1/0 or greater small opacities.

2.3. Indicators of extent of exposure
To put the mortality trends into perspective, we compared them with surrogates of extent of exposure. These differed by outcome and were selected on the basis of the nature of the exposure to the material concerned. For asbestos, the best indicator of extent of exposure available was asbestos consumption, the logical basis being that the more asbestos consumed, the more people are exposed. Data on consumption came from a U.S. Geological Survey report [3]. Silicosis proved more troublesome, in that airborne silica dust often arises in processes incidental to the main product being extracted, made, or maintained. We examined a range of potential data sources, including employment, consumption, and production data available for much of the 20th century for various industries where silica exposure is known to occur. These included metal mining, quarries, sand and gravel, iron and steel production, and construction. For CWP, since few are exposed to coal mine dust apart from those mining the material, the surrogate for extent of exposure was number of underground coal miners. Different patterns of employment and disease occurrence led to us examining the anthracite and bituminous regions of the U.S. separately. In this, the anthracite region was defined as those counties in the eastern third of Pennsylvania, while the bituminous region encompassed all of the other coal fields in Pennsylvania and elsewhere. Employment information on coal mining came from a U.S. Energy Information Administration report [4].

3. Results
The number of asbestosis deaths reported in four age-at-death groups from 1968–2005 together with asbestos consumption is shown in figure 1. Deaths in the oldest age group (75-84) were rising but may have peaked recently. In contrast, deaths in the three younger age groups appear to be declining, with their peaks at earlier years the younger the age at death (142 and 22 for the two youngest age groups in

1986 to 92 and 15 in 2005). In general, the trends in deaths tend to match that for consumption, particularly for the older age groups.

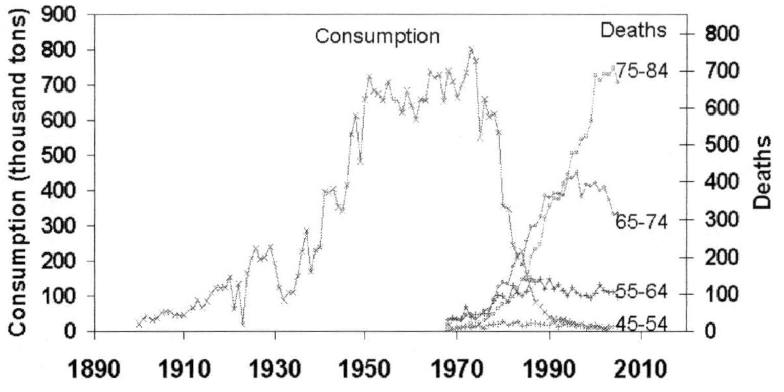

Figure 1. Asbestosis deaths by age at death (1968–2005) and asbestos consumption (1900–2003).

Figures 2 and 3 show silicosis deaths from 1968–2005 with two indicators of extent of exposure: employment in metal mining (figure 2) and employment in construction (figure 3). Fewer deaths over time are seen for all age-at-death groups. The trend in silicosis deaths appears to match employment in metal mining but not that for construction. During 1985–1999 metal mining was reported as the industry on the death certificate for 9% of the deaths, while 12% reported construction work.

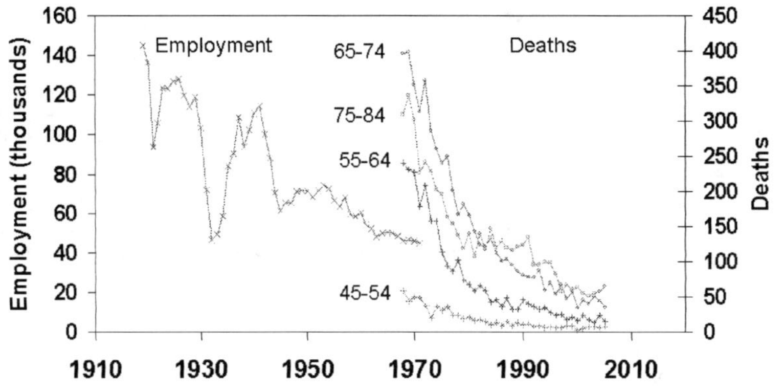

Figure 2. Silicosis deaths by age at death (1968–2005) and employment at metal mines (1919–1971). (9% of all deaths with silicosis from 1985–1999 were associated with metal mining.)

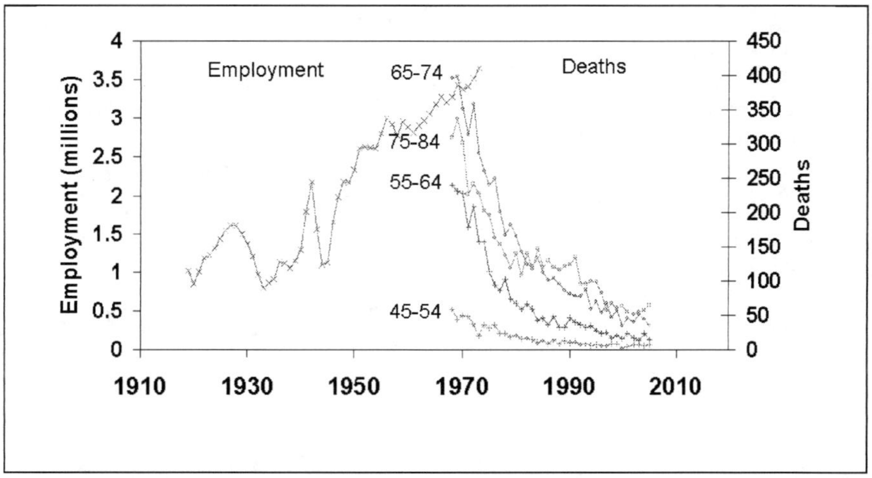

Figure 3. Silicosis deaths by age at death (1968–2005) and employment in construction (1919–1973). (12% of all deaths with silicosis from 1985–1999 were associated with construction.)

Information for CWP deaths in the anthracite region is shown in figure 4. It is clear that coal production has virtually ceased in this area and has been low for the last 30 – 40 years. Mortality in the youngest age-at-death group (45–54) is down to 0.5 cases per year with mention of CWP.

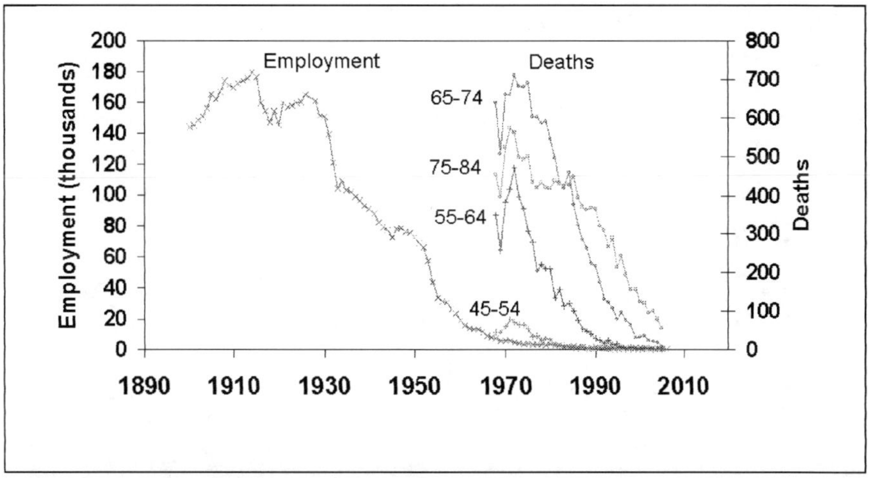

Figure 4. CWP deaths by age at death (1968–2005) and employment (1968–2006): anthracite region.

Figure 5 shows the equivalent picture for the bituminous region. As with the anthracite region, coal mining employment has fallen considerably although, unlike anthracite, there was a rise during the 1970s. Deaths with CWP rose initially in all age groups, and then fell, although there is a disconcerting indication that deaths in the younger age groups are rising.

Unlike the other outcomes studies here, national morbidity data exist for CWP. These data provide CWP prevalence data from 1970 – 2006 (figure 6). Since these are prevalence data, they reflect changes in intensity of exposure to coal mine dust and are independent of the number of miners exposed. They therefore complement the employment data and supply additional information on which to assess the temporal changes in mortality. Overall, the decline in prevalence evident until 1995-1999 appears to reflect and vindicate the reduction in exposure mandated by the 1969 U.S. Coal Mine Health and Safety Act. However, since 1995 there has been disturbing evidence that the prevalence of CWP has been rising (figure 6). Most of these miners would have started work after the introduction of the current federal compliance limit on dust exposure.

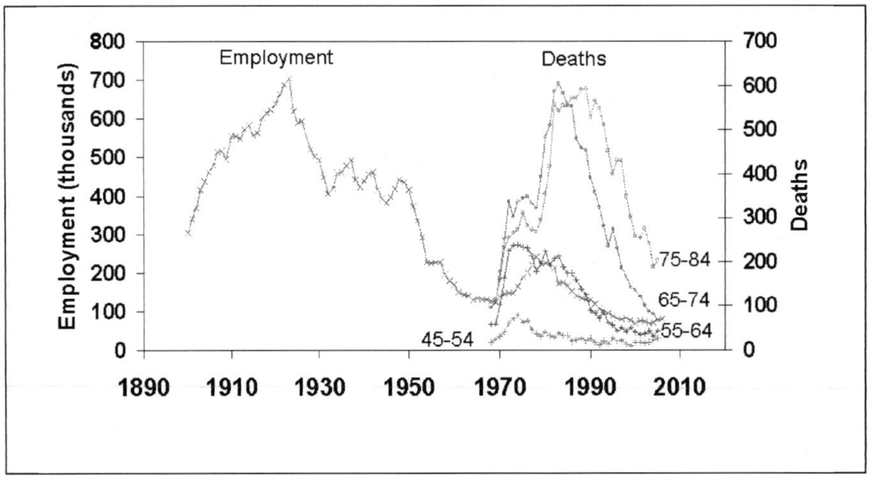

Figure 5. CWP deaths by age at death (1968–2005) and employment (1968–2006): bituminous region.

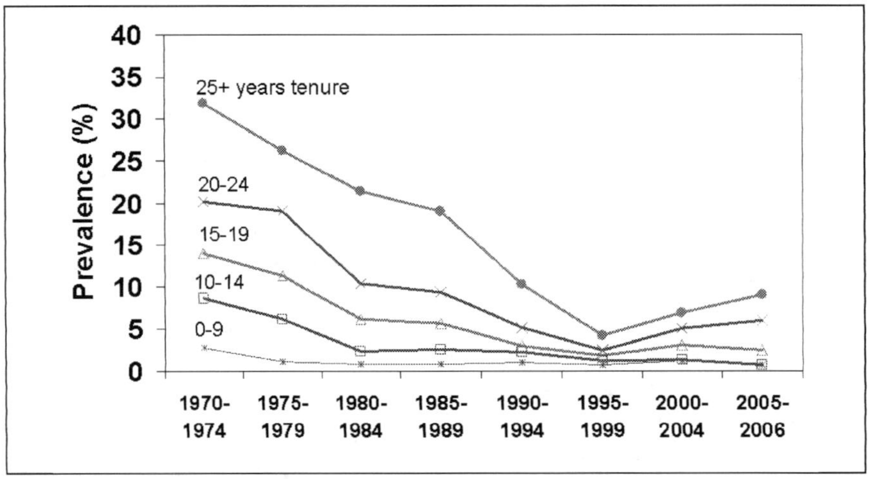

Figure 6. CWP prevalence among working coal miners (1970–2006) by tenure in mining.

No attempt was made to include deaths at age 15-44 for the three outcomes in the above description as they are too few for adequate representation. However, they confirm the general picture for the deaths in the adjacent 45–54 year age group for each outcome – asbestosis deaths appear to be trending lower, bituminous CWP deaths appear to be rising, and silicosis deaths are stationary.

4. Discussion

The correlation observed between deaths with asbestosis and asbestos consumption, coupled with the downward trend in consumption, suggests that the disease will eventually decline in importance and could even disappear. A preliminary tentative analysis undertaken by juxtaposition of the consumption and mortality curves suggests that this might occur around 2050 for the oldest age-at-death group, and before then for the younger age groups. However, this speculation is grounded totally on the assumption that asbestos exposure arises solely from asbestos usage (consumption). Although continued manufacture and application of new asbestos-containing products is still occurring at a low level (with future asbestos consumption being contingent on efforts to ban its use [5]), it could be that the bulk of asbestos exposure to workers in the future will increasingly come from the handling of existing asbestos installations and products. As buildings, factories, and ships that incorporate asbestos in their construction reach the end of their useful life they will be demolished, and those doing that work will be at risk of exposure. Even if buildings are not demolished but are renovated or asbestos is remediated there remains a great risk of exposure. Non-adherence to airborne compliance limits will lead to overexposure and continuing future mortality from asbestosis and other asbestos-related disease. Knowing what we know now about the toxicity of asbestos and how to prevent disease, it would be tragedy if this were ignored and a new epidemic of disease came about.

The apparent correlation between past asbestos consumption and observed mortality observable in Figure 1 supports the derivation of predicted deaths for future years. Deaths for the oldest age-at-death group appear to be peaking. This could coincide with the peak in consumption around 1950. It took 40 years for consumption to decline to negligible levels after 1950, suggesting that asbestosis mortality might continue for another 40 years (i.e., to around 2050) in that age group. Since there is evidence that the peaks in mortality have already occurred for those dying at younger ages, asbestosis mortality might cease to be a major problem before then for those age groups.

Nicholson and colleagues [6] made an intensive study of asbestos exposure and its associated disease outcomes, including a detailed attempt to predict future mortality. In this, the focus was on cancer outcomes, although a small part of the report mentioned asbestosis deaths. They looked only at underlying causes and so their results cannot be compared directly with those presented here. They stated that 200 asbestosis deaths were occurring (presumably around 1980) and that the number would perhaps double during the next two decades (i.e., to around the year 2000) and decline thereafter. It is not exactly clear from where they obtained their claim of 200 deaths, since the actual number around the time the paper was published was about 100 (89 in 1979, 101 in 1980, and 101 in 1981). Nevertheless, they overstated the deaths for 1980, and understated the increase, there being 561 deaths in 2000 – a 5-fold rise. Their statement that asbestosis deaths would decline thereafter is not yet being realized since overall underlying deaths were stationary at around 550 in 2000-2005. However, the overall number is greatly influenced by the deaths occurring at age 75-84. As are seen in figure 1, downward trends in numbers of asbestosis deaths are evident in the younger age groups before 2000.

While there are logical and very pertinent indicators of extent of exposure for asbestosis (consumption) and CWP (employment), silicosis lacks such obvious surrogates. It is often a disease resulting from incidental exposures in the making or processing of other products rather than arising from the production of a silica-based product [7]. While there are obvious industries where silica is the main product (ground silica mining and production, for instance), or where there is a direct link between exposure and product (crushed stone quarrying), the link between the principal product or activity and concomitant silica exposure is less obvious in many other important situations. Such sources of silica exposure are manifold, e.g., sand blasting and road saws in construction; foundry work; ore extraction in metal mining; pottery and clay production; and grinding operations in manufacturing. The variety of exposure situations makes it difficult to arrive at one succinct, pertinent indicator of extent of exposure to silica.

The number of silicosis deaths over time followed the trend in employment in metal mines. However, this relationship could not explain the overall pattern in silicosis deaths since only 9% of deaths were associated with work in metal mining (1985–1999), even though this was the greatest percentage associated with any specific industry. For example, 6% were associated with work in non-metal mining and quarrying. The Construction industry sector, as a whole, accounted for 12% of all deaths, but as is seen in figure 3, employment in construction shows a rise, not a fall, over time. In fact, most indicators of extent of exposure we could locate showed rises in employment or production over time. Many of these did show dips during the Great Depression years. However, anyone working at that time would likely have been born around or before 1900 and so would have died before 1980. So, while this could possible have impacted the trend in deaths for the oldest age group, it could not have impacted the similar downward trends in all of the younger age groups. Nor could all of the silica deaths come from industries for which we might not have obtained exposure surrogates. Of all deaths with silicosis between 1985 and 1999 for which we have information on usual industry, 58% reported working in mining, primary metal industries, construction, and stone, clay, glass, and concrete products [1]. These were industries for which we had information on employment, production, and/or consumption. Ultimately, then, the implication is that the reduction in deaths came about, at least in part, from reductions in worker exposures, probably both prior to and after the imposition of federal compliance limits in the 1970s.

It is possible that some CWP deaths were otherwise reported as silicosis. Some evidence for this can be found in looking at the usual industry and occupation on the death certificate, which mentioned coal mining and coal mining-related jobs in 7% of the deaths reported with silicosis. Whether these are true cases of silicosis, perhaps acquired during work on surface coal mines, or are misdiagnosed/misreported CWP is unknown. However, removal of the silicosis deaths from states where coal mining is prevalent did not eliminate the overall downward trends in deaths. Hence, the observed declines seen in figures 2 and 3 did not simply result from the downward trend in numbers with pneumoconiosis associated with reduced coal miner employment as noted for CWP below. Another possible disease-diagnosis issue relates to the recording of interstitial pulmonary fibrosis (IPF). A number of researchers have examined this issue, the results of which have been summarized by Taskar and Coultas [8]. Their conclusion is that IPF could be an occupational disease and might, in some cases, be simply a misdiagnosis of pneumoconiosis, particularly silicosis. Changes in diagnosis therefore could have impacted the trends in silicosis deaths. However, no data exist with which to evaluate this question. Since the number of deaths with silicosis appears to be currently stable, with no great evidence of continuing decline, silicosis mortality is expected indefinitely into the future. Adoption of the NIOSH recommended limit of 0.05 mg/m3 would aid in reducing this ongoing burden of disease.

For CWP, the reduction in employment in the anthracite region to the two hundred or so individuals currently employed clearly implies that few future cases of CWP will occur. An informal analysis based on past employment of deaths in the two older age groups (where there is the greatest legacy of past exposures), suggests that anthracite CWP deaths will decline to near zero about 2025 and 2037, respectively. This observation is conditional on the continuation of present trends and no revitalization of coal mining in the region. Since deaths appear to be declining faster than past employment in the anthracite industry, it may be that the above predictions are pessimistic and that an estimate of 2030 for the oldest age group is more realistic. With respect to CWP in the bituminous region, it is clear that substantial numbers of miners remain exposed. In addition, in 2008 bituminous coal mining employment was expanding. This alone, ignoring issues concerning the current effectiveness of the current federal dust compliance limit (discussed below), would lead to CWP mortality continuing indefinitely into the future. In fact, this may be the reason that mortality appears to be rising in the younger age groups, its trend following that of the increasing employment in the 1970s. Further follow-up is needed to monitor this situation.

Migration between the anthracite and bituminous regions may impact the findings for CWP. Due to the virtual demise of the anthracite coal industry, it is possible that many anthracite workers migrated to the bituminous coal fields, their deaths there artificially inflating the number of bituminous deaths and likewise underestimating the number of anthracite deaths. Some migration in the opposite direction could have occurred, although it is likely to be less. It is not known to what extent migration took place, as information on the source of work exposures is not given on the death certificate.

The employment data used in the analysis of CWP for both regions includes both surface and underground miners (data giving only underground employment has so far not been identified). The ratio of surface to underground employment has increased over time. Therefore, the number of miners experiencing the higher coal mine dust exposures more prevalent at underground mines decreased rather more rapidly than the employment trends on figures 4 and 5 indicate. Further research on this is needed.

Changes in the cultural understanding of the extent and severity of CWP in the U.S. coalfields have occurred over time. In addition, the ICD system revised its definitions of CWP substantially in the move from ICD-8 to ICD-9. In ICD-8 the principal term used was anthracosilicosis, whereas in ICD-9 the term was coal workers' pneumoconiosis. Overall, this change most likely impacted reporting in the bituminous region since it not only suggests that anthracite miners are affected by their work exposure but also implies that silica exposure is also involved in disease development. Epidemiological evidence emerging in the 1960s and 70s has since clearly shown that CWP does occur from exposure to bituminous coal mine dust, is related to cumulative exposure to that dust, and that silica exposure plays only a minor role in the causation of CWP. Increasing acceptance of the epidemiological findings may have led to the rise in reported deaths with pneumoconiosis in the bituminous region. In the anthracite region, in contrast, with its long history of coal mining (going back into the 18th century) and the higher risk of disease development from anthracite coal [9] it seems likely that CWP was recognized more fully and more promptly than in the bituminous coalfields. The ICD change in definition in 1979 probably came at a time when there was a greater appreciation of the risks that coal miners endure and so did not greatly impact the reporting of CWP in the anthracite region.

As noted above, it is also possible that some CWP was listed as silicosis (and vice versa), since the two diseases can be hard to distinguish medically. Certainly, we have evidence that coal miners have often been reported to have died with silicosis rather than CWP [10]. Possibly some or most of those cases may have worked at surface mines, but this is unknown, and further research is needed. A related issue arises from the reporting of Unspecified Pneumoconiosis. The geographical distribution of deaths with this cause is similar to that for CWP [10]. They may therefore have been CWP deaths, except that the recording physician used the shorthand 'pneumoconiosis' instead of stating CWP explicitly. Again, further research is needed to elucidate this issue.

In terms of predicted mortality for the future, the almost complete decline in the anthracite coal industry indicates that new cases of CWP should be very few and that current mortality arises from the legacy of past exposures. Some tentative extrapolation of the current temporal decline for the oldest age group suggests that mortality with CWP in the anthracite region may virtually disappear by 2030. The situation is much more complex for the bituminous coal region. Here employment appears to be cyclical, depending on the need for coal, with past increases and falls, and a current ongoing increase. Given no impact of other factors, the latter observation suggests that the number of CWP cases should similarly fluctuate in synchrony with past employment. Moreover, the observed increases in disease prevalence and mortality at younger ages do not bode well for the future.

There are several external factors that could have impacted the observed trends in deaths over time. The first is mass screening for asbestosis by legal firms and others intent on obtaining legal compensation for their clients led to the identification of many reported cases of asbestosis. Overall, the number of claims rose sharply in the mid-1990s, but declined abruptly in 2004 [11]. The recognition of the cases by this means might carry over to the later reporting of asbestosis on the death certificate. The mass screening for asbestosis has been followed by similar activities for silicosis. Therefore, the temporal trend in screening activity for compensation for these two diseases could impact the later temporal trend in reported mortality. Unfortunately, reliable data do not exist to assess the magnitude of this effect, if any, on the predictions. Federal compensation for CWP was introduced in the 1969 Coal Mine Health and Safety Act. The immediate effect is that affected individuals applied for, and received compensation, and then retired from coal mining. It is possible that part of the reduction in prevalence seen in the worker monitoring program resulted from the loss of those individuals from the workforce. Leaving coal mining also meant an end to coal mine dust exposure and so a reduction in risk of the disease progressing. However, it seems unlikely that this would impact the reporting of CWP on the death certificate for such individuals. Rather, receipt of compensation probably increased the chance of a report of CWP on the death certificate.

Is it possible to discern whether the U.S. federal compliance dust standards introduced around the 1970s are having an effect on mortality? Clearly, workers who spent time working after the U.S. federal compliance standards were imposed should have realized some benefit, resulting in fewer pneumoconiosis deaths that might have been the case if no standards had been imposed. Is this observable in our results? Of all age-at-death groups, the oldest would be expected to be the least impacted, as their work time under the reduced dust levels is very limited. Examination of the data for that age group for silicosis and CWP shows that they have slower rates of decline compared to the other age groups. For asbestosis, they are the only group whose mortality is still rising. The faster rates of decline of deaths at younger ages compared to the older age group therefore imply that workers have benefited from the imposition of federal dust standards. Overall, the mortality data are not a very satisfactory way to examine this issue. A much better approach is to examine morbidity data, particularly disease prevalence. For CWP, there is convincing evidence that the dust standard has been effective. But, as noted above, the recent increases in CWP prevalence and mortality are disturbing and point to a failure in dust control.

5. Conclusion

This analysis has examined U.S. deaths with asbestosis, silicosis, and coal workers' pneumoconiosis over the period 1968–2005 in relation to past indicators of extent of exposure to asbestos, silica, and coal dust. The findings for asbestosis suggest that due to the legacy of past exposures, older individuals will continue to die with asbestosis until about 2050; deaths among younger individuals may virtually disappear earlier. This prediction is totally contingent upon the effective control of asbestos fiber exposures from demolition and remediation work, as well as in the small remaining manufacturing industry. There has been substantial progress in reducing the burden of mortality from silicosis. Since there is no obvious relationship with diminished employment in at-risk activities, temporal reductions in exposure intensities may well have played a major role in the reduction in mortality. However, deaths with silicosis are still occurring, and the continuing occurrence in younger individuals suggests that intense overexposures remain too frequent. Owing to the virtual cessation of anthracite mining activity, and given no resurgence in the industry, the last few deaths with CWP in that region might well occur around 2030. The situation with respect to CWP in the bituminous coalfields is mixed. There has been a long period of reduced mortality and morbidity, probably due to both reduced mining employment and better dust control. However, the last 8 years has seen a disappointing rise in disease prevalence. At the same time CWP mortality in younger coal miners appears to be rising, rather than dropping as would be expected. Clearly, attention to better dust control is necessary, coupled with a reduction in dust levels to the limit recommended by NIOSH in 1995 [12]. These activities are even more critical when it is understood that bituminous coal mining employment is increasing currently.

References

[1] National Institute for Occupational Safety and Health. National occupational respiratory mortality system (NORMS) http://webappa.cdc.gov/ords/norms.html. Accessed June 06,

2008
[2] National Institute for Occupational Safety and Health. Coal Workers' Health Surveillance Program http://www.cdc.gov/niosh/topics/surveillance/ORDS/CoalWorkersHealthSurvProgram.html. Accessed June 06, 2008
[3] Virta RL 2008 Worldwide asbestos supply and consumption trends from 1900 through 2003. Circular 1298 (Reston, VA: U.S. Geological Survey)
[4] Energy Information Administration 1991 Coal data: A reference. (Washington DC: U.S. Department of Energy)
[5] Asbestos Disease Awareness Foundation 2008 Asbestos has not been banned http://www.banasbestos.us/ Verified June 06, 2008
[6] Nicholson WJ, Perkel G, Selikoff IJ 1982 Am. J. Ind. Med. 3 259-311
[7] National Institute for Occupational Safety and Health 2002 NIOSH Hazard Review. Health effects of occupational exposure to respirable crystalline silica (Cincinnati, OH: National Institute for Occupational Safety and Health)
[8] Taskar VS, Coultas DB. 2006 Proc. Am. Thorac. Soc. 3, 293-298
[9] Attfield MD, Seixas NS 1995 Am. J. Ind. Med. 27 137-51
[10] National Institute for Occupational Safety and Health 2003 Work-related lung disease surveillance report 2002. DHHS (NIOSH) No. 2003-111 (Cincinnati, OH: National Institute for Occupational Safety and Health)
[11] American Academy of Actuaries February 2008 Current issues in asbestos litigation (Washington, DC: American Academy of Actuaries)
[12] National Institute for Occupational Safety and Health 1995 Criteria for a recommended standard: Occupational exposure to coal mine dust. DHHS (NIOSH) Publication No. 95-106 (Washington, DC: National Institute for Occupational Safety and Health)

European and International Standards on health and safety in welding

A Howe

Health and Safety Laboratory, Harpur Hill, Buxton, Derbyshire, SK17 9JN, UK

E-mail: alan.howe@hsl.gov.uk

Abstract. A number of European and International Standards on health and safety in welding have been published in recent years and work on several more is nearing completion. These standards have been prepared jointly by the International Standards Organization (ISO) and the European Committee for Standardization (CEN). The standards development work has mostly been led by CEN/TC 121/SC 9, with excellent technical input from experts within Europe; but work on the revision of published standards, which has recently gathered pace, is now being carried out by ISO/TC 44/SC 9, with greater international involvement. This paper gives an overview of the various standards that have been published, are being revised or are under development in this field of health and safety in welding, seeking to (i) increase international awareness of published standards, (ii) encourage wider participation in health and safety in welding standards work and (iii) obtain feedback and solicit comments on standards that are currently under development or revision. Such an initiative is particularly timely because work is currently in progress on the revision of one of the more important standards in this field, namely EN ISO 10882:2001 *Health and safety in welding and allied processes — Sampling of airborne particles and gases in the operator's breathing zone — Part 1: Sampling of airborne particles.*

1. Background

Welding and allied processes are hazardous and pose various risks to the health and safety of persons who carry them out. In particular, precautions need to be taken to ensure electrical safety; safety from heat, light noise and mechanical hazards; and to avoid harm that can be caused by exposure to the fume and gases that welding processes generate.

The hazards associated with welding and allied processes are more or less identical wherever such processes are carried out, so measures taken to control the risks of harm should be the same in all countries. Ideally, therefore, agreement should be reached internationally on the best ways forward and on methods that can be adopted and applied with equal vigor the world over to help reduce risks and benefit workers and the industry alike. These are the objectives that European and International standardization in the field of health and safety in welding sets out to achieve.

2. Standards organizations

The most important organizations involved in international standards making in the field of health and safety in welding are the International Standards Organization (ISO) and the European Committee for Standardization (CEN). These are not-for-profit organizations that develop voluntary consensus standards to meet the requirements of business and broader needs of society. Such standards only become mandatory when they are given the force of law by citation in laws, regulations, codes, etc.

The International Organization for Standardization (ISO) is the leading producer of international standards in the world and most widely recognized name in the field. It is a network of National Standards Bodies (NSBs) from 158 countries with a Central Secretariat in Geneva, Switzerland that organizes and oversees its activities. NSBs can decide to participate in a particular field of standardization, in which case they are referred to as a P-member and have full voting rights; or they can have observer status, but as an O-member, although they can vote and comment on draft standards, they have no direct role in the standards development process; or of course an NSB can decide to opt out of a given field of work altogether.

The European Committee for Standardization is known as CEN, which is an acronym of the French name of the organization. It is based in Brussels and was founded in 1961 to contribute to the objectives of the European Union (EU) and European Economic Area. CEN, works closely with ISO, often collaborating under the 'Vienna agreement' to produce parallel ISO and CEN standards. CEN also works closely with the EU, carrying out mandated work to produce standards that serve the requirements of European Commission Directives, i.e. one of the EU Directorate Generals funds method validation work to ensure that EU data quality objectives are met.

A third major international organization involved in standards making in the field of health and safety in welding is the American Welding Society (AWS). AWS, which is based in Miami, was founded in 1919 with the goal of advancing the science, technology and application of welding and related joining disciplines. It has developed many recommended practices and guides, following strict American National Standards Institute (ANSI) procedures, for consumption mainly in the USA and Canada, but these are not considered in this paper.

Finally, the role played by Commission VIII of the International Institute of Welding (IIW) should be mentioned. IIW was founded in 1948 by welding institutes or societies in 13 countries who felt the need to make more rapid scientific and technical progress. Its mission is "to act as the world-wide network for knowledge exchange of joining technologies to improve the global quality of life". Whilst not a standards organization per se, IIW has a special relationship with ISO wherein it can prepare and assist in the formulation of International Standards. IW Commission VIII has to date mainly focused on preparing technical papers on the subject of health and safety aspects of welding. However, whilst it has yet to do so, it could be instrumental in promulgating International Standards in the field of health and safety in welding at some point in the future.

3. Standards committees

The committees with the lead role for European and International Standardization in the field of health and safety in welding are CEN/TC 121/SC 9 and ISO/TC 44/SC 9, both of which are entitled 'Health & safety in welding and allied processes'. These two subcommittees have generally collaborated, under the so-called Vienna Agreement (VA), to prepare joint European and International Standards, although each has also prepared a standard in its own right. The VA is an arrangement under which CEN and ISO co-operate on the development of a joint publication, with one or other of the two organizations taking the lead role.

In the past, development of joint European and International Standards in the field of health and safety in welding has been led by CEN/TC 121/SC 9. This subcommittee originally had five working groups, but the first two have now been disbanded and work on revising the standards they published transferred to their sister working groups within ISO/TC 44/SC 9. The working groups concerned are WG 1, which developed three European and International Standards that prescribe laboratory methods for sampling fume and gases generated by arc welding and WG 2, which developed two European and International Standards on sampling airborne particles and gases in the breathing zone of operators carrying out welding and allied processes. Two more CEN working groups have been set dormant: WG 3, which published a European Standards on transparent welding curtains; and WG 5, which developed a European and International Standard on fume data sheets and a European and

Journal of Physics: Conference Series **151** (2009) 012052 doi:10.1088/1742-6596/151/1/012052

International Technical Specification on the identification of thermal-degradation products generated when welding or cutting through products composed wholly or partly of organic material. The only active working groups are WG 4, which is completing work on three European and International Standards on testing and marking of equipment for air filtration, and WG 6 which is developing a European and International Technical Specification on the quantitative determination of fume from resistance spot welding

The parallel international standards committee, ISO/TC 44/SC 9, has not been active until recently, having delegated responsibility for the development of joint European and International Standards in the field of health and safety in welding to CEN/TC 121/SC 9 from the early 1990s until 2004. During that period, ISO/TC 44/SC 9 worked independently on one International Standard on 'Wordless precautionary labels' but, other than that, held only a watching brief. However, since 2004, when the subcommittee became rejuvenated under a new UK Secretariat, ISO/TC 44/SC 9/WG 1 and ISO/TC 44/SC 9/WG 2 have been actively engaged in the revision of the European and International Standards originally developed by CEN/TC 121/SC 9/WG 1 and CEN/TC 121/SC 9/WG 2; and ISO/TC 44/SC 9/WG 5 has been developing a European and International Standard based on the Technical Specification originally prepared by CEN/TC 121/SC 9/WG 5.

Other CEN and ISO technical committees have also prepared standards that prescribe safety requirements for equipment used in welding, e.g. EN 175:1997 'Personal protection — Equipment for eye and face protection during welding and allied processes, safe working procedures', which was prepared by CEN/TC 85 'Eye protective equipment', and product standards that include safety requirements. However, these are not included in the scope of this paper.

Table 1. List of European and International subcommittees and their working groups involved in the development of health and safety in welding standards.

Subcommittee / Working Group	Title
CEN/TC 121/SC 9	Health and safety in welding and allied processes
WG 1 (disbanded)	Laboratory methods for sampling fume and gases
WG 2 (disbanded)	Sampling of airborne particles and gases in the operator's breathing zone
WG 3 (dormant)	Transparent welding curtains
WG 4 (active)	Testing and marking of equipment for air filtration for welding and allied processes
WG 5 (dormant)	Fume data sheets
WG 6 (active)	Procedure for quantitative determination of fume from resistance spot welding
ISO/TC 44/SC 9	Health and safety in welding and allied processes
WG 1 (active)	Laboratory methods for sampling fume and gases
WG 2 (active)	Sampling of airborne particles and gases in the operator's breathing zone
WG 3 (not assigned)	
WG 4 (not assigned)	
WG 5 (active)	Identification of thermal-degradation products generated when welding or cutting through products composed wholly or partly of organic material
WG 6 (not assigned)	

4. Synopsis of published European and International Standards and Technical Specifications

Table 2 gives a synopsis of the eleven European and International Standards and Technical Specifications that have been published in the field of health and safety in welding and allied processes.

One European Standard, EN 1598:1998, was prepared by CEN/TC 121/SC 9, one International Standard, ISO 17846:2004, was prepared by ISO/TC 44/SC 9; and the other eight European and International Standards and one Technical Specifications were prepared jointly by CEN/TC 121/SC 9 and ISO/TC 44/SC 9 under the VA with a CEN lead.

Table 2. List of published European and International Standards and Technical Specifications

Standard or Technical Specification	Title
EN 1598:1997 and amendment A1:2001	Health and safety in welding and allied processes — Transparent welding curtains, strips and screens for arc welding processes
EN ISO 10882-1:2001	Health and safety in welding and allied processes — Sampling of airborne particles and gases in the operator's breathing zone — Part 1: Sampling of airborne particles
EN ISO 10882-2:2000	Health and safety in welding and allied processes — Sampling of airborne particles and gases in the operator's breathing zone — Part 2: Sampling of gases
EN ISO 15011-1:2002	Health and safety in welding and allied processes — Laboratory method for sampling fume and gases generated by arc welding — Part 1: Determination of emission rate and sampling for analysis of particulate fume
EN ISO 15011-2:2003	Health and safety in welding and allied processes — Laboratory method for sampling fume and gases generated by arc welding — Part: 2 Determination of emission rates of gases, except ozone
EN ISO 15011-3:2002	Health and safety in welding and allied processes — Laboratory method for sampling fume and gases generated by arc welding — Part: 3 Determination of ozone concentration using fixed point measurements
EN ISO 15011-4:2006	Health and safety in welding and allied processes — Laboratory method for sampling fume and gases — Part 4: Fume data sheets
CEN/ISO TS 15011-5:2006	Health and safety in welding and allied processes — Laboratory method for sampling fume and gases — Part 5: Identification of fume and gases generated during welding and cutting through products composed wholly and partly of organic materials
EN ISO 15012-1:2004	Health and safety in welding and allied processes — Requirements, testing an marking of equipment for air filtration — Part 1: Testing of the separation efficiency for welding fume
EN ISO 15012-2:2007	Health and safety in welding and allied processes — Requirements, testing an marking of equipment for air filtration — Part 2: Determination of the minimum air volume flow rate of captor hoods and nozzles
ISO 17846:2004	Welding and allied processes — Health and safety — Wordless precautionary labels for equipment and consumables used in arc welding and cutting

5. Synopsis of European and International Standards and Technical Specifications under development, revision or amendment

Table 3 gives a synopsis of the ten European and International Standards and Technical Specifications that are under development, revision or amendment in the field of health and safety in welding and allied processes.

A new proposal (NP) for a work item to revise minor aspects of the methodology described in EN 1598 has recently been circulated for adoption.

Good progress is being made with the revision of Part 1 of EN ISO 10882-1, under an ISO lead, with the work item expected to progress from the working draft (WD) to committee draft (CD) stage before the end of 2008.

Work on the revision of Parts 1, 2 and 3 of EN ISO 15011, which has an ISO lead, is nearly finished. The approval stage is expected to be completed before the end of 2008 for each Final Draft International Standard (FDIS).

A draft amendment to EN ISO 15011-4 is currently available as an FDIS and the approval stage is expected to be completed before the end of 2008.

An approved work item (AWI) has recently been added to the ISO/TC 44/SC 9 work program to develop a full standard based on CEN ISO TS 15011-5.

An approved work item (AWI) has recently been added to the CEN/TC 121/SC 9 work program to develop a new Part 6 of EN ISO 15011.

A preliminary work item (PWI) has recently been proposed to revise EN ISO 15012-1, under the Vienna Agreement with a CEN lead.

Finally, a further PWI, also under the Vienna Agreement with a CEN lead, has recently been proposed to prepare a new Part 3 of EN ISO 15012.

Table 3. List of European and International Standards and Technical Specifications under development, revision or amendment

Standard or Technical Specification	Title
prEN/NP 1598:2008	Health and safety in welding and allied processes — Transparent welding curtains, strips and screens for arc welding processes (revision of EN 1598:1997)
prEN ISO/WD 10882-1:2008	Health and safety in welding and allied processes — Sampling of airborne particles and gases in the operator's breathing zone — Part 1: Sampling of airborne particles (revision of EN ISO 10882-1:2001)
prEN ISO/FDIS 15011-1:2008	Health and safety in welding and allied processes — Laboratory method for sampling fume and gases — Part 1: Determination of emission rate and sampling for analysis of particulate fume (revision of EN ISO 15011-1:2002)
prEN ISO/FDIS 15011-2:2008	Health and safety in welding and allied processes — Laboratory method for sampling fume and gases — Part 2: Determination of emission rates of carbon monoxide (CO), carbon dioxide (CO_2), nitrogen monoxide (NO) and nitrogen dioxide (NO_2) in arc welding, cutting and gouging (revision of EN ISO 15011-2:2003)
prEN ISO/FDIS 15011-3:2008	Health and safety in welding and allied processes — Laboratory method for sampling fume and gases generated by arc welding — Part 3: Determination of ozone emission rate (revision of EN ISO 15011-3:2002)

Standard or Technical Specification	Title
prEN ISO/FDIS 15011-4:2006/DAmd 1	Health and safety in welding and allied processes — Laboratory method for sampling fume and gases — Part 4: Fume data sheets — Amendment 1
prEN ISO/AWI 15011-5:2008	Health and safety in welding and allied processes — Laboratory method for sampling fume and gases — Part 5: Identification of thermal-degradation products generated when welding or cutting through products composed wholly or partly of organic materials using pyrolysis-gas chromatography-mass spectrometry (development of a full standard based on CEN ISO TS 15011-5:2006)
CEN ISO/AWI TS 15011-6:2008	Health and safety in welding and allied processes — Laboratory method for sampling fume and gases — Part 6: Procedure for quantitative determination of fume from resistance spot welding
prEN ISO/PWI 15012-1:2008	Health and safety in welding and allied processes — Requirements, testing an marking of equipment for air filtration — Part 1: Testing of the separation efficiency for welding fume (revision of EN ISO 15012-1:2004)
prEN ISO/PWI 15012-3:2008	Health and safety in welding and allied processes — Requirements, testing and marking of equipment for air filtration — Part 3: Determination of the capture efficiency of on-gun welding fume extraction devices

6. Overview of European and International Standards and Technical Specifications

6.1. EN 1598

EN 1598, which was published in 1997, specifies safety requirements for the transparent welding curtains strips and screens that are used to protect people from the hazardous radiant emissions from welding arcs and from spatter.

The standard was recently confirmed at systematic review, but the responsible CEN technical committee decided that, nevertheless, it should be revised on the basis of comments received. Accordingly, a new work item has recently been proposed to revise the standard.

6.2. EN ISO 10882

6.2.1. *Subdivision of the standard.* EN ISO 10882 is a two-part standard entitled 'Health and safety in welding and allied processes — Sampling of airborne particles and gases in the operator's breathing zone'.

6.2.2. EN ISO 10882-1 . Part 1 of EN ISO 10882-1, which was published in 2001, prescribes a gravimetric method for determination of personal exposure to airborne particles. This involves sampling in the breathing zone, which is considered to extend only behind the welder's face shield, when such personal protective equipment is worn. The standard gives a number of examples of suitable sampler mounting arrangements and includes a recommendation that the user collects separate samples for determination of exposure to welding fume and other airborne particles generated by welding related operations, such as grinding. It also provides guidance on the use of chemical analysis to determine personal exposure to specific chemical agents present to which welders are exposed, but it does not describe analytical methods.

Journal of Physics: Conference Series **151** (2009) 012052 doi:10.1088/1742-6596/151/1/012052

Revision of the standard is now in progress but at present the work item is still at the working draft stage. The new draft includes revised requirements for positioning the sampler and new examples of suitable mounting arrangements including, most notably, a headset-mounted mini sampler[1]. Furthermore, the standard no longer attempts to estimate personal exposure to welding fume by excluding times of heavy grinding from sampling period, as this has been found to be impractical and most countries no longer have a separate limit value for welding fume in any case. The emphasis of the draft is also now very much on sampling, with gravimetric and chemical analysis moved to annexes and given equivalent standing. There is still considerable work to be carried out on the revised standard at this stage, however, so publication is not expected until the end of 2009.

6.2.3. EN ISO 10882-2. Part 2 of EN ISO 10882, which was published in 2000, provides guidance for the determination of personal exposure to gases and organic vapours and, as in Part 1 of the standard, sampling behind the welder's face shield is required when such personal protective equipment is worn.

This standard, which provides no detailed methodology, was confirmed at systematic review in 2005.

6.3. EN ISO 15011

6.3.1. Subdivision of the standard. EN ISO 15011 was originally a three-part standard entitled 'Health and safety in welding and allied processes — Laboratory method for sampling fume and gases generated by arc welding'. However, following its expansion into a six-part series, the title of the standard has been amended to take account of the fact that its scope now includes resistance spot welding as well as arc welding. The new series title is therefore now simply 'Health and safety in welding and allied processes — Laboratory method for sampling fume and gases'.

6.3.2. Rationale. Welding and allied processes generate fume and gases, which, if inhaled, can be harmful to human health. Knowledge of the composition and the emission rates of the fume and gases can be useful to occupational health professionals in assessing worker exposure and in determining appropriate control measures. Absolute exposure is dependent upon factors such as welder position with respect to the plume and draughts and cannot be predicted from emission rate data. However, in the same work situation, a higher emission rate is expected to correlate with a higher exposure and a lower emission rate with a lower exposure. Hence, emission rate data can be used to predict relative changes in exposure that might occur in the workplace under different welding conditions and to identify measures for reducing such exposure, but they cannot be used to calculate ventilation requirements.

6.3.3. EN ISO 15011-1. Part 1 of EN ISO 15011, which was published in 2002, specifies a method for measurement of welding fume emission rate and collection of welding fume for chemical analysis. The procedure set out in the standard involves welding in a test chamber, referred to as a 'fume box', and collecting the fume generated on a suitable filter. The procedure simply prescribes a methodology, leaving selection of the test parameters to the user, so that the effect of different variables can be evaluated. Emission rates vary considerably depending upon the exact test conditions, so test parameters have been prescribed in ISO 15011-4 for the generation of fume emission rate data that can be used for comparing emission rates of welding consumables.

In 2005, work commenced on extensive revision of the standard and, in particular, a more detailed method description and additional examples of suitable fume boxes were added. Publication of the revised standard, which is currently at the approval stage, is expected by end of 2008.

6.3.4. *EN ISO 15011-2.* Part 2 of EN ISO 15011, which was published in 2003, specifies a method for measurement of the emission rates of gases, except ozone. The standard provides guidance on how

Journal of Physics: Conference Series **151** (2009) 012052 doi:10.1088/1742-6596/151/1/012052

to carry out tests using direct reading instruments for inorganic gases and indirect methods for organic vapours, but gives no detailed methodology. Like Part 1 of the standard, welding is carried out in a test chamber that is referred to as a 'fume box'. However, instead of capturing all the emissions, as is the case for fume, emission rates are calculated from the concentration of gases and vapours in the test chamber and the air flow rate through it.

In 2005, work commenced on extensive revision of the standard. Although the equipment used remains essentially unchanged, the standard now refers to the use of a 'hood', rather than a 'fume box', to more correctly describe both its function and physical design. The most significant changes, however, are exclusion of organic vapours from the scope of the standard, due to a lack of evidence to support the applicability of the method to these substances; and comprehensive revision of the normative text to provide a detailed method description. As in EN ISO 15011-1, the procedure simply prescribes a methodology, leaving selection of the test parameters to the user, so that the effect of different variables can be evaluated.

The revised standard, which has been given a new title to specifically identify the gases that are included within its scope, is now at the approval stage. Publication is expected by end of 2008.

6.3.5. *EN ISO 15011-3.* Part 3 of EN ISO 15011, which was published in 2002, originally described a fixed-point measurement method for determination of ozone concentrations at three positions relative to the arc: one intended to corresponds to the typical position of the breathing zone of a welder; one where the ozone concentration was expected to be at a maximum; and a third more remote position intended to correspond to the background concentration in the workplace.

However, doubt about the relevance of results obtained using the method described in the published standard led to the commissioning of the EU project 'WELD-OZONE', the aim of which was to examine alternative methodology for the measurement ozone emissions during welding and allied processes. This work [2] established that fixed-point measurements of ozone concentration do not correlate well with workplace exposure measurements and recommended the drafting of a new standard using a fume box method, which was shown to have much better correlation.

Accordingly, in 2005, work commenced on the extensive revision of EN ISO 15011-3, the title of which was changed to reflect the move away from a method that prescribes the making of fixed-point measurements of ozone concentration to one that determines ozone emission rate. The revised standard defines a method for measuring the emission rate of ozone during arc welding using a hood technique. The procedure simply prescribes a methodology, leaving selection of the test parameters to the user, so that the effects of different variables can be evaluated. Publication of the revised standard, which is currently at the approval stage, is expected by end of 2008.

6.3.6. EN ISO 15011-4. Part 4 of EN ISO 15011, which was published in 2006, prescribes test conditions for generating fume emission rate and chemical composition data for welding consumables using the method prescribed in EN ISO 15011-1. Results obtained may be used on safety data sheets, in risk assessment and/or for classification of welding consumables.

The method was validated in laboratory tests and in two fume emission rate round robins. It was shown [3] to be robust, to exhibit good repeatability (typically ~5% for fume emission rate measurements and in the range 4-10% for chemical composition measurements) and the reproducibility of fume emission rate measurements was deemed to be fit-for-purpose, with most of processes examined having 95% confidence limits in the range ±10% to ±40%. Minor technical changes made to the methodology as a result of the findings of the fume emission rate round robins are the subject of a draft amendment to the standard, which is currently at the approval stage. Publication of the amendment is expected by the end of 2008.

6.3.7. *CEN ISO/TS 15011-5.* Part 5 of EN ISO 15011, which was published in 2006, is a technical specification that describes test procedures for the identification of thermal degradation products generated when welding and cutting through products composed wholly or partly of organic materials.

A wide range of product types is covered, including paints, primers, sealants, adhesives, pressing lubricants, oils etc; and five test methods are described, using: TIG welding; resistance welding; flame cutting with an oxy-acetylene torch; heating in an oven; and pyrolysis.

Since CEN ISO/TS 15011-5 was published, a collaborative project has been undertaken to evaluate the methods described in the technical specification. The conclusion of this work [4] was that pyrolysis-GC-MS is capable of identifying the components of interest in one simple experiment, whilst the other methods require multiple sampling and analysis, which makes them more difficult, time consuming and costly to carry out. The recommendation of the project consortium was therefore that a full standard based on CEN ISO TS 15011-5 should be developed using the pyrolysis method alone. A new work item has recently been approved accordingly, with work due to start on development of the new standard in the second half of 2008.

6.3.8. CEN ISO/TS 15011-6. Part 6 of EN ISO 15011 will be a technical specification that describes a procedure for quantitative determination of fume from resistance spot welding.

A new work item has recently been approved to develop the proposed new technical specification, but work is at an early stage, with the precise methodology to be used still under discussion within the responsible CEN working group. Publication is expected by 2010.

6.4. EN ISO 15012

6.4.1. Subdivision of the standard. EN ISO 15012 is a three-part standard entitled 'Health and safety in welding and allied processes — Requirements, testing an marking of equipment for air filtration'.

6.4.2. *Rationale.* Welding and allied processes generate fume and gases, which, if inhaled, can be harmful to human health. Control is often required to maintain exposure at acceptable levels and this can be achieved by capturing the fume and gases using local exhaust ventilation (LEV), which consists of a capture device, such as a captor hood or nozzle, connected, via ducting, to an exhaust system, or by the use of an on-gun welding fume extraction device. Effective capture of welding fume and gases can only be achieved when the extracted air velocity at the emission point exceeds the resulting velocity of the plume of welding fume and any cross-draught; and effective removal of the fume from the extracted air can only be achieved if associated equipment for air filtration has adequate separation efficiency fro the welding fume. This standard addresses these issues.

6.4.3. EN ISO 15012-1. Part 1 of EN ISO 15012, which was published in 2004, deals with testing of the separation efficiency of equipment for air filtration for welding fume. The method prescribed involves the measurement of welding fume concentrations in ducts upstream and downstream of welding fume separation equipment using isokinetic sampling, collection on a filter and gravimetry. Testing of welding fume separation equipment with an unducted outlet is carried out in a closed test cabin.

The standard was recently confirmed at systematic review, but the responsible ISO and CEN technical committees decided that, nevertheless, it should be revised on the basis of comments received. In particular, there was concern that the classification system for welding fume separation equipment prescribed in the standard was unsatisfactory. Accordingly, a preliminary work item has recently been proposed to revise the standard.

6.4.4. EN ISO 15012-2. Part 2 of EN ISO 15012, which was published in the first half of 2008, addresses determination of the minimum air volume flow rate of captor hoods and nozzles by determination of the air volume flow rate required to achieve an air velocity of 0,4 m s^{-1} at defined test positions in front of the capture device.

6.4.5. EN ISO 15012-3. Part 3 of EN ISO 15012 was to have prescribed a method for determination of the capture efficiency of welding fume extraction devices using a tracer gas method. However, concern about the ability of a tracer gas to adequately mimic the behaviour of welding fume recently led to a decision to delete the existing approved work item from the CEN and ISO work programs and propose a new preliminary work item with a revised scope limited to determination of the capture efficiency of on-gun welding fume extraction devices.

6.5. ISO 17846

ISO 17846, which was published in 2004, lays down requirements for wordless precautionary labels for equipment and consumables used in arc welding and cutting. These are based on recognised hazard and safety symbols which achieved 95% - 100% comprehension rates in tests on 10000 experienced and inexperienced welders. The standard was confirmed at systematic review in 2007.

7. Summary

A number of European and International Standards on health and safety in welding have been published in recent years and work on several more is nearing completion. The standards development work has mostly been led by CEN/TC 121/SC 9, with excellent technical input from experts within Europe; but work on the revision of published standards, which has recently gathered pace, is now being carried out by ISO/TC 44/SC 9, with greater international involvement. This paper gives an overview of the various standards that have been published, are being revised or are under development in this field of health and safety in welding, seeking to (i) increase international awareness of published standards, (ii) encourage wider participation in health and safety in welding standards work and (iii) obtain feedback and solicit comments on standards that are currently under development or revision.

To find out about more about ISO and CEN standards, see the websites of the organizations concerned:

- www.iso.org
- www.cen.eu

To purchase standards, or to get involved in standards work, contact your NSB, e.g.:

- www.bsi-global.com
- www.afnor.org
- www.din.de

References

[1] Lidén G and Surakka J 2008 *Ann. Occ. Hyg.* (in press)

[2] FORCE Technology, Norddeutsche Metall-Berufsgenossenschaft, Finnish Insitute of Occupational Health, TWI Ltd and Instituto de Soldadura e Qualidade 2003 Final Technical Report, EU Project WELD-OZONE Project No: GRD1-1999-20008, Contract No: G6RD-CT-2000-00270 *Examination of the measurement and control of ozone emissions during welding and allied processes* FORCE Technology, Park Allé 345, 2605 Brøndby, Denmark

[3] Howe A and Carter G 2005 *Welding and cutting* **4**(6) 334-40

[4] Pengelly I 2008 *Analysis of weld-through products: Summary and recommendations for EN ISO 15011-5* HSL report AS/2007/26 Health and Safety Laboratory, Harpur Hill, Buxton, Derbyshire, SK17 9JN, UK

Historical exposure levels of inhalable dust in the Polish rubber industry compared to levels in Western Europe.

F de Vocht[1,4], H Kromhout[2], W Sobala[3], and B Peplonska[3]

[1] Occupational & Environmental Health Research Group, School of Translational Medicine, Faculty of Medical and Human Sciences, The University of Manchester, Manchester, UK

[2] Division of Environmental Epidemiology, Institute for Risk Assessment Sciences, Utrecht University, Utrecht, The Netherlands

[3] Department of Occupational and Environmental Epidemiology, NOFER Institute of Occupational Medicine, Lodz, Poland

[4] Correspondence to Dr Frank de Vocht, Occupational & Environmental Health Research Group, School of Translational Medicine, Faculty of Medical and Human Sciences, The University of Manchester, Ellen Wilkinson Building, Oxford Road, Manchester M13 9PL, UK.

Frank.devocht@manchester.ac.uk

Abstract. Although studies have been carried out to assess inhalable dust exposure levels in the rubber manufacturing industry, the levels of exposure in factories in Eastern Europe are less well documented. Routine stationary sampling for compliance testing of inhalable aerosols has however been conducted in a large factory producing tires and tubes in Poland between 1981 and 1996 (N=6,152). This study was conducted to assess historical inhalable aerosol levels in different departments in this rubber plant and to compare the results with estimates based on European data from the United Kingdom, Sweden, the Netherlands and Germany, and also Poland (EXASRUB project). Geometric mean (GM) concentrations in the factory ranged from 2.41 mg/m^3 to 5.82 mg/m^3 and were to a large extent associated to the actual production capacity of the plant and flow of the production process. Whereas 3-4 fold differences between departments existed prior to about 1985, stronger reduction of exposure in the raw materials and finishing departments (-12%/year) compared to other departments (range -5%/yr to -3%/yr), resulted in comparable levels in the 1990s. However, in the pre-treating departments, average concentrations were still about a factor 2-3 higher than in other departments, which could presumably be attributed to the use of anti-tacking agents. GM concentrations have been modelled using (1) stationary measurements collected in the Polish factory only, or (2) all European data collected in the EXASRUB project. Comparison of the estimates showed that these were fairly similar for both datasets. This analysis showed that the levels of inhalable aerosols in the Polish rubber industry have been at least a factor three to four higher than in Western European countries in the 1980s and 1990s, depending on the department, but that these differences were getting smaller in the 1990s. Furthermore, the estimates based on all European data from EXASRUB provides valid estimates compared to factory-specific data.

1. Introduction

The rubber industry as a whole was classified as "entailing exposures that are carcinogenic to humans (group 1)" by the International Agency for Research on Cancer (IARC) as early as 1987 [1,2]. Employment in the rubber manufacturing industry exposes employees to a wide range of chemicals, including n-Nitrosamines, aromatic amines, rubber fumes, rubber process dusts, and more. One the important airborne exposures in this industry is exposure to rubber dusts, which have been shown to be mutagenic [3-5] as well as probably genotoxic [6].

Although studies have been carried out to assess inhalable dust exposure levels in rubber manufacturing [7,8], the levels of exposure in factories in Eastern Europe are less well documented. However, an exposure measurement survey was conducted in a large Polish tire manufacturing factory in the 1980s and '90s in which a large number of airborne chemicals, including inhalable aerosols, were measured [9]. These data were subsequently used to compare Polish exposure levels with those in other European countries in the EXASRUB project (http://exasrub.iras.uu.nl) [10,11] and have also been used to estimate a quantitative job-exposure matrix for this specific factory [12]. The present exercise was conducted to assess the validity of the estimates for average inhalable dust in a Polish factory extrapolated from a "European" statistical model based on data from five European countries compared to estimates from detailed "local" statistical analysis using only measurement data from the Polish factory.

2. Methods

The factory where the measurement survey [9] was conducted was involved in a large cohort study in the Polish rubber industry [13]. Its history and production has been described in detail [12], but in short the factory has been in production since 1950 and it consists of two mixing departments, a car tire production line, a production line for tubes for car tires, and agricultural vehicle tire production line, and a production line for bicycle tubes and tires. Except for the agricultural tires production line, assembly of all types of tires was done using "cold" processes.

Within the measurement survey routine stationary sampling for compliance testing of inhalable aerosols (N=6,152) was conducted between 1981 and 1996 (Table 1). Inhalable aerosol measurements were collected by means of a 42mm sampling head equipped with a 50mm AFPC filter (98%) or 50mm glass fiber filter (2%) at a flow rate of 2L/min (93%) or 3L/min (5%) (2% unknown). The sampling time was 3 hours (79%) or 4 hours (19%), with some 6 hour measurements (2%) as well.

Log-transformed concentrations were analyzed using linear mixed effects models to account for repeated sampling of locations to assess airborne levels and trends in time in the different departments in the factory. Deviations from log-linear trends in time in departments were assessed using penalized smoothing splines and these results were used to improve the estimates from the mixed effects models.

Inhalable dust exposure estimates extrapolated from the "European" analysis were obtained from analyses of the EXASRUB dataset [10]. This dataset included 4,284 personal and 8,463 stationary inhalable aerosol measurements collected in the Netherlands (N=2,285), United Kingdom (N=4,134), Sweden (N=415), Germany (N=185) and Poland (N=6,361) between 1969 and 2003. Estimates of average personal exposure levels for all countries were published previously [10], and these estimates were used to estimate average inhalable dust exposure levels based on stationary sampling using average differences between personal and stationary (57% to 68% lower) samples in different departments from those same statistical models.

Journal of Physics: Conference Series **151** (2009) 012053 doi:10.1088/1742-6596/151/1/012053

3. Results

The results have been summarized in Table 1. Inhalable geometric mean (GM) concentrations in the factory ranged from 2.41 mg/m^3 in the finishing departments to 5.82 mg/m^3 in the mixing and milling departments. The inhalable aerosol levels were determined to a large extent by the production process. More specifically, inhalable aerosols were primarily generated in the first stages of the production process, where crude materials are handled, mixed and milled, and assembled before they are vulcanized in the curing process (GM=5.09 to 5.82 mg/m^3). After curing of products (GM=3.90 mg/m^3), average concentrations drop to 2.41 mg/m^3 in the finishing departments. High inhalable aerosol levels were also found outside the production process for non-process factory workers (GM=4.07 mg/m^3) and in the maintenance and engineering department (4.08 mg/m^3).

Estimated average inhalable dust levels in the different departments between 1980 and 1996 have been shown graphically in Figure 1. Whereas 3-4 fold differences between departments existed prior to about 1985, stronger reduction of exposure in the raw materials and finishing departments (-12%/year) compared to other departments (range -5%/yr to -3%/yr), resulted in comparable average inhalable aerosol levels throughout the factory in the 1990s. Only in the pre-treating departments however, average concentrations were still approximately a factor 2-3 higher than in other departments in the 1990s, which could presumably be attributed to the use of anti-tacking agents like talc and zinc stearate.

Figures 2.1 to 2.6 show the average inhalable dust concentrations in the different departments between 1980 and 1995 estimated using three different methods. Average personal exposure levels have been based on the results from the statistical models using all "European data" collected in the EXASRUB project [10,11] and published in [10]. Average concentration from stationary sampling were based on the same statistical models published in [10] but applying an additional correction factor to account for the difference between personal and stationary measurements, and the third "local data" method was to estimate average concentrations based on stationary measurements collected in the Polish factory only. The models and results of this third method have been published previously in [12]. As shown, personal exposure levels are on average a factor three higher than average concentrations from stationary measurements. However, the estimates of average inhalable dust concentrations based on stationary measurements from the European data are comparable to those based on "local" analysis using measurement data from that specific factory only.

Combining this with the results published in [10] suggests that the inhalable aerosol levels in the 1980s and 1990s in Poland were a factor 3 to 4 higher than in the Netherlands, UK and Sweden, but comparable to those in Germany. However, data from a small field-study assessing the relative performance of the different aerosol samplers to measure rubber dust in the different countries in the EXASRUB project showed that the sampling head used in this Polish survey under-sampled concentrations by approximately 35-50%, depending on the particle size, compared with samplers used in the other countries and to the EU-CALTOOL reference sampler [14]. This implies that in the 1980s and 1990s the differences between countries might even have been somewhat larger and that the average inhalable dust concentrations in Poland were more than twice as high as in western European countries.

4. Discussion

This exercise aimed to assess the validity of the estimates for average inhalable dust in a Polish factory extrapolated from a "European" statistical model, based on data from five European countries, compared to estimates from a detailed "local" statistical analysis using only measurement data from the Polish factory. In general, these data show that the processes across Europe are similar resulting in similar relative ranking of departments in terms of inhalable aerosol exposure levels. Furthermore, the predictions of average concentrations are comparable regardless of whether only "local data" was used or whether predictions are based on a large dataset containing measurement from different countries.

It should be noted that these analyses discuss average inhalable aerosol concentrations in departments and as such do not take into account measurements in specific low- or high-exposure tasks.

The measurement data does suggest that the levels of inhalable rubber aerosols in the Polish rubber industry have been at least a factor three to four times higher than in Western European countries in the 1980s and 1990s, depending on the department, but that the differences were getting smaller in the 1990s.

The reduction in average inhalable dust exposure levels, especially since half-way through the 1980s, can presumably be attributed to major changes in technology and the introduction of more effective exposure control measures in that time period [12]. In the Netherlands, it has been shown that these measures can reduce exposure to aerosols by 34-49% [8]. However, although specific information on local exhaust ventilation and other exposure reduction measures was collected for some locations within the factory but not for other locations,, this information was not used in these analyses to further refine the factory-specific models. Additionally, a reduction in the production level of the factory around the period 1988-1992 [12] might have increased the trend towards lower average exposure levels.

As such, these data suggested that these estimates were comparable to and thus that the "European estimates" for average inhalable aerosol levels in different departments and years in European countries obtained in EXASRUB are also valid at a "local" level.

Acknowledgements
This study was carried out within a large framework sponsored by ECNIS (Environmental cancer risk, nutrition and individual susceptibility), European Union Sixth Framework Programme (Grant number: FOOD-CT-2005-513943), EPITOK (Transfer of Knowledge in Molecular Biology and Epidemiology of Occupational and Environmental Cancer), European Union Fifth Framework Programme (Grant number: MTKD-CT-2004-509829), and EXASRUB (Improved Exposure Assessment for Prospective Cohort Studies and Exposure Control in the European Rubber Manufacturing Industry), Quality of Life and Management of Living Resources Programme, European Union 6th framework (Grant no: QLk4-CT-2001-00160 and QLRT-2001-02786)

Reference List

[1] IARC 1982 Monographs on the evaluation of the carcinogenic risk of chemicals to humans. Volume 28; The rubber industry IARC press Lyon France
[2] IARC 1987 Monographs on the evaluation of the carcinogenic risk of chemicals to humans. Overall evaluation of carcinogenicity: An updating of IARC monographs. Volumes 1 to 42, Supplement 7 IARC Lyon France
[3] Baranski,B, Indulski J, Janik-Spiechowicz E Palus J 1989 Mutagenicity of airborne particulates in the rubber industry J.Appl.Toxicol. **9** 389-393
[4] Vermeulen R Bos R P Kromhout H 2001 Mutagenic exposure in the rubber manufacturing industry: an industry wide survey Mutat.Res **490** 27-34
[5] Vermeulen R Bos R P de,H.J., van,D.H., and Kromhout H 2000a Mutagenic profile of rubber dust and fume exposure in two rubber tire companies Mutat.Res **468** 165-171
[6] Vermeulen R Bos R P Pertijs J and Kromhout H 2003 Exposure related mutagens in urine of rubber workers associated with inhalable particulate and dermal exposure Occup.Environ.Med **60** 97-103
[7] Dost AA Redman D and Cox G 2000 Exposure to rubber fume and rubber process dust in the general rubber goods, tyre manufacturing and retread industries Ann.Occup.Hyg. **44** 329-342
[8] Vermeulen R, de HJ, Swuste P and Kromhout H 2000 Trends in exposure to inhalable particulate and dermal contamination in the rubber manufacturing industry: effectiveness of control measures implemented over a nine-year period Ann.Occup.Hyg. **44** 343-354

[9] Szadkowska-Stanczyk I Wilczynska U Sobala W and Szeszenia-Dabrowska,N 2001 Occupational exposure to chemicals in the manufacture of rubber tires Med.Pr **52** 401-408

[10] de Vocht F Vermeulen R Burstyn I Sobala W Dost A Taeger D Bergendorf U Straif K Swuste P and Kromhout H 2008 Exposure to inhalable dust and its cyclohexane soluble fraction since the 1970s in the rubber manufacturing industry in the European Union Occup.Environ.Me. **65** 384-391

[11] de Vocht F Straif K Szeszenia-Dabrowska N Hagmar L Sorahan T Burstyn I Vermeulen R and Kromhout H 2005 A database of exposures in the rubber manufacturing industry: design and quality control Ann.Occup.Hyg. **49** 691-701

[12] de Vocht F Sobala W Peplonska B Wilczynska U Gromiec J Szeszenia-Dabrowska N and Kromhout H 2008 Elaboration of a quantitative job-exposure matrix for historical exposure to airborne exposures in the Polish rubber industry Am.J.Ind.Med .

[13] Wilczynska U Szadkowska-Stanczyk I, Szeszenia-Dabrowska N Sobala W and Strzelecka A 2001 Cancer mortality in rubber tire workers in Poland. Int.J.Occup Med.Environ Health **14** 115-125

[14] de Vocht F Huizer D Prause M Jakobsson K Peplonska B StraifK and Kromhout H 2006 Field comparison of inhalable aerosol samplers applied in the european rubber manufacturing industry Int.Arch.Occup.Environ.Health **79** 621-629

Table 1. Characteristics of dataset (adapted from (de Vocht F. *et al.*, 2008a). Number of different locations where stationary samples were taken, number of measurements, geometric mean (GM) concentrations and geometric standard deviation (GSD) for inhalable aerosols.

		Inhalable aerosols		
Department	locations	samples	GM (mg/m^3)	GSD
Crude materials	2	242	5.68	1.75
Compounding, mixing and milling	17	2,734	5.82	1.76
Pre-treating	3	494	5.09	1.32
Assembly (tires. Tubes or valves)	7	1,999	5.54	1.60
Curing	2	9	3.90	1.48
Finishing	2	103	2.41	1.60
Storage	1	1	1.70	
Non-process working	12	567	4.07	1.49
Maintenance & Engineering	2	3	4.08	1.18

Figure 1. Average (GM) concentrations in the different factory departments (miscellaneous departments includes curing, storage, non-process workers, and maintenance and engineering departments)

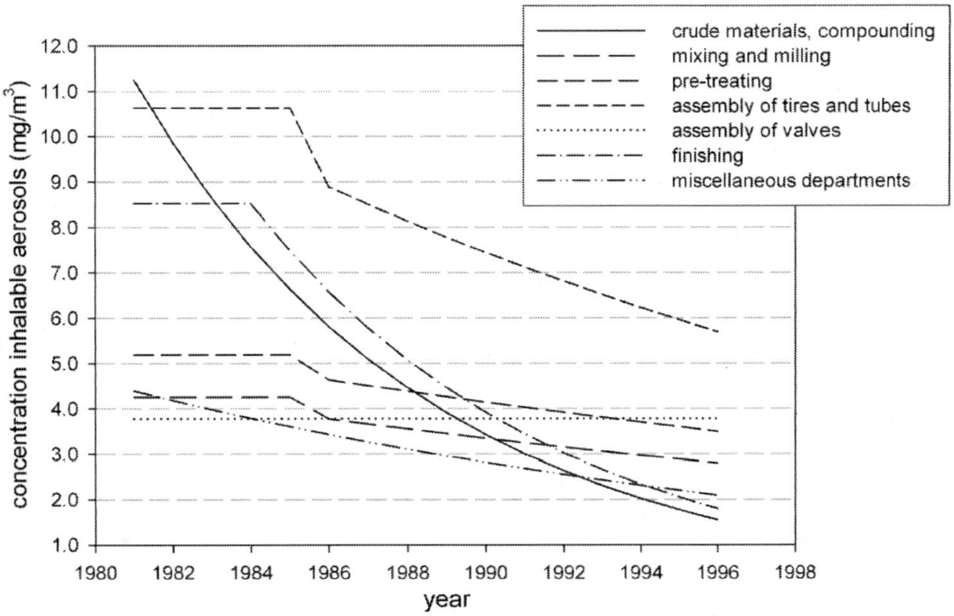

Journal of Physics: Conference Series **151** (2009) 012053 doi:10.1088/1742-6596/151/1/012053

Figures 2.1-2.6. Comparison of estimated average (GM) personal and stationary exposure levels based on the full EXASRUB dataset and average exposure based on stationary measurements based on the Polish data only in different factory departments.

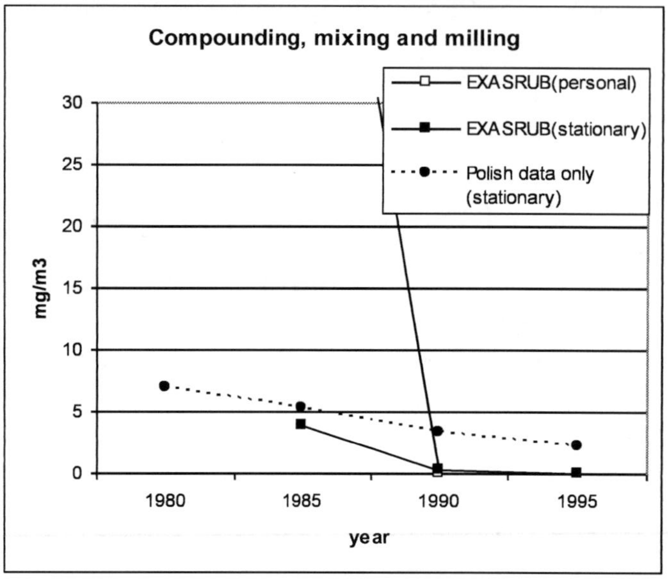

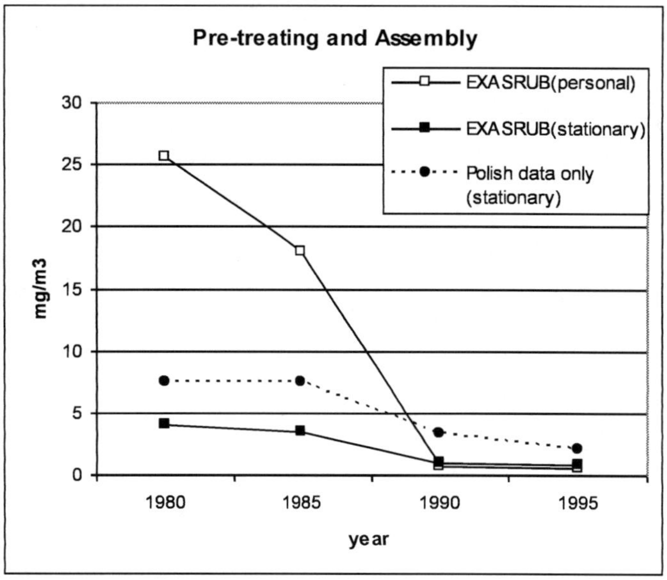

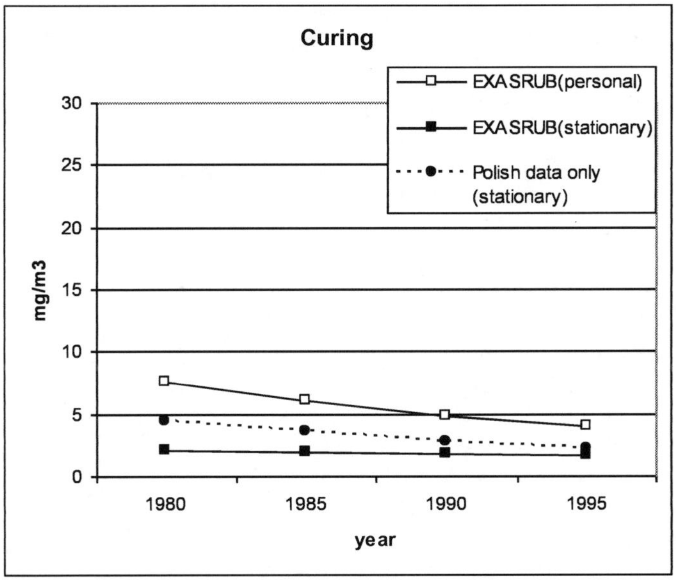
Curing
EXASRUB(personal)
EXASRUB(stationary)
Polish data only (stationary)
mg/m3
30
25
20
15
10
5
0
1980
1985
1990
1995
year

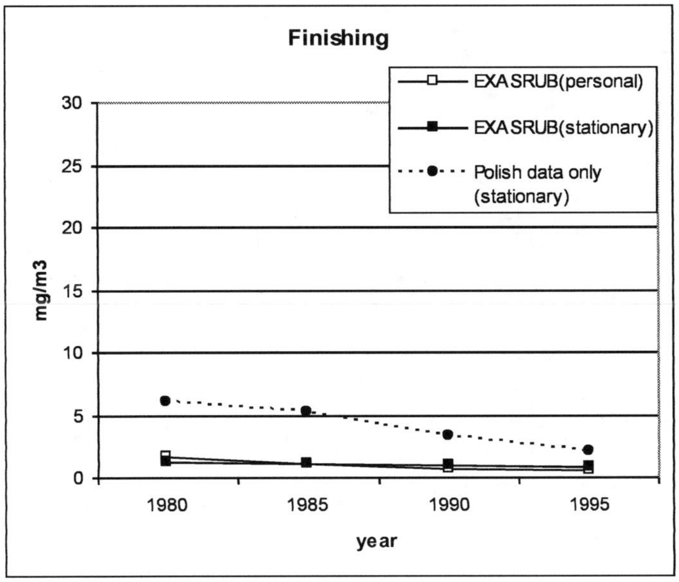
Finishing
EXASRUB(personal)
EXASRUB(stationary)
Polish data only (stationary)
mg/m3
30
25
20
15
10
5
0
1980
1985
1990
1995
year

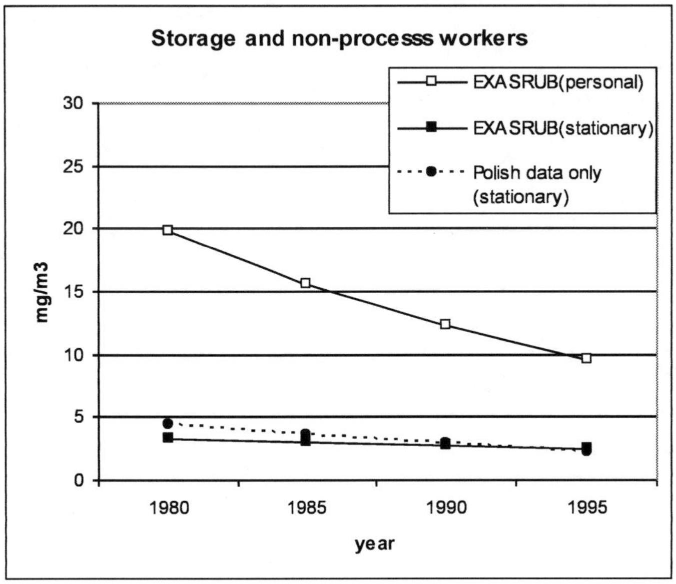
Storage and non-processs workers
EXASRUB(personal)
EXASRUB(stationary)
Polish data only (stationary)
mg/m3
30
25
20
15
10
5
0
1980
1985
1990
1995
year

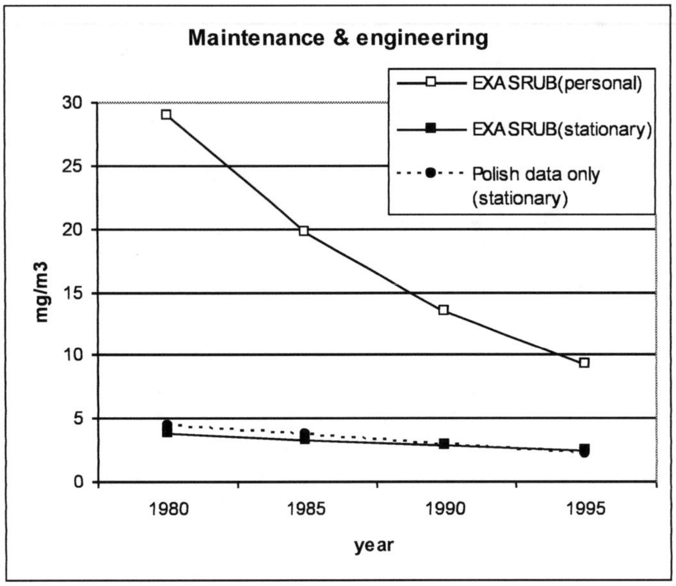
Maintenance & engineering
EXASRUB(personal)
EXASRUB(stationary)
Polish data only (stationary)
mg/m3
30
25
20
15
10
5
0
1980
1985
1990
1995
year

Personal Exposure to Inhalable Cement Dust among Construction Workers

Susan Peters[1], Yngvar Thomassen[2], Edeltraud Fechter-Rink[3] and Hans Kromhout[1]

1. Institute for Risk Assessment Sciences, Environmental Epidemiology Division, Utrecht University, Utrecht, The Netherlands

2. National Institute of Occupational Health, Oslo, Norway

3. TÜV SÜD Industrie Service GmbH, Mannheim, Germany

Corresponding author: s.peters@uu.nl

Abstract. A case study was carried out in 2006-2007 to assess the actual cement dust exposure among construction workers involved in a full-scale construction project and as a comparison among workers involved in various stages of cement and concrete production. Full-shift personal exposure measurements were performed for several job types. Inhalable dust and cement dust (based on analysis of elemental calcium) concentrations were determined. Inhalable dust exposures at the construction site ranged from 0.05 to 34 mg/m3, with a mean concentration of 1.0 mg/m3. For inhalable cement dust mean exposure was 0.3 mg/m3 (range 0.02-17 mg/m3). Reinforcement and pouring workers had the lowest average concentrations. Inhalable dust levels in the ready-mix and pre-cast concrete plants were, on average, below 0.5 mg/m3 for inhalable dust and below 0.2 mg/m3 for inhalable cement dust. Highest dust concentrations were measured in cement production, particularly during cleaning tasks (inhalable dust GM=55 mg/m3; inhalable cement dust GM=33 mg/m3) at which point the workers wore personal protective equipment. Elemental measurements showed highest but very variable cement percentages in the cement plant and very low percentages of cement during reinforcement work and pouring.

Keywords – construction; inhalable cement dust; exposure; workers

1. Introduction

The Health and Safety Executive (HSE, UK) concluded in 1994 that there was no convincing evidence for an increased incidence of any site-specific cancer resulting from cement exposure, but acknowledged that the data available at that time was not consistently and reassuringly negative [1]. In 2005 the HSE reviewed findings in more recently published studies. Among these and earlier studies increased incidences of cancers at several sites (stomach, lungs, colon, head and neck) were noted. A causal association between Portland cement exposure and cancer has however not been established and therefore the uncertainty concerning a possible cancer risk remains [2]. Few studies performed after the HSE review also show some evidence for pharyngeal carcinoma [3], oesophageal adenocarcinoma [4], squamous cell carcinoma of the skin [5] and gastric cancer [6]. Besides a possible cancer risk, occupational exposure to cement dust has been associated with reduced respiratory function. At present evidence is insufficient to establish exposure-response relationships for these effects [2].

Cement dust exposure occurs commonly in the cement production and construction industry. Most studies carried out in these industries, however, focus on silica exposure. There is no measurement data published about cement dust at construction sites at all. On cement dust exposures in cement plants only limited data is available.

The present study was conducted in the Rhein-Neckar area in Germany, where in a recent epidemiological hospital-based case-control study on head- and neck cancer an increased risk of laryngeal carcinoma for construction workers was suggested with cement dust exposure as possible cause [7]. In this context, a case study was carried out in 2006-2007 to assess the actual cement dust exposure of construction workers involved in a full-scale construction project. Job tasks in all stages were included. For comparison, we also incorporated workers involved in the various stages of cement and concrete production. This method is rather unique as we studied the entire process; both the production chain and the total building process. Aim of this study was to assess the actual cement dust exposure among construction workers and workers cement and concrete production.

2. Methods

2.1. Exposure measurements

Full-shift personal inhalable dust measurements in the breathing zones were performed during several tasks in the construction project (i.e. reinforcement work, concrete pouring, concrete repair, bricklaying, floor screed laying, installation work and tile setting) and among workers in a cement plant, five ready-mix concrete plants, and a pre-cast concrete plant. The sampling equipment was in conformity with European standard EN 1232 [8]. A portable pump operating at a flow rate of 3.5 l/min was used, together with a GSP (CIS) sampling head [9], using PVC filters with a 37 mm diameter and a 5-µm pore-diameter. Measurements were all performed after clear instructions of the workers. Daily activities and relevant situations were documented in a structured way. Several field blank GSP filters were collected without passing air through the filter on each sample location. Based on these blanks, the limit of detection (LOD) was estimated to be 0.065 mg/m^3 for inhalable dust for a 7-8 hour sample. The LOD for the cement-content was based on the analytical limit of detection for calcium (0.015 mg) and was estimated to be 0.035 mg/m^3 for 7-8 hour sampling.

2.2. Gravimetrical measurements

Gravimetrical and element measurements were done under continuous temperature and air humidity control. The laboratory balance used was a Sartorius MC Balance, max capacity 5.1 g, stability <1 µg. The mass detection limit using an in house 37 mm PVC filter kept under the above conditions was 0.011 mg (3 times standard deviation). Provisions to discharge the filters prior to weighing were used at all occasions.

2.3. Elemental composition

After weighing, each filter was placed in a Teflon autoclave and 1 ml of *aqua regia*, 1 ml of hydrofluoric acid and a known quantity of beryllium chloride solution were added as an internal spectroscopic standard. The Teflon autoclave was heated in a microwave unit (MLS 1200, Teflon container SV140, 10 Bar, Milestone, Sorisole, Italy); after cooling the digest was added 1 ml of saturated boric acid and reheated before final dilution to 15 ml. The Ca concentration in the solutions were measured by inductively coupled plasma optical emission spectrometry (ICP-OES) using a Perkin-Elmer Optima 3000 spectrometer. Certified NIST reference Portland cement materials were used for quality assurance of the chemical measurements. The cement content of the inhalable dust was based on the Ca-content of the collected samples. Cement coded CEM II/B-S 32.5 R was used as reference. The composition of this cement was as follows: SiO_2 (23.6%), Al_2O_3 (6.67%), Fe_2O_3 (2.54%), CaO (56.8%), MgO (4.23%), MnO (0.14%), K_2O (1.25%), SO_3 (3.05%), Cl (0.078%), S (0.22%), Cr 58.2 ppm.

2.4. Statistical analyses

Exposures were log normally distributed and analyses were consequently performed on log transformed data. Values below limit of detection (LOD) were replaced by LOD/$\sqrt{2}$ (n=1 (0.6%) for inhalable dust and n=10 (6%) for inhalable cement dust). Air concentrations of inhalable dust and cement dust were summarised as arithmetic means (AM), geometric means (GM) and geometric standard deviations (GSD). All analyses were performed using SAS version 9.1 software (SAS Institute Inc.).

3. Results

A total of 180 personal measurements were performed, with an average sampling time of 7 hours and 43 minutes. Eleven measurements lasted less than 6 hours, of which the shortest covered only 2 hours. Concentrations were not standardized to an 8-hour working day.

As shown in table 1, lowest exposures were observed for reinforcement and pouring workers, with a geometric mean (GM) of 0.33 mg/m^3. Personal exposure levels in ready-mix and pre-cast concrete plants were very similar, with GMs of 0.37 and 0.39 mg/m^3, respectively. The highest exposures at the construction site were seen for concrete repairers and tile setters, who had GMs of 2.4 and 3.9 mg/m^3. A similar picture is seen for inhalable cement dust exposure. Cement plant workers had by far the highest inhalable dust exposure. Especially among cleaners it was extremely high. The maximum value of 194 mg/m^3 (in an 8-hour sampling) was measured when a worker performed cleaning tasks, most of the time using a broom.

Table 1. Inhalable dust and inhalable cement dust (mg/m^3), by job type

	D	K	N	AM	GM	GSD	Range	% Cement[a]
Inhalable dust								
Total group	*54*	*55*	*180*	*5.4*	*1.04*	*4.13*	*0.05-194*	*42 %*
All construction workers	*35*	*29*	*120*	*2.0*	*0.97*	*3.20*	*0.05-34*	*37 %*
Reinforcement & pouring	4	10	40	0.46	0.33	2.34	0.05-2.7	23 %
Concrete repair	4	2	8	3.0	2.4	1.96	0.96-6.2	50 %
Bricklaying	5	4	16	0.91	0.76	1.71	0.38-3.7	45 %
Floor screed laying	8	5	16	3.4	1.6	2.36	0.77-34	48 %
Installation work	6	5	20	1.5	1.2	1.90	0.44-6.5	29 %
Tile setting	8	5	20	4.7	3.9	1.85	1.2-15	50 %
Cement plant	6	11	19	7.0	3.0	3.53	0.35-52	63 %
Cement plant (Cleaning)[a]	2	3	6	98	55	3.77	8.8-194	64 %
Ready-mix plants	10	5	10	0.47	0.37	2.17	0.10-1.1	33 %
Pre-cast plant	5	10	25	0.43	0.39	1.61	0.16-1.1	47 %
Inhalable cement dust								
Total group	*54*	*55*	*180*	*3.6*	*0.38*	*5.53*	*0.02-194*	
All construction workers	*35*	*29*	*120*	*0.86*	*0.31*	*4.28*	*0.02-17*	
Reinforcement & pouring	4	10	40	0.09	0.06	2.25	0.02-0.51	
Concrete repair	4	2	8	1.5	1.2	1.96	0.44-3.3	
Bricklaying	5	4	16	0.34	0.32	1.47	0.19-0.62	
Floor screed laying	8	5	16	1.7	0.75	2.45	0.36-17	
Installation work	6	5	20	0.47	0.34	1.94	0.15-3.0	
Tile setting	8	5	20	2.3	1.9	1.80	0.58-7.5	
Cement plant	6	11	19	5.2	1.7	4.72	0.10-49	
Cement plant (Cleaning)[b]	2	3	6	72	33	4.73	3.9-194	
Ready-mix plants	10	5	10	0.19	0.11	3.22	0.02-0.58	
Pre-cast plant	5	10	25	0.21	0.17	1.81	0.07-0.67	

D=number of measurement days; K=number of workers; N=number of samples; AM=arithmetic mean; GM=geometric mean; GSD=geometric standard deviation; [a]AM; [b]These workers wore face pieces as breathing protection the majority of time

From the last column of table 1 - showing cement percentages of inhalable dust - it becomes clear that the actual cement percentage varied extensively from sample to sample, especially in the cement-producing factory where the average percentages in the collected dust lay above 60%. The remaining job types showed much lower percentages. On samples from the ready-mix concrete plants only 33%

of the inhalable dust was cement. Dust in reinforcement and installation work contained even less than 30% cement.

4. Discussion

Personally measured concentrations of inhalable dust and its cement content varied over more than four orders of magnitude. By far the highest concentrations were observed in cement production, particularly during cleaning operations. Personal inhalable dust concentrations in the ready-mix and pre-cast concrete plants were similar to concentrations during reinforcement and pouring work and were on average below 0.5 mg/m^3 for inhalable dust and below 0.2 mg/m^3 for cement. Inhalable dust concentrations at the construction site ranged from 0.05 to 34 mg/m^3, with a geometric mean (GM) concentration of 1.0 mg/m^3. For inhalable cement dust the GM concentration was 0.3 mg/m^3 (range 0.02-17 mg/m^3). Here the highest levels were observed for concrete repairers and tile setters, whereas reinforcement and pouring work had the lowest. Elemental measurements showed highest but very variable cement percentages in the cement factory and very low percentages during reinforcement and pouring work.

The measured concentrations for all cleaners at the cement plant exceeded the current German exposure limit for inhalable dust (10 mg/m^3), except for one sample (8.8 mg/m^3). These workers wore however face pieces as breathing protection while cleaning for the majority of time, probably resulting in a lower actual exposure level. Among the other cement plant workers, concentrations of three out of 19 samples were above the limit. At the construction site only two workers exceeded the limit value: one floor screed layer using a grinding machine and a tile setter. For inhalable cement dust, the same persons and two additional cement plant workers exceeded the German limit value for Portland cement dust exposure (5 mg/m^3).

Inhalable dust exposure has been monitored previously for several tasks in US construction [10]. Exposures during concrete repair (GM 3.6 mg/m^3; with a maximum of 12.9) were slightly higher than levels we found (GM 2.4 mg/m^3; max. 6.2). Inhalable dust concentrations for concrete pouring were with a GM of 2.5 mg/m^3 much higher than exposures we observed for this task (0.33 mg/m^3). The subjects performing concrete pouring in our study were reinforcement workers as well, which is likely to have decreased the overall exposure level in this group. This indicates that it is hard to compare exposure levels of specific tasks between studies, as tasks are difficult to determine exactly. The highest measured concentrations were however more similar for concrete pouring (2.6 mg/m^3 versus 2.7 mg/m^3 in our study). Recess milling is one of the tasks in the construction industry which is known to occur with high dust exposure levels [11,12]. In the current case, most of that type of work was previously done in the pre-cast concrete plant. There was however some recess milling done in this project, performed by installation workers, but only for a limited amount of total work time. Installation workers performed more other tasks with low cement exposure. Recess milling is generally more frequently done in renovation projects. Demolition workers are known to be exposed to high dust levels as well [11]. Measurements of these workers were not conducted in this study though, as no demolition work was done in this building project.

Observed total dust concentrations for production tasks in the cement plant (GM 3.0 mg/m^3; with a maximum of 52) is in the same range as earlier findings among cement plant workers. In the USA a GM of 2.9 mg/m^3 was reported (max 79 mg/m^3) [13], and in a Norwegian study an AM of 7.4 mg/m^3, with exposures up to 54 mg/m^3 [14]. Concentrations in the production areas of a Tanzanian cement factory were somewhat higher: GM=10.6 mg/m^3, with a range of 0.21 to 229 [15].

5. Conclusions

The highest inhalable (cement) dust concentrations at the construction site were observed for concrete repairers, floor screed layers and tile setters. In comparison these exposures were lower than those found in the cement plant, which were particularly high during cleaning tasks.

Acknowledgements
This study was funded by a grant from the European Cement Association (CEMBUREAU www.cembureau.eu).

References

[1] HSE 1994 Portland Cement Dust. Criteria document for an occupational exposure limit. Sudbury.

[2] HSE 2005 Portland Cement Dust. Hazard assessment document. Report No.: EH75/7

[3] Purdue MP, Jarvholm B, Bergdahl IA, Hayes RB and Baris D 2006 Occupational exposures and head and neck cancers among Swedish construction workers. Scand J Work Environ Health. 32(4) 270-5

[4] Jansson C, Johansson AL, Bergdahl IA, Dickman PW, Plato N and Adami J, et al. 2005 Occupational exposures and risk of esophageal and gastric cardia cancers among male Swedish construction workers. Cancer Causes Control. 16(6) 755-64

[5] Mitropoulos P and Norman R 2005 Occupational nonsolar risk factors of squamous cell carcinoma of the skin: a population-based case-controlled study. Dermatol Online J. 11(2) 5

[6] Sjodahl K, Jansson C, Bergdahl IA, Adami J, Boffetta P and Lagergren J 2007 Airborne exposures and risk of gastric cancer: a prospective cohort study. Int J Cancer. 120(9) 2013-8

[7] Dietz A, Ramroth H, Urban T, Ahrens W and Becher H 2004 Exposure to cement dust, related occupational groups and laryngeal cancer risk: results of a population based case-control study. Int J Cancer. 108(6) 907-11

[8] EN1232 1997 Workplace atmospheres. Pumps for personal sampling of chemical agents. Requirements and test methods

[9] Kenny LC, Aitken R, Chalmers C, Fabries JF, Gonzalez-Fernandez E, Kromhout H, et al. 1997 A collaborative European study of personal inhalable aerosol sampler performance. Ann Occup Hyg. 41(2) 135-53

[10] Woskie SR, Kalil A, Bello D and Virji MA 2002 Exposures to quartz, diesel, dust, and welding fumes during heavy and highway construction. AIHA J. 63(4) 447-57

[11] Lumens ME and Spee T 2001 Determinants of exposure to respirable quartz dust in the construction industry. Ann Occup Hyg. 45(7) 585-95

[12] Tjoe Nij E, Hilhorst S, Spee T, Spierings J, Steffens F, Lumens M, et al. 2003 Dust control measures in the construction industry. Ann Occup Hyg. 47(3) 211-8

[13] Abrons HL, Petersen MR, Sanderson WT, Engelberg AL and Harber P 1988 Symptoms, ventilatory function, and environmental exposures in Portland cement workers. Br J Ind Med. 45(6) 368-75

[14] Fell AK, Thomassen TR, Kristensen P, Egeland T and Kongerud J 2003 Respiratory symptoms and ventilatory function in workers exposed to portland cement dust. J Occup Environ Med. 45(9) 1008-14

[15] Mwaiselage J, Bratveit M, Moen B and Yost M 2005 Variability in dust exposure in a cement factory in Tanzania. Ann Occup Hyg. 49(6) 511-9

Journal of Physics: Conference Series **151** (2009) 012055 doi:10.1088/1742-6596/151/1/012055

Silica dust control in small-scale building/structure demolition operations using good work practice guidance

C V Muianga [1,2,3], C H Rice[1] and P Succop[1]

[1]Department of Environmental Health, College of Medicine University of Cincinnati. 3223 Eden Ave., Kettering Laboratory, Cincinnati, OH 45267-0056, USA

[2]Center for Industrial Studies, Safety and Environment, Eduardo Mondlane University, P.o.Box # 257, Maputo, Mozambique

[3]Email: muiangcv@email.uc.edu

Abstract. Work practices can influence exposure, especially in small-scale operations conducted by mobile work crews. This study evaluated the use of information on good work practice in control guidance sheets adapted from UK Silica Essentials guidance sheets by trained workers and supervisors employed in small-scale concrete and masonry demolition operations. A one-page employee silica task-based control guidance sheet for each of four demolition tasks and multiple-page silica control guidance for supervisors were developed. Interactive, hands-on worker training on these task-based good work practice controls was developed. Training was presented to 26 participants from two demolition crews. Feedback on the training and task-based good work practice control guidance sheets was elicited. Observations of work practices were made before and after training. Participants indicated gains in knowledge and checklists were used to document skill attainment. The quality of the training and usefulness of the material/skills was rated high by trainees. Increased use of water to suppress dust and wet cleaning methods on the job were documented following the training. Additional follow-up after training is required to determine long-term impact on sustained changes in work practices, and to evaluate the need for refresher training.

1. Introduction

Small-scale demolition of concrete and masonry structures is a ubiquitous activity that precedes building renovations, remodelling, reconstruction or larger demolition operations. Various researchers have identified demolition of concrete or masonry structures as one of the construction activities with very high exposures to respirable crystalline silica [1]. There is consistency among various professionals that total elimination or substitution of silica from the construction process may not be a feasible option [2]. The use of wet methods, such as sprinkling systems, local exhaust ventilation and high efficiency particulate air (HEPA) filtration systems attached to work tools or equipment, worker's isolation from dust generating sources by tenting off or use of other physical barriers, respiratory protection and work practices may significantly reduce silica dust in many construction activities [3]; however, these controls when used separately or incompletely may fail to reduce exposure to less than established occupational exposure limits [4].

In the late 1990s the Center for Construction Research and Training, responding to the challenge of assessing and controlling occupational health hazards in the construction industry associated with its dynamic nature (highly mobile workforce and perpetually changing worksite, very high exposures for a short working time period, outdoors versus indoors, etc), developed a task-based exposure assessment model. In the model, the task was viewed as a primary building block of an exposure assessment framework. Susi and Schneider, (1995), and Susi et al., (2000) provided detailed description of the model [5]. However, the results are not widely used to control worker exposures. The United Kingdom Health and Safety Executive (UK HSE) developed Control of Substances Hazardous to Health (COSHH) Essentials – a step-by-step occupational risk assessment and risk management tool to provide advice on adequate control measures for supplied chemicals; a number of sector specific and task-based guidance sheets are also provided such as Silica Essentials [6]. Partially as a result of this experience, the European Union (EU) multi-sectorial Negotiation Platform on Silica (NEPSI) developed a Good Practice Guide and detailed task control guidance sheets; these toolkits are widely used in the UK and Europe [7].

Our previous research on silica dust exposure control strategies in small-scale concrete and masonry demolition operations evaluated the usability of the UK HSE Silica Essentials, and the European Union Negotiation Platform on Silica – Good Practice Guide. The results indicated constraints of language differences (English to native languages in industrializing nations), recommended resources could not be accessed (e.g. no Internet access) and absence of exposure controls including local exhaust ventilation, water suppression systems, and respiratory protection [8]. Also, the control guidance sheets included extensive text and technical information on both hazard and controls that may limit their usability by workers and some supervisors as a result of lack of time and literacy. These toolkits were considered valuable resources to address silica hazard prevention and control, but needed adaptation to local working conditions in industrializing countries [7].

The work presented evaluated the implementation of short, task-based good work practice control guidance sheets by employees and supervisors in small-scale concrete and masonry demolition operations to reduce exposure to silica following a training program. Exposure measurements for respirable crystalline silica dust were made. These results will be reported separately.

2. Materials and Methods

Task-based good work practice control guidance sheets were developed to be used by construction supervisors and employees during small-scale demolition of concrete and masonry structures or buildings. To evaluate the effectiveness of the task-based control guidance sheets plus training, data on work practices were collected before and after the training. Human subject review and approval was provided by the University of Cincinnati Institutional Review Board (IRB) and the Social and Behavioral Sciences and the Ethics Committee of the Eduardo Mondlane University, Maputo, Mozambique.

2.1. Development of task-based good work practice control guidance sheets

Task-based good work practice control guidance sheets were developed for four demolition tasks: (i) breaking or dismantling concrete and masonry structure, (ii) segregating, concentrating and stockpiling the debris and scraps and, (iii) removing, transferring or transporting debris for disposal, re-use or recycling and iv) cleaning-up the worksite. The exposure reduction strategy in each guidance sheet was based on the use of:

- wet methods to suppress dust (spray, sprinkle or mop);
- HEPA filtered vacuum cleaner to remove dust;
- natural ventilation by working at an upstream position of the dust generating source and allowing free air circulation through opening doors and windows when appropriate;
- basic personal protective equipment, and,
- basic hygiene and sanitation practice.

Special attention was given to clarity of language and brevity [9].

2.2. Work practice controls walkthrough survey

A walkthrough survey guide was developed for data collectors to document use of work practice controls while workers performed demolition tasks. The walkthrough survey guide was composed of a complete description of the good work practice categories presented on the task-based control guidance sheets plus worksite description, equipment, tools and material used by the demolition crew. A checklist was developed to capture observations on the use of various work practices.

2.3. Training of observers/data collectors

In order to control for information and observation bias [10], two observers were trained to observe and record work practices. They assessed worker training performance by observing participants and completing a performance checklist.

The observers were trained and calibrated by one author (CM) to obtain accurate observations. The calibration was done at the worksite, and at the end of the shift the results were compared. At a random time during each shift the author (CM) visited the worksite and recorded observations for later comparison.

2.4. Worker training

Interactive, hands-on training was developed for the supervision and employees in separate programs so supervisors and workers could learn to use information on the task-based good work practice control guidance sheets to reduce exposure.

The supervisor training was based on structured dialogue [11] in a round table type and at the demolition worksite. The supervisors were challenged to explain a step-by-step small-scale demolition process, tools and material used and details of their work practices/habits. This discussion was followed by a field demonstration of how the small-scale demolition is done, and the names and functions of the tools/equipment. During the training, the supervisors reviewed and provided their feedback on the task-based good work practice control guidance.

The training for crew workers included: (i) silica dust health hazard information, (ii) pre-planning and procedures for applying good work practice controls while performing each task and (iii) step-by-step small-scale demolition processes and reinforcement of key points. Photographic pictures of typical tasks of small-scale concrete and masonry demolition operations in Mozambique and a small-scale concrete and hands-on demolition simulation exercise were used.

The worker training covered the following work practice categories:

- Pre-planning and post self performance evaluation on use of good work practice control guidance
- Use of water to suppress dust (spray, sprinkle until moistened but not wet)
- Use of a high efficiency particulate air (HEPA) filtered vacuum to remove silica dust
- Use of natural ventilation by instructing workers to stay upwind/upstream of the silica dust generating source while performing the task, and allowing free air circulation by opening doors and windows if suitable
- Availability and practice of basic hygiene (hand washing, cleaning working tools and PPE after finishing the work, changing clothes before reaching home) and sanitation (availability of potable water and toilets)
- Availability and usage of basic PPE for the construction industry (hard hat, safety glasses, ear plugs, work gloves, work shoes).

Participant performance was evaluated. The evaluation was composed of two parts. To assess knowledge a three item pre-and-post test was conducted with the group. Participants were asked to raise their hands if they had knowledge in the following areas: 1. What are the hazards of dust in demolition or construction activities? 2. How do dusts generate hazards of silica exposure? 3. How do you keep dust exposure as low as possible during your work? The data were tabulated and the proportions of participants reporting knowledge before training and after training were compared [12]. Secondly, to assess the trainees' skills, performance checklists were completed by the data collectors during a simulation exercise. Any incorrect procedure was discussed with the participants and repeated until correctly performed; thus all participants successfully demonstrated use of skills on the guidance sheets.

2.5. Participant feedback on the training

Written feedback was requested from all participants to better understand whether the training was considered effective and useful. Items on the quality of the training presentation were rated using a four-point Likert scale: 1-strongly disagree, 2-disagree, 3-agree, and 4-strongly agree. The usefulness of the various work practice control methods to reduce dust was rated with a similar scale: 1-not useful at all, 2-somewhat useful, 3-useful, 4-very useful. Participants rated satisfaction with the guidance sheet on a four-point scale: 1-not satisfied at all, 2-somewhat satisfied, 3-satisfied, 4-very satisfied [9].

2.6. Post-training walkthrough survey at worksites

After the training, walkthrough survey observations of work practices were performed by the observers using the same checklist. Work practice changes were obtained by comparing pre-and-post intervention walkthrough survey results for each demolition crew. Pre- and post-training comparisons were made in tabular form, noting the presence or absence of a practice by the crew.

3. Results

3.1. Development of task-based good work practice control guidance sheets

The final version of task-based good work practice control guidance sheets were modification of previous models [8] and all information was captured on a single page for workers, and multiple pages for supervisors. Figure 1 illustrates the Guidance Sheet for Breaking Concrete and Masonry Structures for workers. The supervisor guidance sheet (not shown) included information on: (i) why be concerned about silica dust exposures (regulation compliance requirements, health hazard information and business and economic point of view), (ii) access and premises, (iii) design and equipment, (iv) maintenance, (v) examination and testing, (vi) personal protective equipment, (vii) cleaning, housekeeping and personal hygiene practice, (viii) training and supervision, (ix) further information on control methods for silica dust exposures and useful links.

During the preliminary walkthrough and input from supervisors, three tasks were determined to be relevant for small-scale demolition operations at the participant sites: (i) breaking concrete and masonry structures (walls, floors or ceilings), (ii) cleaning, concentration and removal of demolition debris and scraps, and (iii) transferring and transporting the debris or scraps for salvage and disposal. Therefore, the fourth sheet on cleanup was not used.

Journal of Physics: Conference Series **151** (2009) 012055 doi:10.1088/1742-6596/151/1/012055

Figure 1. Task-based control guidance sheet: Breaking up concrete/masonry structures.

Task-Based Silica Control Guidance Sheet for Small-scale Manual Demolition
Task 1: Breaking concrete and masonry structures (walls, ceilings or floors)

Control Approach: Good Work Practices

IMPORTANT MESSAGE FOR EMPLOYEES

Dust resulting from demolition and other construction activities can be deadly.
The harm from dust depends on:

- What it is made of
- What size it is
- How much is in the air
- How long you breathe it

If you are exposed to dust with silica, **its very fine particles can penetrate deep into your lungs.** If a lot of silica is in the air and you breathe it for a long time you can develop disabling and fatal diseases such as:

- **silicosis**
- **lung cancer**
- **silico-tuberculosis**

The following actions well done will lower silica dust exposure and decrease the risk of diseases.

Preventive actions to reduce silica dust exposures with good work practices	**Answers**	
	Yes	**No**[a]
I take few minutes to pre-plan and assure compliance with good work practices		
I use water to suppress dust (spray, sprinkle or damp cloth)		
- water hose spray		
- sprinkling with my hand or other appropriate resources		
- wet wiping		
I maximize the use of natural ventilation		
- by allowing free air circulation (doors and windows open if possible)		
- working upwind or upstream of the dust-generating source		
I exercise personal hygiene practices by		
- Washing hands before eating, drinking or smoking		
- cleaning and checking working tools after finishing the work		
- I clean my PPE after I finish my work		
- I change clothes before I reach home		
Basic sanitation practice		
- I use the available potable water if needed		
- I use appropriately the available toilet		
I take a few minutes after I perform my task to examine my performance or compliance with these good work practices		

[a]A check mark on "No" needs immediate solution before the task is performed

3.2. Observer training

Comparison of the recorded observations by one of the authors (CM) with the observers during random checks showed complete agreement during all three activities—pre-training, training and post-training. For data analysis, only the recordings made by the observers were used.

3.3. Pre-training work practice observations

Two demolition crews doing renovations in a university setting in Maputo, Mozambique participated in the research. Each crew and its work organization is described below.

3.3.1. *Crew 1.* This crew was composed of four supervisors, four masons and six assistant masons employed by the university. Their daily duties included renovations, remodelling and maintenance of university infra-structures and buildings. Their construction projects always involve some demolition operations of concrete and masonry structures. The crew worked on two sites before the training, and three sites following training. Tools used prior to the training were: sledge hammers, chisels, scrapers, shovels, buckets, pickup truck, wheel barrow and their hands to remove debris. Tools were selected for the task as the work progressed. First, the area to be demolished was marked with a pencil and then the crew used hand tools for demolition, followed by breaking up with a sledge hammer. The cleaning and removal task was conducted to concentrate the debris in one area and the debris was transferred to a container and carried to a disposal or salvage point.

3.3.2. *Crew 2.* This crew was composed of two supervisors, five masons and seven assistant masons employed by an external contractor. The two on-site supervisors were foreign to Mozambique and had limited Portuguese language proficiency. They could communicate with their Mozambican fellows, but not with the trainer; thus, the supervisors did not participate in the training. During the study the crew worked on the same renovation project, which involved some demolition operations before and after training. The crew used the following tools: sledge hammers, chisels, scrapers, circular saw and shovel. They used their hands, push broom, and buckets for debris removal. Tools were selected for the task as the work progressed. This crew worked in teams. One team marked the area with white chalk and used a circular saw to initiate the demolition, another team was assigned to breaking up the concrete and masonry structure with chisel and hammer or sledge hammer, and another team worked on cleaning and removing the debris to a designated point.

3.4. Worker training

Interactive, hands-on worker training on the use of information in the task-based silica good work practice control guidance was delivered to 22 workers and four supervisors, for a total of 26 participants. The evaluation of workers' knowledge of silica dust exposure prevention and controls showed a substantial increase. Prior to training, three of 26 (11.5%) indicated some knowledge on one or more question; following training, 25 of 26 (96%) indicated some knowledge on one or more of the questions. These proportions are significantly different, $p<<0.001$. Skills performance checklists of observations during the simulation of demolition activities showed that all participants (100%) mastered work preplanning, pre-work compliance with good work practices, use of water, maximizing use of natural ventilation, good personal hygiene, no use of compressed air or other dry method to remove dust from clothing and post work review of work practices. No training was conducted using high efficiency particulate air filtered vacuums, as these were not commercially available in the area.

3.5. Participant feedback on the training

Fourteen of the 26 participants (54%) provided some feedback on the training evaluation instruments. This represents 100% of Crew 1. The Crew 2 feedback was not accomplished because of communication difficulties with supervisors and limited time. Twelve of 14 rating the quality of the presentation agreed or strongly agreed that it was at an understandable level. All respondents scored agreement or strong agreement with: Helped me think about solutions to problems in the workplace (n=12) and Encouraged active participation (n=14). Five respondents noted lower scores for following the agenda and remaining flexible to unplanned issues during a session. All scores for feedback from the instructor (n=14) were at the level of agree or strongly agree. All participants completing the evaluation item on effectiveness of the training rated it as effective or very effective overall (n=12).

At the end of training, 13 of 14 participants rated use of water as useful or very useful in suppressing dust at the worksite; only 1 participant rated it as somewhat useful. Supervisors (n=4) and masons (n=4) rated the use of natural ventilation as useful or very useful in reducing dust exposures; assistant masons rated the open windows as useful or very useful (5 of 6) but did not report high confidence in being able to work upstream of a dust source; 3 of 6 rated this approach as not useful.

Participants' overall evaluation of the information in task-based control guidance sheet information to help reduce exposure was very high. Thirteen of 14 (93%) participants reported being very satisfied with the tool; the other participant was satisfied with the tool.

3.6. Post-training work practice observations

Following training, changes in work practices were observed. Table 1 shows components of good work practices used for both crews before and after training. Both crews implemented aspects of the task-based control guidance Sheets. Some differences in the materials used between the crews were observed; for example, crew 1 used plastic sheet to cover dusty materials, acquired a drum to store water, garden hose and watering can to suppress dust, and hand truck for transporting debris. Crew 2 acquired a wet sponge or cloth to clean surfaces and tools, a concrete water tank, plastic bucket and used hands to sprinkle water on surfaces and walls; they used empty cement sacs to manually transport demolition debris from the working site to disposal or salvage point. Neither crew used HEPA filter vacuums, long sleeves and pants, containment/isolation, or local or general exhaust. Vacuum and ventilation is generally not available commercially. High temperatures and high levels of physical exertion influence clothing decisions. Containment and isolation may be useful in some applications; however, these are often coupled with ventilation for best effectiveness.

Table 1: Pre-and-post good work practice controls for crews 1 and crew 2 indicated with – if not used and + if used.

#	Components of good work practice control guidance sheets	Before		After	
		Crew 1	Crew 2	Crew 1	Crew 2
1	Pre-planning	-	-	+	+
	Warning signs, checking if all tools and supplies are available and functioning well	-	-	+	+
2	Water for dust suppression	-	-	+	+
	Water hose spray	-	-	+	-
	Sprinkling by hand or other appropriate resources	-	-	+	+
	Wet wiping	-	-	+	+
3	High efficiency particulate air (HEPA) filter vacuum	-	-	-	-
4	Natural ventilation				
	Free air circulation (doors and windows open if possible)	-	-	+	+
	Working upwind/upstream of the dust-generating source	-	-	+	+
5	Basic personal protective equipment (PPE)				
	Safety glasses with side shields	-	-	+	+
	Hard hat	-	-	+	+
	Safety shoes (boots or steel toe shoes)	-	-	-	-
	Work gloves	-	-	+	+
	Hearing protection	-	-	+	+
	Long sleeves and long pants	-	-	-	-
6	Personal hygiene practices				
	Hand washing facilities	-	-	+	+
	Water for cleaning tools and PPE	-	-	+	+
	Separated space for eating and drinking	-	-	+	+
7	Basic sanitation practice				
	Potable water	+	+	+	+
	Toilet	+	+	+	+
8	Self post-performance evaluation	-	-	+	+
9	Other control measures				
	Containment or isolation	-	-	-	-
	Local exhaust ventilation	-	-	-	-
	General mechanical ventilation	-	-	-	-
	Respiratory protection equipment (N95 mask)	-	-	+	+

4. Discussion

Small scale demolition processes frequently involve generation of dusts. These operations are conducted for short durations of time at temporary work sites. Previous observations showed that the types of engineering controls recommended for dust control generally are not available. More attention to work practices was identified as a method that might be introduced to reduce exposures. This work reported here is part of a larger evaluation of the use of task-based good work practice control guidance sheets with worker training developed especially for small scale demolition operations.

The guidance sheet format presented for each task is much shorter than other models such as the COSHH Essentials. This is appropriate, given the narrowly defined activity of demolition workers. For each task, it is possible to be more limited in the nature of controls in the checklist. Four sheets were developed initially, based on previous work. However, it was found that cleaning and removal was always done at the same time and was followed by transferring and transport. This sequence excludes segregation, concentration and stockpiling of debris which is only done on rare occasions as separated task by the participating crews. This work organization was reported by the supervisors and verified during the completion of initial walkthroughs. It is not known if this combination of tasks is unique to the crews participating in this work, or is now the more usual work organization. In the future, a range of task-specific sheets may be needed for demolition activities. A training program could be structured to include the selection of relevant tasks by the participants. The full complement of sheets would be available, and could be a resource for the participants at future work sites, where work organization might vary.

The observers recorded data consistent with the author during random checks; therefore, inter-observer variability was not relevant. As the data were collected over a four week period, we did not evaluate trends in recording over time.

The work was conducted at one employment site in Mozambique. As English is not the native language, all materials were first developed in English and then translated into Portuguese. Crew 2 had immigrant supervisors who did not read Portuguese and did not participate in the training. The constraint of communicating with supervisors and limited time prevented the trained employees from Crew 2 from completing the training feedback evaluation instruments; their understanding of the goals of the training is demonstrated by the changes implemented by the crew. More time was needed to talk with contractor management who could understand either Portuguese or English and be able to transmit the request to the supervisors in their language and provide time for employees to complete feedback evaluations. If resources were available, an onsite interpreter would be hired to facilitate communication with supervisors. The short time period available to conduct this project prevented long durations between interactions. However, overall, the training program delivery was well-received by participants in the native crew. Time for special training alternatives should be considered.

Knowledge gain was documented by participants raising their hands. Raising hands provided the number of people reporting knowledge before and after training. The assumption was that any worker who would not raise his /her hand, did not have the required knowledge. The instructor asked those who raised their hands to provide a verbal response that was recorded on flipchart. These responses were generally consistent with knowledge of the topic. Although the knowledge gain was assessed using this crude measure it was consistent with the participatory approach. Both employees and supervisors provided very important details throughout the training regarding exposure determinants, tasks generating high dust exposures and measures to reduce exposures while performing demolition task, indicating a high level of engagement. These reports were consistent with previous research that used a task-based exposure assessment model to comprehensively understand tasks, the nature of work, conditions causing or exacerbating exposures and controls that realistically can reduce exposures in construction industry [13]. Alternative knowledge gain assessment methods are the use of a pre-and-post written test or retrospective pre-test [14]. The ability to apply the methods in the task-based control guidance sheets was the main goal and was more rigorously evaluated using observations during simulation exercises and work activities after training.

Before the training, tools and equipment used for small-scale demolition operation included sledge hammers, hammers, chisels, a circular saw, hand scrapers, shovels, and handheld push broom or sweeper, wheelbarrows, use of hands and buckets. After the training, more tools and equipment were added such as the use of water, use of wet/humid clothes or a sponge to remove dust, water reservoir or source (drum or tap water), use of plastic sheet to cover the debris and the use of a hand trailer to transfer the debris from the demolition site to the disposal or salvage point. The workers in these two crews demonstrated high motivation and interest to implement work practice controls. They found sources of water and utilized wet methods and demonstrated awareness for the use of natural ventilation to reduce exposure. Efforts are needed to increase the availability of HEPA vacuum equipment to demolition work crews in this area.

Longer follow-up is needed to determine if changes in work practices can be sustained by the crew members and supervisors at this location, and as they move to other work sites. Refresher training may also be needed, if the implementation of good work practices is documented to decline over time.

5. Conclusions

A program was developed and implemented to train workers in the use of a short Work Practice Control Guidance Sheet to improve work practices that would reduce exposures to dust in small scale demolition. Observations document changes in work practices after training, including preplanning, use of wet methods and end-of-task review.

Acknowledgments

This research was partially supported by the National Institute for Occupational Safety and Health and the Pilot Research Project Training Program of the University of Cincinnati Education and Research Center Grant #T42/OH008432-03. Mr. Muianga received financial support for graduate education through a University of Cincinnati Graduate Assistantship and Grant Number U45ES006184 from the National Institute of Environmental Health Sciences. Preliminary data were collected through the University of Michigan Fogarty International Centre in Occupational and Environmental Health in Southern Africa. The contributions of the faculty and staff at Eduardo Mondlane University, Mozambique, to facilitate data collection are gratefully acknowledged.

References

[1] Raila R 1988 *Ann Occup Hyg* **32**: 215-20

[2] Chisholm, J 1999 *Indoor Built Environ.* **8**:94-106. Lumens M E, Spee T 2001 *Ann Occup Hyg;* **45**: 585-95

[3] Flanagan M E, Seixas N S, Majar M, Camp J E and Morgan M 2003 *Am Ind Hyg Assoc J* **64**:319-28. Flanagan M E, Seixas N S, Becker P, Takacs B, Camp J E and Morgan M 2006 *J Occup Environ Hyg* **3**:144-52. Echt A, Sieber K, Jones E, Schill D, Lefkowitz D, Sugar J and Hoffner K 2003 *Appl Occup Environ Hyg* **18**: 491-95

[4] Croteau G A, Flanagan M E, Camp J E and Seixas N S 2004 *Ann. Occup. Hyg;* **48**:509-18. Lahiri S, Levenstein C, Nelson D I and Rosenberg B J 2005 *Am J Ind Med* **48**:503-14.

[5] Susi P and Schneider S 1995 *Appl Occup Environ Hyg* **10**:394-99. Susi P, Goldberg M, Barnes P and Stafford E 2000 *Appl Occup Environ Hyg* **15**: 26-38

[6] United Kingdom H S E 2006 Publication Series: CN-COSHH Essentials in construction: silica. Internet publication at http://www.hse.gov.uk/pubns/guidance/cnseries.htm (Accessed on February 18, 2007).

[7] The European Network on Silica (EU-NEPSI) 2007 Internet publication. Available at http://www.nepsi.eu/good-practice-guide.aspx (Accessed on February 18, 2007) published by IMA and EU-NEPSI. Garrod A N I, Evans P G and Davy C W 2007 *Journal of Exposure Science and Environmental Epidemiology* **17**:S48–54

[8] Muianga C V and Rice C H 2008 **18th** *Annual Construction Safety Conference*. February 12 – 14, 2008. *Conf. Proc.* Chicago, USA. Muianga C V and Rice C H 2008 *IOHA Newsletter* **16**: 14 – 15

[9] Becker P and Morawetz J 2004 *Am J Ind Med;* **46**:63-70. Caparaz A, Rice C, Graumlich S, Radike M and Morawetz J 1990 *Appl Occup Environ Hyg* **5**:595–603.

[10] Sica G T 2006 *Radiology* **238**: 780-89

[11] Burke L and Hutchins H 2007 *Human Resources Development Review by Sage Publications* **6**: 263-96

[12] Kleinbanum D, Kupper L and Muller K 1998 *Applied Regression Analysis and Other*

Multivariable Methods Kugushev A, Mazow C 3th Edition (Pacific Grove, Brooks/Cole) p 798

[13] Goldberg M 1997 *Am J Ind Med **31**:310-18*

[14] Berger P K, Gunto S J, Rice C and Haley J V 1996 *Appl Occup Environ Hyg* **11**:1198-1203. Szudy E and Arroyo M G 1994 *The Right to Understand: Linking Literacy to Health and Safety Training Darling D (Berkely* Calfornia, Labor Occupational Health Program University of California at Berkeley) p 199

Inhaled Particles X, (23–25 September 2008, Manchester) IOP Publishing
Journal of Physics: Conference Series **151** (2009) 012056 doi:10.1088/1742-6596/151/1/012056

Temporal changes in the variability of respirable mineral dust exposure concentrations

Hans Kromhout[1], Jelle Vlaanderen[1], Richard Jongen[2], Remko Houba[3]

1 Institute for Risk Assessment Sciences, Utrecht University, The Netherlands

2 ArboUnie Occupational Health Service, The Netherlands

3 ArboUnie Expert Centre for Chemical Risk Management, Nijmegen, The Netherlands

Corresponding author: h.kromhout@uu.nl

Abstract: In the last decade a lot of evidence with regard to temporal trends in exposure concentrations in predominantly Western industrial countries has become available. In a recent literature review (Creely et al. 2007) overall percentage of yearly declines up to 32% were presented. To what extent these temporal declines also affect the variability in exposure concentrations is unknown. The main reason is lack of longitudinal data including repeated measurements that would allow evaluating trends in personal and temporal components of exposure variability. The recently elaborated exposure database from the Industrial Minerals Association Dust Monitoring Programme provided an opportunity to study these trends for exposure to respirable mineral dust. This database currently contains more than 11,000 measurements from more than 20 companies and 80 sites throughout Europe. About one-third of the measurement data comprised repeated measurements within a specific site-job-survey combination. Linear mixed models were used to estimate variance components. Variance components were consequently plotted against year of measurements. For 377 groups of workers (with number of workers>2, total number of observations >5 and repeats >1.25) the fold range of the total variability (${}_{t}R_{.95}$) appeared to go down significantly with 3% per campaign (half year): from 34 in summer 2002 to 20 in winter 2005/2006. When the variability was teased apart the fold range for the temporal variability (${}_{ww}R_{.95}$) appeared to decrease significantly as well with 3% per campaign (half year): from 17 in summer 2002 to 10 in winter 2005/2006. The between-worker variability did not show a temporal trend and stayed constant with on average a fold-range (${}_{bw}R_{.95}$) of approximately 4. Downward temporal trends in exposure level of respirable (crystalline silica) dust seem to coincide with downward trends in the size of temporal variability. Fold-ranges of average exposure of individual workers within a job at a particular site appear to be stationary. Implications are that attenuation of exposure-response associations in epidemiogical studies will decline and that fewer measurements will have to be collected to arrive at accurate estimates of long-term exposure to respirable mineral dust in this industry.

1. Introduction

In predominantly Western industrial countries during the last decade a considerable amount of evidence has become available with regard to temporal trends in occupational exposure concentrations. In a recent literature review [1] overall percentage of yearly declines in exposure concentrations of up to 32% annually were presented. To what extent these temporal declines also affect the variability in exposure concentrations is unknown [2]. The main reason is lack of longitudinal data including repeated measurements that would allow an evaluation of trends in personal and temporal components of exposure variability.

2. Materials and methods

The recently elaborated occupational exposure database from the Industrial Minerals Association (IMA-Europe) Dust Monitoring Programme provided an opportunity to study these trends for exposure to respirable mineral dust. This database currently contains more than 10,000 measurements from 24 companies and more than 80 sites throughout Europe, collected from winter 2000/2001 onwards. About one-third of the measurement data comprises repeated measurements within a specific site-job-survey combination. We used linear mixed models to estimate within- and between-worker variance components. Variance components were consequently plotted against campaign (two during each calendar year). The observed trends in fold-ranges (R.95) were analysed with linear mixed models with measurement campaign as the main determinant, adjusted for the number of observations per dataset (fixed effect) and function performed (random effect).

3. Results

Between summer 2002 and winter 2006/2007 359 respirable dust surveys of groups of workers defined by function and site were present that had ≥ 2 measured individuals, a total number of observations ≥ 5, a number of repeated measurements >1.25 and number of observations below limit of detection $\leq$25%). The fold range of the total variability (${}_{t}R_{.95}$) appeared to decline significantly with 4% per campaign (half year): from an estimated median value of 22 in summer 2002 to 15 in winter 2006/2007. When the variability was teased apart the fold range for the within-worker variability (${}_{ww}R_{.95}$) appeared to decrease significantly as well with 5% per campaign (half year): from an estimated median value 13 in summer 2002 to 8 in winter 2006/2007. The between-worker variability did not show a significant temporal trend and stayed constant with a median fold-range (${}_{bw}R_{.95}$) between 3 and 4.

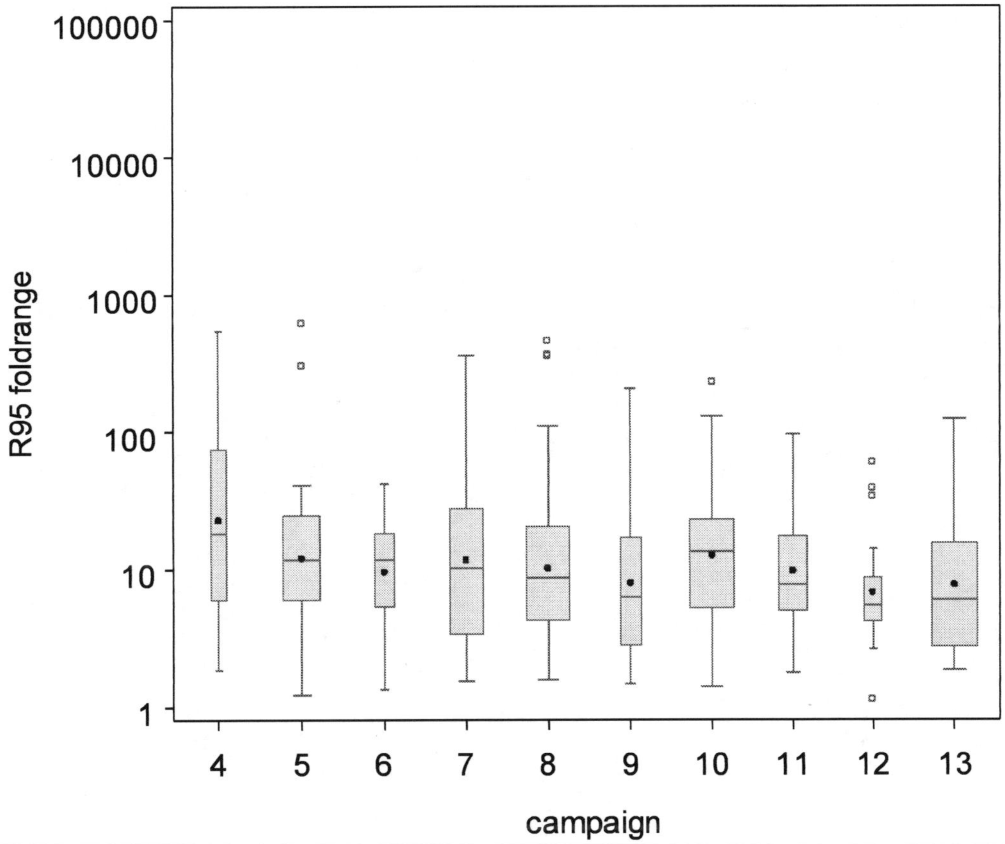

Figure 1 Estimated within-worker variability (presented as fold-range) in respirable dust concentrations for 359 groups of workers defined by function and site from summer 2002 (campaign 4) till winter 2006/2007 (campaign 13)

For respirable quartz trends were quite similar, although with many more measurement results below the limit of detection the number of groups with sufficient data was only 219. The variability in this subset was somewhat increased. The fold range of the total variability (${}_{t}R_{.95}$) appeared to decline significantly with 8% per campaign (half year): from an estimated median value 45 in summer 2002 to 22 in winter 2006/2007. When the variability was teased apart the fold range for the within-worker variability (${}_{ww}R_{.95}$) appeared to decrease significantly with almost 9% per campaign (half year): from an estimated median value 26 in summer 2002 to 10 in winter 2006/2007. The between-worker variability did not show a temporal trend and stayed constant with a median fold-range (${}_{bw}R_{.95}$) between 4 and 6.

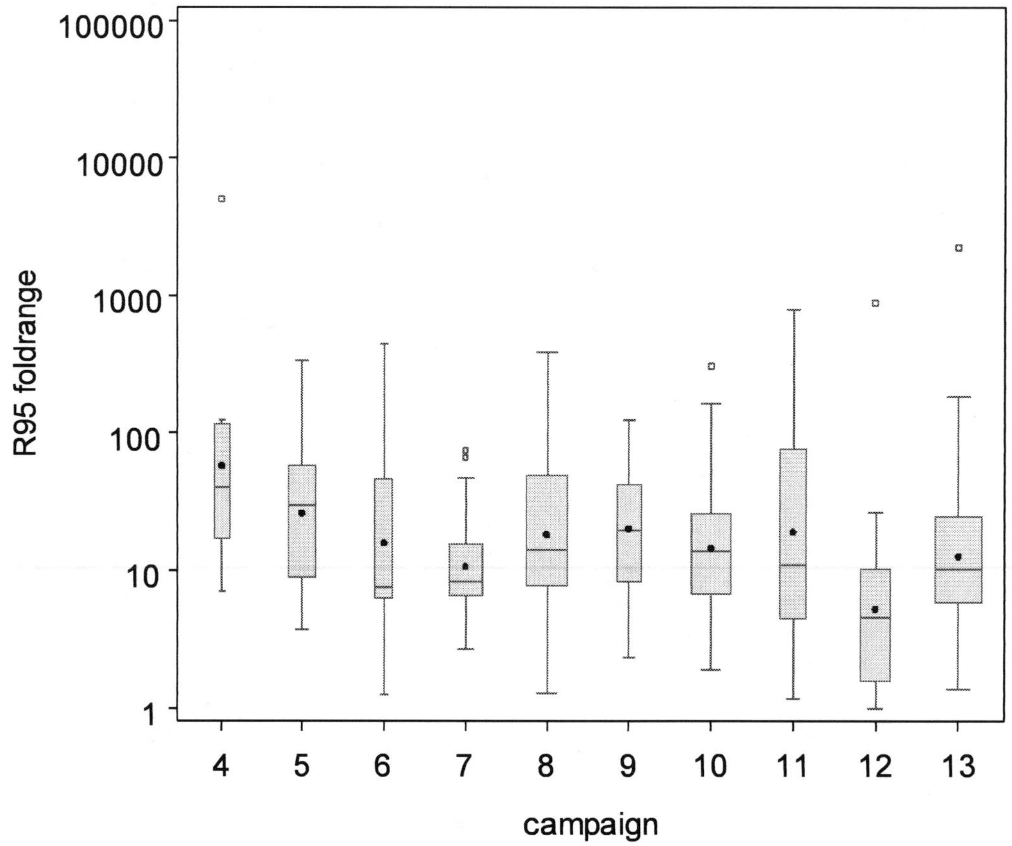

Figure 2 Estimated within-worker variability (presented as fold-range) in respirable quartz concentrations for 220 groups of workers defined by function and site from summer 2002 (campaign 4) till winter 2006/2007 (campaign 13)

4. Conclusions

Downward temporal trends in exposure level of respirable (quartz) dust seem to coincide with downward trends in the size of temporal variability. Fold-ranges of average exposure of individual workers within a function at a particular site appear to be stationary. Implications are that attenuation of exposure-response associations in epidemiogical studies will decline and that fewer measurements will have to be collected to arrive at accurate estimates of long-term exposure to respirable mineral dust in this industry.

References

[1] Creely KS, Cowie H, Van Tongeren M, Kromhout H, Tickner J, Cherrie JW 2007. Trends in inhalation exposure--a review of the data in the published scientific literature. *Ann Occup Hyg.* **51** 665-78.

[2] Kromhout H, Symanski E, Rappaport SM 1993. A comprehensive evaluation of within- and between-worker components of occupational exposure to chemical agents. *Ann Occup Hyg.* **37** 253-70.

Journal of Physics: Conference Series **151** (2009) 012057 doi:10.1088/1742-6596/151/1/012057

A study of the effect of particulate deposit upon fibrous filter efficiency.

Sarah J Dunnett[1], Charles F Clement[2]

[1]Department of Aeronautical and Automotive Engineering, Loughborough University, Loughborough, Leics. LE11 3TU, U.K.

[2]15 Witan Way, Wantage, Oxon, OX12 9EU, U.K.

s.j.dunnett@lboro.ac.uk

Abstract. The aim of this work is to develop numerical methods to model the effects that particle deposit collected by fibrous filters has upon the flow field within the filter and hence upon further deposition. A single fibre model has been developed with the deposit modelled as a porous layer on the fibre surface. Using mathematical techniques the flow field outside and within the porous layer are determined. Once the flow field for a particular deposit has been obtained the equations of motion of the particles are solved and the feedback effects of the deposit upon further deposition investigated. The mechanisms of interception and diffusion are considered. It is found that for the smaller particles the porosity of the deposit has an insignificant effect upon the flow field. However the porosity becomes increasingly important as the relative size of the particle to the fibre increases.

1. Introduction

Fibrous filters are widely used to protect employees from exposure to airborne particulates in the form of face masks or larger scale filtration units. Such filters generally consist of many threadlike fibres which are positioned more or less normal to the direction of the air flow. The air passes through the regions between the fibres and some of the particles within the flow will be deposited onto the fibres surface and hence removed from the airflow. Neglecting the influence of electrical effects the main mechanisms by which particles are removed by the fibres are interception, impaction, diffusion and gravitational settling. The dominant mechanism for any application will depend upon the size of the particles in the flow. The collected particles accumulate forming complex structures which influence the fluid flow and further deposition. Eventually the filter will clog leading to regeneration or replacement.

In order to fully understand the performance of a filter it is necessary to have a detailed understanding of the flow field within the filter. Due to the importance of filters much work has been performed on acquiring this understanding, in particular for clean filters where no significant particle deposition has taken place. See for example, [1-6]. As for the fluid flow, extensive research has been undertaken into particle removal for clean fibres, see for example [7-12]. Due to this earlier work the performances of filters in their clean state are fairly well understood. However, in the case of a loaded filter where significant particle deposition has occurred on the fibres, the situation is more complex and less work has been performed in this area. In this case the deposited material will affect the air flow through the filter and hence affect particle deposition. Previous work which has been performed includes the experimental work of [13,14] where the penetration and pressure drop of various filters were considered under loading. The semi-empirical models of [15] and the studies of the dendritic structures formed on the fibres by the deposited material by [16]. These studies were limited as they did not include the effects of the structures upon the flow field. Work which did take into account the effect the deposited material has upon the flow field includes that of [17-20]. However the models developed were restricted due to their considerable use of computing time and resources. More recently work has been undertaken to study the structure of deposit on single fibres and the effect the deposition has upon efficiency see [21-23].

In the work described here a numerical model has been developed which models the particulate deposit on the fibres as a porous media. The air flow and further particle deposition on the fibres is then investigated. In this way the deposit structure formed as loading increases can be determined. This has been investigated for various values of the relevant parameters. This model is an extension of that developed previously by the authors [24] where in that case the deposit was modelled as a smooth solid layer on the fibre surface.

2. Formulation

2.1 Flow model.

As in [24] the fibres are modelled as infinitely long cylinders with their axis perpendicular to the flow. Hence the flow is two dimensional and the motion takes place in the plane perpendicular to the cylinder axis. A single fibre is considered and a cell model adopted. This is different from an isolated fibre as the presence of the other fibres is accounted for in the boundary conditions on the cell. The size of the cell is determined by the packing fraction of the filter considered, c, where the packing fraction is defined as the fraction of the perceived volume of the filter that is actually occupied by fibres. It is assumed that the Reynolds number of the flow, Re, is small and creeping flow is modelled. Where Re is defined by, $\mathrm{Re} = \frac{d\rho U_0}{\mu}$, d is the diameter of the fibres making up the filter, ρ is the density of the air, U_0 is the air velocity and μ the air viscosity. This is a reasonable assumption as generally for flow through a fibrous filter Re<1. The particulate deposit collected on the fibre is modelled as a porous media and hence the situation modeled is shown in figure 1. Due to symmetry only half the fibre is considered.

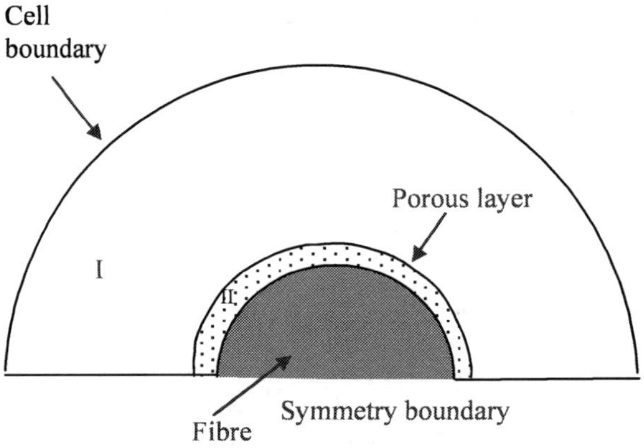

Figure 1. Cell model.

In the model there are two regions to consider, region I the area in the filter between fibres where the air flows and region II which is the porous region made up of deposited particles. In region I the flow is described by the biharmonic equation for the stream function ψ, i.e.

$$\nabla^4 \psi = 0 \tag{1}$$

In region II the flow motion is described by Darcy's equation:

$$\nabla p = -\frac{\mu}{k} \underline{U} \tag{2}$$

Where k is the intrinsic permeability of the porous medium.

In terms of ψ equation (2) becomes

$$\nabla^2 \psi = 0 \tag{3}$$

Using the same approach as [24] the Boundary Element Method (BEM) was adopted to solve for ψ in regions I and II. In order to do this an extension of the Beavers-Joseph condition developed for fluid flow adjacent to a porous medium was adopted, see [25],

$$\left(\frac{\partial u_T}{\partial N} + \frac{\partial u_N}{\partial T}\right)_I = \frac{\alpha_{BJ}}{\sqrt{k}}\left((u_T)_I - (u_T)_{II}\right) \tag{4}$$

Where N and T are the normal and tangent to the boundary, u_N and u_T are the components of fluid velocity in the normal and tangential directions respectively and the subscripts I and II denote the regions in which the terms are taken. The term α_{BJ} is a dimensionless parameter which depends on the material parameters that characterize the structure of the permeable material.

Once ψ can be obtained anywhere in regions I and II the flow velocities in those regions can be determined using

$$u_r = -\frac{1}{r}\frac{\partial \psi}{\partial \theta} \quad \text{and} \; u_\theta = -\frac{\partial \psi}{\partial r} \tag{5}$$

Once the flow velocities are known the motion of particles within the flow can be determined.

2.2 Particle Motion.

In this work we are considering the motion of smaller particles as these are the particles of most interest when considering health effects. The major mechanism by which such particles are captured is diffusional deposition, in which the combined effect of the air motion and the Brownian motion of the particle brings it into contact with a fibre. In this case, for filtration problems it can be shown that, see [24], the particle concentration, n, is described by

$$u_r \frac{\partial n}{\partial r} + \frac{u_\theta}{r}\frac{\partial n}{\partial \theta} = \frac{2}{Pe}\left(\frac{\partial^2 n}{\partial r^2} + \frac{1}{r}\frac{\partial n}{\partial r}\right) \tag{6}$$

Where all terms in the equation are non-dimensional and Pe is the Peclet number which is a measure of the relative magnitude of the diffusional motion of the particles and the convective motion of the air past the fibre, given by;

$$Pe = \frac{U_0 d}{D} \tag{7}$$

Where D is the coefficient of diffusion of the particles. The particle concentration has been non-dimensionalised with respect to n_0, the particle concentration at the cell boundary.

Equation (6) was solved numerically to give values of n at finite points in the domain of interest, the region outside the surface formed by previously deposited particles. It is assumed that particles are attached to the porous boundary if they impact with it. The interception of particles with fibres has been accounted for by using the appropriate boundary conditions

The rate at which particles are deposited on a surface by diffusion is proportional to the concentration gradient $\partial n / \partial N$ on the surface. Hence in this work $\partial n / \partial N$ on the deposit surface is investigated for various situations. Also once $\partial n / \partial N$ is known the new surface formed by depositing particles can be determined for a certain time interval. This new surface then defines the boundary of the porous layer and the flow field can be recalculated using the process described in the previous section. In this way the shape of the surface can be investigated as deposition increases.

3. Results

It has been found in earlier work, [24], that for small particles and for a certain range of the parameters κ and Pe, the surface formed by the early stages of deposit on the fibres can be approximated by

$$r = r_0\left[1 + \lambda\left(1 - a\beta^p\right)^b\right] \tag{8}$$

where r_0 is the radius of the clean fibre, λ is the thickness of the deposit on the fibre front divided by r_0 and a, p and b are fitted constants. The angle β is measured from the front of the fibre. The parameter κ is the ratio of particle and fibre diameter and Pe is the Peclet number given by equation (7). Hence in this work equation (8) has been taken to define the boundary between the porous deposit and the air flow.

In order to determine the flow field it is necessary to impose a value for α_{BJ}, the empirical constant in the boundary condition (4). From earlier work [25] and [26], for the situations considered here, it appears that α_{BJ} will lie in the range $2 \le \alpha_{BJ} \le 4$. From an investigation of the flow field it appears that the effects of varying α_{BJ} within the relevant range are insignificant hence in all results shown here α_{BJ} has been taken to be 3.

In equation (2) the parameter k is introduced which is a measure of the permeability of the particulate deposit. This has been shown to be related to the porosity of the deposit, φ, by equation (9), see [25]

$$k = \frac{\varphi^3 d_p^2}{180(1-\varphi)^2} \tag{9}$$

Where d_p is the characteristic diameter of a particle in the porous deposit. In non-dimensional terms this becomes

$$k' = \frac{k}{r_0^2} = \frac{\varphi^3 \kappa^2}{45(1-\varphi)^2} \tag{10}$$

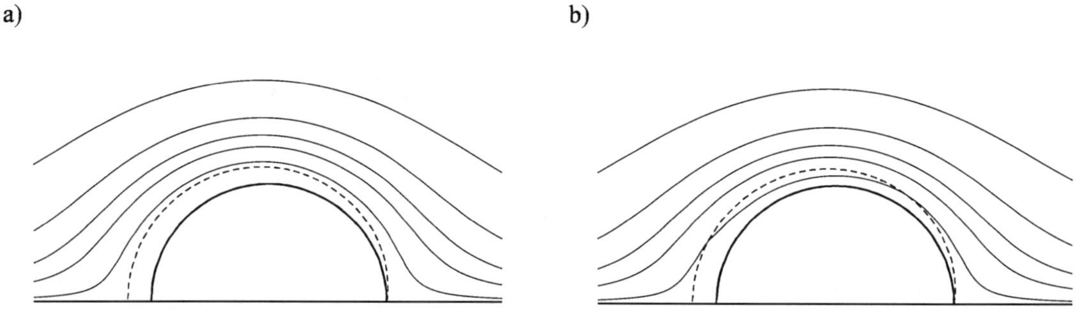

Figure 2. Streamlines of the flow for a) k'=0.01 and b) k'=0.05

Considering the flow field past the fibre for different values of k', figure 2 shows the flow streamlines for k'=0.01 and 0.05. In the results shown a typical shape for the deposit, as determined in earlier work, [24], has been assumed and the non-dimensional thickness of the deposit layer at the front of the fibre has been taken to be 0.2. Due to symmetry, only half of the fibre is shown and the porous region is enclosed within the dotted curve. Only the flow in the region close to the fibre with deposit is shown. As can be seen, in the case of low permeability of the porous layer, which corresponds to the smaller value of k', the streamlines are shifted away from the deposit with very little flow passing through. As k' increases more flow passes through the layer, as seen in figure 2b with the porous layer offering less resistance to the flow. Considering the corresponding values of κ and φ to the values of k' taken in figure 2, as small particles are being considered κ is restricted to the range $\kappa \leq 0.05$. Hence for k'=0.01, for κ=0.05 then φ=0.933 and for κ=0.01 then φ=0.985. For k'=0.05, shown in figure 2b) for κ=0.05 then φ=0.968 and for κ=0.01 then φ=0.993. It appears therefore that for the creeping flow through filters, for these small particles considered when diffusion is the main mechanism of deposition, the porosity of the deposit must be high for any significant flow in pass through the deposit. In figure2, in general, over 90% of the volume of the deposit is void space.

In order to consider the effect deposited material has upon further deposit the value of $\partial n/\partial N$ on the surface of the porous deposit is investigated, as the rate at which more particulate is deposited is proportional to this term. The way this has been achieved here is by, for a particular value of Pe and κ:

i) determining the flow field for a clean fibre and solving for n. This enables $\partial n/\partial N$ to be determined on the fibre surface.

ii) The coefficients a,b and p which give the best fit to the equation $\frac{\partial n}{\partial N} = \left[\frac{\partial n}{\partial N}\right]_{\beta=0} \left(1 - a\beta^p\right)^b$. are then determined.

iii) As the rate at which particles deposit is proportional to $\partial n/\partial N$ the non-dimensional equation of the surface of the porous material on the fibre is written as $r = 1 + \alpha \frac{\partial n}{\partial N} = 1 + \lambda\left(1 - a\beta^p\right)^b$

iv) The new flow field and particle concentration can then be obtained and $\partial n/\partial N$ on the porous surface determined.

v) Step ii) is then repeated and more deposit added with the new surface of deposited porous material given by $r_{new} = r_{old} + \alpha \frac{\partial n}{\partial N} = r_{old} + \lambda\left(1 - a\beta^p\right)^b$

Steps iv) and v) are then repeated as deposit grows.

In figure 3 $\partial n/\partial N$ is shown as a function of the angle β for Pe=5891 and κ=0.0025, β=0 correspond to the front of the fibre and $\beta=\pi$ is at the back. The amount of deposited material added before the flow field was recalculated was taken to be such that the non-dimensional width of the porous layer added was 0.2 at the front of the fibre. In figure 3a) the deposit has been assumed to form a solid layer on the fibre and hence no flow passes through it, in figure 3b) the deposit is assumed to be highly porous with φ=0.95. The values of Pe and κ shown correspond to a fibre of 20μm, particles of size 0.05 μm and a flow velocity of 0.1m/s. As can be seen for these parameters, even when the deposit has a high porosity, the effect upon $\partial n/\partial N$, and hence further deposit is insignificant.

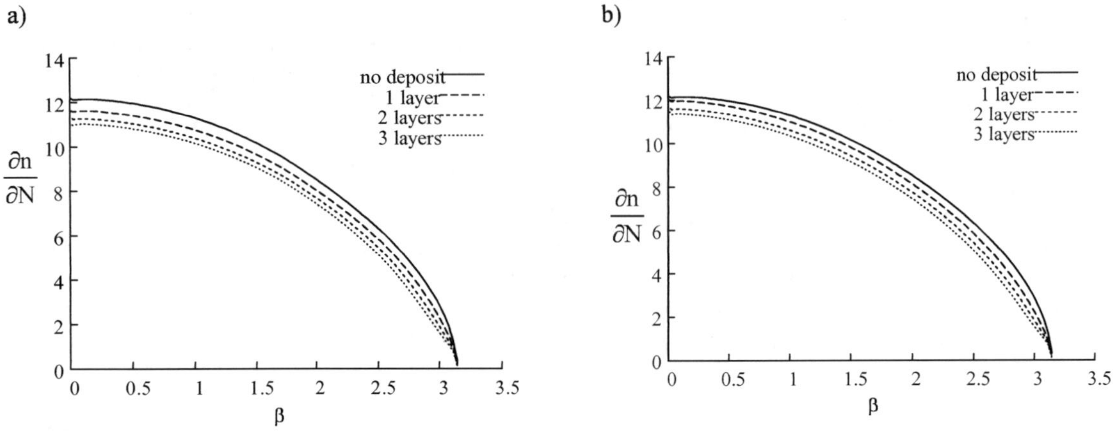

Figure 3. $\frac{\partial n}{\partial N}$ as a function of β for Pe=5891, κ=0.0025 when deposit is assumed to be a) solid, b) porous, φ=0.95.

Considering Pe=1473, κ=0.05 figure 4 shows $\partial n/\partial N$ as a function of β for a) solid and b) porous, φ=0.9, deposit. In this case the parameters correspond to a fibre of 1 μm diameter, particles of size 0.05 μm diameter and a flow velocity of 0.5m/s. It can be seen that for this larger value of the relative size of particles considered, κ, the porosity of the deposit does have a significant effect upon the results. As can be seen the effect of the deposit is to increase the value of $\frac{\partial n}{\partial N}$, and hence the amount of deposit, at the front of the fibre, around β=0, and decrease it at the back. This effect continues as the amount of deposited material increases, as seen by the curves for 2, 3 and 4 layers of deposit.

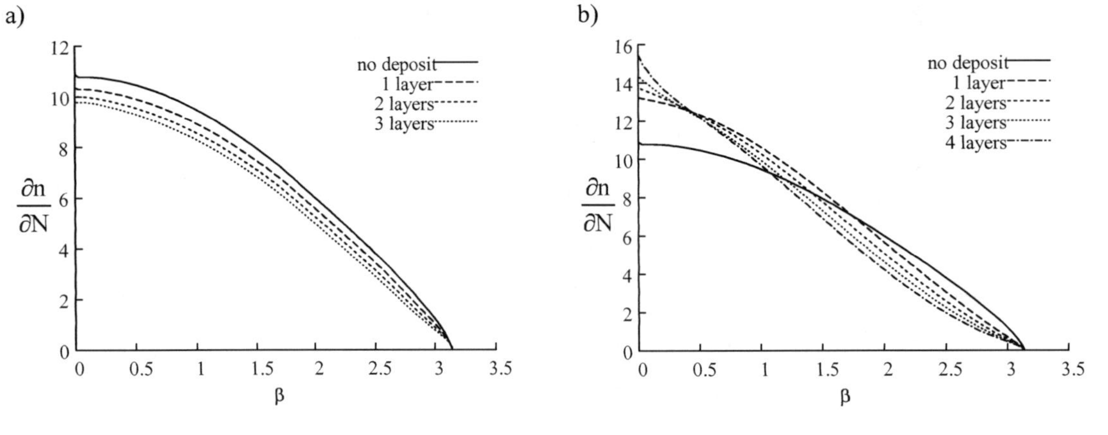

Figure 4. $\frac{\partial n}{\partial N}$ as a function of β for Pe=1473, κ=0.05 when deposit is assumed to be a) solid, b) porous, φ=0.9.

An important characteristic used to evaluate the performance of a filter is the particle collection efficiency; η, which is the measure of the efficiency of a single representative fibre in the filter. This is defined to be the ratio of the number of particles that are actually removed by the fibre to the number entering the 'cell'. As shown by [24] this can be written as

$$\eta = \frac{2I\sqrt{c}}{Pe} \tag{11}$$

where I is the integral

$$I = \int_{\theta=0}^{\theta=\pi} \left[\frac{\partial n}{\partial N} \right]_{r=r_1} ds \tag{12}$$

r_1 is the interception radius which is given by $r_s+\kappa$, where r_s is the non-dimensional distance from the fibre centre to the surface formed by the fibre with deposit. This radius will change as deposition of particles occurs. As an example of the effect of deposit, and its porosity, upon efficiency figure 5 shows η as a function of the number of layers of deposit for Pe=1473 and κ=0.05. The cases of a solid deposit and the two values of porosity 0.8 and 0.9 have been considered. As can be seen the effect of deposit is to increase efficiency as the area upon which particles can deposit increases. Also as the porosity of the deposit increases and more flow is able to pass through it, the efficiency of the filter also increases. Other values of Pe and κ were considered and the same behaviour was found to occur.

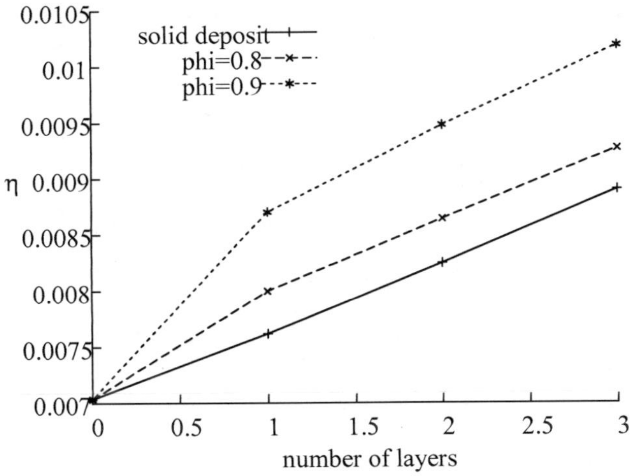

Figure 5. Filter collection efficiency as a function of layers of deposit for Pe=1473, κ=0.05.

4. Conclusions

When using fibrous filters to protect people from exposure to particulate it is important to understand how the filters will perform. One area in which we do not have a detailed understanding is how the filters perform when they are loaded with particulate deposit. The aim of the work, some of which is described here, is to develop a numerical model of a filter which describes the loading process and the effects the porosity of the deposit has upon filter efficiency. Such knowledge could then be used to understand the changes occurring in real filters due to particle deposition. In developing the model certain assumptions have been made, such as the neglect of electrostatic forces. These assumptions do not affect the results obtained describing the effects of the particulate deposit.

In this work a relatively straightforward mathematical model has been described that considers the particulate deposit on the fibres to be a porous medium. The model enables the airflow around and through the porous medium to be obtained in an efficient manner. In the preliminary results obtained it appears that for many of the small particles, which are of interest for occupational health, the porosity of the deposit does not have a significant effect upon the flow behaviour within the filter. In this case a simpler model with a solid deposit appears adequate. However for larger particles the effect of the porosity is more significant causing the particle build up to increase on the front of the fibres and decrease on the back.

In future work these effects will be investigated in more detail and larger size particles will be considered where the mechanism of impaction becomes important for deposition.

5. References

[1] Kuwabara, S 1959 The forces exerted by randomly distributed parallel circular cylinders or spheres in a viscous flow at small Reynolds numbers, *J. Phy. Soc. Japan* **14** 527-32.

[2] Happel, J 1959 Viscous flow relative to arrays of cylinders, *A.I.Ch.E. J.* **5** 174-7.

[3] Rao, N and Faghri, M 1988 Computer modelling of aerosol filtration by fibrous filters, *Aerosol Sci. and Tech.* **8** 133-56.

[4] Dhaniyala, S and Liu, B Y H 1999 An asymmetrical, three-dimensional model for fibrous filters, *Aerosol Sci. and tech.* **30** 333-48.

[5] Hildyard, M L, Ingham, D B, Heggs, P J and Kelmanson, M A 1985 Integral equation solution of viscous flow through a fibrous filter *Boundary Elements VII,* ed C A Brebbia and G Maier, (Berlin: Springer-Verlag)

[6] Ingham, D B, Hildyard, M L and Heggs, P J 1991 The particle collection of an array of cylinders using the Boundary Element Method, *Eng. Analysis with Boundary Elements* **8** 36-44.

[7] Fardi, B and Liu, B Y H 1994 Flow field and pressure drop of filters with rectangular fibres, *Aerosol Sci. and tech.* **17** 36-44.

[8] Ingham, D B, Hildyard, M L and Heggs, P J 1989 The particle collection efficiency of a cascade of cylinders, *The Canadian J. of Chem. Eng.* **67** 545-53.

[9] Pich, J 1973 Theory of gravitational capture of particles in fibrous aerosol filters, *J. Aerosol Sci.* **4** 217-26.

[10] Stechkina, I B and Fuchs, N A 1966 Studies on fibrous aerosol filters –I. Calculations of diffusional deposition of aerosols in fibrous filters, *Ann. Occup. Hyg.* **9** 59-64

[11] Zhu, C, Lin, C H and Cheung, C S 2000 Inertial impaction-dominated fibrous filtration with rectangular or cylindrical fibers, *Powder tech.* **112** 149-62.

[12] Asgharian, B and Chenh, Y S 2002 The filtration of fibrous aerosol. *Aerosol Sci. and Tech.* **36** 10-7

[13] Japuntich, D A, Stenhouse, J I T and Liu, B Y H 1994 Experimental results of solid monodisperse particle clogging of fibrous filters. *J.Aerosol Sci.* **25** 385-93.

[14] Thomas, D, Penicot, P, Contal, P, Leclerc, D and Vendel, J 2001 Clogging of fibrous filters by solid aerosol particles, Experimental and modelling study. *Chem. Eng. Sci.* **56** 3549-61.

[15] Stenhouse, J I T, Japuntich, D A and Liu, B Y H 1992 The behaviour of fibrous filters in the initial stages of filter loading. *J.Aerosol Sci.* **23** 761-4.

[16] Payatakes, A C and Gradon, L 1980 Dendritic deposition of aerosols by convective Brownian motion for small, intermediate and high Knudsen numbers. *A.Ch.E.J.* **26** 443-54.

[17] Jung, Y and Tien, C 1993 Simulation of aerosol deposition in granular media. *Aerosol Sci. and Tech.* **18** 418-40.

[18] Biggs, M J, Humby, S J, Buts, A and Tuzun, U 2003 Explicit numerical simulation of suspension flow with deposition in porous media: influence of local flow field variation on deposition processes predicted by trajectory methods. *Chem. Eng. Sci.* **58** 1271-88.

[19] Karadimos, A and Ocone, R 2003 The effect of the flow field recalculation on fibrous filter loading: a numerical simulation. *Powder tech.* **137** 109-19.

[20] Przekop,R, Moskal, M, and Gradon, L 2003 Lattice-Boltzmann approach for description of the structure of deposited particulate matter in fibrous filters. *J. Aerosol Sci.* **34**, 133-47

[21] Hoferer, J, Lehmann, M J, Hardy, E H, Meyer, J and Kasper, G 2006 Highly resolved

determination of structure and particle deposition in fibrous filters by MRI. *Chem. Eng. Technol.* **7** 816-9.

[22] Schollmeier, S, Meyer, J and Kasper, G 2007 Measurement of the collection efficiency of single dust-loaded filter fibers, *Proc. European Aerosol Conf.*.

[23] Schollmeier, S, Meyer, J and Kasper, G 2007 Characterization of deposit structures in single dust-loaded filter fibers, *Proc. European Aerosol Conf.*

[24] Dunnett, S J and Clement, C F 2006 A numerical study of the effects of loading from diffusive deposition on the efficiency of fibrous filters. *J. Aerosol Sci.* **37** 1116-39.

[25] Nield, D A and Bejan, A 1992 Convection in porous media. (New York: Springer Verlag).

[26] Sahraoui, M and Kaviany, M 1992 Slip and no-slip velocity boundary conditions at interface of porous, plain media. *Int. J. Heat Mass Transfer* **35** 927-43.

Size selective isocyanate aerosols personal air sampling using porous plastic foams

Cong Khanh Huynh, Trinh Vu Duc

Institut Universitaire Romand de Santé au Travail (IST), 21 rue du Bugnon - CH-1011 Lausanne, Switzerland

chuynh@hospvd.ch

Abstract. As part of a European project (SMT4-CT96-2137), various European institutions specialized in occupational hygiene (BGIA, HSL, IOM, INRS, IST, Ambiente e Lavoro) have established a program of scientific collaboration to develop one or more prototypes of European personal samplers for the collection of simultaneous three dust fractions: inhalable, thoracic and respirable. These samplers based on existing sampling heads (IOM, GSP and cassettes) use Polyurethane Plastic Foam (PUF) according to their porosity to support sampling and separator size of the particles. In this study, the authors present an original application of size selective personal air sampling using chemical impregnated PUF to perform isocyanate aerosols capturing and derivatizing in industrial spray-painting shops.

1. Introduction

Polyurethane-based paints are widely used in the car industry owing to their outstanding technical features such as durability, color stability and resistance to abrasion, chemicals and weather extremes. Isocyanates (NCO) are important reactive chemicals, the so-called hardener of two-component polyurethane-based paints, and a leading cause of occupational asthma in industrialized countries [1-3]. Exposure to NCO can occur by inhalation and by dermal contact [4]. Sampling and analysis of airborne NCO is not easy. NCO occurs in a variety of chemical forms, such as, monomers, oligomers and mixtures of all these forms. NCO oligomers and prepolymers are commonly used in car spray-painting paints, as they are less volatile than the monomers. NCO occurs in a variety of physical form, for example, vapor, aerosol and liquids. A sampling method for NCO in the workplace should be suitable for all physical forms. Contrary to the American Conference of Governmental Industrial Hygienists (ACGIH) who recommended a threshold limit value (TLV) of 34 $\mu g.m^{-3}$ (5 ppb) only on HDI monomer and nothing about prepolymer form [5], European countries like the UK Health and Safety Executive (HSE) has set a long-term workplace exposure limit (WEL, 8 h time weighted average reference period) of 20 $\mu g.m^{-3}$ of total isocyanate –NCO group and a short-term limit (15 min) of 70 $\mu g.m^{-3}$. Swiss regulation also adopts a short-term limit (15 min) of 20 $\mu g.m^{-3}$ of total isocyanate –NCO group [6].

Because of the reactivity of NCO, sampling in the workplace is commonly carried out by chemical reaction of NCO with derivatization reagent to produce stable derivative that is easier to trap on the supports. In general, the sampling device recommended in the UK method MDHS 25/3 [7] depends upon the physical form of the isocyanate. Isocyanate vapor can be sampled by an impregnated filter and aerosols are most effectively sampled by a combination of impinger containing 1-(2-

methoxyphenyl)piperazine (MP) solution backed up by a MP impregnated filter. However, it is widely acknowledged that impingers are difficult to use for personal monitoring and several groups are currently researching the limitations and applicability of filter only sampling methods because they are easier to use.

The exposure of workers to aerosols and particles is a major concern for occupational health. Polyurethane Plastic Foam (PUF) provides a low-cost method for separating dust into health-related fractions in accordance with recognized sampling conventions (ISO/TR/7708 [8], EN481 [9]). As part of a European project (SMT4-CT96-2137), various European institutions specialized in occupational hygiene (BGIA, HSL, IOM, INRS, IST, Ambiente e Lavoro) have established a program of scientific collaboration to develop one or more prototypes of European personal samplers for the collection of particles according to the penetration curves. These samplers based on existing sampling heads (IOM, GSP and cassettes) use PUF according to their porosity to support sampling and size separation of the particles. Conventional applications of size selective PUF samplers were focused on general occupational dust monitoring [10-11], welding fumes [12], mining dust, metals, bio-aerosols [13-14], etc.

The health effect of isocyanate exposure in the gas phase and in different size fractions are not well understood and only a few papers report the results of field situation, depending spray-painting or thermal degradation operation, using denuder and cascade impactor sampler [15-18]. In addition, it is not know whether the selection characteristics of these methods comply with the CEN/ISO convention for inhalable, thoracic and respirable aerosol. Impinger, though offering an immediate solubilizing and derivatizing environment by the use of solvent, suffer from the inability to capture particles much smaller than 2 μm. In contrary, filters collect more effectively particulate, but because the lack of solvent, shown a less efficiency for the vapor phase reaction. In field situation of spray-painting shops and focus on the monitoring of commercial two-component paints based on isocyanates prepolymers, only less than 0.5% of monomer HDI remains in the conventional paint composition. In fact, impregnated filter method was sufficiently robust for field evaluation [17]. Concerning impregnated filter method, the only sensitive problem was the necessity to dissolve the derivatized isocyanates collected on the filters in acetonitrile (ACN) solution immediately after sampling and sampling time do not excess one hour.

In this study, an original application of size selective personal air sampling using chemical impregnated PUF to perform isocyanate aerosols capturing and derivatizing in paint shops, comply with the CEN/ISO convention, was presented.

2. Methodology

2.1. Reagents

1-(2-methoxyphenyl)piperazine (MP) was purchased from Aldrich (Chemie Brunschwig, Basel, Switzerland) and Hexamethylen diisocyante (HDI) monomer from Eastman Kodak (Rochester, N.Y.). Desmodur N-100, which contains 99.3% of prepolymers, predominately in the form of biuret trimer of HDI, and Desmodur N-75, which contains also more than 99.2% of prepolymers, predominately in the form of isocyanurate trimer of HDI were obtained from Bayer Chemical (Bayer AG, Zürich, Switzerland). Both isocyanate reagents contain less than 0.5% of HDI. Local, commercial, two-component paints based on Desmodur N-75 (PolyJal, Jallut SA, Lausanne, Switzerland) were used for the generation of the aerosol. Glass fibre filters (GF/A) were obtained from Whatman and filter holders, Swinnex-type, 25-mm diameter were from Millipore.

2.2. Impregnation of filter, PUF and sample preparation

An Andersen impactor (Ambient Particle Sizing Sampler, model 2000, Andersen Inc., USA) equipped with glass fiber filters (Whatmann filter 934AH Ø 81mm) impregnated with MP solution (3 ml of solution of 1 $mg.ml^{-1}$ of MP in anhydrous toluene, which corresponds to about 3 mg of reagent on the filter) was used at 28.3 $l.min^{-1}$ (1 Cubic Feet per minute or CFM) for the collecting and the weighing

Journal of Physics: Conference Series **151** (2009) 012058 doi:10.1088/1742-6596/151/1/012058

of dust according to the size of particles. Conventional size-selective samplers such as cyclone (GS-3 Respirable Dust Cyclone, SKC), at flowrate 2.75 $l.min^{-1}$ for 50% cut-point of 4.0 µm, and two-stage virtual impactor (Respicon Particle Sampler, Model 8522, TSI), inhalable fraction, cut point at 100 µm, thoracic fraction, cut point at 10 µm, respirable fraction, cut point at 4 µm, at flow rate of 3.11 $l.min^{-1}$, were also used with the same derivatizing MP impregnation treatment. Porous polyurethane foam plugs of nominal porosity, 30 Pores Per Inch (PPI), black foam and 90 PPI white foam at different diameters (15 and 29.5 mm) were delivered by the coordinator to participants during the European project. The PUF plugs were ready to be adapted to conventional sampling heads such as IOM and BGIA GSP samplers. PUF plugs were first washed with detergent and water, then with distillated methanol by Sohxlet over night, dried in an oven at 100°C during 1 hour and finally impregnated with MP solution, similar to filter impregnation.

2.3. PUF sampling heads

The PUF sampling heads are schematically shown in Figure 1. It consists of a commercial glass fibre filter of 25 mm diameter (Gelman, Skan AG), which is mounted on a glass tube containing two PUF plugs at different porosity (35 and 90 PPI). These prototypes are designed based on the characteristics of conventional sampling heads cassette, IOM and GSP.

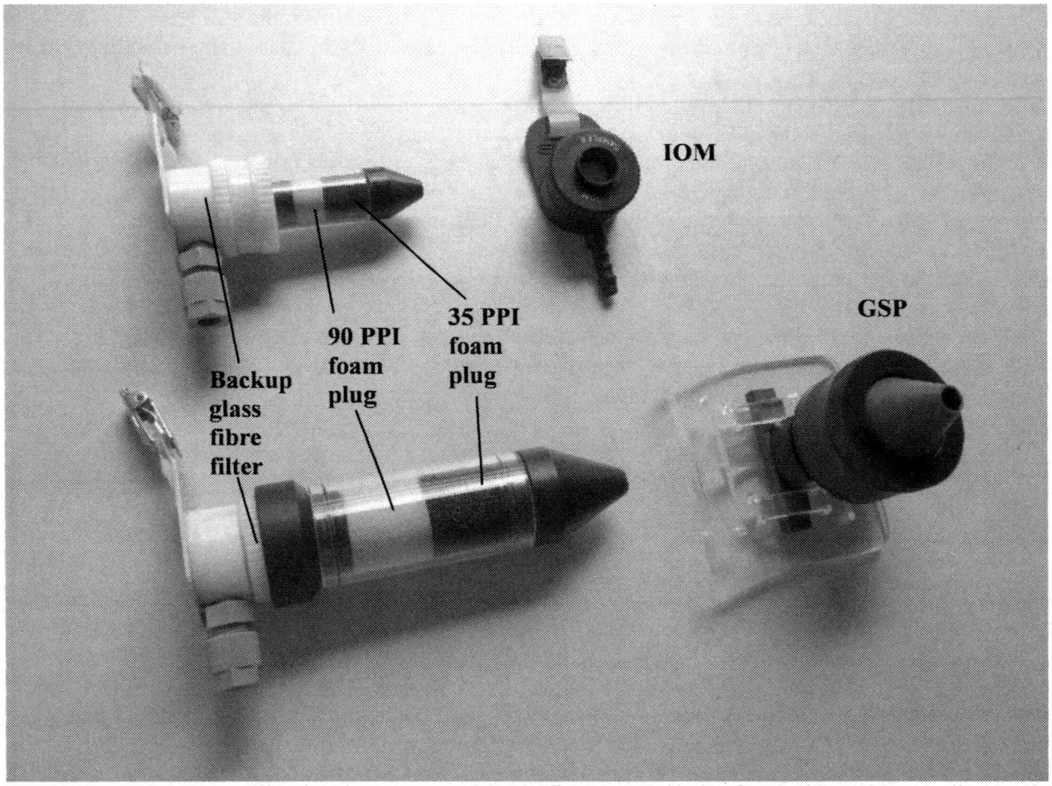

Figure 1: PUF sampling heads prototype 1(top), flow rate 2 l/min, foam plugs (15 mm diameter): 35 PPI (black), thickness 9 mm and 90 PPI (white), thickness 12.5 mm, inlet cone orifice 6 mm diameter and PUF sampling heads prototype 2 (bottom), flow rate 3.5 l/min, foam plugs (29.5 mm diameter): 35 PPI (black), thickness 25 mm and 90 PPI (white), thickness 25 mm, inlet cone orifice 8 mm diameter. The conventional IOM and BGIA GSP samplers were placed at the right of the picture for the comparison.

The diameters and thicknesses of foam plugs are set according to the sampling flow rate:

1. Prototype 1 : flow rate 2 l/min, foam plugs (15 mm diameter): 35 PPI, thickness 9 mm and 90 PPI, thickness 12.5 mm, inlet cone orifice 6 mm diameter.
2. Prototype 2 : flow rate 3.5 l/min, foam plugs (29.5 mm diameter) : 35 PPI, thickness 25 mm and 90 PPI, thickness 25 mm, inlet cone orifice 8 mm diameter.

The pore size, dimensions of foam plugs, aerosol penetration characteristics and applications in particle size-selective sampling were already investigated in previous studies [10-14], and the entry characteristics of practical workplace aerosol samplers in relation to the health-related aerosol fraction was reported based on measurements of aerodynamic optical particle counter [19]. In this study, the authors prefer to adopt the pragmatic solutions [20] of inlet cone orifice to standard GSP sampler, with the inlet diameter was adapted to flow rate. Side-by-side comparisons of samplers were made using "inhalable mass fraction" of aerosols by weight.

2.4. Analysis of HDI and prepolymers

Analysis of derivatized HDI and prepolymers was based on MDHS 25/3 method [7, 17, 21, 22] by HPLC with UV and electrochemical detection (EC), the reference procedure for evaluation of the developed solid PUF samplers. Inter-method comparisons to appraise the performances and limits of the reference method were made within the context of previously published papers [21-23]. The deviations of analytical HPLC method used in this laboratory were 3%. Calculation of the total equivalent NCO group for the HDI prepolymers in terms of mass per unit volume of air was done by using the approach proposed by Silk et al. [24]. Essentially, the trimer structure of prepolymers contains three –N=C=O groups (major peak) and each prepolymers peak surface area, obtained from data system, was calculated as HDI-equivalent [17].

2.5. Generation of commercial aerosol HDI prepolymers

To study the size distribution of the isocyanate particles according to the spray-painting operations, we simulated these operations, first in a vertical aerosol test chamber [17], and second in our experimental chamber of 10 m^3 of volume. Briefly, isocyanate-based aerosols generated by a low airflow in an all-glass nebulizer were introduced downstream at the top of the vertical aerosol test chamber. Spatial distribution and temporal stability of the aerosols concentration in the test section of the chamber were measured and the coefficients of variation are between 5-8%. Performance of PUF samplers was also tested in real field situations in industrial spray-painting shops.

3. Results and discussions

3.1. Penetration of particles through PUF plugs

The penetration of particles through foam plugs with selected geometry and porosity in prototype sampler was measured in laboratory tests and complied sufficiently well with the required EN 481 thoracic and respirable curve [25]. The experimental 50% penetration particle diameters (D_{50}) were 4.3 µm for respirable fraction (foam 90 PPI) and 10 µm for thoracic fraction (foam 35 PPI). This means the collected isocyanate aerosols on the first stage 35 PPI foam plug were assigned to extra-thoracic fraction (>10 µm). Thoracic fraction was the results the sum of the second 90 PPI foam plug and the backup filter and the respirable fraction <4.3 µm was the mass collected on the backup filter. Summing the three collected results (foam 35 PPI, foam 90 PPI and backup filter) provided the total inhalable fraction. Mass collections on different fractions were also tested using polydisperse dust (glass microspheres) in a wind tunnel and a calm air chamber [25, 26].

3.2. Penetration of isocyanate-based prepolymer aerosols through PUF plugs

Table 1 summarized the experimental conditions of different samplers used during this comparative study. Two PUF prototypes, three cyclones, two-stage virtual Respicon impactor, Andersen impactor and impingers were used at their respective sampling flow rate and different inlet geometries for comparison.

Table 1 : Summary of experimental conditions

Sampler	Flow rate ($l.min^{-1}$)	Stage	Range (µm)	50% Cut-point	Impregnation Volume (ml) 1 $mg.ml^{-1}$ MP	Extraction Volume (ml ACN)
Impactor Andersen 2000 Whatmann filter 934AH Ø 81mm	28	0	8.3-17.0	11.0	3 ml	10
		1	5.2-10.5	7.0		10
		2	3.5-7.3	4.7		10
		3	2.7-5.0	3.3		10
		4	1.5-3.1	2.1		10
		5	0.75-1.6	1.1		10
		6	0.45-1.0	0.6		10
		7	0.3-0.68	0.4		10
		F	<0.4	0.2		10
Cyclone GS-3 GF/B Filter Ø 25mm	2.75	Filter	Respirable fraction	4.0	0.3	3
Respicon 8522 GF/B Filter Ø 37mm with hole	3.11	Top	<4.0	4.0	0.3	3
		Middle	4.0-10.0	10.0		3
		Bottom	>10.0	100		3
Prototype 1 foam plugs Ø 15 mm, inlet cone orifice Ø 6 mm	2.0	35 PPI PUF Black Foam	7.0-11.0	10.0	2	3
		90 PPI PUF White Foam	3.3-11.0	4.3	3	3
		GF/B Ø 25mm	<3.3	0.2	0.3	3
Prototype 2 foam plugs Ø 29.5 mm, inlet cone orifice Ø 8 mm	3.5	35 PPI PUF Black Foam	7.0-11.0	10.0	3	10
		90 PPI PUF White Foam	3.3-11.0	4.3	9	10
		GF/B Ø 25mm	<3.3	0.2	0.3	3

In practice, as reported Kenny [20], the only factor distinguishing the sample designs tested by Mark and Vincent [27, 28], is the orifice size and hence the inlet velocity. In contrast to the conventional IOM sampler (which has an orifice of 15 mm, a flow of 2 $l.min^{-1}$ and an inlet velocity of 0.2 $m.s^{-1}$), the manikin's mouth has an elliptical orifice measuring ~6 x 30 mm, an average flow rate of 20 $l.min^{-1}$ and an average inlet velocity of ~4.7 $m.s^{-1}$. On the another hand, the GSP sampler has an orifice of 8 mm, a flow rate of 3.5 $l.min^{-1}$ and an inlet velocity of 1.16 $m.s^{-1}$, similar to Andersen impactor (inlet orifice 25.6 mm, flow rate 28 $l.min^{-1}$, inlet velocity of 0.91 $m.s^{-1}$, and PAS-6 personal sampler, University of Wageningen (inlet orifice 6 mm, flow 2 $l.min^{-1}$, inlet velocity 1.18 $m.s^{-1}$). Finally, for inter-comparison facility, the authors prefer to adapt the inlet entry for two PUF prototypes with inlet velocity approaching the value of GSP, ranked by Kenny as the first most precise sampler at

0.5 - 1 m.s^{-1} velocity of wind tunnel [29], and Andersen values, about 1 m.s^{-1}, this means for the IOM like PUF sampler, the inlet cone orifice of the 6 mm diameter gave 1.18 m.s^{-1} at an airflow of 2 l.min^{-1}.

To illustrate the performance of size-selective PUF samplers, three rounds were performed in real situation, during spray-painting in a shop. Total NCO equivalent mass fraction expressed in mg.m^{-3} collected on different samplers: impactors, cyclone, PUF sampling heads and impingers were used and results reported in table 2.

Table 2 : Summary of all results of different samplers in field situation, spray-painting shop with two-component HDI based paint, expressed in total NCO equivalent [mg.m^{-3}].

Round	Sampler	Respirable fraction < 4.3 μm	Thoracic fraction <10 μm	Extra-Thoracic fraction >10 μm	Total Inhalable fraction
1	Prototype 1	0.412	0.748	2.322	3.070
25 min		0.425	0.714	1.716	2.430
		0.394	0.740	1.572	2.313
	Prototype 2	0.561	1.073	1.925	2.998
		0.482	1.013	1.782	2.795
		0.512	1.077	1.479	2.556
	Cyclone	0.659	-	-	-
		0.604	-	-	-
		0.608	-	-	-
	Respicon	0.748	1.187	2.742	3.015
	Andersen	0.757	0.967	1.262	2.018
	Impinger	-	-	-	4.475
		-	-	-	5.002
		-	-	-	4.839
2	Prototype 1	0.519	0.852	2.162	3.014
25 min		0.460	0.868	2.424	3.291
		0.462	0.828	2.211	3.039
	Prototype 2	0.503	0.860	1.732	2.592
		0.511	0.896	1.856	2.752
		0.532	0.985	1.990	2.974
	Cyclone	0.980	-	-	-
		0.875	-	-	-
		0.938	-	-	-
	Respicon	1.002	1.657	3.370	3.904
	Andersen	0.789	1.018	1.112	1.901
	Impinger	-	-	-	4.377
		-	-	-	5.148
		-	-	-	4.187
3	Prototype 1	0.090	0.206	0.397	0.740
20 min		0.084	0.167	0.440	0.607
		0.110	0.204	0.489	0.693
	Prototype 2	0.110	0.206	0.402	0.607
		0.095	0.200	0.418	0.617
		0.107	0.208	0.391	0.599
	Cyclone	0.273	-	-	-
		0.243	-	-	-
		0.308	-	-	-
	Respicon	0.325	0.509	0.427	0.794
	Andersen	0.133	0.185	0.168	0.301
	Impinger	-	-	-	0.984

Remarks:

1. Prototype 1 : PUF plugs Ø 15 mm, inlet cone orifice Ø 6 mm
2. Prototype 2 : PUF plugs Ø 29.5 mm, inlet cone orifice Ø 8 mm
3. Respicon : extra-thoracic fraction = (inhalable – thoracic) x 1.5
4. Andersen : Respirable fraction = sum of stages 2 to 7 and backup filter
   Thoracic fraction = sum of stages 1 to 7 and backup filter
   Inhalable fraction = sum of all stages and backup filter
5. Swiss Occupational Exposure Limits (OEL-VME) : 0.02 mg.m^{-3} of total NCO equivalent

We observed very reproducible intra-PUF samplers results and acceptable comparative results of different samplers. For example at round 1, the total concentrations of NCO equivalent mass fraction of the respirable fraction (< 4.3 µm), were 0.394 to 0.561 mg.m^{-3} for PUF prototypes (calculated as NCO collected on the backup filter), 0.064 to 0.659 for cyclones, 0.748 for Respicon and 0.757 for Andersen impactor (calculated as the sum of stages 2 to 7 and backup filter). The thoracic fraction (<10 µm) was 0.714 to 1.170 mg.m^{-3} for PUF prototypes (calculated as the sum of 90 PPI foam plug and backup filter), 1.187 for Respicon and 0.967 for Andersen impactor (calculated as the sum of stages 1 to 7 and backup filter). No thoracic data were available with cyclone samplers, because loss inside the cyclone. Aerodynamic distribution of paint aerosols is about 10 µm, and the composition was in majority HDI isocyanurate prepolymers, with a small part of HDI monomer (<0.5%). In this case, size-selective PUF samplers were useful for personal monitoring, without the need of a heavy impactor or denuder devices. Compared to cyclones, PUF personal samplers give more details on different fractions, to comply with CEN/ISO convention: respirable, thoracic, extrathoracic and inhalable fractions.

4. Concluding remarks

Porous PUF size-selective samplers for isocyanates aerosols in spray-painting survey are presented. Compared with other conventional size-selective samplers such as Andersen impactor, cyclone and the two-stage virtual impactor Respicon, the respective corresponding fractions (inhalable, extra-thoracic, thoracic, and respirable), were very similar in term of mass fraction of equivalent total NCO isocyanates. During spraying of isocyanate-based paints, large particle-borne isocyanate and prepolymers fractions dominate. Vapor HDI in commercial two-component paints tend to be a negligible part of total emission (lower than 0.5%) and denuder techniques are not necessary. In term of regulation of occupation isocyanate exposure, more attention is needed to focus on the mass fraction of large prepolymer isocyanate aerosols, less on monomer HDI vapor. Survey of isocyanate occupational exposure with there proposed size-selective personal samplers can be useful to understand the mechanism of occupation asthma developed in spray-painting shops.

Acknowledgements

The authors wish to thank Mr. Philippe Boiteux and Ferdinand Storti for their appreciated technical assistance.

References

[1] Vandenplas O, Malo J L, Saetta M, Mapp C E and Fabbri L M **1993** *Br J of Ind Med* **50**:213-228
[2] Chan-Yeung M and Malo J T **1995** *N Engl J Med* **333**:107-11
[3] Woskie SR, Sparer J, Gore R J, Stowe MH, Bello D, Liu Y, Young F, Redlich CA, Eisen E and Cullen M **2004** *Ann Occup Hyg* **48**:393-403
[4] Bello D, Redlich CA, Stowe M H, Sparer J, Woskie S R, Streicher R P, Hosgood H D and Liu Y **2008** *Ann Occup Hyg* **52**:117-124
[5] ACGIH **2007** Threshold Limit Values for chemical substances and physical agents
[6] SUVA **2007**: Valeur limites d'exposition aux postes de travail (Permissible Exposure Limit at working places), Switzerland
[7] HSE **1999** Methods for the determination of hazardous substances #25/3 organic isocyanates in air (MDHS 25/3). Sudbury, UK: Health and Safety Executive
[8] International Standard Organisation (ISO) **1995** ISO 7708, CH-1211, Geneva, Switzerland
[9] European Committee for Standardization **1993** EN 481, B-1050 Bruxelles, Belgium
[10] Vincent J H, Aitken R J and Mark D **1993** *J Aerosol Sci* **24**:929-944
[11] Kenny L C, Aitken R J, Beaumont G and Görner P **2001** *J Aerosol Sci* **32**:271-285
[12] Chung K Y K, Aitken R J and Bradley D R **1997** *Ann Occup Hyg* **41**:355-372
[13] Kenny L C, Stancliffe J D, Crook B, Stagg S, Griffiths W D, Stewart I W and Futter S J **1998** *Amer Ind Hyg Ass J* **59**:831-841
[14] Kenny L C, Bowry A, Crook B and Stancliffe J D **1999** *Ann Occup Hyg* **43**:393-404
[15] Dahlin J, Spanne M, Karlsson D, Dalene M and Skarping G **2008** *Ann Occup Hyg* advance access published on april 30, 2008
[16] Huynh C K, Vu Duc T and Savolainen H **1989** *Ann Occup Hyg* **33**: 573-581
[17] Huynh C K, Vu Duc T and Savolainen H **1992** *Am Ind Hyg Assoc J* **53**:157-162
[18] Rudzinski W E, Yin J, England E, Carlton E, Key-Schwartz R and Lesage J **2001** *Am Ind Hyg Assoc J* **62**:246-250
[19] Vincent J H and Mark D **1990** *Ann Occup Hyg* **34**:249-262
[20] Kenny L **2003** *Ann Occup Hyg* **47**:437-440
[21] Bagon D A, Warwick C J and Brown R H **1984** *Am Ind Hyg Assoc J* **45**:39-43
[22] White J **2006** *Ann Occup Hyg* **50**:15-27
[23] HSL **1993** *Report* #IR/L/SP/93/01, UK: Health and Safety Executive
[24] Silk S J and Hardy H L **1983** *Ann Occup Hyg* **27**:333-339
[25] Möhlmann C, Aitken R J, Kenny L C, Görner P, Vuduc T and Zambelli G **2002** *Ann Occup Hyg* **46** (suppl 1):386-389
[26] Huynh C K and Vuduc T **2002** *Arch Mal Prof et Méd Trav* **63**:586-588
[27] Mark D and Vincent J H **1986** *Ann Occup Hyg* **30**:89-120
[28] Vincent J H and Mark D **1990** *Ann Occup Hyg* **34**:249-262
[29] Kenny L C, Aiken R, Chalmers C, Fabriès J F, Gonzales-Fernandez E, Kromhout H, Lidén G, Mark D, Riediger G and Prodi V **1997** *Ann Occup Hyg* **41**:135-153

Journal of Physics: Conference Series **151** (2009) 012059 doi:10.1088/1742-6596/151/1/012059

Characteristics of Welding Fume Aerosol Investigated in Three Swedish Workshops

C Isaxon[1], J Pagels[1], A Gudmundsson[1], C Asbach[2], A C John[2], T A J Kuhlbusch[2], J E Karlsson[3], R Kammer[3], H Tinnerberg[3], J Nielsen[3] and M Bohgard[1]

1. Division of Ergonomics and Aerosol Technology (EAT), Lund University, Faculty of Engineering, PO Box 118, SE-221 00, Lund, Sweden
2. Institute of Energy and Environmental Technology, (IUTA), Duisburg, Germany
3. Occupational and Environmental Health, Lund University, University Hospital, Lund, Sweden

Abstract: Potentially high human exposures to nanometer sized airborne particles occur due to welding and other thermal processes in industrial environments. Detailed field measurements of physical and chemical particle characteristics were performed in three work-shops in Sweden. Measurements were performed both in the plume 5-20 cm above the welding point and in the background air (more than 5 m away from the nearest known particle source). Particle number and mass concentrations were measured on-line. A low pressure impactor was used for size-resolved chemical particle composition. The in-plume measurements generated the chemical signatures for different welding processes. These signatures were then used to identify contributions from various processes to the particle concentrations in different size classes. The background number and mass concentrations increased by more than an order of magnitude during intense activities in the work-shops compared to low activities during breaks.

1. Introduction

In industrial environments nano sized particles occur at much higher concentrations than in ambient and indoor air. Welding is one of the more important sources for fine and ultrafine particles in industrial environments worldwide. Only in the United States there are nearly 0.5 million people performing welding and cutting operations full-time, and additional 1-2 million workers who weld intermittently [1]. The total amount of smoke emitted from the welding industry is estimated to 5000 tons/year [2]. Particles from welding processes are enriched in metals, which from a health perspective makes the study even more relevant.

The focus of our project is mild-steel welding, particularly Gas Metal Arc Welding (GMAW), since it is one of the most frequently used indoor welding methods. This method is often referred to as Metal Inert Gas (MIG) or Metal Active Gas (MAG), depending on the composition of the shielding gas. The gas is used for protecting the molten metal from oxidation by creating a microclimate around the welding point, thus shielding it from the oxygen in the surrounding atmosphere. The choice of shielding gas depends on the substrate; however a welding process is often optimized by using a mixture of the two types of gases. Also the composition of the electrode is substrate dependent, so that a welding joint with mechanical properties similar to the substrate is obtained.

Journal of Physics: Conference Series **151** (2009) 012059 doi:10.1088/1742-6596/151/1/012059

It has been shown that the fume composition is reasonably independent of the substrate material itself and that the welding fume is mainly composed by species from the electrode material [3]. The substrate only seems to influence the fume composition if covered with oil, paint or any similar substance [4].

The aim of this study is to determine detailed emission characteristics of welding aerosols in three welding workshops in southern Sweden to improve the knowledge of nanoparticle exposure, particularly to welding fumes in real-world work-places. This knowledge will also be used to generate relevant welding fume for controlled human chamber exposures in the laboratory at Ergonomics and Aerosol Technology at Lund University.

2. Methods

Workshop 1 produces wind mills and railway wagons, and resides in the spacious open halls of a former shipyard. Workshop 2 is placed in a foundry, where several activities other than welding takes place, such as carbon arc gouging, chiseling and grinding. In the third workshop the main activity is welding, only interrupted by a few other operations where the welding joints are being smoothed.

Samples were taken at two different points in each of the three workshops. The first sampling point was at 2 m height, at least 5 m distance from the nearest known source of fine and ultrafine particles, using a PM10 inlet. The purpose of this was to sample the workshop background air. The second sampling point was in a freshly generated welding plume at a distance 5-20 cm above the welding point using a sampling probe immediately connected to a two stage ejector dilution system, which provided a fume dilution factor of 1:150. The dilution probe was introduced to reduce the magnitude of aerosol dynamics processes, particularly coagulation and to cool and dilute the sample to the range of the measurement systems. The in-plume measurements provided the signature size distribution and composition of welding fumes.

The total sampled airflow of 16.7 lpm was led to a flow splitter which distributed the flow to the different instruments, which were as follows: fine and ultrafine particle size distributions were determined using a Scanning Mobility Particle Sizer (SMPS, 10-1000 nm, time resolution of several minutes) and a Fast Mobility Particle Sizer (FMPS, 5-560 nm, 1 s time-resolution), coarse particle number size distributions were determined with an Aerodynamic Particle Sizer (APS, 0.5-20 um). The PM_{10} mass concentration was determined using a Tapered Element Oscillating Microbalance (TEOM), the results of this instrument were compared to a photometer (DustTrak). The lung deposited surface area concentration was determined using a Nanoparticle Surface Area Monitor (NSAM). Two low pressure impactors (LPI) were used, a Hering LPI for collecting samples for Transmission Electron Microscopy (TEM) and a Dekati LPI to collect size-segregated samples (13 stages, 30-10000 nm) for chemical analysis using Particle Induced X-ray Emission (PIXE). To estimate personal particle exposure, selected welders carried personal filter samplers incorporating respirable fraction cyclone pre-collectors. Relative humidity and temperature were also monitored.

3. Results and discussions

3.1. Number- and mass concentrations

A representative sample of the large time-fluctuations in the background measurements are shown in figure 1 for total number concentration (15-1000 nm) and for PM10 mass concentration. It can be seen that the activities increase the exposure both in terms of number and mass by more than an order of magnitude, compared to the level during breaks. A clear correlation between increases in number and mass could be found. The PM_{10} measurements in the "background" zone ranged from < 100 $\mu g/m^3$ during longer breaks up to 3000 $\mu g/m^3$ during intense activities. Correlations between PM_{10} determined with the TEOM and the photometer are being investigated. Personal exposures measured as respirable dust ranged from 600-3400 $\mu g/m^3$.

From the in-plume measurements the signature number size distributions from welding could be determined and typically yielded a single mode with geometric mean diameter of 100-150 nm. It was found that the high time resolution of the FMPS instrument was a clear advantage compared to the SMPS in these measurements with strong concentration fluctuations over time (not shown).

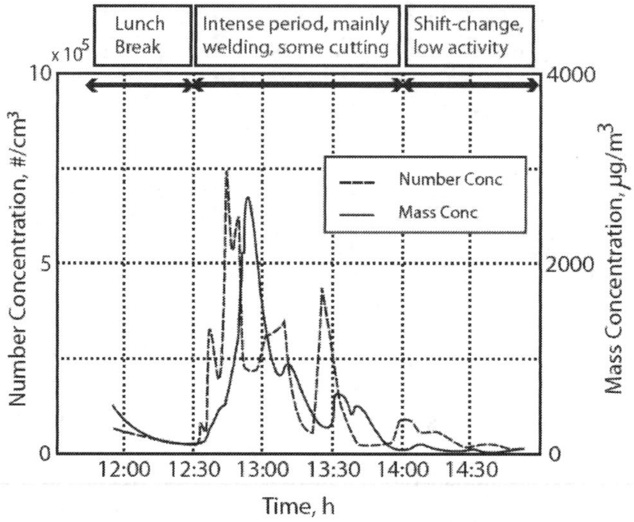

Figure 1. Number concentrations measured with SMPS and mass concentration measured as PM10 with the TEOM at the background measurement point during the afternoon at workshop 3.

3.2. Mass size distributions and chemical composition

Normalized mass size distributions based on the elements detected with PIXE are given in figure 2 for the in-plume measurements and in figure 3 for the background measurements. It can be seen that the signature size distributions of MAG-welding appears as a single mode with mass median aerodynamic diameter (MMAD) of 200-300 nm, while the experiment with stick welding generated larger particles. The stick electrodes used (Elga, Maxeta 22, 6 x 600 mm) were of high output type. They were used to generate filling in larger cracks (5-10 cm) in the components being manufactured.

The elemental composition as a function of particle size was calculated for health relevant and major elements. In figure 4, an example is given for in-plume measurements of MAG-welding in mild steel. The particle composition is dominated by Iron, while the Manganese fraction increases with size. This suggests that Manganese condenses onto existing particles at lower temperatures in the formation process. Further, the fraction of Nickel and Chromium is very low, as expected for welding in mild steel.

In figure 5, a sample of the background air in the same workshop is given. It can be seen that the fraction of Ni and Cr is about an order of magnitude higher compared to the mild steel welding signature in figure 4. This indicates the presence of work in stainless steel within the workshop. The fraction of these compounds is strongly elevated for particles smaller than 150 nm. This either reflects that welding in stainless steel generates smaller particles, or that another operation in stainless steel, for example cutting is dominating in this size range. In the size-range between 150 and 500 nm, the fractions of Fe and Mn are in the same range as in figure 4 and the Ni and Cr concentrations are low. This suggests that MAG welding in mild steel is the main source for these sizes. In the background measurement, the fraction of Mn decreases towards a few percent for coarse particles, which is in the same range as the raw material (mild steel). This would be expected for an aerosol being generated by mechanical processes, in contrast to the evaporation-condensation processes which are responsible for the fine mode. Further, the fractions of Ni and Cr increases with increasing particle size above 300 nm. This indicates the presence of mechanical processes in stainless steel as well.

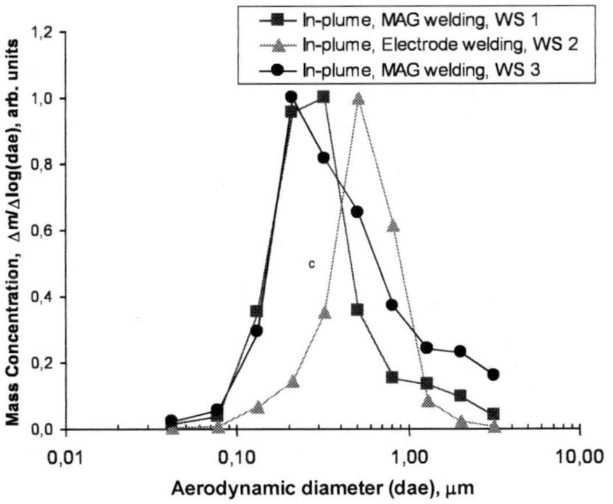

Figure 2. Normalized mass size distributions from in-plume measurements in work shops.

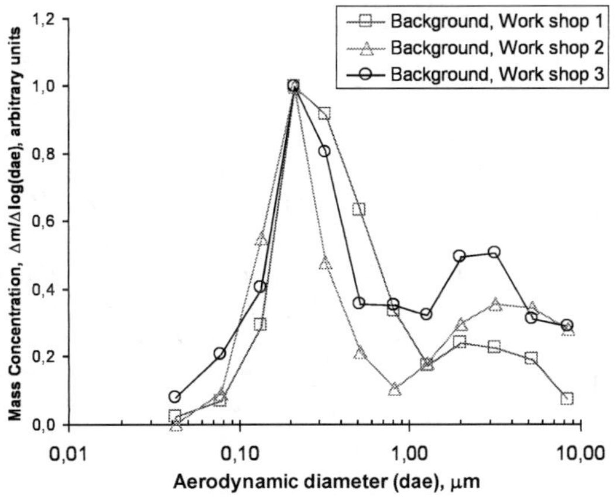

Figure 3. Normalized mass size distributions from background measurements in work shops

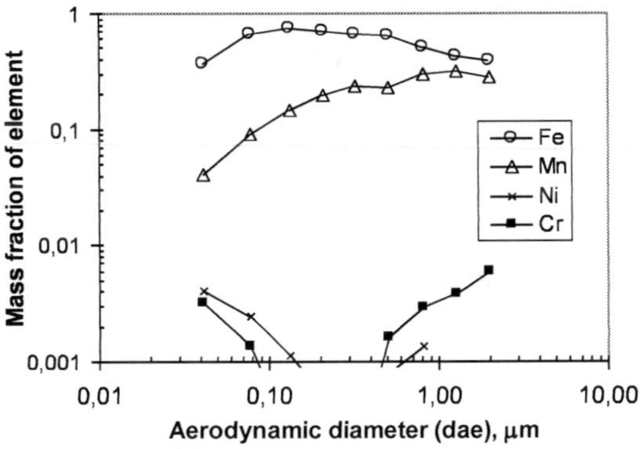

Figure 4. Size-fractionated elemental composition of welding fumes from an in-plume measurement in workshop 3, MAG-welding in mild steel.

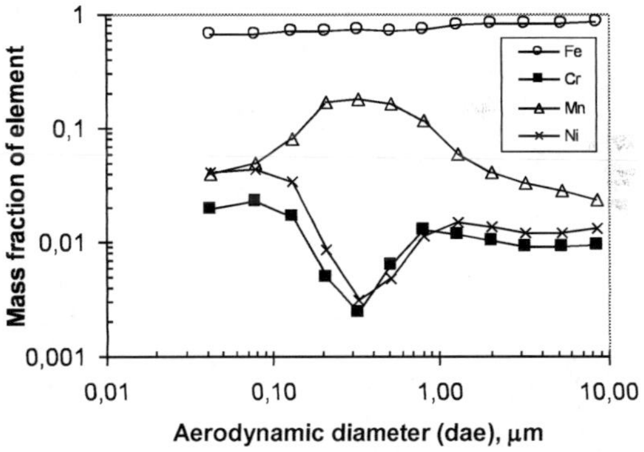

Figure 5. Size-fractionated elemental composition of particles sampled during a background measurement in workshop 3.

4. Conclusions

Using a large range of aerosol measurement techniques, representing the state of the art, exposure levels and particles characteristics have been identified in field measurements in real work environments. Further the combination of techniques can be used to identify detailed knowledge of the contributions from separate sources to the airborne particle concentrations.

Acknowledgements:
The Swedish society for working life and social research is acknowledged for financial support.

References

[1] Jenkins N. T., *Chemistry of airborne particles from metallurgical processing*, PhD thesis, Massachusettes Institute of Technology, 2003.

[2] Redding C. J., *Fume model for Gas Metal Arc Welding*, Welding Journal, p 95-103, 2002

[3] Antonini J. M., Afshari A. A., Stone S., Chen B., Schwegler-Berry D., Fletcher W. G., Goldsmith W. T., Vandestouwe K. H., McKinney W., Castarova V., Frazer D. G., *Design, construction and characterization of a novel robotic welding fume generator and inhalation exposure system for laboratory animals*, Journal of Occupational and Environmental Hygiene 3 (4), p 194-203, 2006.

[4] Quimby B. J. & Ulrich G. D., *Fume formation Rates in Gas Metal Arc Welding*, Welding Journal 78 (4), p 142-149, 1999.

Journal of Physics: Conference Series **151** (2009) 012060 doi:10.1088/1742-6596/151/1/012060

Parallel particle impactor – novel size-selective particle sampler for accurate fractioning of inhalable particles

S Trakumas[1] **and E Salter**[2]

[1] SKC Inc., 863 Valley View Road, Eighty Four, PA 15330
[2] SKC Ltd., 11 Sunrise Park, Higher Shaftesbury Road, Blandford Forum, Dorset, DT11 8ST

E-mail: saulius@skcinc.com

Abstract. Adverse health effects due to exposure to airborne particles are associated with particle deposition within the human respiratory tract. Particle size, shape, chemical composition, and the individual physiological characteristics of each person determine to what depth inhaled particles may penetrate and deposit within the respiratory tract. Various particle inertial classification devices are available to fractionate airborne particles according to their aerodynamic size to approximate particle penetration through the human respiratory tract. Cyclones are most often used to sample thoracic or respirable fractions of inhaled particles. Extensive studies of different cyclonic samplers have shown, however, that the sampling characteristics of cyclones do not follow the entire selected convention accurately. In the search for a more accurate way to assess worker exposure to different fractions of inhaled dust, a novel sampler comprising several inertial impactors arranged in parallel was designed and tested. The new design includes a number of separated impactors arranged in parallel. Prototypes of respirable and thoracic samplers each comprising four impactors arranged in parallel were manufactured and tested. Results indicated that the prototype samplers followed closely the penetration characteristics for which they were designed. The new samplers were found to perform similarly for liquid and solid test particles; penetration characteristics remained unchanged even after prolonged exposure to coal mine dust at high concentration. The new parallel impactor design can be applied to approximate any monotonically decreasing penetration curve at a selected flow rate. Personal-size samplers that operate at a few L/min as well as area samplers that operate at higher flow rates can be made based on the suggested design. Performance of such samplers can be predicted with high accuracy employing well-established impaction theory.

1. Introduction

Currently, a target specification for size-selective sampling instruments is determined by inhalable, thoracic, and respirable sampling conventions agreed upon internationally [1, 2, 3]. Various particle inertial classification devices are available to fractionate airborne particles to approximate one of the mentioned size-selective sampling criteria. The Button Sampler (SKC Inc., Eighty Four, PA), IOM Sampler (SKC Inc., Eighty Four, PA), GSP Sampler (Strohlein GmbH, Kaarst, Germany), and the Seven-Hole Sampler (Casella CEL Ltd., Bedford, UK) are examples of devices where the inhalable convention is followed well through the appropriate combination of inlet size, shape, and sampling flow rate [4, 5, 6]. Cyclones are the samplers most often used to separate thoracic or respirable fractions of inhalable particulates. A number of respirable and thoracic cyclones operating at flow rates ranging from 1.0 L/min to 10.0 L/min are available from different manufacturers. Extensive studies of different cyclonic samplers have shown, however, that the sampling characteristics of cyclones do not follow the entire ACGIH/CEN/ISO-defined respirable or thoracic sampling convention accurately [7, 8, 9, 10, 11, 12]. In most cases, cyclones oversample smaller particles and undersample larger ones as compared to the above mentioned conventions. Additionally, the performance of some personal respirable cyclones may be altered by particle build-up on the inner walls of the cyclone and particle re-entrainment and bounce-off [8, 13].

The use of porous polyurethane foam (PUF) as an alternative to cyclones has been proposed and is an inexpensive method for particle size-selective sampling [13, 14, 15, 16]. However, the particle separation using foams does not provide more accurate results when compared to those obtained using cyclones. Another alternative to traditional cyclones is a virtual cyclone – this device is based on a non-impact particle separation [17]. A respirable sampler employing the virtual cyclone concept was designed and shown to be able to closely follow the respirable convention [18].

In general, conventional inertial impactors, the most well known and studied inertial particle separators, cannot be used for thoracic or respirable particle collection due to the impactor's sharp cut-off characteristics. In 1978, Marple suggested the method to fractionate particles according to predetermined convention using a number of conventional impactors with different cut-offs [19]. A sampler proposed by Marple contained a number of inertial impactors arranged in parallel where each impactor simulated part of the predetermined curve so that overall performance of the sampler followed the entire selected curve. Different cut-offs were achieved using sets of differently sized nozzles. The aerosol was divided equally among different nozzle sets using an appropriate number of nozzles in each set. For example, to match the ACGIH respirable curve that was valid at that time, Marple used one 2.4-mm, eight 0.87-mm, and fifty-three 0.33-mm nozzles to provide 5.8, 3.5, and 2.2 μm 50% cut-offs at 2.0 L/min [19]. A similar approach can be used to design samplers for nearly any flow rate, with penetration characteristics that approximate any monotonically decreasing penetration curve [20]. Several single-stage two- and three-cut-off personal impactors with respirable aerosol penetration characteristics were designed and tested [21, 22]. Sampler prototypes with two cut-offs were found to work well, however, the smallest cut-off nozzles did not work properly in the three-cut-off sampler [22, 23].

Jones [23] suggested respirable and thoracic samplers that included a number of separate single-inlet impactors with different cut-offs arranged in parallel. In this design, each impactor had its own collection substrate and filter. In addition, each impactor had an independent flow control valve to maintain the appropriate flow rate through each impactor. Therefore, the sampler proposed by Jones was bulky and inconvenient to use [23]. Other researchers also used the approach of dividing a specific curve into increments: John [24] described a universal impactor for particle sampling within selected criteria and Chen et al. [14] used foams of different porosity placed in parallel to simulate the respirable curve.

As discussed above, several impactors arranged in parallel can be employed to build a sampler capable of simulating a predetermined curve accurately. However, none of the above mentioned samplers using the parallel impactor technique found its application in industrial hygiene. One reason for this, in the authors' opinion, was the lack of reliable and convenient design. This paper presents a novel sampler design [25] that contains a number of impactors arranged in parallel. During this study, several personal sampler prototypes were built, tested, and shown to be in good agreement with the sampling criteria for which they were designed.

2. Parallel particle impactor design

Performance of an inertial impactor is defined in terms of its 50% cut-off size, d_{50}. This means that 50% of cut-off-size particles penetrate through the impactor and the other 50% are collected on the impaction plate. The d_{50} can be found using the following equation [26, 27]:

$$d_{50} = \left(\frac{9\mu W Stk_{50}}{\rho_p V_o C} \right)^{1/2}, \tag{1}$$

where μ is air viscosity, W is the width or diameter of the impactor nozzle, Stk_{50} is the Stokes number corresponding to a 50% particle cut-off, ρ_p is the particle density, V_0 is average air velocity in the nozzle, and C is the size-dependent Cunningham slip correction factor. The Stk_{50} depends on the Reynolds number of the flow, *Re*, jet-to-plate distance, *S*, and impactor nozzle throat length, *T*. The Stokes number may change significantly if porous material is used as a collection substrate [28, 29].

Typically, particle penetration through the conventional impactor decreases sharply from 100% to 0% near the 50% cut-off point. The solid line in Figure 1 represents a penetration curve for an impactor with a d_{50}=4.0 μm. The dashed curve in Figure 1 shows the respirable convention which has the same 50% cut-off of 4.0 μm as the single impactor. Despite the fact that both curves have a d_{50}=4.0 μm, it is clear that the conventional impactor (solid curve in Figure 1) will significantly oversample particles smaller than 4.0 μm and undersample particles larger than 4.0 μm compared to the respirable convention (dashed curve in Figure 1). As discussed, the curve of any predetermined shape may be approximated by combining several impactors in parallel [19, 21]. The square symbols in Figure 1 show the shape of a penetration curve of a hypothetical sampler containing two impactors arranged in parallel where one of the impactors has a 50% cut-off of 2.95 μm and the other has a d_{50} of 5.35 μm. The selected cut-off sizes correspond to 75% (midpoint of 50-100) and 25% (midpoint of 0-50) penetration efficiency of the respirable convention [19]. These numbers were determined based on the condition that the flow rate through each impactor is the same and equal to half of the total flow rate through the sampler. The curves in Figure 1 representing penetration through samplers containing four and six impactors (triangular symbols and circular symbols respectively) were constructed in a manner similar to that described above. The airflow through the individual impactors in these samplers is equal as it was in the sampler containing two impactors. The flow through a single impactor is one-fourth of the total flow for the sampler containing four impactors and one-sixth in the case of the six-impactor sampler. The curves presented in Figure 1 show that the respirable convention can be simulated rather closely using a sampler containing four or more impactors arranged in parallel.

The above described procedures can be applied to model a sampler with characteristics simulating the shape of any predetermined curve. The first step is to decide the number of impactors to be used and the overall flow rate through the sampler. The second step is to determine the cut-off sizes for each impactor according to the shape of the curve the sampler is to simulate and the flow rate through the individual impactor chosen. With this information, the inlet nozzle sizes of individual impactors can be defined easily using well-established impaction theory [27].

Control of the airflow through the individual impactors should be addressed when designing a sampler containing a number of impactors arranged in parallel. Marple's design was based on the assumption that the pressure drop across the impactor nozzle, ΔP, is approximately equal to the dynamic pressure of the air jet in the nozzle [19]:

$$\Delta P = \frac{1}{2} \rho V_0, \tag{2}$$

where ρ is air density and V_0 is the average air velocity in the nozzle. The amount of air for each cut-off was then controlled by the appropriate size and number of nozzles for each cut-off assuming that the air velocity in each nozzle of the sampler was approximately the same. Jones [23] simply placed valves beneath each impactor to control the amount of air passing though each impactor.

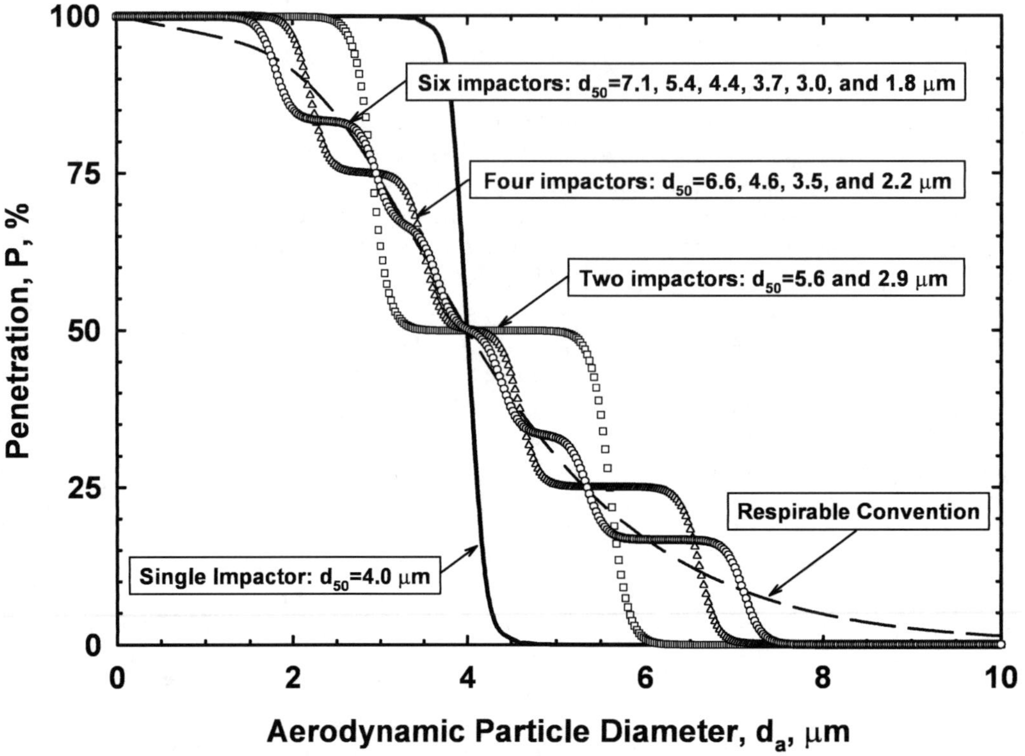

Figure 1. Comparison of the respirable convention curve with theoretically predicted performance of samplers containing a different number of inertial impactors arranged in parallel.

In the new sampler design, flow through each impactor is controlled using an appropriately sized exit orifice. Figure 2 shows schematically a sampler containing two impactors arranged in parallel. Both impactors are completely separated and each has an inlet nozzle, W_{In}, and exit orifice, W_{Out}. Since all inlets of the sampler are at the same atmospheric pressure and all outlets are connected to a single pump, the pressure drop across each impactor, ΔP_i, is the same and equal to the overall pressure drop across the whole sampler, ΔP_S:

$$\Delta P_S = \Delta P_1 = \Delta P_2. \tag{3}$$

Pressure drop across an individual impactor, ΔP_i, is a sum of pressure drops across the inlet nozzle, ΔP_{iIn}, and outlet orifice, ΔP_{iOut}:

$$\Delta P_i = \Delta P_{iIn} + \Delta P_{iOut}\,. \tag{4}$$

Using Equations 2, 3, and 4, the following equation may be written:

$$\frac{1}{2}\rho V_{1In}^2 + \frac{1}{2}\rho V_{1Out}^2 = \frac{1}{2}\rho V_{2In}^2 + \frac{1}{2}\rho V_{2Out}^2\,. \tag{5}$$

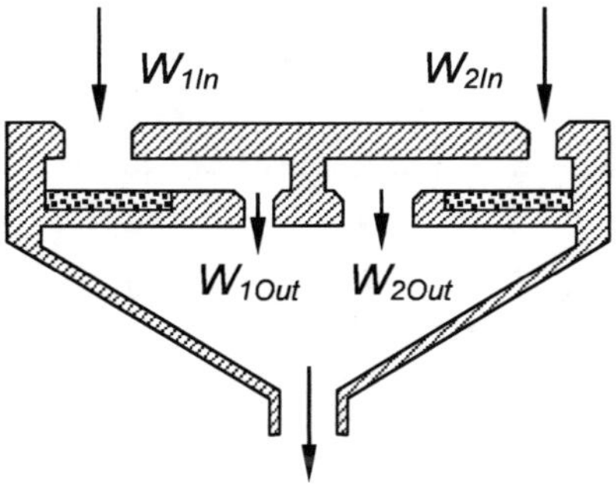

Figure 2. Schematic of sampler comprising two inertial impactors arranged in parallel.

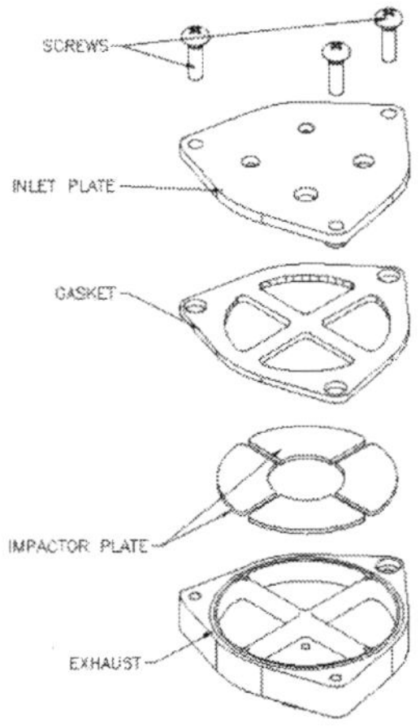

Figure 3. Schematic of prototype containing four impactors arranged in parallel impactor.

Equations 3, 4, and 5 can be extended for any number of impactors incorporated in the sampler. Thus, taking into account that airjet velocity in the nozzle $V=Q/(\pi(W/2)^2)$, Equation 5 for the sampler containing N impactors in parallel would be:

$$Q_1^2\left(\frac{1}{W_{1In}^4}+\frac{1}{W_{1Out}^4}\right)=Q_2^2\left(\frac{1}{W_{2In}^4}+\frac{1}{W_{2Out}^4}\right)=.....=Q_N^2\left(\frac{1}{W_{NIn}^4}+\frac{1}{W_{NOut}^4}\right). \quad (6)$$

As indicated earlier, the size of the inlet nozzle for each impactor is selected depending on the shape of the curve the sampler is going to simulate, the overall sampling flow rate, the desired particle size cut-offs, and the number of impactors in the sampler. The sizes of outlet orifices then can be calculated using Equation 6. Such a sampler can be designed for an overall flow that is equally distributed among all impactors ($Q_1=Q_2=...\ =Q_N$) or a flow rate through each impactor may be set individually.

A schematic of a prototype sampler that includes four impactors arranged in parallel is shown in Figure 3. The prototype sampler has an inlet plate with impaction nozzles and an exhaust that is divided in four compartments. Each compartment includes collection substrate and an exit orifice. A gasket is used between the inlet plate and exhaust to prevent air leakage from outside and among individual impactors. Collection plates were cut out of a porous plastic support pad (Cat. No. 225-2902, SKC Inc., Eighty Four) and soaked with a few drops of silicone oil to minimize particle bounce and re-entrainment. Three prototype samplers were tested during this study: two respirable parallel particle impactors, RPPI2 and RPPI4, were designed to follow the respirable convention when operated at 2.0 and 4.0 L/min flow rates respectively. Thoracic parallel particle impactors (TPPI2) was designed to follow the thoracic curve when operated at 2.0 L/min. Parameters of these samplers are presented in Table 1. Nozzle sizes were calculated using value of Stokes number different from that suggested by impaction theory [27]. Lower value of Stokes number was based on other studies that show how porous collection substrate affects the performance of impactors [28, 29, 30] and our own experience using porous impaction substrate [31].

Table 1. Specifications of prototype samplers tested. Each sampler comprises four inertial impactors arranged in parallel.

Respirable Impactor, RPPI2 Q_S=2.0 L/min			Respirable Impactor, RPPI4 Q_S=4.0 L/min			Thoracic Impactor, TPPI2 Q_S=2.0 L/min		
D_{50}, µm	W_{In}, mm	W_{Out}, mm	D_{50}, µm	W_{In}, mm	W_{Out}, mm	D_{50}, µm	W_{In}, mm	W_{Out}, mm
6.5	3.45	1.68	6.5	2.59	1.30	17.5	5.10	2.15
4.6	2.71	1.72	4.6	2.06	1.33	11.9	4.00	2.18
3.5	2.26	1.80	3.5	1.73	1.40	8.9	3.25	2.25
2.3	1.68	3.45	2.3	1.30	2.59	4.8	2.15	5.10

3. Experimental methods

The performance of three parallel impactor prototypes was evaluated by measuring aerosol concentration upstream and downstream of the sampler using an Aerodynamic Particle Sizer (APS, Model 3320, TSI Inc., St. Paul, MN). A similar technique has been used widely by other researchers and is proven to allow accurate and rapid evaluation of particle samplers [13, 18, 22, 32, 33, 34]. Figure 4 shows a schematic of the experimental setup used in the study. The aerosol chamber was made of a clear plexiglass cylinder approximately four feet high and one foot in diameter. Test particles and HEPA-filtered dry air were introduced and mixed at the top of the chamber. Aerosol passed through a honeycomb flow straightener before reaching the test area where two identical sampling lines were installed. Air was expelled from the chamber through a 2.5-inch diameter hose to a biological safety cabinet (Model SG603, The Baker Company, Sanford, MN). The overall airflow through the chamber combining clean air and test aerosol ranged from 40 to 80 L/min resulting in a vertical air velocity in the chamber, V_{Ch}, from 10 to 20 mm/s. Thus, calm air conditions were simulated in the test chamber.

Three different aerosol generators were employed depending on the type of particle used. A six-jet Collison Nebulizer (BGI Inc., Waltham, MA) and a Sono Tek Ultrasonic Atomizer (Sono Tek Corporation, Milton, NY) were used to generate Potassium Sodium Tartrate (PST, Spectrum Chemical Mfg. Corp., Gardena, CA) and Dioctyl Phthalate (DOP, Spectrum Chemical Mfg. Corp., Gardena, CA). Glass spheres (GS, Spheriglass solid spheres, 10000E, Potters Industries Inc., Carlstandt, NJ) were aerosolized using a Fluidized Bed Aerosol Generator (Model 3400A, TSI Inc., St. Paul, MN). The charge on the generated particles was reduced to Boltzman equilibrium by passing test aerosol through five Polonium 210 Alpha source neutralizers (Model 2U500, NRD LLC, Grand Island, NY).

One of two identical sampling lines was used to measure aerosol size distribution in the chamber, $N_{Ch}(d_{ae})$. The second sampling line, on which the test sampler was mounted, was employed to register particle concentration downstream of the sampler, $N_{Down}(d_{ae})$. Ball valves were used to connect the required sampling line to the inlet of an Aerodynamic Particle Sizer (APS) placed beneath the chamber. In normal operating condition, the APS is sampling at 5.0 L/min. From this flow, 4.0 L/min is used as a sheath air and aerosol is sampled through the sensor at 1.0 L/min. In the setup, both sampling lines were connected directly to the 1.0 L/min aerosol sampling inlet using an air splitter which also enabled the addition or subtraction of makeup air so that flows through the sampler and the APS were maintained at required levels. The APS drew sheath air directly from the room.

During a single test, a set of six measurements lasting 60 seconds each was performed. $N_{Ch}(d_{ae})$ and $N_{Down}(d_{ae})$ were measured three times each alternating appropriate sampling lines to the APS inlet. Particle penetration $P(d_{ae})$ through the sampler was calculated using the following equation:

$$P(d_{ae}) = \frac{N_{Down}(d_{ae})}{N_{Ch}(d_{ae})} \times 100\% . \quad (7)$$

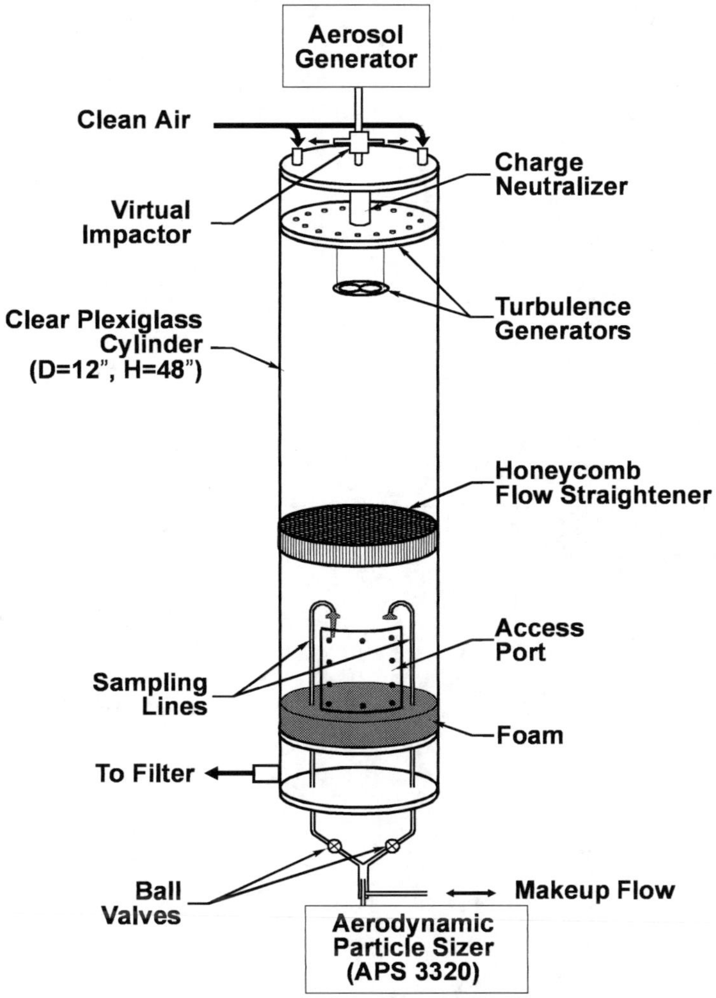

Figure 4. Schematic of experimental setup.

It is known that when using APS Model 3320 false large-particle counts may occur due to small particle recirculation in the optics chamber. To minimize this effect, overall particle concentration was kept below 80 cc^{-1}. A virtual impactor was employed, when needed, to minimize the number of particles smaller than 1.0 μm. APS calibration was checked periodically using NIST-traceable polymer microspheres of 1.0, 3.0, 4.0, 7.0, and 10.0 μm (Particle Counter Size Standards, Duke Scientific Corporation, Palo Alto, CA). The calibration of the APS was found to be stable and in good agreement with values provided by the manufacturer. In addition, the validity of the test technique was confirmed through testing several commercially available samplers. Data obtained for the respirable Higgins-Dewell Cyclone (Model BGI4CP, BGI, Inc., Walthman, MA) and thoracic cyclone (Model GK2.69, BGI, Inc., Walthman, MA) were in good agreement with previously reported results for similar samplers. Sampling characteristics measured for the Personal Environmental Monitor (PEM) (Model PEM 200-2-10, MSP Corporation, Shoreview, MN) also agreed well with manufacturer specifications.

4. Results and discussion

Individual impactors of the newly designed samplers were tested to confirm the theoretically predicted cut-off sizes. This was accomplished by closing three out of the four inlets so that air was pulled only through the inlet of the impactor to be tested. Single impactors were tested at a flow rate equal to one-fourth of the overall flow through the sampler because all sampler prototypes tested in this study were designed to have the overall sampling flow rate split equally among the individual inpactors of the sampler.

Data presented in Figure 5 show penetration as a function of particle aerodynamic size measured for each impactor composing prototype sampler RPPI2. PST test particles were used during these experiments. The symbols in the graph represent an average of at least three measurements and the error bars show standard deviation (the same applies for all other data presented in this paper). Solid lines represent the modeling of experimental data using the following sigmoid function:

$$P(d_{ae}) = \frac{a}{1+\exp\left\{-\frac{(d_{ae}-c)}{b}\right\}}. \quad (8)$$

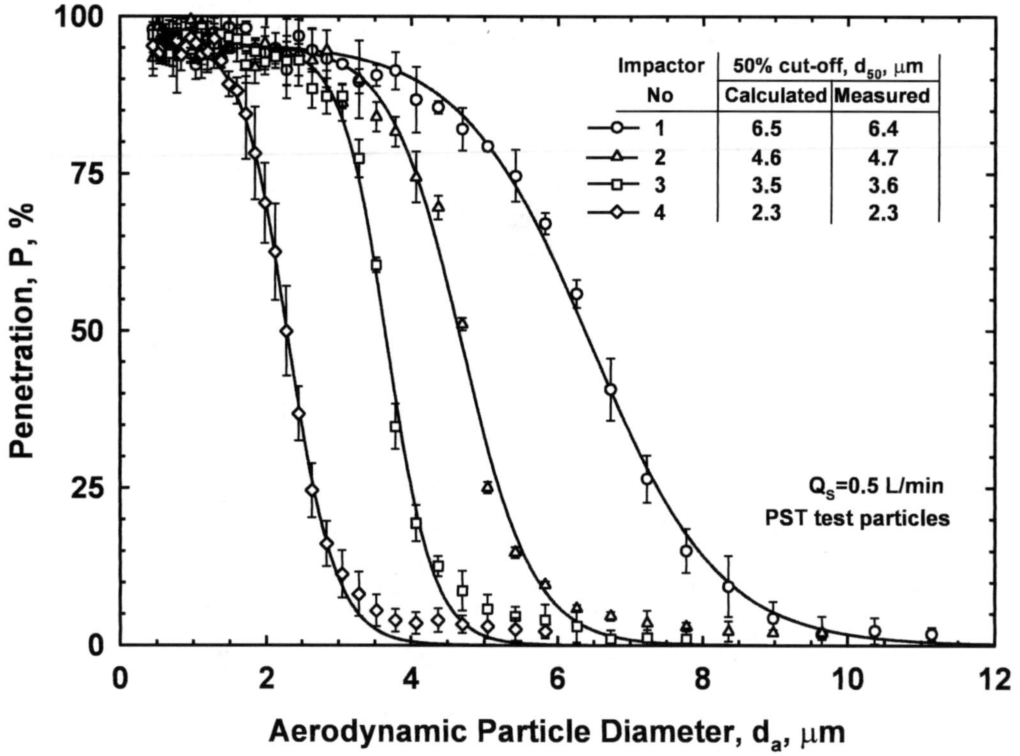

Figure 5. Particle penetration through individual impactors composing the RPPI2 respirable impactor.

The sigmoid function (Equation 8) was found to fit well with data obtained for the individual impactors tested. The same function was used by Kwon et al. [35] to fit collection efficiency data measured for a five-stage cascade impactor. Once constants *a*, *b*, and *c* were found for the best fit, Equation 8 was used to calculate d_{50}. The table in the corner of Figure 5 includes calculated and measured 50% cut-off sizes for each impactor incorporated in the RPPI2 sampler. The data shows that the measured d_{50} are in very good agreement with theoretically predicted values. Agreement was similarly good between the design and the experimental d_{50} for the impactors composing samplers RPPI4 and TPPI2.

Figure 6 shows the penetration characteristics of prototype sampler RPPI2. The open circles represent the initial performance of the sampler, i.e., the sampler was tested with new, clean collection substrates installed. The solid triangles show the penetration measured for sampler RPPI2 after it was exposed to coal mine dust for six hours. Figure 6 also includes a curve representing the respirable convention (solid line) and a curve mathematically constructed using data obtained for individual impactors (dashed line). The latter simulates the respirable curve more smoothly than was theoretically predicted for the sampler containing four impactors (Figure 1). The theoretical prediction was based on the assumption that all four impactors have sharp penetration characteristics as shown in Figure 1 for a single impactor. However, the data (Figure 5) indicate that the actual shape of the penetration curves for all individual impactors is not as steep as the one employed during mathematical simulation. Thus, the combined performance of the four impactors arranged in parallel is smoother (Figure 6, dashed line) when compared to the predicted characteristics (Figure 1, triangles). The initial overall penetration measured for the RPPI2 sampler (circles) follows closely the respirable curve (solid line) and the curve constructed using individual impactor data (dashed line). Thus, data presented in Figure 6 indicate that the performance of the RPPI2 sampler is in good agreement with the entire respirable convention, and, at the same time show the validity of the new design.

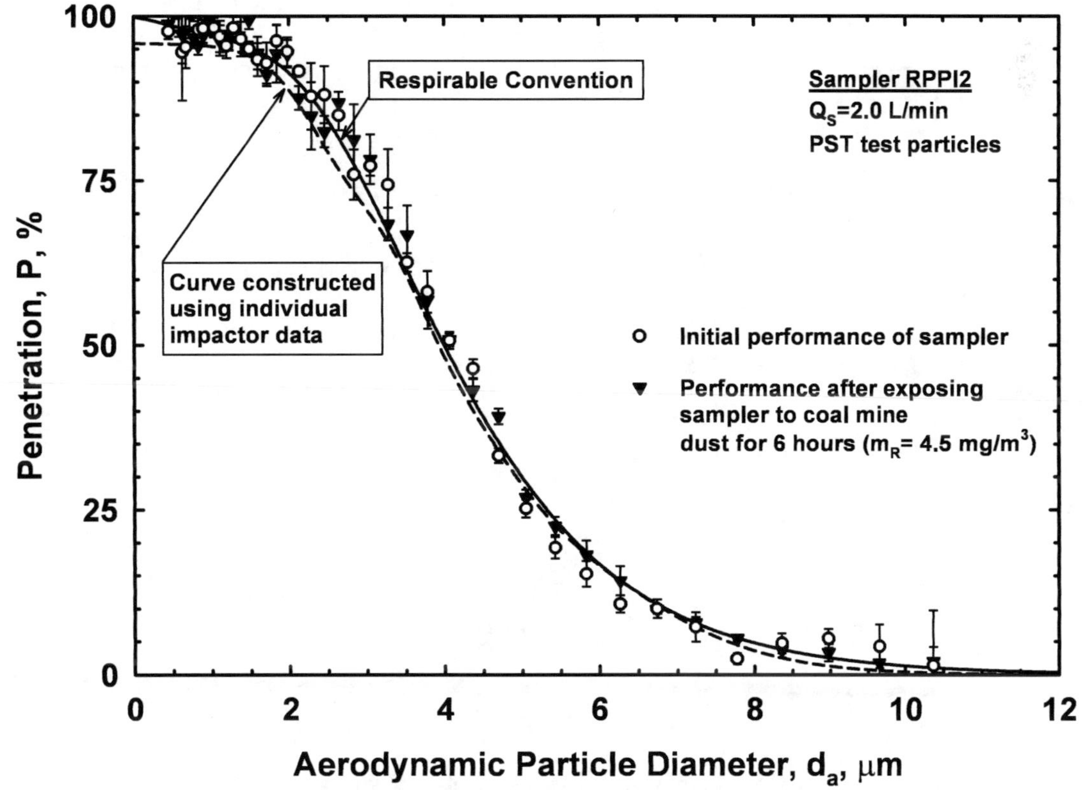

Figure 6. Comparison of performance of RPPI2 sampler with respirable convention.

To assess loading effect on the performance of the RPPI2 and RPPI4 samplers, they were operated for six hours in a Marple chamber at the NIOSH Pittsburgh Research Laboratory (Pittsburgh, PA). The mass of respirable fraction of Keystone coal dust measured using a Higgins-Dewell (HD) respirable cyclone was 4.76 mg/m^3. During an earlier study using the same Marple chamber under similar conditions [36], it was established that the size distribution of Keystone coal dust had mass median aerodynamic diameter (MMAD) equal to 3.91 μm and a geometric standard deviation (GSD) equal to

3.05. It can be shown that the mass of respirable fraction for the mentioned size distribution would make approximately 50 % of the total mass. Thus, we can conclude that the collection plates of RPPI2 and RPPI4 samplers were loaded with approximately 3.4 and 6.8 mg of coal dust respectively. As shown in Figure 6 (solid triangles), such load did not affect the performance of the RPPI2 sampler: neither particle bounce-off nor blow-off of collected particles was observed. The same was found to be valid for the RPPI4 sampler. These findings are in agreement with the stable performance of oil-impregnated porous impaction plates as reported by Marple and McCormack [18].

It is known that solid particles tend to be more bouncy than liquid particles. For this reason, sampler performance is often assessed using both types of test particles. Therefore, in addition to the potassium sodium tartrate (PST) particles recommended by ASTM standard D6061-01 [37], the parallel impactor samplers were also tested using glass microspheres (GS) and dioctyl phthalate (DOP) test particles. Figure 7 shows how three different types of test particles penetrate through the RPPI2 sampler. There is no significant difference in sampler performance obtained using PST, GS, or DOP test particles. In all cases, sampling characteristics stayed close to the respirable convention – the curve the sampler was designed to follow. Thus, it can be concluded that the sampler containing four impactors arranged in parallel collects liquid and solid particles similarly and that there is no indication of increased particle bounce even for smooth-surface, high-density glass spheres.

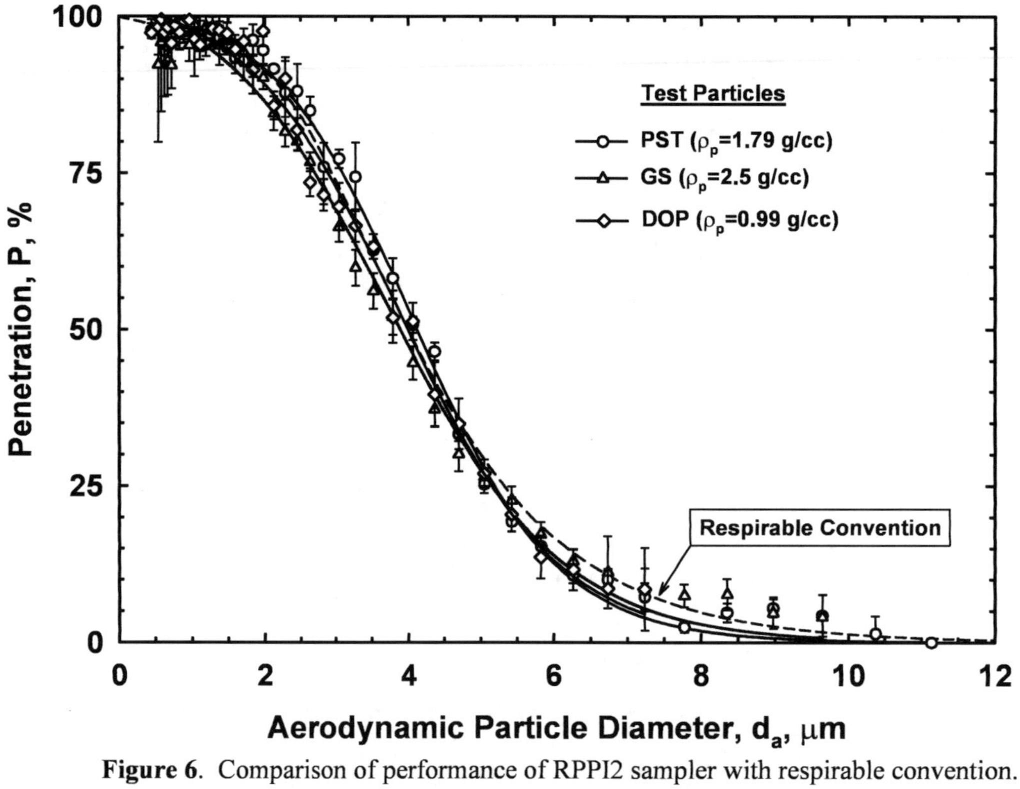

Figure 6. Comparison of performance of RPPI2 sampler with respirable convention.

To assess loading effect on the performance of the RPPI2 and RPPI4 samplers, they were operated for six hours in a Marple chamber at the NIOSH Pittsburgh Research Laboratory (Pittsburgh, PA). The mass of respirable fraction of Keystone coal dust measured using a Higgins-Dewell (HD) respirable cyclone was 4.76 mg/m^3. During an earlier study using the same Marple chamber under similar conditions [36], it was established that the size distribution of Keystone coal dust had mass median aerodynamic diameter (MMAD) equal to 3.91 µm and a geometric standard deviation (GSD) equal to of a similar geometry. Although neither of these two studies tested the cyclone at a 2.2 L/min sampling flow rate, data presented in both studies suggest that at this particular flow the cyclone will match the respirable convention best. Nevertheless, the HD Cyclone oversamples smaller particles and undersamples larger ones (Figure 8., solid squares) when compared to the respirable convention as do many other cyclonic samplers [8, 9, 10, 12]. The results presented in Figure 8 clearly indicate that both newly designed parallel impactor samplers follow the respirable convention more closely than the respirable cyclone tested.

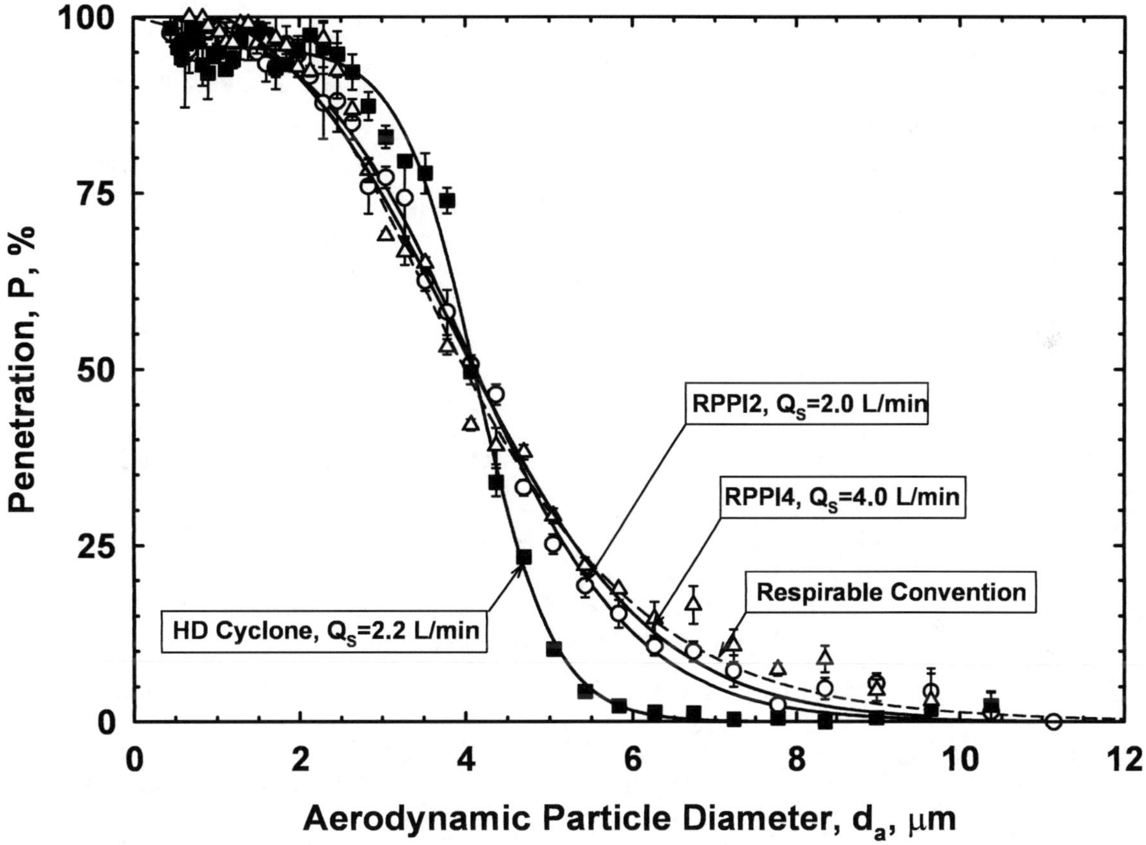

Figure 8. Comparison of penetration characteristics measured for the RPPI2 and RPPI4 parallel impactor samplers and HD Cyclone.

Comparison of the bias maps plotted for the RPPI4 sampler and the HD Cyclone (Figure 9) leads to a similar conclusion: the parallel impactor outperforms the cyclonic sampler. Bias mapping is a widely accepted way to present the performance of a sampler [7, 9, 37]. As discussed earlier, the mathematical model (Equation 8) was used to fit experimental data. Once the model for the sampler in question is established, it can be used to calculate the mass concentration sampled by this sampler, C_{Samp}, for any chosen aerosol size distribution. If C_R is the concentration measured by an ideal respirable sampler, then mean relative bias, Δ, is defined as follows [7, 37]:

$$\Delta = \frac{C_{Samp} - C_R}{C_S} \tag{9}$$

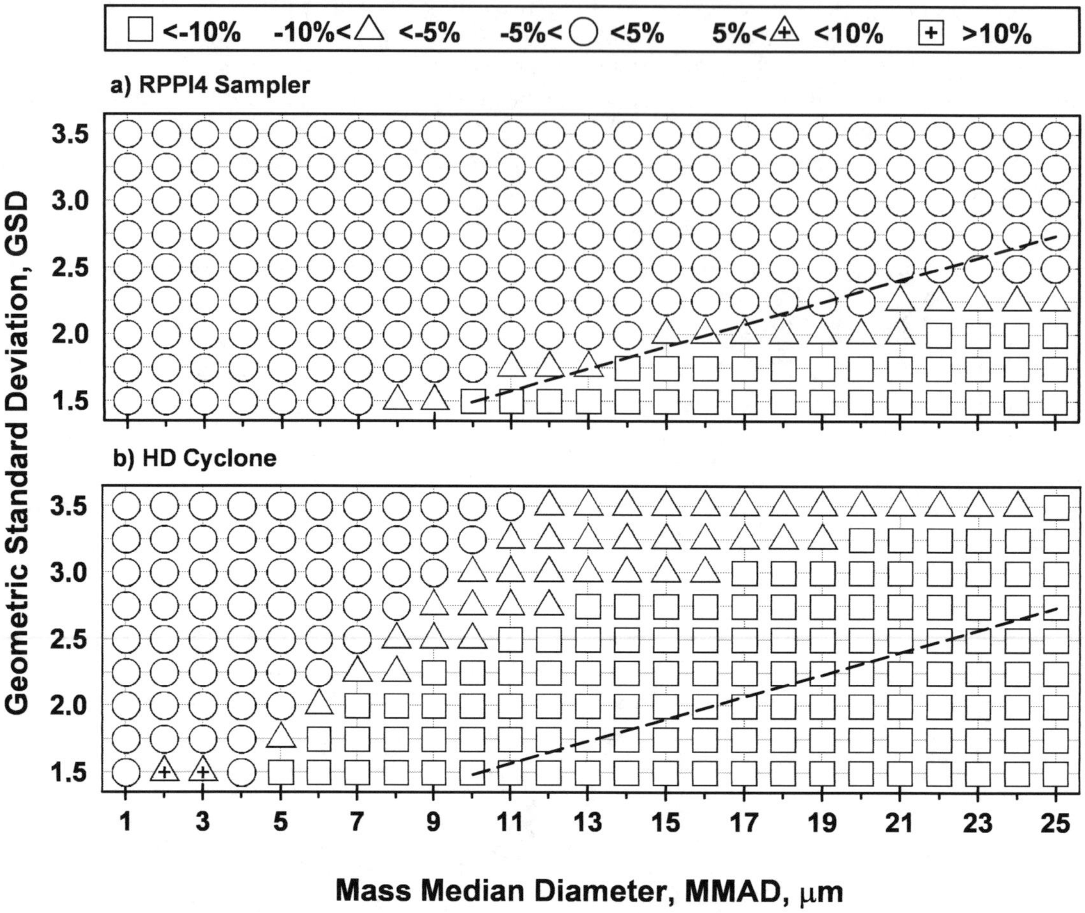

Figure 9. Bias maps plotted for the RPPI4 respirable sampler and the HD Cyclone.

Figure 9 includes the RPPI4 sampler and HD Cyclone bias charts for lognormal aerosol size distributions with geometric standard deviations between 1.5 and 3.5 and mass median diameters of < 25 μm. Data show that for a wide range of aerosol size distribution, particulate mass measured using the RPPI4 sampler will not differ by more than ± 5 % from the mass collected using an ideal respirable sampler. This difference falls below –10 % only for the rare narrow distribution of large aerosol (small GSD and large MMAD) which can be excluded from sampler evaluation. Per ASTM D60061-01 [37], the respirable sampler would only be evaluated at aerosol size distributions where the respirable fraction is greater than 5 % of a total aerosol. This omits sizes below the dashed line in Figures 9a and 9b defined by: (MMAD,GSD) = (10 μm, 1.5) to (25 μm, 2.75) [37]. Data provided in Figure 9 shows that the HD Cyclone will provide less accurate results compare to the RPI4 sampler.

Figure 10 includes penetration data for the newly designed thoracic parallel impactor sampler (TPPI2), Personal Environmental Monitor (PEM 10, MSP Corporation, Shoreview, MN), and GK2.69 Cyclone (BGI, Inc., Waltham, MA). Figure 10 also shows a curve representing the thoracic convention. As indicated earlier, the TPPI2 sampler was designed using the parallel impactor concept to meet the thoracic convention at a 2.0 L/min sampling flow rate. According to manufacturer specifications, the PEM 10 is designed to have a 10.0 μm 50% cut-off at a sampling flow rate of 2.0 L/min and the GK2.69 Cyclone will conform to the thoracic curve at 1.6 L/min.

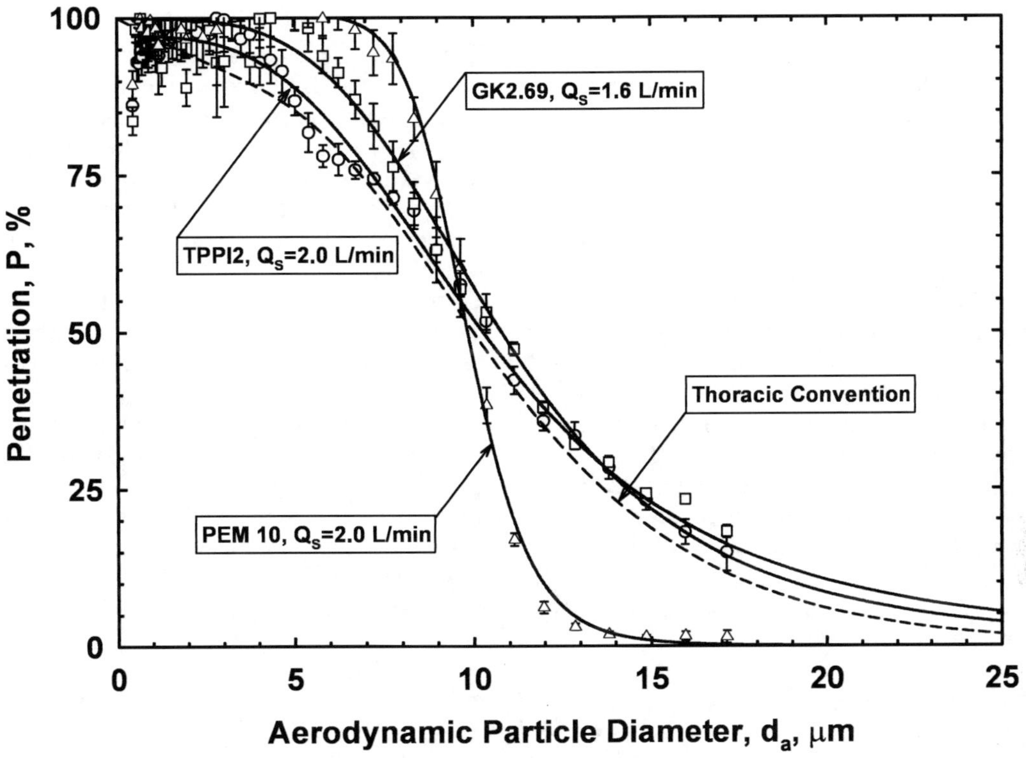

Figure 10. Comparison of penetration measured for TPPI2 sampler, GK2.69 cyclone, and PEM 10.

Penetration of PST test particles measured for the TPPI2 sampler (Figure 10, circles) demonstrates good agreement with the thoracic convention. It should be noted that the tests did not include a thorough evaluation of inlet aspiration efficiency which may influence results for particles larger than 10 μm. Also, reliable data was not produced for particles larger than approximately 17 μm due to the low particle count above that size. Experimental data were fitted and performance of the TPPI2 sampler was extrapolated for larger particle sizes using the following model:

$$P(d_{ae}) = \frac{a}{1+\left(\frac{d_{ae}}{b}\right)^{c}}. \tag{10}$$

The same equation was used to model penetration data obtained for the GK2.69 Cyclone (squares in Figure 10).

Experimental data for the PEM 10 (triangles in Figure 10) show good agreement with manufacturer specifications. Penetration characteristics measured for the GK2.69 Cyclone (squares in Figure 10) conform well with sampler specifications and with results reported by Maynard [33]. These facts indicate that the test system used in this study provides accurate results in the larger particle range despite the limitations mentioned earlier. Thus, it can be concluded that the newly designed TPPI2 sampler closely follows the entire thoracic convention.

5. Conclusion

The novel design of a sampler comprising a number of inertial impactors arranged in parallel was applied to manufacture two respirable samplers operating at a 2.0 and 4.0 L/min sampling flow rate and one sampler approximating the thoracic convention at 2.0 L/min. Each sampler incorporated four impactors arranged in parallel. Test results for all samplers showed good agreement with predicted characteristics. Penetration characteristics of both respirable samplers followed the entire respirable convention more closely when compared to the performance of the commercially available and widely used Higgins-Dewell respirable cyclone. The newly designed thoracic sampler also performed as predicted and showed good agreement with the entire thoracic convention.

Using solid and liquid test particles, it was shown during this study that performance of the newly designed samplers does not depend on the type of particles collected. Experiments revealed also that an approximate load of 6.4 mg of coal mine dust on the sampler collection plates did not affect the performance of the new sampler.

The suggested parallel impactor arrangement can be used to design samplers that will accurately fractionate inhalable particles according to predetermined characteristics including the ACGIH/CEN/ISO-defined respirable and thoracic conventions or any other monotonically decreasing penetration curve. The new design can be applied to fabricate personal samplers operating at a few liters per minute or area samplers operating at higher flow rates. Parameters of samplers with parallel impactors can be predicted with high accuracy. The suggested parallel impactor design insures consistent sampler performance for different types of aerosol and a wide range of particle concentrations.

Acknowledgements

The authors would like to thank Jon Volkwein and Bob Vinson of NIOSH Pittsburgh Research Laboratory (Pittsburgh, PA) for their help with the particle loading experiment.

References

[1] ISO 1995 ISO 7708:1995(E) Air Quality – Particle Size Fraction Definitions for Health-related Sampling (Geneve: ISO)

[2] CEN 1993 CEN Standard EN 481, Workplace Atmospheres: Size Fraction Definitions for Measurement of Airborne Particles in the Workplace (Brussels: CEN)

[3] ACGIH 2003 TLVs® and BEIs® (Cincinnati:ACGIH)

[4] Aizenberg V, Grinshpun S A, Willeke K, Smith J and Baron P A 2000 Measurement of the Sampling Efficiency of Personal Inhalable Aerosol Samplers Using a Simplified Protocol J. Aerosol Sci. **31** 169-179

[5] Kenny L C, Aitken R, Chalmers C, Fabries J F, Gonzales-Fernandez E, Kromhout H, Liden G, Mark D, Riediger G and Prodi V 1997 A Collaborative European Study of Personal Inhalable Aerosol Sampler Performance Ann. occup. Hyg **41** 135-153

[6] Kenny L C, Aitken R, Baldwin P E J, Beaumont G C and Maynard A D 1999 The Sampling Efficiency of Personal Inhalable Aerosol Sampler in Low Air Movement Environments J. Aerosol Sci. **30** 627-638

[7] Bartley D L, Chen C C, Song R and Fischbach T J 1994 Respirable Aerosol Sampler Performance Testing Am. Ind. Hyg. Assoc. J. **55** 1036-1046

[8] Chen C C and Huang S H 1999 Shift of Aerosol Penetration in Respirable Cyclone Samplers Am. Ind. Hyg. Assoc. J. **60** 720-729

[9] Gorner P, Wrobel R, Micka V, Skoda V, Denis J and Fabries J F 2001 Study of Fifteen Respirable Aerosol Samplers Used in Occupational Hygiene Ann. Occup. Hyg. **45** 43-54

[10] Liden G and Kenny L C 1993 Optimization of the Performance of Existing Respirable Dust Samplers Appl. Occup. Environ. Hyg. **8** 386-391

[11] Maynard A D and Kenny L C 1995 Performance Assessment of Three Personal Cyclone

Models, Using an Aerodynamic Particle Sizer J. Aerosol Sci. **26** 671-684
[12] Trakumas S and Hall P 2003 Performance Assessment of Personal Respirable Cyclone Samplers Abstracts American Industrial Hygiene Conference & Exposition (Dallas) p 45
[13] Chen C C, Lai C Y, Shih T S and Hwang J S 1999 Laboratory Performance Comparison of Respirable Samplers Am. Ind. Hyg. Assoc. J. **60** 601-611
[14] Chen C C, Lai C Y, Shih T S and Yeh W Y 1998 Development of Respirable Aerosol Samplers Using Porous Foams Am. Ind. Hyg. Assoc. J. **59** 766-773
[15] Gorner P, Wrobel R, Mohlmann C, Aitken R J, Kenny L C and Fabries J F 2002 Performances of European Multifractional Aerosol Samplers Abstracts 21th Annual AAAR Conference (Charlotte) p 35
[16] Kenny L, Chung K, Dilworth M, Hammond C, Wyn Jones J, Shreeve Z and Winton J 2001 Applications of Low-cost, Dual-fraction Dust Samplers Ann. occup. Hyg. **45** 34-42
[17] Torczynski J R and Rader D J 1997 The Virtual Cyclone: A Device for Nonimpact Particle Separation Aerosol Sci. Techno. **26** 560-573
[18] Chen C C, Huang S H, Lin W Y, Shih T S and Jeng F T 1999 The Virtual Cyclone as a Personal Respirable Sampler Aerosol Sci. Techno. **31** 422-432
[19] Marple V A 1978 Simulation of Respirable Penetration Characteristics by Inertial Impaction J. Aerosol Sci. **9** 125-134
[20] Marple V A, Rubow K L and Olson B.A 1993 Inertial, Gravitational, Centrifugal, and Thermal Collection Techniques Aerosol Measurement: Principles, Techniques, and Applications ed K Willeke and P A Baron (New York: Van Nostrand Reinhold) pp 206-232
[21] Marple V A and McCormack J E 1983 Personal Sampling Impactor with Respirable Aerosol Penetration Characteristics Am. Ind. Hyg. Assoc. J. **44** 916-922
[22] Baron P A 1983 Sampler Evaluation with an Aerodynamic Particle Sizer Aerosols in the Mining and Industrial Work Environment ed V A Marple and B Y H Liu (Ann Arbor: Ann Arbor Science) pp 861-877
[23] Jones G W 1987 Design and Calibration of a Multi-Purpose Aerosol Sampler Doctoral dissertation. Graduate School of Public Health (Pittsburgh: University of Pittsburgh)
[24] John W 1994 Universal Impactor for Particle Collection within Sampling Criteria Abstracts 13th Annual AAAR Conference (Los Angeles) p 929
[25] U.S. Patent 7,073,402
[26] Marple V A and Willeke K 1976 Impactor Design Atmos. Envir. **10** 891-896
[27] Rader D J and Marple V A 1985 Effect of Ultra-Stokesian Drag and Particle Interception on Impaction Characteristics Aerosol Sci. Techno. **4** 141-156
[28] Kavouras I G and Koutrakis P 2001 Use of Polyurethane Foam as the Impaction Substrate/Collection Medium in Conventional Inertial Impactors Aerosol Sci. Techno **34** 46-56
[29] Marjamaki M and Keskinen J 2004 Effect of Impaction Plate Roughness and Porosity on Collection Efficiency J. Aerosol Sci. **35** 301-308
[30] Demokritou P, Gupta T, Ferguson S and Koutrakis P 2002 Development and Laboratory Performance Evaluation of a Personal Cascade Impactor J. Air & Waste Manage. Assoc. **52** 1230-1237
[31] Trakumas S and Hall P 2002 Modification and Validation of Diesel Particulate Matter Cassette for PM-1.0 Sampling at Different Flow Rates Abstracts 2002 American Industrial Hygiene Conference & Exposition (San Diego) p 19
[32] Maynard A D, Kenny L C and Baldwin E J 1999 Development of a System to Rapidly Measure Sampler Penetration up to 20 μm Aerodynamic Diameter in Calm Air, Using the Aerodynamic Particle Sizer J. Aerosol Sci. **30** 1215-1226
[33] Maynard A D 1999 Measurement of Aerosol Penetration Through Six Personal Thoracic Samplers Under Calm Air Conditions J. Aerosol Sci. **30** 1227-1252
[34] Trunov M, Trakumas S, Willeke K, Grinshpun S A and Reponen T 2001 Collection of

Bioaerosol Particles by Impaction: Effect of Fungal Spore Agglomeration and Bounce Aerosol Sci. Techno. **35** 617-624

[35] Kwon S B, Lim K S, Jung J S, Bae G N and Lee K W 2003 Design and Calibration of a 5-stage Cascade Impactor (K-JIST Cascade Impactor) J. Aerosol Sci. **34** 289-300

[36] Volkwein J C, Tuchman D P and Vinson R P 2002 Performance of a Prototype Personal Dust Monitor for Coal Mine Use Mine Ventilation ed E De Souza (Lisse: A.A.Balkema Publishers) pp 633-639.

[37] ASTM 2003 D6061-01: Standard Practice for Evaluating the Performance of Respirable Aerosol Samplers Annual Book of ASTM Standards (West Conshohocken: ASTM)

High efficiency CIP 10-I personal inhalable aerosol sampler

P Görner, R Wrobel, X Simon

INRS, Rue du Morvan, CS 60027, 54519 Vandoeuvre les Nancy, France

E-mail: peter.gorner@inrs.fr

Abstract. The CIP 10 personal aerosol sampler was first developed by Courbon for sampling the respirable fraction of mining dust. This respirable aerosol sampler was further improved by Fabries, then selectors for sampling thoracic and inhalable aerosols were designed. Kenny *et al.* evaluated the particle-size dependent sampling efficiency of the inhalable version in a large-scale wind tunnel using a life-size dummy. The authors found that the overall sampling efficiency decreases more rapidly than the CEN-ISO-ACGIH target efficiency curve. Görner and Witschger measured the aspiration efficiency of the CIP 10 omni-directional inlet. They found that the aspiration efficiency was high enough for inhalable aerosol sampling. This result led to the conclusion that the low sampling efficiency is due to some internal losses of the aspirated particles before they reach the final sampling stage, namely the CIP 10 rotating filter. Based on the assumption that the inhalable particles are selected at selector aspiration level, an experimental research project was conducted to improve particle transmission to the collection stage of the sampler. Two different inhalable selectors were designed by Görner and tested in a laboratory wind tunnel. The transmission efficiency of both models was measured by Roger following an experimental protocol described by Witschger. The T-shaped air flow circuit was finally adopted to draw the aspirated particles into the final collection stage of the CIP 10. Actually, in this selector, the almost horizontally aspirated particles should be conducted vertically to the rotating cup. In two previous prototypes, particles could be deposited in certain places by inertia (where the aerosol was forced to deviate drastically) or by sedimentation (where the aerosol decelerated). The aerodynamic behaviour of the adopted solution causes the particles to accelerate radially between two horizontal plates before they enter a vertical tube. This acceleration avoids the particles being deposited on the lower horizontal plate. At the beginning of the vertical tubing, the mutually opposing particle trajectories limit particle wall deposition by virtual impaction effect. The inner selector walls are polished to avoid particles being stopped by eventual surface asperities. Particle size-dependent sampling efficiency was measured in the laboratory wind tunnel. The experimental aerosol was composed of polydisperse glass micro-spheres. The size analysis of the particles collected was done by the Coulter Counter technique. The transmission efficiency (reciprocal to wall losses) was found to be close to 100 % for the entire range of particle sizes, and indicated no particle loss. The overall sampling efficiency was measured using a rotating bluff body at an external wind speed of 1 m/s. The rotating bluff body represents a scaled torso of an operator. The "high efficiency" CIP 10-I (I for inhalable) responds fairly well to the conventional CEN-ISO-ACGIH criteria for sampling the inhalable health-related aerosol fraction.

1. Introduction

The CIP 10 personal aerosol sampler was first developed as an individual version of the static coal-mine respirable dust sampler CPM3 [1]. The sampler is based on a rotating cup principle described by Quinot [2]. The efficiency of the rotating filter and its flow rate as a function of rotating speed were studied by Görner *et al.* [3]. The respirable aerosol selector of the CIP 10 was further improved [4], and the performance of the CIP 10 was laboratory tested by [5] along with fourteen other respirable samplers used in occupational hygiene all over the world. A selector for sampling the thoracic fraction of aerosols was then designed for the CIP 10 sampler [6, 7]. The sampling of the inhalable fraction by the CIP 10 was achieved simply by directing the aspirated particles directly into the rotating cup with no additional selection unit. This resulting in the CIP 10 sampler being able to mesure any of the three conventional health-related aerosol fractions [8, 9, 10]. Finally, a special rotating cup was developed for biological aerosol sampling into a liquid medium [11, 12]. The CIP 10 sampler is operated at a flow rate of 10 $L.min^{-1}$, except for the thoracic version which is operated at 7 $L.min^{-1}$.

Kenny *et al.* [13, 14] evaluated the particle-size dependent sampling efficiency of the original version of the inhalable selector of the CIP 10-I (I for inhalable) in a large-scale wind tunnel using a full-scale dummy. A series of quasi monodisperse aluminium oxide (Aloxite) particles was used to generate the test aerosol. The authors found that the sampling efficiency decreased more rapidly than the CEN-ISO-ACGIH [8, 9, 10] target efficiency curve (figure 1).

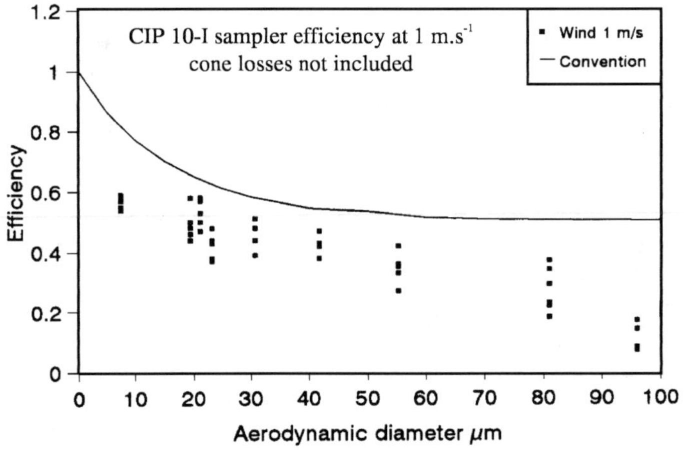

Figure 1. Overall sampling efficiency of the original CIP 10-I inhalable aerosol sampler as a function of particle size, measured in a large wind tunnel with rotating full-size mannequin.

A hypothesis was made that the low efficiency could be due to wall losses inside the selector (figure 3a).

Görner *et al.* [15] measured the aspiration efficiency of the CIP 10 omni-directional inlet using a polydisperse aerosol of glass microspheres. They found that the entry efficiency of the aspiration slot was sufficiently high for inhalable aerosol sampling (figure 2). This result led to the conclusion that the low sampling efficiency is actually due to the internal losses of the aspirated particles in the original selector.

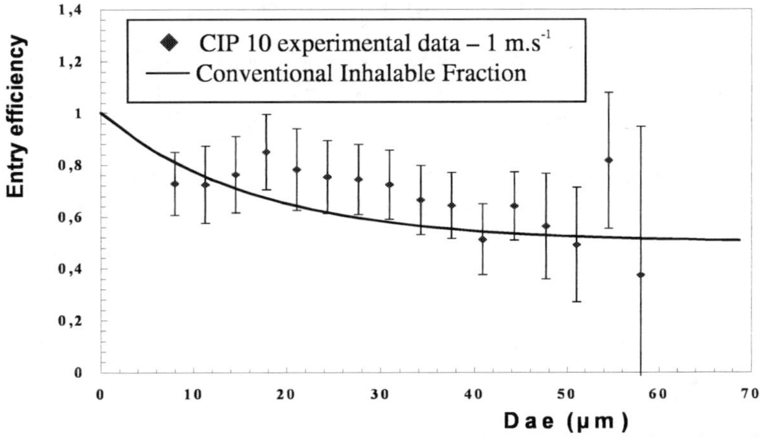

Figure 2. Experimental entry efficiency of the CIP 10 aspiration slot as a function of particle size. Sampler in isolated static position (with no bluff body).Sampler flow rate: 10 $L.min^{-1}$ - Tunnel wind speed: 1 $m.s^{-1}$

2. Design of "high efficiency" inhalable selector

The aerosol is aspirated through an omni-directional aspirating slot shielded by a protective cap. Based on the assumption that the inhalable particles are selected at aspiration slot level, an experimental research project was conducted to improve particle transmission to the collection stage of the sampler. First, we looked for the main position of the particle deposit inside the selector. The deposit was formed by impaction in flow deviation under the protective cap near the six inlet orifices of the selector body (as represented in figure 3a). A second inhalable selector with a modified air circuit was designed (figure 3b). However, this second prototype still exhibited an unsatisfactory transmission efficiency, and the deposits of particles were mainly observed in the enlarged conical section, probably due to flow turbulences and sedimentation in decelerated flow.

The third prototype designed uses an annular aspiration slot with two circular horizontal plates connected to a vertical cylindrical tubing (figure 3c).

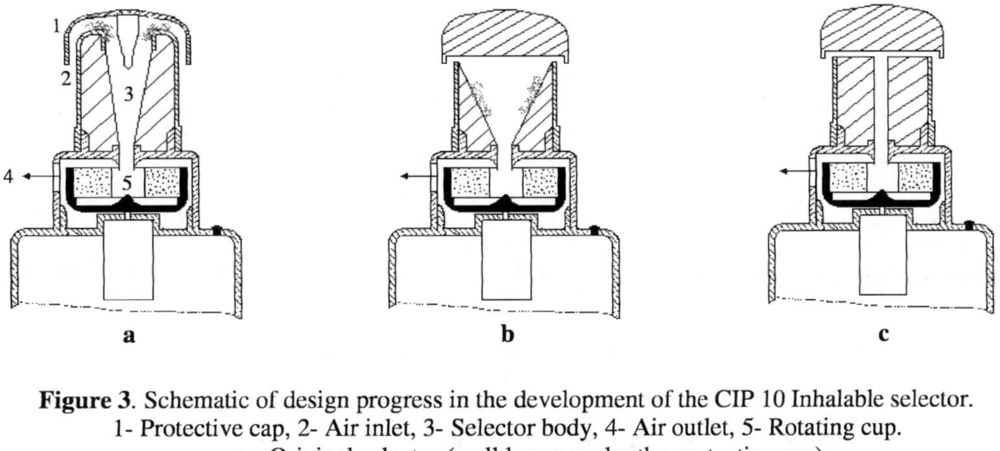

Figure 3. Schematic of design progress in the development of the CIP 10 Inhalable selector.
1- Protective cap, 2- Air inlet, 3- Selector body, 4- Air outlet, 5- Rotating cup.
a – Original selector (wall losses under the protective cap)
b – Second selector (wall losses in large conical duct)
c – Final, T-shape selector (no wall losses)

This geometric configuration was studied earlier by Witschger *et al.* [16] and Roger *et al.* [17], but not in connection with the CIP 10 sampler [18]. The solution in figure 3c avoids the six orifices of the original selector and the large conical section of the second prototype, where the particle losses were concentrated. In the final prototype, the air velocity is accelerated in a radial path through the annular slot, as indicated in figure 4. The acceleration is due to the decreasing section of the slot toward its centre (provided the height of the sampling slot is constant, as shown in figure 3c). The horizontal acceleration as well as the following downward vertical flow minimise the possibility of particle deposit between the aspiration slot and the rotating cup, which is the final particle collector.

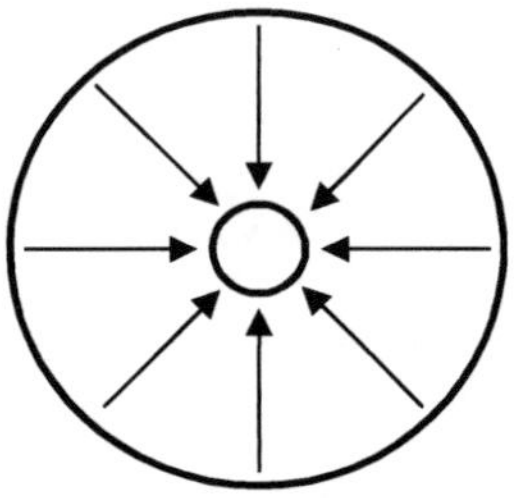

Figure 4. Schematic of radial air flow acceleration between the sampling slot and the central cylindrical duct.

The transmission efficiency of the particles, defined as the difference between aspiration and overall sampling efficiency, was measured by Roger for all three models following an experimental protocol described by Witschger *et al.* [19] using a polydisperse glass sphere aerosol. The particle size-dependent transmission efficiency of the two first selectors from figures 3a and 3b was situated approximately between 60 and 80 %, meaning that the inner wall losses represent about 20-40 % of the aspirated aerosol. The transmission efficiency of the final selector (figure 3c) is close to 1 for all particle sizes, and clearly higher than those of the two previous prototypes (figure 5). This means that the wall losses were minimised and that almost 100 % of the aspirated particles of any size reach the collecting stage, i.e. the rotating cup.

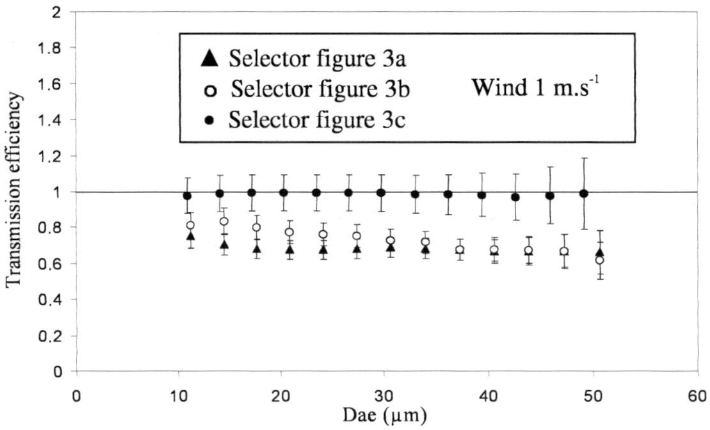

Figure 5. Transmission efficiency (reciprocal to wall losses) of the three inhalable aerosol selectors from figures 3a,b,c.
Sampler in isolated static position (with no bluff body).
Sampler flow rate: 10 L.min^{-1} - Tunnel wind speed: 1 m.s^{-1}

3. Measurement of overall sampling efficiency

The particle-size dependent sampling efficiency of the final version of the CIP 10-I was studied in an experimental wind tunnel using glass beads to generate the aerosol. In the case of the inhalable fraction, the wind conditions in the vicinity of the sampler should be taken into account because of their influence on particle aspiration. The sampling efficiency of personal aerosol samplers is influenced by the operator wearing the sampler. Therefore, the personal samplers are tested attached to a mannequin modelling the operator's body. To simulate a moving operator, the sampler wearing mannequin rotates. This means that the sampler is exposed to the experimental aerosol at different angles. An optimal experimental method to measure sampler efficiency is described in the EN 13205 [20] standard. A large-scale experimental wind tunnel containing a full-sized mannequin is required. Consequently, high flow rates are necessary and some problems with particle size and concentration homogeneity over the entire measuring zone can be expected. Several authors have worked to reduce this by scaling the dimensions of the experimental equipment without deteriorating the experimental results [21-23]. Promising results were obtained by Paik and Vincent [24] using a rectangular, three-dimensional bluff body. We checked that our experimentation with a cylindrical bluff body 110 mm in diameter and 55 mm in width, rotating at 2 rpm, led to results similar to those of Paik and Vincent.

A detailed description of the wind tunnel used and its flow parameters, along with their spatial and time stability, were given by Witschger *et al.* [19]. Polydisperse aerosol generation and determination of the aerodynamic particle diameter (Dae) using the Coulter-Counter technique are described in Görner *et al.* [18].

The overall sampling efficiency of a sampler is defined as the ratio of the size-resolved particle concentration sampled on the final support versus the free stream particle concentration measured by a reference probe. The thin-walled, sharp-edged cylindrical reference probe (10 mm inner diameter) was designed in accordance with the geometric criteria of Belyaev and Levin [25] for unbiased sampling when operated in isokinetic isoaxial conditions. In the case of sequential sampling, used in our case, the sampler on test and the reference probe are placed consecutively one after the other right in the tunnel axes. This solution was preferred to parallel sampling because of better time stability than the space stability of the aerosol concentration in the tunnel [19].

4. Results

The resulting overall sampling efficiency of the final CIP 10-I inhalable aerosol sampler, measured at a wind speed of 1 $m.s^{-1}$, is reported in figure 6. The sampling efficiency of the well-known IOM sampler was also measured and is reported to compare the results of both samplers in the same experimental conditions. The particle size-resolved efficiencies are represented by experimental points along with their error bars corresponding to $\pm$ 2 SD at 95 % confidence level. (SD being Standard Deviation from 3 experimental runs). The CEN-ISO-ACGIH conventional Inhalable penetration is also reported.

The sampling efficiencies of the two samplers are close to each other and decrease around the conventional penetration curve. The final version of the CIP 10 Inhalable aerosol sampler performs much better (figure 6) than the original version of this sampler (figure 1).

To quantify the sampling performance of the sampler placed in various polydisperse aerosols situations, the bias and accuracy maps were elaborated according to the EN 13205 specifications. The goal of this numerical calculation is to simulate the inhalable mass concentration assumed to be measured by the sampler and to compare it to the conventional inhalable concentration (target aerosol fraction), for any given aerosol with a known particle mass distribution. The calculation were made for a large range of log-normally distributed polydisperse aerosols.

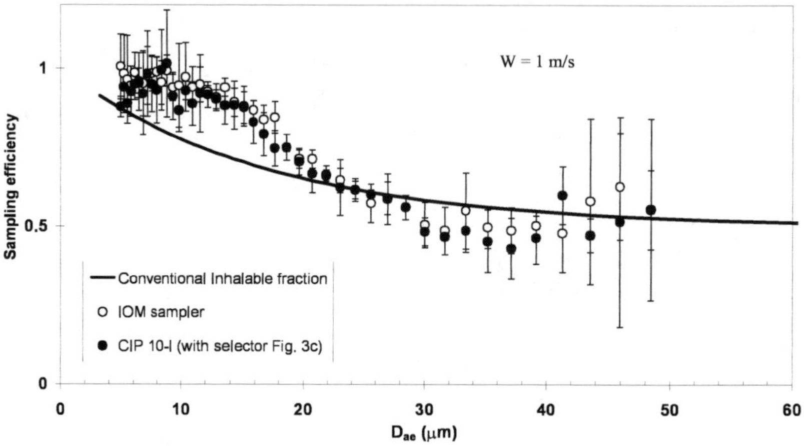

Figure 6. Overall sampling efficiency of the CIP 10-I "high efficiency" inhalable sampler (figure 3c) along with the efficiency of the IOM sampler, measured in the same experimental conditions. Samplers attached to a rotating bluff body. Sampler flow rate: CIP 10-I 10 $L.min^{-1}$ IOM 2 $L.min^{-1}$ – Tunnel wind speed: 1 $m.s^{-1}$

The experimental efficiency values were fitted by a polynomial model of second degree to obtain an analytical expression of the sampling efficiency as a function of aerodynamic diameter of particles. The Newton-Gauss least-squares algorithm was used to optimise the parameters of the model [26]. The model fits quite well the sampler efficiency in the interval of experimental data (0 – 50 µm) but it was found that it idealises the sampler efficiency in the extrapolated interval (50 – 100 µm), keeping the efficiency close to a horizontal line at a level of about 45%. It is obvious that the real sampler efficiency decreases with increasing particle size. The more realistic, decreasing, model was reached by minimising the first fitted parameter of the polynomial model. The resulting sampling efficiency function "E_s" used in bias and accuracy calculations is as follows:

$$E_s(D_{ae}) = 1.10^{-12} . D_{ae}^{2} - 0.0138 . D_{ae} + 1.028$$

The detailed method for bias and accuracy calculations is fully described elsewhere [6, 27]. The bias and accuracy maps of the CIP 10-I "high efficiency" inhalable aerosol sampler are represented in figures 7 and 8. Bias and accuracy are expressed in percentage of sampled mass concentration compared to the concentration which would be measured by a device with the efficiency following exactly the CEN-ISO-ACGIH conventional penetration curve, for each aerosol distribution.

The bias and accuracy are calculated for a series of aerosol distributions with the MMAD (Mass Median Aerodynamic Diameter) within an interval from 1 to 25 µm (step of 1 µm) and the GSD (Geometric Standard Deviation) from 1.5 to 3.5 (step of 0.25). The EN 13205 standard require the bias within −10% to +10% and accuracy within 0 to +30% for a first quality sampler. The bias and accuracy performance criteria were defined by Görner *et al.* [28] as a percentage of the aerosols in the map for which the EN 13205 requirement is met. More the criterion approaches 100%, more versatile is the sampler, i.e. its using is suitable for any aerosol distribution.

It is possible to use the bias and accuracy maps to predict the sampler behaviour in workplace in the case of respirable aerosol measurement. In the case of inhalable samplers, high caution is recommended. Actually, the sampling of smaller particles (respirable aerosol fraction) is less influenced by the external wind speed and by the operator in the vicinity of the sampler. A difference between experimental conditions and workplace conditions has low impact on sampling efficiency of the sampler. In the case of inhalable fraction, coarse particles having big inertia are involved in sampling procedure and the sampler efficiency become extremely sensible to the external flow conditions. For this reason the bias and accuracy maps of inhalable aerosol samplers are less predictive in workplace conditions which can be different from the conditions of experimental run. The experimental measurement of the CIP 10-I sampling efficiency was made at a wind speed of 1 m.s^{-1}, the sampler rotating with a scaled bluff body. The reported bias and accuracy maps (figures 7 and 8) correspond to such flow and operating conditions.

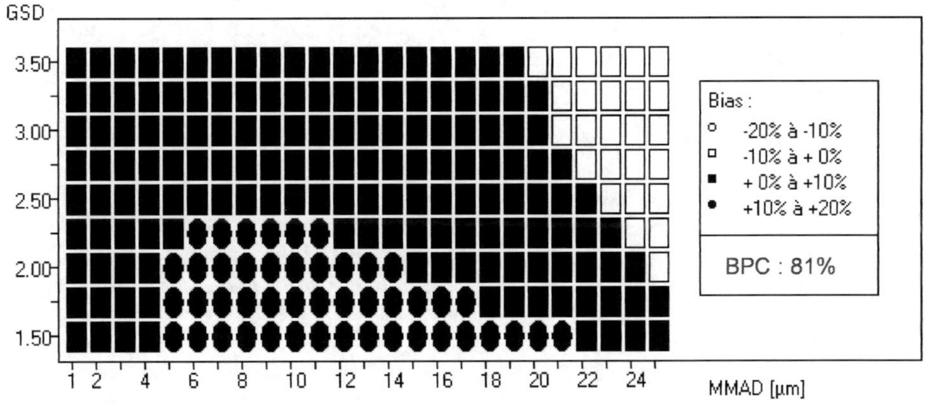

Figure 7. Bias map of the "high efficiency" CIP 10-I.
BPC: Bias Performance Criterion

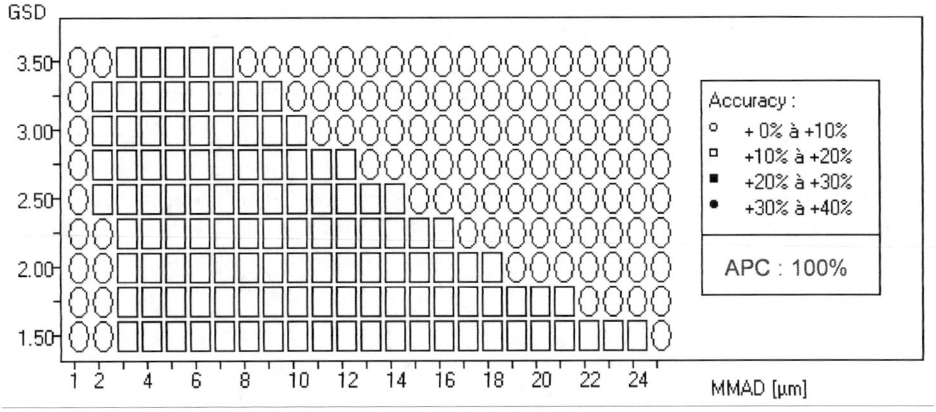

Figure 8. Accuracy map of the "high efficiency" CIP 10-I.
APC: Accuracy Performance Criterion

5. Conclusion

The three-dimensional T-shaped air flow circuit inside the CIP 10-I selector was adopted (figure 3c) to draw the aspirated particles into the final collecting stage of the sampler [29]. In the two previous prototypes, particles could be deposited in certain places due to inertia (where the aerosol was forced to deviate drastically) or sedimentation (where the aerosol decelerated). This kind of wall loss brings down the overall sampling efficiency of the sampler. The aerodynamic behaviour of the adopted solution (figure 3c) causes the particles to accelerate radially between the two horizontal plates before they enter a vertical tube. This acceleration avoids the particles being deposited on the lower horizontal plate. At the beginning of the vertical tube, the mutually opposing particle trajectories limit particle wall deposit by virtual impaction effect. The inner selector walls are polished to avoid particles being stopped by eventual surface asperities.

The "high efficiency" CIP 10-Inhalable aerosol sampler meets fairly well the conventional CEN-ISO-ACGIH criteria for sampling the inhalable health-related aerosol fraction.

References

[1] Courbon P 1984 Mesure des empoussièrages et CPM 3 *Industrie Minérale Technique* **4** 307
[2] Quinot E 1968 Study of a dust sampling instrument with cyclonic airflow: The Turbocaptor - Report n° 92 June 1968 Centre d'études Médicales Minières
[3] Görner P, Wrobel R and Fabriès JF 1990 Air filtration by rotating porous media. *In Proceedings of the 5th World Filtration Congress (Nice, France, 5-8 June 1990)* pp 165-167
[4] Courbon P, Wrobel R and Fabriès JF 1988 A new individual respirable dust sampler: The CIP 10 *Annals of Occupational Hygiene* **32** 129
[5] Görner P, Wrobel R, Micka V, Skoda V, Denis J and Fabriès JF 2001 Study of fifteen respirable aerosol samplers used in occupational hygiene *Annals of Occupational Hygiene* **45** 43
[6] Fabriès JF, Görner P, Kauffer E, Wrobel R and Vigneron JC 1998 Personal thoracic CIP 10-T sampler and its static version CATHIA-T *Annals of Occupational Hygiene* **42** 453
[7] Fabriès JF, Wrobel R and Görner P 1988 Capteur individuel portatif pour recueillir une fraction déterminée des suspensions solides d'une atmosphère *Patent n° 88 15974 (Paris, 6 December 1988)*
[8] CEN EN481 1993 Workplace atmosphere - Specification for conventions for measurement of suspended matter in workplace atmospheres *Comité Européen de Normalisation (Brussells)*
[9] ISO IS7708 1995 Air Quality - Particle size fractions definitions for the health-related sampling *International Standards Organisation (Geneva)*
[10] ACGIH 1994-1995 Threshold limit values for chemical substances and physical agents and biological exposure indices. *American Conference of Governmental Industrial Hygienists (Cincinnati, OH)*
[11] Görner P, Courbon P, Champion C, Wrobel R, Witschger O, and Fabriès JF 2003 Coupelle d'échantillonnage des particules microbiologiques en suspension dans une atmosphère et dispositif portatif d'échantillonnage pourvu d'une telle coupelle *Patent n° 03 02323 (Paris, 26 February 2003)*
[12] Görner P, Fabriès JF, Duquenne P, Witschger O and Wrobel R 2006 Bioaerosol sampling by a personal rotating cup sampler CIP 10-M *Journal of Environmental Monitoring* **8** 43
[13] Kenny LC 1995 Pilot study of CEN protocols for the performance testing of workplace aerosol sampling instruments *Final report of the European contract MATI-CT92-0047 (HSL, Sheffield)*
[14] Kenny LC, Aitken R, Chalmers C, Fabriès JF, Gonzalez-Fernandez E, Kromhout H, Lidén G, Mark D, Riediger G and Prodi V 1997 A collaborative European study of personal inhalable aerosol sampler performance *Annals of Occupational Hygiene* **41** 135
[15] Görner P, Witschger O and Fabriès JF 1996 Annular aspiration slot entry efficiency of the CIP-10 aerosol sampler *The Analyst* **121** 1257
[16] Witschger O, Fabriès JF and Görner P 1997 Particle entry efficiency of an annular slot aerosol sampler *Journal of Aerosol Science* **28** S679
[17] Roger F, Fabriès JF, Görner P, Wrobel R and Renoux A 1999 Numerical study of the behaviour of an annular aerosol sampling slot in laminar and turbulent flow *Journal of Aerosol Science* **30** S899
[18] Görner P, Witschger O, Roger F, Wrobel R and Fabriès JF 2008 Aerosol sampling by annular aspiration slots *Journal of Environmental Monitoring* DOI:10.1039/B802881H
[19] Witschger O, Wrobel R, Fabriès JF, Görner P and Renoux A 1997 A new experimental wind tunnel facility for aerosol sampling investigations *Journal of Aerosol Science* **28** 833
[20] CEN EN 13205 1998 Workplace atmospheres - Assessment of performance of instruments for measurement of airborne particle concentrations *Comité Européen de Normalisation (Brussels)*

[21] Witschger O, Willeke K, Grinshpun SA, Aizenberg V, Smith J and Baron PA 1998 Simplified method for testing personal inhalable aerosol samplers *Journal of Aerosol Science* **29** 855

[22] Kenny LC, Beaumont G, Gudmundssen A and Koch W 2000 Small-scale aerosol sampler testing systems *Journal of Aerosol Science* **31** S406

[23] Kennedy NJ, Tatyan K and Hinds WC 2001 Comparison of a simplified and full-sized mannequin for the evaluation of inhalable sampler performance *Aerosol Science and Technol*ogy **35** 564

[24] Paik SY and Vincent JH 2004 The orientation-averaged aspiration efficiency of IOM-like personal aerosol sampler mounted on bluff bodies *Annals of Occupational Hygiene* **48** 3

[25] Belyaev SP and Levin LM 1974 Techniques for collection of representative aerosol samples *Journal of Aerosol Science* **5** 325

[26] Fabriès JF 1990 Simulation of particle size-selective air samplers placed in polydisperse aerosols *Aerosol Science and Technology* **12** 673

[27] Görner P, Fabriès JF 1996 Industrial aerosol measurement according to the new sampling conventions *Occupational Hygiene* **3** 361

[28] Görner P, Wrobel R, Micka V, Denis J and Fabriès JF 1998 Performance testing of several respirable aerosol samplers *Advances in the Prevention of Occupational Respiratory Diseases (Excerpta Medica International Congress Series* vol 1153) ed K Chiyotani, Y Hosoda, Y Aizawa, ELSEVIER Tokyo pp 1013-1018

[29] Görner P, Wrobel R, Witschger O, Roger F and Fabriès JF 2003 Sélecteur de particules en suspension dans une atmosphère et dispositif d'échantillonnage pourvu d'un tel sélecteur *Patent n° 03 02322 (Paris, 26 February 2003)*

Inhalability for aerosols at ultra-low windspeeds

Darrah K. Sleeth and James H. Vincent

Department of Environmental Health Sciences, School of Public Health, University of Michigan, Ann Arbor, MI, 48109, U.S.A.

jhv@umich.edu

Abstract. Most previous experimental studies of aerosol inhalability were conducted in wind tunnels for windspeeds greater than 0.5 ms^{-1}. While that body of work was used to establish a convention for the inhalable fraction, results from studies in calm air chambers (for essentially zero windspeed) are being discussed as the basis of a modified criterion. However, information is lacking for windspeeds in the intermediate range, which – it so happens – pertain to most actual workplaces. With this in mind, we have developed a new experimental system to assess inhalability – and, ultimately, personal sampler performance – for aerosols with particle aerodynamic diameter within the range from about 9 to **90** μm for ultra-low windspeed environments from about 0.1 to 0.5 ms^{-1}. This new system contains an aerosol test facility, fully described elsewhere, that combines the physical attributes and performance characteristics of moving air wind tunnels and calm air chambers, both of which have featured individually in previous research. It also contains a specially-designed breathing, heated, life-sized mannequin that allows for accurate recovery of test particulate material that has been inhaled. Procedures have been developed that employ test aerosols of well-defined particle size distribution generated mechanically from narrowly-graded powders of fused alumina. Using this new system, we have conducted an extensive set of new experiments to measure the inhalability of a human subject (as represented by the mannequin), aimed at filling the current knowledge gap for conditions that are more realistic than those embodied in most previous research. These data reveal that inhalability throughout the range of interest is significantly different based on windspeed, indicating a rise in aspiration efficiency as windspeed decreases. Breathing flowrate and mode of breathing (i.e. nose versus mouth breathing) did not show significant differences for the inhalability of aerosols. On the whole however, the data obtained here are within the range of inhalability data that exist from the large body of the previous experimental work performed at the higher windspeeds. These latest findings are an important contribution to the ongoing discussion in international standards-setting bodies about the possible adjustment of the quantitative definition of what constitutes the inhalable fraction.

1. Introduction

The entry of airborne particles into the noses and mouths of human subjects during breathing is strongly dependent on a number of factors, primarily particle aerodynamic diameter (d_{ae}), defined as the diameter of an equivalent spherical particle of density 10^3 kg/m^3 that has the same falling speed in air as the particle in question, the windspeed (U), orientation of the subject with respect to the wind, and breathing flowrate and pattern. Similar factors are influential in how airborne particles are drawn into aerosol sampling devices. These scenarios and their relevance to human exposures to aerosols in working and living environments have been discussed extensively in the literature[1]. The relationship between the efficiency of aerosol aspiration during inhalation and the variables mentioned provide the basis of criteria by which to define what has become widely known as the *inhalable fraction*. This fraction provides a quantitative definition of the aerosol that can enter the human body through the nose and/or mouth during breathing, and so becomes a primary ingredient in health-related standards. It is therefore an important yardstick against which the performances of aerosol sampling instruments may be compared.

The development of health-related particle size-selective criteria for occupational aerosol sampling and standards remains of considerable interest to international standards-setting bodies, notably the International Standards Organization (ISO), the Comité Européen de Normalisation (CEN) and the American Conference of Governmental Industrial Hygienists (ACGIH), as well as to national regulatory bodies throughout the world. Previous experiments to characterize the *aspiration efficiency* (A) of the human head were carried out mostly in wind tunnels with windspeeds in excess of 0.5 ms^{-1}, based on the experiments with life-sized, breathing mannequins for orientation with respect to the wind averaged over all possible orientations [2, 3, 4]. Using this body of work from several laboratories, the experimentally-obtained data for aspiration efficiency were used to develop a convention for what has become known as '*inhalability*' (I). In its most general expression it is given by the empirical form [5].

$$I(d_{ae}) = 0.5\{1 + \exp(-0.06 d_{ae})\} + \mathrm{f}(d_{ae}, U) \qquad (1)$$

where the term on the right hand side may be expressed as

$$f(d_{ae}, U) \equiv pU^{q} \exp(r d_{ae}) \qquad (2)$$

in which d_{ae} is in [µm] and the windspeed U is in [ms^{-1}], and where the coefficients $p = 1 \times 10^{-5}$, $q = 2.75$ and $r = 0.055$ for windspeeds in the range $1 \leq U \leq 9$ ms^{-1}. Thus it is seen that Equation (1) contains a windspeed dependency for large particle sizes and for higher windspeeds. But for lower windspeeds ($U < 4$ ms^{-1}), Equation (1) reverts to the simpler (and more familiar)

$$I(d_{ae}) = 0.5\{1 + \exp(-0.06 d_{ae})\} \qquad (3)$$

which is the form that appears in the ACGIH definition of the inhalable fraction considered to this day as being applicable to most workplaces.[6]

Since the adoption of Equation (3) as *the* definition of inhalable aerosol applicable to workplaces, questions have been posed about *what is* the appropriate range of windspeeds in working environments. Measurements of actual windspeeds prevailing in a wide range of workplaces have revealed that actual workplace windspeeds tend to be much lower than originally thought, typically in the range from about 0.05 to 0.5 ms^{-1} [7, 8]. This has produced interest in re-defining the inhalable fraction for what are now regarded as more realistic windspeeds. In one body of work, new experiments to measure inhalability with a breathing mannequin were carried out in a calm air chamber,[9] and revealed some systematic differences with what had been observed at the higher windspeeds of the previous studies. However, there has so far been no systematic research to determine the aspiration efficiency of the human head as a function of windspeed within the important low – or, as we call it, '*ultra-low*' – range of windspeeds that have been identified as most relevant to most workplaces. The current research therefore has set out to acquire new data with which this knowledge gap may be filled.

2. Methods

The challenge in conducting research at the very low windspeeds mentioned was to develop an experimental arrangement in which sufficiently uniform aerosol distribution could be achieved in a test system in order to allow valid measurements of aspiration efficiency for the human head. This was achieved by means of what might best be described as a 'hybrid' ultra-low windspeed facility, combining elements of both a wind tunnel and a calm air chamber. A full description of the facility that was designed and built in our laboratory has been described in detail elsewhere[10] and so the details will not be reproduced here. It is sufficient to refer to the schematic diagram in Figure 1. Figure 1a identifies the main features, indicating the provision to introduce test aerosol from upstream (as in a wind tunnel) and from above (as in a calm air chamber). In principle, each aerosol injection location may be adjusted with respect to the other to provide uniform spatial distribution of the aerosol in the working section. In particular, Figure 1a illustrates the higher windspeed situation where particle settling velocity is relatively negligible, so that all the airborne particles in the test section arrive there by horizontal convection. Figure 1b illustrates the calm air situation where all the airborne particles in the test section arrive there vertically by gravitational settling. Figure 1c is the most interesting in the context of the current work, showing how aerosol in the test section receives contributions from both upstream and above. It indicates the horizontal and vertical velocity vectors for particles of a given size, revealing the non-horizontal net trajectory. In principle it is thus possible to maintain uniform spatial distribution of both concentration and particle size distribution. As the work progressed, just one modification was made to the experimental system as shown in Figure 1.

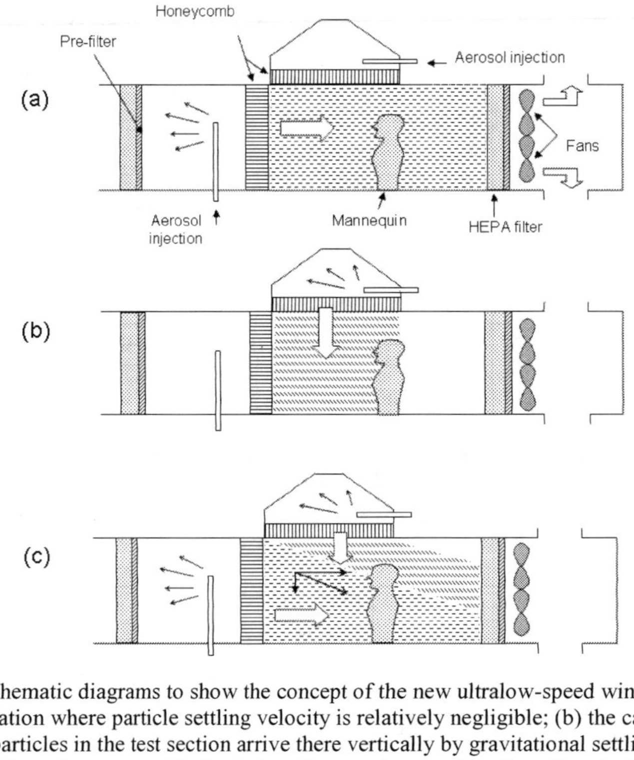

Figure 1. Schematic diagrams to show the concept of the new ultralow-speed wind tunnel: (a) the higher windspeed situation where particle settling velocity is relatively negligible; (b) the calm air situation where all the airborne particles in the test section arrive there vertically by gravitational settling; and (c) the ultra-low windspeed situation where aerosol in the test section receives contributions from both upstream and above.[10]

In the earliest experiments it was found that, under certain conditions, there were significant losses of particles during transport into the working section. To overcome this, the honeycomb flow straighteners at the entrance to the working section were replaced by a perforated metal screen, and this was the experimental arrangement used for most of the results that will be described below. It suffices to say here that the modification produced no observable impact on the results for inhalability.

Test aerosol, mechanically generated from the same narrowly-graded powders of fused alumina that were featured widely in previous such research, was dispersed into the working section of the facility, which measured approximately 3 m in length and 1.2 m by 1.2 m in cross section. The particle size distribution was measured using 5 modified Marple-type cascade impactors fitted with foam plugs to extend the size collection range[11] and the aerosol concentration was measured using IOM inhalable samplers acting as static samplers. It was shown that spatial uniformity – for both particle size distribution and concentration – was achieved to within ±10% across the range of windspeeds of interest. However, unlike in the previous work, where it was possible to assume the original calibrations of particle size distribution first tabulated by Mark *et al.* (1985) [12], the enhanced losses of particles by gravitational effects at the very low windspeeds in the present work required that particle size distributions of the test aerosol needed to be actually measured for each and every combination of starting powder grade and windspeed. The results are summarized in Table 1, and provide the desired information about particle size for each set of experimental conditions. Here it is important to note that, for many of the experimental conditions of interest, the particle size distribution was significantly modified in comparison to what might have been expected for higher windspeeds, in terms of both mass median aerodynamic diameter (*MMAD*) and geometric standard deviation (σ_g). In fact, for the latter, the experimentally-measured σ_g-values were significantly greater than for the original 'parent' aerosol.

Table 1. Particle size distributions measured using modified Marple-type cascade impactors in ultra-low windspeed environments, as represented by mass median aerodynamic (*MMAD* in μm) and geometric standard deviation (σ_g).

Powder Grade	Windspeed (ms^{-1})						Nominal Size	
	0.10		0.24		0.42			
	MMAD	σ_g	*MMAD*	σ_g	*MMAD*	σ_g	*MMAD*	σ_g
F1200	9.6	1.28	9.5	1.32	9.3	1.34	6.0	1.36
F800	13.9	1.49	12.8	1.47	12.4	1.56	13.0	1.38
F500	28.8	1.62	32.7	1.71	28.7	1.93	26.0	1.30
F400	37.7	1.62	44.3	1.59	40.0	1.74	34.0	1.20
F280	74.0[a]	1.19[a]	62.4	1.42	66.9	1.45	74.0	1.19
F240	89.5[a]	1.29[a]	60.1	1.45	63.0	1.49	89.5	1.29

[a] Nominal value used for this condition

As already mentioned, the subject of measurement is the *aspiration efficiency* of the human head, as represented by the life-sized mannequin shown in Figure 2. This mannequin was designed to incorporate not only breathing through the nose and/or mouth (in whatever combination – in through the nose/out through the mouth, in through the nose/out through the nose, etc. – is desired to be tested) but also body heat. For the latter, however, previous research described by Schmees *et al.* (2008) [13] showed that elevated body temperature had no appreciable effect on the motion of air near the human body, even at the very low windspeeds of interest. So this aspect will not feature further in this paper.

In our experiments, inhaled aerosol was collected on the 47-mm glass fibre filter that was located just inside the nose and/or mouth entry, and on other inside surfaces between the entry and the filter. The latter deposits were recovered at the end of each experiment by swabbing with isopropyl alcohol-impregnated cotton balls. It is also important to note that special care was taken to design the air pathways inside the mannequin such that expired air did not pass through the same filter, or over the same surfaces, as during inspiration, thus eliminating any risk of re-entraining particulate material already collected. For each experiment, both the filter and cotton were conditioned overnight in a desiccator to stabilize the mass, after which it was then weighed. After sample collection, the filter and cotton balls were similarly conditioned again and re-weighed. The resultant filter and wall particulate masses were combined to provide the total inhaled aerosol mass. From knowledge of the corresponding total volume of air sampled, the inhaled aerosol mass concentration (c_I) was obtained directly.

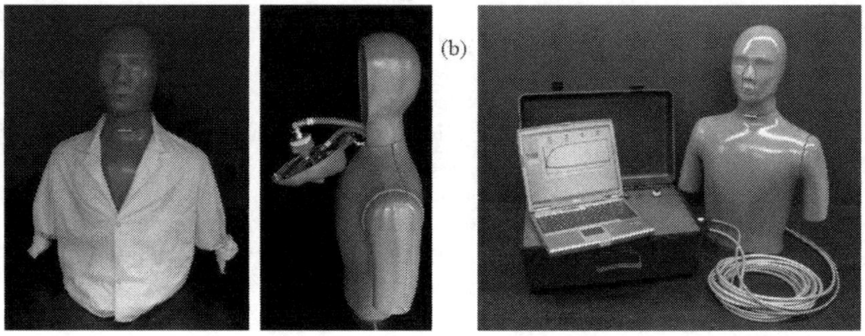

Figure 2. Photographs of the mannequin: (a) both assembled and disassembled (face manifold dropped down) to reveal aerosol collection filter and internal connections; and (b) the mannequin system, including the breathing machine and control lap-top computer.

The reference concentration was obtained from samples taken 0.75 m upstream of the mannequin using thin-walled isokinetic samplers. Here, provided that conditions were truly isokinetic, the particulate material entering through the plane of the sampler was indeed representative of the total airborne concentration. Similarly, as for the mannequin, the aspirated particulate material was recovered from both the filter and the inside walls of the thin-walled sampling tube, and combined to provide the reference concentration (c_0). Finally, aspiration efficiency (A) was determined from the ratio c_I/c_0.

Experiments were conducted for particle size (as represented by *MMAD*) in the range from approximately 9 to 90 μm (represented by 6 different powder grades), windspeed values of 0.10, 0.24 and 0.42 ms^{-1}, breathing flowrate values (as represented by minute volume, V) of 6 and 20 lmin^{-1}, and breathing modes of 'mouth-only' (M), 'nose-only' (N), and 'in through nose/out through mouth' (NM) (with 'nose-only' breathing not tested for the higher breathing flowrate). Each sampling run lasted for 20 minutes and employed the breathing mannequin heated to typical skin temperature (33 °C) while dressed in a laboratory coat. Each set of conditions was tested twice, for a total of 180 experiments.

3. Results

The experimental results shown here for the aspiration efficiency of the human head (A) represent an important contribution to the ongoing discussion in the international community about how to define the inhalable fraction. Figures 3, 4 and 5 show a summary of the results, all plotted graphically as A versus d_{ae} for each windspeed, breathing flowrate and mode of breathing, respectively. Each data point represents a mean for the pooled data at that aerosol particle size, containing between 2 and 18 values, across all experimental conditions. The data shown include those for the earlier experimental set-up as well as those for the later arrangement (and which, as already mentioned, exhibited no observable differences). It should also be noted that the values for d_{ae} represent the mass median aerodynamic diameter and so were obtained as part of a distribution, the spread of which can be assessed by the σ_g values shown in Table 1. The current inhalable convention, as adopted by ACGIH, ISO and CEN, is also displayed on the graphs for comparison. Overall, Figures 3, 4 and 5 reveal a very comprehensive picture, providing aspiration efficiency data for the full range of conditions mentioned above.

ANOVA results indicated that aspiration efficiency in the lowest windspeed (0.10 ms^{-1}) was significantly different from aspiration efficiency measured at both higher windspeeds (p-value <0.0001). By contrast, the results for different breathing flowrates (p-value = 0.066) and modes of breathing (p-value = 0.155) showed no statistically significant differences. Overall, it is evident that the results all appear to be close – within the scatter – to the inhalability curve that was derived from the original bodies of experimental work at higher windspeeds.

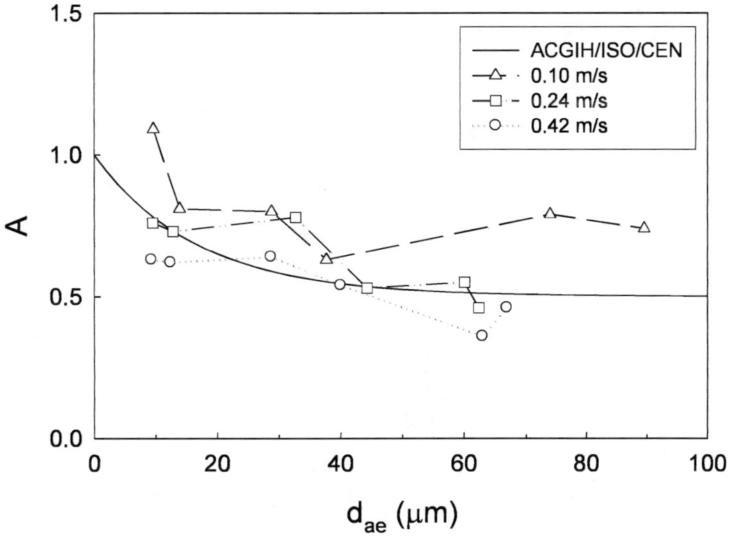

Figure 3. Mean human aspiration efficiency (A) as a function of particle aerodynamic diameter (d_{ae}) at different windspeeds, across all breathing flowrates and modes. Also shown is the current inhalable convention as adopted by ACGIH, ISO and CEN.

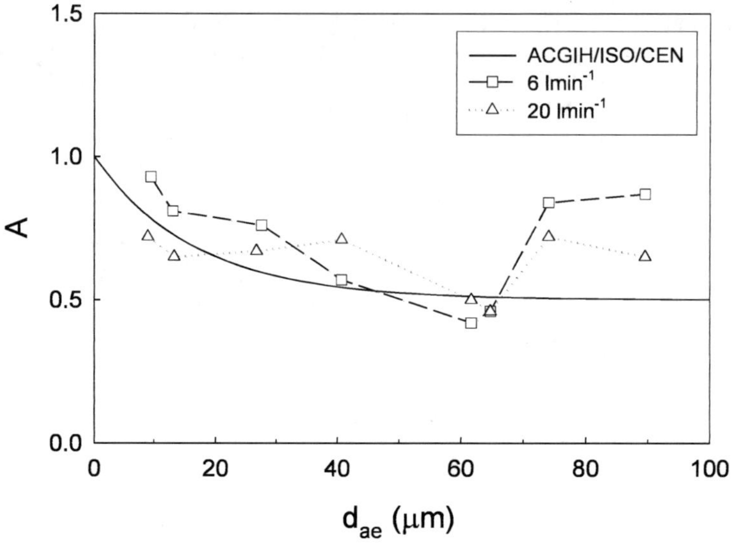

Figure 4. Mean human aspiration efficiency (A) as a function of particle aerodynamic diameter (d_{ae}) at different breathing flowrates, across all windspeeds and modes of breathing. Also shown is the current inhalable convention as adopted by ACGIH, ISO and CEN.

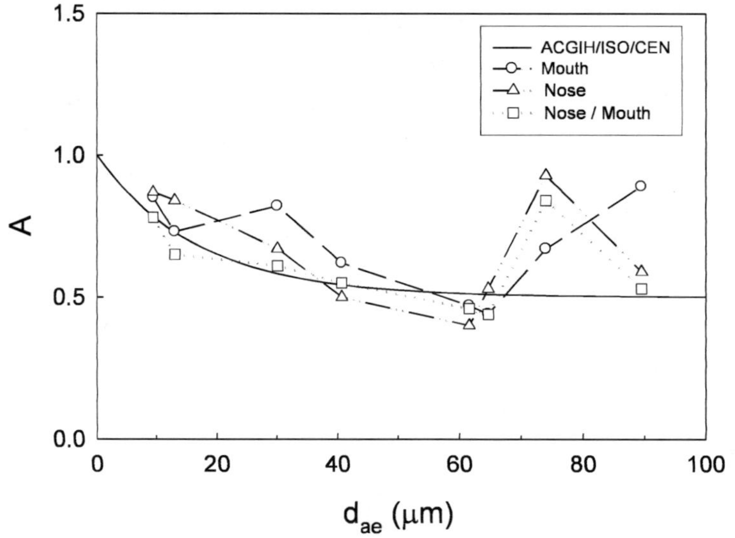

Figure 5. Mean human aspiration efficiency (A) as a function of particle aerodynamic diameter (d_{ae}) at different modes of breathing, across all windspeeds and breathing flowrates. Also shown is the current inhalable convention as adopted by ACGIH, ISO and CEN.

4. Discussion

The aspiration efficiency of the human head (i.e. inhalability) has been successfully assessed at ultra-low windspeeds. The difficulty in performing experiments at such low windspeeds was overcome by the development of novel facilities and methods, and thus enabled the inhalability of aerosols to be assessed under realistic working conditions – at windspeeds between 0.1 and 0.5 ms^{-1}. While mannequin breathing flowrate and mode of breathing did not appear to be important factors in determining inhalability, windspeed did have an impact on the inhalability of the coarse aerosols tested here.

It should be noted that the current inhalable convention was established based on general trends in the large body of work that was performed for aspiration efficiency in fast moving air, without regard to rigorous statistical models. Bearing in mind that those earlier data were all obtained for windspeeds ranging from 0.5 to 9 ms^{-1}, it is interesting to observe that the set of inhalability data obtained in this research at the ultra-low windspeeds indicated lies within the range of those values previously measured.

As mentioned earlier, studies performed in essentially calm air have shown that inhalability in those environments is typically somewhat higher than what was seen in the previous data for higher windspeeds[9]. Taking a closer look at the importance of windspeed as it might relate to human aspiration efficiency will therefore be instructive. In the comprehensive analyses of workplace windspeeds discussed before, it was observed that approximately 85% of measured windspeeds were below 0.3 ms^{-1}, with 50% of values lower than 0.1 ms^{-1} [8]. When one considers the widespread use of ventilation systems in most working environments, especially where inhalable hazards are known to be present, it is difficult to imagine many workplaces having windspeed conditions close to 'calm'. In addition, workers who are not stationary will create additional air movements that would further limit the applicability of actual calm air results. Although our results were limited to environments with windspeeds at 0.1 ms^{-1} and greater, the similarities seen between the set of data obtained here and those data used to set the inhalable convention, coupled with the limited applicability of true calm air models, windspeed – even ultra-low – does not appear to be a significant factor in determining the inhalability of aerosols.

Overall, these data provide an important assessment of human aspiration efficiency under realistic workplace environments and will be important for filling the gap between calm air and moving air with respect to human inhalability. It should also be noted that the preliminary analyses performed here have been purely statistical. The results of a more rigorous analysis, also incorporating ideas taken from the physical theories of aerosol aspiration, will be reported in a later paper.

5. Acknowledgements

We would like to thank the US-NIOSH for their continued grant support (5-RO1-OH002987-09), Engineering Laboratory Design (Lake City, MN, USA) for its contributions to designing and building the wind tunnel, and Measurement Technology Northwest (Seattle, WA, USA) for its contributions to designing and building the mannequin system.

References

[1] Vincent J H 2007 *Aerosol Sampling: Science, Standards, Instrumentation and Applications* (Chichester: Wiley & Sons)
[2] Ogden T L and Birkett J L 1977 The human head as a dust sampler In: *Inhaled Particles IV* (Oxford: Pergamon Press) 93-105
[3] Armbruster L and Breuer H 1982 Investigations into defining inhalable dust In: *Inhaled Particles, Vol. V* (Oxford: Pergamon Press) 21-32
[4] Vincent J H and Mark D 1982 Application of blunt sampler theory to the definition and measurement of inhalable dust In: *Inhaled Particles V* (Oxford: Pergamon Press) 3-19

[5] Vincent J H, Mark D, Miller B G, Armbruster L and Ogden T L 1990 Aerosol inhalability at higher windspeeds *Journal of Aerosol Science* **21** 577-586

[6] Vincent J H 1999 *Particle Size-Selective Sampling for Particulate Air Contaminants* (Cincinnati: American Conference of Governmental Industrial Hygienists)

[7] Berry R D and Froude S 1989 An investigation of wind conditions in the workplace to asses their effect on the quantity of dust inhaled *U.K. Health and Safety Executive Report IR/L/DS/89/3* (London: Health and Safety Executive)

[8] Baldwin P E J and Maynard A D 1998 A survey of windspeeds in indoor workplaces, *Annals of Occupational Hygiene* **20** 303-313

[9] Aitken R J, Baldwin P E J, Beaumont G C, Kenny L C and Maynard A D 1999 Aerosol inhalability in low air movement environments *Journal of Aerosol Science* **30**, 613-626

[10] Schmees D K, Wu Y-H and Vincent J H 2008 Experimental methods to determine inhalability and personal sampler performance for aerosols in ultra-low windspeed environments *Journal of Environmental Monitoring* DOI:10.1039/b806431h

[11] Wu Y-H and Vincent J H 2007 A modified Marple-type cascade impactor for assessing particle size distributions in workplaces *Journal of Occupational & Environmental Hygiene* **4** 798-807

[12] Mark D, Vincent J H, Gibson H and Witherspoon W A 1985 Applications of closely graded powders of fused alumina as test dusts for aerosol studies *Journal of Aerosol Science* **16** 125-131

[13] Schmees D K, Wu Y-H and Vincent J H 2008 Visualization of the air flow around a life-sized, heated, breathing mannequin at ultra-low windspeeds *Annals of Occupational Hygiene* **52** 351-360

Journal of Physics: Conference Series **151** (2009) 012063 doi:10.1088/1742-6596/151/1/012063

Modelling exposure to pharmaceutical agents

J W Cherrie [1], A T Gillies [2], A Sleeuwenhoek [1], M van Tongeren [1], P McDonnell [3], M Coggins [3], S R Bailey [4]

1 Institute of Occupational Medicine, Edinburgh, EH14 4AP, UK.

2 Gillies Associates Ltd, 34 Holman Road, Aylsham, Norfolk NR11 6BZ, UK.

3 National University of Ireland, Department of Physics, Galway, Ireland.

4 GlaxoSmithKline, Southdownview Way, Worthing, BN14 8NQ, UK.

Email: john.cherrie@iom-world.org

Abstract. Aerosol exposure to the active ingredients of pharmaceuticals arises in research, development and manufacturing. In most instances it is only possible to make small numbers of measurements of exposure and given the inter- and intra-individual variability it is often difficult to obtain sufficient objective data to make reliable decisions about the appropriateness of control measures. This paper describes the development and validation of an exposure model with the potential to predict airborne exposure from both new and existing operations. The model could be used to more efficiently target exposure measurement resources.

1. Introduction

There is widespread handling of powders in the pharmaceutical industry. Manufacturing is divided into two major stages: the production of the active pharmaceutical ingredients (APIs) and secondary processing, involving the addition of other substances to the active ingredients and conversion into medicines. The manufacturing process is largely "powder technology" and the control of aerosol emissions from these processes into the surrounding work environment poses an ongoing challenge [1].

While the effects of pharmaceutical agents may be beneficial for the individuals to whom they are prescribed, it is not desirable or acceptable for workers manufacturing such products to be excessively exposed to these APIs. Heron and Pickering [2] reviewed the published evidence for the presence of adverse occupational health effects from APIs and categorized them as: acute pharmacological effects; chronic effects from potent compounds; respiratory sensitisation and bronchoconstriction; and skin sensitisation. Also drug exposures in the workplace can add to, be synergistic with, or even potentiate drugs the employees take therapeutically [3]. New APIs are increasingly potent, with many having exposure limits less than $1 \mu g/m^3$

Many pharmaceutical companies develop in-house occupational exposure limits (OELs) for the APIs in their products. For new pharmaceuticals, it is difficult to specify OELs because of limited toxicity data and this has led to the development of "banding systems" that provide performance-based occupational exposure limits (PB-OEL) systems [4]. These approaches offer systematic methods of assigning compounds to a particular Occupational Hazard Category (OHC) based on the potency, pharmacological and toxicological effects of the API. An Exposure Control Approach (ECA) is then chosen to achieve the performance specification, taking into account exposure factors such as physical form, dustiness, energy input, and volume of material. This is illustrated in Table 1, which summarises the classifications used by GlaxoSmithKline (GSK).

Table 1. An example of a Banding System

OHC	Performance Band ($\mu g/m^3$)		ECA	Summary Description of Control Options
1	$> 1000 \leq 5000$		A	Room ventilation
2	$> 100 \leq 1000$		B	Local extract ventilation
3	$> 10 \leq 100$	Assessment	C	Partial enclosure
4	$> 1 \leq 10$		D	Enclosed processing and isolation
5	≤ 1		E	Enclosed processing plus additional containment

Exposure measurements show substantial variation within- and between-workers. Kromhout and coworkers investigated data from about 14,000 measurements from more than 1,500 workers in 165 occupational groups in general industry, defined by job title and factory [5]. Only 25% of the occupational groups had 95% of the individual mean exposures within a factor of 2, while almost 30% of the groups had greater than a 10-fold range in individual mean exposures. In most cases the within-worker variability exceeded the between-worker variability, and the within-worker exposures were much greater for intermittent processes, such as those typical of the pharmaceutical industry, when compared to continuous processes. This variability makes it difficult to reliably characterize exposure in a specific scenario with a reasonable number of personal exposure samples, e.g. less than about 5 to 10.

An alternative approach to assess exposure is to use modeling techniques based on the underlying determinants of exposure. There are many benefits that can arise from exposure models, including prediction of exposures before the process has been commissioned, maximizing the benefit of existing data by combining with modelled exposures, and reducing the number of occupational hygiene samples required to characterize and monitor exposure. This type of model could also be useful in implementing the new European REACH regulations for chemicals.

In this paper we describe the adaptation of an existing deterministic model to predict the mean inhalation exposure to an API in a given scenario and describe a series of exposure assessments from scenarios in the pharmaceutical industry that have been used to validate the model output.

2. Methods

Estimates of exposures for pharmaceutical scenarios were based on the methods originally described by Cherrie and his co-workers [6-7], with a small number of adaptations to reflect the approach to risk management taken in this industry.

The exposure estimates were based on a simple theoretical model, which comprises a source term that is dependent on the *intrinsic properties* of the contaminant (ε_i), e.g. the dustiness of a solid and the proportion of API in the mixture, the way the material is *handled (h)*, e.g. careful scooping of a powder, and finally, the efficiency of *local controls* ($1-\eta_{lv}$). The three parameters (i.e. ε_i, h and ($1-\eta_{lv}$)) were multiplied together to provide the active emission of the source (e_a). We assume that the proportion of the API in the bulk product will be the same as is found in the aerosol, and therefore multiply the total emission by the percent API present.

Journal of Physics: Conference Series **151** (2009) 012063 doi:10.1088/1742-6596/151/1/012063

Three further parameters were incorporated into the basic model. These are the passive or fugitive emission (ε_p), the fractional time the source was active (t_a) and the efficiency of any respiratory protection ($1-\eta_{ppe}$). So for a single source close to a worker, i.e. in their near-field, the estimated exposure level (C) would be:

$$C = (\varepsilon_i \, . \, h \, . \, (1 - \eta_{lv}) \, . \, t_a + \varepsilon_p) \, . \, (1 - \eta_{ppe}) \quad (1)$$

The passive emission term is shown as an additive term unrelated to the active source.

The model simplifies the dispersion of contaminants away from sources using two notional spatial regions: the *near-field*, which comprises a volume of $8m^3$ (a 2m x 2m x 2m cube) around the worker whose exposure is being investigated; and the *far-field*, which comprises the remainder of the work environment. General ventilation in a workroom will have an impact on the contaminant concentration in both the near-field and the far-field. Equation 1 should therefore more correctly be written with suffixes for the near-field, i.e. "NF" and where the source is in the far-field with "FF", as in equations 2 and 3. The term d_{gv} accounts for the dilution effect of general ventilation in the work room or area.

$$C_{NF} = ((\varepsilon_i \, . \, h \, . \, (1 - \eta_{lv}))_{NF} \, . \, t_{a,NF} + \varepsilon_{p,NF}) \, . \, (1 - \eta_{ppe}) \, . \, d_{gv,NF} \quad (2)$$

$$C_{FF} = ((\varepsilon_i \, . \, h \, . \, (1 - \eta_{lv}))_{FF} \, . \, t_{a,FF} + \varepsilon_{p,FF}) \, . \, (1 - \eta_{ppe}) \, . \, d_{gv,FF} \quad (3)$$

In this scheme the intrinsic and passive emissions have concentration units (e.g. $\mu g/m^3$). This would correspond to the airborne concentration generated with a certain "standardised" handling. The other terms in these equations are dimensionless.

To take account of some of the specific issues associated with pharmaceutical manufacturing we subdivided the model terms of handling activities and local controls to make more explicit some of the contributing factors. For handling we had factors for the energy input during the activity, the quantity being handled per hour and the likelihood that the workers behaviour could influence their actual exposure. For local control we assigned the control factor illustrated in Table 2. In addition if the information suggested that the design of the controls was better or worse than the general requirements for that level of control the assessor could apply an adjustment local control factor (3, or 0.3).

Table 2. Control factors used in the model

ECA	Summary Description of Control Options	$(1 - \eta_{lv})$
A	Room ventilation	1
B	Local extract ventilation	0.1
C	Partial enclosure	0.01
D	Enclosed processing and isolation	0.001
E	Enclosed processing plus additional containment	0.0001

To estimate exposure levels, the assessor must assign numeric values to each of the parameters. These parameters were based on the descriptive information obtained during exposure measurements in the exposure scenarios of interest. The information was summarised by one of the authors (AG) and any clarification was obtained by telephone or email correspondence. The reconstructions were undertaken by three independent assessors (JWC, AS and PMcD) who were blinded to the results from the exposure measurements at the time of the assessments. Guidance on the values of the parameters was developed by JWC, who trained AS in the methodology, who subsequently trained PMcD. The parameters were then adjusted by the assessors as they completed the set of scenarios to ensure that the allocation of numeric values to parameters was consistent across all situations.

Little quantitative data is available about the dustiness characteristics of the powders being handled in the pharmaceutical industry. In the modelling work described in this paper we used a single factor for intrinsic emission for almost all products.

Data for the validation exercise was abstracted from the occupational hygiene records of GSK. All of the measurements were collected during the scenario rather than over a full-shift. In addition to the measurement results details of the process, work tasks, control measures, materials being handled, the size of the workroom and other relevant contextual information was summarised. Where available photographs of the operator carrying out their work tasks were obtained. The exposure scenarios were deliberately selected to cover a wide range of handling operations and containment levels, and hence represent a broad range of exposure levels. Abnormal operating conditions (e.g. mixer with and without leaks) were also included in the list of scenarios. Also, where possible the same or similar jobs from different sites were included.

3. Results

Data were obtained for 27 scenarios across five pharmaceutical manufacturing sites. These ranged from dispensing gram quantities of API in a negative pressure glove box through to dispensing about a tonne of API into 25kg quantities in kegs. Most of the operators wore respiratory protective equipment (RPE) although all samples were collected on the lapel of the operator and so for the purposes of this exercise we did not take account of the protection afforded by the RPE. There were between two and 31 measurements of exposure available for each scenario; in some cases only the arithmetic mean measured exposure and the range was available, while in others all of the measurement results were retrieved. The scenario mean exposure levels ranged from <0.005 to 20,000 $\mu g/m^3$. Six scenarios had an average exposure level less than 100 $\mu g/m^3$, eight were between 100 and 1,000 $\mu g/m^3$, 10 scenarios were between 1,000 and 10,000 $\mu g/m^3$ and three above 10,000 $\mu g/m^3$.

Table 3 summarises the geometric mean level for all of the scenarios from the three assessors along with the bias, i.e. the ratio of the geometric mean level from the assessor to the geometric mean of the scenario means and the correlation between the log-transformed assessments and scenario means.

Table 3. Summary data from the exposure assessments and measurements

	Exposure measurements	JWC	AS	PMcD
Geometric mean (GM) of scenario estimates ($\mu g/m^3$)	278	360	220	400
Bias, i.e. ratio of assessor GM to data GM	-	1.3	0.8	1.4
Correlation between log-transformed estimates from assessor and exposure measurements	-	0.97	0.93	0.88
Correlation between log-transformed estimates from assessor and exposure measurements (excluding data below 100 $\mu g/m^3$)	-	0.84	0.76	0.80

The assessors showed a small positive bias (between 30 and 40% high on average) or a small negative bias (20% low). There was an excellent correlation between the log-transformed estimates and the log-transformed means from the corresponding scenario (between 0.88 and 0.97). Limiting the correlation analysis to data above 100 $\mu g/m^3$ caused a slight decrease in the association.

Figure 1 shows a scatter plot of the assessed exposure levels in relation to the mean measured exposure for each scenario.

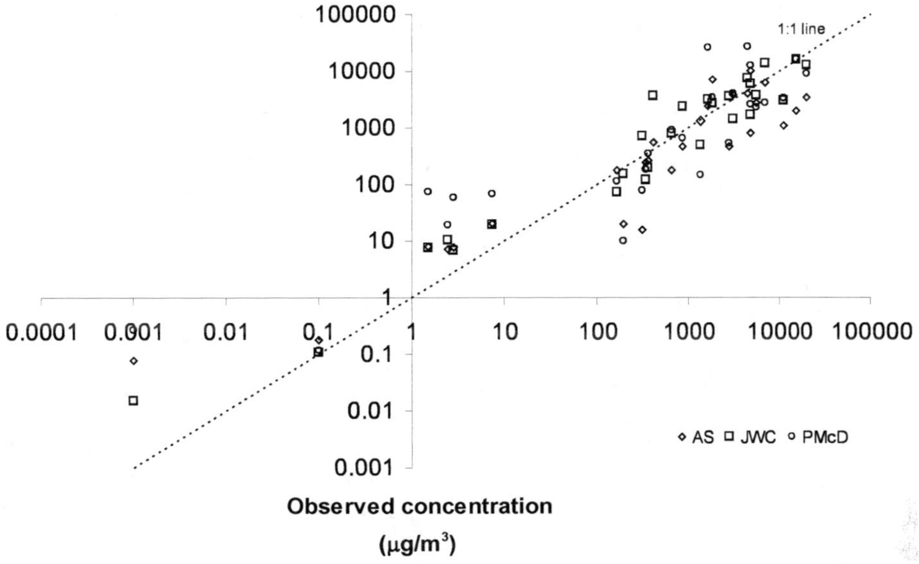

Figure 1. Scatter plot of assessed exposure level in relation to the scenario mean exposure level.

The results from this validation exercise are comparable to that which was obtained when this methodology was validated for non-pharmaceutical scenarios [7-8]. Results suggest that the exposure estimates are of comparable accuracy and precision to the available measurement data. Individual estimates from the assessors were generally no more than an order of magnitude different from the average measured value, which was comparable to the variation in the measurement data from the same scenario. There is some possible overestimation of exposure at lower exposure levels, but this may be partly related to potential bias in the measured exposure data, particularly because of the relatively small number of measurements available for most scenarios.

4. Discussion

The exposure model appears to provide a useful way of estimating the mean exposure to airborne APIs in pharmaceutical manufacturing operations using descriptive information on key contextual variables. It could certainly be used to more effectively target and prioritise monitoring efforts to focus on scenarios where there is a greater probability that exposure could be high in relation to the relevant occupational exposure limit.

The validation exercise produced results that are comparable to those previously produced for non-pharmaceutical agents. Cherrie and Schneider [7] reported bias between 0.47 and 2.86 and correlation coefficients between the log-transformed data of zero to 0.93 for asbestos, toluene man-made mineral fibre, respirable dust and styrene. They noted that the correlation coefficients between the assessments and the measurements were greater when the span of the data (ratio of the highest to lowest mean scenario measurement) was relatively large, i.e. greater than about 100. In the present study we also noted that the correlation between the measurements and assessments decreased when the lower exposure level scenarios were excluded from the comparison, although this did not substantially reduce the association. Semple and colleagues [8] quoted the mean correlation coefficients between the scenario mean measured and estimated levels for a variety of agents of 0.82 with a range of 0.73 to 0.85 for five assessors.

Preliminary data that we have obtained suggests that the range of dustiness of the pharmaceutical products included in this evaluation may be relatively small, which may explain the good correlation obtained without taking variation in dustiness into account. Additional data on the dustiness of pharmaceutical powders, the generation of airborne dust from pharmaceutical processes and the effectiveness of control measures used in the industry would further aid the accuracy of the model predictions. Boundy and co-workers [9] developed a small-scale apparatus for testing the dustiness of pharmaceutical powders, which would minimize the mass of API handled. However, data on the magnitude of the dustiness of pharmaceutical powders was not provided by the authors.

It would be advantageous to develop a user-friendly interface for the model and to remove some of the professional judgment needed in assigning the model input parameters. This model could be implemented as an integrated exposure assessment tool by extending it to incorporate a probabilistic approach using Monte Carlo simulation and then to use Bayesian statistics to update the modelled estimates with measurements contained in a linked exposure database, as suggested previously [10]. Importantly, this approach would allow an iterative exposure assessment process, whereby an initial appraisal is obtained using the exposure model results and when further data becomes available the model estimates can be updated to account for the increased understanding of the exposure in the scenario.

References

[1] Wood J 2001 Why Contain? Then and Now *Containment in the Pharmaceutical Industry* vol 8, ed J P Wood (New York: Marcel Dekker Inc) p1

[2] Heron R and Pickering, F 2003 Health effects of exposure to active pharmaceutical ingredients (APIs) *Occup. Med.* **53** 357-62

[3] Heidel D 2001 Industrial hygiene aspects of pharmaceutical manufacturing *Containment in the pharmaceutical industry* vol 8, ed JP Woods (New York: Marcel Dekker Inc) p 22

[4] Nauman N, Sargent E, Starkman B, Fraser W, Becker G and Kirk D 1996 Performance-based exposure control limits for pharmaceutical active ingredients *Am. Ind. Hyg. Assoc. J.* **57** 33-42

[5] Kromhout H, Symanski E and Rappaport M 1993 A comprehensive evaluation of within- and between-worker components of occupational exposure to chemical agents *Ann. Occup. Hyg.* **37**(3) 253-70

[6] Cherrie J, Schneider T, Spankie S and Quinn M 1996 A new method for structured, subjective assessments of past concentration *Occup. Hyg.* **3** 75-83

[7] Cherrie J, and Schneider T 1999 Validation of a new method for structured subjective assessment of past concentrations *Ann. Occup. Hyg.* **43**(4) 235-45

[8] Semple S, Proud L, Tannahill S, Tindall M and Cherrie J 2001 A training exercise in subjectively estimating inhalation exposures *Scand. J. Work Environ. Health* **27**(6) 395-401

[9] Boundy M, Leith D and Polton T 2006 Method to Evaluate the Dustiness of Pharmaceutical Powders *Ann. Occup. Hyg.* **5**(5) 453-58

[10] Tielemans E, Warren N, Schneider T, Tischer M, Ritchie P, Goede H, Kromhout H, Van Hemmen J, Cherrie JW 2007 Tools for regulatory assessment of occupational exposure: development and challenges. *J. Expo. Sci. Environ. Epidemiol.* **17** (supplement 1) 572-580

An analysis of employee exposure to organic dust at large-scale composting facilities.

P Sykes, J A Allen, J D Wildsmith and K P Jones

University of Wales Institute Cardiff (UWIC), Cardiff School of Health Sciences, Western Avenue, Cardiff, CF5 2YB.

psykes@uwic.ac.uk

Abstract. The occupational health implications from exposure to dust, endotoxin and 1-3 β Glucan at commercial composting sites are uncertain. This study aims to establish employee exposure levels to inhalable and respirable dust, endotoxin and 1-3 β Glucan during various operational practices in the composting process. Personal samples were collected and the inhalable and respirable dust fractions were determined by gravimetric analysis. Endotoxin concentrations were determined using a Limulus Amebocyte Lysate assay (LAL). 1-3 β Glucan levels were estimated using a specific blocking agent to establish the contribution that these compounds gave to the original endotoxin assay. Employees' exposure to dust was found to be generally lower than the levels stipulated in the Control of Substances Hazardous to Health Regulations (COSHH) 2002 (as amended), (median inhalable fraction 1.08 mg/m^3, min 0.25 mg/m^3 max 10.80 mg/m^3, median respirable fraction 0.05 mg/m^3, min 0.02 mg/m^3, max 1.49 mg/m^3). Determination of the biological component of the dust showed that employees' exposures to endotoxin were elevated (median 31.5 EU/m^3, min 2.00 EU/m^3, max 1741.78 EU/m^3), particularly when waste was agitated (median 175.0 EU/m^3, min 2.03 EU/m^3, max 1741.78 EU/m^3). Eight out of 32 (25%) of the personal exposure data for endotoxin exceeded the 200 EU/m^3 temporary legal limit adopted in the Netherlands and thirteen out of 32 (40.6%) exceeded the suggested 50 EU/m^3 guidance level suggested to protect workers from respiratory health effects. A significant correlation was observed between employee inhalable dust exposure and personal endotoxin concentration ($r = 0.728$, $p<0.05$) and also personal endotoxin exposure and 1-3 β Glucan concentration ($r = 0.817$, $p<0.05$). Further work is needed to explore the possibility of using inhalable dust concentration as a predictor for personal endotoxin exposure. The general dust levels stipulated in the COSHH Regulations 2002 (as amended) are inadequate for managing the potential health risks associated with endotoxin exposure at composting sites. Employee exposure levels and dose-response disease mechanisms are not well understood at this present time. Consequently, in light of this uncertainty, it is recommended that a precautionary approach be adopted in managing the potential health risks associated with inhalation of organic dusts at composting sites.

1. Introduction

The European Landfill Directive (EC/31/1999) and the implementation of subsequent more stringent waste strategies in the UK require at least 45% of biodegradable waste to be diverted from the traditional practice of landfill, by 2015. The proliferation of commercial composting is already underway in the UK and composting rates have risen from 6% in 1995/1996 to 31% in 2006/07 [1,2].

There is also a trend appearing for larger, centralized composting facilities such as in-vessel composters (IVC) and Mechanical Biological Treatment (MBT) technologies to be utilized in an attempt to achieve these diversion targets [2]. These newer composting technologies are most commonly sited indoors to minimize environmental impacts such as odours, bioaerosol emissions and the attraction of vermin. The occupational health implications of conducting composting operations indoors are poorly understood. One particular area of concern is employees' exposure to bioaerosols which at this present time is poorly characterised. It is unclear whether indoor composting processes have the potential to expose site workers to higher concentrations of organic dust than more traditional 'open – windrow' processes that have more commonly been used.

Bioaerosols are defined as a collection of aerosolised biological particles such as bacteria, fungi and protozoa, cell components and fragments of cells [3-10].

The concern regarding bioaerosols from composting activities arises because of their potential to cause adverse health effects in employees and the public living in close proximity to such facilities. These adverse effects can potentially occur in susceptible individuals from exposure to the micro-organisms associated with the composting process and can elicit an adverse response by infection, allergy or an adverse response to toxins [11].

The effects of exposure to organic dust on respiratory health may lead to or exacerbate a number of distinct identifiable conditions including aspergillosis in immunocompromised individuals [6,7]; allergic rhinitis and asthma [12-14]; extrinsic allergic alveolitis (Farmer's Lung) [15] where prolonged (usually occupational) exposure occurs; chronic obstructive pulmonary disease (COPD) [16-20], toxic pneumonitis /Organic Dust Toxic Syndrome (ODTS) [17,21], upper airway irritation/ mucous membrane irritation [20, 22-24] and an accelerated decline in Forced Vital Capacity (FVC%) [20].

The biological hazards associated with bioaerosol liberation from composting activities are fairly well defined and include fungi, bacteria, in particular actinomycetes, endotoxin and 1–3 β glucans [25].

1.1. Fungi

Fungi, in particular *Aspergillus fumigatus* have the ability to degrade cellulose and hemicelluloses and are therefore particularly important to the composting process. It is however an allergenic fungus and can cause aspergillosis in immuno-compromised individuals [17,22]. 1-3 β Glucan present in the cell wall of fungi has been associated with an increased prevalence of atopy and decreased lung function [26,27].

1.2. Bacteria

Whilst respiratory infection from bacteria is unlikely to be a significant hazard [28]. Endotoxin, the lipopolysaccharide portion of the outer layer of the cell walls of gram-negative bacteria, is known to cause acute illness such as flu-like symptoms and Organic Dust Toxic Syndrome (ODTS), chronic exposure has been associated with chronic pulmonary inflammation leading to chronic bronchitis and reduced lung function [29]. Actinomycetes are gram positive bacteria and thermophilic species thrive in wet compost at temperatures between 30 and 60 °C. Thermophilic actinomycetes are known respiratory allergens and are primarily responsible for occupational diseases such as Farmer's Lung and Mushroom Workers lung [7].

Even though the organisms that predominate throughout the composting process have been identified, it is extremely difficult to obtain a concise characterisation of bioaerosol emissions from composting activities. The existence and concentration of individual species associated with the composting process is highly variable and heavily dependant on a number of factors. These include, the nature of the material being composted, individual bioaerosol properties, the temperature and moisture content of the compost, process design, site operations such as turning, shredding and screening, geographical, topographical and meteorological conditions and whether the process is enclosed or carried out in the open air [28,30].

At present the literature available to determine the potential risks to employees' health from bioaerosol exposure is not sufficiently exhaustive or conclusive. A large proportion of employee exposure studies have utilised static air sampling methodologies which have limited relevance to actual employee exposures in the context of the current regulatory framework for inhalation of inhalable and respirable dust. More significantly, a large number of the studies undertaken to date have focused solely on the enumeration of viable organisms liberated from the composting activities. It has been suggested that the culturable or viable component of bioaerosol emissions may be less than 10% of the total bioaerosol release [31-33]. Therefore, the assessment of viable organisms alone, provides a considerable under-estimation of personal exposure and the potential risks to employees' health as cell components and fragments are known to elicit adverse health outcomes in exposed individuals [24]. Bacterial endotoxin and 1-3 β Glucan have been associated with the development of respiratory symptoms in a number of occupational environments [34] and have been measured in order to provide an indication of the biological composition of employees' exposure to organic dust at composting facilities.

2. Managing the occupational risks from organic dust exposure

Regulation 2 of the Control of Substances Hazardous to Health Regulations (COSHH) 2002 (as amended) define 'dust' as a 'substance hazardous to health' when present in concentrations in excess of 10mg/m^3 for inhalable dust and 4mg/m^3 for respirable dust. It has been suggested that these general dust levels are unsuitable for managing the occupational risks posed by dusts containing biological material because of their increased toxic potential [35].

Currently in the UK there are no Workplace Exposure Limits for endotoxin and 1-3 β Glucan. Guidelines for 'no-effect' levels for environmental endotoxin have been suggested by Rylander [35] and are presented in Table 1.

Table 1. Guidelines for 'no-effect' levels for environmental endotoxin.

Disease	ng/m^3 [a]
Toxic pneumonitis	200
Airways inflammation	10
Systemic effects	100

[a] 1ng/m^3 ≡ 10 EU/m^3

Rylander [35] also reports that a level for 1-3 β Glucan of 10 ng/m^3 and above may give rise to health effects similar to those witnessed with endotoxin exposure such as airway inflammation, fatigue and headache.

The Dutch Expert Committee on Occupational Standards of the National Health Council [36] proposed a temporary legal limit of 200 EU/m^3 and a guidance limit of 50 EU/m^3 over an 8 hour exposure period for endotoxin. No similar limit has been suggested for 1-3 β Glucan.

This study investigates the impact of operational practices on bioaerosol liberation and employee exposure in four large-scale commercial composting sites. Work on this project is ongoing and the results presented in this paper detail the work conducted to date.

3. Methods

Monitoring was undertaken at two open windrow sites, an in-vessel composting facility and an enclosed windrow facility in order to establish the impact that indoor composting activities have on employee exposure compared to those undertaken in the open air.

Bioaerosol liberation from composting activities are influenced by waste composition, moisture content and meteorological conditions [30,37]. Consequently, monitoring was undertaken at each site three times per year during January –March, April – June and July – September to determine the impact of seasonal variation on employee exposure. No monitoring was undertaken during October – December as site operations were limited at this time.

Job Exposure Matrices (JEM) were completed for each employee and the predominant work activity was coded into one of the following predominant work activities;

1)Manual sorting of waste, 2)Shredding, 3)Turning, 4)Screening. Analysis was undertaken to investigate the impact of these operational practices on employee exposure.

A total of 32 personal samples were taken and the inhalable and respirable fraction, endotoxin and 1-3 β Glucan was determined. 67 static samples were also taken to evaluate the working environment of employees and analysed for the inhalable fraction, endotoxin and 1-3 β Glucan.

Inhalable and respirable dust was sampled in accordance with MDHS 14/3 [38]. Three weighing room blanks were used as laboratory controls for the gravimetric analysis and one additional set of inhalable and respirable samples were left unexposed as field controls. Sampling was performed using Casella Apex Pro constant flow pumps, set to run for 4 hours, in combination with IOM inhalable and Higgins-Dewell cyclonic respirable sampling heads, set at 2.0 l min^{-1} and 2.2 l min^{-1} respectively and fitted with Whatman GF/A glass fibre filters. Pumps were calibrated before and after sampling and the results extrapolated to an 8 –hour time weighted average (TWA) to enable comparisons with the regulatory levels to be made. The dust concentrations were determined by gravimetric analysis in accordance with MDHS 14/3 [38]. Following gravimetric analysis of the dust, extraction and analyses for endotoxin was performed as previously described [24]. Extraction for endotoxin was performed in 10ml of 0.05% Tween-20 in pyrogen free water. Each sample tube was then placed on a Stuart SB3 Rotator and spun at 25 rev/min for one hour. Following the washing process the sample tubes were spun in a centrifuge at 3000 rpm for 10 minutes. Endotoxin levels were determined, in supernatant, by a chromogenic kinetic Limulus Amebocyte Lysate (LAL). 1-3 β-Glucan levels were estimated using a specific blocking agent to establish the contribution that these compounds gave to the original endotoxin assay [39].

4. Results

4.1. Dust

Employees' exposure to dust was found to be generally low (median inhalable fraction 1.08 mg/m^3, min 0.25 mg/m^3 max 10.80 mg/m^3, median respirable fraction 0.05 mg/m^3, min 0.02 mg/m^3, max 1.49 mg/m^3). Only one of the 32 personal samples exceeded the 10mg/m^3 level stated in the COSHH Regulations and none of the respirable samples exceeded the 4 mg/m^3 COSHH level. Employees' exposures were increased when waste was being agitated but no statistically significant difference was observed between the operational practices of sorting, shredding, turning and screening.

The static dust samples showed extensive variation (median inhalable fraction 0.66 mg/m^3, min 0.01 mg/m^3 max 172.42 mg/m^3). Indoor static inhalable dust levels were significantly higher than outdoor static levels ($p<0.05$ Mann-Whitney *U*-Test).

No statistically significant difference was observed in inhalable dust levels during waste sorting, shredding, turning and screening. Despite this, final product screening and manual sorting of waste were identified as areas of concern as high dust concentrations were experienced on occasion and further work is needed to investigate employee exposures whilst undertaking these activities.

4.2. Endotoxin

Despite the relatively low dust levels, determination of the biological component of the dust indicated that employees' exposure to endotoxin was elevated (median 31.5 EU/m^3, min 2.00 EU/m^3, max 1741.78 EU/m^3), particularly when waste was agitated during sorting, shredding and screening operations (median 178.1 EU/m^3, min 6.02 EU/m^3, max 1741.78 EU/m^3).

The static sampling results showed large variations in endotoxin concentrations, range 0.65 to 584.02 EU/m^3. Personal exposure levels for endotoxin were found to be higher than those obtained from static sampling, range 2.03 to 1741.78 EU/m^3. Personal endotoxin levels were higher when operatives were working indoors, particularly when the waste is agitated.

Figure 1 illustrates that, although not statistically significant, differences in employee exposures were seen in employees working indoors compared to outdoors and between personal sampling and static sampling results

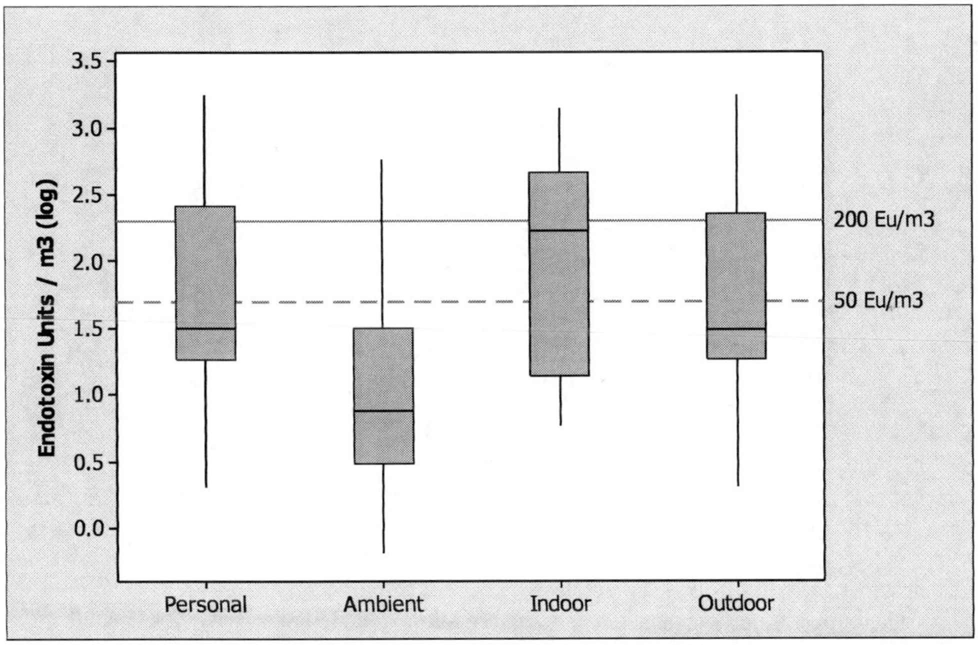

Figure 1. Median and Inter-quartile range (log scale) of endotoxin exposure concentrations by location.

As shown in Figure 2, operational practices such as waste sorting, shredding, turning and screening was seen to increase bioaerosol liberation and had an increased effect on employee exposure to endotoxin. Eight out of 32 (25%) of the personal exposure data exceeding the 200 EU/m^3 temporary legal limit

Journal of Physics: Conference Series **151** (2009) 012064 doi:10.1088/1742-6596/151/1/012064

adopted in the Netherlands and thirteen out of 32 (40.6%) exceed the 50 EU/m^3 guidance level suggested to protect workers from respiratory health effects [36].

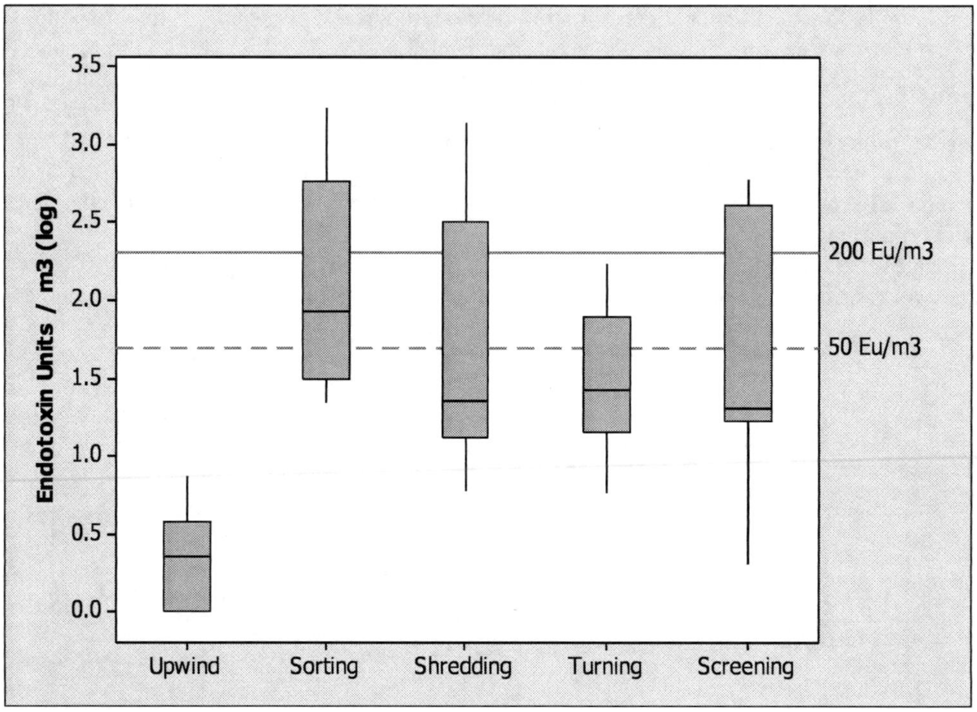

Figure 2. Median and Interquartile range (log scale) of personal endotoxin exposure concentrations by predominant work activity.

4.3. Glucan

Generally, employees' exposure to 1-3 β Glucan was low (median 0.52 ng/m^3, min 0.00 ng/m^3, max 28.72 ng/m^3). Five out of 32 (15.6%) exceeded the 10ng/m^3 level believed to be related to clinical symptoms [35].

There was a statistically significant correlation between employees' exposure to inhalable dust and personal endotoxin concentration (Spearman rho correlation coefficient $r = 0.728$, $p<0.05$) and also between personal endotoxin exposure and 1-3 β Glucan concentration (Spearman rho correlation coefficient $r = 0.817$, $p<0.05$).

5. Discussion

The findings of this study are consistent with others in that employees of composting sites are exposed to high levels of bacterial endotoxin [20,23,24] even though in most cases, their personal dust exposure is below regulatory levels.

The results obtained from this study are highly variable due to a number of influencing factors such as site operations and practices, waste composition and meteorological conditions. This has been highlighted in previous studies [30] and consequently the determination of 'typical exposures' from composing operations such as sorting, shredding, turning and screening is problematical.

The results indicate that employees' exposure to dust and endotoxin was higher at indoor composting facilities. This finding requires further investigation and employees' exposures to dust and endotoxin needs to be extensively characterised as indoor facilities are becoming more prevalent within the waste management industry.

The agitation of waste during various site activities resulted in generally higher employee exposures to dust and endotoxin. This increase however was not found to be statistically significant. Waste shredding, turning and screening is fundamental to the composting process and consequently practical risk mitigation measures are required to reduce employees' exposure as far as reasonable practicable during these activities. The efficacy of any control measures adopted should be investigated to determine their actual impact on reducing employee exposure to dust and endotoxin.

There is general agreement that organic dusts containing bacterial endotoxin are associated with important adverse health effects and that this potential risk should not be considered as a general dust [35]. It is well established that endotoxin concentration is highly dependant on the process generating the dust and considerable variation in the levels of endotoxin concentrations have been described in a number of different industrial sectors such as composting, swine farming, grain handling and agriculture [40]. Consequently the general dust levels stipulated in the Control of Substances Hazardous to Health Regulations 2002 (as amended), as currently applied, are inadequate in effectively managing the potential risks specifically arising from the inhalation of endotoxin.

At present, endotoxin monitoring and analysis is unlikely to be widely adopted in the waste management industry due to cost and technical constraints in conducting the assay [3,25]. Consequently, if an occupational exposure standard for endotoxin was developed, it would be difficult to comply with in practice. This study does however indicate a significant correlation between inhalable dust and endotoxin concentration. Further work is needed to explore this relationship in order to establish whether dust may be used as a reliable proxy measure for endotoxin exposure. This may form the basis of a pragmatic, yet evidence-based way to regulate and manage employees' exposure to endotoxin.

The potential risks from exposure to endotoxin at composting sites should be managed with caution at this present time. Site operators need to comprehensively assess the risk to site operatives from exposure to bioaerosols and implement personal monitoring, health surveillance, screening and immunological testing, where applicable, to manage these potential risks in a proactive and responsible manner.

6. Conclusion

Compost workers are exposed to airborne concentrations that may, in conjunction with other factors contribute to the development of respiratory symptoms. This study demonstrates that employees on composting sites are commonly exposed to levels of bacterial endotoxin believed to be associated with respiratory illness. Employee exposure to 1-3 β Glucan was generally below levels considered to be related to adverse effects. There is however considerable uncertainty regarding the body burdens experienced by site operatives and the extent of dose-related effects from bioaerosol exposure at composting sites.

Employee exposure to endotoxin at composting sites is generally high, especially during operations that involve the agitation of waste. A significant correlation was found between the inhalable dust fraction and endotoxin concentration and further work is needed to explore the possibility of using inhalable dust concentration as a predictor for personal endotoxin concentration. The general dust levels that define a dust as a 'substance hazardous to health' as stipulated in the Control of Substances Hazardous to Health Regulations 2002 (as amended) are inadequate for managing the potential health risks associated with endotoxin exposure at composting sites. At present, no statutory or reference levels exist in the UK for occupational exposure to endotoxin.

References

[1] DEFRA 2007 *Waste strategy for England 2007* (Norwich: TSO)
[2] Composting Association 2006 *The state of composting and biological waste treatment in the UK 2004/05.* (Northampton : The Composting Association)
[3] Sykes P, Jones K and Wildsmith J D 2007 Managing the potential public health risks from bioaerosol liberation at commercial composting sites in the UK: An analysis of the evidence base *Resources, Conservation and Recycling* **52** 410-24
[4] Cox C S and Wathes C M, ed 1995 *Bioaerosols handbook.* (Florida: CRC Press Ltd.)
[5] Epstein E 1994 Composting and bioaerosols. *BioCycle* **35(1)** 51–7
[6] Millner P 1995 Bioaerosols and composting. *BioCycle* **36(1)** 48–54
[7] Millner P D, Olenchock S A, Epstein E, Rylander R, Haines J,Walker J, Ooi BL, Horne E and Maritato M 1994 Bioaerosols associated with composting facilities. *Compost. Sc.i. Util.* **2** 6–57
[8] Fischer G, Muller T, Ostrowski R and Dott W 1999 Mycotoxins of *Aspergillus fumigatus* in pure culture and in native bioaerosols from compost facilities. *Chemosphere* **38(8)** 1745–55
[9] Gilbert J and Ward C 1998 Monitoring bioaerosols at composting facilities. *BioCycle* **39(10)**75–6
[10] Van der Werf P 1996 Bioaerosols at a Canadian composting facility. *BioCycle* **37(9)** 78–83
[11] Environment Agency 2005 *Health impact assessment of waste management: Methodological aspects and information sources.* Science *report P6-011/1/SR1* (Bristol : Environment Agency)
[12] Cockcroft D W, Ruffin R E and Dolovich J 1977 Allergen-induced increase in non-allergic bronchial reactivity. *Clin Allergy* **7** 503–13
[13] Newman Taylor A J 1987 The lung and the work environment.. *Current perspectives in the immunology of respirtory disease.* ed Kay A B and Goetzl E J (Edinburgh: Churchill Livingstone)
[14] Zuskin E, Schachter E N, Kanceljak B, Mustajbegovic J and Wiltek T 1994 Immunological and respiratory reactions in workers exposed to organic dusts. *Int. Arch. Occup. Environ. Health.* **66** 317–24
[15] Flannigan B, McCabe E M and McGarry F 1991 Allergenic and toxigenic micro-organisms in houses. *J. Appl. Bacteriol.* **70**:61–73
[16] Clapp W D, Becker S, Quay J, Watt J L, Thorne P S, Frees K L, Zhang X, Koren H S, Lux C R and Schwartz D A 1994. Grain dust-induced airflow obstruction and inflammation of the lower respiratory tract. *Am. J. Resp. Crit. Care. Med.* **150** 611–7
[17] Lacey J and Crook B 1988 Fungal and actinomycete spores as pollutants of the workplace and occupational allergens. *Ann. Occup. Hyg.* **32** 513–33
[18] Matheson M C, Benke G, Raven J, Sim M R, Krombout H, Vermeulen R, Johns D P, Walters E H and Abramson M J 2005 Biological dust exposure in the workplace is a risk factor for chronic obstructive pulmonary disease. *Thorax* **60** 645–51
[19] Olenchock S A, Christiani D C, Mull J C, Ye T T and Lu P L 1990 Airborne endotoxin concentrations in various work areas within two cotton textile mills in the People's Republic of China. *Biomed. Environ. Sc.i.* **3** 443–51
[20] Bunger J, Schlappler-Scheele B, Hilgers R and Hallier E 2007 A 5-year follow-up study on respiratory disorders and lung function in workers exposed to organic dust from composting plants. *Int. Arch. Occup. Environ. Health.* **80** 306-312

[21] Chan-Yeung M, Enarson D A and Kennedy S M 1992 The impact of grain dust on respiratory health. *Am. Rev. Respir. Dis* .**145** 476–87

[22] Dutkiewicz J 1997 Bacteria and fungi in organic dust as potential health hazard. *Ann. Agric. Environ .Med..* **4** 11–6

[23] Herr C E W, zur Nieden A, Jankofsky M, Stilianakis N I, Boedeker R-H and Eikmann T F 2003 Effects of bioaerosol polluted outdoor air on airways of residents: a cross sectional study. *Occup. Environ. Med.* **60** 336–42

[24] Wouters I M, Spaan S, Douwes J, Doekes G and Heederik D 2006 Overview of personal occupational exposure levels to inhalable dust, endotoxin, β (1–3) glucan and fungal extracellular polysaccharises in the waste management chain. *Ann. Occ. Hyg.* **50** 39–53

[25] Douwes J, Thorne P, Pearce N and Heederik D 2003 Bioaerosol health effects and exposure assessment: progress and prospects. *Ann. Occup. Hyg.* **47** 187–200

[26] Douwes J, Dubbeld H, van Zwieten L, Wouters I, Doekes G, Heederik D and Steerenberg P 1997 Work related acute and (sub-)chronic airways inflammation assessed by nasal lavage in compost workers. *Ann. Agric. Environ. Med.* **4**149–51

[27] Wouters I M, Hilhorst S K M, Kleppe P, Doekes G, Douwes J, Peretz C and Heederik D 2002 Upper airway inflammation and respiratory symptoms in domestic waste collectors. *Occup. Environ. Med.* **59** 106–12

[28] Swan J R M, Crook B and Gilbert E J 2002 Microbial emissions from composting sites. *editor. Issues in Env Sci & Tech.* ed Harrison R (Cambridge: Royal Soc Chemistry)

[29] Jacobs R R, Heederik D, Douwes J and Zahringer U 1997 Endotoxin structure. *In.t J. Occ. Environ. Health.* **97;3** s6–7

[30] Taha M P M, Drew G H, Longhurst P J, Smith R and Pollard S J T 2006 Bioaerosol releases from compost facilities: evaluating passive and active source terms at a green waste facility for improved risk assessments. *Atmos. Environ* .**40** 1159–69

[31] Amman R, I Ludwig W and Schleifer K H 1995 Phylogenetic identification and in situ detection of individual microbial cells without cultivation. *Microbiol. Rev* .**59** 143–69

[32] Blomquist G 1994 Sampling of biological particles. *Analyst* **119** 53–6

[33] Pillai S and Ricke S 2002 Bioaerosols from municipal and animal wastes: background and contemporary issues. *Can. J. Microbiol.***48** 681–96

[34] Douwes J, McLean D, Van der Maarl E, Heederik D and Pearce N 2000 Worker Exposures to Airborne Dust, Endotoxin and β(1-3) Glucan in Two New Zealand Sawmills. *Am. J. Ind. Med.* **38** 426-30.

[35] Rylander R 1997 Evaluations of the risks in endotoxin exposures. *Int. J. Occ. Environ. Health* .**3** s32–6

[36] Heederik D and Douwes J 1997 Towards an occupational exposure limit for endotoxins. *Ann. Agric. Environ. Med.***4** 17–9

[37] Wheeler P A, Stewart I, Dumitrean P and Donovan B 2001 *Health effects of composting—a Study of three compost sites and review of past data. Environment Agency R&D Technical Report P1-315/TR* (Bristol: Environment Agency)

[38] Health and Safety Executive 2000 *MDHS 14/3 General methods for sampling and gravimetric analysis of respirable and inhalable dust.* (Norwich: HSE Books)

[39] Hryhorczuk D, Curtis L, Scheff P, Chung J, Rizzo M, Lewis C, Keys N and Moomey M 2001. Bioaerosol emissions from a suburban yard waste composting facility. *Ann. Agric. Environ. Med.* **8** 177–85

[40] Health and Safety Executive 2007 *The HSE Grain Dust Study – workers exposure to grain dust contaminants, immunological and clinical responses*. RR 540 (Norwich: HSE Books)

Journal of Physics: Conference Series **151** (2009) 012065 doi:10.1088/1742-6596/151/1/012065

Occupational rhinitis and occupational asthma; one airway two diseases?

M J Seed, M Gittins, F De Vocht and R M Agius

Occupational and Environmental Health Research Group, University of Manchester.

Correspondence to Dr Martin Seed, Occupational & Environmental Health Research Group, School of Translational Medicine, Faculty of Medical & Human Sciences, University of Manchester, Ellen Wilkinson Building, Oxford Road, Manchester M13 9PL, UK. Martin.seed@manchester.ac.uk

Abstract. The concept of 'one airway, one disease' refers to the frequent comorbidity of asthma and rhinitis. However, only limited research has been done on this association for the diverse range of occupational respiratory sensitisers. The relative frequency of rhinitis was determined for the 15 respiratory sensitisers reported to cause at least 10 cases of rhinitis or asthma to The Health and Occupation Reporting (THOR) network between 1997 and 2006. Of 1408 cases, 1190 were sole diagnoses of asthma, 138 sole diagnoses of rhinitis and in 80 cases asthma coexisted with rhinitis. The six sensitisers for which rhinitis featured in over 15% of cases were all particulates and known to cause release of mast cell mediators, either directly or through IgE antibodies. Four of the other nine sensitisers often exist as vapours and only two have been consistently associated with IgE-mediated disease mechanisms. Particle size did not appear to correlate with the relative frequency of rhinitis. Despite its limitations this study would support the hypothesis that there are at least two mechanistic categories of respiratory sensitisation with rhinitis being relatively more common where the mechanism is IgE-mediated. Particulate nature may be another important factor to consider in future studies.

1. Introduction

The frequent comorbidity of rhinitis and asthma in the general population is thought to be due to common IgE-mediated allergic pathology in atopics in addition to an increased risk of asthma developing in individuals with perennial rhinitis regardless of atopic status. [1-3] This 'one airway one disease' hypothesis might also be expected to apply to occupational rhinitis and occupational asthma. Indeed the prevalence of occupational rhinitis in cases of occupational asthma has been estimated at between 76 and 92%. [4]

However, the diverse range of occupational respiratory sensitisers differ with respect to several parameters including molecular weight, particle type and size as well as pathophysiological mechanism. It is possible therefore that the relative frequency with which they cause rhinitis compared to asthma varies as a function of one or more of these factors. We therefore set out to examine variation in the relative incidences of rhinitis and asthma caused by different respiratory sensitisers using data from cases reported to The Health and Occupation Reporting (THOR) network in the UK. [5]

Journal of Physics: Conference Series **151** (2009) 012065 doi:10.1088/1742-6596/151/1/012065

2. Methods

Cases of rhinitis or asthma were taken from those reported to THOR by either chest physicians through the Surveillance of Work-related and Occupational Respiratory Disease (SWORD) [6] scheme or by occupational physicians through the Occupational Physicians Reporting Activity (OPRA).[7] All the individual cases of rhinitis and/or asthma reported during the decade 1997-2006 were firstly grouped according to the suspected causative agent. After cases had been excluded for which more than one suspected causative agent had been reported, each substance that had caused more than 10 cases in total was identified. Cases of asthma thought to be caused by irritant mechanisms or the Reactive Airways Dysfunction Syndrome (RADS) were excluded. However, the reporting cards did not ask for information on the mechanism of occupational rhinitis hence reported cases could be caused by either irritant or allergic pathology. For each respiratory sensitiser causing ten or more total cases, the proportion of cases in which rhinitis had been reported was determined and 95% confidence intervals calculated for this 'rhinitis:asthma' ratio.

To assess a possible association between the rhinitis:asthma ratio and particle size, these causative agents were a priori categorised as gases, vapours, mists or particles. Fumes were included in the category of particles, which was further subdivided in having primarily respirable or primarily inhalable particles based on the size distribution of the airborne exposure resulting from 'typical' jobs or tasks. In practice however, the size distribution of the different airborne substances can vary between different jobs and tasks caused by differences in handling of the parent compounds. Therefore we have based this classification on a combination of expert judgement and a weighing of the size distributions for each substance in different occupational scenarios reported in published literature.

The medical literature was then searched using the Medline database for evidence of the molecular size of each respiratory sensitiser and its pathophysiological mechanism. The cut-point used to distinguish high from low molecular weight compounds was 1kDa. Evidence for the presence of specific IgE antibodies to the causative sensitiser or its hapten-protein conjugate was sought in published case series in addition to individual case reports. For the purpose of this study a sensitiser was deemed to cause disease through IgE-mediated mechanisms if specific IgE had been demonstrated by either serological immunoassay or skin prick testing in at least 50% of published cases.

3. Results

In total 2039 cases of asthma were reported to SWORD during the study period for which coexistent rhinitis was reported in 104 cases. 184 cases of rhinitis without asthma were reported. Thus the rhinitis:asthma ratio was 13% for cases of respiratory sensitisation reported to SWORD. For OPRA during this period 317 cases of asthma were reported for which coexistent rhinitis was reported in 7 cases. Rhinitis was reported to OPRA without coexistent asthma in 73 cases and the overall rhinitis:asthma ration for OPRA cases was 21%. When the SWORD and OPRA data were combined 201 cases of asthma or rhinitis were identified for which two or more suspected causes were named. These suspect causative agents were diverse but the most frequent dual combinations were laboratory animals with latex (6 cases of asthma, 1 case of rhinitis) and isocyanates with formaldehyde (5 cases of asthma).

For the remaining cases with only one named suspect cause fifteen causative agents were identified for which there were ten or more cases of asthma due to sensitisation and/or rhinitis reported to THOR. Of the 1408 cases of respiratory sensitisation caused by the 15 selected substances, 1190 were sole diagnoses of asthma, 138 sole diagnoses of rhinitis and in 80 cases asthma coexisted with rhinitis. For each of these agents the number of cases for which rhinitis or asthma was reported alone or both diagnoses coexisted are shown in table 1. The two groups of agents with the highest number of total cases demonstrated a large difference in the relative frequency of rhinitis and asthma. The rhinitis:asthma ratio for isocyanates, with 431 total cases, was only 1% whereas for laboratory animals, with 221 total cases, the rhinitis:asthma ratio was 45%.

Table 1. Breakdown of cases by specific diagnoses for the fifteen leading causes of rhinitis or asthma reported to THOR 1997-2006.

	Total cases	Rhinitis alone	Rhinitis & asthma	Asthma alone
Isocyanates	431	0	4	427
Laboratory animals	221	72	28	121
Flour	186	20	14	152
Latex	109	15	14	80
Colophony	101	3	1	97
Glutaraldehyde	89	9	2	78
Wood dust	77	4	3	70
Chromium compounds	48	3	0	45
Acrylics & acrylates	40	2	1	37
Cobalt	28	0	0	28
Seafood allergens	22	0	0	22
Platinum salts	21	7	1	13
Morphine	13	0	11	2
Acid anhydrides	12	0	0	12
β-lactams	10	3	1	6

The rhinitis:asthma ratios for each of the 15 selected sensitisers are illustrated in figure 1 with 95% confidence intervals for this proportion. Information concerning particle nature and size, molecular weight and evidence for IgE-mediated mechanism is recorded below the corresponding rhinitis:asthma ratios. .

All of the six substances with rhinitis:asthma ratio greater than 15% are particulates but whether or not they were considered predominantly inhalable or predominantly respirable did not appear related to relative frequency of rhinitis. All of these six substances have consistent evidence in the literature of either IgE-mediated disease mechanisms or, in the case of morphine, cause direct degranulation of mast cells. However, these six substances had an even distribution of high and low molecular weight classification.

Figure 1. Proportion of respiratory sensitisation cases that included a diagnosis of rhinitis for individual agents, THOR 1997-2006. (with 95% confidence intervals)

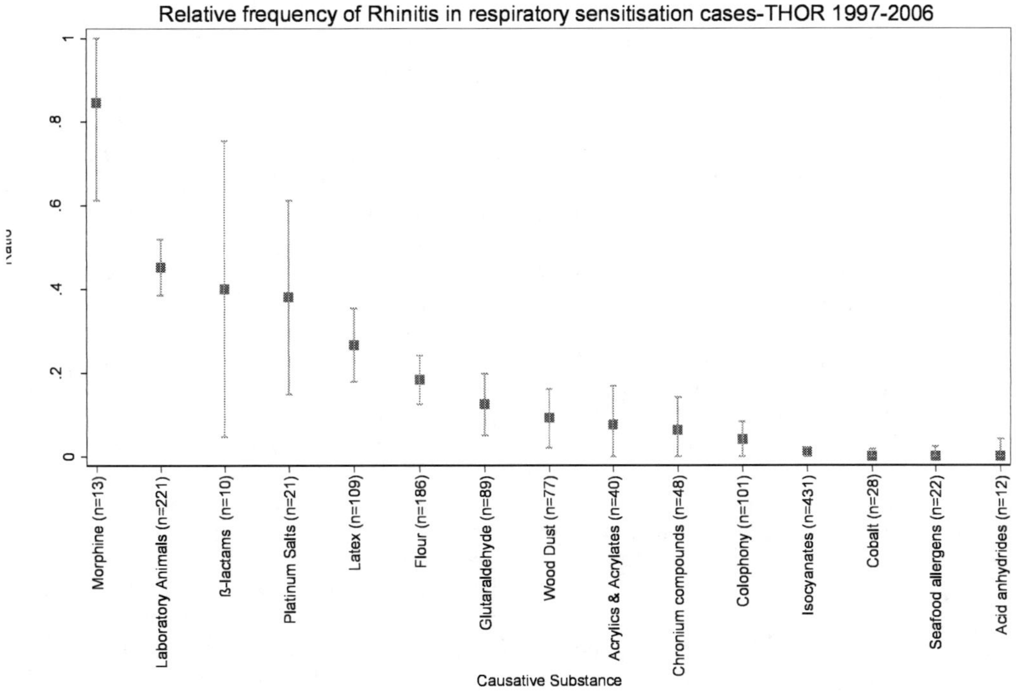

Particle	R	I	R	R	I	I	V	I	R	V	R	R/ V	R	R	V
Molecular weight	L	H	L	L	H	H	L	L	L	L	L	L	L	H	L
IgE	(+)	+	+	+	+	+	-	-	-	-	-	-	-	+	+

R = respirable particle, I = inhalable particle, V = vapour
L = low molecular weight, H = high molecular weight
+ = consistent evidence for IgE-mediated mechanism
- = limited evidence for IgE-mediated mechanism

4. Discussion

The extent to which conclusions can be drawn from this THOR data relating to rhinitis and asthma has limitations. There might be biases in the collection of data relating to variation in physician reporting behaviour. It has been shown that limited diagnostic consensus exists between physicians for both occupational asthma [8] and occupational rhinitis. [9] The relative frequency with which occupational rhinitis is reported to THOR might also be influenced by the fact that rhinitis could present to otorhinolaryngologists or allergists, neither of whom currently report to the THOR scheme. In contrast, occupational asthma is more likely to be diagnosed by chest and occupational physicians and be reported to the SWORD and OPRA schemes. Another reason why there might be relative underreporting of occupational rhinitis is that following the diagnostic work-up resulting in a diagnosis of occupational asthma, the physician might omit to report coexistent rhinitis, particularly if it does not cause disabling symptoms. The reporting threshold for symptoms of rhinitis is likely to vary between physicians who might see proportionately more cases caused by a particular sensitiser.

Nevertheless the data suggest differences in the relative incidences of reported asthma and rhinitis due to individual respiratory sensitisers. No clear effect of particle or size distribution on the rhinitis:asthma ratio was found by grouping sensitisers into basic categories. However, particle size might be expected to affect the relative distribution of a given respiratory sensitiser between the nose and lower airways upon inhalational exposure. Many of the particulate sensitisers have a heterogeneous distribution of particle sizes and it might be that the predominant type does not necessarily govern the rhinitis:asthma ratio for that substance. Further toxicokinetic explanations for such differences might include variation in water and fat solubility, which have not been examined here.

Whilst no clear effect of molecular weight was observed, five of the six sensitisers with the highest rhinitis:asthma ratios have been consistently demonstrated to have immunoglobulin E (IgE) mediated mechanisms. [10-12] The sixth substance, morphine is known to cause direct mast cell degranulation.[13] By contrast IgE mechanisms have been firmly demonstrated for only two of the nine respiratory sensitisers that have lower rhinitis:asthma ratios; seafood allergens [14] and acid anhydrides.[15] The other seven substances are LMW respiratory sensitisers and their relatively low rhinitis:asthma ratios could reflect their predominant mechanism being non-IgE-mediated. Some of the cases of rhinitis caused by the latter agents might have irritant mechanisms. The non-IgE-mediated mechanisms of asthma caused by many LMW respiratory sensitisers have yet to be fully characterised.

It is likely that a combination of toxicokinetic and toxicodynamic factors explain the relative causation of rhinitis and asthma for a given respiratory sensitiser. A larger prospective study using consistent diagnostic criteria might help to further characterise their relative importance.

References

[1] Leynaert B, Neukirch F, Demoly P and Bousquet J 2000 Epidemiologic evidence for asthma and rhinitis comorbidity *J. Allergy Clin. Immunol.* **106**(5 Suppl) S201-5.

[2] Bachert C, van Cauwenberge P and Khaltaev N 2002 World Health Organization. Executive summary. Allergic rhinitis and its impact on asthma. In collaboration with the World Health Organization *Allergy* **57** 841-55

[3] Cruz A A, Popov T and Pawankar R 2007 Common characteristics of upper and lower airways in rhinitis and asthma: ARIA update, in collaboration with GALEN *Allergy* **62** (Suppl.84) 1-41

[4] Siracusa A, Desrosiers M and Marabini A 2000 Epidemiology of occupational rhinitis: prevalence, aetiology and determinants *Clin. Exp. Allergy* 2000;**30**:1519-34.

[5] The University of Manchester. The Health and Occupation Reporting Network. Centre for Occupational & Environmental Health, University of Manchester. Available at: http://www.medicine.manchester.ac.uk/coeh/thor/ (Accessed on 30 August 2008).

[6] McDonald J C, Chen Y, Zekveld C and Cherry N M 2005 Incidence by occupation and industry of acute work related respiratory diseases in the UK, 1992-2001 *Occup. Environ. Med.* **62** 836-42

[7] Cherry N M, Meyer J D, Holt D L, Chen Y and McDonald JC 2000 Surveillance of work-related diseases by occupational physicians in the UK: OPRA 1996-1999 *Occup. Med.* **50** 496-503

[8] Fishwick D, Bradshaw L, Henson M, Stenton C, Hendrick D, Burge S, et al 2007 Occupational asthma: an assessment of diagnostic agreement between physicians. *Occup. Environ. Med.* **64** 185-90

[9] Castano R and Theriault G 2006 Defining and classifying occupational rhinitis *J. Laryngol. Otol.* **120** 812-17

[10] Wild L G and Lopez M 2003 Occupational asthma caused by high-molecular-weight substances *Immunol. Allergy Clin. North Am.* **23** 235-50

[11] Baker D B, Gann P H, Brooks S M, Gallagher J and Bernstein I L 1990 Cross-sectional study of platinum salts sensitization among precious metals refinery workers *Am. J. Ind. Med.* **18** 653-64

[12] Blanca M, Romano A, Torres M J, Demoly P and DeWeck A 2007 Continued need of appropriate betalactam-derived skin test reagents for the management of allergy to betalactams *Clin. Exp. Allergy* **37** 166-73

[13] Nasser S M and Ewan P W 2001 Opiate-sensitivity: clinical characteristics and the role of skin prick testing *Clin. Exp. Allergy* **31** 1014-20

[14] Jeebhay M F, Robins T G, Lehrer S B and Lopata A L 2001 Occupational seafood allergy: a review *Occup. Environ. Med.* **58** 553-62

[15] Bernstein D I, Gallagher J S, D'Souza L and Bernstein I L 1984 Heterogeneity of specific-IgE responses in workers sensitized to acid anhydride compounds *J. Allergy Clin. Immunol.* **74** 794-801

Respiratory Tract Deposition of Particles from Biomass Combustion

Jakob Löndahl[1], Erik Swietlicki[1], Joakim Pagels[2], Andreas Massling[1], Christoffer Boman[3], Jenny Rissler[1], Anders Blomberg[4] and Thomas Sandström[4]

1. Department of Physics, Lund University, PO Box 118, SE-221 00, Lund, Sweden

2. Division of Aerosol Technology (EAT), Lund Institute of Technology, PO Box 118, SE-221 00, Lund, Sweden

3. Energy Technology and Thermal Process Chemistry, Umeå University, SE-901 87 Umeå, Sweden

4. Department of Respiratory Medicine and Allergy, University Hospital, SE-901 85 Umeå, Sweden

E-mail: jakob.londahl@nuclear.lu.se

Abstract. The respiratory tract deposition of particles from two types of biomass combustion was measured for 10 healthy subjects. The aerosol was extensively characterised. Model calculations of the deposition were made based on the particle properties. The results show that particle water absorption has substantial impact on deposition. The particles from biomass combustion obtained a size in the respiratory tract at which the deposition probability is close to its minimum.

1. Introduction

There is an ongoing discussion about the relative importance of particle number, surface and mass concentration in health effects of an aerosol. Air quality guidelines usually regulate the mass concentration. It is possible that the differences in health effects of various aerosols could be partly explained by a potentially large variation in the respiratory tract deposited dose of the particles. In this study the respiratory tract deposition has been measured and calculated for particles from biomass combustion. A full report of the study is published elsewhere (Löndahl et al. 2008 [1]).

Smoke from biomass combustion is one of the major environmental threats to human health – especially in the poor part of the world (Lopez et al. 2006 [2]). It is also of increasing interest in search for renewable fuel to reach a sustainable development. One key parameter to fully understand the health effects of the particles is to estimate their probability to deposit in the lungs.

Biomass combustion particles are often hygroscopic and will thus change size in the lungs by absorption of water vapor. The relative humidity (RH) in the lungs is estimated to 99.5%. The common dose models do not usually take particle water absorption into account. Hence they need to be modified in order to make a correct estimate of the deposition biomass combustion particles. To best of the authors' knowledge there are no previous measurements reported on the deposition of hygroscopic real-world particles which could be used to validate the models.

2. Methods

The respiratory tract deposition of particles from particles from biomass combustion was measured for healthy human subjects with a novel instrument named RESPI (Löndahl et al. 2006 [3]). Data fulfilling our quality assurance protocol was obtained for 10 subjects (4f/6m). A modern residential wood pellet burner fired with a typical softwood pellet fuel was used for generating the emissions. Two entirely different operation conditions were applied: 1) efficient "complete" combustion at high load (~10 kWfuel) with an intensive/turbulent and hot (900-1100°C) flame and 2) low temperature (600-800°C) incomplete combustion at ~5 kWfuel simulating "wood smoke" by burning moist fuel with reduced air/fuel mixing. The combustion aerosols were cooled and diluted using a three stage dilution system. In addition the deposition was measured for nebulised hydrophobic DEHS-oil particles.

Biomass particles usually contain a large fraction of salts and therefore grow by absorption of water in a humid environment such as the respiratory tract. Thus, deposition is expected to change. To verify this, growth by water absorption was measured at 90% relative humidity (RH) with an H-TDMA. Based on these data and on the measured chemical composition, the particle hygroscopic growth factors at the RH prevailing in the lung were estimated. Thereby, the well-established ICRP dose model (ICRP Publication 66 1995 [4]) could be adjusted for particle water absorption. This was done in order to show consistency between measured deposition and theory.

3. Results and Discussion

There were considerable differences in deposition between the three studied aerosols (Figure 1). Average tidal volume for the 10 subjects was 0.68 ± 0.18 L and breathing frequency 12.4 ± 2.1 min^{-1}. The measured deposition fraction (DF) of the particles emitted from efficient combustion had a minimum around 60-70 nm. For particles from low temperature combustion the minimum was stretched out from 60-300 nm. On average, 21%-24% of the number, surface and mass of the inhaled particles were deposited for both studied biomass combustion aerosols. The DF for the reference oil higher but similar in shape to the deposition estimated from the common ICRP model (Figure 4).

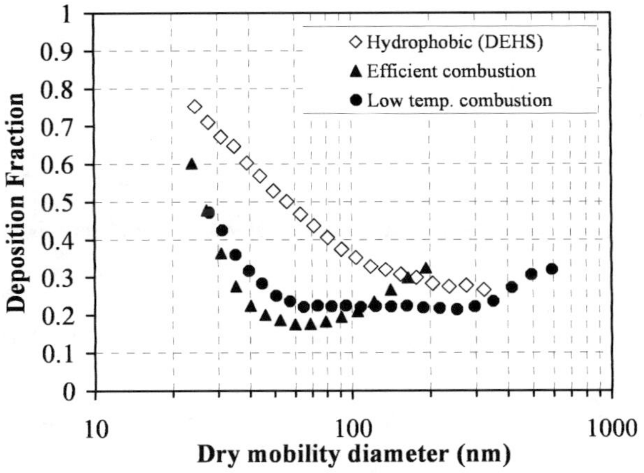

Figure 1. Measured respiratory tract deposition of particles from two types of biomass combustion (efficient and low temperature) and the deposition of hydrophobic reference particles.

The two biomass combustion aerosols had different size distributions, chemical composition and ability to absorb water – properties of fundamental importance for the respiratory tract deposition (Figure 2 and 3). Figure 2 shows the hygroscopic growth factor measured at 90% (left) and extrapolated to 99.5% RH (right). Figure 3 shows how this growth changes the size distribution in the lungs at 99.5% RH. The efficient combustion aerosol had a number concentration of $1.46 \pm 0.06 \times 10^5$ cm^{-3}, a count median mobility diameter, CMD, of 81 nm and a geometric standard deviation, σ_g, of 1.5. The low temperature combustion aerosol had a number concentration of $3.8 \pm 1.2 \times 10^4$ cm^{-3}, CMD of 137 nm and σ_g of 1.8. PM_1 samples of efficient combustion particles consisted by mass to 96% of soluble alkali salts, primarily K_2SO_4, KCl and K_2CO_3. The soluble volume fraction of KCl was 0.51, as derived from the HTDMA measurements. The fraction of K_2SO_4, which was only partly in solution at 90% RH, was 0.38 according to the chemical analysis. The increase in hygroscopic growth caused by inclusion of K_2SO_4 as it goes into solution at higher RH is shown in Figure 3 (right). The total carbon content was <1%. In contrast, the particles from low temperature combustion were more complex and consisted of 85% carbonaceous material (organic compounds and soot) and 14% alkali salts. The fraction of insoluble material in these particles, calculated from the HTDMA data, increased with size and particles above 400 nm were nearly hydrophobic.

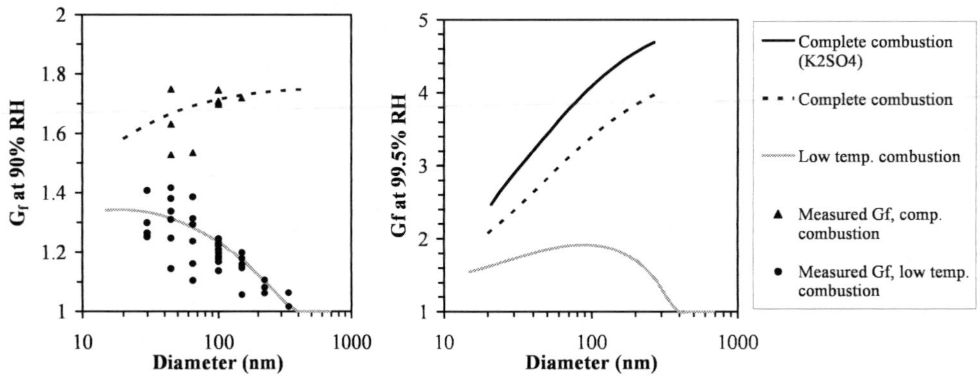

Figure 2. The size-dependent particle growth factor (Gf) measured at 90% RH (left) and extrapolated to 99.5% RH (right).

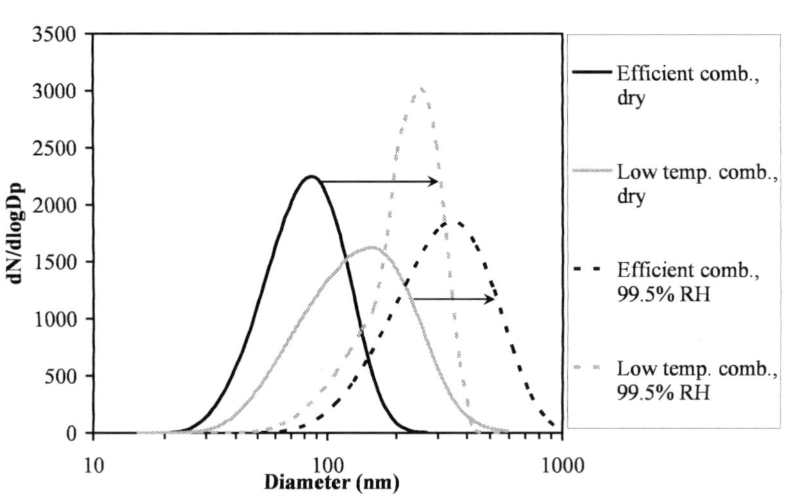

Figure 3. The particle number size distributions of the two biomass combustion aerosols at dry conditions and at 99.5% RH.

The experimentally determined DF agreed largely with the predicted DF when hygroscopic growth was incorporated in the model calculations (Figure 1 and 4). The modeled DF curve was shifted to the left towards smaller dry diameters, for both combustion aerosols, although most significantly for the efficient combustion due to higher growth factors. For instance, the "measured" growth factor for 100 nm particles from efficient combustion was 4.1 (Figure 3). Hence, these particles were assumed to grow to a size of 410 nm in the respiratory tract and thus deposit with the same probability as hydrophobic 410 nm particles.

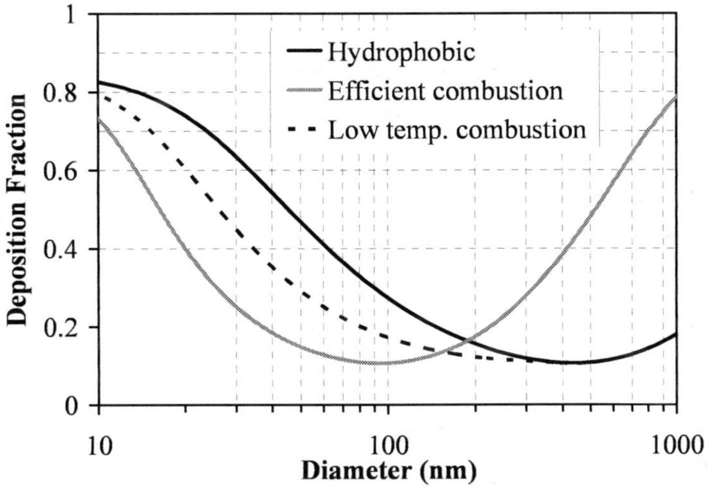

Figure 4. The deposition fraction according to the ICRP model after adjustment for particle water absorbtion.

4. Conclusion

This study experimentally verifies that the ability of real-world particles to absorb water sometimes substantially alter the deposition in the respiratory tract. This has previously only been shown by theoretical calculations and a limited number of experiments whereof most with NaCl particles. The majority of the particles from the two studied and completely different types of biomass combustion aerosols obtain a size in the respiratory tract (after water up-take) which coincides with the minimum in deposition probability. This was evident from both the measurements on human subjects and from model calculations, where particle characteristics were included. The respiratory tract deposited fraction of particles did thus become low (0.21-0.24) compared to what could be expected from for example traffic exhaust particles which are both smaller in size and more hydrophobic.

References

[1] Löndahl J, Pagels J, Massling A, Boman C, Swietlicki E, Rissler J, Blomberg A, Sandström T. 2008. Deposition of Biomass Combustion Aerosol Particles in the Human Respiratory Tract. Inhalation Toxicology 20:923-933.

[2] Lopez AD, Mathers CD, Ezzati M, Jamison DT, Murray CJL. 2006. Global and regional burden of disease and risk factors, 2001: systematic analysis of population health data. Lancet 367:1747-1757.

[3] Löndahl J, Pagels J, Swietlicki E, Zhou JC, Ketzel M, Massling A, Bohgard M. 2006. A set-up for field studies of respiratory tract deposition of fine and ultrafine particles in humans. Journal of Aerosol Science 37:1152-1163.

[4] ICRP Publication 66. 1995. Human Respiratory Tract Model for Radiological Protection, 66.International Comission on Radiological Protection.

Journal of Physics: Conference Series **151** (2009) 012067 doi:10.1088/1742-6596/151/1/012067

Rat inhalation test with particles from biomass combustion and biomass co-firing exhaust

B Bellmann, O Creutzenberg, H Ernst, and H Muhle

Fraunhofer Institute of Toxicology and Experimental Medicine, Nikolai-Fuchs-Str.1, 30625 Hannover, Germany

E-mail: bernd.bellmann@item.fraunhofer.de

Abstract. The health effects of 6 different fly ash samples from biomass combustion plants (bark, wood chips, waste wood, and straw), and co-firing plants (coal, co-firing of coal and sawdust) were investigated in a 28-day nose-only inhalation study with Wistar WU rats. Respirable fractions of carbon black (Printex 90) and of titanium dioxide (Bayertitan T) were used as reference materials for positive and negative controls. The exposure was done 6 hours per day, 5 days per week at an aerosol concentration of 16 mg/m^3. The MMAD of all fly ash samples and reference materials in the inhalation unit were in the range from 1.5 to 3 µm. The investigations focused predominantly on the analysis of inflammatory effects in the lungs of rats using bronchoalveolar lavage (BAL) and histopathology. Different parameters (percentage of polymorphonuclear neutrophils (PMN), interleukin-8 and interstitial inflammatory cell infiltration in the lung tissue) indicating inflammatory effects in the lung, showed a statistically significant increase in the groups exposed to carbon black (positive control), C1 (coal) and C1+BM4 (co-firing of coal and sawdust) fly ashes. Additionally, for the same groups a statistically significant increase of cell proliferation in the lung epithelium was detected. No significant effects were detected in the animal groups exposed to BM1 (bark), BM2 (wood chips), BM3 (waste wood), BM6 (straw) or titanium dioxide.

1. Introduction

In the current study, the health effects of six different fly ash samples from biomass combustion and co-firing plants were investigated in a 28-day nose-only inhalation study in Wistar WU rats. Respirable fractions of carbon black (Printex 90) and of titanium dioxide (Bayertitan T) were used as reference materials for positive and negative controls, as long-term experience with these materials has been gained at the Fraunhofer ITEM [1] [2] and has been reported in literature.

The health effects of other relevant ambient air particles, like diesel exhaust particles (DEP), have been investigated in the past decades in various studies including animal studies, *in vitro* cell tests, epidemiological studies, and controlled short-term human exposure. These studies have shown two major endpoints: induction of lung tumors [1] [3], and cardiovascular effects in a sensitive human subpopulation [4]. For both endpoints, inflammatory effects in the lungs were considered to be early indicators in short-term studies. Therefore, investigation of the aerosols from biomass combustion in the present study was focused predominantly on the analysis of inflammatory effects in the lungs of rats by bronchoalveolar lavage (BAL) and histopathology.

2. Materials and Methods

2.1. Test and reference materials

The fly ash samples BM1 (bark) and BM3 (waste wood) have been sampled during pilot-scale test runs at a Mawera combustion unit from the clean flue gas (downstream multi-cyclone) by applying a sampling device consisting of a cyclone and a fabric filter. Fly ash sample BM6 (straw) was sampled directly from the filter fly ash at a grate-fired Elean power station (UK), whereas the fly ash samples C1 (Polish coal) and C1+BM4 (co-firing of coal and sawdust) were sampled at the smallest precipitation stage of the ESP (electrostatic precipitator) during large scale test runs performed at the Dolna Odra power station (Poland). All mentioned samples were subsequently sieved at 40 µm. The fly ash sample BM2 (wood chips) was produced at a small-scale test run and applying a high-temperature sampling device. The chemical composition of the fly ash samples is listed in table 1

Titanium dioxide was purchased from Bayer AG, Krefeld, Germany (lot # 85/13928). By chemical analysis, the material was 99.5% TiO_2; annealing loss 0.22%; 99.5% of the TiO_2 was rutile. Its density was 4.26 g/cm^3.

The carbon black type Printex® 90 was purchased from Degussa AG. Its density was 1.8-1.9 g/m^3. This reference item had been used in two chronic inhalation studies in rats at the Fraunhofer ITEM [1] [5]. In these studies, lung tumors had been detected after exposure to Printex 90.

Table 1. Chemical composition of test materials

Sample	Ca	Si	Mg	K	Na	Zn	Mn	Pb	S	Cl	P	Ti	Fe	Al	Ba	TOC	TIC
Bark (BM1)	18.0	1.2	1.5	16.4	0.6	3.9	0.7	0.2	8.0	5.8	0.7	0.0	0.5	0.4	0.2	1.0	0.9
Wood chips (BM2)	0.6	0.0	0.1	37.5	1.4	0.8	0.0	0.0	9.3	3.4	0.1	0.0	0.1	0.0	0.0	0.0	3.0
Waste wood (BM3)	5.7	1.3	1.0	12.1	3.2	12.2	0.2	8.2	2.7	32.9	0.3	0.2	0.3	0.2	0.1	0.6	0.1
Coal+ sawdust (C1+BM4)	3.4	19.3	1.9	2.5	1.5	0.0	0.1	0.0	0.7	0.0	0.7	0.7	4.6	13.9	0.0	3.1	0.1
Coal (C1)	3.0	20.2	1.8	2.4	1.3	0.0	0.1	0.0	0.5	0.0	0.4	0.7	3.8	13.5	0.0	5.8	0.1
Straw (BM6)	12.7	2.1	0.2	31.1	0.2	0.1	0.0	0.0	4.3	28.5	1.1	0.0	0.0	0.0	0.0	2.7	0.3

Values are presented in mass % of elements (oxygen not shown)

TOC: total organic carbon; TIC: total inorganic carbon

2.2. Animal model

Female Wistar rats (strain Crl:WU) were purchased from Charles River Deutschland, Sulzfeld, Germany. At the start of exposure, the animals were approx. 9 weeks of age. For a period of 3 weeks prior to exposure, the animals were trained to become accustomed to nose-only tubes.

2.3. Aerosol Generation and Exposure

For each exposure group, the aerosol was generated by dispersing the test item. Dispersion was achieved by a feeding system and a high-pressure, high-velocity pressurized air dispersion nozzle developed at the Fraunhofer ITEM [6]. For each nose-only exposure unit, the aerosol was generated by a high-pressure pneumatic disperser. Air flow, temperature, and relative humidity were measured continuously and recorded as 20-minute means. Limits were set at 22 °C $\pm$ 4 °C for temperature and 50% $\pm$ 20% for relative humidity. The aerosol was monitored (A) by an aerosol photometer (on-line) and (B) by determination of gravimetric concentration (weighing of filter samples once a week). The aerosol photometer signal was used in the computerized feedback loop to control the test item feeding rate. By this procedure an exposure concentration of 16 mg/m^3 was generated for test items and reference items.

The animals were restrained in Batelle type polycarbonate tubes; with this system animals in the supply tubes keep their noses close to the airflow at the opening of the tubes. The exposure of animals was performed in identical exposure chambers of cylindrical shape, each housing up to 32 animals. Control animals were exposed in an identical unit to filtered air only. The duration of the exposure was 6 hrs/day on 5 days per week for 4 weeks.

2.4. Gross Pathology/Necropsy

All animals were subjected to complete necropsy, including careful examination of the external surface of the body, all orifices, and the cranial, thoracic and abdominal cavities and their contents. The rats were anesthetized with an overdose of pentobarbital sodium (Narcoren®) and killed by cutting the vena cava caudalis (for BAL of rats, see 2.54). Rats used for histopathological investigations (see 2.6) were killed with an overdose of carbon dioxide. The abdominal cavity was opened and the diaphragm was cut carefully allowing the lungs to collapse. Heart, esophagus, upper half of the trachea, thymus, and lung-associated lymph nodes (LALN) were removed from the lung convolute.

2.5. Bronchoalveolar Lavage (BAL)

Bronchoalveolar lavage was performed in 6 rats per group after end of exposure. The method of Henderson et al. [7] was used with minor modifications. Following preparation, the lungs were lavaged with saline using two lavages of 4 ml each. The lavage fluid was collected in calibrated tubes and the collected volume was recorded. Until processing the lavage fluid was kept on ice. Leukocyte concentration of the lavagate was determined using a counting chamber and two cytoslides were prepared with a cytocentrifuge (Shandon Co., Frankfurt, Germany) for differential cell count (macrophages, neutrophils, lymphocytes). After centrifugation of the lavage fluid, biochemical indicators relevant for diagnosis of lung damage were determined in the supernatant (lactic dehydrogenase - LDH, β-glucuronidase, total protein). These three parameters were analyzed according to routine clinical chemistry protocols using a Randox daytona device (Roche Co., Grenzach, Germany).

The bronchoalveolar lavage fluids were analyzed for tumor necrosis factor-α (TNF-α) and CINC-1 concentrations. Cytokine-induced neutrophil chemoattractant-1 (CINC-1) represents the equivalent molecule for interleukin-8 in the rat system, acting through rat CXCR2, a receptor which shows a > 85% homology to human CXCR2. CXCR2 was formerly termed Interleukin-8 receptor b. Therefore, a commercially available ELISA system for rat CINC-1 was used in this study.

Measurements were performed using the following commercially available ELISA systems:

TNF-α: Rat TNF-α Duoset, R&D Systems GmbH, Wiesbaden (catalog number: DY510). The lower limit of quantification was 62.5 pg/ml.

CINC-1: Rat CINC-1 Duoset, R&D Systems GmbH, Wiesbaden (catalog number: DY515). The lower limit of quantification was 15.2 pg/ml.

The bronchoalveolar samples were measured in duplicates using the respective ELISA system following the technical manual of the supplier.

2.6. Histopathology/Cell Proliferation

For 6 animals per group histopathology of lung lobes, including bronchi and the lung-associated lymph nodes (LALN) was performed. A semi-quantitative scoring of fibrotic changes was included.

Lungs were fixed in buffered formalin (10%), embedded in paraffin, sectioned, and stained with hematoxylin and eosin (H & E). Masson trichrome was applied as special stain for diagnosis of fibrotic changes.

Formalin-fixed tissue of the terminal bronchioles and lung parenchymal cells were examined for cell proliferation using the sensitive S-phase response method. Proliferating cells were labeled by 5-bromo-2'-deoxyuridine (BrdU; 20 mg/ml in phosphate-buffered saline), which was administered to the animals by a minipump (Alzet type 2 ML1; 10 μl/h) 5 days prior to sacrifice. The lung section slides were prepared according to histological routine procedures and stained immunohistochemically following denaturation of the DNA (antibody technique). The antibody to BrdU was labeled by the chromogen DAB, by which the nuclei of proliferating cells are marked brown. The nuclei of non-proliferating cells were labeled blue by hemalaun staining.

The slides were evaluated by analyzing an appropriate number of airway cells and cells of the proximal regions of the pulmonary parenchyma per rat. For at least 4 terminal bronchioles, the number of positive cells and the total length of the analyzed bronchiolar epithelium were measured. From these data, the unit length labelling index was calculated in cells per mm. In four regions around each of these terminal bronchioles, the number of brown coloured cell nuclei (proliferating cells) and the number of blue coloured cell nuclei of the non-proliferating epithelial cells were counted. From these data, the labelling index of the lung parenchymal cells was calculated as percentage of positive cells.

3. Results and Discussion

The MMAD (mean mass aerodynamic diameter) of all fly ash samples and reference materials was in a similar range between 1.5 and 3 µm. Therefore, the deposition of these particles in the respiratory system was similar for all samples.

For different parameters (percentage of polymorphonuclear neutrophils (PMN), Interleukin-8 and interstitial inflammatory cell infiltration in the lung tissue) indicating inflammatory effects in the lung, a significant increase was observed in the groups exposed to carbon black (positive control), C1 (coal) and C1+BM4 (co-firing of coal and sawdust) fly ashes (see all results in table 2). Additionally, for the same groups a significant increase of cell proliferation in the lung epithelium was detected. These effects which were observed already after 4 weeks of exposure could be early indicators for the development of lung tumours. In a 2-year inhalation study with rats using a similar exposure concentration as in the present study carbon black as well as diesel exhaust particles induced lung tumours.

Table 2. Summary of results from the in-vivo study

Parameter	Control	BM1	BM2	BM6	BM3	C1+ BM4	C1	TiO_2	Carbon black
Lung weight	100	105	100	102	98	103	101	100	**120
BAL results									
PMN	100	73	138	188	120	**1,099	**1672	68	**6,203
LDH	100	86	84	88	110	151	161	88	**396
Total Protein	100	103	106	97	99	**143	**148	95	**218
GSH	100	93	94	76	126	*136	121	94	128
Interleukin-8	100	107	106	86	113	**318	**529	89	**1413
Histopathology									
Lung inflammatory cell infiltration[a]	17	50	67	67	50	67	*83	17	**100
Interstitial fibrosis[a]	0	0	0	0	0	17	33	0	**100
Cell proliferation lung parenchyma	100	98	102	121	96	*137	**172	86	**155

Values are presented in % of controls

[a] Values are shown as percentual incidence of lesion

Statistics: Comparison to control group by Dunnett's test: * p < 5%; ** p < 1%

The common characteristic of carbon black, C1 and C1+BM4 is the high content of insoluble particles whereas the other biomass fly ash samples BM1, BM2, BM3 and BM6 mainly consist of soluble salts. After inhalative exposure these salts are dissolved very fast, can reach the capillary system and are removed by renal clearance. Therefore, the concentration of soluble fly ash particles is very low in the lung after long term inhalation whereas insoluble particles accumulate in the lung. For the insoluble particle samples C1 and C1+BM4 the effects on inflammation and cell proliferation in the lung were statistically significant but lower than the effects induced by the carcinogenic carbon black or diesel exhaust particles. For the soluble fly ash samples no adverse health effects were detected in the present study.

This means, in this rat inhalation study the percentage of insoluble particles in the fly ash samples was the important parameter which influences the potency for the induction of adverse health effects.

It has to be taken into account that the particle samples used in this study originated from state-of-the-art biomass combustion plants respectively from real-scale coal as well as biomass and coal co-firing power plants. In these installations combustion technology and combustion control are highly advanced and therefore, the gas phase burnout is almost complete and the TOC concentrations in the total ashes are <5 wt% d.b.. As a consequence only minor amounts of TOC emissions (<2 mg/Nm^3) could be determined and therefore, the fine particulate emissions almost exclusively consist of inorganic salts. Moreover, particle precipitation devices such as baghouse filters or ESPs are applied in order to reduce particulate emissions. In old biomass residential heating systems on the other side, TOC emissions and consequently also soot and organic aerosol emissions can be significantly higher. In this case, the health effects caused by the fine particulate emissions might be more comparable with those of the carbon black samples applied in this study than with the samples from well controlled biomass combustion. Additionally, also polyaromatic hydrocarbons (PAHs) and dioxins may be emitted, which could induce additional health effects.

However, an important aim for future research on this topic should be to investigate the influence of the combustion quality on the health effects of the particulate emissions with special respect to biomass based residential heating systems.

Acknowledgements
The work was supported by the EC through the SES-CT-2003-502679 project BIOASH.

References

[1] Heinrich U, Fuhst R, Rittinghausen S, Creutzenberg O, Bellmann B, Koch W, Levsen, K 1995 Inhal. Toxicol. **7** 533.
[2] Muhle H, Bellmann B, Creutzenberg O, Dasenbrock C, Ernst H, Kilpper R, MacKenzie JC, Morrow P, Mohr U, Takenaka S, Mermelstein R, 1991 Fundam. Appl. Toxicol. **17** 280
[3] Garshick E, Laden F, Hart JE, Rosner B, Smith TJ, Dockery DW, Speizer FE 2004 Health Perspect. **112** 1539
[4] Mills NL, Tornqvist H, Robinson SD, Gonzalez M, Darnley K, MacNee W, Boon NA, Donaldson K, Blomberg A, Sandstrom T, Newby DE, 2005 Circulation **112** 3930
[5] Heinrich U, Dungworth DL, Pott F, Peters L, Dasenbrock C, Levsen K, Koch W, Creutzenberg O, Schulte A 1994 Ann. Occup. Hyg. **38** (Suppl. 1) 351
[6] Koch W 1998 Application of aerosols, in: S Uhlig, AE Taylor (Eds.), Pulmonary Research. Birkhäuser, Basel-Boston-Berlin.
[7] Henderson RF, Mauderly JL, Pickerell JA, Hahn RF, Muhle H, Rebar AH, 1987 Exp. Lung Res. **13** 329

AUTHOR INDEX

AUTHOR INDEX

AUTHOR INDEX

AUTHOR INDEX